AF519002

NATURAL PRODUCTS: *Research Reviews*

— Volume 3 —

The Editor

Dr. Vijay Kumar Gupta, Ph.D., FLS, London (born 1953-) Ex. Chief Scientist, CSIR- Indian Institute of Integrative Medicine, Jammu, India. He did his M.Sc. (1975) and Ph.D. (1979) in Zoology both from University of Jammu, Jammu-India. His research capabilities are substantiated by his excellent work on histopathology, ecology and reproductive biology of fishes, turtles, birds and mammals, which has already got recognition in India and abroad.

Dr. Gupta has to his credit more than 150 scientific publications and review articles which have appeared in internationally recognized Indian and foreign journals. Founder fellow, life member and office bearer of many national societies, academies and associations. He has successfully completed a number of research/consultancy projects funded by government, private and multinational agencies. His current areas of interest are histopathology, toxicology, pre-clinical safety pharmacology, reproductive efficacy studies of laboratory animals and biodiversity.

He is the Series Editor of the recently published multi-volume set of books, "**Comprehensive Bioactive Natural Products (Vols. 1-8)**", published by M/S Studium Press, LLC, USA. He is also Editor-in-Chief of the books, "**Utilisation and Management of Medicinal Plants (Vols. 1-3)**", "**Medicinal Plants: Phytochemistry, Pharmacology and Therapeutics (Vols. 1-4)**", "**Traditional and Folk Herbal Medicine (Vols. 1-3)**", "**Natural Products: Research Reviews (Vols. 1-4)**", "**Bioactive Phytochemicals: Perspectives for Modern Medicine (Vols. 1-3)**", "**Perspectives in Animal Ecology and Reproduction (Vols. 1-10**)" and "**Animal Diversity, Natural History & Conservation (Vols. 1-5)**". The Editor-in-chief of the American Biographical Institute, USA, has appointed him as *Consulting Editor* of *The Contemporary Who's Who*. Dr. Gupta also appointed as Nominee for the *Committee for the Purpose of Control and Supervision of Experiments on Animals* (CPCSEA, Govt. of India). The *Linnaean Society of London, U.K.* has awarded fellowship to him in November 2009 in recognition of his contribution towards the cultivation of knowledge in Science of Natural History. Recently, Modern Scientific Press, USA has nominated Dr. Gupta as the Editor of the *International Journal of Traditional and Natural Medicine*.

NATURAL PRODUCTS: *Research Reviews*

— *Volume 3* —

Editor
V.K. Gupta
Chief Scientist (Ex)
CSIR-Indian Institute of Integrative Medicine
Canal Road, Jammu – 180 001,
India

2016
Daya Publishing House®
A Division of
Astral International Pvt. Ltd.
New Delhi – 110 002

Cataloging in Publication Data--DK
Courtesy: D.K. Agencies (P) Ltd.
<docinfo@dkagencies.com>

Natural products : research reviews / editor, V.K. Gupta.
volume 3 cm
Includes bibliographical references and index.
ISBN 978-93-5124-685-5 (Hardbound)
ISBN 978-93-5130-882-9 (International Edition)
ISBN 978-81-7035-774-2 (Series)
ISBN 978-93-5130-104-2 (Series) (International Edition)

1. Natural products. 2. Botanical chemistry. 3. Materia medica, Vegetable. I. Gupta, V. K. (Vijay Kumar), 1953- editor.

QK861.N38 2016DDC 572.2 23

Published by : **Daya Publishing House®**
A Division of
Astral International Pvt. Ltd.
– ISO 9001:2008 Certified Company –
4760-61/23, Ansari Road, Darya Ganj
New Delhi-110 002
Ph. 011-43549197, 23278134
E-mail: info@astralint.com
Website: www.astralint.com

Laser Typesetting : **Classic Computer Services,** Delhi - 110 035

Printed at : **Sanat Printers**

PRINTED IN INDIA

Editorial Board

Prof. Ian Fraser Pryme

Department of Biomedicine,
University of Bergen,
Jonas Lies vei 91, N-5009 Bergen, NORWAY
Tel: +47-55-586438
Fax: +47-55-586360
E-mail: ian.pryme@biomed.uib.no

Dr. George Qian Li

Herbal Medicines Research and Education Centre,
Faculty of Pharmacy, The University of Sydney,
NSW 2006, AUSTRALIA
Tel: 612 9351
Fax: 9351 8638
E-mail: george.li@sydney.edu.au

Dr. Julia Serkedjieva

Institute of Microbiology,
Bulgarian Academy of Sciences,
26, Academician Georgy Bonchev St.,
1113 Sofia, BULGARIA
Tel: +359 2 979 31 85
Fax: +359 2 870 01 09
E-mail: jserkedjieva@microbio.bas.bg

Bulgarian Academy of Sciences
The Stefan Angeloff Institute of Microbiology
26 G. Bonchev str., 1113 Sofia, Bulgaria
Tel: (+3592) 9793195
E-mail: nina@microbio.bas.bg

Foreword

Natural products, particularly from microbial and plant origin have been used for centuries by people as a source of an unlimited range of chemically diverse structures. The interest for new, natural compounds is renewed as a result of the growing resistance of microorganisms to commonly used drugs. Thus, nosocomial infections have become great public health problems. The selection of plants is a first research step based on certain criteria such as ethnobotanical and ethnopharmacological approaches, including information about the traditional use of plants to treat specific diseases. Also, many studies explore chemotaxonomic approach, which relay on the study of plants from the same family or genus of a species from which active compounds have been already isolated.

The volume 3 of **Natural Products: Research Reviews** edited by Dr. V. K. Gupta includes twenty two chapters devoted to the identification and isolation of bioactive components with antimicrobial and antioxidant properties and also chapters related to the potential application of natural products in cardiovascular, cancer and neurodegenerative disorders. Drawing upon the collective wisdom of traditional medical systems and modern science, this book identifies the revered herbs from around the world that can comprehensively inhibit pathologic conditions. It discusses the challenges of using traditional medicines for diseases where access to modern medicine is limited, and the research areas needed to improve quality, safety, and efficacy for enhancing healthcare. It is a pleasure for me to recommend this volume to all researchers in the field of natural products and I believe that the book will attract the attention of great number of readers.

Prof. Nina Ivanovska

Preface

Natural products have been investigated and utilized to alleviate disease since early human history. Medicine among the ancient civilizations was a blend of religion and science. Nature produces a large number of chemical compounds whose structure and properties are of great interest and fascination for the majority. These are organic compounds formed by living system and are emerging from the cocoon of organic chemistry. The term 'natural products' is applied to materials derived from plants, microorganisms, invertebrates and vertebrates, which are the fine biochemical factories for biosynthesis of these metabolites and over 40 per cent of medicines, have origin from these compounds. Natural products are the cornerstone of modern therapeutics. Building on the discovery of morphine from poppy seeds in the 1800s and penicillin from mold in the 1920s, natural products have been a key source of new therapeutics. Natural products are currently used to treat cancer, high levels of cholesterol, fungal infections, bacterial infections, and parasitic diseases. As of 1990, nearly 80 per cent of all drugs in use were natural products or natural product derivatives. In developing countries, the search for biologically active compounds obtainable from locally available plants is also important to reduce public health.

Despite the great successes already achieved, we have barely begun to tap the potential of our molecular diversity and just 5-15 per cent of the 2,50,000 species of higher terrestrial plants in existence have been pharmacologically investigated in systematic fashion. The percentage of insects, marine organisms and microbes investigated is far lower still. In the case of microbes, it is estimated that 95 to 99 per cent of existing species are currently not even known, never mind analyzed. Even with all the challenges facing drug discovery, natural products isolated from natural sources can be predicted to remain an essential component in the research for new medicines.

It was with this approach that this volume "***Natural Products: Research Review Vol. 3***" has been compiled. The contributions made by some of the worlds leading scientists in the field should prove of wide interest to the readers. This is a timely and interesting collection of review articles covering a very wide range of biologically active natural products. The book is produced to the high standard that we have come to expect of this volume. It contains a good subject index. There is a great deal of information in these chapters which should be of interest not just to the natural product chemist but also to those in the pharmaceutical and agrochemical industries who are interested in potential leads for structure activity studies. The book can be recommended to all who are interested in biologically active natural products.

V.K. Gupta

Contents

Natural Products: Research Reviews Vol. 3 (2016) *Pages* **1–33**
Editor: **V.K. Gupta**
Published by: **DAYA PUBLISHING HOUSE, NEW DELHI**

1

Use of Garlic, Ginkgo, Ginseng, and St. John's Wort and Potential Interactions with Cardiovascular Drugs: A Brief Review

M.G. Miguel*[1] and M.A.P. Silva[1]

ABSTRACT

Cardiovascular drugs concomitantly administered with herbal supplements may be responsible for unexpected adverse effects, particularly in the elderly. The majority of this population presents physical and mental weakness and some have depression. Ginkgo (to treat dementia and for memory improvement), ginseng (adaptogen), garlic (antihypertensive, hypolipidemic, antiatherogenic) and St. John' wort (antidepressant) are indicated for these situations. In the present work, a very brief review was made on interaction between these herbal supplements and some cardiovascular drugs.

Keywords: Alternative medicine, Cardiovascular drugs, Herb-drug interaction, St. John' wort, Ginkgo, Ginseng, Adaptogen.

1. Introduction

In developing countries more than of 80 per cent of population uses herbal medicines due to the fact that they only have this type of remedies as their primary source of healthcare. In developed countries, the consumption of herbal remedies has increased in the last decades, because such products are generally considered harmless and without adverse side effects. About 30 per cent of adults in these countries use herbal remedies and natural supplements because they consider that

1 Faculdade de Ciências e Tecnologia, Departamento de Química e Farmácia, Edifício 8, Universidade do Algarve, Campus de Gambelas, 8005-139 Faro, PORTUGAL

* *Corresponding author*: E-mail: mgmiguel@ualg.pt

such type of products are less toxic than chemical drugs, underestimating that herbs can exert important toxicities (Izzo, 2004; Effert and Kaina, 2011; Na *et al.*, 2011).

It is quite often that patients and consumers utilize herbal medicines or plant-based therapeutic products as complementary therapy in the treatment and management of chronic ailments such as diabetes, tuberculosis, hypertension, cancer, HIV/AIDS and other diseases, because such herbs have been used since ancient times in traditional medicine to treat diseases and ailments (Effert and Kaina 2011; Obodozie, 2012). This does not mean that people does not take their conventional drugs, only they use both therapies in the treatment of their diseases. However, many times, patients or consumers do not report to their physicians the co-intake of conventional drugs and plant-based-therapeutic products. This is a real problem due to the possible interactions between chemical and herbal preparations beyond to the absence of the guarantee of quality, safety and efficacy of these products because they are generally registered as dietary supplements (Effert and Kaina, 2011; Na *et al.*, 2011).

The interaction among drugs may alter their pharmacological response, either in intensity or duration, or an adverse reaction or a totally new side effect may appear, which would not be observed if administered alone (Rahal *et al.*, 2013). Considering that herbal preparations present great variations in composition and dosages, the risk of interactions among therapeutic drugs and these plant-based-therapeutic products is even higher, and extremely dangerous when co-administered with narrow therapeutic index drugs (Effert and Kaina, 2011).

2. General Aspects of Mechanisms for Drug Interaction with Herbal Supplements

The mechanisms proposed for herb-drug interaction are based on the same pharmacokinetic and pharmacodynamic mechanisms reported for drug-drug interactions (Izzo, 2004). Nevertheless, herbal supplements usually contain more than one pharmacologically active ingredient and this fact significantly increases the possibilities of interactions (Shi, 2012).When an herbal supplement changes the absorption, distribution, metabolism, or excretion of a therapeutic drug there is pharmacokinetic interaction, which results in altered levels of the drug or its metabolites on blood stream. Pharmacodynamic interactions can affect organ systems, receptor sites or enzymes because they are related to the pharmacology action of the interacting agents (Renuka *et al.*, 2011).

Pharmacokinetic herb-drug interactions may be attributed to the induction or inhibition of hepatic and intestinal drug-metabolizing enzymes, namely cytochrome P450 (CYP) and/or drug transporters such as P-glycoprotein (P-gp) (Izzo, 2004; Effert and Kaina, 2011; Renuka *et al.*, 2011). Nevertheless other enzymes may metabolize herbal drugs which may influence the therapeutic drugs biotransformation. Such enzymes include epoxide hydrolase, carboxylesterase and amidase, peroxidise, alcohol/aldehyde dehydrogenase, monoamine oxidase, nicotinamide adenine dinucleotide phosphate (NADPH), quinone reductase, all of them of Phase I biotransformation reactions. In Phase II conjugation reactions, UDP-glucuronosyl transferases (UGTs) that catalize glucuronidation of xenobiotics, and sulfotransferases (SULTs) that catalize sulfonation of drugs and/or herbs may alter their disposition and clearance when co-administered. Nevertheless, the interactions mediated by UGTs, SULTs, and the remaining enzymes of Phase I appear to be fewer than those mediated by the CYPs (Choi *et al.*, 2011; Na *et al.*, 2011).

Drug-herb interaction is also influenced by genetic differences in patients. Specific polymorphisms may lead different responses to drugs and consequently cause herb drug interactions, which do not occur in the majority of other patients (Tomlinson *et al.*, 2008; Effert and Kaina, 2011). There is evidence for polymorphisms in CYP1A2, CYP2A6, CYP2D6, CYP2C19 and CYP3A. With the increasing consumption of herbal medicines or supplements in the Western world, where they are often co-administered with conventional medicines, it is possible that constituents in herbal preparations may be inhibitors or inducers of CYPs with the consequent impact on the pharmacokinetics of any co-administered drugs metabolized by these systems (Zhou *et al.*, 2003).

P-gp is one of the most prevalent efflux transporters, playing an important role in the intestinal absorption of many compounds that are its substrates. For example, inhibition of P-gp leads to the improvement of oral bioavailability (Yang *et al.*, 2012). The expression and function of P-gp can also be modified by genetic polymorphisms of the multidrug resistance 1 gene (MDR1 gene). Several MDR1 single-nucleotide polymorphisms have been identified associated with alteration of P-gp expression and/or function. These polymorphisms may affect drug disposition (Tandon *et al.*, 2006).

A systematic review made by Hermann and Richter (2012) in which herb-drug interaction were evaluated, showed that at commonly recommended doses, none of the herbs under study (garlic, gingko, and gingseng) act as potent or moderate inhibitors or inducers of CYP enzymes or P-glycoprotein (ABCB1).

A study undertaken by some authors about the utilization of chronic medications and herbal use during anaesthetic preoperative assessment, showed that most commonly used treatments were angiotensin-converting enzyme inhibitors, angiotensin-II receptor blockers, β-blockers, and platelet inhibitors. Some of the patients also used alone or concomitantly with those medicines, herbal remedies such as valeriane, ginseng, ginkgo, St. John's wort, echinacea and ephedra (Baillard *et al.*, 2007). According the study made by these authors, women and patients aged 40-70 years old were those whom preferentially took herbal products.

In the present work, the interactions between some herbs and supplements (St. John's wort, *Ginkgo biloba*, ginsengs and garlic) and therapeutic drugs for cardiovascular diseases will be preferentially reported. Such choice is due to the fact that herbal remedies are preferably used by patients from the age of 40, which generally coincides with the use of cardiovascular therapies.

3. Herbal Products in Patients with Cardiovascular Diseases

Aloe vera, bilberry, garlic, ginkgo, ginseng, green tea, grapefruit juice, St. John's wort among other herbals and supplements are used with some frequency by populations, particularly by senior citizens, whom are also generally medicated with cardiovascular therapies, such as hypertension regulators, among much more therapeutic drugs. The demand for herbal drugs by these populations is remarkably due to the fact that they need to alleviate some symptoms characteristics of the elderly: pain symptoms, anxiety, depression and prevalence of chronic disorders, coupled with the belief that natural products are harmless. Ara Tachjian *et al.* (2010) summarized the herbal products to be avoided by patients with cardiovascular diseases as well as the cardiac adverse effect of interaction. Izzo *et al.* (2005) in a review reported interactions between cardiovascular therapies and herbal medicines.

4. Cardiovascular Therapy

Some of the major types of commonly prescribed cardiovascular medications include anticoagulants, angiotensin-converting enzyme (ACE) inhibitors, angiotensin II receptor blockers (or inhibitors), β-blockers, calcium channel blockers, diuretics, digitalis preparations and statins. They are used since for decreasing the clotting (coagulating) ability of the blood, preventing harmful clots forming in the blood vessels, to lower LDL cholesterol (and triglyceride levels), raise HDL cholesterol and lower arterial pressure.

According to a systematic review made by Posadziki *et al.* (2012), serious herb–drug interactions were noted for *Hypericum perforatum* and *Viscum album* and moderately severe interactions were noted for *Ginkgo biloba, Panax ginseng, Piper methysticum, Serenoa repens* and *Camellia sinensis.* The most severe interactions reported by those authors included transplant rejection, delayed emergence from anaesthesia, cardiovascular collapse, renal and liver toxicity, cardiotoxicity, bradycardia, hypovolaemic shock, inflammatory reactions with organ fibrosis and death.

5. Potential Interaction between Drugs and Herb Supplements

5.1. *Hypericum perforatum* (St. John's wort)

Flavonol glycosides, biflavones, tannins, xanthones, phloroglucinols are some compounds present in *Hypericum perforatum* (St. John's wort) (Figure 1.1). Some of them, particularly hyperforin (phloroglucinol) inhibits monoamine oxidase, the uptake of serotonin, noradrenaline, dopamine and possess γ-aminobutyric acid-A receptor affinity, being therefore used as antidepressant (Henderson *et al.,* 2002; Nadkarni *et al.,* 2012).

Animal, preclinical and clinical studies have suggested that *H. perforatum* may modulate CYP activity. A number of clinical trials have demonstrated that *H. perforatum* induces CYP3A4, CYP2E1, and CYP2C19, with no effect on CYP1A2 of males, CYP2D6 or CYP2C9 (Borrelli and Izzo, 2009).

Table 1.1: Summary of Herb-Cardiovascular Drugs Interaction.

Evaluated Herb	*Drugs*	*Possible Mechanism of Interaction*	*References*
Garlic	Acetylsalicylic acid	Additive effect	German *et al.*, 1995; Tachjian *et al.*, 2010
	Cilostazol	None	Mateen *et al.*, 2011
	Warfarin	Additive effect	German *et al.*, 1995; Tachjian *et al.*, 2010
	Atorvastatin	Increase of plasma concentration inrats by inibition of CYP3A4 and P-gp. Decrease of plasma concentration in humans by induction of CYP3A4.	Reddy *et al.*, 2012; Markowitz *et al.*, 2003; Thomas *et al.*, 2012
	Lovastatin	Decrease of plasma concentration in humans by induction of CYP3A4.	Markowitz *et al.*, 2003; Thomas *et al.*, 2012
	Simvastatin	Decrease of plasma concentration in humans by induction of CYP3A4.	Markowitz *et al.*, 2003; Thomas *et al.*, 2012
	Propanolol	Increase of plasma concentration in rats. The mechanism is unknown.	Asdaq and Inamdar 2011
	Metoprolol	Potencial sinergism ascardioprotective.	Kshirsagar and Asdaq 2014
	Verapamil	Decrease of plasma concentration by induction of CYP3A4.	Markowitz *et al.*, 2003; Thomas *et al.*, 2012
	Nifedipine	Decrease of plasma concentration by induction of CYP3A4.	Nowack *et al.*, 2008; Chan *et al.*, 2010
Ginkgo	Atorvastatin	Slightly decrease of plasma concentration by induction of CYP3A4, UDP-glucoronosyltranferae 1A1 and UGT1A3.	Guo *et al.*, 2012
	Cilostazol	Additive effect	Ryu *et al.*, 2009
	Digoxin	None	Mauro *et al.*, 2003
	Diltiazem	Decrease of plasma concentration in rats by induction of CYP3A4.	Ohnishi *et al.*, 2003
	Nicardipine	Decrease of plasma concentration in rats by induction of CYP3A2.	Shinozuka *et al.*, 2002
	Nifedipine	No pharmacokinetic interaction but possible pharmacodynamic interaction.	Yoshioka *et al.*, 2004
	Talinolol	None	Fane t *al.*, 2009 a,b.
	Thiazide diuretic	Possible metabolic inibition.	Chen *et al.*, 2011; Hu *et al.*, 2005.
	Ticlopidine	None	Kim *et al.*, 2010
	Warfarin	Inibition of platelet activating factor by ginkgolides. No pharmacokinetic ineraction.	Gurley *et al.*, 2005; Jiang *et al.*, 2005.

Contd...

Table 1.1–*Contd...*

Evaluated Herb	*Drugs*	*Possible Mechanism of Interaction*	*References*
Ginseng	Nifedipine	Increase of plasma concentration.The mechanism is unknown.	Smith *et al.*, 2001
	Warfarin	None with use of *P. ginseng.*Decrease of plasma concentration by induction of CYP2C9 with use of American ginseng.	Lee *et al.*, 2008; Garley *et al.*, 2005; Yuan *et al.*, 2004
St. John's wort	Atorvastatin	Decrease of plasma concentration.The mechanism was not reported bythe authors.	Andrén *et al.*, 2011
	Clopidogrel	Increase of plasma concentration ofthe active metabolite by induction of CYP3A4.	Chen *et al.*, 2011
	Digoxin	Decrease of plasma concentration by induction of P-gp.	Anonymous, 2004; Gurley *et al.*, 2008
	Ivabridine	Decrease of plasma concentration by induction of CYP3A4.	Portolés *et al.*, 2006
	Nifedipine	Decrease of plasma concentration by induction of CYP3A4.	Tannergren *et al.*, 2004; Wang *et al.*, 2009
	Phenprocoumon	Decrease of plasma concentration by induction of CYP3A4.	Borrelli and Izzo, 2009
	Prasugrel	Increase of plasma concentration ofthe active metabolite by inducion of CYP3A4.	Chen *et al.*, 2011
	Pravastatin	None	Sugimoto *et al.*, 2001
	Simvastatin	Decrease of plasma concentration by induction of CYP3A4.	Sugimoto *et al.*, 2001
	Talinolol	Increase of clearance by indution of P-gp.	Scharz *et al.*, 2007
	Ticlopidine	Increase of plasma concentration ofthe active metabolite by inducion of CYP3A4.	Chen *et al.*, 2011
	Verapamil	Decrease of plasma concentration by induction of CYP3A4.	Tannergren *et al.*, 2004; Wang *et al.*, 2009
	Warfarin	Decrease of plasma concentration by induction of CYP3A4, CYP1A2 and CYP2C9.	Jiang *et al.*, 2004

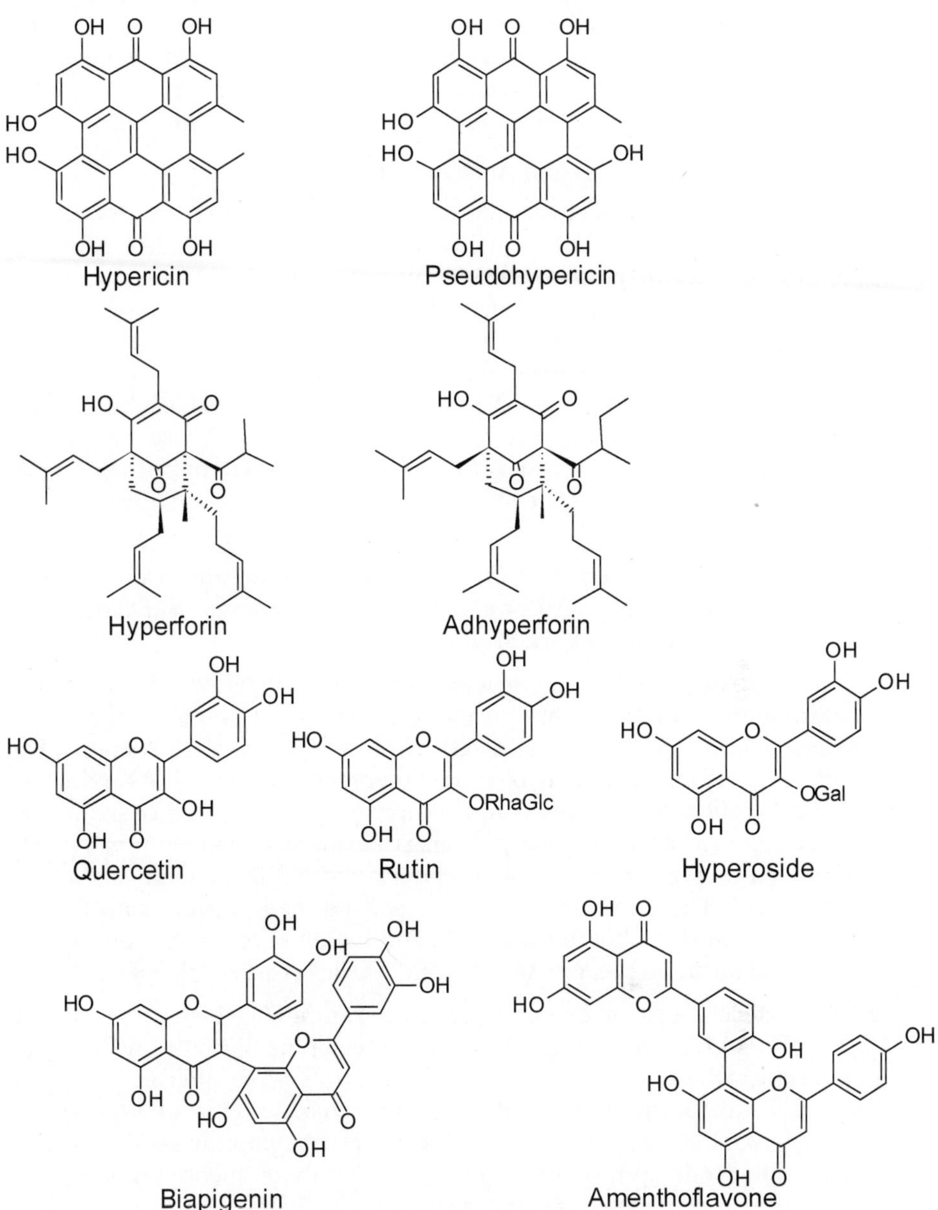

Figure 1.1: Structures of some Components Present in *H. perforatum*.

H. perforatum has been shown to induce P-gp expression in intestinal isolated cells as well as in the human intestine in healthy volunteers which suggest an active role in drug elimination and absorption. This fact leads to decrease plasma concentration of some P-gp substrates, such as digoxin, talinol, among much more drugs after long treatment (more than ten days) (Borrelli and Izzo, 2009). Co-administration of St. John's wort and therapeutic drugs for lesser number of days is generally scarce or even absent according to the same authors whom made a revision about this subject. Hyperforin seems to be the potent CYP3A4 and P-gp transporter inducer, causing reduced oral bioavailability and increased clearance of co-administered drugs which are substrates of CYP3A4 and P-gp (Chen *et al.*, 2011).

5.1.1 *Hypericum perforatum* and Anticoagulants

St. John's wort induces the metabolism of both warfarin enantiomers (Figure 1.2), which are mediated by induction of CYP1A2, CYP2C9, and CYP3A4, resulting in a significant reduction in the pharmacological effect of this drug (Jiang *et al.*, 2004). Nevertheless, one case of severe bleeding diathesis, manifested by hematemesis and melena was reported by Bayramiçli *et al.* (2011). The authors attributed such unexpected effect to one or more combinations of active compounds present in the extract of *H. perforatum* (hyperforin, flavonols, flavonol glycosides, biflavones, naphthodianthrones, acylphloroglucinols, and phenylpropanes). These components that may affect the metabolism of warfarin in sensitive individuals by potentiating its effect on the clotting cascade, in spite of the utilization of standard extracts (Obach, 2000).

St. John's wort by induction of CYP3A4 and possibly other isoforms also decreased plasma levels of phenprocoumon (Figure 1.2) (a coumarin anticoagulant chemically related to warfarin) (Borrelli and Izzo, 2009).

The pharmacokinetics of hyperforin is age-dependent. Hyperforin clearance may be reduced in elderly subjects probably resulting from reduced liver blood flow and/ or diminished liver weight and not necessarily from alterations of CYP activity. For this reason, elderly consumers may be more vulnerable to the CYP3A4 inductive effects of *H. perforatum*, because small doses of hyperforin may produce significant induction of CYP3A4 activity. The great problem is that the St John's wort supplements are standardized to hypericin and this compound did not present any ability for inducing CYP3A4. These standardized supplements may present considerable variability with regard to hyperforin content and therefore render St. Jonh's wort extracts as potent inducers of CYP3A4 in elderly (Gurley *et al.*, 2005).

The anti-platelet drug clopidogrel is a prodrug (Figure 1.2) which is metabolized by CYP2C9, CYP3A4/5 and CYP2B6 to its active metabolite. With the induction of CYP3A4 through the co-administration of St. John's wort, there is a higher biotransformation of the prodrug with the consequent release of the active drug. The herbal drug is also able to enhance the inhibitory effect of ticlopidine and prasugrel (Figure 1.2), both prodrugs belonging to the same family of thienopyridine ADP-receptor antagonists on platelet aggregation (Chen *et al.*, 2011).

5.1.2. *Hypericum perforatum* and Statins

St. John's wort significantly increased serum level of LDL cholesterol in patients treated with a stable dose of atorvastatin (Figure 1.3) compared with control and increased the total cholesterol level (Andrén *et al.*, 2007). The mechanism involved in these clinical results was not reported by the authors.

St. John's wort decreased plasma concentration of simvastatin (Figure 1.3), probably to the fact that this statin is metabolized by CYP3A4 in the intestinal wall

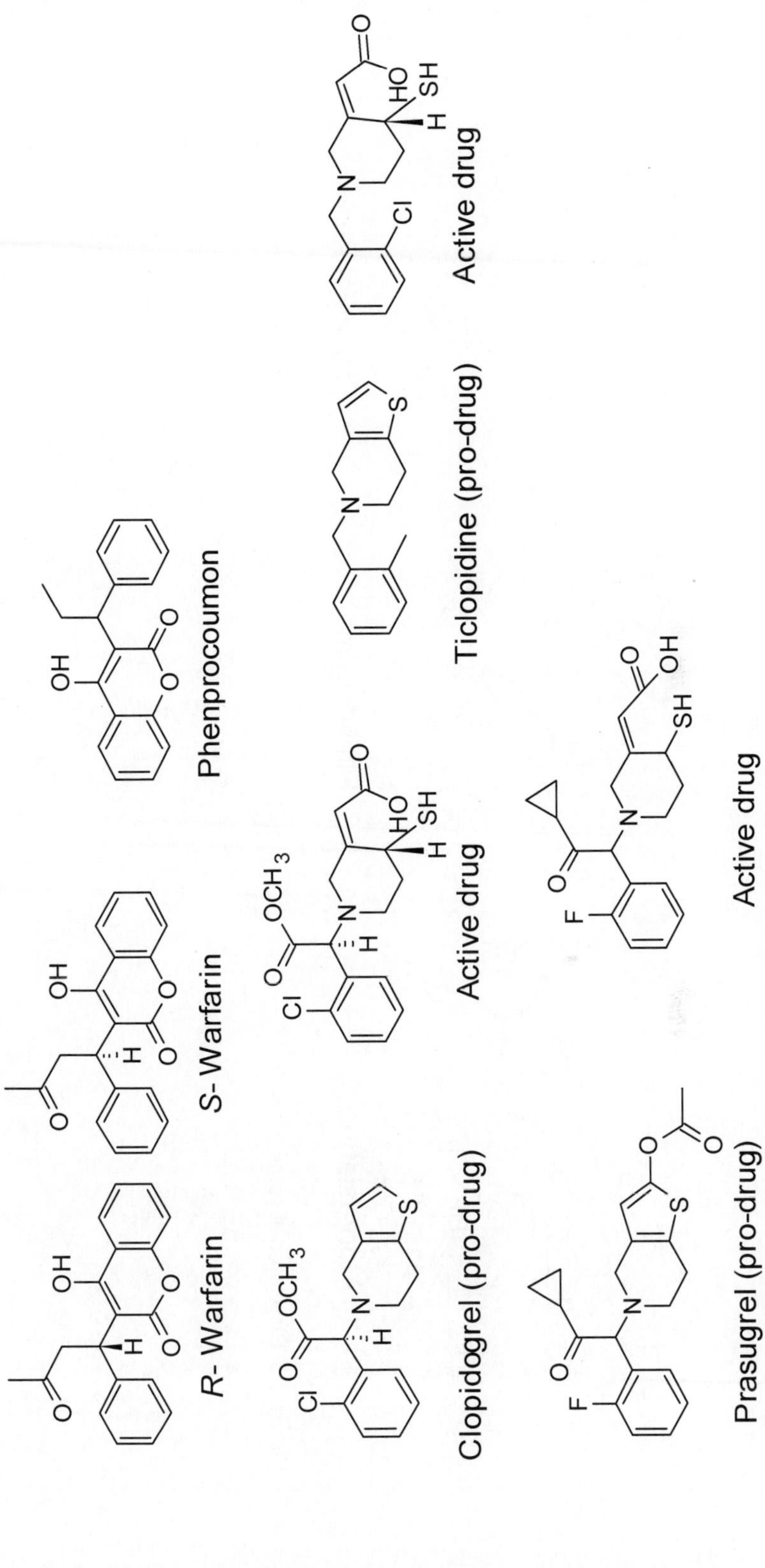

R- Warfarin

S- Warfarin

Phenprocoumon

Clopidogrel (pro-drug)

Active drug

Ticlopidine (pro-drug)

Active drug

Prasugrel (pro-drug)

Active drug

Figure 1.2: Structures of some Anticoagulants/Antiplatelets.

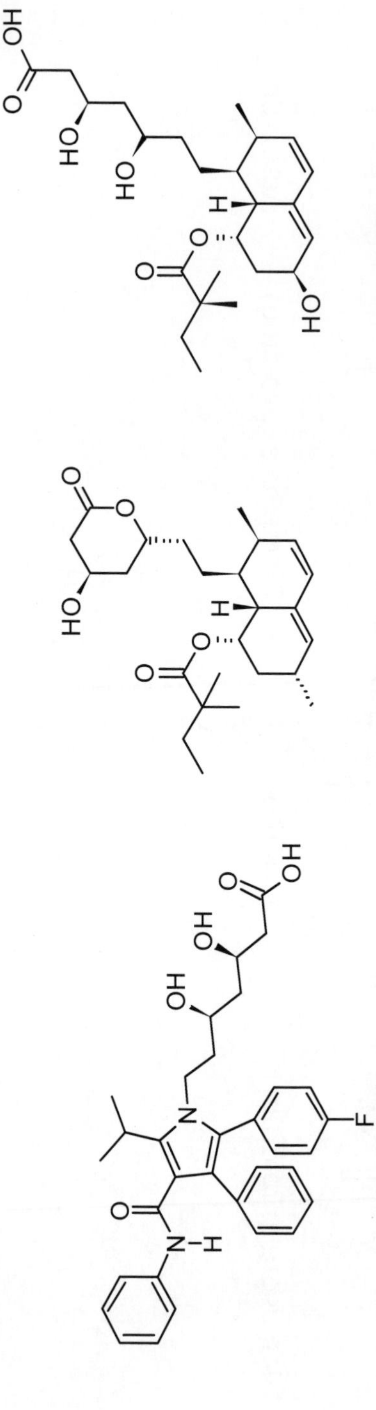

Atorvastatine

Simvastatin

Pravastatine

Figure 1.3: Structures of some Statins.

and liver, which is induced by St. John's wort (Sugimoto *et al.*, 2001). Pravastatin (Figure 1.3) did not decrease because it is not a substrate for CYP3A4.

5.1.3 *Hypericum perforatum* and β-blockers

Talinolol (Figure 1.4) is a β1-adrenergic blocker used in essential hypertension and also as an antiarrhythmic drug. The clearance of this drug increased when associated with St. John's wort. An increased MDR1 messenger ribonucleic acid (mRNA) as well as P-gp in the duodenal mucosa was observed due to the inductive effect of that herbal drug on intestinal P-gp (Scharz *et al.*, 2007).

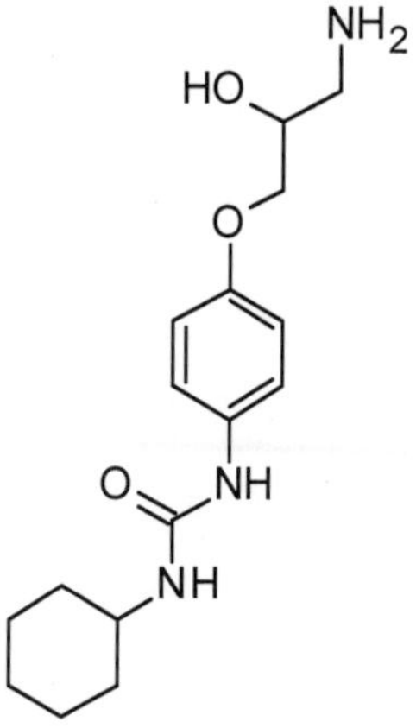

5.1.4 *Hypericum perforatum* and Calcium Channel Blockers

Nifedipine and verapamil (Figure 1.5) are calcium blockers used in the treatment of hypertension, among other indications. Both are metabolized by CYP3A4, which is induced by *H. perforatum*. For this reason, there is an expected decrease of bioavailability of verapamil enantiomers and nifedipine (Tannergren *et al.*, 2004). Nevertheless, some authors showed that different haplotype of pregnant X receptor (*PXR*) (steroid and xenobiotic receptor and pregnane-activated receptor) would result in PXR variant proteins with distinct binding properties to *CYP3A4* gene and consequently having different inducing ability by *H. perforatum* (Wang *et al.*, 2009).

Tanilolol

Figure 1.4: Structure of Tanilolol.

5.1.5 *Hypericum perforatum* and Antianginal Drugs

Ivabridine (Figure 1..6) is an antianginal drug and is metabolized by intestinal and hepatic CYP3A4 which is inducible by *H. perforatum*. This effect is responsible for the lower plasma concentration of this antianginal drug, which can fall up to half (Portolés *et al.*, 2006).

5.1.6 *Hypericum perforatum* and Cardiac Ionotropic Drugs

Digoxin (Figure 1.7) used in the treatment of heart failure has a narrow therapeutic index and is a substrate for P-gp. Study performed by several enterprises interested in the area of St. John's wort demonstrated that the 10-day coadministration of digoxin and *H. perforatum* reduced the levels of digoxin in plasma due to an intestinal induction of P-gp by the *H. perforatum* (Anonymous, 2004). In spite of this result, and according to the same authors, no therapeutic interactions between St. John's wort and digoxin have been reported to date. The interaction of St John's wort and digoxin varies within St John's wort preparations and doses and seems to be correlated with the dose, particularly of hyperforin (Anonymous, 2004).

The effects of St. John's wort on the pharmacokinetic of digoxin in healthy young adults, a recognized P-gp substrate, were studied by Gurley *et al.* (2008). These authors found that St. John's wort products containing sufficient amount of hyperforin were equivalent to a positive control used in the assay, rifampin, in terms of P-gp induction.

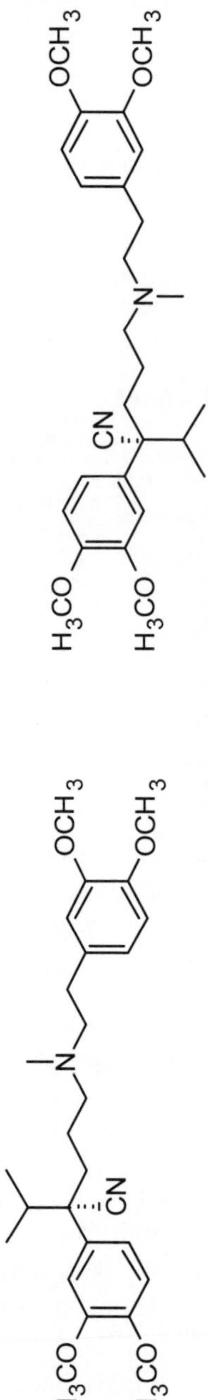

Varapamil enantiomers

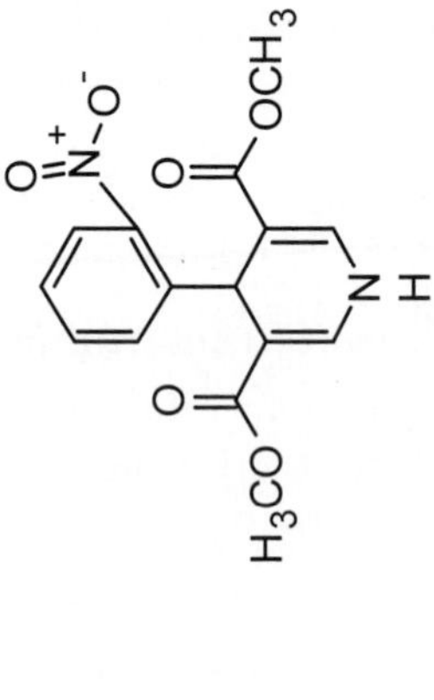

Nifedipin

Figure 1.5: Structures of some Calcium Channel Blockers.

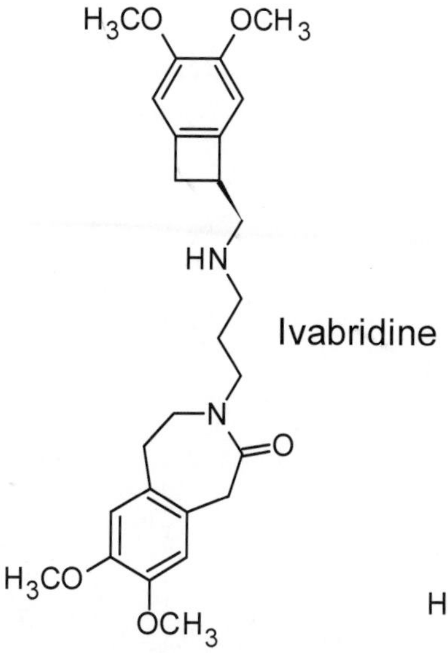

Figure 1.6: Structure of Ivabridine.

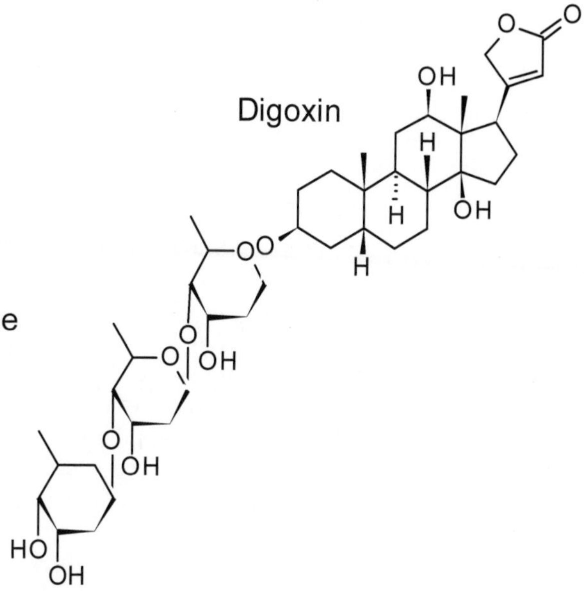

Figure 1.7: Structure of Digoxin.

Through this mechanism, St. John's wort may reduce efficacy of digoxin and make a patient a no responder, whereas increased toxicity may be anticipated after withdrawal of the herb (Izzo *et al.*, 2005).

5.2. *Ginkgo biloba*

Ginkgo biloba (ginkgo) is most commonly used to treat dementia, for memory improvement, cerebrovascular and arterial insufficiency, tinnitus, vertigo, asthma, allergies and many other diseases (Posadzki *et al.*, 2012). These properties of *Ginkgo biloba* extract (GbE) have been attributed to flavonoids (quercetin, kaempferol and isorhamnetin) (Figure 1.8) and terpene lactones (ginkgolides and bilobalide) (Figure 1.8) which have been reported as being able to remove free radicals, protect the endothelial cells of blood vessels, block platelet activating factors and improve brain circulation (Jiang *et al.*, 2013). A well defined extract, GbE 761 is produced from leaves which contain 24 per cent w/w (weight of active compounds/total weight of extract) flavone glycosides and 6 per cent w/w terpene lactones. This extract is marketed as a food supplement (Birks and Grimley, 2009). According to a review made by these authors they concluded that GbE 761 appears to be safe in use but evidence that the extract as clinically significant benefit for people with dementia or cognitive impairment was inconsistent and unreliable.

Other studies demonstrated that GbE, namely ginkgolide A, was able to activate human PXR and capable of increasing the expression of P-gp. These findings permit to suggest that such extract has the capacity for enhancing the clearance of brain β-amyloid in the Alzheimer's disease (Lau *et al.*, 2010).

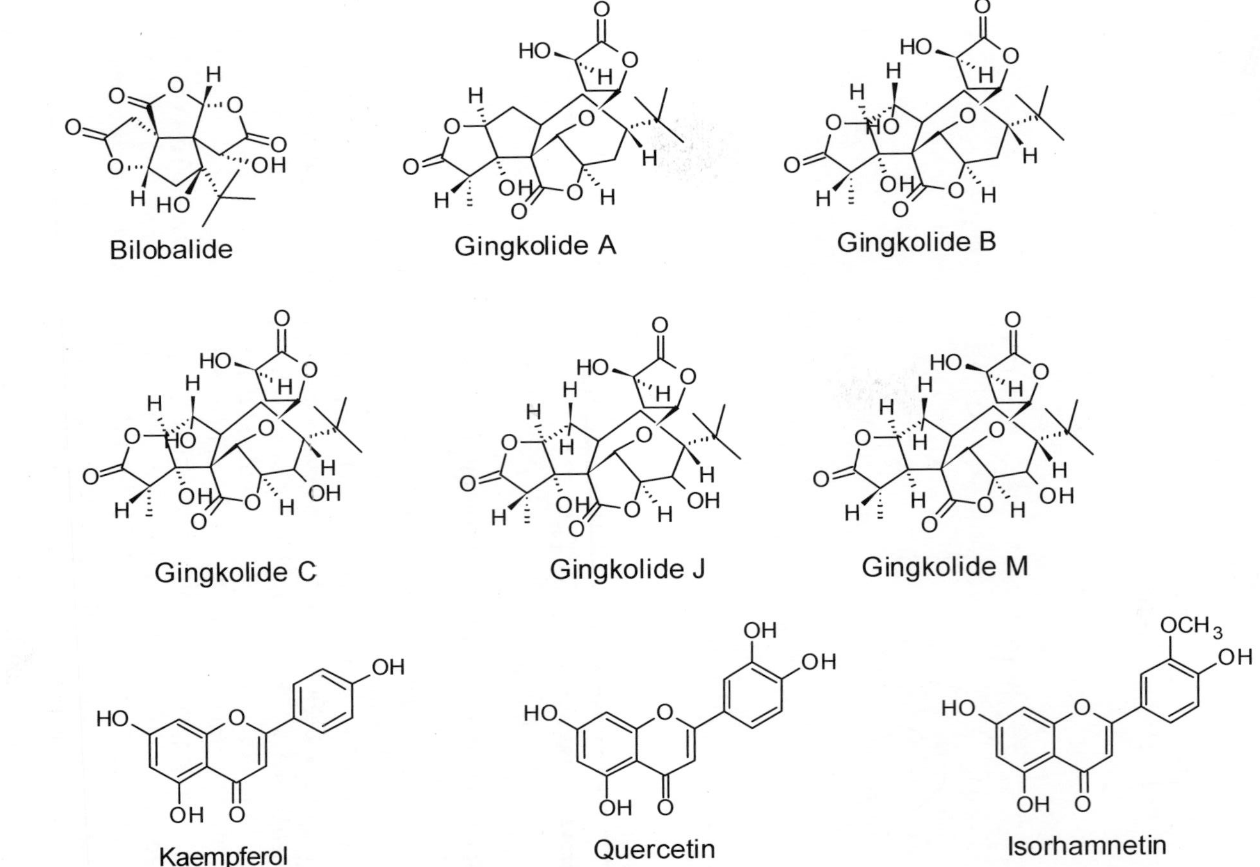

Figure 1.8: Some Structures of Gingkolides, Bilobalide and Flavonoids Present in Gingko.

5.2.1. *Ginkgo biloba* and Anticoagulants

GbE has been found to increase bleeding risks according to some authors, when administered concomitantly with anticoagulants or antiplatelet drugs (Cordier and Steenkamp, 2010). Nevertheless, an open-label, randomized, crossover study demonstrated that there was no difference in bleeding times or platelet aggregation between ticlopidine (Figure 1.2) and ticlopidine/GbE treatment groups (Table 1.1) (Kim *et al.*, 2010).

The antiplatelet effect can be potentiated with the association of GbE and cilostazol without prolongation of the bleeding time or coagulation time. The authors even consider that this combination could be a potential new antithrombotic therapy (Ryu *et al.*, 2009).

A possible interaction between GbE and warfarin has been reported, particularly relevant to elderly patients on anticoagulant therapy. This interaction seems to be attributed to the inhibition of platelet activating factor by diverse ginkgolides and not the result of phytochemical-mediated effects on CYP-isoforms (Gurley *et al.*, 2005). In another study, Jiang *et al.* (2005) reported that recommended doses of GbE did not affect the apparent clearance of warfarin enantiomers, which could mean that GbE did not significantly influence CYP1A2, CYP3A4 and CYP2C9. Recently, and in a review article, Wanwimolruk and Prachayaasittikul (2014) reported the evidence indicating that, at commonly recommended doses, *Ginkgo biloba* and other selected herbs including echinacea, garlic, goldenseal and milk thistle do not act as potent or moderate inhibitors or inducers of CYP enzymes.

5.2.2. *Ginkgo biloba* and β-Blockers

A single oral dose of GbE did not affect the pharmacokinetics of talinolol (Figure 1.4) in healthy Chinese volunteers; nevertheless repeated ingestion of GbE increased the talinolol maximum plasma concentration probably by inhibition of P-gp-mediated drug efflux and/or other drug transporters (Fan *et al.*, 2009a,b).

5.2.3. *Ginkgo biloba* and Cardiac Ionotropic Drugs

The concomitant use of GbE and digoxin (Figure 1.7) did not appear to have any significant effect on the pharmacokinetics of orally administered digoxin in healthy volunteers according to a study developed by Mauro *et al.* (2003).

5.2.4. *Ginkgo biloba* and Statins

In some patients with Alzheimer's disease, GbE and statins are frequently used at the same time. GbE when co-administered with atorvastatin slightly decreased the plasma atorvastatin (Figure 1.3) concentrations and did not affect atorvastatin cholesterol-lowering efficacy. In the same study, the authors found that GbE may only have a little effect on CYP3A4 *in vivo* as well as in UDP-glucuronosyl-transferase1A1 and UGT1A3, which mediate interconversion from the acid to the lactone form of atorvastatin (Guo *et al.*, 2012).

5.2.5. *Ginkgo biloba* and Calcium-Channel Blockers

GbE with nifedipine (calcium-channel blocker used in hypertension and angina) co-admininstered in healthy volunteers did not significantly affect the pharmacokinetic parameters of nifedipine and its metabolite, despite in two adults the maximal plasma nifedipine concentrations were approximately doubled by GbE and also presenting severe and longer-lasting headaches with GbE than without this extract (Yoshioka *et al.*, 2004). In addition, the authors reported that in general the mean value for heart rate after oral administration of nifedipine and GbE tended to be faster than that without the GbE supplementation. For these observations, the authors considered the existence of a pharmacodynamic interaction and advice that GbE and nifedipine or other similar calcium-channel blockers should not be at concomitantly ingested as much as possible.

Nevertheless in previous studies made by Ohnishi *et al.* (2003) in rats, the authors reported a pharmacokinetic interaction between GbE and other calcium channel blocker (diltiazem) (Figure 1.9), which is a typical probe of CYP3A4. Such conclusion was based on their results in which the concomitant use of GbE in rats increased the bioavailability of diltiazem by inhibiting both intestinal and hepatic metabolism, at least in part by inhibition of CYP3A4. In contrast, Shinozuka *et al.* (2002) reported that GbE induced CYP3A2, attenuating therefore the hypotensive effects of nicardipine (Figure 1.9) in rats.

GbE administered to rats for 4 weeks concomitantly with the calcium channel blocker, nicardipine, significantly reduced the hypotensive effect of this drug that was reported to be metabolized by CYP3A2. The authors suggested that GbE reduced the therapeutic potency of nicardipine through the enhancement of cytochrome P-450 expression (Shinozuka *et al.*, 2002).

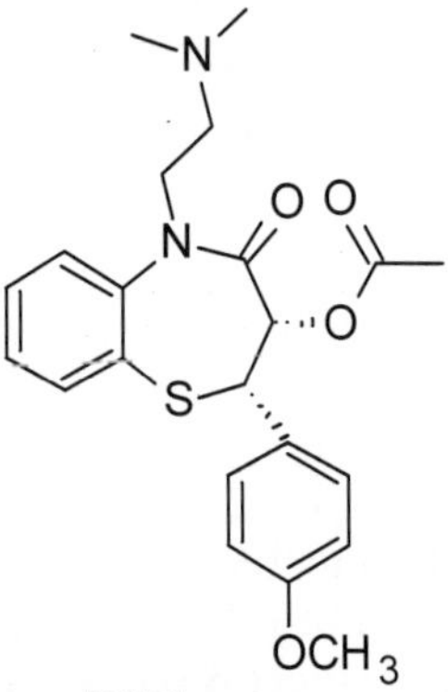

Diltiazem

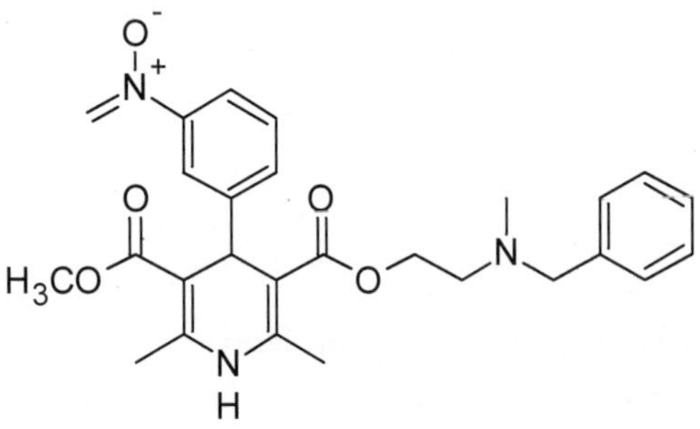

Nicardipine

Figure 1.9: Structures of Diltiazem and Nicardipine.

5.2.6. *Ginkgo biloba* and Diuretics

Case reports (one elderly patient) have demonstrated that ginkgo could elevate blood pressure when combined with a thiazide diuretic (Chen *et al.*, 2011). Metabolic inhibition of the thiazide diuretic by ginkgo is a possible mechanism (Hu *et al.*, 2005).

5.2.7. *Ginkgo biloba* and others Therapeutic Drugs

Some interactions between GbE and other therapeutic drugs with controversial pharmacokinetics and pharmacodynamics parameters found in diverse works are compiled in several review articles (Miller, 1998; Izzo and Ernst, 2001; Izzo, 2004; Izzo *et al.*, 2005; Yang *et al.*, 2006; Zhou *et al.*, 2007; Gardiner *et al.*, 2008; Hellum *et al.*, 2008; Gardiner *et al.*, 2008; Chen *et al.*, 2011; Na *et al.*, 2011; Vieira and Huang, 2012; Tsai *et al.*, 2012). In such works the general conclusions were always the necessity to confirm the clinical impact of the combinations and the risk versus benefit profiles should be thoroughly considered prior to initiating new therapies.

5.3. Ginsengs

Panax ginseng CA Mayer (commonly referred to as Asian, Chinese or Korean ginseng) and *P. quinquefolius* L (American or North American ginseng) are the two most recognized ginseng herbs worldwide. There are also *Panax notoginseng* (Burk.) F. H. Chen, *Panax japonicus* C.A. Meyer (Japanese ginseng) and *Panax vietnamensis* Ha et Grushv (Vietnamese ginseng) but less known (Qi *et al.*, 2011a,b). Ginsenosides or ginseng saponins are the main components responsible for the biological properties of roots and rhizomes of *P. ginseng* after 3-6 years of cultivation, although other components can also show pharmacological properties such as polysaccharides, peptides, polyacetylenic alcohols and fatty acids (Attelle *et al.*, 1999). Ginsenosides (except only one) belong to a family of steroids named steroidal saponins, but also named ginsenoside saponins, triterpenoid saponins, or dammarane derivatives (Attelle *et al.*, 1999; Wang *et al.*, 2005).

Ginsenosides (Figure 1.10) are generally divided in two groups, namely the protopanaxadiol and protopanaxatriol ginsenoside groups. The chemical structures of ginsenosides are different from each other in the number, linkage position and type of sugar (Peng *et al.*, 2012). Protopanaxadiol and protopanaxatriol ginsenoside groups may present different bioactivities, even opposite effects (Kang *et al.*, 1995).

The quantity and quality of ginsenosides in ginseng may vary depending on the species, genetics, age, part of plant, season of harvest, method of cultivation, and means of preservation (Lim *et al.*, 2005; Wang *et al.*, 2005; Lee *et al.*, 2011; Peng *et al.*, 2012).

Ginseng is an adaptogen that purportedly increases humans' resistance to physical, chemical, and biological stress and builds up general vitality, physical, and mental capacity for work (Shao and Jia, 2013).

Also have been reported antioxidant, anti-inflammatory, and immuno-stimulatory activities for Asian and American ginseng. Ginseng has been reported as possessing beneficial properties on central nervous system disorders; aging; neurodegenerative diseases; some cardiovascular diseases, acting as anti-ischemic, anti-arrhythmic and anti-hypertensive; antidiabetic, namely of type 2; anticancer chemoprevention and inhibition of tumour growth (Jia *et al.*, 2009; Qi *et al.*, 2011c).

The most common side effects of ginseng are nervousness and excitability when administered at relative high doses. In inadequate uses, the most common side-effects

Ginsenosides

Ginsenoside	R_1	R_2	R_3
Ra	glc(2-1)glc	H	glc(6-1)glc
Rb1	glc(2-1)glc	H	glc(6-1)glc
Rb2	glc(2-1)glc	H	glc(6-1)ara(p)
Rc	glc(2-1)glc	H	glc(6-1)ara(f)
Rd	glc(2-1)glc	H	glc
Re	H	–O-glc(2-1)rha	glc
Rf	H	–O-glc(2-1)glc	H
Rg1	H	–O-glc	glc
Rg2	H	–O-glc(2-1)rha	H
Rg3	glc(2-1)glc	H	H
Rh1	H	–O-glc	H
Rh2	glc	H	H

Figure 1.10: Structures of Ginsenosides in Ginseng.

are the inability to sleep and high blood pressure, nausea, diarrhoea, euphoria, headache, epistaxis, mastalgia, eruptions and vaginal bleeding (Jia *et al.*, 2009).

The pharmacokinetic profile of ginseng is not well understood owing to the heterogeneous chemical structures of different ginsenosides (Qi *et al.*, 2011a).

Studies made by Yu *et al.* (2005) demonstrated that there was no evidence of the induction of CYP2B1, CYP3A23, or CYP1A2 gene expression by *P. ginseng* and *P. quinquefolius* extracts in primary cultures of rat hepatocytes and adult rats. In other studies ginseng and ginkgo did not show any effect on CYP3A4, CYP1A2, CYP2E1 and CYP2D6 of young healthy volunteers (Gurley *et al.*, 2002), nevertheless in elderly persons, CYP2D was slightly inhibited (Gurley *et al.*, 2005).

5.3.1. Ginsengs and Anticoagulants

In a randomized, open-label, controlled study investigating the interaction between warfarin and *P. ginseng* by observing the prothrombin time and the international normalized ratio in ischemic stroke patients who did not have a history of taking warfarin did not influence the pharmacologic action of this therapeutic drug (Table 1.1) (Lee *et al.*, 2008).

Studies in young volunteers indicate that *P. ginseng* effect on CYP activity is minimal, and therefore ginseng supplementations did not present significant interactions with drugs that are substrates for CYP1A2, CYP2D6, CYP2E1, or CYP3A4 (Garley *et al.*, 2005). However, some other CYP isoforms may be influenced by ginsengs. For example, some authors demonstrated that American ginseng reduced warfarin plasma concentrations in young, healthy volunteers (Yuan *et al.*, 2004). Being CYP2C9 the principal isoform involved in the biotransformation of the active S-warfarin, the results reported by those authors suggest a ginseng-mediated induction of this enzyme (Garley *et al.*, 2005; Greenblatt and Moltke, 2005).

According to the clinical evidences which were compiled by Shi and Klotz (2012) suggest that the potential for Asian and American ginsengs-drug interactions is low.

5.3.2. Ginsengs and Calcium Channel Blockers

Smith *et al.* (2001) in a clinical trial verified an increase of levels of nifedipine in plasma of those taking concomitantly gingseng. As a consequence, there was an increase of side effects of nifedipine, such as headache, constipation and ankles oedema. The mechanism for this fact is unknown.

5.4. *Allium sativum* (Garlic)

Garlic is widely used around the world for its pungent flavour as a seasoning or condiment. In addition, garlic is one of the most sold herbal supplement in developed countries due to the number of its clinically established pharmacological properties – antihypertensive, hypolipidemic, antiatherogenic, antitumorigenic, fibrinolytic, anti-carcinogenic, immunomodulatory, antimicrobial and hypoglycemic activities (Wanwimolruk and Prachayaasittikul, 2014). Garlic has been reported as possessing antiplatelet activity, either *in vitro* studies (Apitz-Cástro *et al.*, 1983; Srivastava, 1986) and in healthy humans and patients with coronary artery disease, being lower doses more effective in long-term periods of administration instead high doses in one single dose (Bordia *et al.*, 1996).

Whole garlic typically contains alliin (Figure 1.11), together with (1)-*S*-methyl-L-cysteine sulfoxide (methiin) and (1)-*S*-(*trans*-1-propenyl)-L-cysteine sulfoxide. *S*-(2-Carboxypropyl)glutathione, γ-glutamyl-*S*-allyl-Lcysteine, γ-glutamyl-*S*-(*trans*-1-propenyl)-L-cysteine and γ-glutamyl-*S*-allyl-mercapto-L-cysteine are also present in garlic cloves. In crushed garlic and garlic essential oil were identified sulfide, diallyl disulfide, diallyl trisulfide, methyl allyl disulfide, methyl allyl trisulfide, 2-vinyl-1,3-dithiin, 3-vinyl-1,2-dithiin (Fenwick and Hanley 1985) and *E*,*Z*-ajoene (Block *et al.*, 1984). According to these authors, *E*,*Z*-ajoene possesses antithrombotic activity. Cutting, crushing, chewing or dehydration, alliin is rapidly lised by the vacuolar enzyme, alliinase, in alkyl alkane thiosufinates such as the intermediate allicin. In turn, these alkyl alkane thiosufinates including allicin are rapidly decomposed in those compounds reported above which have been described as beneficial in human health (Amagase *et al.*, 2001).

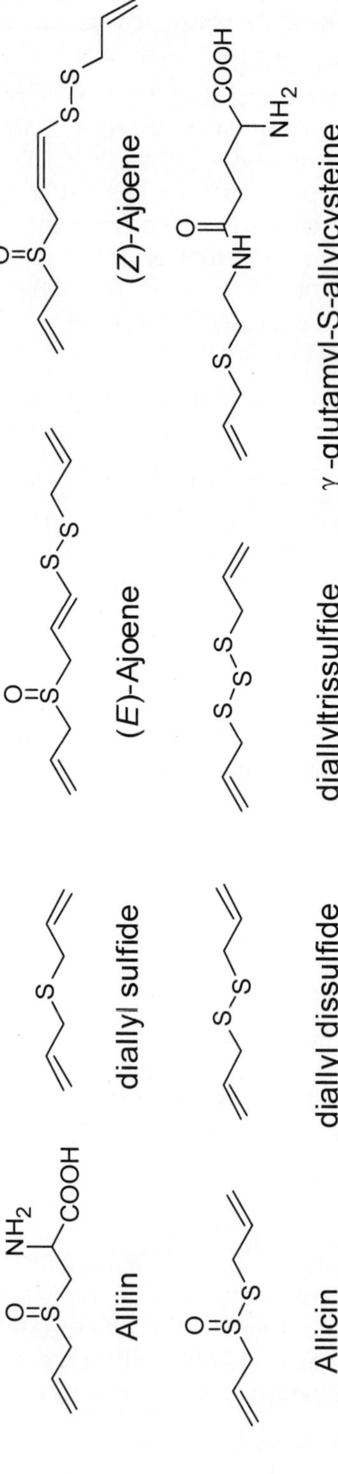

Figure 1.11: Structures of some Components of Garlic.

Srivastava (1986) found that garlic reduced the formation of thromboxane from exogenous arachidonate; inhibited the phospholipase activity; inhibited the formation of thromboxane and lipoxygenase products formed in platelets prelabelled with arachidonate; and garlic inhibited the incorporation of arachidonate into platelet phospholipids. These effects may explain, in part, inhibition of platelet aggregation. The same authors also reported that garlic was also effective in inhibiting aggregation induced by calcium ionophore A23187 which effect may be related to intraplatelet mobilization of calcium. However, other in vitro studies showed that platelets incubated with the inhibitors after incorporation of radiolabeled arachidonate, although did not aggregate, produced, after thrombin activation, similar amounts of radiolabeled thromboxane X2 and lipoxygenase products as the controls. Electron microscopy of inhibited platelets, in the presence of thrombin, showed no degranulation but an increase of spherical forms. These findings suggest that the effects found might be mediate by a perturbation of the physico-chemical properties of the plasma membrane rather than by affecting arachidonate or calcium metabolism in the cells (Apitz-Cástro *et al.*, 1983). Chemical structures of active components could be diallytrisulfide, 2-vinyl-1,3-dithiene, and most probably allyl 1,5-hexadienyltrisulfide (Apitz-Cástro *et al.*, 1983).

Diallytrisulfide and diallydisulfide also showed antiplatelet activity and inhibited platelet thromboxane formation in vitro studies carried out by Bordia *et al.* (1998). The same authors, in in vivo studies found that garlic, administered in a daily dose of 2 x 2 capsules (each capsule containing ethyl acetate extract from 1 g peeled and crushed raw garlic) reduced significantly total serum cholesterol and triglycerides, and increased significantly HDL-cholesterol and fibrinolytic activity (Bordia *et al.*, 1998). Other authors proposed that adenosine was the component responsible for the antiplatelet activity of garlic (Makheja and Mailey, 1990).

Other active components present in garlic have been described such as flavonoids, sapogenis and saponins, selenium compounds and fructosamines (Berginc and Kristl, 2012).

Dosages generally recommended in the literature for adults are 4 g (one to two cloves) of raw garlic per day, one 300 mg dried garlic powder tablet (standardized to 1.3 per cent allicin or 0.6 per cent allicin yield) two to three times per day, or 7.2 g of aged garlic extract per day (Stanger *et al.*, 2012).

Pharmaceutical products developed from garlic bulbs/cloves are made available in the forms of powders, oily preparations, or aqueous-alcoholic extracts of fresh or aged garlic. The preparations obtained by different manufacturing processes result in products of different chemical composition and pharmaceutical properties, and therefore with diverse safety, and bio-availability characteristics (Wanwimolruk and Prachayaasittikul, 2014).

Garlic preparations such as garlic oil among others have been shown to be efficient inhibitors of CYP2E1, and inducer of CYP1A, 2B and 3A. Garlic has also been reported as be able to induce some phase II enzymes such as UDP-glucuronosyltransferase, glutathione S-transferase, and NAD(P)H: quinine oxidoreductase (Guyonnet *et al.*, 1999; Le Bon *et al.*, 2003; Zhou *et al.*, 2003). However, the effects of garlic constituents on various CYP isoforms *in vivo* were dependent on the species (Zhou *et al.*, 2003).

5.4.1. Garlic and Anticoagulants

The true risks of these interactions and effects are difficult to characterize due to the limited number and nature of existing reports. In a review, Stanger *et al.* (2011) reported that garlic had the capacity for reducing ADP-induced platelet aggregation, thromboxane and clotting time, and therefore, bleeding has been associated with oral garlic use in several studies and case reports (Ulbricht *et al.*, 2008). Nevertheless, it appears that garlic supplements when taken at recommended doses who are not taking anticoagulant medication do not have anticoagulant properties. Ulbricht *et al.* (2008) reported from a cross-sectional, point-of-care survey of almost 2000 patients, 25 cases were identified as potential clinically significant interactions between garlic and anticoagulants, although some authors consider that diverse reports contain inadequate information (Izzo *et al.*, 2005). Nevertheless, those authors suggest that garlic should be administered cautiously in patients taking warfarin and other anticoagulants.

Some authors suggest that garlic supplements should be discontinued about 10 days before elective surgical procedures, manly in those taking acetylsalicylic acid or warfarin (Table 1.1) (German *et al.*, 1995; Tachjian *et al.*, 2010).

Studies on the genetic variability in vitamin K epoxide reductase subunit 1 gene (*VKORC1*) have demonstrated to be involved in warfarin resistance. According to Abdul *et al.* (2008) co-administration of garlic did not significantly alter warfarin pharmacokinetics or pharmacodynamics, nevertheless subjects who carry the *VKORC1* variant type (CT and TT alleles) were more prone to interactions with warfarin, requiring a significantly lower warfarin dose compared with people with wild-type (CC allele). However, the same authors also consider some limitations in their work because a relative small sample size for each genotype was used.

Co-administration of aged garlic extract and cilostazol (anti-platelet) (Figure 1.12) did not enhance the anti-platelet activity compared to individual drugs, nevertheless the authors considered that further studies must be carried out to evaluate the possible interaction of garlic in higher doses and in combination with other anti-platelet activity drugs (Mateen *et al.*, 2011).

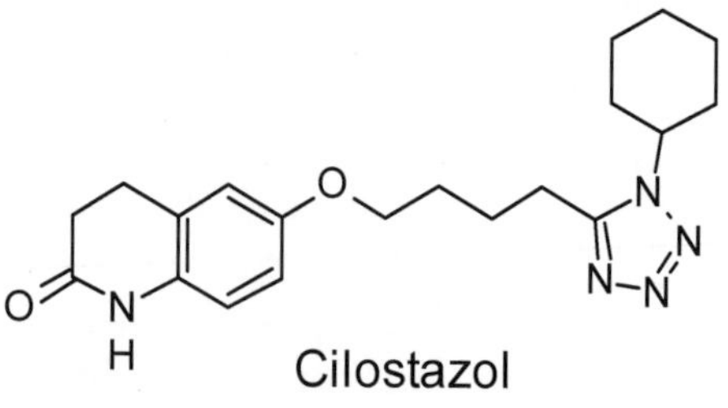

Figure 1.12: Structure of Cilostazol.

5.4.2. Garlic and Statins

Studies regarding the pharmacokinetic interactions of garlic with atorvastatin in dyslipidemic *Sprague Dawley* rats demonstrated that it increased the concentration of atorvastatin and other pharmacokinetic parameters, also reducing the elimination rate constant, suggesting a synergistic pharmacokinetic profile (Reddy *et al.*, 2012). The authors attributed these results to the possible inhibitory effect of garlic on CYP3A4 and P-P-gp, which may increase oral availability of atorvastatin due to inhibition of first-pass metabolism. In fact, previous studies *in vitro* made on human cytochrome P450 had already reported that diverse types of garlic samples (fresh extracts, freeze dried) were able to inhibit CYP2C19, 3A4, 3A5 and 3A7 and very low to moderate P-gp interaction when compared to the positive control (Foster *et al.*, 2001). In contrast, Markowitz *et al.* (2003) in other studies made in healthy volunteers reported that CYP3A4 were induced by garlic, which resulted in clinically important decreases in concentrations of drugs metabolized by this enzyme, such as lovastatin, simvastatin and atorvastatin. Allicin was reported as being responsible for this capacity (Markowitz *et al.*, 2003; Thomas *et al.*, 2012). However, the experiments have demonstrated that the modulation of CYP enzyme activity and expression are highly dependent on the type and chemical structure of garlic constituents, dose regimen, animal species and tissue, and source of garlic (dosing regimen of garlic constituents appeared to influence the modulation of CYP isoforms (Zhou *et al.*, 2003; Gurley *et al.*, 2005).

Reddy *et al.* (2010) have also reported that the use of atorvastatin at high doses along with garlic resulted in renal damage of dyslipidaemic *Sprague dawley* rats. According to Reddy *et al.* (2012), such results may be attributed to the fact that diallyl sulphide, present in garlic, has been reported to decrease the mouse hepatic Organic Anion Transport Proteins (OATP) expression. OATPs are reported to be responsible for transport of large number of endogenous and xenobiotic compounds across cell membranes (Cheng *et al.*, 2005). OATP1B transporters are responsible for major hepatic uptake systems for atorvastatin and its metabolites (Reddy *et al.*, 2012).

5.4.3. Garlic and β-blockers

Pharmacokinetic and pharmacodynamic interactions of garlic preparations and propranolol (Figure 1.13) in rats submitted to chronic prophylactic administration of garlic homogenate were evaluated by Asdaq and Inamdar (2011). They reported that garlic could cause increase of bioavailability and half life along with decrease in the clearance and elimination rate constant of propranolol when taken orally. These findings may have a negative implication due to the possible propranolol accumulation and consequent toxicity. Nevertheless and the same authors also consider that addition of garlic in moderate doses may have beneficial effects during treatment of hypertension in animals with myocardial stress. The same conclusion was reported by Kshirsagar and Asdaq (2014) when their findings suggest that combined therapy of garlic aqueous extract and metoprolol (Figure 1.13) possess synergistic potential as cardioprotective remedy for combating myocardial stress in rats.

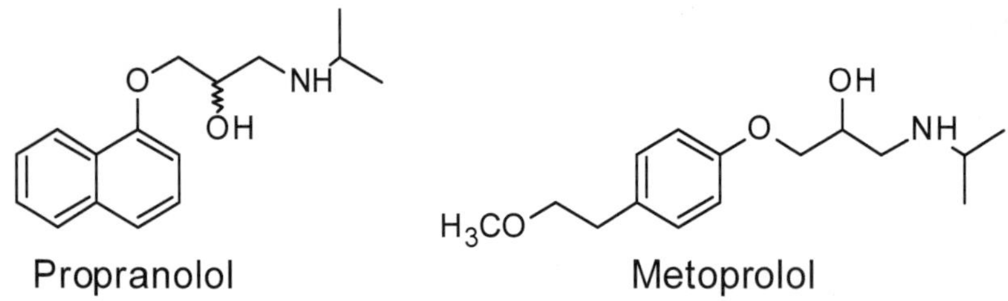

Propranolol

Metoprolol

Figure 1.13: Structures of Propranolol and Metoprolol.

Garlic and Calcium Channel Blockers

Allicin has been found to induce the enzyme CYP3A4 in humans, which may result in clinically important decreases in concentrations of drugs metabolized by this enzyme such as verapamil (Markowitz *et al.*, 2003; Thomas *et al.*, 2012). Garlic has been described as inhibitor of CYP3A5 and therefore with great probability to interfere with drugs, such as verapamil and nifedipine, which are also metabolized by the same CYP isoenzyme (Nowack *et al.*, 2008; Chan *et al.*, 2010).

Conclusion

Complexes mixtures with diverse concentrations of active compounds constitute the great majority of herbal drugs used as supplements. Even those which are standardised as for example, extracts of *H. perforatum*, or the well defined extract GbE 761, it is not yet known with certitude which are the active compounds for Alzheimer's disease. In addition, the action of the extracts is unpredictable, particularly those not standardized, as well as their pharmacokinetic and pharmacodynamic interactions with conventional medicines. These factors, along with the prevalence of consumption by the elderly who may be taking herb supplements concomitantly with conventional medicines with narrow therapeutic index, such as the case of digitalis medication, may be responsible for a very dangerous association between conventional and herbal drugs. This is also true when multiple drugs and multiple supplements are ingested simultaneously. A considerable attention over the last years has been given to herb-drug interactions with many scientific works developed by worldwide researchers but the review of literature shows that the available clinical evidence is still limited and sometimes inconclusive.

Moreover, the majority of patients who use herbal products do not reveal this fact to their physicians or pharmacists. Such fact increases the probability that herb-drug interaction occurs that, in turn, cannot be identified and resolved in a timely manner. For this reason it is advisable that these health professionals question their patients about the possible herb supplements dietary that they are taking and properly monitor the concomitant use of herbal drugs and conventional medicines or discourage this simultaneous administration.

Physicians, pharmacists, nurses and other health professionals have to be well prepared to deal with these issues whom in collaboration with researchers will contribute to the clarification of herb-drug interactions. With this collaboration, scientists in the field of drug development will be also able to design structures with a minimal potential interaction with herbs, having always in mind that any new drug which is identified to be CYP and/or P-gp substrates has a potential for interactions with many commonly used herbal medicines that can modulate CYP and P-gp expression and consequently activity (Yang *et al.*, 2006; Chen *et al.*, 2011).

References

Abdul, M.M., Jiang, X., Williams, K.M., Day, R.O., Roufogalis, B.D., Liauw, W.S., Xu, H., and Mclachlan, A.J. (2008). Pharmacodynamic interaction of warfarin with cranberry but not with garlic in healthy subjects. *British Journal of Pharmacology*, **154**: 1691-1700.

Amagase, H., Petesch, B.L., Matsuura, H., Kasuga, S., and Itakura, Y. (2001). Intake of garlic and its bioactive components. *Journal of Nutrition*, **131(Suppl)**: S955–S962.

Andrén, L., Andreasson, A., and Eggertsen, R. (2007). Interaction between a commercially available St. John's wort product (Movina) and atorvastatin in patients with hypercholesterolemia. *European Journal of Clinical Pharmacology*, **63**: 913-916.

Anonymous (2004). Effect of St John's wort dose and preparations on the pharmacokinetics of digoxin. *Clinical Pharmacology and Therapeutics*, **75**: 546-557.

Apitz-Cástro, R., Cabrera, S., Cruz, M.R., Ledezma, E., and Jain, M.K. (1983). Effects of garlic extrac and of three pure components isolated from it on human platelet aggregation, arachidonate metabolism, release reaction and platelet ultrastructures. *Thrombosis Research*, **15**: 155-169.

Ara Tachjian, M.D., Viqar Maria, M.B.B.S., and Arshad Jahangir, M.D. (2010). Use of herbal products and potential interactions in patients with cardiovascular diseases. *Journal of the American College of Cardiology*, **55**: 515-525.

Asdaq, S.M.B., and Inamdar, M.N. (2011). Pharmacodynamic and pharmacokinetic interactions of propranolol with garlic (*Allium sativum*) in rats. *Evidence-based Complementary and Alternative Medicine*, article ID824042, 11 pages, doi: 10.1093/ecam/neq076.

Attelle, A.S., Wu, J.A., and Yuan, C.-S. (1999). Ginseng Pharmacology, multiple constituents and multiple actions. *Biochemical Pharmacology*, **58**: 1685-1693.

Baillard, C., Bianchi, A., Gehan, G., Sitbon, P., Denantes, C., Benhamou, D., Cupa, P., and Samama, C.-M. (2007). Anaesthetic preoperative assessment of chronic medications and herbal medicine use: a multicenter survey. *Annales Françaises d'Anesthésie et de Réanimation*, **26**: 132-135.

Bayramiçli, U.O., Kalkay, M.N., Oskay Bozkaya, E., Köse, E.D., Iyigün, Ö., Görük, M., and Sezgín, G. (2011). St. John's wort (*Hypericum perforatum*) and warfarin: dangerous liaisons! *Turkish Journal of Gastroenterology*, **22**: 115.

Berginc, K., and Kristl, A. (2012). The effect of garlic supplements and phytochemicals on the ADMET properties of drugs. *Expert on Opinion Drug Metabolism and Toxicology*, **8**: 295-310.

Birks, J., and Grimley, E.J. (2009). *Ginkgo biloba* for cognitive impairment and dementia (review). The Cochrane Collaboration, John Wiley and Sons, Ltd.

Block, E., Ahmad, S., Jain, M.K., Crecely, R.W., Apitz-Castro, R., and Cruz, M.R. (1984). (*E,Z*)-Ajoene: a potent antithrombic agent from garlic. *Journal of the American Chemical Society*, **106**: 8295–8296.

Bordia, A., Verma, S.K., and Srivastava, C. (1996). Effect of garlic on platelet aggregation in humans: a study in healthy subjects and patients with coronary artery disease. *Prostaglandins, Leukotrienes, and Essential Fatty Acids*, **55**: 201-205.

Bordia, A., Verma, S.K., and Srivastava, K.C. (1998). Effect of garlic (*Allium sativum*) on blood lipids, blood sugar, fibrinogen and fibrinolytic activities in patients with coronary artery disease. *Prostaglandins, Leukotrienes, and Essential Fatty Acids*, **58**: 257-263.

Borrelli, F., and Izzo, A.A. (2009). Herb-drug interactions with St. John's wort (*Hypericum perforatum*): an update on clinical observations. *The AAPS Journal*, **11**: 710-727.

Chan, E., Tan, M., Xin, J., Sudarsanan, S., and Johnson, D.E. (2010). Interactions between traditional Chinese medicines and western therapeutics. *Current Opinion in Drug Discovery and Development*, **13**: 50-65.

Chen, X.-W., Serag, E.S., Sneed, K.B., Liang, J., Chew, H., Pan, S.-Y., and Zhou, S.-F. (2011). Clinical herbal interactions with conventional drugs: from molecules to maladies. *Current Medicinal Chemistry*, **18**: 4836-4850.

Cheng, X., Maher, J., Dieter, M.Z., and Klaassen, C.D. (2005). Regulation of mouse organic anion-transporting polypetides (OAPTS) in liver by prototypical microsomal enzyme inducers that activate distinct transcription factor pathways. *Drug Metabolism and Disposition*, **33**: 1276-1282.

Choi, Y.H., Chin, Y.-W., and Kim, Y.G. (2011). Herb-drug interactions: focus on metabolic enzymes and transporters. *Archives of Pharmacal Research*, **34**: 1843-1863.

Cordier, W., and Steenkamp, V. (2012). Herbal remedies affecting coagulation: a review. *Pharmaceutical Biology*, **50**: 443-452.

Effert, T., and Kaina, B. (2011). Toxicities by herbal medicines with emphasis to traditional Chinese medicine. *Current Drug Metabolism*, **12**: 989-996.

Fan, L., Mao, X.Q., Tao, G.Y., Jiang, F., Chen, Y., Li, Q., Zhang, W., Lei, H.P., Hu, D.L., Huang, Y.F., Wang, G., and Zhou, H.H. (2009b). Effect of *Schisandra chinensis* extract and *Ginkgo biloba* extract on the pharmacokinetics of talinolol in healthy volunteers. *Xenobiotica*, **39**: 249-254.

Fan. L,, Tao, G.Y., Wang, G., Chen, Y., Zhang, W., He, Y.J., Li, Q., Lei, H.P., Jiang, F., Hu, D.L., Huang, Y.F., and Zhou, H.H. (2009a). Effects of *Ginkgo biloba* extract ingestion on the pharmacokinetics of talinolol in healthy Chinese volunteers. *Annals of Pharmacotherapy*, **43**: 944-949.

Fenwick, G.R., and Hanley, A.B. (1985). The genus *Allium*. Part 2. *Critical Reviews in Food Science and Nutrition*, **22**: 273–377.

Foster, B.C., Foster, M.S., Vandenhoe, K.S., Krantis, A., Budzinski, J.W., Arnason, J.T., Gallicano, K.D., and Choudri, S. (2001). An *in vitro* evaluation of human cytochrome P450 3A4 and P-glycoprotein inhibition by garlic. *Journal of Pharmacy and Pharmaceutical Sciences*, **4**: 176.14.

Gardiner, P., Phillips, R., and Shaughnessy, A.F. (2008). Herbal and dietary supplement-drug interactions in patients with chronic illness. *American Family Physician*, **77**: 73-78.

German, K., Kumar, U., and Blackford, H.N. (1995). Garlic and the risk of TURP bleeding. *British Journal of Urology*, **76**: 518.

Greenblatt, D.J., and von Moltke, L.L. (2005). Interaction of warfarin with drugs, natural substances, and foods. *Journal of Clinical Pharmacology*, **45**: 127-132.

Guo, C.-X., Pei, Q., Yin, J.-Y., Peng, X.-D., Zhou, B.-T., Zhao, Y.-C., Wu, L.-X., Meng, X.-G., Wang, G., Li, Q., Ouyang, D.-S., Liu, Z.-Q., Zhang, W., and Zhou, H.-H. (2012). Effects of *Ginkgo biloba* extracts on pharmacokinetics and efficacy of atorvastatin based on plasma indices. *Xenobiotica*, **42**: 784-790.

Gurley, B.J., Gardner, S.F., Hubbard, M.A., Williams, D.K., Gentry, W.B., Cui, Y., and Ang, C.Y.W. (2002). Cytochrome P450 phenotypic ratios for predicting herb-drug interactions in humans. *Clinical Pharmacology and Therapeutics*, **72**: 276-287.

Gurley, B.J., Gardner, S.F., Hubbard, M.A., Williams, K., Gentry, W.B., Cui, Y., and Ang, C.Y.W. (2005). Clinical assessment of botanical supplementation on cytochrome P450 phenotypes in the elderly: St. John's wort, garlic aicd, *Panax ginseng*, and *Ginkgo biloba*. *Drugs Aging*, **22**: 525-539.

Gurley, B.J., Swain, A., Williams, D.K., Barone, G., and Battu, S.K. (2008). Gauging the clinical significance of P-glycoprotein-mediated herb-drug interactions: comparative effects of St. John's wort, Echinacea, clarithromycin, and rifampin on digoxin pharmacokinetics. *Molecular Nutrition and Food Research*, **52**: 772-779.

Guyonnet, D., Siess, M.H., le Bon, A.M., and Suschetet, M. (1999). Modulation of phase II enzymes by organosulfur compounds from allium vegetables in rat tissues. *Toxicology and Applied Pharmacology*, **154**: 50-58.

Hellum, B.H., and Nilsen, O.G. (2008). *In vitro* inhibition of CYP3A4 metabolism and P-plycoprotein-mediated transport by trade herbal products. *Basic and Clinical Pharmacology and Toxicology*, **102**: 466-475.

Henderson, L., Yue, Q.Y., Bergquist, C., Gerden, B., and Arlett, P. (2002). St John's wort (*Hypericum perforatum*): drug interactions and clinical outcomes. *British Journal of Clinical Pharmacology*, **54**: 349-356.

Hermann, R., and von Richter, O. (2012). Clinical evidence of herbal drugs as perpetrators of pharmacokinetic drug interactions. *Planta Medica*, **78**: 1458-1477.

Hu, Z., Yang, X., Ho, P.C.L., Chan, S.Y., Heng, P.W.S., Chan, E., Duan, W., Koh, H.L., and Zhou, S. (2005). Herb-drug interactions, a literature review. *Drugs*, **65**: 1239-1282.

Izzo, A.A. (2004). Herb-drug interactions: an overviewvof the clinical evidence. *Fundamental and Clinical Pharmacology*, **19**: 1-16.

Izzo, A.A., di Carlo, G., Borrelli, F., and Ernst, E. (2005). Cardiovascular pharmacotherapy and herbal medicines: the risk of drug interaction. *International Journal of Cardiology*, **98**: 1-14.

Izzo, A.A., and Ernst, E. (2001). Interactions between herbal medicines and prescribed drugs. *Drugs*, **61**: 2163-2175.

Jia, L., Zhao, Y., and Liang, X.-J. (2009). Current evaluation of the millennium phytomedicine-ginseng (II) collected chemical entities, modern pharmacology, and clinical applications emanated from traditional Chinese medicine. *Current Medicinal Chemistry*, **16**: 2924-2942.

Jiang, L., Su, L., Cui, H., Ren, J., and Li, C. (2013). *Ginkgo biloba* extract for dementia: a systematic review. *Shangai Archives of Psychiatry*, **25**: 10-21.

Jiang, X., Williams, K.M., Liauw, W.S., Ammit, A.J., Roufogalis, B.D., Duke, C.C., Day, R.O., and McLachlan, A.J. (2004). Effect of St. John'S wort and ginseng on the pharmacokinetics and pharmacodynamics of wafarin in healthy subjects. *British Journal of Clinical Pharmacology*, **57**: 592-599.

Jiang, X., Williams, K.M., Liaw, W.S., Ammit, A.J., Roufogalis, B.D., Duke, C.C., Day, R.O., and McLachlan, A.J. (2005). Effect of ginkgo and ginger on the pharmacokinetics and pharmacodynamics of warfarin in healthy subjects. *British Journal of Clinical Pharmacology*, **59**: 425-432.

Kang, S.Y., Schini-Kerth, V.B., and Kim, N.D. (1995). Ginsenosides of the protopanaxatriol group cause endothelium-dependent relaxation in the rat aorta. *Life Sciences*, **56**: 1577-1586.

Kim, B.-H., Kim, K.-P., Lim, K.S., Kim, J.-R., Yoon, S.H., Cho, J.-Y., Lee, Y.-O., Lee, K.-H., Jang, I.-J., Slin, S.-G., and Yu, K.S. (2010). Influence of *Ginkgo biloba* extract on the pharmacodynamic effects and pharmacokinetic properties of ticlopidine: an open-label, randomized, two period,, two treatment, two sequence, single-dose crossover study in healthy Korean male volunteers. *Clinical Therapeutics*, **32**: 380-390.

Kshirsagar, V.V., and Asdaq, S.M.B. (2014). Studies on pharmacodynamic interaction of garlic and metoprolol in rats with chronic myocardial disfunction and metabolic derangments. *International Journal of Experimental Pharmacology*, **4**: 8-13.

Lau, A.J., Yang, G., Rajaraman, G., Baucom, C.C., and Chang, T.K.H. (2010). Human pregnance X receptor agonism by *Ginkgo biloba* extract: assessment of the role of individual ginkgolides. *Journal of Pharmacology and Experimental Therapeutics*, **335**: 771-780.

Le Bon, A.M., Vernevaut, M.F., Guenot, L., Kahane, R., Auger, J., Arnault, I., Haffner, T., and Siess, M.-H. (2003). Effect of garlic powders with varying alliin contents on hepatic drug metabolizing enzymes in rats. *Journal of Agricultural and Food Chemistry*, **51**: 7617-7623.

Lee, M.-J., Choi, J.-S., Cha, S.-W., Lee, K.-S., Lee, Z.-W., Hwang, G.-S., Lee, S.H., Kamal, A.H.M., Jung, Y.-A., Seung, N.-S., and Woo, S.-H. (2011). Variation in the ginsenoside profiles of cultivated ginseng (*Panax ginseng* C.A. Meyer) landraces in Korea. *Process Biochemistry*, **46**: 258-264.

Lee, S.-H., Ahn, Y.-M., Ahn, S.-Y., Doo, H.-K., and Lee, B.-C. (2008). Interaction between wafarin and *Panax ginseng* in ischemic stroke patients. *Journal of Alternative and Complementary Medicine*, **14**: 715-721.

Lim, W., Mudge, K.W., and Vermeylen, F. (2005). Effects of population, age, and cultivation methods on ginsenoside content of wild American ginseng (*Panax quinquefolium*). *Journal of Agricultural and Food Chemistry*, **53**: 8498-8505.

Makheja, A.N., and Bailey, J.M. (1990). Antiplatelet constituents of garlic and onion. *Agents and Actions*, **29**: 360-363.

Markowitz, J.S., Devane, C.L., Chavin, K.D., Taylor, R., Ruan, Y., and Donovan, J.L. (2003). Effects of garlic (*Allium sativum* L.) supplementation on cytochrome P450 2D6 and 3A4 activity in healthy volunteers. *Clinical Pharmacology and Therapeutics*, **74**: 170-177.

Mateen, A.A., Rani, P.U., Naidu, M.U.R., and Chandrashekar, E. (2011). Pharmacodynamic interaction study of Allium sativum (garlic) with cilostazol in patients with type II diabetes mellitus. *Indian Journal of Pharmacology*, **43**: 270-274.

Mauro, F., Mauro, L.S., Kleshinski, J.F., Khuder, S.A., Wang, Y., and Erhadt, P.W. (2003). Impact of *Ginjgo biloba* on the pharmacokinetic digoxin. *American Journal of Therapy*, **10**: 247-251.

Miller, L.G. (1998). Herbal medicines, selected clinical considerations focusing on known or potential drug-herb interactions. *Archives in International Medicine*, **158**: 2200-2211.

Na, D.H., Ji, H.Y., Park, E.J., Kim, M.S., Liu, K.-H., and Lee, H.S. (2011). Evaluation of metabolism-mediated herb-drug interaction. *Archives of Pharmacal Research*, **34**: 1829-1842.

Nadkarni, A., Oldham, M.A., Howard, M., and Berenbaum, I. (2012). Drug-drug interactions between warfarin and psychotropics: updated review of the literature. *Pharmacotherapy*, **32**: 932-942.

Nowack, R. (2008). Review article: Cytochrome P450 enzyme, and transport protein mediated herb-drug interactions in renal transplant patients: grapefruit juice, St John's Wort - and beyond! *Nephrology*, **13**: 337-347.

Obach, R.S. (2000). Inhibition of human cytochrome P450 enzymes by constituents of St. John's wort, an herbal preparation used in the treatment of depression. *Journal of Pharmacology and Experimental Therapeutics*, **294**: 88-95.

Obodozie, O.O. (2012). Pharmacokinetics and drug interactions of herbal medicines: a missing critical step in the phytomedicine (drug development process). In: Readings in Advanced Pharmacokinetics – Theory, Methods, and Applications. Ed.: Ayman Noreddin, ISBS: 978-953-51-0533-6, In Tech.

Ohnishi, N., Kusuhara, M., Yoshioka, M., Kuroda, K., Soga, A., Nishikawa, F., Koishi, T., Nakagawa, M., Hori, S., Matsumoto, T., Yamashita, M., Ohta, S., Takara, K., and Yokoyama, T. (2003). Studies on interactions between functional foods or dietary supplements and medicines. I. Effects of *Ginkgo biloba* leaf extract on the pharmacokinetics of diltiazem in rats. *Biological and Pharmaceutical Bulletin*, **26**: 1315-1320.

Peng, D., Wang, H., Qu, C., Xie, L., Wicks, S.M., and Xie, J. (2012). Ginsenoside Re: its chemistry, metabolism and pharmacokinetics. *Chinese Medicine*, **7**: 2 (doi10.1186/1749-8546-7-2.

Peng, L., Sun, S., Xie, L.-H., Wicks, S.M., and Xie, J.-T. (2011). Ginsenoside Re: Pharmacological effects on cardiovascular system. *Cardiovascular Therapeutics*, **30**: e183-188.

Portolés, A., Terleira, A., Calvo, A., Martínez, I., and Resplandy, G. (2006). Effects of *Hypericum perforatum* on ivabradine pharmacokinetics in healthy volunteers: na open-label, pharmacokinetic interaction clinical trial. *Journal of Clinical Pharmacology*, **46**: 1188-1194.

Posadzki, P., Watson, L., and Ernst, E. (2012). Herb-drug interactions: an overview of systematic reviews. *British Journal of Clinical Pharmacology*, **75**: 603-618.

Qi, L.-W., Wang, C.-Z., Du, G.-J., Zhang, Z.-Y., Calway, T., and Yuan, C.-S. (2011a). Metabolism of ginseng and its interactions with drugs. *Current Drug Metabolism*, **12**: 818-822.

Qi, L.-W., Wang, C.-Z., and Yuan, C.-S. (2011b). Isolation and analysis of ginseng: advances and challenges. *Natural Product Reports*, **28**: 467-495.

Qi, L.W., Wang, C.-Z., and Yuan, C.-S. (2011c). Ginsenosides from American ginseng: chemical and pharmacological diversity. *Phytochemistry*, **72**: 689-699.

Rahal, A., Ahmad, A.H., Kumar, A., Mahima, and Verma, A.K. (2013). Clinical drug interactions: a holistic review. *Pakistan Journal of Biological Sciences*, **16**: 751-758.

Reddy, G.D., Reddy, A.G., Rao, G.S., Haritha, C., and Jyothi, K. (2010). Interaction study on garlic and atorvastatin with reference to nephrotoxicity in dyslipidaemic rats. *Toxicology International*, **17**: 90-93.

Reddy, G.D., Reddy, A.G., Rao, G.S., and Kumar, M.V. (2012). Pharmacokinetic interaction of garlic and atorvastatin in dyslipidemic rats. *Indian Journal of Pharmacology*, **44**: 246-252.

Renuka, G., Thiruvengadarajan, V.S., Amruthkumar, N., Malesh, K., and Chetty, C.M. (2011). A review of herb-drug interactions. *International Journal of Pharmaceutical Research and Development*, **3**: 136-153.

Ryu, K.H., Han, H.Y., Lee, S.Y., Jeon, S.D., Im, G.-J., Lee, B.Y., Kim, K., Lim, K.-M., and Chung, J.-H. (2009). *Ginkgo biloba* extract enhances antiplatlet and antithrombotic effects of cilostazol without prolongation of bleeding time. *Thrombosis Research*, **124**: 328-334.

Scharz, U.I., Hanso, H., Oertel, R., Miehlke, S., Kuhlisch, E., Glaeser, H., Hitzl, M., Dresser, G.K., Kim, R.B., and Kirch, W. (2007). Induction of intestinal P-glycoprotein by St. John's wort reduces the oral bioavailability of talinolol. *Clinical Pharmacology and Therapeutics*, **81**: 669-678.

Shao, J., and Jia, L. (2013). Potential serious interactions between nutraceutical ginseng and warfarin in patients with ischemic stroke. *Trends in Pharmacological Sciences*, **34**: 86.

Shi, S., and Klotz, U. (2012). Drug interactions with herbal medicines. *Clinical Pharmacokinetics*, **51**: 77-104.

Shinozuka, K., Umegaki, K., Kubota, Y., Tanaka, N., Mizuno, H., Yamauchi, J., Nakamura, K., and Kumitomo, M. (2002). Feeding of *Ginkgo biloba* extract (GbE) enhances gene expression of hepatic cytochrome P-450 and attenuates the hypotensive effect of nicardipine in rats. *Life Sciences*, **70**: 2783-2792.

Smith, M., Lin, K.M., and Zheng, Y.P. (2001). An open trial of nifedipine-herb interactions: nifedipine with St John's wort, ginseng or *Ginkgo biloba*. *Clinical Pharmacology and Therapeutics*, **69**: 86.

Srivastava, K.C. (1986). Evidence for the mechanism by which garlic inhibits platelet aggregation. *Prostaglandins Leukotriens and Medicine*, **22**: 313-321.

Stanger, M.J., Thompson, L.A., Young, A.J., and Lieberman, H.R. (2012). Anticoagulant activity of select dietary supplements. *Nutrition Reviews*, **70**: 107-117.

Sugimoto, K., Ohmori, M., Tsuruoka, S., Nishiki, K., Kawaguchi, A., Harada, K., Arakawa, M., Sakamoto, K., Masada, M., Miyamori, I., and Fujimura, A. (2001). Different effects of t. John wort on the pharmacokinetics of simvastatin and provastatin. *Clinical Pharmacology and Therapeutics*, **70**: 518-524.

Tachjian, A., Maria, V., and Jahangir, A. (2010). Use of herbal products and potential interactions in patients with cardiovascular diseases. *Journal of the American College of Cardiology*, **55**: 515-525.

Tandon, V.R., Kapoor, B., Bano, G., Gupta, S., Gillani, Z., Gupta, S., and Kour, D. (2006). P-glycoprotein: pharmacological relevance. *Indian Journal of Pharmacology*, **38**: 13-24.

Thomas, A.S., Varughese, P., Shirwaikar, A., and Shirwaikar, A. (2012). Herb-drug interactions: a review. *Hygeia: Journal of Drugs and Medicines*, **4**: 33-40.

Tomlinson, B., Hu, M., and Lee, V.W. (2008): *In vivo* assessment of herb-drug interactions: possible utility of a pharmacogenetic approach? *Molecular Nutrition and Food Research*, **52**: 799-809.

Tsai, H.-H., Lin, H.-W., Pichard, A.S., Tsai, H.-Y., and Mahady, G.B. (2012). Evaluation of documented drug interactions and contraindications associated with herbs and dietary supplements: a systematic literature review. *International Journal of Clinical Practice*, **66**: 1056-1078.

Ulbricht, C., Chao, W., Costa, D., Seamon-Rusie, E., Weissner, W., and Woods, J. (2008). Clinical evidence of herb-drug interactions: a systematic review by the natural standard research collaboration. *Current Drug Metabolism*, **9**: 1063-1120.

Vieira, M.L.T., and Huang, S.-M. (2012). Botanical-drug interactions: a scientific perpective. *Planta Medica*, **78**: 1400-1415.

Wang, A., Wang, C.-Z., Wu, J.A., Osinski, J., and Yuan, C.-S. (2005). Determination of major ginsenosides in *Panax quinquefoluis* (American ginseng) using high performance liquid chromatography. *Phytochemical Analysis*, **16**: 272-277.

Wang, X.-D., Li, J.-L., Su, Q.-B., Guan, S., Chen,.J, Du, J., He, Y.-W., Zeng, J., Zhang, J.-X., Chen, X., Huang, M., and Zhou, S.-F. (2009). Impact of the haplotypes of the human pregnane X receptor gene on the basal and St. John's wort-induced activity of cytochrome P4503A4 enzyme. *British Journal of Clinical Pharmacology*, **67**: 255-261.

Wanwimolruk, S., and Prachayaasittikul, V. (2014). Cytochrome P450 enzyme mediated herbal drug interactions (Part 1). *EXCLI Journal*, **13**: 347-391.

Yang, X.-X., Hu, Z.-P., Duan, W., Zhu, Y.-Z., and Zhou, S.-F. (2006). Drug-herb interactions: eliminating toxicity with hard drug design. *Current Pharmaceutical Design*, **12**: 4649-4664.

Yang, Z., Wang, J.-R., Niu, T., Gao, S., Yin, T., You, M., Jiang, Z.-H., and Hu, M. (2012). Inhibition of P-glycoprotein leads to improved oral bioavailability of compound K, an anticancer metabolite of red ginseng extract produced by gut microflora. *Drug Metabolism and Disposition*, **40**: 1538-1544.

Yoshioka, M., Ohnishi, N., Koishi, T., Obata, Y., Nakagawa, M., Matsumoto, T., Tagagi, K., Takara, K., Ohkuni, T., Yokoyama, T., and Kuroda, K. (2004). Studies on interactions between functional foods or dietary supplements and medicines. IV. Effects of *Ginkgo biloba* leaf extract on the pharmacokinetics and pharmacodynamics of nifedipine in healthy volunteers. *Biological and Pharmaceutical Bulletin*, **27**: 2006-2009.

Yu, C.T., Chen, J., Teng, X.W., Tong, V., and Chang, T.K.H. (2005). Lack of evidence for induction of CYP2B1, CYP3A23, and CYP1A2 gene expression by *Panax ginseng* and *Panax quinquefolius* extracts in adult rats and primary cultures of rat hepatocytes. *Drug Metabolism and Disposition,* **33**: 19-22.

Yuan, C.-S., Wei, G., Dey, L., Karrison, T., Nahlik, L., Maleckar, S., Kasza, K., Ang-Lee, M., and Moss, J. (2004). American ginseng reduces warfarin's effect in healthy patients. A randomized, controlled trial. *Annals of Internal Medicine*, **141**: 23-27.

Zhou, S., Gao, Y., Jiang, W., Huang, M., Xu, A., and Paxton, J.W. (2003). Interactions of herbs with cytochrome P450. *Drug Metabolism Reviews*, **35**: 35-98.

Zhou, S.-F., Zhou, Z.-W., Li, C.-G., Chen, X., Yu, X., Xue, C.C., and Herington, A. (2007). Identification of drugs that interact with herbs in drug development. *Drug Discovery Today*, **12**: 664-673.

Natural Products: Research Reviews Vol. 3 (2016) *Pages* **35–63**
Editor: **V.K. Gupta**
Published by: **DAYA PUBLISHING HOUSE, NEW DELHI**

2

Polyene Macrolide Antibiotics, Microbial Derived Natural Bioactive Products: Fermentation, Pharmacological and Chemotherapeutic Importance

Suresh R. Naik[1*]

ABSTRACT

Polyene macrolide antibiotics are produced by soil actinomycetes; the majority of them belong to members of the genera Streptomyces, Streptoverticillium, Actinosporangium and others. They are isolated from the biomass and exhibit antifungal activity against wide spectrum of fungal organisms. However, some of them also elicit antibacterial activity. They are classified on the basis of number conjugated double bonds as triene, tetraene, pentanene, hexanae, heptanea, octaene, oxopolyenes and mixed polyenes. They are not soluble in water, poorly absorbed,toxic and contains amino sugar joined to macrolide ring through glycoside bond. Besides, antifungal, antibacterial activity, some of them also showed hypolipidemic, hypoprostatic, antitumor, antiviral and anti-inflammatory activity. Some of the antifungal antibiotics are incorporated into liposomal preparations which are found to be clinically effective in the treatment of systemic mycosis/deep seated mycosis (like Aspergillosis, Candidiasis and Cryptococcosis). Polyene macrolide antibiotics form an important chemotherapeutic category of drugs in the treatment of a wide variety of fungal infections.

Keywords: Polyene macrolide antibiotics, Actinomycetes, Chemotherapeutic, Fungal infections.

1 Department of Pharmacology, Sinhgad Institute of Pharmaceutical Sciences, S. No. 309/ 310, Kusgaon (Bk), Off Mumbai Pune Expressway, Lonavala, Pune – 410 401, India.

* *Corresponding author*: E-mail: srnaik5@rediffmail.com

Introduction

Fungi have been recognized as causative agents of human, veterinary and plant fungal diseases. Fungi causing favus (*Trichophyton scholeini*) and thrush (*Candida albicans*) have been described as early as 1839. The study of pathogenic fungi had received scant attention in comparison with the study of other pathogens, probably due to the relative benign nature of the common mycotic diseases. The types of mycosis described by Ananthnarayan and Panikar (1971) are: 1) superficial (Dermatophytes, Pityriasis, Piedra, various types of Tinea affecting skin, hair and nails), 2) cutaneous (Candidiasis), 3) subcutaneous (Mycotic mycetoma, Chromoblasto mycosis, Sporotrichosis, subcutaneous Phycomycosis etc), 4) systemic mycosis (Cryptocococcosis, blastomycosis, histoplasmosis, paracoccidioidomycosis, coccidioidomycosis etc.).

Today, 17 per cent of the world population is suffering from fungal infection and antifungal market projected and was valued around $ 3-4 billion and currently representing around 8 per cent of the total anti-infective. In India, antifungal represent 10 per cent of the total pharmaceutical sales.

Mycophenolic acid was the first antifungal antibiotic isolated in pure form from Moldy rice by Gosio in 1896, while investigating the cause of pellagra Korzybski and Kurylowiez in 1961 found that it was active against large number of fungi which are pathogenic to plants and humans too, and less active against bacteria. A range of chemical structures are available for screening the prospectus of elaborating secondary metabolites in a single batch of submerged culture, however, the paucity of availability of research tools and sensitive methods for creating novel screens that have the capacity to detect novel chemicals species for new therapeutic application are very few.

Isolation and screening methods are the crucial steps in the search and discovery of previously unknown natural products from microorganism. The "**golden era**" of Antibiotics is aimed at major clinical problems that have been less amenable to rational design. In proposing the search for microbial products for disease modifying drugs, the question of the rationale for supporting the existence of such molecules shall also be stressed. The present review focused on some important polyene macrolide antibiotics and, especially polyene antibiotics discovered at the research and development division of Hindustan Antibiotics Ltd., Pimpri, India.

Polyene macrolides antibiotics comprise a large group of antifungal antibiotics produced by soil actinomycetes; the majority of them belongs to genera *Streptomyces, Streptoverticillium, Actinosporangium,* Chinia (Omura and Tanaka, 1984). Since 1950 more than 300 polyene macrolide have been reported. They elicit mainly antifungal activity, albeit some of them also elicit antibacterial (Prave *et al.*, 1972), antiprotozoal (Chuenkova *et al.*, 1982), antiviral (Jordan and Seet, 1978), antitumor, hypo prostatic, hypolipidemic, anti-inflammatory etc. (Aszalos *et al.*, 1968; Schaffner and Gordan, 1968; Naik *et al.*, 1981, 1987).

Polyene macrolide are characterized by large (C 20-44 member) lactone ring containing 3-8 conjugated carbon-carbon bonds called double bonds in Trans form, usually combined with one sugar moiety. Some compounds contain aliphatic side chain with an additional aromatic moiety (Tereshin, 1976).Other highly characteristic feature of polyene antibiotics is large number of hydroxyl groups present on the molecule varying from 6 to 14.

Polyenes are extremely unstable in the presence of oxidants and UV rays, extreme pH, stable between 5-6 pH and thermo labile (prolonged heating above 50°C). Polyene contains carbon, hydrogen, oxygen and in some instances nitrogen. The amino group present in some of the polyene is associated with an amino sugar moiety that is connected to the macrolide ring through glycoside bond. Polyenes are classified on the basis of number of conjugated double bonds (Berdy, 1980). They are: triene, tetraene, pentaene, hexaene, heptaene, octaene, oxopolyene and mixed polyene groups. Different sophisticated spectroscopy techniques, like field desorption, fast atom bombardment (Rinehart, 1982) and secondary ion mass spectroscopy (Harada *et al.*, 1982), high resolution ^{1}H NMR and, X-ray structure analysis (Mechlinski *et al.*, 1970) and two dimensional ^{13}C NMR actonide analysis (Rychnovsky *et al.*, 1997) were employed in polyene macrolide antibiotics structure elucidation.

Trienes

The trienes polyene antibiotics showed some atypical properties. Aurantinin and proticin produced by *Bacillus* sp., showed antibacterial activity against Gram (+ve and –ve) bacteria (Omura *et al.*, 1976; Nishikiori *et al.*, 1978). Resistaphylin produced by *Streptomyces* showed strong Gram +ve activity (Aizawa *et al.*, 1971). The triene showed antitumor activity (Aszalos *et al.*, 1968). Rapamycin and demethoxy rapamycin are structurally proved atypical trienes (Findlay, 1982) and chromatrienin

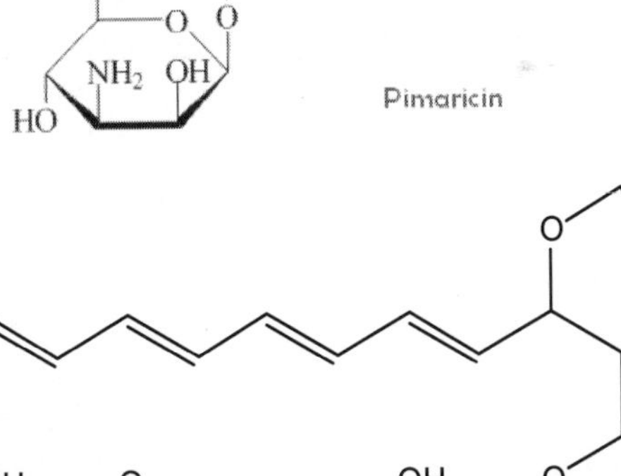

Pimaricin

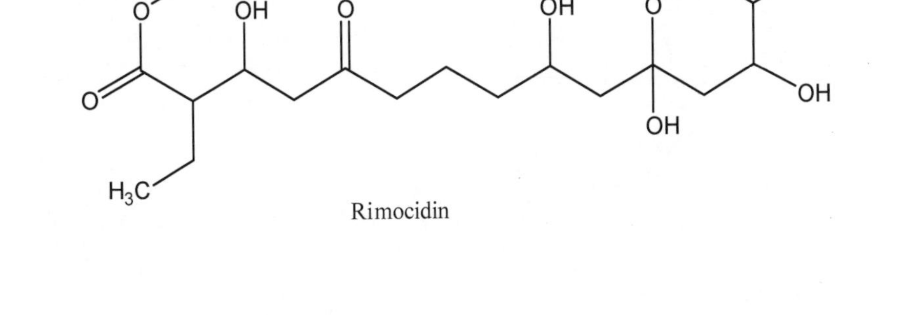

Rimocidin

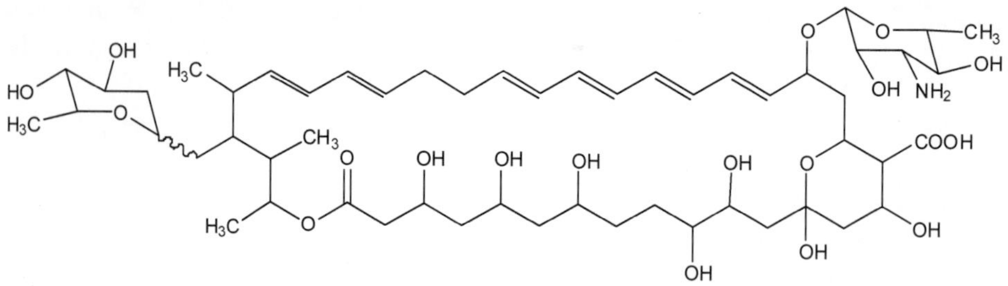

Nystatin A3

produced by *Actinimyces, chromogenes* var *trienicus* showed promising antifungal activity (Vezina *et al.*, 1975). Triene polyene antibiotics showed very good antifungal activity against wide varieties of fungal organisms including clinical isolates. Structures of some members of this group have been elucidated and shown to contain mycosamine moiety. Among them whose structures are small (26-28 member) and large (38- member) ring compounds have been documented. The antibiotics belonging to this group are: nystatin, pimaricin, rimocidin, tetramycin A and B etc. (Golding *et al.*, 1966; Radics *et al.*, 1982). Among these antibiotics, nystatin is widely used for topical applications in clinical practice. Nystatin exerts fungicidal activity by forming non-permeable channel with the cell membrane.

Hence, by interaction of nystatin with phosphatidyl choline, liposomes of varied size (large and small unilamellar vesicles) were developed for targeted antifungal delivery systems, especially, for the treatment of systemic mycosis (Sedlak *et al.*, 2007 a,b).

Pentaenes

This class of pentaenes includes two subgroups, methyl pentaenes and nitrogen containing pentaenes. The subgroup consists of neutral, nitrogen and sugar free compounds having 26 or 28 member macrolide lactone ring, such as Aurenin (26-member lactone ring), fungichromin (28- member lactone ring) (Taig *et al.*, 1969; Pandey

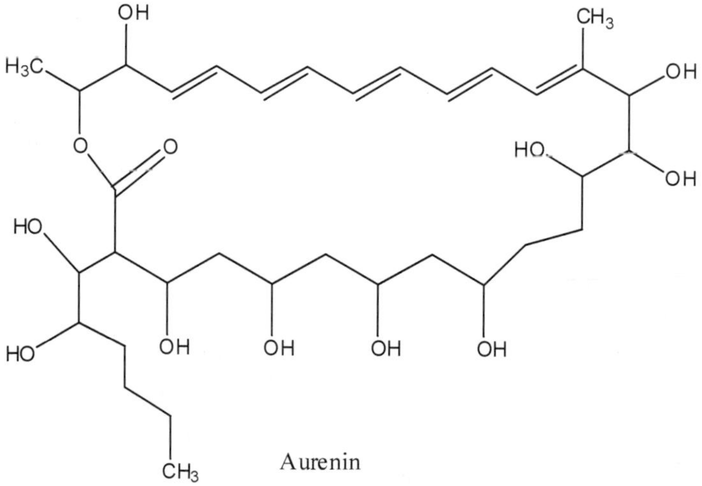

Aurenin

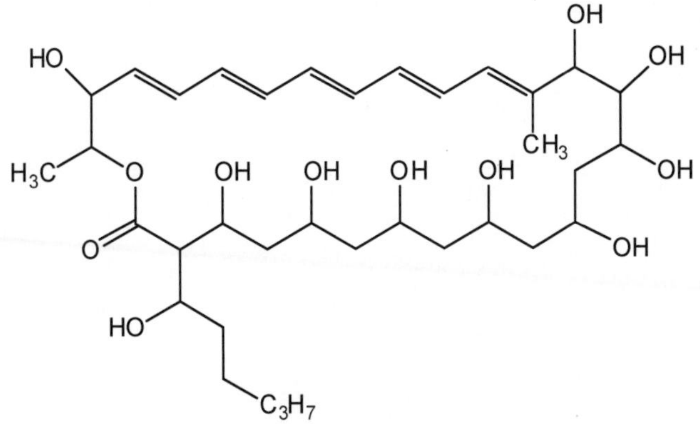

Fungichromin

et al., 1982). The 3-methyl pentaenes included were: fungichromin, lagosin and gomycin. The nitrogen containing pentaenes are: rectilavendomycin, eurocidins A and B, gannibamycin (Mitskevich *et al.*, 1975; Horii *et al.*, 1970; Shenin *et al.*, 1981).

Hexaenes

This group is relatively small and the most important antibiotic in this group is dermostatin.Generally hexaenes are produced together with tetraenes, *e.g.*: candicidin A and B, flavacid, dermostatin etc. (Martin and Mack Daniel, 1974; Rychnovsky *et al.*, 1997; Pandey *et al.*, 1980).

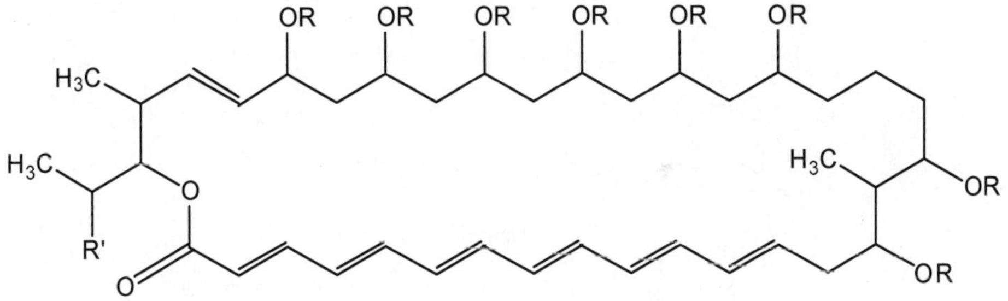

1 A: R=H, R'=CH3
1 B: R=H, R'=CH2CH3

2 A : R=Ac, R'=CH3
2 B : R=Ac, R'=CH2CH3

Dermostatins A and B

Heptaenes

This class of polyenes is the largest and include important antibiotics that are used widely in clinical practice. Members that are characterized structurally showed to possess 38 member lactone ring. Heptaenes are further subdivided into non aromatic (amphotericin B and candicidin etc) (Mechlinski *et al.*, 1970) and aromatic hamycin,

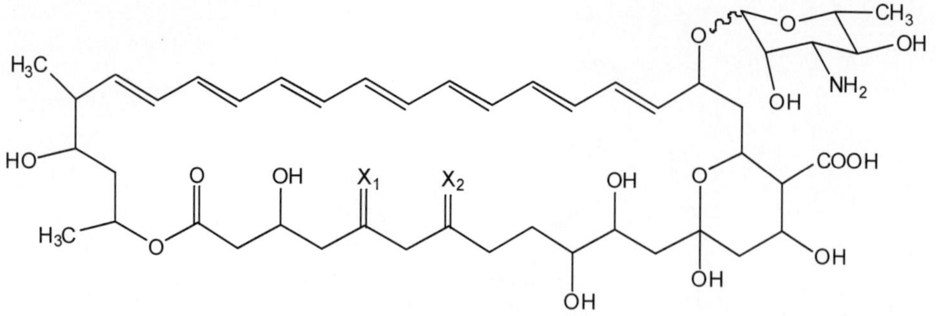

	X1	X2
Candidin	H,OH	=O
Mycoheptin	=O	H,OH

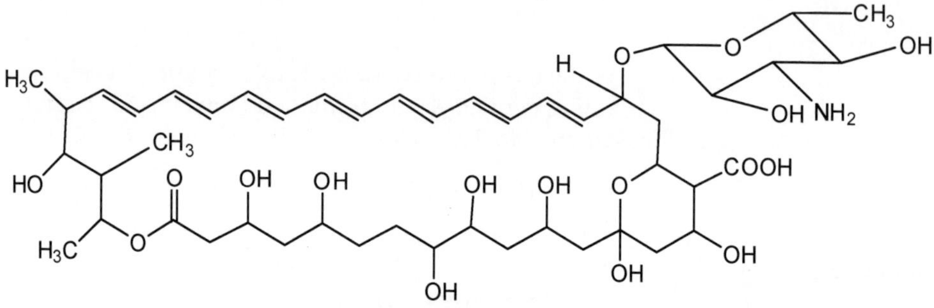

Amphotericin B

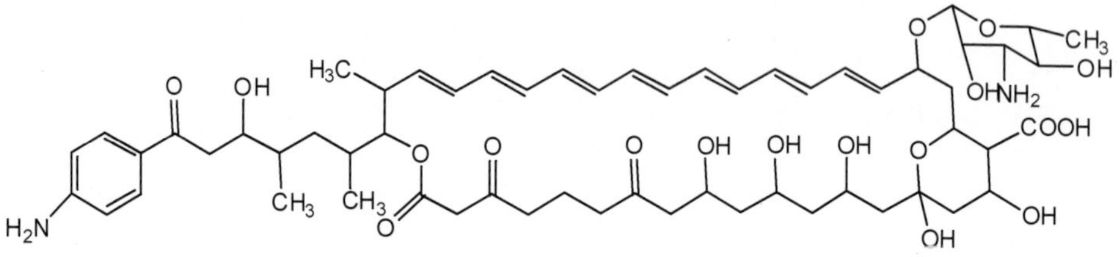

Levorin A_2 (Candicidin D)

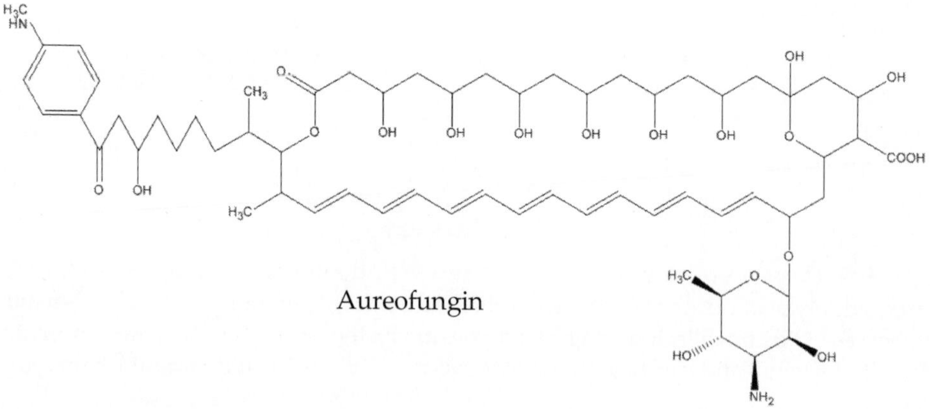

Aureofungin

aureofungin A and B, candicidin etc (Thirumalachar *et al.*, 1961, 1964; Oroshuik *et al.*, 1955). These aromatic heptaenes are always produced as a complex of several components/isomers.

Hamycin

The first Indian antibiotic, Hamycin was discovered in 1961 (Thirumalchar *et al.*, 1961) at research and development division of Hindustan Antibiotic Ltd. Pimpri, Pune. It is a polyene (heptaene) antibiotic isolated as the metabolic product of *Streptorynces pimprina* Thirum. The physio-chemical properties of hamycin are outlined below in the Table 2.1.

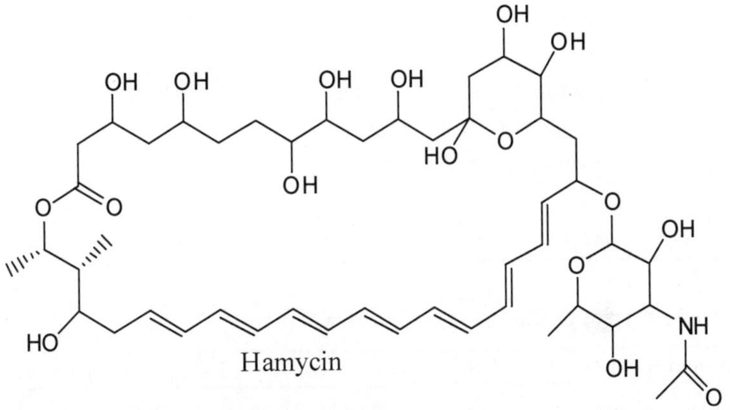

Hamycin

Table 2.1: Physio-chemical Properties of Hamycin

Colour	Golden yellow, amorphous powder
Chemical nature	Amphoteric
Melting Point (MP)	No definite MP, decomposes slowly above 160ºC
Analysis	Found C- 58.2, H- 8.3, N- 2.1 per cent
Optical rotation	$[\alpha]_D^{25}$ + 213 (C-0.2 per cent pyridine)
UV absorption in (80 per cent methanol)	345,363,383,406,413,635,916,875
Major Infrared bands	3.0,3.45,4.35,5.8,6.1,6.3,6.85,7.25,8.5,8.83,9.35,9.6,9.96,11.8 μ
Solubility (28ºC)	60 per cent methanol 0.54, 60 per cent ethanol 2.06, n-propanol 3.08, 50 per cent isopropanol 3.22, soluble in pyridine and dimethyl formamide

Others

The octaene, ochramycin produce by *Streptoverticilliun orinoci* showed Gram(+ve) antibacterial activity and octamycin produced by *Actinoplanes iantinogenes* sub sp. octamycin showed antitumor activity (Cassinelli *et al.*, 1967; Brazhnikova *et al.*, 1979). The oxopolyene is a polyene bearing oxoconjugated polyene system. These antibiotics showed a single broad peak in UV spectrum. Members of this group have 32 or 36 member lactone ring and no sugars which do not exist in other polyene groups (dermostatin A and B, roflamycoin, HA-1-92 etc.) (Rychnovsky *et al.*, 1997; Swamy *et al.*, 1994; Naik *et al.*, 2001; Naik and Gupte, 2007).

Table 2.2: Activities (*In vitro*) of Hamycin, Amphotericin B and Nystatin

Organisms	*Minimal Inhibitory Concentration (MIC), µg/ml*		
	Hamycin	*Nystatin*	*Amphotericin B*
C. parapsilosis	1.0	–	–
C. albicans (SC 1539)	0.010	1.1	0.04
C. tropicalis (SC 2620)	0.014	0.54	0.013
C. krusei (SC 2613)	0.015	1.7	0.16
Sachharomyces cerevisiae (SC 1600)	0.010	0.55	0.012
Aspergillus niger (SC 2828)	0.44	0.76	1.11
Penicillium notatum (SC 2827)	2.4	2.1	2.0
Microsporium canis (SC 2627)	0.012	0.50	0.05
Trichophyton mentagrophtes (SC 2637)	3.9	2.8	0.66
Staphylococoaus aureus	>50.0	>50.0	>50.0
Escherichia coli (SC 2975)	>50.0	>50.0	>50.0

Table 2.3: Acute Toxicity Studies of Hamycin in Rodents

Species and Route	*LD_{50} (mg/kg)*	*Behavioural Symptoms*
Mouse, Intraperitoneally (ip)	1.12 (1.04-1.2)	Sluggish, hind limb paralysis, ataxia bradycardia; depressed respiration, clonic and tonic convulsion. Cyanosis and death due to respiratory arrest.
Mouse, Intravenously (iv)	0.62 (0.55-0.73)	
Rat Intravenously (iv)	0.91 (0.80-1.05)	
Rabbit Intravenously (iv)	1.41 (1.28-1.54)	

Table 2.4: Activity (*In vitro*) of Hamycin against Fungi Causing Systemic Mycosis

Test Organism	*MIC (µg/ml)*
Blastomyces dermatitidis	0.005
Histoplasma capsulation	0.05-0.78
H. farciminosium U.MH 2869U.MH 2870	0.005-0.0060.001
Aspergillus fumigatus	1.56-6.25
Cryptococcus neoformans	0.05-0.06
Nocardia asteroides	0.01-0.39
Cephalosporium madurae	20.0
C. ifestans	20.0
Madurella mycetomi	5.0
Glenospora semoni	5.0

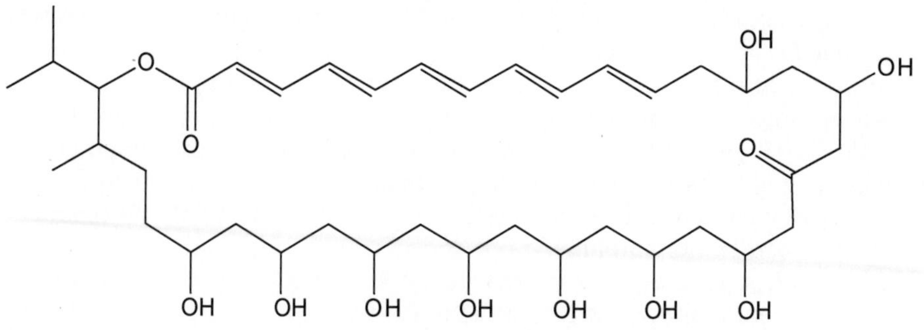

Roflamycoin

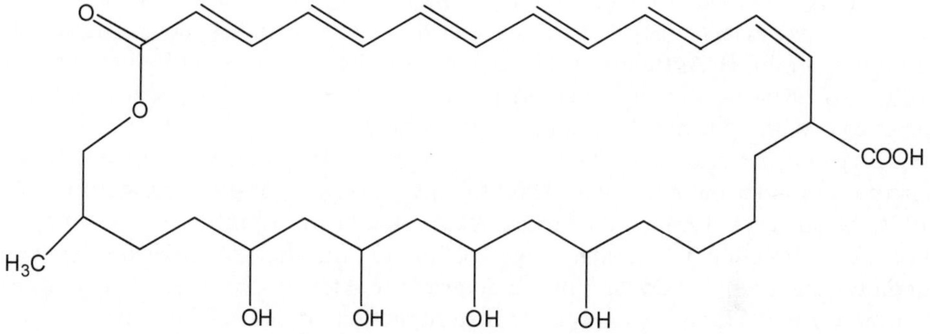

HA-1-92

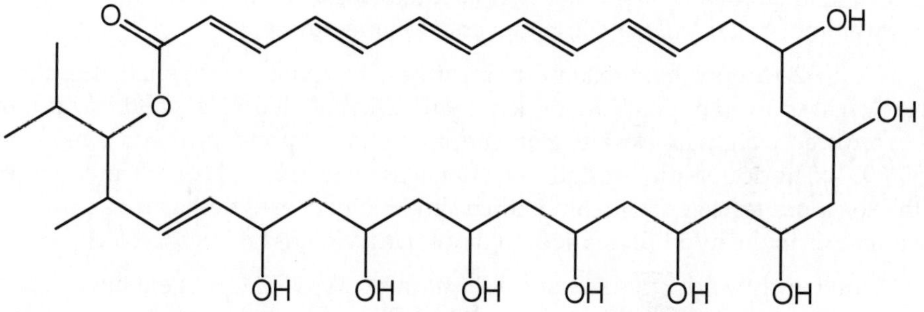

Mycoticin

Biology of Polyene Antibiotics

The biology of polyene has been described in the excellent reviews (Naik, 1981; Santra and Chandra,1981).Polyenes are known to exert various degrees of inhibition against many species of fungi such as yeast (*e.g. Candida, Cryptococcus*),dimorphic fungi (*e.g.*, Histoplasma, Blastomyces and Coccidioides), dermatophytes (*e.g. Trichophyton, Microsporum* and *Epidermophyton*) and moulds (*e.g.*, *Aspergillus* and *Penicillium*) (Medoff and Kobayashi,1980). The effect of nystatin, aureofungin, amphotericin B and hamycin on the growth of certain pathogenic keratinophilic fungi *viz. Nannizia incurvata* (+) and (-) strains, *Molbranchea aurantiaca* and *Botryotrichum keratinophylum* have been established. Among these antibiotics, amphotericin B and aureofungin causes 100 per cent growth inhibition at 500 ppm concentration. Pulmonary infections with *Aspergillus fumigatus* can be prevented by inhalation of aerosols containing amphotericin B, nystatin or pimaricin. Antifungal effect of flucytocine is increased when used in combination with amphotericin B against *Cryptococcus neoforman* and various species of *Candida* (Medoff *et al.*, 1972).Similarly combined treatment of amphotericin B and rifampicin increases effect against *C. albicans, S. cervisiae, Aspergillus* sp., *H. capsulatum* and *B. dermatitidis* (Medoff *et al.*, 1972). Flavopentin showed antiviral activity (*in vivo*) in chick embryos and in chorioallantoic fragments from chick embryos (*in vitro*).

Lipid enveloped viruses containing cholesterol in their coat showed sensitivity to the polyene macrolides *e.g., Vascular stomatitis, Vaccinia sindbis* and *Herpes simplex* (Jordan *et al.*, 1978). Antiviral activity of amphotericin B and its derivatives against influenza virus has been documented. Polyene antibiotics are reported to be toxic to protozoan's of medical importance such as *Trichomonads, Entamoeba* sp., *Naegleria* sp., *Trypanosomes* and *Leishmania donavani* (Bialasiewicz *et al.*, 1982).Polyenes have also been found to be active against algae, *Chlorella vulgaris* (Malewicz and Borowski, 1979). Martin *et al.* (1994) studied the effect of cytokines and antifungal agents on the phagocytic function of human monocytes. Potentiating effect of fungicidal capacity of monocytes seem to be an important determinant factor for therapeutic efficacy of amphotericin B. Hamycin showed cardio tonic activity in isolated heart preparation possibly through increased potassium content. It stimulates smooth muscle (rat uterus, rat ileum) *in vitro* at low concentration, and at higher concentration showed depressant effect (Arora, 1962). It has been reported that hamycin treatment clinically cured vaginal monilliasis and other vaginal infections, Athletes' foot, *Tinea captis* etc. (Bhagwat and Anjaneyulu, 1972; Ganju and Anjaneyulu, 1966).

HA-1-92 (a new oxohexaene macrolide), hamycin and JA 20 (heptaene macrolides) showed hypolipidemic activity ((Naik *et al.*, 1981; Naik and Harindran, 2000) reduced weight of prostate gland and secretion of prostatic fluid (Naik, 1981, 1983, 1984). The decrease in prostatic secretion was reversible and completely restored with exogenous testosterone treatment. The involvement of hypothalamus-pituitary-gonadal axis in the hypo-prostatic activity of hamycin has been suggested.

Hamycin showed a marked anti-*M.ovale* activity, which was enhanced in the presence of divalent cations like Cu^{++} and Zn^{++}. It is suggested that hamycin alone or along with Cu^{++} may be useful clinically in the treatment of dandruff and seborrhoic dermatitis (Gupte *et al.*, 1999a).

SJA-95

The strain (*Streptomyces* sp. *S24*) was isolated from the soil collected from Japan. Submerged fermentation studies on *Streptomyces* sp. *S-24* were carried out at 28°C for 120 h in a 1000L fermentor with aeration (4:1, v/v) and agitation (300 rev/min) and back pressure (0.5 kg/cm^2) for production of SJA-95 (Figure 2.1).

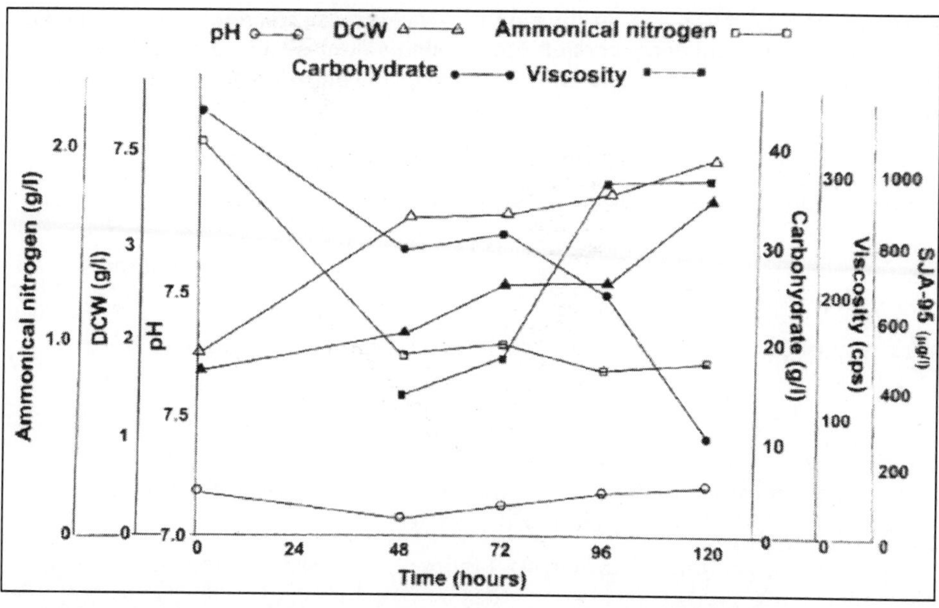

Figure 2.1: Time Course of Submerged Fermentation of *Streptomyces* sp. S24.

Table 2.5: Physio-chemical Properties of SJA-95

Test	*Observation*
Colour	Golden yellow
Nature	Amorphous
Melting point	Decomposed above 190°C
Solubility	Soluble in methanol, ethanol, butanol, dimethyl formamide, DMSO, pyridineInsoluble in water, ether chloroform, CCL_4
pH (25µg/ml)	4.35
Optical rotation $[\alpha]^{23}D$	66.60 (c 0.3, DMSO)
Functional groups	
(a) Molisch reagent	Negative (Sugar absent)
(b) Benedict's reagent	Negative (Sugar absent)
(c) Ninhydrin test	Positive (free amino group present)
(d) Sulphuric acid test	Positive (unsaturation present)
(e) Ferric chloride test	Negative (phenolic OH absent)
(f) Silver mirror test	Positive (aldehyde or ketone present)
(g) KI/CH_3COOH	Negative (epoxide group absent)
Elemental analysis	
Observed	C = 62.73 per cent, H = 7.82 per cent, N-2.48 per cent
Calculated	C = 62.47 per cent, H = 7.71 per cent, N = 2.51 per cent
UV λ max nm (E1 per cent 1cm)	400, 380, 358
IR í max nm (K Br) cm-1	3350, 2940, 2860, 1710, 1640, 1600, 1560, 1530, 1410, 1320, 1260, 1180, 1070, 1000, 800
Retention time (HPLC)*	2.197 (major) and 1.728 (minor)
R_f value of pure SJA 95	0.58

* Solvent system 0.05M sodium citrate buffer, (pH 5.3): acetonitrile (60:40), Column C-18(25x0.4), flow rate 1 ml/min, Detection 380 nm. TLC solvent system: butanol: acetic acid: water: tetrahydrofuran (6:2:2:1v/v)

Flow sheet for Isolation and purification of SJA-95.*these fractions were spotted on TLC (preparative) and each fraction gave a single spot with R_f 0.77, 0.58 and 0.51, respectively. The second fraction was found to be the major fraction and heptaenic in nature by U.V. spectrum.

The fermented broth filtered using 1 per cent super cells

Biomass/mycelium 2kg (washed with H_2O to make it free from broth colour)

↓ Extract with 95 per cent butanol

2 kg mycelium in 4L of BuOH and stirred with centrifugal stirrer Butanol extract

↓

Concentration under vacuum flash evaporator to its 1/10th volume chilled at 4°C overnight

Filtration (Whatman no. 1 Filter paper, pore size 10µm)

↓

Extract suspended in ethyl acetate stirred and filtered Ethyl *acetate (After* evaporation –concentration (solid residue). Suspended in hexane, maintained for15 min. and then filtered.

↓

Hexane extract ion under vacuum flash evaporator Crude SJA-95 (3g)

↓ Silica gel column chromatography
Glass column (5.5x2.5 cm)
Flow rate (2ml/min)

Methanol -Benzene gradient (10:90 to 90:10 elution) Active 3 fractions collection and concentration*

↓

Preparative TLC (Precoated GF264); Solvent system: Butanol-acetic acid-water-tetrahydrofuran (6:2:2:1)

↓

SJA-95 pure (1.4 g)

Acute toxicity (LD_{50}) of SJA-95 was carried out in mice by different routes. The LD_{50} observed were: 1.95 (1.8-2.1)mg/kg i.v., 2.4 (2.24-2.56) mg/kg i.p. and 230(164.5-322.0) mg/kg p.o. The behavioural symptoms observed were: rigidity, respiratory distress, diarrhoea, anorexia, mortality from 2nd day onwards.

Antifungal activity (*in vitro*) of SJA-95, hamycin and amphotericin B are outlined in Table 2.6.

Table 2.6: Comparative Antifungal Activity (*in vitro*) of SJA-95 and other Polyene Antibiotics

Organism	*MIC (µg/ml)*			
	SJA-95 A	*Hamycin*	*Amphotericin B*	*Griseofulvin*
Yeast				
Candida albicans ATCC 4463	0.02	0.01	0.05	>100
Saccharomyces cerevisiae ATCC 9763	0.03	0.02	0.0015	>85
Cryptococcous neoformans	0.06	0.06	0.05	>100
Trychophyton mentagrophytes	50	43	42	>100
Aspergillus fumigatus	5	4.5	3.9	–
Epidermophyton floccosum	>50	>50	–	>100
Filamentous fungi				
Paecilomyces varioti (HA-1585)	0.5	0.25	0.01	>100
Trichoderma virdi (NCIM 992)	5	6.25	7	>100
Aspergillus awamor	1.5	1	0.85	–
Aspergillus niger	0.5	0.3	3.85	>100
Aspergillus ochraecus (ATCC 28706)	0.5	0.8	4.8	–
Fusarium semitactrum	0.5	0.75	5	–
Fusarium udum	3.5	2.9	4.3	–
Westerdykella multispora	0.5	1.5	2.9	>100
Westerdykella cylindrica	50	–	–	–
Westerdykella dispersa	>100	–	–	–
Westerdykella purpurea	>100	–	–	–
Plant pathogen				
Fusarium fugikori	5	8	11.2	2.5
Collatotrichum gleosporodies	>100	–	–	>100
Helminthosporium oryzea	0.5	0.8	1.5	50
Alternaria macrophomina	10	15	–	–
Alternaria brassica	10	18	25	–
Macrophomina phaseoli	10	–	–	–
Rhizopus sp.	5	4.5	7.8	>100

Spectroscopic Studies of SJA-95

SJA-95 is a new aromatic heptaene polyene antibiotic having preferential binding towards ergosterol of fungal cell membrane. Treatment with SJA-95 both in aspergillosis, cryptocococcosis, progressively prolonged survival time and decreased the fungal load in vital organs (liver, spleen, kidney) respectively. Liposomal incorporated SJA-95 showed reduced toxicity and prolonged the survival time and decreased the fungal load (effective microbiological clearance) and seem to be a promising therapeutic agent in systemic fungal infections (Naik *et al.*, 2007; Desai and Naik, 2008; 2009) and immune modulation activity(Naik *et al.*, 2011).

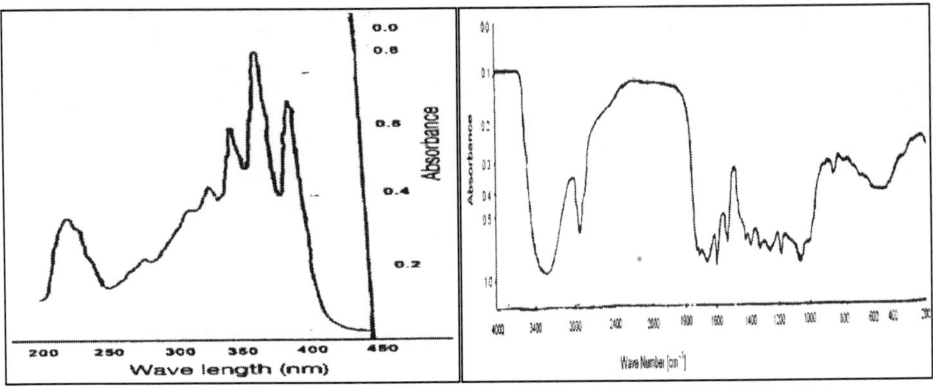

Figure 2.2(a): IR Spectrum of SJA-95. **Figure 2.2(b)**: UV Visible Spectrum of SJA-95.

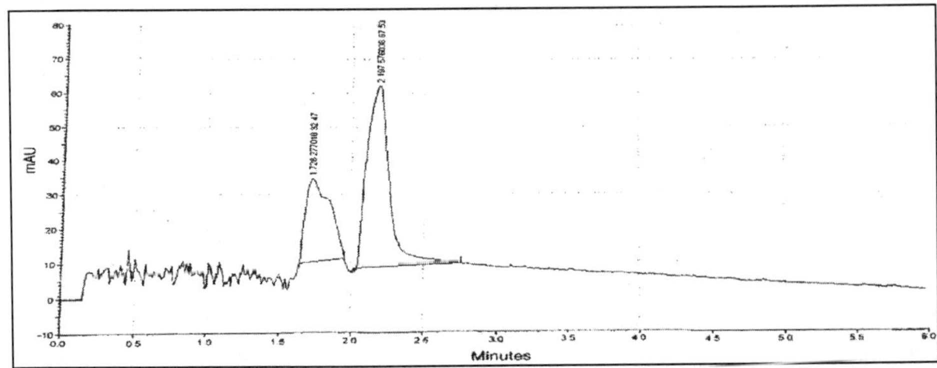

Figure 2.2(c): HPLC Analysis of SJA-95 on C-18 Column Detection, Flow Rate 1ml/min, Solvent System: Citrate Buffer (0.05M): Acetonitrile (60:40) Adjusted to pH 5.3 with Injection Volume 100 µl.

HA-2-91

During soil screening program, an isolate from Ukraine soil, Uk-10 has shown promising antifungal activity and nature of antibiotic was found to be tetraene type polyene macrolide antibiotic. The strain Uk-10 was identified by morphological, cultural biochemical, physiological and chemotaxonomic studies as *S. arenae* var *ukrainiana*. The antibiotic produced intracellularly by *S. arenae* var *ukrainiana* was designated as HA-2-91(Gupte and Naik, 1999). The fermentation profile of HA-2-91 production is depicted in Figure 2.3. The isolation of HA-2-91 from the harvested mycelia is outlined in the form of flow diagram.

Time Course of HA-2-91 Production by *Streptomyces arenae var ukrainiana*

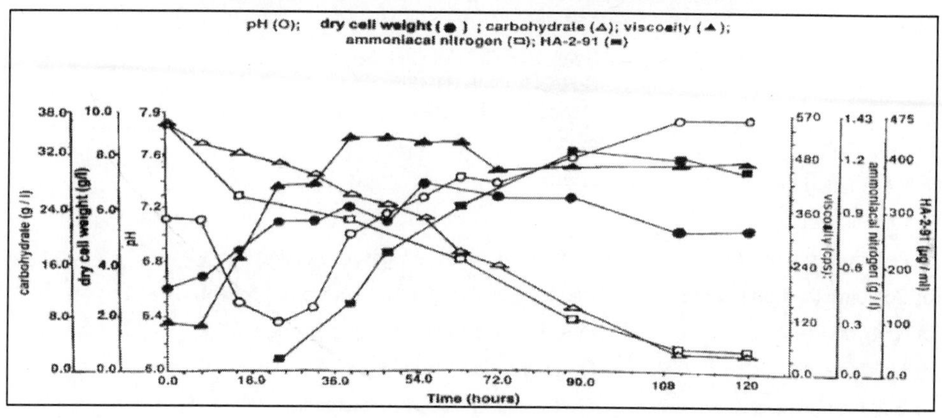

Figure 2.3: HA-2-91 Production by *Streptomyces arenae* var. *ukrrainiana* in Fermentor (450L).

Flow diagram for isolation and purification of HA-2-91

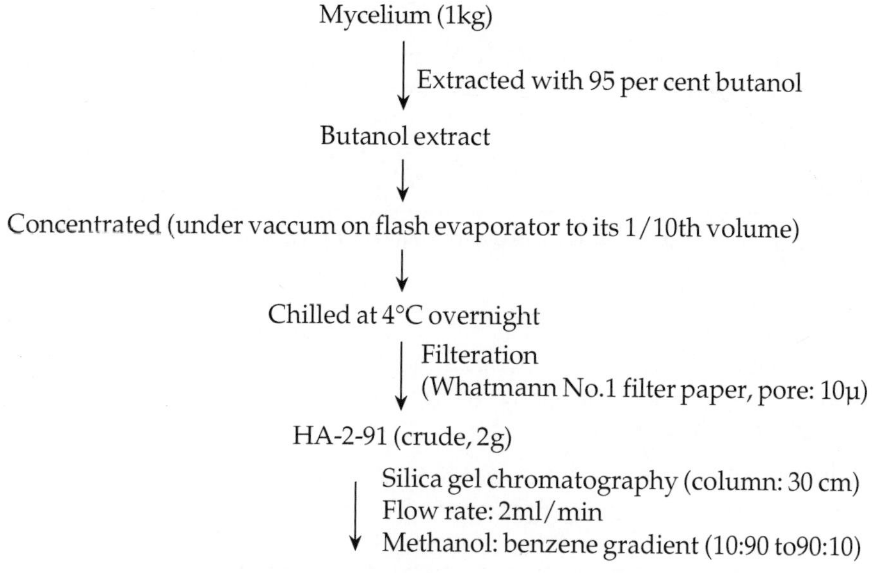

Mycelium (1kg)

↓ Extracted with 95 per cent butanol

Butanol extract

↓

Concentrated (under vaccum on flash evaporator to its 1/10th volume)

↓

Chilled at 4°C overnight

↓ Filteration (Whatmann No.1 filter paper, pore: 10µ)

HA-2-91 (crude, 2g)

↓ Silica gel chromatography (column: 30 cm)
Flow rate: 2ml/min
Methanol: benzene gradient (10:90 to90:10)

Active fractions collected; Prep. TL
(butanol- acetic acid-water-dioxone:: 4:1:2:1 v/v)

↓

Pure HA-2-91 (1.45 g)

Table 2.7: Physio-chemical Properties of HA-2-91.

Test	*Observation*
Colour	Pale Yellow
Nature	Amorphous
Melting point	154-160 °C (decomposition)
Solubility	Soluble in CH_3OH,C_2H_5OH, C_4H_9OH, DMF, DMSO. Insoluble in H_2O,$CHCl_3$, n-C_6H_{14}, $(CH_3)_2CO$.
pH (10µg/ml)	6.33
Optical rotation $[\alpha]^{23}D$	+0.077
Functional groups	
(a) Fehling's solution test	Positive (aldohexose group present)
(b) Sodium bicarbonate test	Weakly positive (carboxylic group present)
(c) Ninhydrin test	Positive (free amino group present)
(d) Molisch test	Positive (sugar present)
(e) Sulphuric acid test	Bluish violet colour (unsaturation)
(f) Ferric chloride test	Negative (phenolic OH absent)
(g) KI/CH_3COOH	Negative (epoxide group absent)
Retention time (HPLC)*	8.58 min
R_f value (TLC)**	0.63
Elemental analysis (Observed)	(C),58.37; (H),7.28;(N),1.97 per cent
$C_{39}H_{63}O_{15}N.H_2O$ (Calculated)	(C),58.28; (H),8.09;(N),1.74 per cent
UV λ max nm (E1 per cent 1cm)	290(453), 305(566), 318(449)
FTIR ν max nm (K Br) cm^{-1}	3500-3250, 2900, 2800, 1715, 1640, 1578, 1510, 1400, 1260, 1170, 1060, 840, 800
FAB-MS (M^+)	783 ($C_{39}H_{63}O_{15}N$)

* Solvent system: CH_3CN-0.05M aq. Sodium citrate buffer pH 5.3 (40:60)**Solvent system: C_4H_9OH-CH_3OH-H_2O-dioxane

Based on all physio-chemical and spectroscopic analysis the structure of HA-2-91 was found to be a tetraene having 28 member macrocyclic lactone rings with a side chain C_4H_8 at C-27 position and a mycosamine sugar moiety attached to C-17 position through ether linkage.

Table 2.8: Acute Toxicity (LD_{50}) Studies of HA-2-91 in Mice

Route of Administration	*LD_{50} (mg/kg)*	*Observation*
Oral	> 2000 mg/kg	Normal behaviour
Intraperitoneal	74(62.71 – 87.32) mg/kg	Ataxia followed by ptosis, stomach distension, diarrhoea, cause of death is respiratory arrest
Intravenous	62(58.22 – 66.03) mg/kg	Ataxia followed by depression, asphexysia, cause of death is respiratory arrest.

The acute toxicity of HA-2-91 in mice by various routes and antifungal activity profile against wide variety pathogenic and non-pathogenic yeasts and filamentous fungi (Table 2.8 and 2.9).

Table 2.9: Antimicrobial Activity Profile of HA-2-91

Organism	*MIC (μg/ml)*
Yeast	
Candida albicans HA 4463	**1.5**
Saccharomyces cerevisiae ATCC 9763 0.4	0.4
Pityrosporum ovale	8
Cryptococcus neoformans	2
Filamentous fungi	
Paecilomyces varioti HA 1585	0.6
Aspergillus niger	15
Aspergillus awamori	8
Aspergillus ochracious ATCC 28706	7.5
Penicillium chrysogenum (Toyo Jozo)	15.5
Penicillium chrysogenum (PAN LAB)	30
Acremonium chrysogenum	7
Westerdykella cylindrical	25.5
Trichoderma viride NCIM 99	27.5
Fusarium udum	15
Plant pathogens	
Helminthosporium sp. (rice)	5
Fusarium oxysporium (wilt of cotton)	1000
Macrophomina phaeseolina (rot of safflower)	10
Alternaria macrosporum (leaf spot of cotton)	5
Scleroticum sp. (collar rot of cluster bean)	5
Colletotrichicum gleosporides (rot of mango)	5
Alternaria macrosporum	10
Alternaria brassica	10
Clinical pathogens	
Aspergillus flavus	25
Aspergillus fumigatus	7.5
Epidermophyton floccosum	8
Trichophyton mentagrophytes	1.8

The probable mechanism(s) of antifungal activity of HA-2-91 found to be through binding both ergo sterol and cholesterol present on fungal or mammalian cell membranes. However, it showed preferential binding towards ergosterol than cholesterol. It was also observed that the interaction of HA-2-91, seem to be reversible process. With such preferential degree of binding of HA-2-91on memberane sterol, the effect will be greater on fungal cell membrane than mammalian cell membrane. Hence, adverse reactions like haemolysis may not occur with HA-2-91 treatment (Naik and Gupte, 2009).

Table 2.10: A Comparative Data on HA-2-91, Nystatin, Pimaricin and Rimocidin

Biological Profiles	*Tetraenes*			
	HA-2-91	*Nystatin*	*Pimaricin*	*Rimocidin*
Producing organism	*S. arenae* var *ukrainiana*	*S. noursei*	*S. natalensis*	*S. rimosus*
Other producing organism	–	*S. albulus*	*S. chattanoogenesis* *S. gilveosporus*	–
Anti-fungal activity *in-vitro* (µg/ml)				
Cryptococcus neoformans	2.0	0.3	5-10	5
Candida albicans	1.8	0.5	6	5
Aspergillus niger	1.5	>50	>50	5
Trichophyton sp.	1.5	12.5	50	5
Acute toxicity (LD_{50}) in mice (mg/kg i.v.)	62	3	30	–

Its structure has been determined on the basis of spectral data (UV, IR, ^{1}H and ^{13}C NMR, Mass).

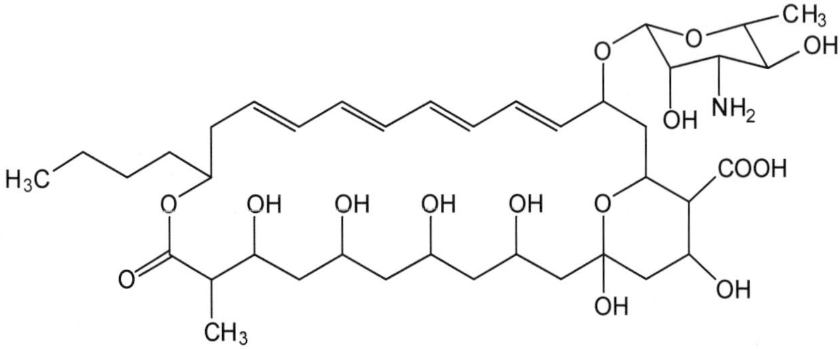

Proposed structure of HA-2-91

HA-1-92

During soil screening program, in search of new antifungal antibiotic, a *Streptomyces* culture obtain from CDRI, Lucknow, India named CDRIL-312,was found to produce a polyene macrolide presumably belonging to the oxohexaene group and designated as HA-1-92. The strain, CDRIL was cultured in a submerged condition and fermentation was carried out for 120h at 28°C with back pressure (0.7 kg cm^{-2}), aeration (1:1) and agitation (200 rev/min) in a 41 L fermentor (chemap, Sweden).

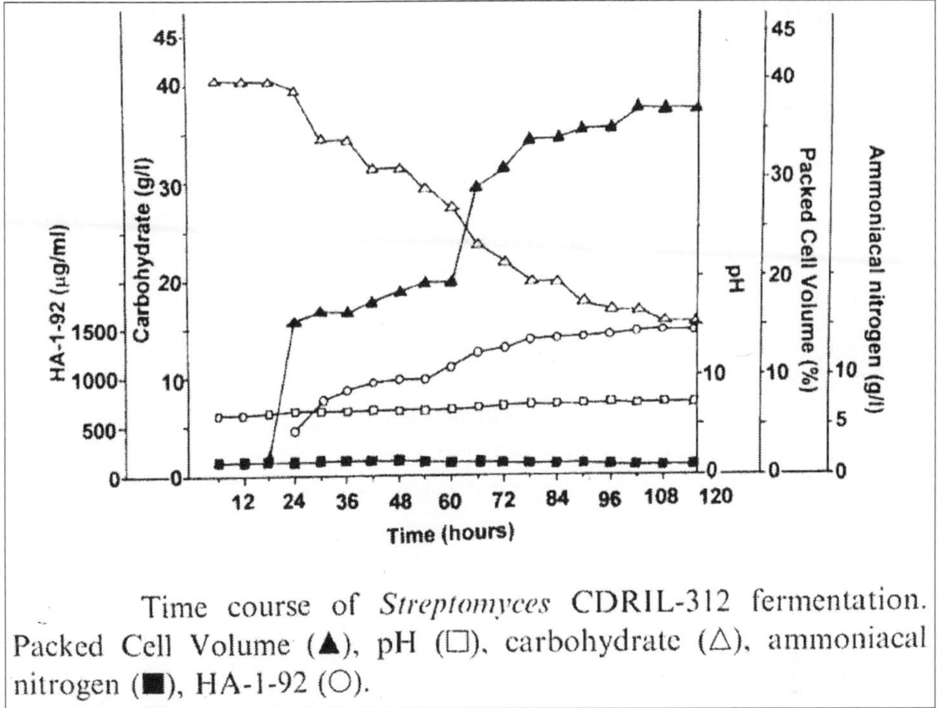

Time course of *Streptomyces* CDRIL-312 fermentation. Packed Cell Volume (▲), pH (□), carbohydrate (△), ammoniacal nitrogen (■), HA-1-92 (○).

Figure 2.4: HA-1-92 Production by *Streptomyces CDRIL-312* in a Fermentor (41L).

During fermentation, various parameters were monitored (Figure 2.4). The harvested broth was filtered and mycelial cake was suspended in butanol (1:2) stirred and then filtered under vacuum. The extract was concentrated to $1/10^{th}$of the initial volume in a rotary vacuum evaporator and chilled at 4°C for 16 h to allow HA-1-92 to precipitate. The residue was washed 3 times with hexane followed by ethyl acetate and dried at 60°C. The crude HA-1-92 was purified by silica gel (100-200 µm particle size) column chromatography. Gradient elution was carried out using benzene: methanol (90:10 to 10:90).The active fraction showing the desired UV absorption (at 404,382,361 and 432 nm) were pooled and subjected to preparative TLC (precoated silica gel,GF_{254}Merck) plates using solvent system (butanol: acetic acid: water, 4:1:5 v/v). The major active component was recovered with methanol and concentrated under vacuum at 40°C. The purity was confirmed using different solvent systems and activity was determined by microbiological assay using *Candida albicans* HA4463 as the test organism.

The physio-chemical properties of HA-1-92 were determined by spectroscopic and chromatographic techniques (UV spectrum, FTIR spectrum and others including HPLC chromatography) (Harindran *et al.*, 1999) (Table 2.11).

Table 1.11: Physio-chemical Properties of HA-1-92

Test	*Observation*
Colour	Yellow
Nature	Amorphous
Melting point	215-217 °C (decomposition)
Solubility	Soluble in methanol, ethanol, butanol, dimethyl formamide, dimethyl sulfoxide, acetoneand pyridine. Insoluble in water, ether, chloroform, carbon tetrachloride.
pH (10µg/ml)	4.39
Optical rotation $[\alpha]^{23}D$	–89.29 (c 0.2, methanol)
Functional groups	
(a) Molisch reagent	Negative (Sugar absent)
(b) Benedict's reagent	Negative (Sugar absent)
(c) Tollen's reagent	Negative (diazotization absent)
(d) Ninhydrin test	Negative (free amino group absent)
(e) Sulphuric acid test	Positive (unsaturation present)
(f) Ferric chloride test	Negative (phenolic OH absent)
(g) Silver mirror test	Positive (aldehyde or ketone present)
Elemental analysis	
Observed:	C (66.43 per cent), H (7.86 per cent), O (25.87 per cent)
Calculated:	C (67.67 per cent, H (8.29 per cent), O (24.26 per cent)
UV λ max nm ($E^{1\ per\ cent}$ 1cm)	404 (868), 382(1023), 361(772),342 (542) (Sh)
IR ν max nm (K Br) cm^{-1}	3800-3500, 3000, 2850, 2700, 1720-1700, 1650, 1600-1590, 1550, 1500, 1450, 1357-1350, 1200, 1100, 1050, 850, 750-710, 700, 650
Retention time	3.23 (major), 3.8, 4.81, 5.62 and 6.47
(HPLC)*	
R_f value of pure HA-1-92	(a) 0.32 (b) 0.52 (c) 0.61

* Solvent system, 0.05M Sodium citrate buffer, pH 5.3: acetonitrile (60:40); Column, C-18 (25x0.4 cm); Flow rate, 1 ml/min; Detection, 380nm (a) Solvent system, butanol: acetic acid: water (4:1:5 v/v). (b) Solvent system, pyridine: ethyl acetate: acetic acid: water (5:5:1:3 v/v)

Acute toxicity (LD_{50}) was carried out in mice (Swiss strain) by different routes by the method of Litchfield and Wilcoxon (1949). The LD_{50} of HA-1-92 in mice by oral found to be 1000 (900-1100) mg/kg, 7.25 (7.05-7.45) mg/kg by intravenous and 90 (75-105) mg/kg by the intra-peritoneal routes. These results suggest HA-1-92 is less toxic in mice than other known oxohexaenes *viz.*, dermostatin A 50 mg/kg ip., mycotocin A 15 mg/kg, ip., roflamycoin 24 mg/kg ip. (Berdy *et al.*, 1987).The antimicrobial activity (*in-vitro*) against yeast and filamentous fungi including plant pathogens and clinical isolates (Harindran *et al.*, 1999) are outlined in Table 1.12.

Table 1.12: Comparative Antimicrobial Activity (*in-vitro*) of HA-1-92, Hamycin, Aureofungin, Amphrotericin B and Griseofulvin

Organism	*MIC (μg/ml)*				
	HA-1-92	*Hamycin*	*Aureo-fungin*	*Ampho-tericin B*	*Griso-fulvin*
Bacteria					
Bacillus subtilis ATCC 6633	75	50	–	–	–
Staphyloccocus aureus P209	25	50	–	>50	–
Sarcina lutea ATCC 9341	75	50	–	–	–
Escherichia coli ATCC10536	10	50	–	>50	–
Pseudomonas aeruginosa NCTC 6750	5	–	–	–	–
Klebseilla pneumonia ATCC 10031	7.5	–	–	–	–
Yeast					
Candida albicans HA4463	0.05	0.01	0.05	0.04	>100
Saccharomyces cerevisiae NCTC 9763	2	0.01	0.4	0.012	–
Filamentous fungi					
Paecilomyces varioti HA 1056	0.1	0.25	5.5	0.01	–
Plant pathogens					
Helminthosporium sp.	0.5	4.5	2.5	–	–
Fusarium oxysporium	0.9	–	2	–	–
Macrophomina phaeseolina	1.5	50	2.5	–	–
Alternaria macrosporum	1	11	2.25	–	–
Scleroticum sp.	1.25	6.5	1.5	–	–
Colletotrichicum gleosporides	1.75	4.5	3.5	–	–
Curvularia lunata	0.5	5	1	–	–
Clinical pathogens					
Cryptococcus neoformans (clinical isolate)	5	0.005	0.05	0.06	>100
Aspergillus fumigatus (clinical isolate)	0.1	>20	10	3.12	>100

The liposomal incorporated HA-1-92 showed decreased toxicity and also improved pharmacokinetic profile in rats, when administered to aspergillosis and cryptococcosis – infected Balb/c mice. Liposomal HA-1-92 showed increased antifungal activity, compared with that of free HA-1-92, along with improved survival rate and decreased colony forming units in lung, liver, spleen and kidney. Hence it may be useful in controlling experimental systemic mycosis (Harindran *et al.*, 1999).However, more research work is warranted.

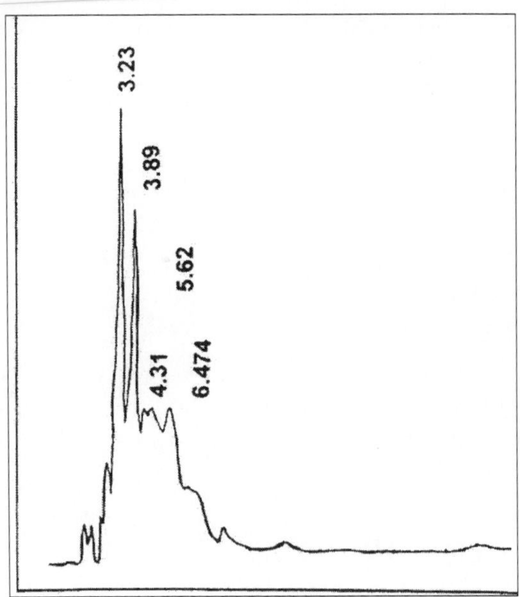

Figure 2.5: HPLC of HA-2-91.

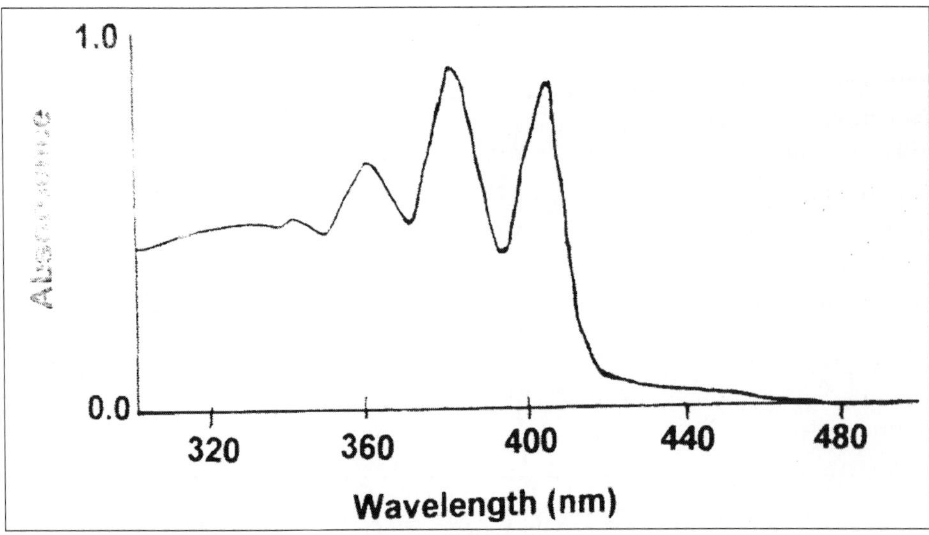

Figure 2.6: UV Spectrum of HA-1-92 in 95 per cent Methanol.

HA-94

A novel *Strepto verticillium* sp. *G-55* was isolated from a soil sample (collected from Panjim, Goa) which produces scerotia under specific conditions, both in liquid and solid media. The strain was further identified by taxonomic studies as *Strepto verticillium cinnamoneum* var. *scleroticum*. The species produced pentaene polyene

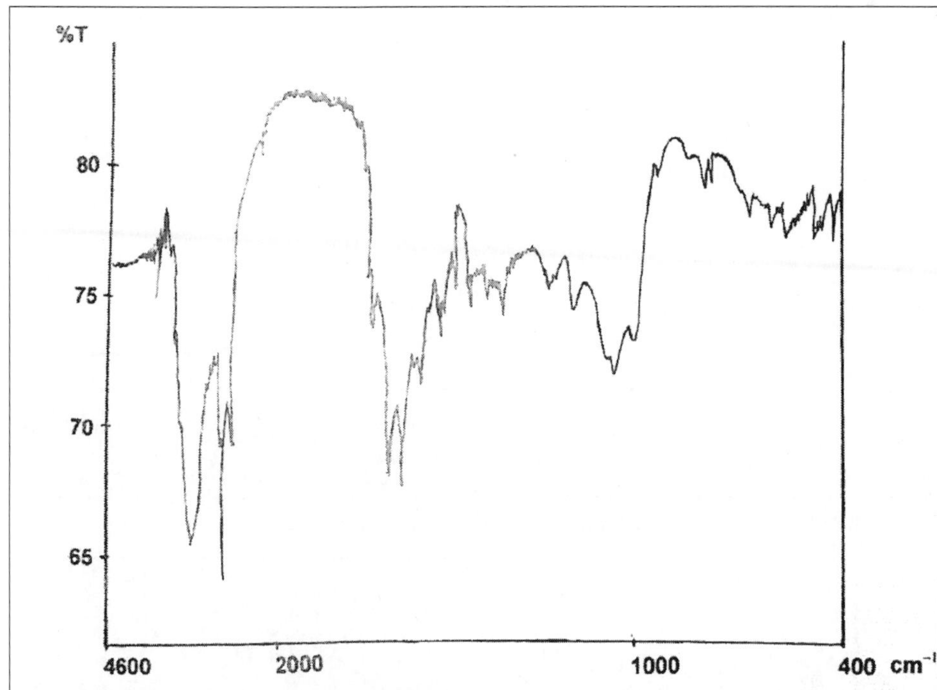

Figure 2.7: FTIR Spectrum of HA-1-92.

Table 1.13: Physico-chemical and Biochemical Characteristics of Strain G-55

Character	Description	Character	Description
Spore chain morphology	monoverticillate type with 4–5 whorls at regular intervals	crystal violet	rendering medium colourless after 10 days
Spore surface	smooth surfaced to slightly rough	sodium azide (0.02%)	no growth
Other morphological features	production of sclerotial bodies under certain environmental conditions	(0.01%)	no growth
		phenol	no growth
		lecithinase	moderate growth
Colour of spore mass	white, later on buff coloured		
Diffusible pigment	brown on certain standard media	*Sensitivity and resistance to antibiotics* (μg/ml)	
Optimum temperature	28 °C (25 ° ~ 29 °C)	Resistant to:	ampicillin (10)
Coagulation of milk	negative		carbenicillin (30)
Peptonization of milk	negative		colistin (10)
Nitrate reduction	negative		erythromycin (15)
H_2S production	negative		amphotericin B (100)
			nystatin (100)
Degradation of:			hamycin (100)
guanine	negative (moderate growth)	Sensitive to:	cephalexin (30)
tyrosine	positive		rifampicin (5)
xanthine	negative		kanamycin (30)
keratin	negative (moderate growth)		neomycin (30)
gelatin	negative		tetracycline (30)
Hydrolysis of:			
chitin	moderate	*Acid production from sugars (sugar 1% w/v):*	
pectin	negative	D-galactose	negative
starch	positive	inositol	positive
		D-fructose	negative
Growth in the presence of inhibitory compounds:		D-ribose	positive
sodium chloride (4%)	slight growth	D-trehalose	negative
(7%)	no growth		

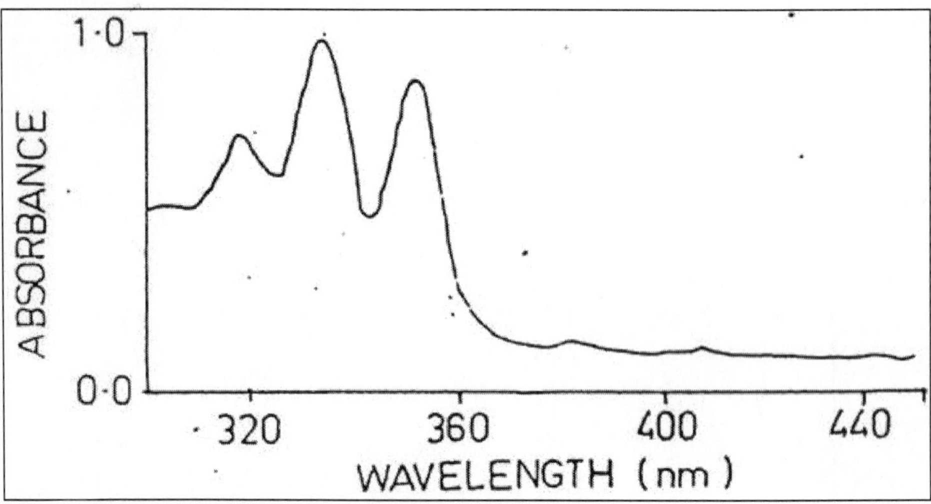

Figure 2.8: UV-Visible Spectrum of Antibiotic HA-94 in Butanol.

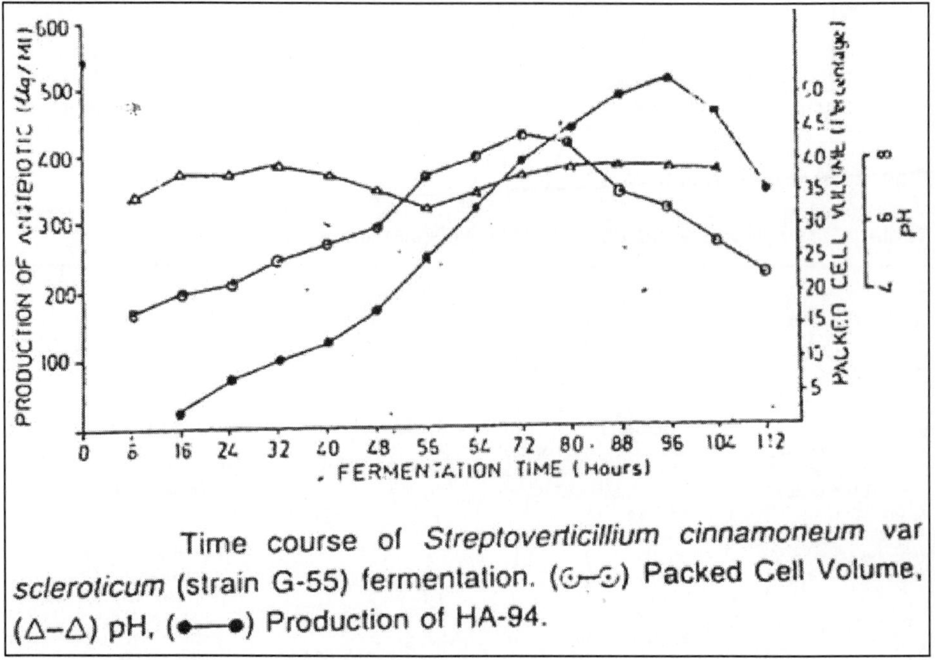

Time course of *Streptoverticillium cinnamoneum* var *scleroticum* (strain G-55) fermentation. (○–○) Packed Cell Volume, (Δ–Δ) pH, (●—●) Production of HA-94.

Figure 2.9: Fermentative Production of HA-94.

macrolide antibiotic (HA-94) under submerged culture conditions, which showed promising antifungal and antibacterial activity (*in vitro*) (Paradkar *et al.*, 1998).

Conclusion

Polyene antifungal antibiotics are heterogeneous compounds with diverse chemical structures. Furthermore, polyene antibiotics also reduce benign prostatic hyperplasia, lower serum lipid levels besides antifungal activity. Despite such claims today, we have very few polyene macrolide antibiotics namely (amphotericin-B, hamycin, candicidin, pimericin, and nystatin) that are largely active topically because of their poor absorption and toxicity. The development of new therapeutic modalities involves the use of drug carriers such as liposomes which can modify pharmacokinetics, biodistribution and also target specific delivery. However, so far, clinically curative antifungals especially for systemic or deep mycosis are not available. Therefore, a rigorous search for polyene antibiotics from microbial sources need to be explored with the improved screening models coupled with modern technology and knowledge based approach. Alternatively, development of synthetic derivatives of microbial metabolites need to be undertaken to achieve better biodistribution, improved pharmacokinetics pattern with reduced toxicity profile.

Acknowledgements

I am thankful to my ex-doctoral students, Dr J.Harindran, T.E.Gupte, S.K.Desai and postgraduate, Mr Rishabh Singh for their technical support in preparing the review article. I am also thankful to Dr. S.B. Bhise (Principal) for his encouragement.

References

Aizawa, S., Shibuya, M., and Shirato, S. (1971). Resistaphylin, a new antibiotic. Production, isolation and properties. *J.Antibiot.*, **24**: 393-396.

Aszalos, A., Robinson, R.S., Lemanski, P., and Berk, B. (1968). Trienine: an antitumor triene antibiotic. *J. Antibiot.*, **21**: 611-615.

Ananthnarayan, R., and C.K.Jay Ram Panikar. (1981). *Medical mycology*. In *textbook of microbiology*. (2nd Ed.). *Orient Longman*. pp. 555-571.

Arora, H.R.K. (1962). Cardiotnic action of *Hamycin. J. Pharm and Pharmacol*., **14**: 320

Berdy, J., Aszalos, A., Bastian, M., and McNitt, K. L. (1980). *Handbook of Antibiotics Compounds II* (CRC press, Florida), pp.165.

Berdy, J., Aszalos, A., Bostian, M., and McNitt, K.L. (1987). *Macrocyclic Lactone Antibiotics.* In CRC Handbook of Antibiotic Compounds, Vol II. Pp307-311. Boca Raton, Florida: CRC Press. ISBN 0-84933436-5.

Bhagwat, S. K., and Anjaneyulu, R. (1972). Hamycin in the treatment of Vaginitis. *J. Obstet.Gyaecol.*, **22**: 69-75

Brazhnikova, M.G., Borisova, V.N., Lysenkova, L.N., Rubasheva, L.M., Fedorova, G.B., and Bazhanov, V.S. (1979). Isolation and physio-chemical properties of components of the antitumor antibiotics 4041 produced by *Actinoplanes ianthino* genes subspecies octamycini sub species nova. *Antibiotiki.*, **24**: 582.

Cassinelli,G., Grein, A., Orezzi, P., Penella, P., and Sanfilippo, A. (1967). New antibiotics produced by *Streptoverticillium orinoci*, n sp. *Arch. Microbiol.*, **55**: 358-368.

Chuenkova,M.V., Sukhareva,N.M., Lyubimova,L.K., and Egorov, N.S.(1982). Study of trypanosome sensitivity to polyene antibiotics. *Antibiotiki.*, **27**: 424-430.

Findlay,J.A. Liu, J.S., Burnell, D.J., and Nakashima, T.T. (1982). The structure of demethoxyrapamycin. *Can. J. Chem.*, **60**: 2046-2047.

Ganju, Durga and Anjaneyulu. R.(1966). Leucorrhoea due to *Trichomonas vaginalis* cliunical trials with hamycin. *Hind. Antibiot. Bul.*, **9**: 751-769

Golding, B. T., Richard, R.W., Meyer,W.E., Patrick, J.B., and Barber, M. (1966). The structure of the macrolide antibiotic pimaricin.*Tetrahedron Lett.*, **7**: 3551-3557.

Gupte,T.E., and Naik, S. R.(1999).Isolation, taxonomic andfermentation studies on new strain of *Streptomyces arnae* var *ukrainiana* producing a tetraene antibiotic. *World J Microbiol and Biotechnol.*, **15**: 545-552.

Gupte,T.E., Gaikwad, U.V., and Naik, S. R.(1999a).Experimental studies (*in vitro*) on polyene macrolide antibiotics with special reference to hamycin against *Malassezeia ovale. Comp. Immunol. Microbiol. Infect. Dis.*, **22**: 93-102

Harindran,J., Gupte, T.E., and Naik, S.R. (1999). HA-1-92, A new antifungal antibiotic produced by *Streptomyces* CDRIL-312: Fermentation, Isolation, Purification and Biological activity. *World J Microbiol and Biotechnol.*, **15**: 425-430.

Harada, K., Suzuki, M., Takeda, N., and Tatematsu, A. (1982). Molecular secondary ion mass spectra of kanamycins. *J.Antibiot.*, **35**: 102-105.

Harindran J. (1996). *Fermentation, Chemistry and Pharmacology of new microbial products of soil origin. Ph D. Thesis* submitted to the University of Pune, Maharashtra, India.

Horii, S., Shima, T., and Ouchida, A. (1970). Partial structure of eurocidin complex. *J. Antibiot.*, **23**: 102-104.

Jordan, G.W., and Seet, E.C.(1978).Antiviral effects of amphotericin B methyl ester. *Antimicrobial Agents and Chemotherepy*, **13**: 199-204.

Korzybski, T., and Kurylowicz, W. (1961). Antibiotica; Herkunft, Arten, Eigenschaften. *Fischer Jena.*, 232-241.

Litchfeild, J.T., and Wilcoxon, F. (1949). A simplified method of evaluating dose-effect experiments. *J Pharmacol Exp Ther.*, **96**: 99-113.

Malewicz, B., and Borowski, E. (1979). Energy dependence and reversibility of membrane alternations induced by polyene antibiotics in *Chorella vulgaris. Nature*, **281**: 80-82.

Martin J.F., and McDaniel, L.E. (1974). Isolation, purification and properties of the hexane macrolides. Candihexin I and Candihexin II.J. *Antibiot.*, **27**: 610-619.

Martin, E., Stuben, A., Gorz, A., Weller, V., and Bhakdi, S. (1994). Novel aspects of Amphotericin B action: Accumulation in human Monocyte potentiates killing of phagocytosed *C. albicans. Antimicrob. Agents Chemother.*, **38**: 13-22.

Mechlinski, W., Schaffner, C.P., Ganis, P., and Avitable, G. (1970). Structure and absolute configuration of the polyene macrolide antibiotic amphotericin B. *Tetrahedron Lett.*, **11**: 3873-3876.

Medoff, G., and Kobayashi, GA. (1980). *The polyenes. Antifungal Chemotherepy*. Ed. By D.C.E. Speller. A Wiley-Interscience Publication.pp. 3-34.

Medoff, G., Kobayashi, G.S., Kwan, C.N., Schlessinger, D., and Venkov, P. (1972). Potentiation of rifampin and 5-fluorocytosine as antifungal antibiotics by amphotericin B. *Proc. Natl. Acad. Sci.*, **69**: 196-1999.

Mitskevich,V.M., Tsyganov, V.A., Lvovich, T.A., and Yu. D. Shenin. (1975). *Actinomyces rectila vendulae* var *pantaenicus*, an organism producing rectilavendomycin, a new pentaenic antibiotic. *Antibiotiki.*, **20**: 202.

Naik S R. (1983). Effect of Hamycin on prostate secretion and serum testosterone levels in the rabbits. *Hind Antibiot Bull.*, **25**: 125-130.

Naik, S. R., Raman, P. H., Jalit, R. N. and Karnick, A. K. (1981). Hypolipidemic and hypocholesterolemic activity of JA-20, a polyene macrolide. *Hind Antibiot Bull.*, **23**: 12-17.

Naik, S.R., Desai, S.K., Nanda, R.K., and Narayan, M.S. (2007). Fermentation, isolation, purification and biological activity of SJA-95, a heptaenepolyene macrolide antibiotic produced by *Streptomyces* sp. Strain S-24. *Arzneim Forsch (Drug Res).*, **57**: 171-179.

Naik, S.R., and Harindran, J. (2000). Hypolipidemic, hypoprostatic and pharmacological activity of oxohexanepolyene antibiotic (HA-1-92). *Indian J. Exp. Biol,.* **38**, 727-729.

Naik, S.R., Harindran, J., and Nanda, R. K. (2001). Structural studies on aoxohexaene antibiotic (HA-1-92) produced by *Streptomyces CDRIL-312. Ind. J.Chem.*, **40B**: 1183-1186.

Naik, S.R., Thakare, V.N., Desai, S.K., and Rahalkar, P.R. (2011). Study of immunological aspects of Aspergillosis in mice and effect of polyene macrolide antibiotic (SJA 95) and IFN- γ: A possible role of IFN- γ as an adjuant in antifungal therepy. *Immunol Lett.*, **141**: 68-73.

Naik S.R., and Gupte T.E. (2009). Probable mechanism(s) of antifungal activity of HA-2-91; A tetraene antibiotic. *J. Biotechnol.*, **8**: 91-96

Nishikiori, T., Masuma, R., Oiwa, R., Katagiri, M., Awaya, J., Iwai, Y., and Omura, S. (1978). Aurantinin, a new antibiotic of bacterial origin. *J. Antibiot.*, **31**: 525-532.

Oroshnik,W., Vining, L.C., Mebane, A.D., and Taber, W.A. (1955). Polyene antibiotics. *Science*, **121**: 147-149.

Omura, S., and Tanaka, H. (1984). Production, structure and antifungal activity of polyene macrolide. *In Macrolide Antibiotics: Chemistry, Biology and Practice*, ed. Omura S. pp. 351-404. Boca Raton, Florida: Academic Press. ISBN 0-12-526450-X.

Omura, S., Nishikiori, T., Oiwa, R., Iwai, Y., Masuma, R., and Katagri, M. (1976). A new antibiotic KM-214 produced by a strain of *Bacillus*. *J. Antibiot.*, **26**: 477-478.

Pandey, R.C., Guenther, E.C., Aszolos, A.A., and Brajtburg, J. (1982). Physio-chemical and biological composition of polyene macrolide antibiotics fungichromin, lagosin and cogomysin. *J. Antibiot.*, **35**: 988-996.

Pandey, R.C., Rinehart Kenneth, L. Jr. and Millington Davis, S.(1980). Dermostatin's chemical structure. *Hind.Antibiot. Bull.*, **22**: 4-15

Paradkar U.R., Gupte T.E., Joshi, A.P., and Naik S.R. (1998). A novel *Streptoverticillium cinnamomeum* var *sclerotium* producing a polyene antibiotic. *World J.Microbiol.Biotechnol.*, **14**: 705-709.

Prave, P., Sukatsch, D., and Vertesyn, L. (1972). Protocidin. A new phosphorus-containing antibiotic. I. Taxanomy: Fermentation, Isolation and biological properties *J. Antibiot.*, **25**: 1-3.

Radics, L., Incze, M., Dornberger, K., and Thrum, H. (1982). Tetramycin B, a new polyene macrolide antibiotic.The structure of Tetramycins A and B as studied by high- field NMR spectroscopy. *Tetrahedron*, **38**: 183-189.

Rinehart, K.L., Jr. (1982). Fast atom Bombardment Mass Spectroscopy. *Science*, **218**: 254-260.

Rychnovsky, S.D., Richardson, T.I., and Rogers, B.N. (1997). Two dimentional NMR analysis of Acetonide Derivatives in the stereochemical assignment of polyol chains. The absolute configuration of Dermostatin A and B. *J. Org. Chem.*, **62**: 2925-2934.

Santra, S.C., and Chandra, A.L. (1981). Chemical and biological studies of polyenes-a review. *Trans. Bose.Res.Inst.*, **44**: 65 -73.

Schaffner, C.P., and Gordon, H.W. (1968). The hypocholesterolemic activity of orally administered polyene macrolides. *Proc. Natl. Acad. Sci*, U.S.A., **61**: 36-41.

Sedlak, M., Pravda, M., Kubicova, L., Mikulcikova, P., and Ventura, K. (2007a). Synthesis of pH-sensitive amphotrecin B-poly (ethylene glycol) - b-poly (l-lysine) conjugate. *Bio org Med Chem Lett.*, **17**: 2554-2557.

Sedlak, M., Pravda, M., Staud, F., Kubicova, L., Tycova, K., and Ventura, K. (2007b). Synthesis of pH-sensitive amphotrecin B-poly (ethylene glycol) conjugates and study of their controlled release *in vitro*. *Bio org Med Chem Lett.*, 1**5**: 4069-4076.

Shenin, Yu. D., Andreeva, L.B., and Kruglikova, L.F. (1981). Aglycon structure of gannibamycin, a polyene antibiotic. *Antibiotiki.*, **26**: 19-23.

Swamy, M.B., Sastry, M.K., and Nanda, R.K.(1994). Dermostatin A and B: Chromatography,Structure and Configural studies using HPLC, CCD,^{13}C (125 MHz) and 1H (500 MHz) NMR spectroscopy. *Hind Antibiot Bull.*, **36**: 6-20

Taig, M.M., Solovieva, N.K., and Braginskaya, P.S. (1969). Characteristics of aurenin-producing organism. *Antibiotiki.*, **14**: 873-876.

Tereshin, I.M. (1976). Sources and biosynthesis of polyene antibiotics.In, *polyene* antibiotics- Present and Future. Univ. of Tokyo press, Tokyo. pp. 7-62.

Thirumalachar, M.J., Menon, S.K., and Bhatt, V.V. (1961). Hamycin: a new antifungal antibiotic I. Discovery and biological studies. *Hind Antibiot Bull.*, **3**: 136-138.

Thirumalachar, M.J., Rahalkar, P.W.,Sukapure,R.S., and Gopalkrishnan, K.S. (1964). Aureofungin : a new heptane antibiotic. I. Microbiological studies. *Hind Antibiot Bull.*, **6**: 108-111.

Vezina, C., Kudelski, A., and Sehgal, S.N. (1975). Rapamycin (AY-22,989), a new antifungal antibiotic. I. Taxonomy of the producing Streptomycetes and isolation of the active principle. *J.Antibiot.*, **28**: 721-726.

Natural Products: Research Reviews Vol. 3 (2016) *Pages* **65–74**
Editor: **V.K. Gupta**
Published by: **DAYA PUBLISHING HOUSE, NEW DELHI**

3

Pongamia pinnata: A Review on Pharmacological and Therapeutic Properties

Disha B. Menpara[1] and Sumitra V. Chanda[1*]

ABSTRACT

Many Indian plants have been used from time immemorial to treat various diseases and infections in traditional medicinal systems of India. Traditional system of medicine consists of large number of plants with various medicinal and pharmacological importance and hence represents a priceless tank of new bioactive molecules. Therapeutic approaches with natural products provide a fruitful source for searching safe, effective and relatively inexpensive new remedies for treatment of various ailments. Pongamia pinnata Pierre (Fabaceae) is a medicinal plant native to Western Ghats and chiefly found in tidal forests of India. It is commonly known as Karanj and has been recognized in different traditional system of medicines for the treatment of varied diseases of human beings. It has been used as folk medicinal plant, particularly in Ayurveda and Siddha systems of Indian medicine. All parts of this plant have been used as a crude drug for the treatment of tumours, piles, skin diseases, itches, abscess, painful rheumatic joints wounds, ulcers, diarrhoea etc. This review is an attempt to examine and compile the information on recent investigations regarding documented pharmacological activities.

Keywords: *Pongamia pinnata*, Pharmacological activity, Medicinal plant.

Introduction

Pongamia pinnata Pierre (Fabaceae) (Figure 3.1) is a medicinal plant (native to Western Ghats and chiefly found in tidal forests of India (Krishnamurthi, 1969). It is

1 Phytochemical, Pharmacological and Microbiological Laboratory, Department of Biosciences (CAS), Saurashtra University, Rajkot-360005, Gujarat, India

* *Corresponding author*: E-mail: svchanda@gmail.com

Figure 3.1: *Pongamia pinnata*.

widely distributed throughout the greater part of India, especially at low levels in wet places. It is locally known as Karanja. It is a medium size glabrous tree attaining height of round18 m (Chopra, 1986; Simin, 2002). It is also called as *Pongamia glabra*, *Derris indica*.

Botanical Classification

Kingdom	:	Plantae
Division	:	Magnoliophyta
Class	:	Magnoliopsida
Order	:	Fabales
Family	:	Fabaceae
Genus	:	*Pongamia*
Species	:	*pinnata*

Traditional Uses

In traditional system of medicine leaves are used in eczema, scabies, leprosy, piles, ulcers (Nadkarni, 1976) and rheumatic pain (Kirtikar and Basu, 2004). This plant has been used in traditional medicine for bronchitis, whooping cough, treatment of rheumatic joints and to quench dipsia in diabetes (Kirtikar and Basu, 1995). The bark is used in piles. Roots are used for cleaning gums, teeth, and ulcers also effective in fistulous sores and gonorrhea (Rastogi and Malhotra, 2001; Chauhan and Chauhan, 2002). Seed extract has hypotensive effects and produces uterine contractions. Powdered seed is used in bronchitis, chronic fever, whooping cough, chronic skin diseases and painful rheumatic joints (Ingredient guide). Seed oil is used in scabies, leprosy, piles, ulcers, chronic fever and lever pain. Its oil is a source of biodiesel and it is also used as fuel for cooking and in lamps (Mahli *et al.*, 1989). In a study conducted by Bhat *et al.* (2014) *P. pinnata* was found to have highest use value in the coastal parts of Central Western Ghats, Karnataka, India.

Reported Activities

Antimicrobial Activity

Medicinal plants have a rich repository of antimicrobial compounds. They are sources of many potent and powerful drugs. Kesari *et al.* (2010) reported that *Pongamia pinnata* seed oil showed inhibition against the tested fungal and bacterial cultures. Unsaturated fatty acids in seed oil were attributed to be responsible for this activity. Aqueous seed extract of *P. pinnata* was evaluated for *in vitro* antiviral properties using Vero cells against herpes simplex virus type- 1 (HSV- 1) and type- 2 (HSV-2) (Elanchezhiyan *et al.*, 1993). They found inhibition of growth of HSV-1 and HIV-2 with no toxic effect on Vero cells.

According to Kagithoju *et al.* (2012) flower extracts showed antibacterial activity with 80 per cent acetone in water and petroleum ether as solvents against enteric pathogens which are the major bacterial agents causing serious infections. Simin *et al.* (2002) isolated a new rotenoid Pongarotene along with Karanjin from seeds of *P. pinnata*. Pongarotene was strongly active against *Escherichia coli* and *Staphylococcus aureus* and karanjin was strongly active against *Coynebacterium diphtheriae*.

Antibacterial property of leaf of *P. pinnata* was reported by Arote *et al.* (2009). Antibacterial activity was studied by disc diffusion method against certain enteric bacterial pathogens using petroleum ether extract, chloroform extract, ethyl acetate extract and methanol extracts of leaf. Out of all these extracts, methanol extracts had wide range of antibacterial activity on these bacterial pathogens.

In a study by Sajid *et al.* (2012) different solvent extracts from bark, leaves and seeds were used for antimicrobial activity against a set of bacterial and fungal strains. The strongest antimicrobial activity with the largest inhibition zone and lowest minimum inhibitory concentration (MIC) was found with the bark extract.

In vitro antimicrobial activity of cycloart-23- ene-3β, 25-diol (called as B2) isolated from stem bark of *P. pinnata* was investigated by Badole *et al.* (2011). Antimicrobial activity of B2 was determined by cup plate method in different concentration range of 10-100μg/mL. B2 exhibited broad spectrum activity against bacteria and strong activity against yeast type of fungi.

Antioxidant Activity

Anuradha *et al.* (2011) reported that methanolic extract from the flowers of *P. pinnata* was evaluated against lead acetate induced hepatic damage in male albino rats. The decreased enzymatic antioxidants and non-enzymatic antioxidants were restored towards normalization by methanolic extract at a dose level of 150 mg/kg b.wt/day, administered orally daily once for 90 days. This indicated that methanolic extract of *P. pinnata* has got a potent antioxidant action against lead acetate induced hepatic damage in rats.

Badole *et al.* (2011) reported *in vitro* antioxidant activity of Cycloart-23- ene-3β, 25-diol isolated from stem bark of *P. pinnata* by using various antioxidant methods like DPPH radical scavenging, hydroxyl radical scavenging, hydrogen peroxide scavenging, nitric oxide radical scavenging and metal chelating activity.

Anti-diabetic Activity

The current diabetes epidemic is a global concern. Current therapeutics for diabetes is often associated with undesirable side effects. This plant has been used in traditional medicine to quench dipsia in diabetes.

The effect of pongamol, obtained from the fruits of *P. pinnata*, on glucose uptake and GLUT4 translocation in skeletal muscle cells was studied by Tamrakar *et al.* (2011). L6-GLUT4myc myotubes were treated with pongamol. This significantly promoted both glucose transport and GLUT4 translocation to the cell surface in a concentration-dependent manner. It was concluded that pongamol stimulated glucose uptake by increasing surface GLUT4 level in skeletal muscle cells.

A study was conducted by Vadivel and Biesalski (2011) to analyze antioxidant and type II diabetes related enzyme inhibition properties of methanolic extract of raw and traditionally processed *P. pinnata* seeds. The study was aimed to promote seeds as a dietary ingredient in the supplementary foods to be formulated with therapeutic value. It was found that methanolic extract contains appreciable levels of total free phenolics with promising antioxidant and type II diabetes related enzyme inhibition properties. Indigenous processing methods like soaking and cooking had resulted in substantial loss of total free phenolics, antioxidant and type II diabetes related enzyme inhibition properties.

Gastroprotective Activity

Giri *et al.* (2010) used hydroalcoholic extract of leaves of *P. pinnata* to treat acute gastric ulceration in rats produced by oral administration of various noxious chemicals including aspirin, ethanol, indomethacin and pylorus-ligated technique. Extract was administered at the dose of 400 mg/kg orally in all experiments. The anti-ulcer effect was further confirmed histologically. The extract showed significant protection against gastric ulcers in all the models.

In another study conducted by Singh *et al.* (2011), acute gastric ulceration was induced in Wistar Albino rats by oral administration of ethanol or indomethacin and pylorous ligation and cold restraint stress. Methanolic extract of bark was used to treat acute gastric ulceration. The extract was found to decrease the ulcer index thereby increasing the percentage of ulcer protection.

Prabha *et al.* (2009) reported ulcer protective and healing effects of methanolic extract of seeds of *P. pinnata* in rats. The gastric ulcers were induced by 2 h cold restraint stress. Dose dependent ulcer protective effects were seen with the extract. The ulcer protective effects were attributed to the flavonoids present in the extract.

Gastroprotective properties of Karanjin from *P. pinnata* seeds was reported by Vismaya *et al.* (2010). Karanjin, a furano-flavonoid from seeds was evaluated for anti-ulcerogenic property in adult male albino rats. Karanjin could inhibit 50 and 74 per cent of ulcers induced by swim stress at 10 and 20 mg/kg b.w., respectively, protect gastric mucin and inhibit oxidative stress. Also the Karanjin treated rats indicated no adverse effects on major organs.

Others

The seed oil of Karanj has been found to have anti-proliferative properties that inhibit proliferation of human cancer cell lines MCF-7 and HeLa (Arulvasu *et al.*, 2012). The mechanism of inhibitory action was attributed to DNA fragmentation. In another study by Behera *et al.* (2012b) hepatoprotective activity of hydroalcoholic leaf extract of *P. pinnata* against ischemia/reperfusion (I/R) induced hepatic injury was assessed in Wistar Albino rats. Treatment with hydro-alcoholic leaf extract reversed all biochemical parameters as well as histological alterations induced by I/R and a significant correlation existed between concentration of the extract and alteration in the biochemical and histological parameters. Analgesic activity of ethanolic extract of *P. pinnata* leaves was reported by Yadav *et al.* (2011). *In vivo* analgesic activity was evaluated by using the Eddy's Hot plate method in mice. A significant analgesic activity was elicited in a dose-dependent manner.

Conclusion

In traditional system of medicine, *Pongamia pinnata* has been widely used as remedial agent for variety of ailments. This investigation therefore justifies its ethnomedicinal use and can thus be used to discover bioactive natural products that will lead to the development of new pharmaceuticals. Owing to its versatile characteristics, this plant can be considered as multipurpose tree with immense medicinal value and can be used in the future for the formulation of many herbal products. A detailed part wise reported activity of *P. pinnata* has been given in Table 3.1.

Table 3.1: Reported Activities of *P. pinnata*

Sl.No.	*Part Used*	*Activity*	*Extract*	*Reference*
1.	Leaf	Antioxidant	Hydroalcoholic	Sagar *et al.*, 2013
2.	Leaf	Antioxidant	Hydroalcoholic	Behera *et al.*, 2012a
3.	Leaf	Hepatoprotective	Hydroalcoholic	Behera *et al.*, 2012b
4.	Leaf	Antioxidant, Antimicrobial	Leaf essential oils	Reddy *et al.*, 2012
5.	Leaf	Antioxidant, Antimicrobial	ME, ET, AC, AQ, AQME, AQET, AQAC	Sajid *et al.*, 2012
6.	Leaf	Antibacterial	PE, CH	Biswal *et al.*, 2011a
7.	Leaf	Antifungal	CH	Biswal *et al.*, 2011b
8.	Leaf	Antioxidant	ME	Gupta and Sharma, 2011
9.	Leaf	Antioxidant	ME	Sagwan *et al.*, 2011
10.	Leaf	Antixanthomonas	Cow urine extract	Shanthi *et al.*, 2011
11.	Leaf	Antimicrobial	CH, AC	Chandrashekhar and Prasanna 2010

Contd...

Table 3.1–*Contd...*

Sl.No.	Part Used	Activity	Extract	Reference
12.	Leaf	Gastroprotective	Hydroalcoholic	Giri *et al.*, 2010
13.	Leaf	Antioxidant	ET	Govindaraj *et al.*, 2010
14.	Leaf	Antibacterial, Larvicidal	CH	Mondal *et al.*, 2010
15.	Leaf	Antibacterial	PE, CH, EA, ME	Arote *et al.*, 2009
16.	Leaf	Antibacterial	PE, CH, EA, ME	Dahikar *et al.*, 2008
17.	Leaf	Antioxidant	ET	Essa and Subramanian, 2007
18.	Leaf	Antimicrobial	Crude decoction	Brijesh *et al.*, 2006
19.	Seed	Antibacterial	ME, ET	Rani *et al.*, 2013
20.	Seed	Anticancer	Oil	Arulvasu *et al.*, 2012
21.	Seed	Antibacterial	Cow urine extract, HE, CH, EA, ET, ME, AQ	Murugan *et al.*, 2012
22.	Seed	Insecticidal	ME	Singh and Sinha, 2012
23.	Seed	Antioxidant, Antimicrobial	ME, ET, AC, AQ, AQME, AQET, AQAC	Sajid *et al.*, 2012
24.	Seed	Antimicrobial	Pongarotene and Karanjin	Simin *et al.*, 2012
25.	Seed	Antioxidant	ME	Sagwan *et al.*, 2011
26.	Seed	Antioxidant	ME	Vadivel and Biesalski, 2011
27.	Seed	Gastroprotective	Karanjin	Vismaya *et al.*, 2010
28.	Seed	Antimicrobial	Oil	Kesari *et al.*, 2009
29.	Seed	Antibacterial, Larvicidal	CH	Mondal *et al.*, 2010
30.	Seed	Antiulcer	ME	Prabha *et al.*, 2009
31.	Seed	Antibacterial	PE	Baki *et al.*, 2007
32.	Seed	Antimicrobial	Oil	Wagh *et al.*, 2007
33.	Seed	Antibacterial	ME, EA	Simin *et al.*, 2012
34.	Seed	Antiviral	AQ	Elanchezhiyan *et al.*, 1993
35.	Flower	Antibacterial	PE, AC	Kagithoju *et al.*, 2012
36.	Flower	Antioxidant	ME	Anuradha and Krishna-moorthy 2011
37.	Flower	Antioxidant	ET, EA, BT, AQ	Babu and Rao, 2010
38.	Flower	Antihyperglycemic, Antilipidperoxidative	AQ	Punitha *et al.*, 2006
39.	Stem	Antioxidant, Antimicrobial	Cycloart-23-ene-3β, 25-diol	Badole *et al.*, 2011
40.	Stem	Antioxidant	ME	Sagwan *et al.*, 2011
41.	Stem	Gastroprotective	ME	Singh *et al.*, 2011
42.	Stem	Antibacterial,Larvicidal	CH	Mondal *et al.*, 2010
43.	Stem	Antidiabetic	PE, AQ	Badole and Bodhanker, 2009

Acknowledgements

The authors thank Prof. S. P. Singh, Head, Department of Biosciences, Saurashtra University, Rajkot, Gujarat, India for providing excellent research facilities. One of the authors, Disha Menpara is thankful to DSA, UGC, New Delhi, India for providing financial support (Junior Research Fellowship).

References

Anuradha, R., and Krishnamoorthy, P. (2011). Antioxidant activity of methanolic extract of *Pongamia pinnata* on lead acetate induced hepatic damage in rats. *African Journal of Biochemistry Research*, **5**: 348-351.

Arote, S.R., Dahikar S.B., and Yeole, P.G. (2009). Phytochemical screening and antibacterial properties of leaves of *Pongamia pinnata* Linn. (Fabaceae) from India. *African Journal of Biotechnology*, **8**: 6393-6396.

Arulvasu, C., Suppriya, S.V., and Bapu, G. (2012). Anticancer activity of *Pongamia glabra* V. seed oil extract against selected human cancer cell lines. *International research Journal of Pharmacy*, **3**: 131-134.

Babu, D.R., and Rao, G.N. (2010). *In vitro* studies on extracts of *Pongamia pinnata* (L.) Pierre flowers as a potent antioxidant. *International Journal of Agriculture and Food Science Technology*, **1**: 7-11.

Badole, S.L., and Bodhankar, S.L. (2009). Investigation of antihyperglycaemic activity of aqueous and petroleum ether extract of stem bark of *Pongamia pinnata* on serum glucose level in diabetic mice. *Journal of Ethnopharmacology*, **123**: 115–120.

Badole, S.L., Zanwar, A.A., Khopade, A.N., and Bodhankar, S.L. (2011). *In vitro* antioxidant and antimicrobial activity cycloart-23-ene-3β, 25- diol (B2) isolated from *Pongamia pinnata* (L. Pierre). *Asian Pacific Journal of Tropical Medicine*, **4**: 910-916.

Baki, M.A., Sadik, G., Shahadat Hossain Mondal, K.A.M., Mosaddik M.A., and Rahman, M.M. (2007). Methylkarranjic acid and pongamol from *Derris indica* seeds and their antibacterial activity. *Dhaka University Journal of Pharmaceutical Sciences*, **6**: 9-13.

Behera, S., Manohar Babu, S., and Roja Ramani, Y. (2012a). Phytochemical investigation and study on antioxidant properties of *Pongamia pinnata* hydro-alcoholic leaf extract. *Plant Sciences Feed*, **2**: 74-78.

Behera, S., Manohar Babu, S., Roja Ramani, Y., and Choudhury, P.K. (2012b). Studies on hepatoprotective activity of hydroalcoholic leaf extract of *Pongamia pinnata* against i/r induced hepatic reperfusion injury. *International Journal of Pharmacy and Biological Sciences*, **2**: 15-30.

Bhat, P., Hegde, G.R., Hegde, G., and Mulgund, G.S. (2014). Ethnomedicinal plants to cure skin diseases–An account of the traditional knowledge in the coastal parts of Central Western Ghats, Karnataka, India. *Journal of Ethnopharmacology*, **151**: 493–502.

Biswal, B., Saha, D., Jena, A., Jana, S.B., Koley, A., Sur, D., and Mohanty, J.C. (2011a). Antibacterial activity of leaf extract of *Derris indica*. *International Journal of Research in Pharmaceutical and Biomedical Sciences*, **2**: 271-273.

Biswal, B., Saha, D., Koley, A., Sur, D., Jana, S.B., Jena, A., and Sarkar, A. (2011b). Antifungal activity of leaf extract of *Derris indica*. *International Journal of Applied Biology and Pharmaceutical Technology*, **2**: 30-32.

Brijesh, S., Daswani, P.G., Tetali, P., Rojatkar, S.R., Antia, N.H., and Birdi, T.J. (2006). Studies on *Pongamia pinnata* (L.) Pierre leaves: understanding the mechanism(s) of action in infectious diarrhea. *Journal of Zhejiang University*, **7**: 665-674.

Chandrashekar, K.S., and Prasanna K.S. (2010). Antimicrobial activity of *Pongamia pinnata* leaves. *International Journal of Medicinal Research*, **1**: 18-20.

Chauhan, D., and Chauhan, J.S. (2002). Flavonoid glycosides from *Pongamia pinnata*. *Pharmaceutical Biology*, **40**: 171-174.

Chopra, R.N., Nayar, S.L., and Chopra, I.C. (1986). Glossary of Indian Medicinal Plants (Including the Supplement). Council of Scientific and Industrial Research, CSIR Publications, New Delhi. C.S.I.R. (Council of Scientific and Industrial Research). 1948–1976. The Wealth India 11 vols. New Delhi.

Dahikar, S. B., Arote S. R., and Yeole, P. G. (2008). Phytochemical screening and antibacterial properties of leaves of *Pongamia pinnata* Linn. (Fabaceae) from India. *Continental Journal of Microbiology*, **2**: 5 - 10.

Elanchezhiyan, M., Rajarajan, S., Rajendran, P., Subramanian, S., and Thyagarajant, S.P. (1993). Antiviral properties of the seed extract of an Indian medicinal plant, *Pongarnia pinnata* Linn., against Herpes Simplex viruses: *in vitro* studies on Vero cells. *Journal of Medical Microbiology*, **38**: 262-264.

Essa, M.M.M., and Subramanian, P. (2007). Effects of *Pongamia pinnata* on lipid peroxidation products and antioxidants in hyperammonemic rats: with reference to circadian variations. *Iranian Journal of Pharmacology and Therapeutics*, **6**: 119-123.

Giri, M.A., Bhalke, R.D., and Pal, S.C. (2010). Gastroprotective effect of hydroalcoholic leaves extract of *Pongamia pinnata*. *International Journal of Pharma and Bio Sciences*, **1**: 1-6.

Govindaraj, Y., Karthikeyan, V., Sivakumar, G., Suresh, A., and Nema, R.K. (2010). Antioxidant activity of the combined ethanolic extract of *Derris brevipes* and *Derris indica* against carbon tetrachloride induced oxidative stress in Wistar rats. *International Journal of Pharmaceutical and Clinical Research*, **2**: 54-57.

Gupta, V., and Sharma, M. (2011). *In vitro*, antioxidant studies on methanolic extracts of *Pongamia Pinnata* (L.) Pierre: A great versatile leguminous plant. *Journal of Pharmacy Research*, **4**: 2241-2244.

Kagithoju, S., Godishala, V., Pabba, S.K., Kurra, H.B., and Swamy, N.R. (2012). Anti bacterial activity of flower extract of *Pongamia pinnata* Linn. An elite medicinal plant. *International Journal of Pharmacy and Pharmaceutical Sciences*, **4**: 130-132.

Kesari, V., Das, A., and Rangan, L. (2009). Physico-chemical characterization and antimicrobial activity from seed oil of *Pongamia pinnata*, a potential biofuel crop. *Biomass and Bioenergy*, **34**: 108 – 115.

Kirtikar, K.R., and Basu, B.D. (1995). Indian Medicinal Plants, Vol. 1., 2nd ed. International Book Distributors, Dehradun, India.

Kirtikar, K.R., and Basu, B.D. (2004). Indian Medicinal Plants. Ist ed., Periodic Experts Book Agency, 380-382.

Krishnamurthi, A. (1969). The Wealth of India. Vol. VIII. Publication and Information Directorate, CSIR, New Delhi, India.

Mondal, O.A., Haque M.E., and Islam, N. (2010). Antibacterial and larvicidal potentials of *Derris indica* (Lamk.) Bennet. extractives. *University Journal of Zoology Rajshahi University*, **29**: 17-22.

Murugan, A.M., Shanthi, S., Arunachalam, C., Sivakumar, N., Elamathy, S., and Rajapandian, K. (2012). Study on cow urine and *Pongamia pinnata* Linn. seed in farmyard: A natural, cost effective, ecofriendly remedy to bacterial leaf blight (BLB) of paddy. *African Journal of Biotechnology*, **11**: 9591-9598.

Nadkarni, K. M. (1976). Indian Materia Medica. Vol. 1, 1001-1004.

Mahli, S.S., Basu, S.P., Sinha, K.P., and Banerjee, N.C. (1989). Pharmacological effects of karanjin and pongamol from seed oil of *Pongamia pinnata*. *Indian Journal of Animal Science*, **59**: 657-660.

Prabha, T., Dorababu, M., Goel, S., Agarwal, P.K., Singh, A., Joshi, V.K., and Goel, R.K. (2009). Effect of methanolic extract of *Pongamia pinnata* Linn seed on gastro duodenal ulceration and mucosal offensive and defensive factors in rats. *Indian Journal of Experimental Biology*, **47**: 649-659.

Punitha, R., Vasudevan, K., and Manoharan, S. (2006). Effect of *Pongamia pinnata* flowers on blood glucose and oxidative stress in alloxan induced diabetic rats. *Indian Journal of Pharmacology*, **38**: 62-63.

Rani, M.S., Dayanand, C.D., Shetty, J., Vegi, P.K., and Kutty, A.V.M. (2013). Evaluation of antibacterial activity of *Pongamia pinnata* Linn. on pathogens of clinical isolates. *American Journal of Phytomedicine and Clinical Therapeutics*, **1**: 645-651.

Rastogi, R.P., and Malhotra, B.N. (2001). Compendium on Medicinal Plants, Central drug Research Institute Lucknow and National Institute of Science Communication, New Delhi, India, 522 – 523.

Reddy, L.J., Gopu, S., Jose, B., and Jalli, R.D. (2012). Evaluation of antibacterial and DPPH radical scavenging activities of the leaf essential oils of *Pongamia Pinnatta* and *Eucalyptus maculata*. *Asian Journal of Biochemical and Pharmaceutical Research*, **2**: 25-32.

Sagar, P.K., and Paliwal, R.K. (2013). Phytochemical screening and antioxidant potential of *Pongamia pinnata* (vent) hydro-alcoholic root and leaves extracts. *International Journal of Natural Product Science*, **3**: 19-29.

Sagwan, S., Rao, D.V., and Sharma, R.A. (2011). *In-vitro* and *In-vivo* antioxidant activity and total phenolic content of *Pongamia pinnata* (L.) Pierre: An important medicinal plant. *International Journal of Biotechnology*, **4**: 568-574.

Sajid, Z.I., Anwar, F., Shabir, G., Rasul, G., Alkharfy, K.M., and Gilani, A. (2012). Antioxidant, antimicrobial properties and phenolics of different solvent extracts from bark, leaves and seeds of *Pongamia pinnata* (L.) Pierre. *Molecules*, **17**: 3917-3932.

Shanthi, S., Elamathy, S., Panneerselvam, A., and Radha, N. (2011). Antixanthomonas activity of *Pongamia pinnata* linn leaves – cow urine extract- a natural cost effective ecofriendly remedy to bacterial leaf blight of paddy (BLB). *Journal of Pharmacy Research*, **4**: 650-652.

Simin, K., Ali, Z., Khaliq-Uz-Zaman, S.M., and Ahmad, V.U. (2002). Structure and biological activity of a new rotenoid from *Pongamia pinnata*. *Natural Product Letters*, **16**: 351-357.

Singh, K.D., Sherief, S.H., Sengottuvelu, S., Lalitha, V., and Sivakumar, T. (2011). Gastroprotective activity of *Pongamia pinnata* stem bark in Wistar albino rats. *International Journal of Research in Ayurveda and Pharmacy*, **2**: 655-659.

Singh, R.R., and Sinha, B.K. (2012). Evaluation of *Pongamia pinnata* seed extracts as an insecticide against American bollworm *Helicoverpa armigera* (Hubner). *International Journal of Agriculture Sciences*, **4**: 257-261.

Tamrakar, A.K., Jaiswal, N., Yadav, P.P., Maury, R., and Srivastava, A.K. (2011). Pongamol from *Pongamia pinnata* stimulates glucose uptake by increasing surface GLUT4 level in skeletal muscle cells. *Molecular and Cellular Endocrinology*, **339**: 98– 104.

Vadivel, V., and Biesalski H.K. (2011). Contribution of phenolic compounds to the antioxidant potential and type II diabetes related enzyme inhibition properties of *Pongamia pinnata* L. Pierre seeds. *Process Biochemistry*, **46**: 1973–1980.

Vismaya., Belagihally, S.M., Rajashekhar, S., Jayaram, V.B., Dharmesh, S.M., and Thirumakudalu, S.K. (2010). Gastroprotective properties of Karanjin from Karanja (*Pongamia pinnata*) seeds; role as antioxidant and H, K-ATPase inhibitor. *Evidence Based Complementary and Alternative Medicine*, **2011**: 1-10.

Wagh, P., Rai, M., Deshmukh, S.K., and Durate, M.C.T. (2007). Bio-activity of oils of *Trigonella foenum-graecum* and *Pongamia pinnata*. *African Journal of Biotechnology*, **6**: 1592-1596.

Yadav, R.D., Jain, S.K., Alok, S., and Sharma, S. (2011). Analgesic activity of ethanolic extract of *Pongamia pinnata* Linn. leaves. *Der Pharmacia Lettre*, **3**: 179-182.

Natural Products: Research Reviews Vol. 3 (2016) *Pages* **75–93**
Editor: **V.K. Gupta**
Published by: **DAYA PUBLISHING HOUSE, NEW DELHI**

4

Tricin: A Phytochemical with Potential Therapeutic Benefits Against Cancer

Abdullah Al-Nahain[1], Rownak Jahan[2] and Mohammed Rahmatullah[1*]

ABSTRACT

Tricin is an O-methylated flavone and belongs to the flavonoid group of compounds. The compound can be obtained from a number of plants and has been reported from common cereal plants like paddy, wheat, barley and maize. Recent studies have shown that tricin and tricin-derivatives have important pharmacological activities, including anti-cancer activity against different cancer cell lines and forms of cancer. Since the compound does not appear to have the usual adverse side-effects associated with most anti-cancer drugs, tricin can be of immense potential in forming the new generation of anti-cancer compounds. This review explores the different plant sources from which tricin and tricin derivatives have been isolated and explores the mechanism of action of the anti-cancer activity of these compounds.

Keywords: Tricin, Ttricin derivatives, Anti-cancer, Phytochemical, Flavones.

Introduction

Tricin (IUPAC name: 5,7-dihydroxy-2-(4-hydroxy-3,5-dimethoxyphenyl)-4H-chromen-4-one, also known as 5,7,4'-trihydroxy-3',5'-dimethoxyflavone) is an O-methylated flavone and belongs to the flavonoid group of compounds. Flavonoids are a class of plant secondary metabolites. Among the diverse pharmacological activities being studied with different flavonoids, anti-cancer activity studies have

1 Department of Pharmacy

2 Department of Biotechnology and Genetic Engineering,
University of Development Alternative, Dhanmondi, Dhaka-1209, Bangladesh.

* *Corresponding author*: E-mail: rahamatm@hotmail.com

occupied a predominant position, not only because cancer is a difficult disease to cure, but also because most anti-cancer drugs that are available on the market are costly and may have serious adverse side-effects. From that view point, flavonoids represent a promising class of a new generation of anti-cancer agents, and a number of flavonoids and plant extracts containing flavonoid compounds are being intensively studied towards that aspect. A few such examples of such flavonoid compounds apart from tricin are given below, while tricin will be discussed in details in the remaining portion of the chapter.

The inhibitory effect of fisetin (3,7,32,42-tetrahydroxyflavone), a dietary flavonoid (found in Fabaceae family plants like *Acacia greggii* and *Acacia berlandieri*), has been investigated against bladder cancer through an autochthonous rat model of bladder cancer induced by intravesical N-methyl-N-nitrosourea (MNU). It has been observed that fisetin induced apoptosis in bladder cancer, which was mediated via up-regulation of p53 and down-regulation of NF-κB pathway, thus causing changes in the ratio of pro- and anti-apoptotic proteins (Li *et al.*, 2014). Deguelin is another flavonoid compound, which can be found in *Derris trifoliata* or *Mundulea sericea*. This compound has also been shown to exhibit anti-tumorigenesis and anti-proliferative activity in various types of cancer both *in vivo* and *in vitro* (Wang *et al.*, 2013). The compound mainly acts through promoting apoptosis or cell cycle arrest or as an anti-angiogenesis agent.

The flavonoid, isoquercitrin (found in the plant), has been reported to inhibit the progression of human liver cancer *in vivo* and *in vitro* by inhibiting the proliferation of liver cancer cells, promoting the apoptosis of liver cancer cells, and blocking the cell cycle in the G1 phase (Huang *et al.*, 2014). Apigenin, another plant flavonoid present in plants like parsley (*Petroselinum crispum*), has been found to be effective against chronic myeloid leukemia, as observed with two cell lines, namely, K562 and K562/IMA3. Apigenin induced G2/M arrest in K562 cells and S phase arrest in K562/IMA3 cells (Solmaz *et al.*, 2014). It may be noted that a drug, Imatinib (as well as other tyrosine kinase inhibitors), is available for treatment of chronic myeloid leukemia (CML) therapy, but newer drugs are necessary for treatment because of development of resistance, intolerance, and side effects. From that view point, tricin also is a plant flavonoid with excellent potential as a therapeutic for several forms of cancer.

Plant Sources of Tricin and Tricin Derivatives

Tricin and tricin derivatives have been reported from a number of plants. As shown in Table 4.1, at least 57 plants reportedly contain tricin and tricin derivatives. The Poaceae family appears to be an excellent source for tricin, for 20 of the 57 plants have been reported to contain the compound and which contain tricin. Poaceae family plants are one of the largest groups of plants present throughout the world and are a large and nearly ubiquitous family of monocotyledonous flowering plants. Poaceae family plants are also known as Gramineae or true grasses. The Poaceae family plants comprise of many of the most important cereal crops such as maize (corn), wheat, rice, barley, and millet, and tricin as well as tricin derivatives have been reported to be present in virtually all of these plants. While tricin has been

Table 7.1: Reported Isolation of Tricin and Tricin Derivatives from Various Plants

Sl.No.	Plant Name (Family)	Plant Part from which Isolated	Name of Isolated Compound	Reference(s)
1.	*Agelaea pentagyna* (Lam.) Baill. (Connaraceae)	Whole plant	Tricin	Kuwabara *et al.*, 2003.
2.	*Agropyrons repens* (L.) Beauv. (Poaceae)	Shoot	Tricin	Weston *et al.*, 1987.
3.	*Arenaria kansuensis* Maxim. (Caryophyllaceae)	Whole plant	Tricin, tricin-7-O-β-D-glucopyranoside, tricin 4'-O-β-guaiacylglyceryl ether	Lei *et al.*, 2007.
4.	*Arnica japonica* (L.) (Asteraceae)	Flower	Tricin	Merfort and Wendisch, 1988.
5.	*Artemisia copa* Phil. (Asteraceae)	Aerial part	Tricin	Moscatelli *et al.*, 2006.
6.	*Artemisia rupestris* L. (Asteraceae)	Whole plant	Tricin	Cui *et al.*, 2013.
7.	*Atriplex centralasiatica* Iljin (Amaranthaceae)	Seed	Tricin	Zhang and Zhao, 2005.
8.	*Avena sativa* L. (Poaceae)	Whole plant	Three flavonolignans derived from tricin	Wenzig *et al.*, 2005.
9.	*Centipeda minima* (L.) A.Braun et Asch. (Asteraceae)	Whole plant	Tricin	Pu *et al.*, 2009.
10.	*Cirsium rivulare* (Jacq.) All. (Asteraceae)	Flower	Tricin	Nazaruk and Gudej, 2003.
11.	*Chrysanthemum indicum* L. (Asteraceae)	Flower	Tricin	Wang *et al.*, 2010b.
12.	*Chrysopogon aciculatus* (Retz.) Trin. (Poaceae)	Whole plant	Tricin	Shen *et al.*, 2012a.
13.	*Cyperus rotundus* L. (Cyperaceae)	Aerial part	Tricin	Sayed *et al.*, 2007.
14.	*Dendrobium aphyllum* (Roxb.) C.E.C. Fisch. (Orchidaceae)	Stem	4'-methoxyl-tricin, tricin (2), 7, 3', 5'-tri-O-methyl-tricetin	Shao *et al.*, 2008.
15.	*Deschampsia japonicas* E. Desv. (Poaceae)	Whole plant	Tricin	Webby and Markham, 1994.
16.	*Echinochloa colona* L. (Poaceae)	Shoot	Tricin	Hegab *et al.*, 2013.
17.	*Echinochloa utilis* Ohwi and Yabuno (Poaceae)	Grain	Tricin	Watanabe, 1999.
18.	*Elephantopus tomentosus* L. (Asteraceae)	Whole plant	Tricin	Li *et al.*, 2013.
19.	*Emilia sonchifolia* DC. (Asteraceae)	Aerial part	Tricin-7-O-β-D-glucopyranoside	Shen *et al.*, 2012(b).

Contd...

Table 7.1–*Contd...*

Sl.No.	*Plant Name (Family)*	*Plant Part from which Isolated*	*Name of Isolated Compound*	*Reference(s)*
20.	*Epimedium acuminatum* Franchet (Berberidaceae)	Aerial part	Tricin	Liang *et al.*, 1993.
21.	*Epimedium hunanense* (Hand.-Mazz.) Hand.-Mazz. (Berberidaceae)	Aerial part	Tricin	Liang *et al.*, 1997.
22.	*Epimedium pubescens* Maxim. (Berberidaceae)	Leaf	Tricin	Zhang *et al.*, 2013.
23.	*Fargesia robusta* Pingwu (Poaceae)	Leaf	Tricin-5-O-glucopyranoside	Van Hoyweghen *et al.*, 2010.
24.	*Ficus hirta* Vahl (Moraceae)	Root	Tricin	Zheng *et al.*, 2013.
25.	*Ficus stenophylla* Standl. (Moraceae)	Root	Tricin	Zhang *et al.*, 2008b.
26.	*Hordeum vulgare* L. (Poaceae)	Seed	Tricin, (2RS)-dihydrotricin 7-O-β-D-glucopyranoside, (2RS)-dihydrotricin	Nakano *et al.*, 2011.
27.	*Hyparrhenia hirta* (L.) Stapf. (Poaceae)	'Leaves	Tricin 4'-O-(erythro-β-guaiacylglyceryl) ether; tricin4'-O-(threo-β-guaiacylglyceryl) ether together with their 7-O-glucosides	Bouaziz *et al.*, 2002.
28.	*Lespedeza virgata* (Thunb.) DC. (Fabaceae)	Aerial parts	Tricin	Chen *et al.*, 2008.
29.	*Leucas cephalotes* Spreng. (Lamiaceae)	Whole plant	Tricin	Miyaichi *et al.*, 2006.
30.	*Livistona australis* (R.Br.) Mart. (Arecaceae)	Leaf	Tricin 7-O-β-glucopyranoside-23-sulphate sodium salt	Kassem *et al.*, 2012.
31.	*Livistona chinensis* (Jacq.) R. Br. ex Mart. (Arecaceae)	Fruit	Tricin	Yao *et al.*, 2012.
32.	*Lygodium japonicum* (Thunb.) Sw. (Lygodiaceae)	Whole plant	Tricin-7-O-β-D-glucopyranoside	Zhang *et al.*, 2008a.
33.	*Meconopsis quintuplinervia* Regel (Papaveraceae)	Whole plant	Tricin	Cheng *et al.*, 2011.

Contd...

Table 7.1–*Contd...*

Sl.No.	*Plant Name (Family)*	*Plant Part from which Isolated*	*Name of Isolated Compound*	*Reference(s)*
34.	*Medicago sativa* L. (Fabaceae)	Aerial part, leaf	7-O-[2'-O-sinapoyl-β-D-glucuronopyranosyl(1→2)-O-β-D-glucuronopyranoside] tricin; 7-O-[2'-O-feruloyl-β-D-glucuronopyranosyl(1→2)-O-β-D-glucuronopyranoside]tricin; 7-O-[2'-O-p-coumaroyl-β-D-glucuronopyranosyl(1→2)-O-β-D-glucuronopyranoside] tricin; 7-O-[2'-O-feruloyl-[β-D-glucuronopyranosyl(1→3)]-O-β-D-glucuronopyranosyl(1→2)-O-β-D-glucuronopyranoside] tricin; 7-O-β-D-glucuronopyranoside tricin; 7-O-[β-D-glucuronopyranosyl(1→2)-O-β-D-glucuronopyranoside] tricin; tricin	Stochmal *et al.*, 2001; Gatouillat *et al.*, 2014.
35.	*Medicago truncatula* Gaertn. (Fabaceae)	Aerial part, leaves	Tricin 7-O-β-D-glucuronopyranosyl-4'-O-glucopyranoside; tricin7-O-[2'-O-feruloyl-β-D-glucuronopyranosyl-(1→2)-O-β-D-glucopyranoside]; tricin7-O-{2'-O-p-coumaroyl-[β-D-glucuronopyranosyl-(1→3)]-O-β-D-glucuronopyranosyl(1→2)-O-β-D-glucuronopyranoside}; 7-O-[2'-O-sinapoyl-β-D-glucuronopyranosyl(1→2)-O-β-D-glucuronopyranoside] tricin; 7-O-[2'-O-feruloyl-β-D-glucuronopyranosyl (1→2)-O-β-D-glucuronopyranoside]tricin; 7-O-[2'-O-p-coumaroyl-β-D-glucuronopyranosyl(1→2)-O-β-D-glucuronopyranoside] tricin; 7-O-[2'-O-feruloyl-[β-D-glucuronopyranosyl(1→3)]-O-β-D-glucuronopyranosyl(1→2)-O-β-D-glucuronopyranoside] tricin; 7-O-β-D-glucuronopyranoside tricin; 7-O-[β-D-glucuronopyranosyl(1→2)-O-β-D-glucuronopyranoside] tricin; tricin glucuronide	Kowalska *et al.*, 2007; Marczak *et al.*, 2010.

Contd...

Table 7.1–*Contd...*

Sl.No.	*Plant Name (Family)*	*Plant Part from which Isolated*	*Name of Isolated Compound*	*Reference(s)*
36.	*Pleioblastus amarus* (Keng) Keng f.	Leaf	Tricin	Wang *et al.*, 2004.
37.	*Polygala stenopetala* Klotzsch (Polygalaceae)	Root	Tricin	Ahmed *et al.*, 1999.
38.	*Oryza sativa* L. (Poaceae)	Bran, whole brown rice, hull, defatted rice bran	Tricin, tricin 4'-O-(erythro-β-guaiacyl-glyceryl) ether, tricin 4'-O-(threo-β-guaiacylglyceryl) ether	Hudson *et al.*, 2000; Chung *et al.*, 2005; Devi and Arumughan, 2007; Mohanlal *et al.*, 2011
39.	*Ranunculus japonicus* Thunb. (Ranunculaceae)	Whole plant	Tricin	Zheng *et al.*, 2006; Liang *et al.*, 2008.
40.	*Saccharum officinarum* L. (Poaceae)	Stem	Tricin derivatives; tricin-7-O-β-(6"-methoxycinnamic)-glucoside	Maurício Duarte-Almeida *et al.*, 2006; Duarte-Almeida *et al.*, 2007.
41.	*Salsola collina* Pallas (Chenopodiaceae)	Whole plant	Tricin; tricin -7-O-β-D-glucopyranoside; tricin-4'-O-β-D-apioside	Xiang *et al.*, 2007.
42.	*Sasa albomarginata* (Franch. and Sav.) (Poaceae)	Whole plant	Tricin,	Sakai *et al.*, 2008.
43.	*Sasa borealis* (Hack.) Makino and Shibata (Poaceae)	Whole plant, leaf	Tricin; tricin 7-O-β-D-glucopyranoside	Jeong *et al.*, 2007; Park *et al.*, 2007.
44.	*Sasa kurilensis* var. *gigantea* (Rupr.) (Poaceae)	Leaf	Tricin-4'-O-β-d-glucopyranoside; tricin-5-O-β-d-glucopyranoside	Hasegawa *et al.*, 2008.
45.	*Sasa senanensis* (Franch. and Sav.) Rehder (Poaceae)	Leaf	Tricin	Matsuta *et al.*, 2011.
46.	*Setaria viridis* (L.) P. Beauv. (Poaceae)	Aerial part	Tricin; tricin-7-O-β-D-glucoside	Kwon *et al.*, 2002.
47.	*Simaba multiflora* A. Juss. (Simaroubaceae)	Wood	Tricin	Arisawa *et al.*, 1985.
48.	*Spartina cynosuroides* (L.) Roth (Poaceae)	Aerial parts	Tricin	Bhattacharyya *et al.*, 1978.
49.	*Spergularia diandra* (Gussone) Heldreich (Caryophyllaceae)	Aerial parts	Tricin	El-Dien *et al.*, 2013.

Contd...

Table 7.1–*Contd...*

Sl.No.	*Plant Name (Family)*	*Plant Part from which Isolated*	*Name of Isolated Compound*	*Reference(s)*
50.	*Spergularia marina* (L.) Griseb. (Caryophyllaceae)	Aerial parts	Tricin	El-Dien *et al.*, 2013.
51.	*Trigonella foenum-graecum* L. (Fabaceae)	Seed, whole plant	Tricin	Shang *et al.*, 1998; Wang *et al.*, 2010a.
52.	*Triticum aestivum* L. (Poaceae)	Leaf, husk	Tricin	Moheb *et al.*, 2013a; Moheb *et al.*, 2013b.
53.	*Triticum durum* Desf. (Poaceae)	Leaf and stem	Tricin derivatives present exclusively as O-glycosides	Cavaliere *et al.*, 2005.
54.	*Valeriana laxiflora* DC. (Caprifoliaceae)	Above ground biomass, root	Tricin	Gu *et al.*, 2004.
55.	*Vernonia remotiflora* L.C. Rich (Asteraceae)	Aerial parts	Tricin	Jacobs *et al.*, 1986.
56.	*Wickstroemia indica* (L.) C.A. Mey (Thymelaeaceae)	Whole plant, stem bark	Tricin	Wang *et al.*, 1998; Geng *et al.*, 2006.
57.	*Zea mays* L. (Poaceae)	Cob	Tricin	Ashour *et al.*, 2013.

Figure 4.1: Chemical Structures of Tricin and some of its Derivatives.

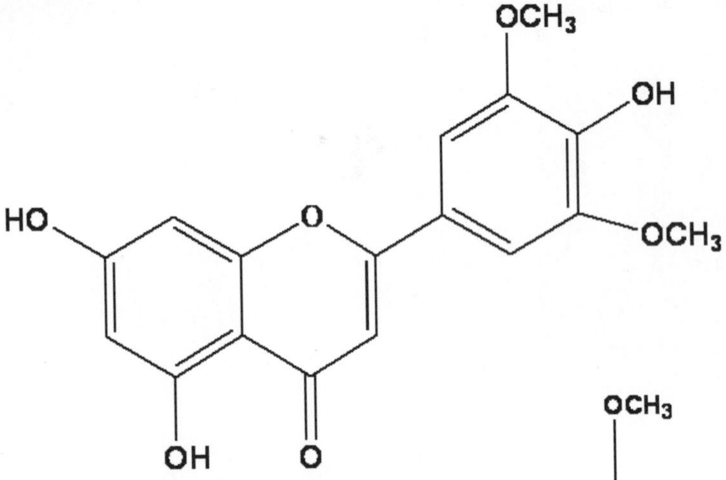

Tricin (5,7,42-trihydroxy-32,52-dimethoxyavone)

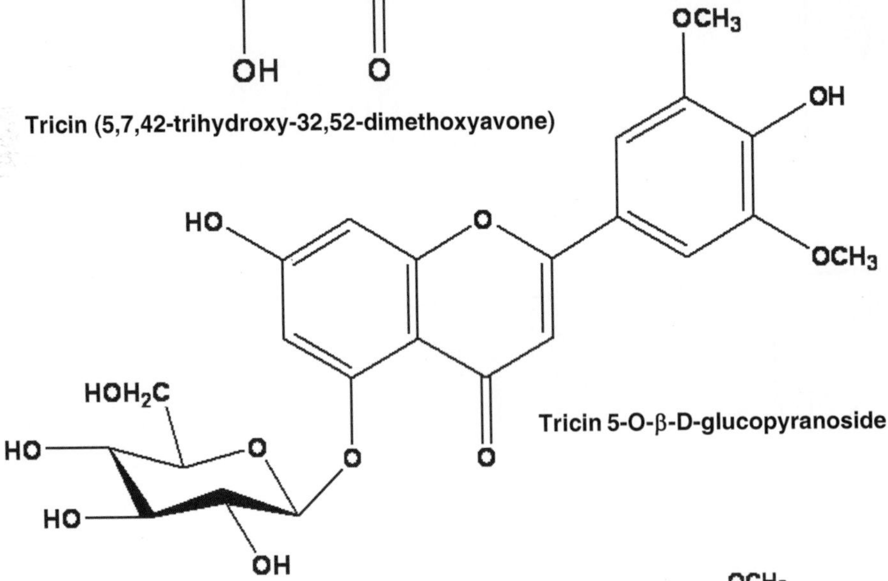

Tricin 5-O-β-D-glucopyranoside

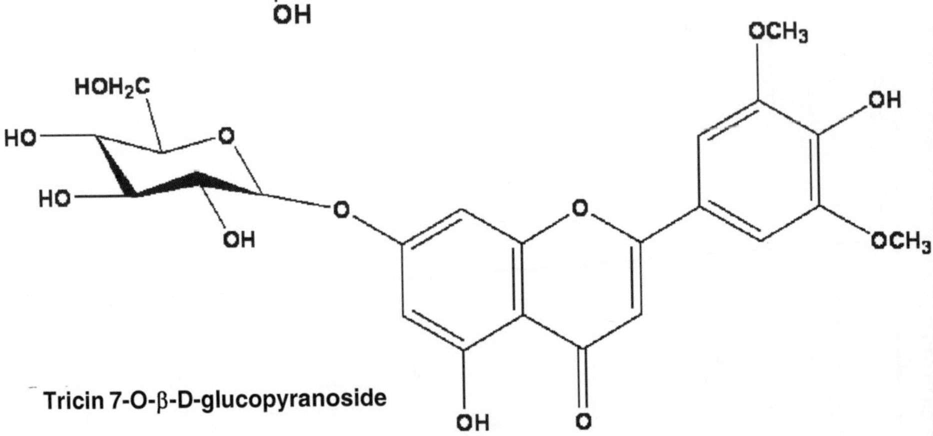

Tricin 7-O-β-D-glucopyranoside

Contd...

Figure 4.1–*Contd...*

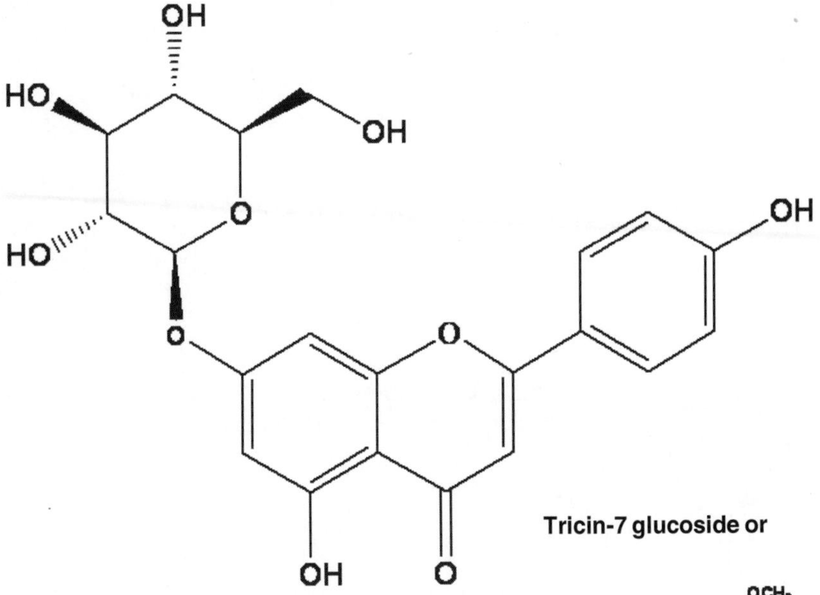

Tricin-7 glucoside or

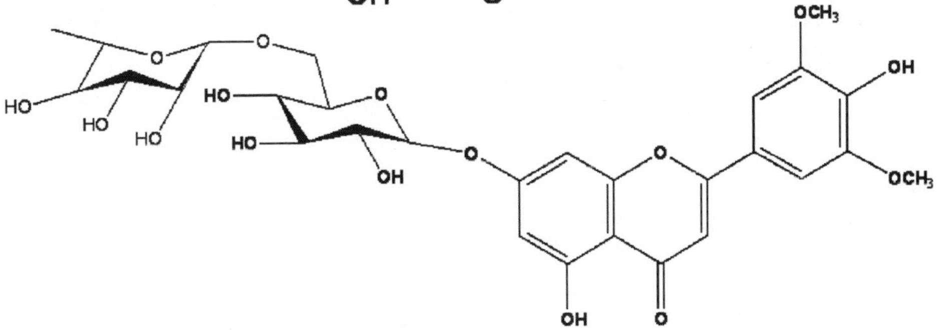

Tricin-7-O-routinoside

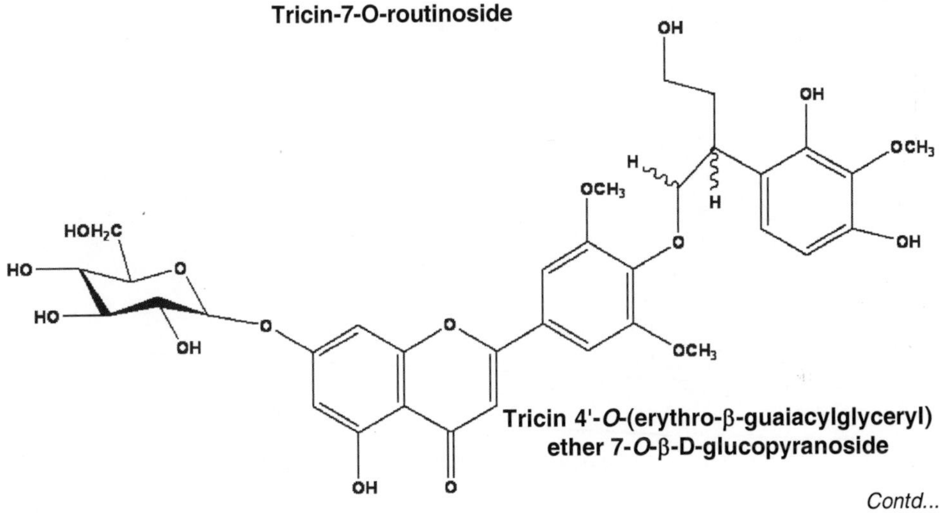

Tricin 4'-*O*-(erythro-β-guaiacylglyceryl) ether 7-*O*-β-D-glucopyranoside

Contd...

Figure 4.1–*Contd...*

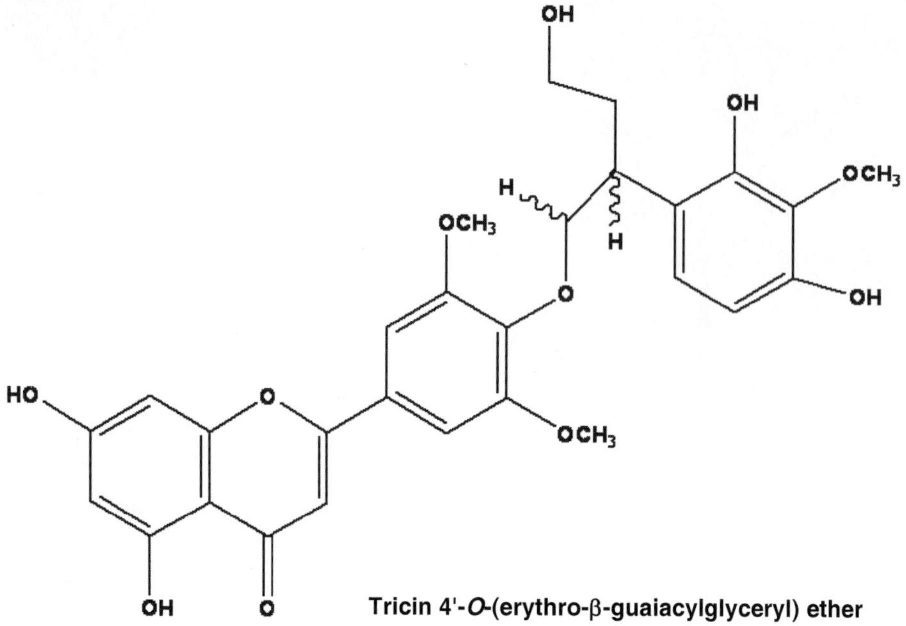

Tricin 4'-*O*-(erythro-β-guaiacylglyceryl) ether

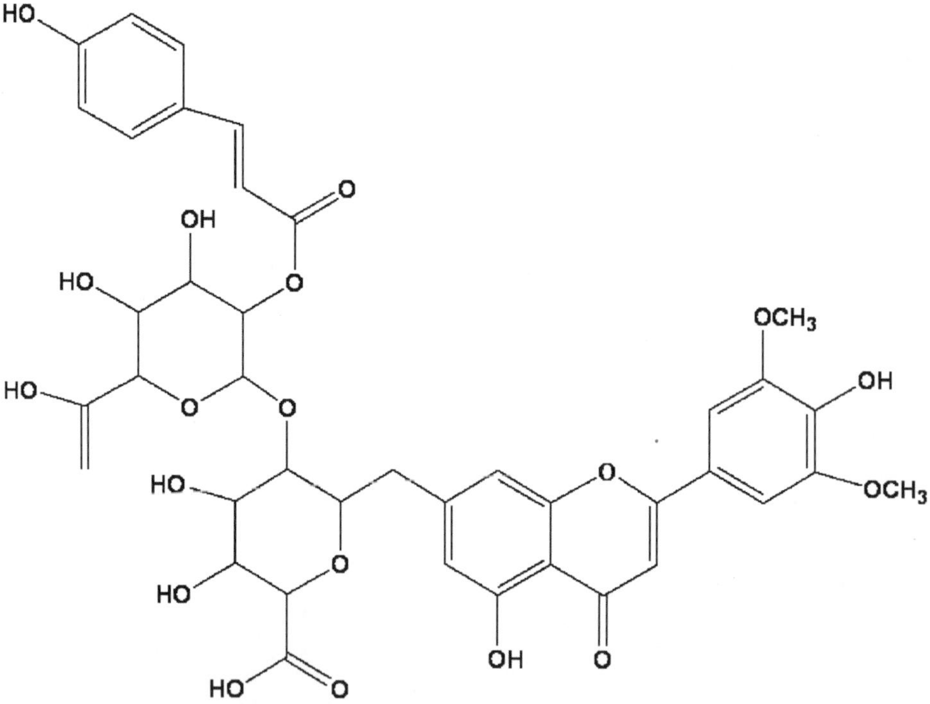

Tricin 7-[p-coumaroyl-(→2)-glucuronyl-(1→2)-glucuronide]

Contd...

Figure 4.1–*Contd...*

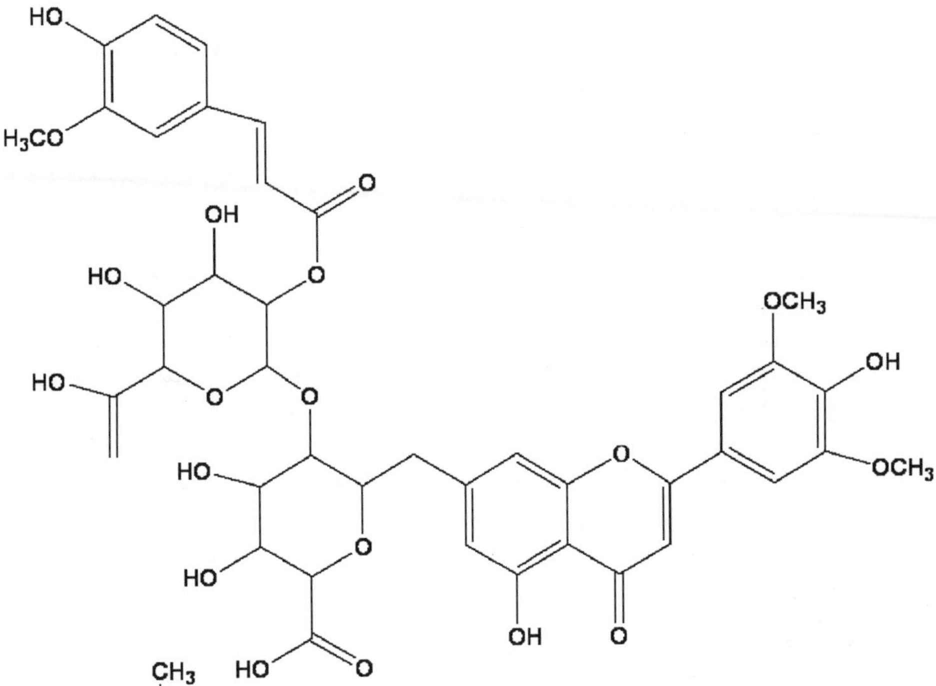

Tricin 7-[feruloyl-(→2)-glucuronyl-(1→2)-glucuronide

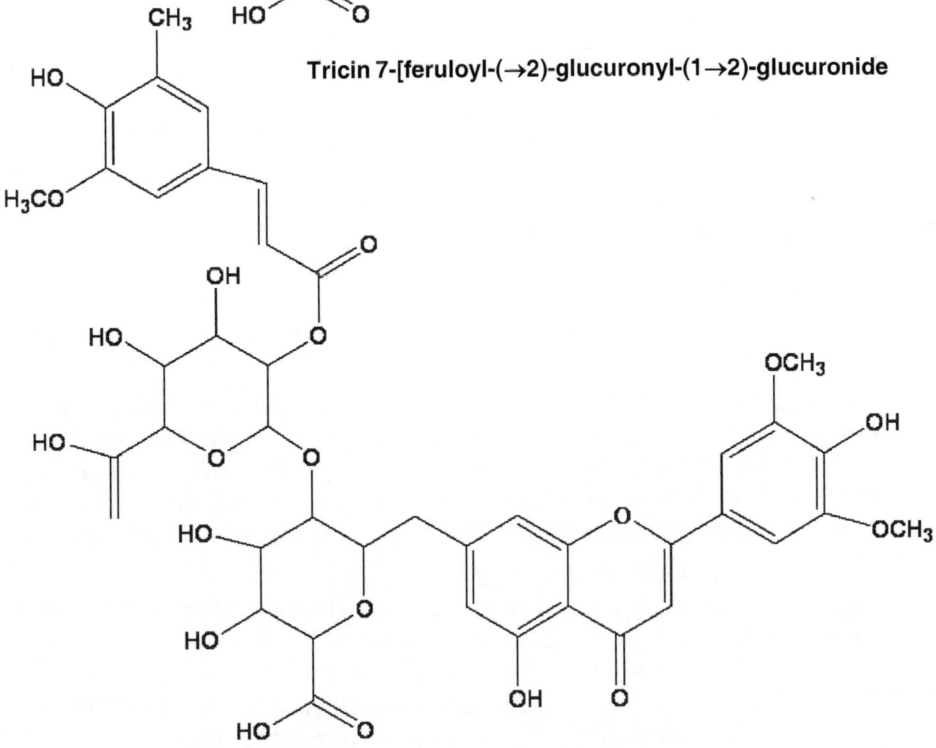

Tricin 7-[sinapoyl-(→2)-glucuronyl-(1→2)-glucuronide]

reported to be present in both rice seed (paddy) husk and bran, in maize it is present in corn cob (Table 4.1 for more details).

Anti-Cancer Activity of Tricin and Tricin Derivatives

A number of reports are available on the anti-cancer activity of tricin and tricin derivatives isolated from various plant sources.

Moheb and his group reported potential anti-cancer effect of tricin isolated from wheat (*Triticum aestivum*) selectively on rat pancreatic cells (INS832/13) and human hepatocellular carcinoma cells (HepG2) (Moheb *et al.*, 2013).

Anti-carcinogenesis effect of tricin was studied by administrating dietary tricin in azoxymethane-induced colon cancer male Crj:CD-1 mice. Colonic adenomas and adenocarcinomas were significantly inhibited. Beside this, significant reduction in the expression of tumor necrosis factor (TNF)-α was observed, which was the main reason behind inhibition of adenocarcinomas (Oyama *et al.*, 2009].

Anti-cancer effect of tricin was demonstrated by Cai and his group in human-derived breast cancer cells *in vitro* and in nude mice *in vivo* (Cai *et al.*, 2004).

Recent study by Mohanlal *et al.* (2013) demonstrated significant antitumor activities of two rare flavonolignans, tricin 4'-*O*-(erythro-β-guaiacylglyceryl) ether and tricin 4'-*O*-(threo-β-guaiacylglyceryl) ether isolated from 'Njavara' black in three cancer cell lines, namely, colon adenocarcinoma cell line HCT 116, ovarian cancer cell line SKOV3, and breast cancer cell line MCF-7. They showed that these two potent compounds exhibited anti-cancer effects by apoptosis and targeting cancer cells through mitochondrial membrane potential loss and subsequent chromatin condensation.

Tricin inhibited platelet-derived growth factor (PDGF)-BB-induced cell proliferation by blocking cell cycle progression and cell migration in the human HSC line LI90 and culture-activated HSCs. It also reduced the phosphorylation of PDGF receptor β and the downstream signaling molecules ERK1/2 and Akt, which might be due to its tyrosine kinase inhibitor properties rather than inhibition of the direct binding between PDGF-BB and its receptor. The study suggested that tricin might be beneficial in HSC-targeting therapeutic or chemopreventive applications for hepatic fibrosis (Seki *et al.*, 2012).

Chemoprotective effect of tricin isolated from brown rice was confirmed by Hudson *et al.* (2004), where they investigated eight isolated phenolic compounds from brown rice (Hudson *et al.*, 2000), and among them tricin showed significant inhibition in the growth of human breast and colon cancer cell line. In their next study they demonstrated the mechanism of tricin as a chemoprotective agent (Hudson *et al.*, 2004).

Significant antitumor and chemoprotective activities by tricin-7-*O*-b-(600-methoxycinnamic)-glucoside isolated from sugarcane juice was observed against different cell lines including breast (MCF-7), multidrug resistant breast (NCI-ADR), prostate (PC-03), ovary (OVCAR03), lung, non-small cell (NCI 460) and colon (HT-29) cell lines melanoma (UACC-62) and kidney (786-0) cell lines depending on concentration (Duarte-Almeida *et al.*, 2007).

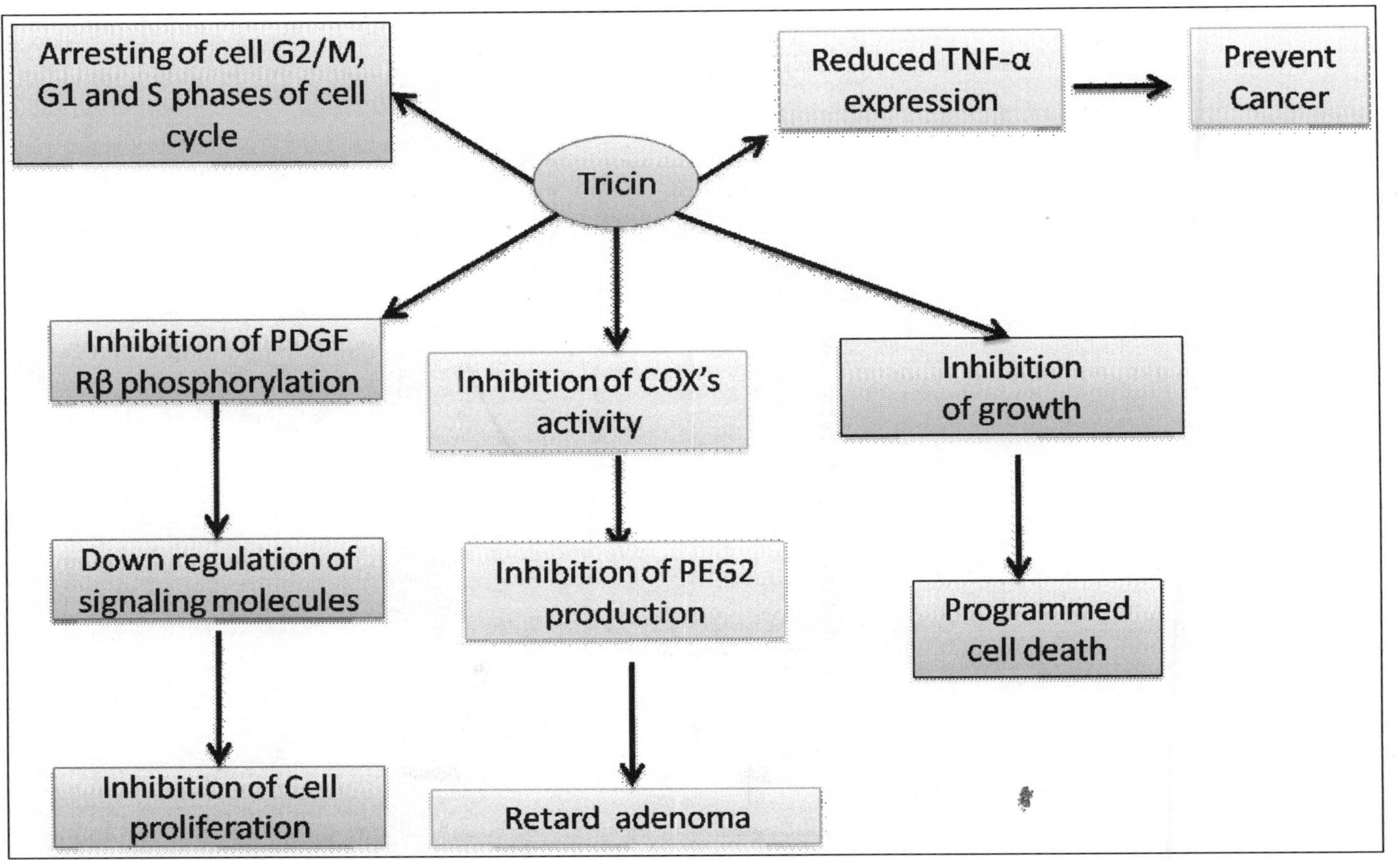

Figure 4.2: Mechanism of Tricin as an Anticancer Agent.

Antiluekemic activity by tricin has been reported by Bhattacharyya (Bhattacharyya *et al.*, 1978).

Reported Anti-Cancer Mechanism of Tricin

Molecular anti-cancer or anti-proliferative effects by tricin and its derivatives are not clearly understood. Study by Cai *et al.*, reported that tricin showed that tricin inhibited cell growth and arrested cell cycle especially G2/M phase (Cai *et al.*, 2004). Another study showed tricin suppressed the expression of TNF-α and therefore inhibited inflammation-related mouse colon carcinogenesis as well as the proliferation of adenocarcinomas (Oyama *et al.*, 2009). Seki and his group found that tricin blocked tyrosine phosphorylation of platelet derived growth factor binding receptor beta (PDGF-Rβ) which further blocked signaling pathway. Therefore, tricin showed significant cell inhibition activity against human hepatic stellate cells (Seki *et al.*, 2012). It has been reported that tricin retarded adenoma by inhibiting the activity cyclooxygenase enzymes which subsequently reduced the production of prostaglandin E2 (PGE 2) (Cai *et al.*, 2005). Tricin induced apoptosis through mitochondrial potential loss and subsequently condensation of chromatin (Mohanlal *et al.*, 2013)

Cytotoxicity to Normal Cell

Recent studies by different groups have confirmed that tricin and its metabolites are not toxic to normal cells. Tricin was totally non-toxic to normal cell when the toxicity was invested in NIH 3T3 (Moheb *et al.*, 2013). Non toxic properties of tricin were also observed by (Verscholve *et al.*, 2006). Beside this, metabolite of tricin is stable in cell (Cai *et al.*, 2011).

Conclusion

Due to great demand of obtaining new potential therapeutic agent against cancer, tricin can be a strong candidate which has already been proved by carried out a large number of scientific investigations. The available potential sources and the nontoxic propertiy of tricin are also great advantages. Though, significant anticancer activities have been reported, find out the molecular mechanism behind this is a great field for further development to enhance the potency and efficacy of tricin as an anticancer agent.

References

Ahmed, B., Hamburger, M., and Hostettmann, K. (1999). Isolation of flavones from *Polygala stenopetala*. *Pakistan Journal of Pharmaceutical Sciences*, **12**: 17-19.

Arisawa, M., Fujita, A., Morita, N., Kinghorn, A.D., Cordell, G.A., and Farnsworth, N.R. (1985). Plant Anticancer Agents XXXV. Further Constituents of *Simaba multiflora*. *Planta Medica*, **51**: 348-349.

Ashour, A., Amer, M., Marzouk, A., Shimizu, K., Kondo, R., and El-Sharkawy, S. (2013). Corncobs as a potential source of functional chemicals. *Molecules*, **18**: 13823-13830.

Bhattacharyya, J., Stagg, D., Mody, N.V., and Miles, D.H. (1978). Constituents of *Spartina cynosuroides*: isolation and 13C-NMR analysis of tricin. *Journal of Pharmaceutical Sciences*, **67**: 1325-1326.

Bouaziz, M., Veitch, N.C., Grayer, R.J., Simmonds, M.S., and Damak, M. (2002). Flavonolignans from *Hyparrhenia hirta*. *Phytochemistry*, **60**: 515-520.

Cavaliere, C., Foglia, P., Pastorini, E., Samperi, R., and Laganà A. (2005). Identification and mass spectrometric characterization of glycosylated flavonoids in *Triticum durum* plants by high-performance liquid chromatography with tandem mass spectrometry. *Rapid Communications in Mass Spectrometry*, **19**: 3143-3158.

Chen, Y., Deng, H.Z., and Liang, L. (2008). Contents of flavonoids in the extract of *Lespedeza virgata*. *Nan Fang Yi Ke Da Xue Xue Bao*, **28**: 858-860.

Cheng, F., Guo, M., Zhang, Y., Yu, X.H., and Wang, Z.W. (2011). Tracking the active component of Tibetan medicine *Meconopsis quintuplinervia* from Gansu. *Zhong Yao Cai*, **34**: 69-71.

Chung, I.M., Hahn, S.J., and Ahmad, A. (2005). Confirmation of potential herbicidal agents in hulls of rice, *Oryza sativa*. *Journal of Chemical Ecology*, **31**: 1339-1352.

Cui, F.X., Zhang, C., Jiang, Y., and Tu, P.F. (2013). Chemical constituents from ethyl acetate extract of *Artemisia rupestris*. *Zhongguo Zhong Yao Za Zhi*, **38**: 1757-1759.

Devi, R., and Arumughan, C. (2007). Antiradical efficacy of phytochemical extracts from defatted rice bran. *Food and Chemical Toxicology*, **45**: 2014-2021.

Duarte-Almeida, J.M., Negri, G., Salatino, A., de Carvalho, J.E., and Lajolo, F.M. (2007). Antiproliferative and antioxidant activities of a tricin acylated glycoside from sugarcane (*Saccharum officinarum*) juice. *Phytochemistry*, **68**: 1165-1171.

El-Dien, O.G., Shawky, E., Aly, A.H., Abdallah, R.M., and Abdel-Salam, N.A. (2013). A validated high-performance thin-layer chromatography (Hptlc) method for the quantitative determination of tricin in two *Spergularia* species. *American Journal of Analytical Chemistry*, **4**: 668-673.

Gatouillat, G., Alabdul Magid, A., Bertin, E., Okiemy-Akeli, M.G., Morjani, H., Lavaud, C., and Madoulet, C. (2014). *Nutrition and Cancer* [Epub ahead of print].

Geng, L.D., Zhang, C., and Xiao, Y.Q. (2006). A new dicoumarin from stem bark of *Wikstroemia indica*. *Zhongguo Zhong Yao Za Zhi*, **31**: 43-45.

Gu, J.Q., Wang, Y., Franzblau, S.G., Montenegro, G., Yang, D., and Timmermann, B.N. (2004). Antitubercular constituents of *Valeriana laxiflora*. *Planta Medica*, **70**: 509-514.

Hasegawa, T., Tanaka, A., Hosoda, A., Takano, F., and Ohta, T. (2008). Antioxidant C-glycosyl flavones from the leaves of *Sasa kurilensis* var. *gigantea*. *Phytochemistry*, **69**: 1419-1424.

Hegab, M.M., Abdelgawad, H., Abdel Hamed, M.S., Hammouda, O., Pandey, R., Kumar, V., and Zinta, G. (2013). Effects of tricin isolated from jungle rice (*Echinochloa colona* L.) on amylase activity and oxidative stress in wild oat (*Avena fatua* L.). *Allelopathy Journal*, **31**: 345-354.

Huang, G., Tang, B., Tang, K., Dong, X., Deng, J., Liao, L., Liao, Z., Yang, H., and He, S. (2014). Isoquercitrin inhibits the progression of liver cancer *in vivo* and *in vitro* via the MAPK signalling pathway. *Oncology Reports*, [Epub ahead of print].

Hudson, E.A., Dinh, P.A., Kokubun, T., Simmonds, M.S., and Gescher, A. (2000). Characterization of potentially chemopreventive phenols in extracts of brown rice that inhibit the growth of human breast and colon cancer cells. *Cancer Epidemiology, Biomarkers and Prevention*, **9**: 1163-1170.

Jacobs, H., Bunbury, M., and McLean, S. (1986). Tricin from *Vernonia remotiflora*. *Journal of Natural Products*, **49**: 1164-1164.

Jeong, Y.H., Chung, S.Y., Han, A.R., Sung, M.K., Jang, D.S., Lee, J., Kwon, Y., Lee, H.J., and Seo, E.K. (2007). P-glycoprotein inhibitory activity of two phenolic compounds, (-)-syringaresinol and tricin from *Sasa borealis*. *Chemistry and Biodiversity*, **4**: 12-16.

Kassem, M.E., Shoela, S., Marzouk, M.M., and Sleem, A.A. (2012). A sulphated flavone glycoside from *Livistona australis* and its antioxidant and cytotoxic activity. *Natural Product Research*, **26**: 1381-1387.

Kowalska, I., Stochmal, A., Kapusta, I., Janda, B., Pizza, C., Piacente, S., and Oleszek, W. (2007). Flavonoids from barrel medic (*Medicago truncatula*) aerial parts. *Journal of Agricultural and Food Chemistry*, **55**: 2645-2652.

Kuwabara, H., Mouri, K., Otsuka, H., Kasai, R., and Yamasaki, K. (2003). Tricin from a Malagasy connaraceous plant with potent antihistaminic activity. *Journal of Natural Products*, **66**: 1273-1275.

Kwon, Y.S., Kim, E.Y., Kim, W.J., Kim, W.K., and Kim, C.M. (2002). Antioxidant constituents from *Setaria viridis*. *Archives of Pharmacal Research*, **25**: 300-305.

Lei, N., Du, S.S., Li, L., Zheng, Y.F., Zhang, W.S., and Wang, Y.Y. (2007). Chemical constituents from a Tibetan medicine *Arenaria kansuensis* I. *Zhongguo Zhong Yao Za Zhi*, **32**: 918-920.

Li, Y., Zhang, C.Y., Lin, T., Wang, G.H., Zeng, D.Q., Guo, Z.J., Zou, X.H., and Chen, H.F. (2013). Chemical constituents from *Elephantopus tomentosus*. *Zhongguo Zhong Yao Za Zhi*, **38**: 1751-1756.

Li, J., Qu, W., Cheng, Y., Sun, Y., Jiang, Y., Zou, T., Wang, Z., Xu, Y., and Zhao, H. (2014). The inhibitory effect of intravesical fisetin against bladder cancer by induction of p53 and down-regulation of NF-κB pathways in a rat bladder carcinogenesis model. *Basic and Clinical Pharmacology and Toxicology*, [Epub ahead of print].

Liang, H.R., Yan, W.M., Li, L., Jiang, L., and Dai, L. (1993). Studies on the chemical constituents of *Epimedium acuminatum* Franch. *Zhongguo Zhong Yao Za Zhi*, **18**: 677-679.

Liang, H.R., Vuorela, P., Vuorela, H., and Hiltunen, R. (1997). Isolation and immunomodulatory effect of flavonol glycosides from *Epimedium hunanense*. *Planta Medica*, **63**: 316-319.

Liang, Y.F., Chen, Z.T., and Liu, L.H. (2008). Studies on chemical constituents of *Ranunculus japonicas*. *Zhongguo Zhong Yao Za Zhi*, **33**: 2201-2203.

Marczak, L., Stobiecki, M., Jasiñski, M., Oleszek, W., and Kachlicki, P. (2010). Fragmentation pathways of acylated flavonoid diglucuronides from leaves of *Medicago truncatula*. *Phytochemical Analysis*, **21**: 224-233.

Matsuta, T., Sakagami, H., Satoh, K., Kanamoto, T., Terakubo, S., Nakashima, H., Kitajima, M., Oizumi, H., and Oizumi, T. (2011). Biological activity of luteolin glycosides and tricin from *Sasa senanensis* Rehder. *In Vivo*, **25**: 757-762.

Maurício Duarte-Almeida, J., Novoa, A.V., Linares, A.F., Lajolo, F.M., and Inés Genovese, M. (2006). Antioxidant activity of phenolics compounds from sugar cane (*Saccharum officinarum* L.) juice. *Plant Foods for Human Nutrition*, **61**: 187-192.

Merfort, I., and Wendisch, D. (1988). Flavonoid Glucuronides from the Flowers of *Arnica montana*. *Planta Medica*, **54**: 247-250.

Miyaichi, Y., Segawa, A., and Tomimori, T. (2006). Studies on Nepalese crude drugs. XXIX. Chemical constituents of Dronapuspi, the whole herb of *Leucas cephalotes* SPRENG. *Chemical and Pharmaceutical Bulletin (Tokyo)*, **54**: 1370-1379.

Mohanlal, S., Parvathy, R., Shalini, V., Helen, A., and Jayalekshmy, A. (2011). Isolation, characterization and quantification of tricin and flavonolignans in the medicinal rice Njavara (*Oryza sativa* L.), as compared to staple varieties. *Plant Foods for Human Nutrition*, **66**: 91-96.

Mohanlal, S., Maney, S. K., Santoshkumar, T. R., and Jayalekshmy, A. (2013). Tricin 4'-O-(erythro-β-guaiacylglyceryl) ether and tricin 4'-O-(threo-β-guaiacylglyceryl) ether isolated from Njavara (Oryza sativa L. var. Njavara), induce apoptosis in multiple tumor cells by mitochondrial pathway. *Journal of Natural Medicines*, **67**: 528-533.

Moheb, A., Agharbaoui, Z., Kanapathy, F., Ibrahim, R.K., Roy, R., and Sarhan, F. (2013a). Tricin biosynthesis during growth of wheat under different abiotic stresses. *Plant Science*, **201-202**: 115-120.

Moheb, A., Grondin, M., Ibrahim, R.K., Roy, R., and Sarhan, F. (2013b). Winter wheat hull (husk) is a valuable source for tricin, a potential selective cytotoxic agent. *Food Chemistry*, **138**: 931-937.

Moscatelli, V., Hnatyszyn, O., Acevedo, C., Megías, J., Alcaraz, M.J., and Ferraro, G. (2006). Flavonoids from *Artemisia copa* with anti-inflammatory activity. *Planta Medica*, **72**: 72-74.

Nakano, H., Kawada, N., Yoshida, M., Ono, H., Iwaura, R., and Tonooka, T. (2011). Isolation and identification of flavonoids accumulated in proanthocyanidin-free barley. *Journal of Agricultural and Food Chemistry*, **59**: 9581-9587.

Nazaruk, J., and Gudej, J. (2003). Flavonoid compounds from the flowers of *Cirsium rivulare* (Jacq.) All. *Acta Poloniae Pharmaceutica*, **60**: 87-89.

Park, H.S., Lim, J.H., Kim, H.J., Choi, H.J., and Lee, I.S. (2007). Antioxidant flavone glycosides from the leaves of *Sasa borealis*. *Archives of Pharmacal Research*, **30**: 161-166.

Pu, S., Guo, Y., and Gao, W. (2009). Chemical constituents from *Centipeda minima*. *Zhongguo Zhong Yao Za Zhi*, **34**: 1520-1522.

Sakai, A., Watanabe, K., Koketsu, M., Akuzawa, K., Yamada, R., Li, Z., Sadanari, H., Matsubara, K., and Muroyama, T. (2008). Anti-human cytomegalovirus activity of constituents from *Sasa albo-marginata* (Kumazasa in Japan). *Antiviral Chemistry and Chemotherapy*, **19**: 125-132.

Sayed, H.M., Mohamed, M.H., Farag, S.F., Mohamed, G.A., and Proksch, P. (2007). A new steroid glycoside and furochromones from *Cyperus rotundus* L. *Natural Product Research*, **21**: 343-350.

Shang, M., Cai, S., Han, J., Li, J., Zhao, Y., Zheng, J., Namba, T., Kadota, S., Tezuka, Y., and Fan, W. (1998). Studies on flavonoids from Fenugreek (*Trigonella foenum graecum* L.). *Zhongguo Zhong Yao Z Zhi*, **23**: 614-616, 639.

Shao, L., Huang, W.H., Zhang, C.F., Wang, L., Zhang, M., and Wang, Z.T. (2008). Study on chemical constituents from stem of *Dendrobium aphyllum*. *Zhongguo Zhong Yao Za Zhi*, **33**: 1693-1695.

Shen, C.C., Cheng, J.J., Lay, H.L., Wu, S.Y., Ni, C.L., Teng, C.M., and Chen, C.C. (2012a). Cytotoxic apigenin derivatives from *Chrysopogon aciculatus*. *Journal of Natural Products*, **75**: 198-201.

Shen, S.M., Shen, L.G., Lei, Q.F., Si, J.Y., Liu, C.M., and Lu, H. (2012b). Chemical constituents contained in aerial parts of *Emilia sonchifolia*. *Zhongguo Zhong Yao Za Zhi*, **37**: 3249-3251.

Solmaz, S., Gokbulut, A.A., Cincin, B., Ozdogu, H., Boga, C., Cakmakoglu, B., Kozanoglu, I., and Baran, Y. (2014). Therapeutic potential of apigenin, a plant flavonoid, for Imatinib-sensitive and resistant chronic myeloid leukemia cells. *Nutrition and Cancer*, [Epub ahead of print].

Stochmal, A., Simonet, A.M., Macias, F.A., and Oleszek, W. (2001). Alfalfa (*Medicago sativa* L.) flavonoids. 2. Tricin and chrysoeriol glycosides from aerial parts. *Journal of Agricultural and Food Chemistry*, **49**: 5310-5314.

Van Hoyweghen, L., Karalic, I., Van Calenbergh, S., Deforce, D., and Heyerick, A. (2010). Antioxidant flavone glycosides from the leaves of *Fargesia robusta*. *Journal of Natural Products*, **73**: 1573-1577.

Wang, H.K., Xia, Y., Yang, Z.Y., Natschke, S.L., and Lee, K.H. (1998). Recent advances in the discovery and development of flavonoids and their analogues as antitumor and anti-HIV agents. *Advances in Experimental Medicine and Biology*, **439**: 191-225.

Wang, H.B., Yao, H., Bao, G.H., Zhang, H.P., and Qin, G.W. (2004). Flavone glucosides with immunomodulatory activity from the leaves of *Pleioblastus amarus*. *Phytochemistry*, **65**: 969-974.

Wang, G.R., Tang, W.Z., Yao, Q.Q., Zhong, H., and Liu, Y.J. (2010a). New flavonoids with 2BS cell proliferation promoting effect from the seeds of *Trigonella foenum-graecum* L. *Journal of Natural Medicines*, **64**: 358-361.

Wang, J., Chen, D., Liang, L., Xue, P., and Tu, P. (2010b). Chemical constituents from flowers of *Chrysanthemum indicum*. *Zhongguo Zhong Yao Za Zhi*, **35**: 718-721.

Wang, Y., Ma, W., and Zheng, W. (2013). Deguelin, a novel anti-tumorigenic agent targeting apoptosis, cell cycle arrest and anti-angiogenesis for cancer chemoprevention. *Molecular Clinical Oncology*, **1**: 215-219.

Watanabe, M. (1999). Antioxidative phenolic compounds from Japanese barnyard millet (*Echinochloa utilis*) grains. *Journal of Agricultural and Food Chemistry*, **47**: 4500-4505.

Webby, R.F., and Markham, K.R. (1994). Isoswertiajaponin 2''-O-β-arabinopyranoside and other flavone-C-glycosides from the Antarctic grass *Deschampsia antarctica*. *Phytochemistry*, **36**: 1323-1326.

Wenzig, E., Kunert, O., Ferreira, D., Schmid, M., Schühly, W., Bauer, R., and Hiermann, A. (2005). Flavonolignans from *Avena sativa*. *Journal of Natural Products*, **68**: 289-292.

Weston, L.A., Burke, B.A., and Putnam, A.R. (1987). Isolation, characterization and activity of phytotoxic compounds from quackgrass [*Agropyron repens* (L.) Beauv.]. *Journal of Chemical Ecology*, **13**: 403-421.

Xiang, Y., Li, Y.B., Zhang, J., Li, P., and Yao, Y.Z. (2007). Studies on chemical constituents of *Salsola collina*. *Zhongguo Zhong Yao Za Zhi*, **32**: 409-413.

Yao, H., Chen, Y., Shi, P., Hu, J., Li, S., Huang, L., Lin, J., and Lin, X. (2012). Screening and quantitative analysis of antioxidants in the fruits of *Livistona chinensis* R. Br using HPLC-DAD-ESI/MS coupled with pre-column DPPH assay. *Food Chemistry*, **135**: 2802-2807.

Zhang, Y., and Zhao, Y.M. (2005). Studies on the chemical constituents in seeds of *Atriplex centralasiatica*. *Zhongguo Zhong Yao Za Zhi*, **30**: 679-681.

Zhang, L.H., Fan, C.L., Ye, W.C., and Li, Y.P. (2008a). Study on flavonoiels and phenolic acids from the herb of *Lygodium japonicum*. *Zhong Yao Cai*, **31**: 224-226.

Zhang, X.Q., Jiang, W.W., Wang, Y., Li, Y.L., and Ye, W.C. (2008b). A new phenylpropanoic acid derivatives from the roots of *Ficus stenophylla*. *Yao Xue Xue Bao*, **43**: 281-283.

Zhang, H.F., Yan, L.H., Zhang, Q.W., and Wang, Z.M. (2013). Flavonoids from leaves of *Epimedium pubescens*. *Zhongguo Zhong Yao Za Zhi*, **38**: 1942-1946.

Zheng, W., Zhou, C.X., Zhang, S.L., Weng, L.J., and Zhao, Y. (2006). Studies on the chemical constituents in herb of *Ranunculus japonicus*. *Zhongguo Zhong Yao Za Zhi*, **31**: 892-894.

Zheng, R.R., Ya, J., Wang, W.J., Yang, H.B., Zhang, Q.W., Zhang, X.Q., and Ye, W.C. (2013). Chemical studies on roots of *Ficus hirta*. *Zhongguo Zhong Yao Za Zhi*, **38**: 3696-3701.

Natural Products: Research Reviews Vol. 3 (2016) *Pages* **95–145**
Editor: **V.K. Gupta**
Published by: **DAYA PUBLISHING HOUSE, NEW DELHI**

5

A Review: Therapeutical Applications on Phytochemicals of Plants Belongs to *Celosia* Genus

P. Rajesh[1*] and V. Rajesh Kannan[2]

ABSTRACT

The main goal of the present comprehensive review is to present the research carried out with species of Celosia *genus, widely spread in the world, in order to organize the data produced as resource for reference piece. This review was summarized the phytoconstituents, its chemical structure and the corresponding pharmacological activities reported in various species of the* Celosia *genus.* Celosia *is a medicinally important genus belonging to the family of Amaranthaceae consisting of about 158 scientific plant namesin the genus* Celosia. *From the above, 51 are accepted species rather 20 species were infra specific genus belonging to* Celosia. *About, the* Celosia *genus 65 species are cute familiar and distributed in the region of Africa, America and Asia continents. Among the 65 species, some of the species only explored for the medicinal purposes with its phytochemicals. Such that, the plant* Celosia *known as edible and ornamental plants; in ancient literatures are also recommended for treatment of blood diseases, eye problems, diarrhea, dressings for boils, Intestinal worms, mouth sores, and sores. In the past decades, the research study of the* Celosia*genus has also been reported to have anti-oxidant, anti-mitotic, anti-bacterial, ant-iinflammatory, hepatoprotective, immunological activity. The plant phytochemicals (secondary metabolites)*

---

1 Assistant Professor, Department of Biotechnology, KSR College of Arts and Science, Tiruchengode – 637 215, Tamil Nadu, India.

2 Assistant Professor, Department of Microbiology, Bharathidasan University, Tiruchirappalli – 620 024, Tamil Nadu, India.

* *Corresponding author*: E-mail: raju19marc@gmail.com

have been reported to have flavonoids, phenolic glycoside, pigments, peptides, steroids as presence as 90 different compounds and its derivatives. Further, the chapter deals with chemical compounds, its isolation, characterization methods and therapeutical applications were reviewed here.

Keywords: Amaranthaceae, *Celosia*, Celosin, Celogentin, Moroidin.

Introduction

A larger number of higher plants possessed and reported that pharmacologically potent healing activities with its medicinal properties through phytochemicals. The applications of medicinal plants with pharmaceutical production have received and augmented interest nowadays in both homeopathic and allopathic branches. However allopathic drugs have side effects and people are going back to the nature with hope of safety and security.

Herbal traditional methods have been developed through many experiences (analytical methods) of many generations (past decades) (Zingare, 2012). Herbal medicines are easily available, safe, cheaper, without any side effect and adverse reactions. Many valuable herbal drugs have been discovered by knowing that particular plant was used by ancient folk healers for the treatment of some kind of ailments; nowadays, the folk medicines phytochemical as secondary metabolites were proven through research and development (Ekka and Dixit, 2007).

Medicinal plants play an important role in public health, especially in developed and developing countries and leads to examination for therapeutic and pharmacological applications (Mossi*et. al.*, 2009). In ethno-botany, information collection regarding medicinal and economic importance of plants from local people is of great importance. Ethno-botany provides us profound understanding and appreciation of the richness and intimacy of relationship between nature and human (Jagtap *et al.*, 2013).

About, the phytochemistry or plant chemistry is concerned with the enormous variety of organic substances that are elaborated and accumulated through the chemical structured of these substances, biosynthesis, metabolism, natural distribution and its biological function.

Medicinal plants contain some organic compounds, which provide definite physiological action in human body and these bioactive substances include alkaloids, carbohydrates, flavonoids, steroids, tannins and terpenoids. These compounds are synthesized by primary or rather secondary metabolism of plants. Secondary metabolites are chemically and taxonomically extremely diverse compounds with obscure function. They are regarded as secondary metabolites because the plants that manufacture or release for its functional uses. They are naturally synthesized in all parts of the plant such as bark, leaves, stem, root, flower, fruits, seeds, *etc.* The quantity and quality of phytochemicals present in plant parts may differ from one part to another.

Till now, the *Celosia* genus was unexplored for its therapeutical applications and no more studies were reports evaluated critically. Since, there was pity reports have been researched and submitted worldwide for their significant ailment properties to the various debilitating diseases through *in vivo* and *in vitro* research studies. The plant genus *Celosia* consisting of about 65 species in the family Amaranthaceae (Caryophyllales) is native in subtropical and temperate zones of Africa, America and Asia. The seeds, young leaves and inflorescences are used as a vegetable in China and other countries (Palada and Crossman, 1999) and also the dried leaves, inflorescences and seeds are used in traditional Chinese medicine (Wong, 1994 and Xu, 1996). About 65 species of *Celosia*, some common species only used in folk medicines and research development such as *C. aristata, C. cristata, C. isertii, C. laxa, C. nitida, C. leptostachya, C. palmeri, C. plumose, C. polygonoids* and *C. trigyna etc.*

The present chapter attempt and deals with many therapeutical applications of the above species such as anti-mitotic, anti-cancer, anti-inflammatory, anti-diabetic *etc.*, through various phytochemicals reported that about 90 compounds (primary and secondary metabolites) such as celosins, celogentins, stigmasterol, β-sitosterol, moroidin, isorhamnetin-3-O-β-D-glucopyranoside,quercetin-3-O-β-D-glucopyranoside and rhamnazin-3-O-β-D-glucopyranoside *etc.* were reviewed here.

Phytochemicals and Therapeutical Applications of the *Celosia* Genus

Celosia argentea

The aqueous, ethanolic, petroleum ether, ethyl acetate seeds extract of *Celosia argentea* showed and reported that alkaloids, carbohydrate, flavonoids, glycosides, proteins, saponin, steroids and tannins (Patel *et al.*, 2010).The aqueous and methanolic extracts of *C. argentea* were showed and reported to the presence of alkaloid, amino acids, carbohydrates, flavonoids, glycosides, gums, mucilage, proteins, steroidsand triterpenoids (Gautam *et al.*, 2011). Also, the methanolic extract of *C. argentea* of showed the presence of carbohydrates, flavonoids, glycosides, phenolic compounds, saponin and tannin (Urmila *et al.*, 2013).

The complete pharmacognosy includes microscopy of the *C. argentea* of the leaves showed the thick, cylindrical and collateral vascular strand were reported. The preliminary phytochemical screening indicated the presence of carbohydrates, glycosides, flavonoids, steroids and tannins in the aerial parts such as seed and root extracts which were reported through chromatographic techniques such as TLC (Priyanka *et al.*, 2012). The flavonoids, glycosides, steroids and tannins were presented and reported in leaves extracts of *C. argentea* and the compounds reported for the abnormal vaginal discharge (Alqasim, 2013).The root, seed, flower of *C. argentea* used for redness, swelling of the eyes, photophobia, lacrimation headache, colic, gonorrhea and eczema. Further, reported to possess alkaloids, flavonoids, phenolic compounds, saponins, steroids and tannin (Jadhao, 2013).In the preliminary phytochemical analysis, protein, fat, flavonoids, reducing sugar, saponins, starch, sugar and tannins presence were reported in root, stem, leaf extract of aqueous and ethanolic extracts of *C. argentea* (Rajni and Deokule, 2013).

Figure 5.1: Phytochemical Compounds of *Celosia* Genus (**Chemical structures drawn by ISIS Chem draw).

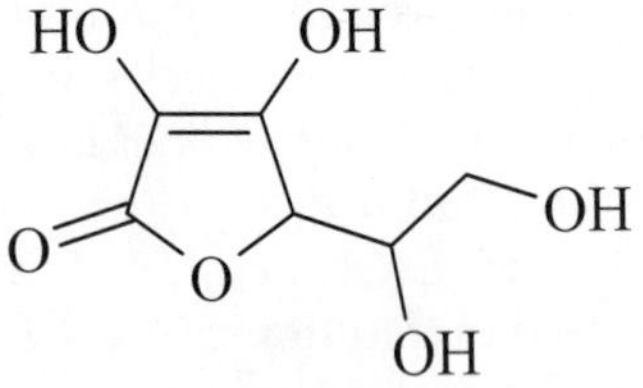

(1) Vitamin C

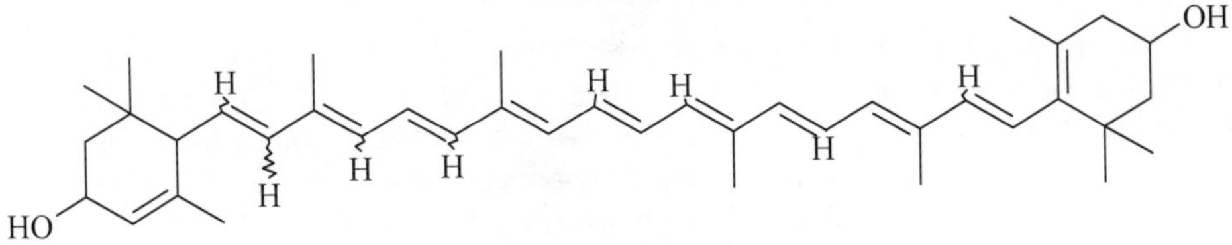

(2) Carotenoid

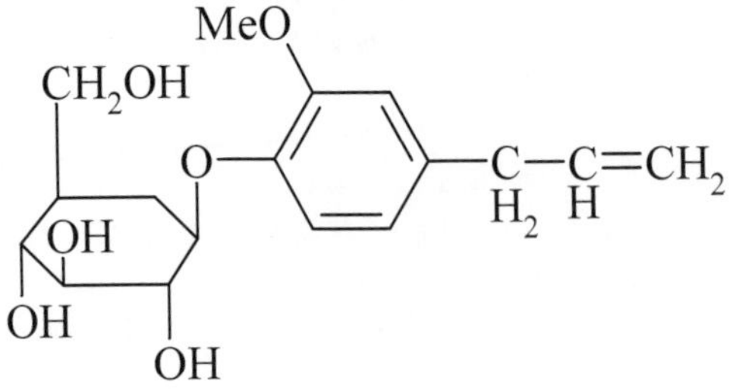

(3) 1-(4-O-β-glucopyranosyl-3-methoxyphenyl) propane-2-ene (Citrusin C)

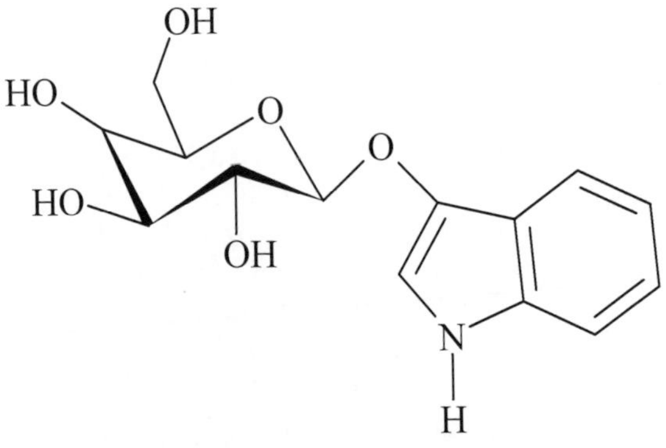

(4) 3-O-β-glucopyranosyl-1H-indole (Indican)

Contd...

Figure 5.1–*Contd...*

O
Glc_6-Rham

(5) (3Z)-hexenyl-1-O-α-rhamnopyranosyl-β-glucopyranoside

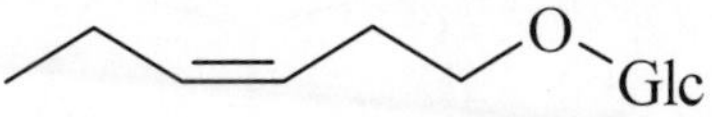

O
Glc

(6) (3Z)-hexenyl-1-O-β-glucopyranoside

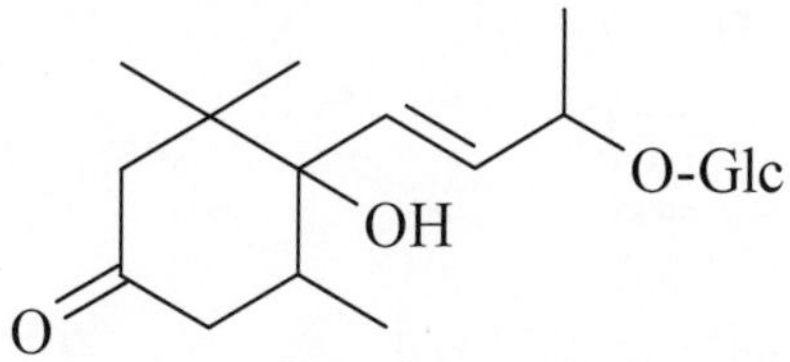

OH
O-Glc
O

(7) (7E)-6,9-dihydromegastigma-7-ene-3-one-9-O-β-glucopyranoside

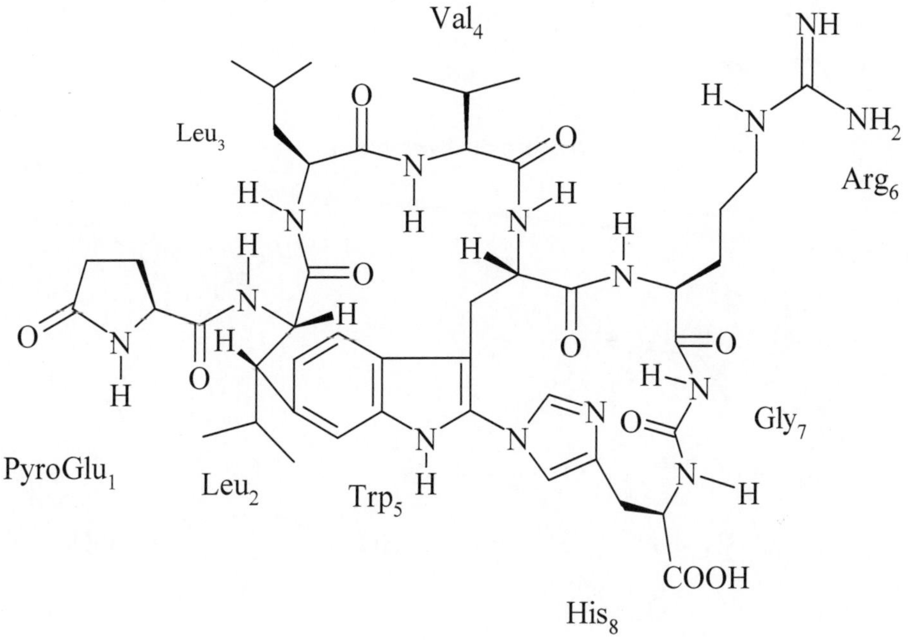

Val_4
Leu_3
Arg_6
$PyroGlu_1$
Leu_2
Trp_5
Gly_7
His_8
COOH

(8) Moroidin

Contd...

Figure 5.1–*Contd...*

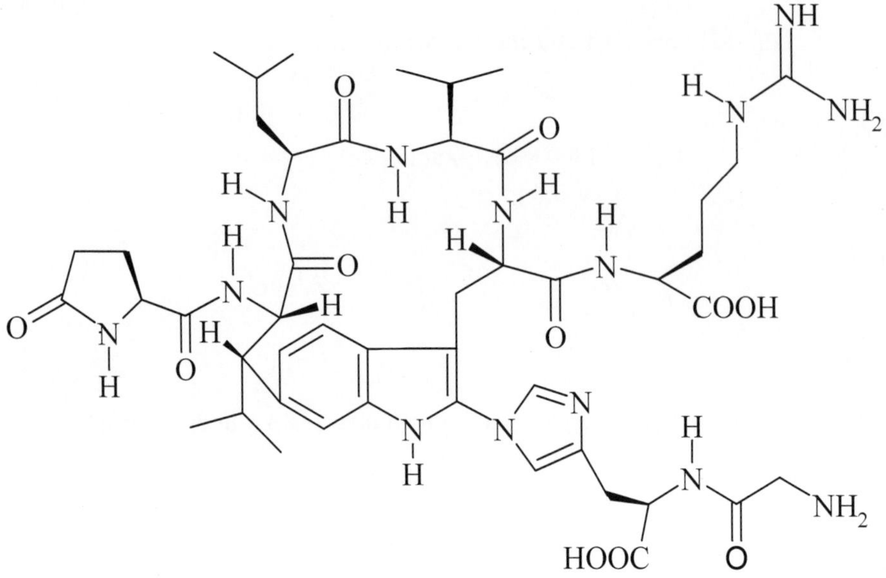

(9) Moroidin hydrolysate

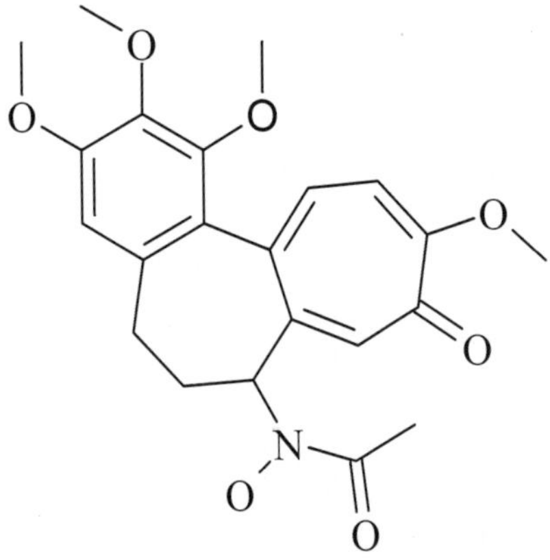

(10) Colchicines

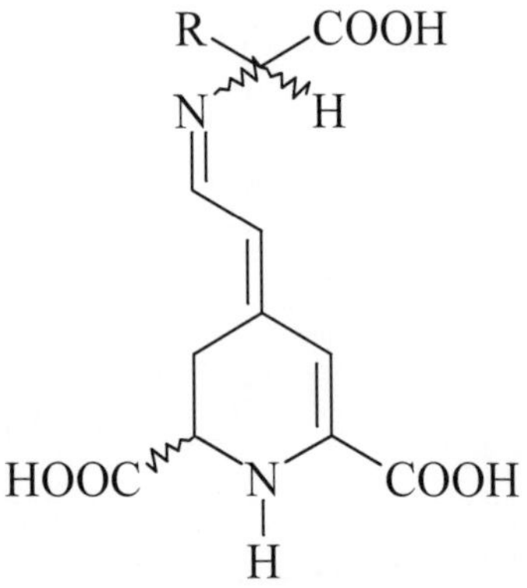

(11) Betaxanthin

Contd...

Figure 5.1–*Contd...*

(12) Amaranthin

(13) Isoamaranthin

Contd...

Figure 5.1–*Contd...*

(14) Betalamic acid

(15) Dopamine

(16) 3-methoxytyramine

(17) (S)-tryptophan

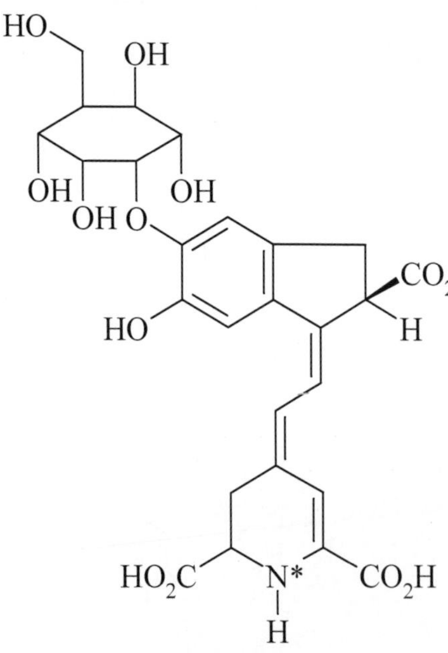

(18) Betacyanin

(19) 2-descarboxy-betanidin

Contd...

Figure 5.1–*Contd...*

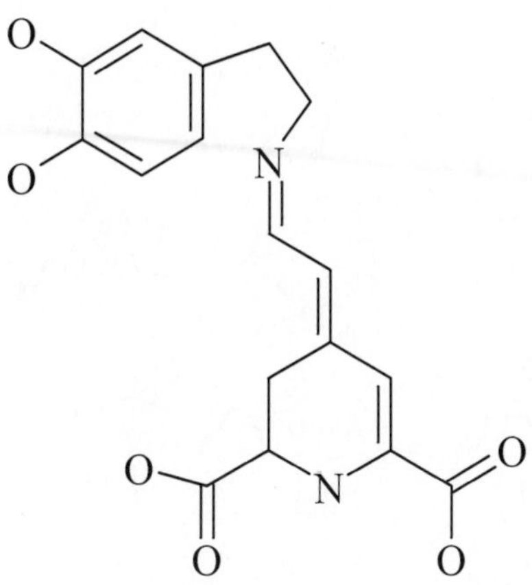

(20) Miraxanthin V

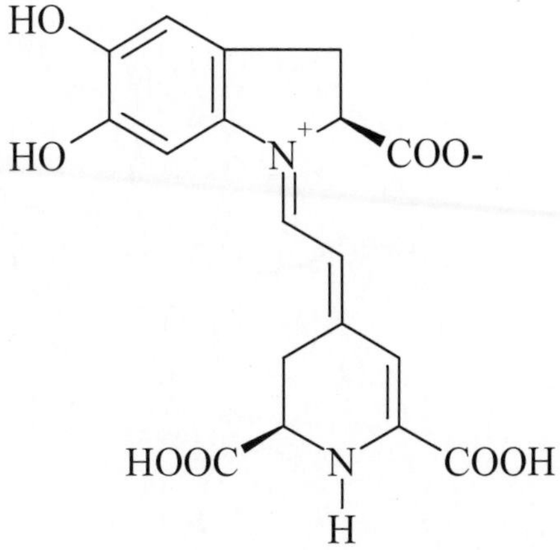

(21) Betanidin

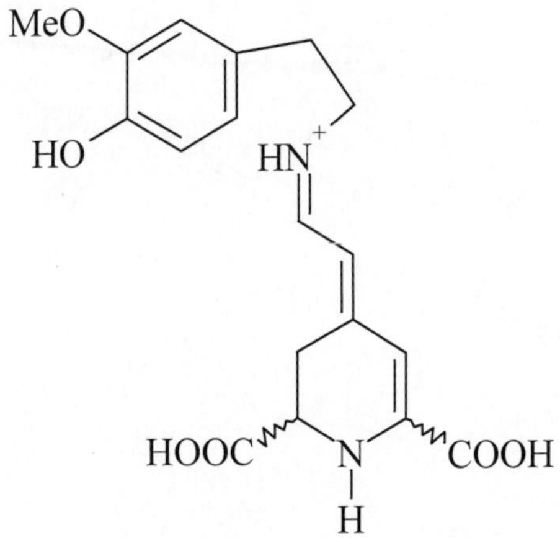

(22) 3-methoxytyramine-betaxanthine

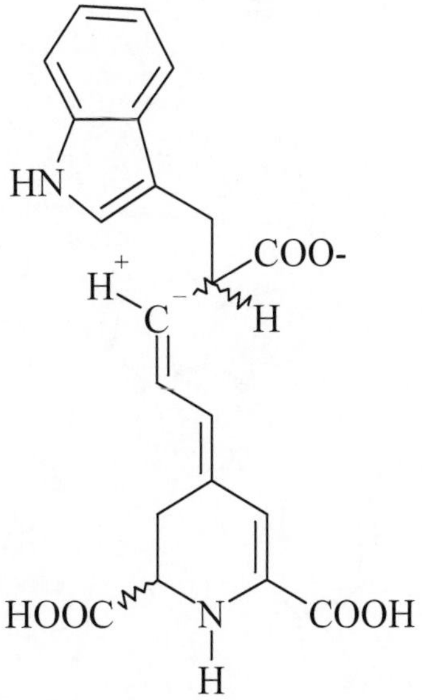

(23) (S)-tryptophan-betaxanthine

Contd...

Figure 5.1–*Contd...*

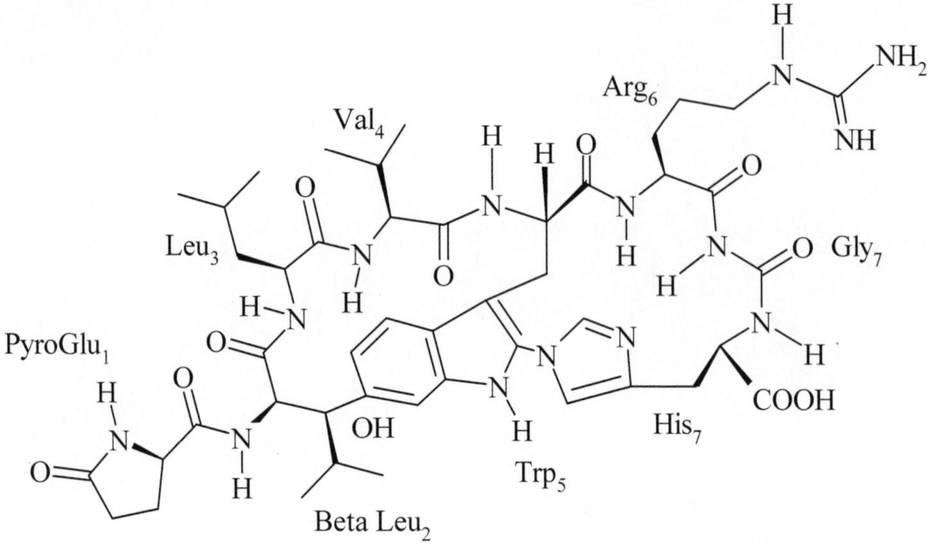

(24) Celogentin A

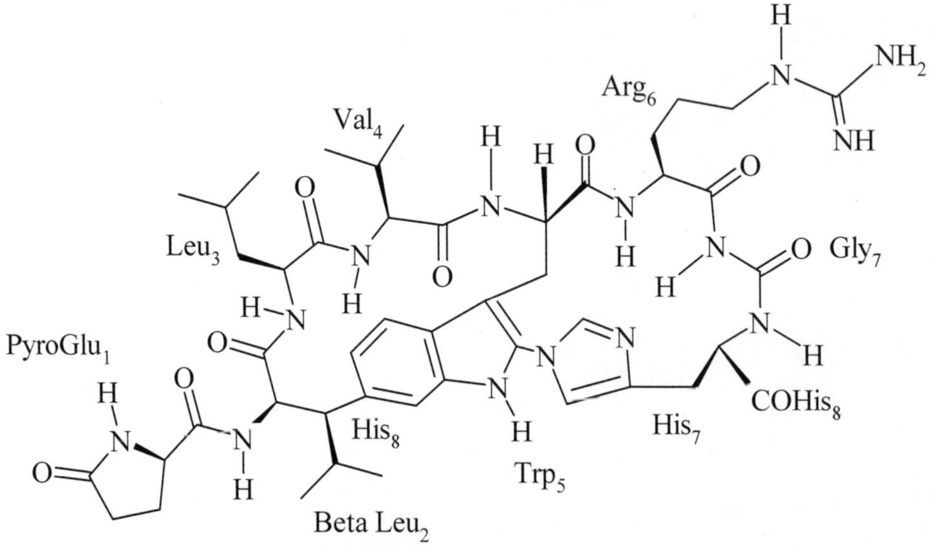

(25) Celogentin B

Contd...

Figure 5.1–*Contd...*

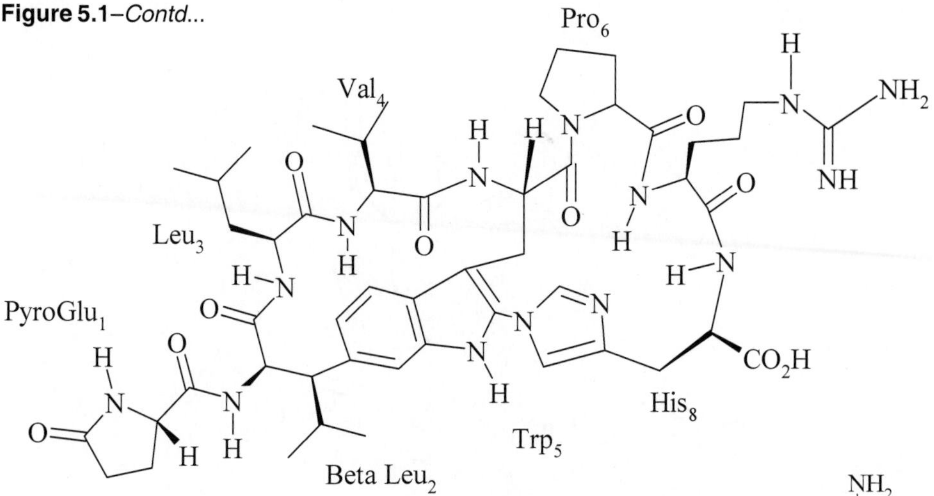

(26) Celogentin C

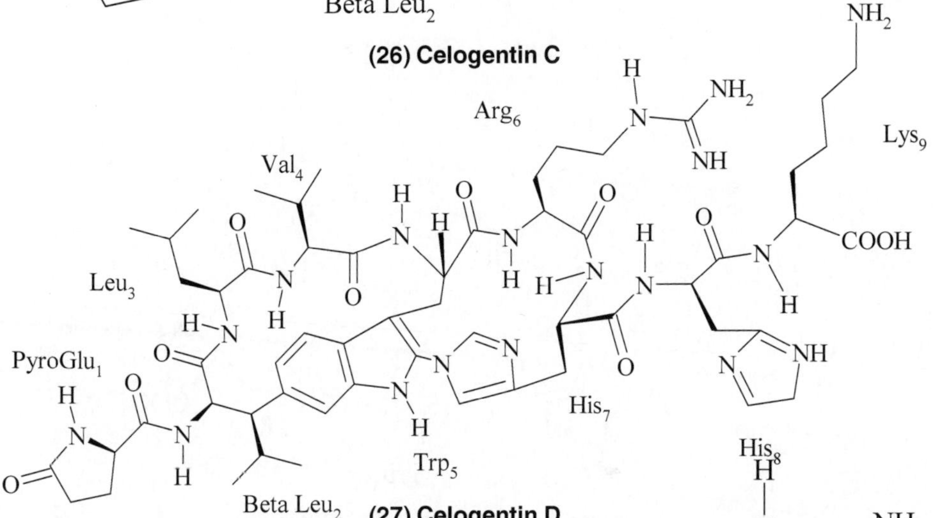

(27) Celogentin D

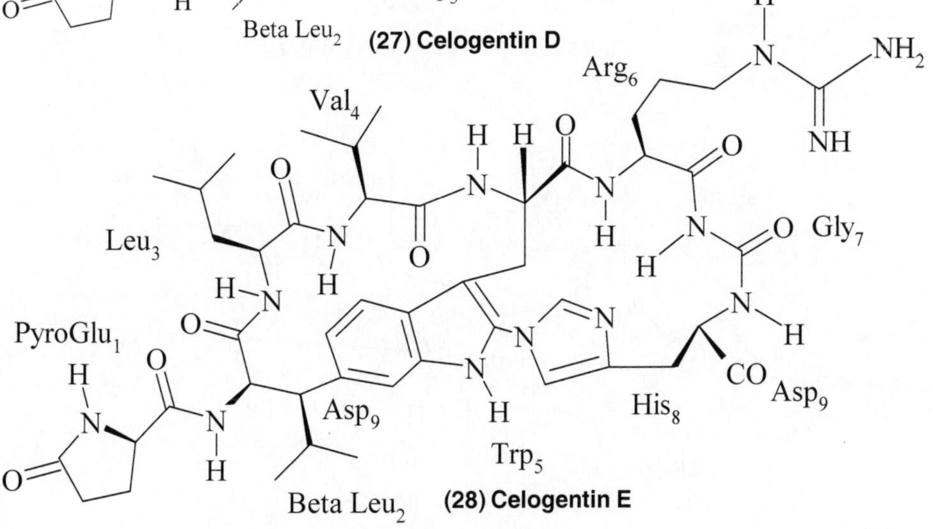

(28) Celogentin E

Contd...

Figure 5.1–*Contd...*

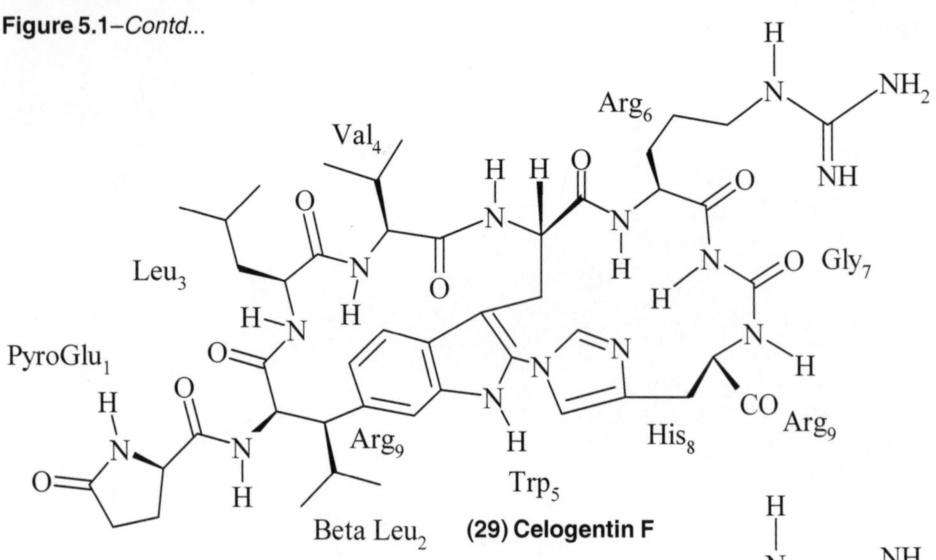

(29) Celogentin F

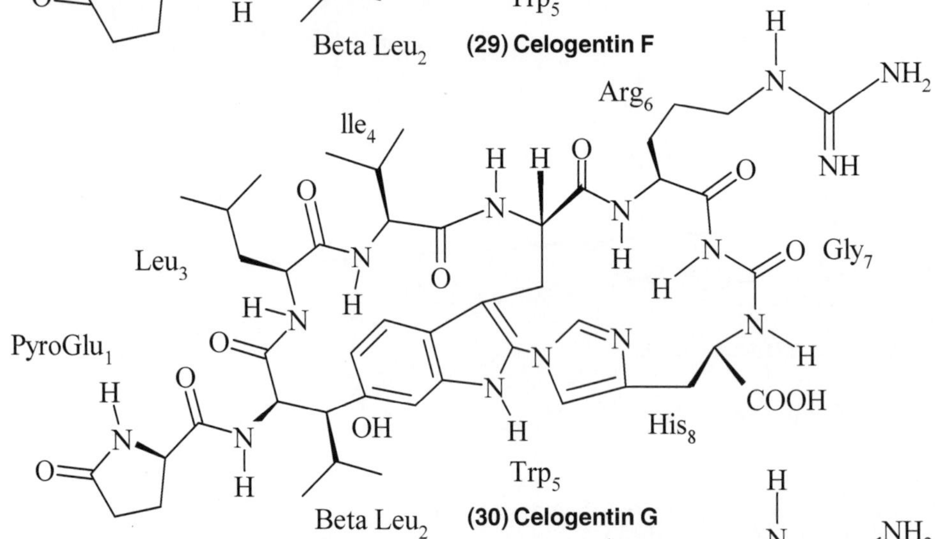

(30) Celogentin G

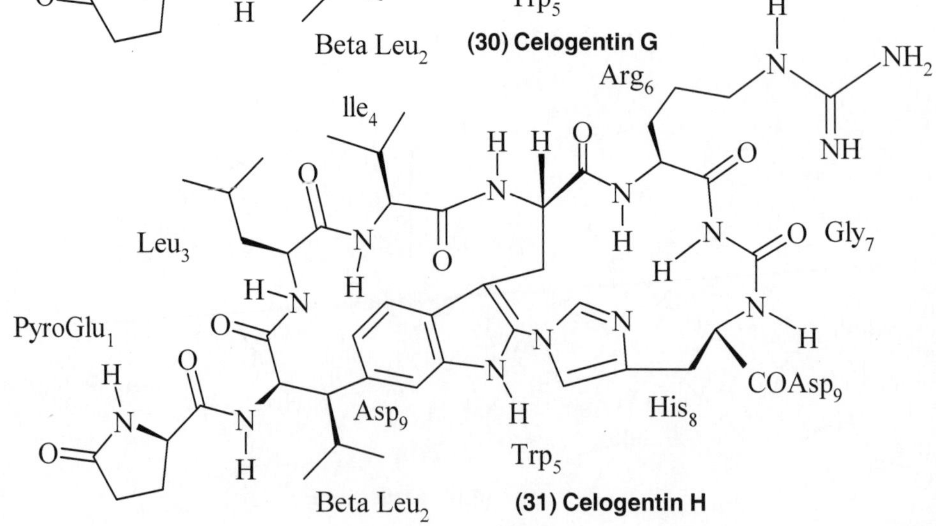

(31) Celogentin H

Contd...

Figure 5.1–*Contd...*

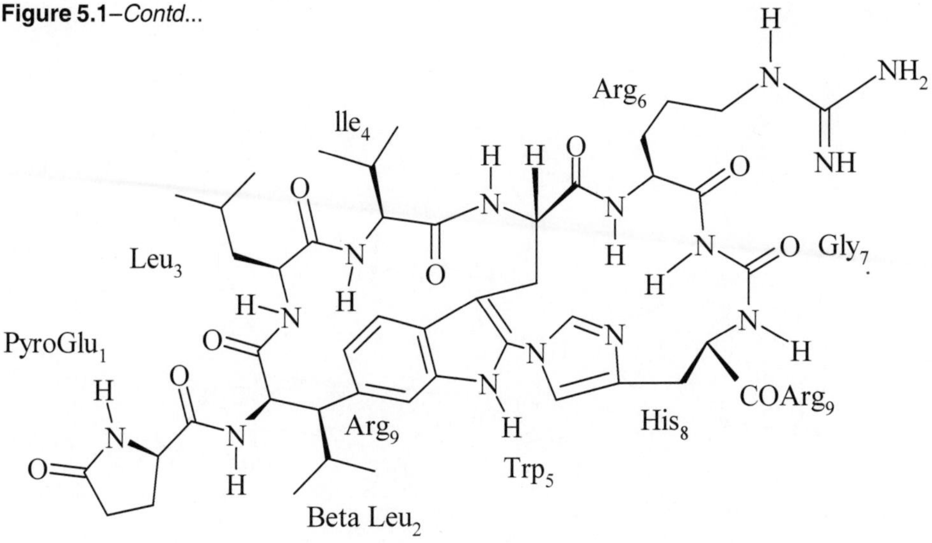

(32) Celogentin J

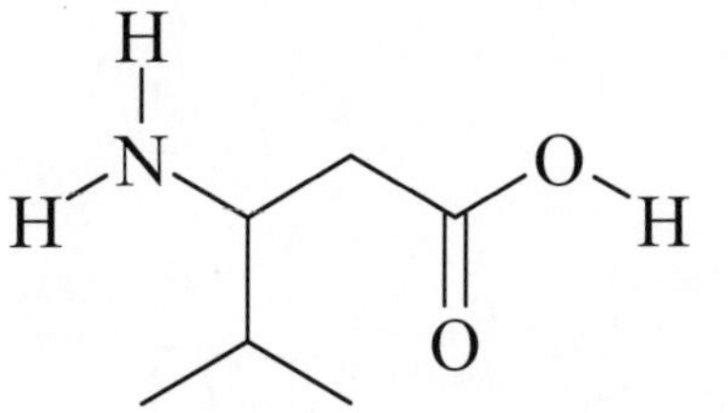

(33) β^s-leucine

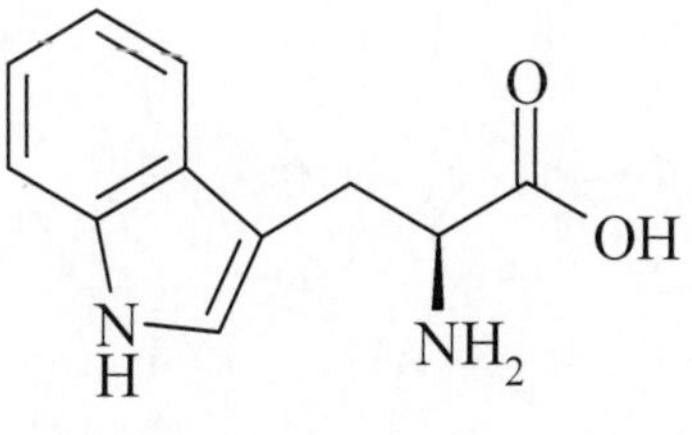

(34) Tryptophan

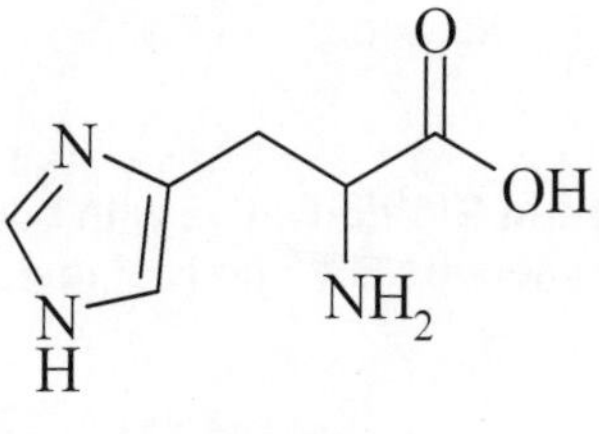

(35) Histidine

Contd...

Figure 5.1–*Contd...*

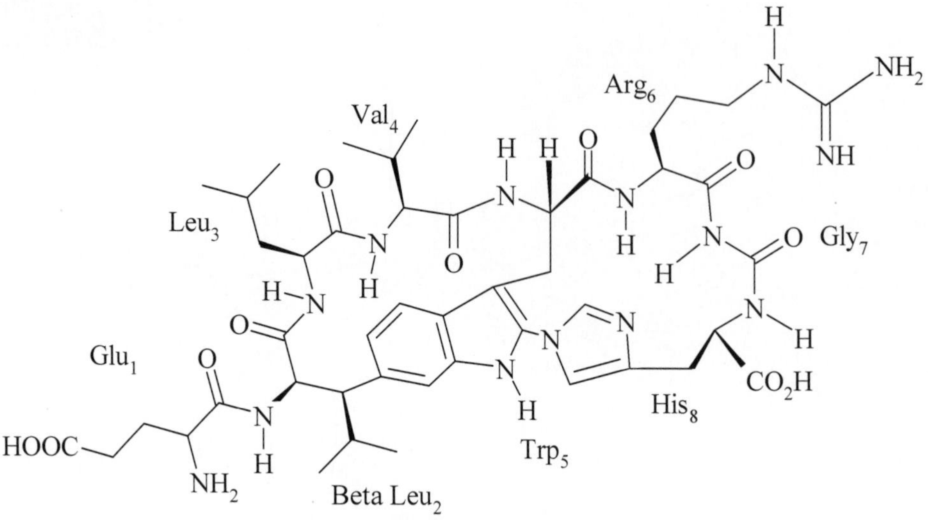

(36) Celogentin Glu[1] derivative

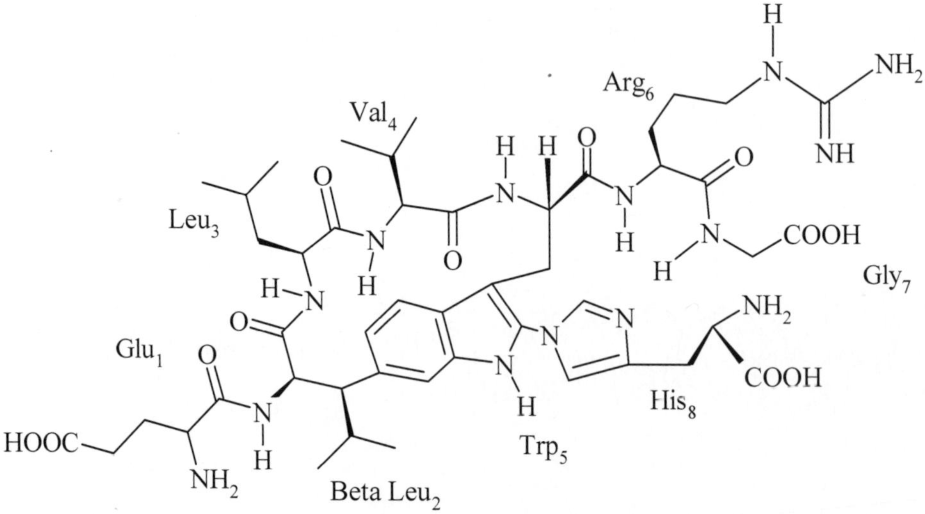

(37) Celogentin Glu[1] derivative with bond cleavage between the Gly[7] and His[8] residues

Contd...

Figure 5.1–*Contd...*

(38)Celogentin Glu^{1}derivative with bond cleavage between the $Arg^{6}$and Gly^{7} residues

(39) Celogentin derivative without the $PyroGlu^{1}$ residue

(40) Oligogalacturonic acid

Contd...

Figure 5.1–*Contd...*

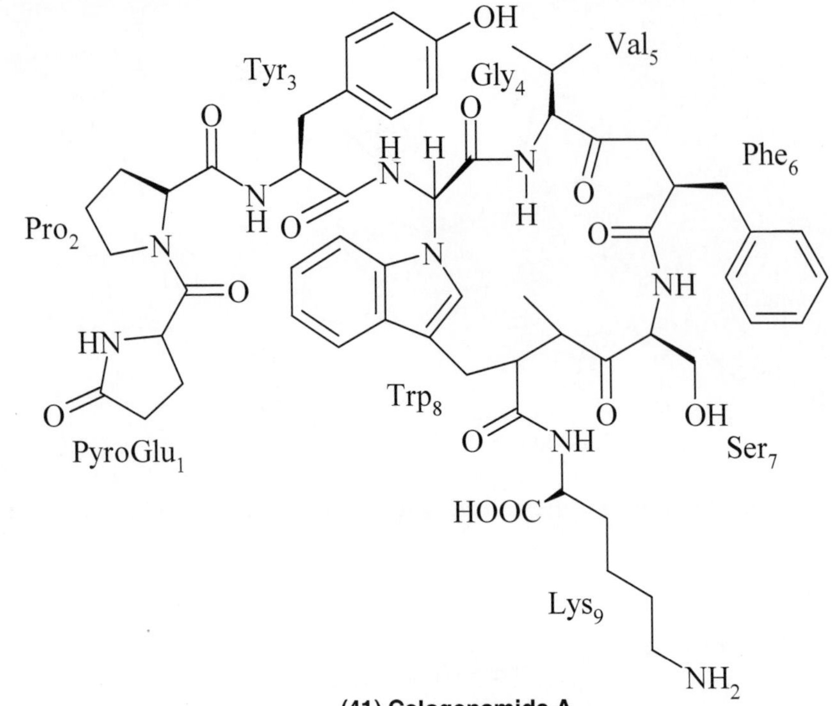

(41) Celogenamide A

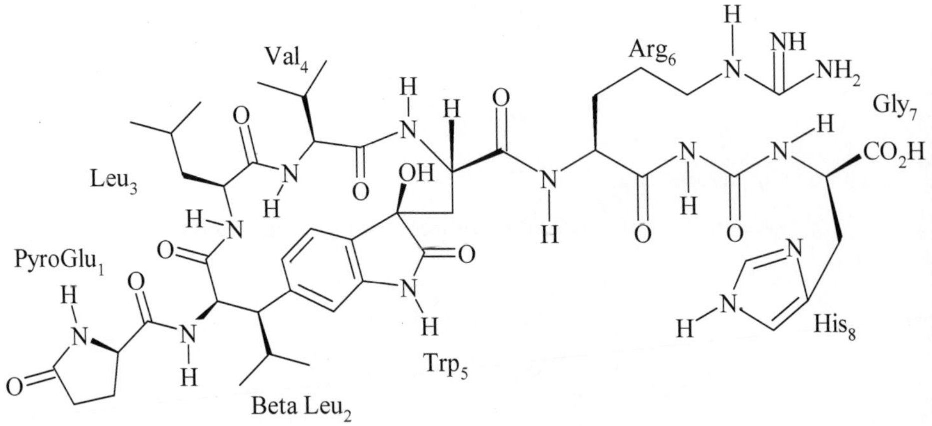

(42) Celogentin K

Contd...

Figure 5.1–*Contd...*

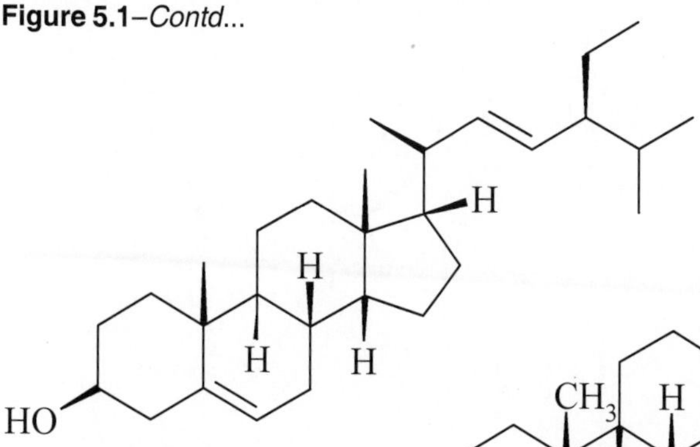

(43) Stigmasterol

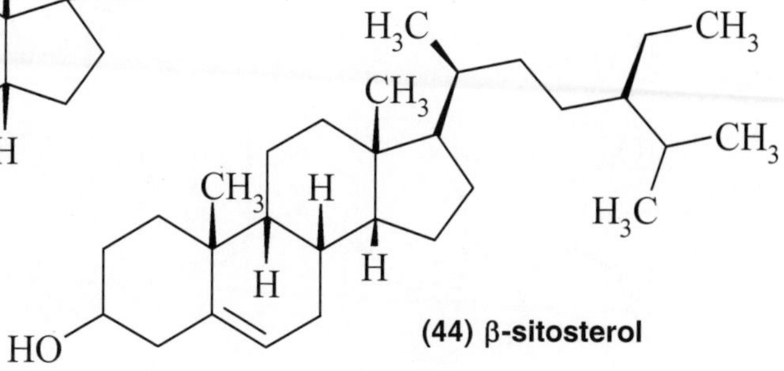

(44) β-sitosterol

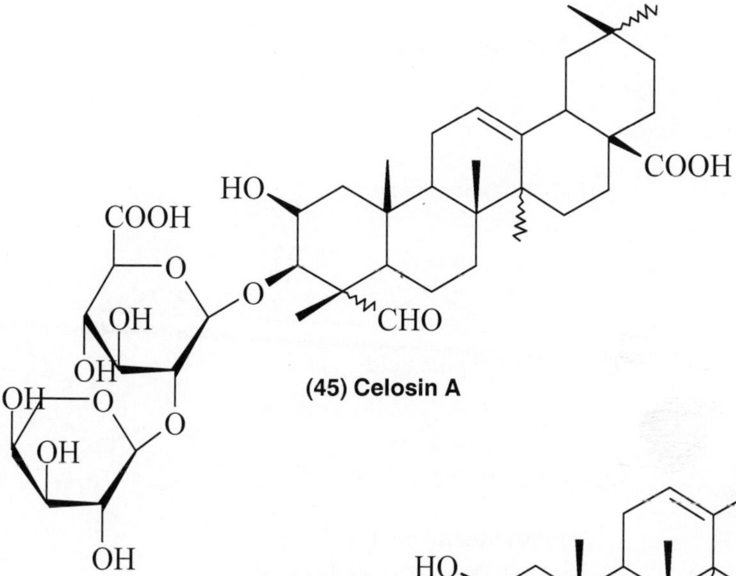

(45) Celosin A

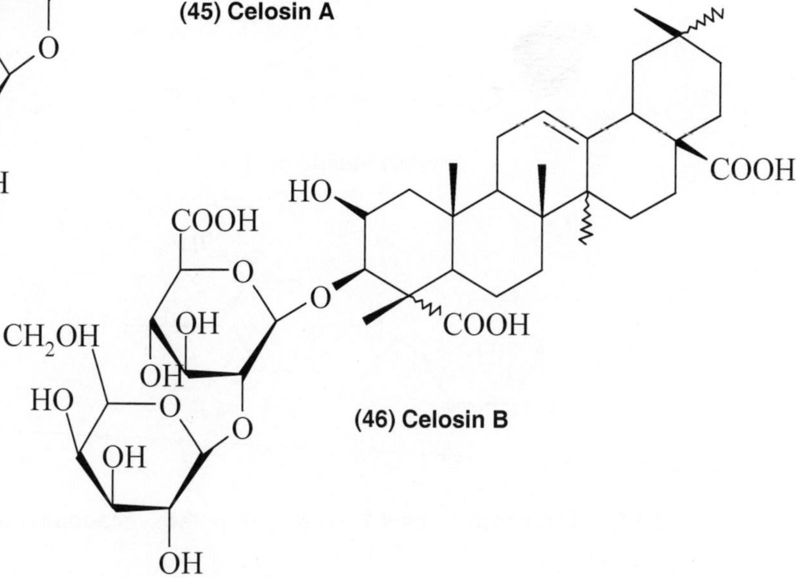

(46) Celosin B

Contd...

Figure 5.1–*Contd...*

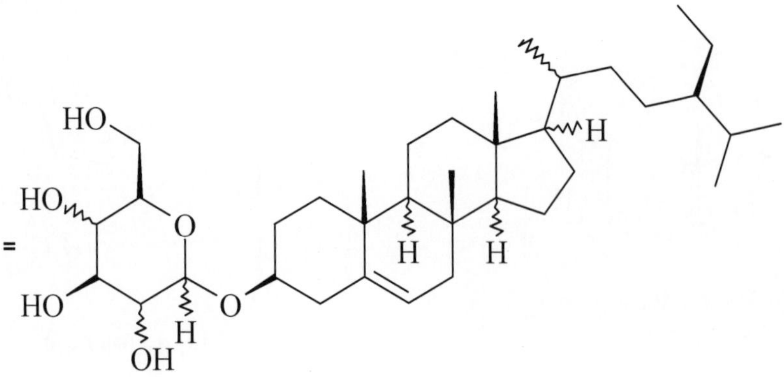

=

(47) β-daucosterol

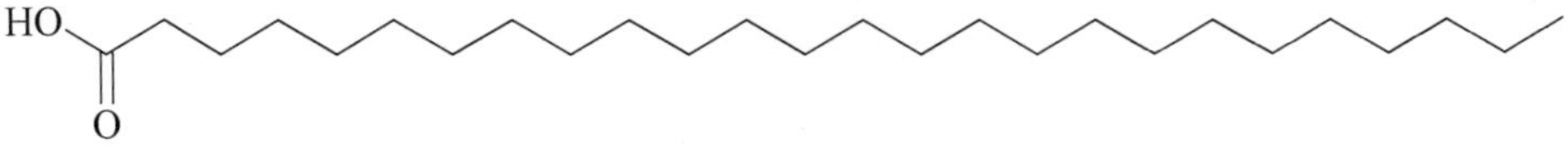

(48) n-hexacosnic acid

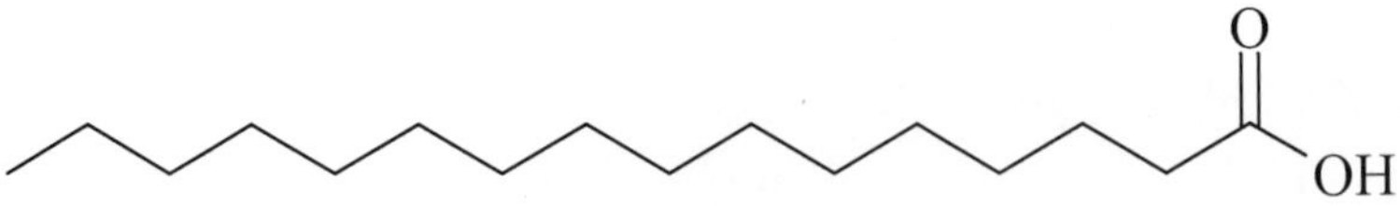

(49) Palmitic acid

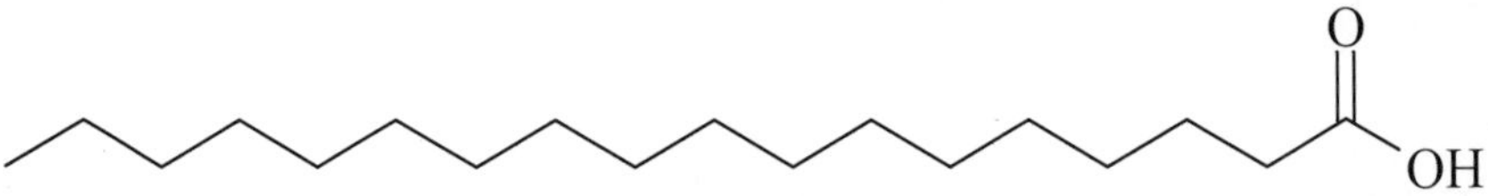

(50) Stearic acid

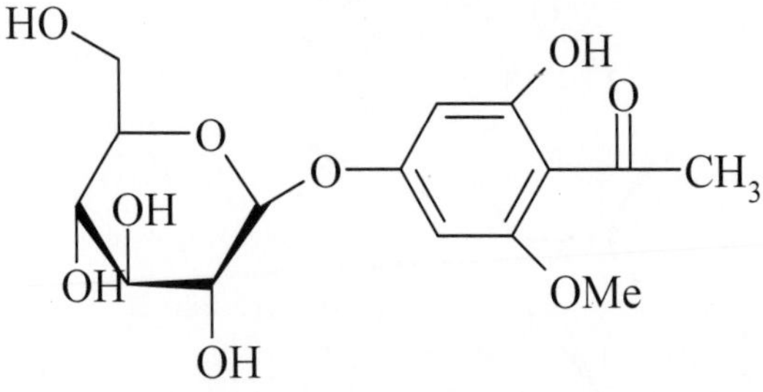

(51) 4-O-β-D-glucopyranosyl-2-hydroxy-6-methoxyace-tophenone

Contd...

Figure 5.1–*Contd...*

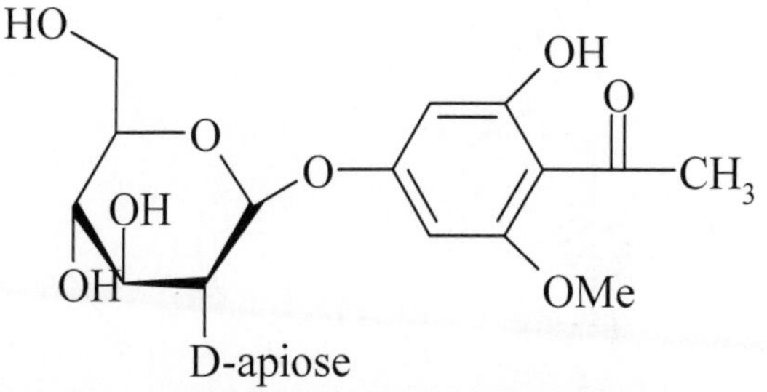

(52) 4-O-β-D-apifuranosyl-(1→2)-β-D-glucopyranosyl-2-hydroxy-6-methoxyacetophenone

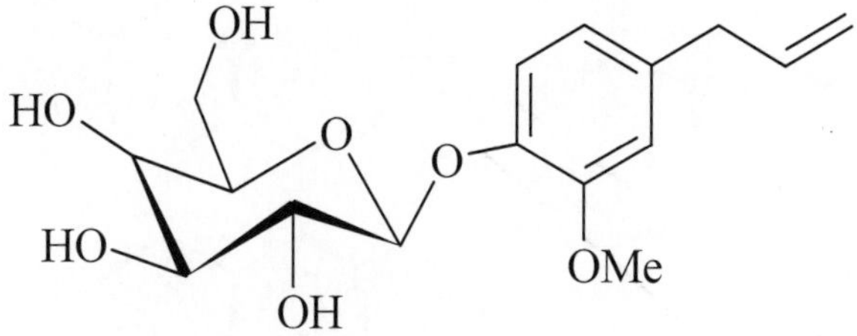

(53) Eugenyl-O-β-D-glucopyranoside

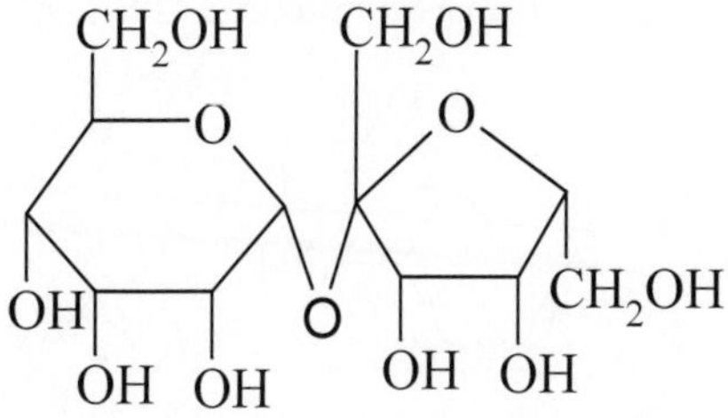

(54) Sucrose

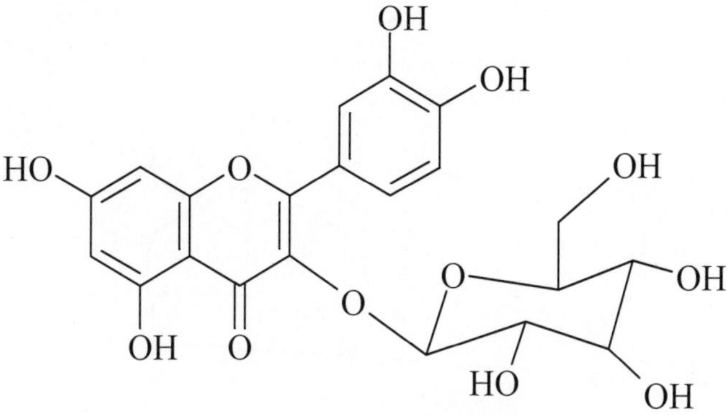

(55) Quercetin-3-O-β-D-glucopyranoside

Contd...

Figure 5.1–*Contd...*

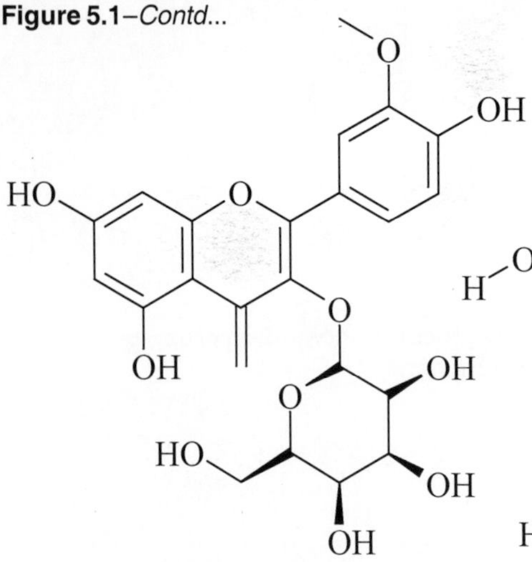

(56) Isorhamnetin-3-O-β-D-glucopyranoside

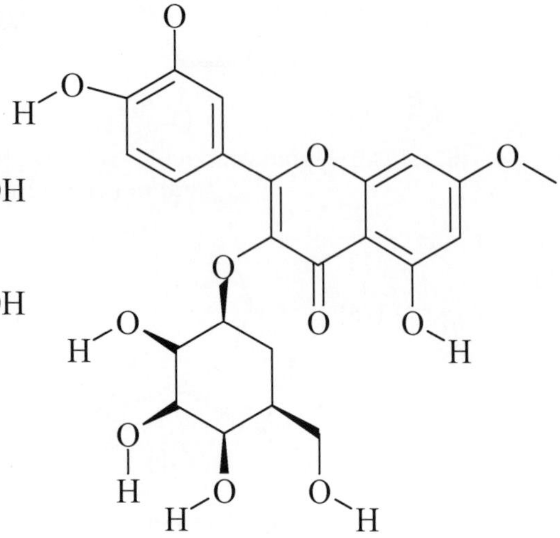

(57) Rhamnazin-3-O-β-D-glucopyranoside

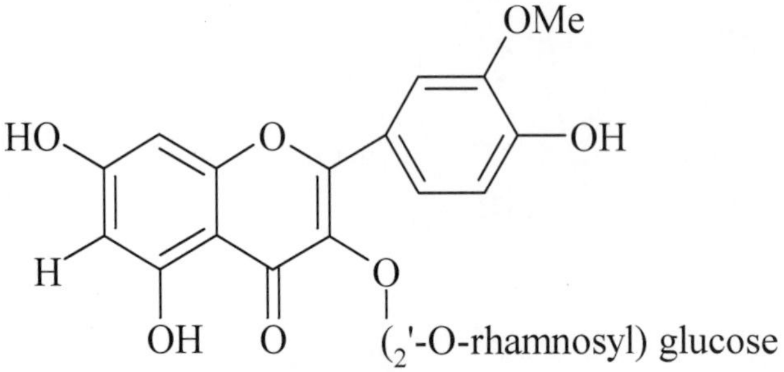

(58) Isorhamnetin-3-O-α-L-rhamnopyranosyl-(1→2)-β-D-glucopyranoside

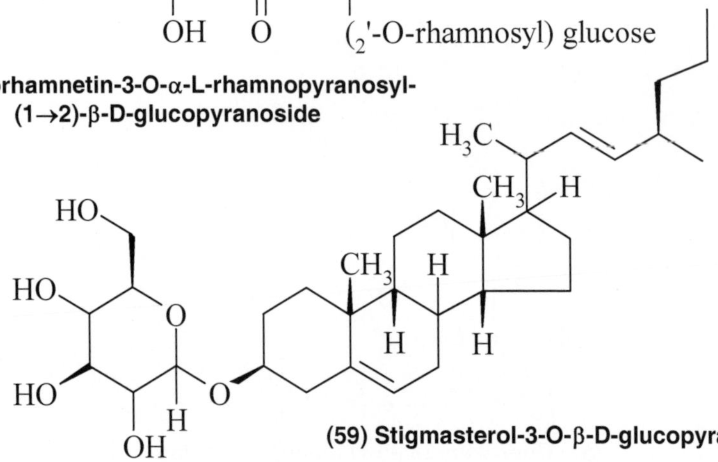

(59) Stigmasterol-3-O-β-D-glucopyranoside

Contd...

Figure 5.1–*Contd...*

(60) Oleanolic acid

(61) Glycerides

(62) Sterols

(63) Vitamin A

(64) Vitamin D

Contd...

Figure 5.1–*Contd...*

(65) Vitamin E

(66) Vitamin K

(67) Cristatain

(68) Celosin C

Contd...

Figure 5.1–*Contd...*

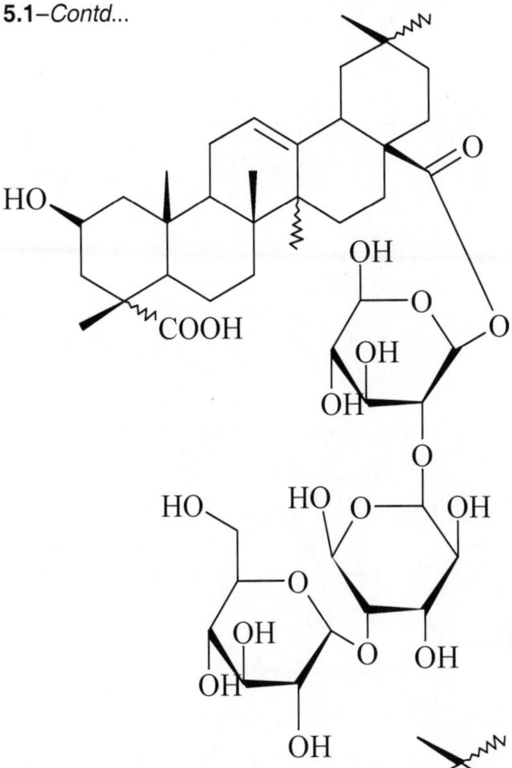

(69) Celosin D

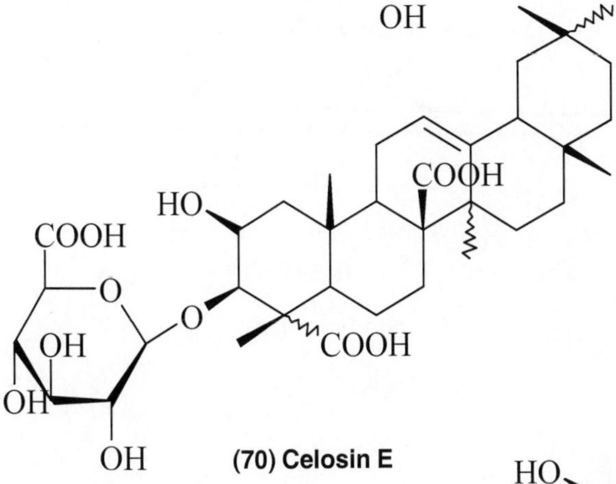

(70) Celosin E

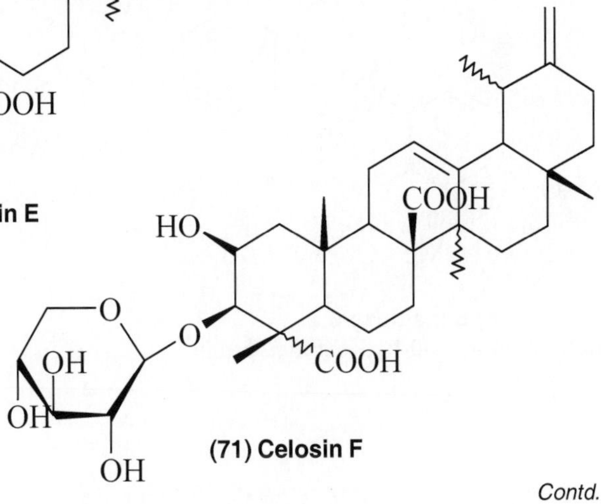

(71) Celosin F

Contd...

Figure 5.1–*Contd...*

(72) Celosin G

(73) Giberellic acid

(74) 5-methoxy-6,7-methylenedioxy-2-hydroxyisoflavone

(75) Tlatlancuayin

Contd...

Figure 5.1–*Contd...*

(76) Trans-ferulic acid

(77) Hyaluronic acid

(78) Betanin

Contd...

Figure 5.1–*Contd...*

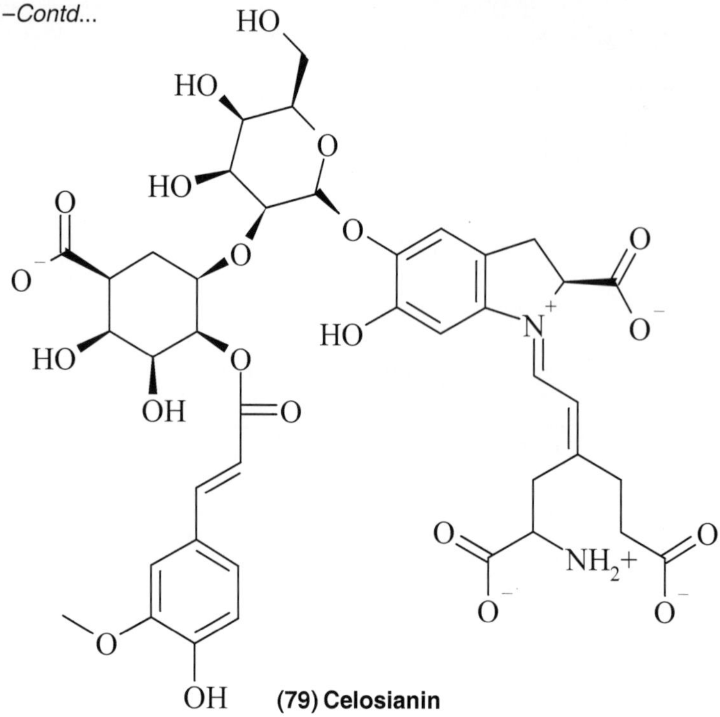

(79) Celosianin

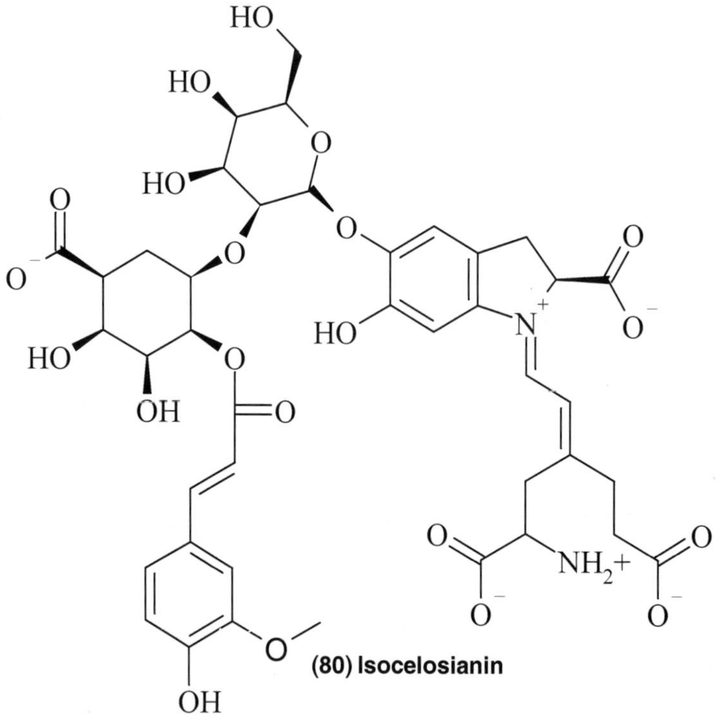

(80) Isocelosianin

Contd...

Figure 5.1–*Contd...*

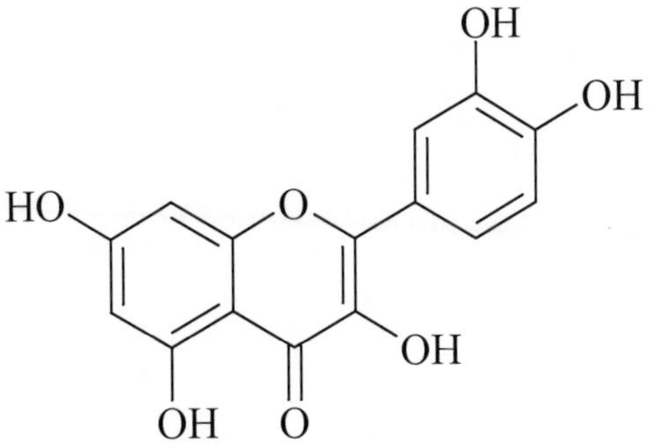

(81) Quercetin

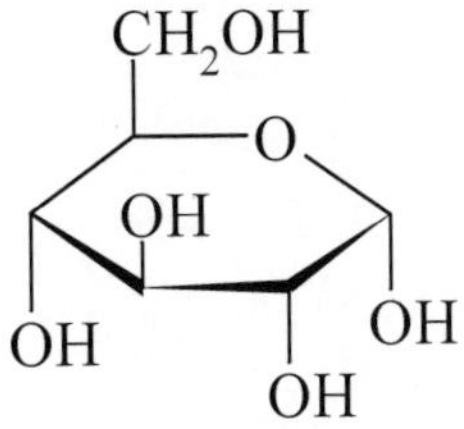

(82) Sugar

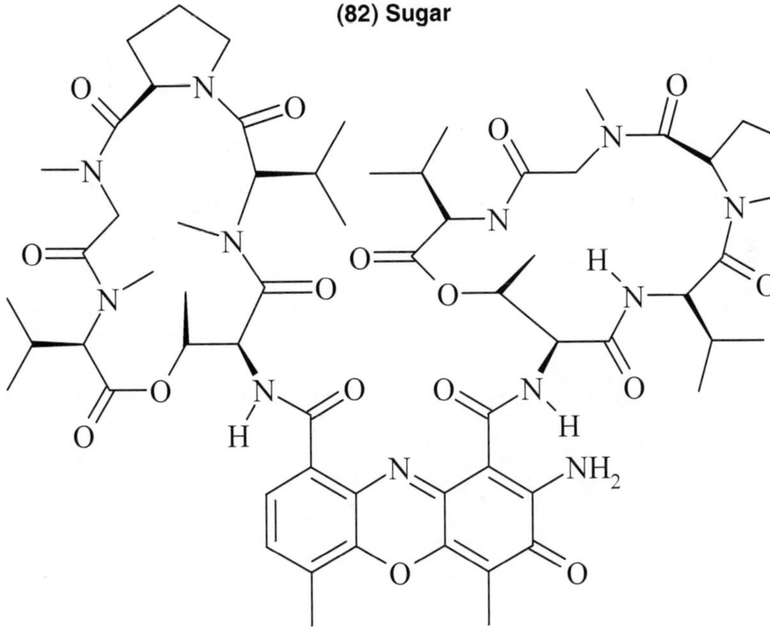

(83) Actinomycin D

Contd...

Figure 5.1–*Contd...*

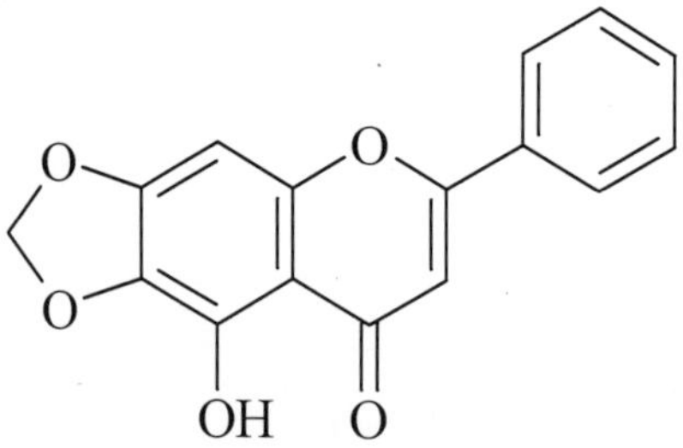

(84) Cochliophilin A (5-hydroxy-6,7-methylenedioxyflavone)

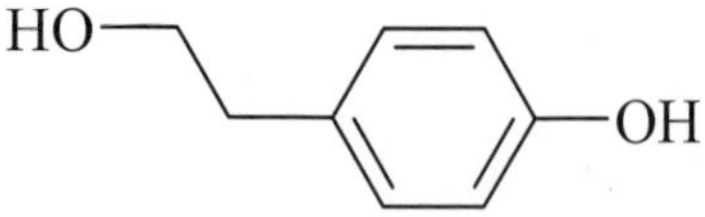

(85) 4-hydroxyphenethyl alcohol

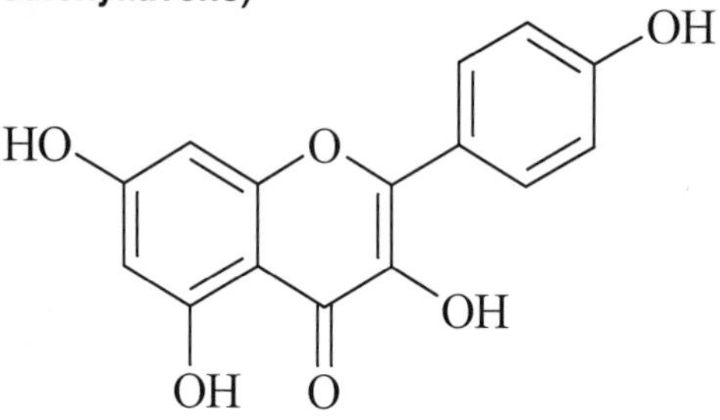

(86) Kaempferol

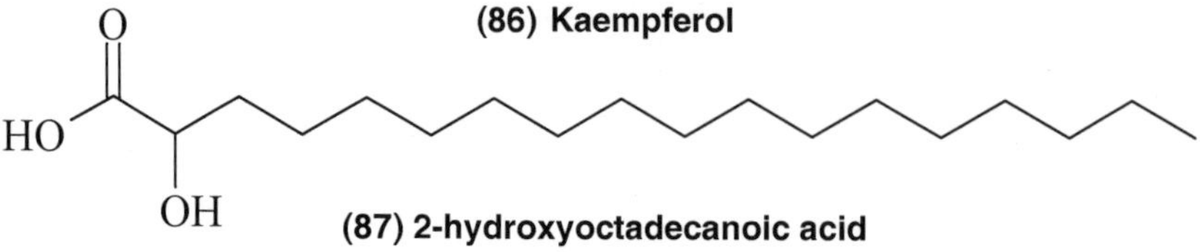

(87) 2-hydroxyoctadecanoic acid

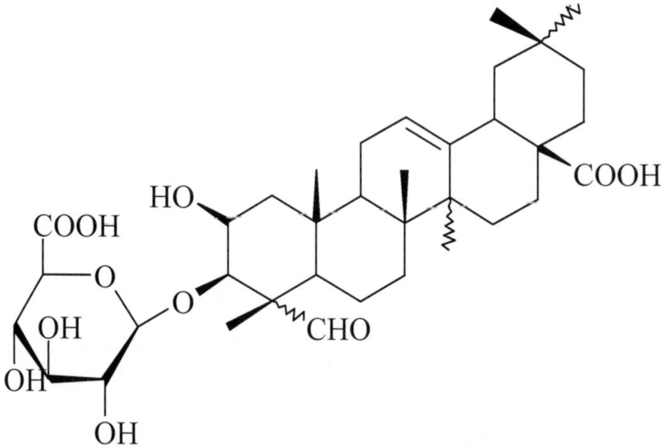

(88) Semenoside A (Terpenoid saponin)

Contd...

Figure 5.1–*Contd...*

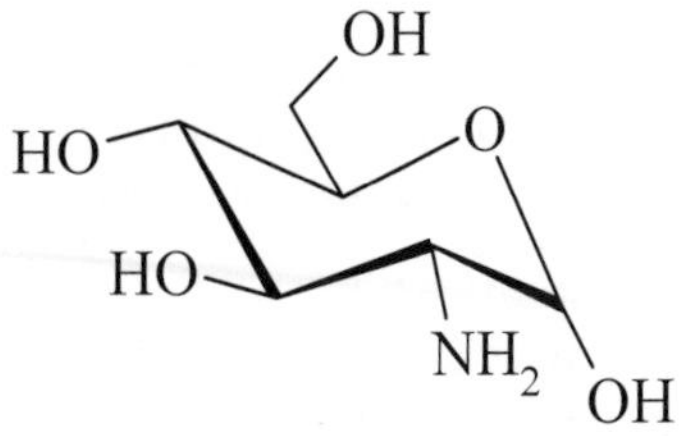

(89) Hexosamine

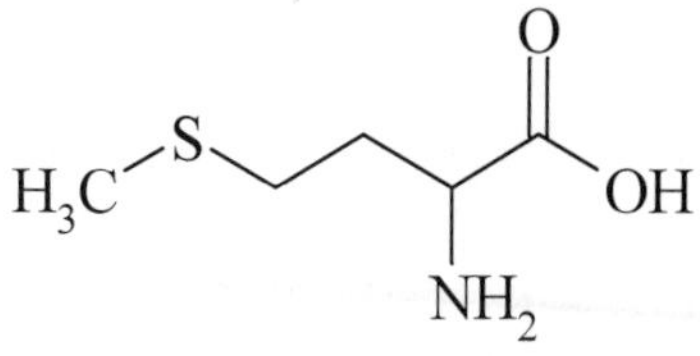

(90) Methionine

About twenty two lines/accessions belonging to 5 species of *Celosia* were analyzed and reported for their nutritional and anti-nutritional composition. The nutrient contents, vitamin C**(1)** varied from carotenoid**(2)**, protein whiles the anti-nutritional factors, nitrate, oxalate and moisture. Variation of these parameters with leaf position and the amino acids profile were also evaluated and reported (Prakash *et al.*, 1995).

Five glycosides were isolated from the leaves of *C. argentea* and their structures were determined based on UV, MS, ^{1}H NMR and ^{13}C NMR spectroscopic data as 1-(4-O-β-glucopyranosyl-3-methoxyphenyl) propane-2-ene (citrusin C)**(3)**, 3-O-β-glucopyranosyl-1H-indole (indican)**(4)**, (3Z)-hexenyl-(1-O-α-rhamnopyranosyl-β-glucopyranoside)**(5)**, (3Z)-hexenyl-1-O-β-glucopyranoside**(6)** and (7E)-6,9-dihydromegastigma-7-ene-3-one-9-O-β-glucopyranoside**(7)**. The above compounds were detected and reported in leaves of *C. argentea* for the first time in this study (Akiyoshi *et al.*, 1999).

A unique bicyclic peptide, moroidin **(8)**,moroidin hydrolysate **(9)** and colchicines **(10)**, from the seeds of *C. argentea* strongly inhibited and reported the polymerization of tubulin. Amongst the above compounds, stereo structure of moroidin **(8)** was reinvestigated and reported by spectroscopic data, chemical degradation and molecular dynamics simulation through ^{1}H and ^{13}C NMR. The above data was clearly indicated and reported the phytochemicals were possesses the anti-mitotic activity (Hiroshi *et al.*, 2000).

The chemical stability and colorant properties of three betaxanthin **(11)** was identified and reported from *Celosia argentea*. The three betaxanthin **(11)**had slightly higher pigment retention than amaranthin **(12)** and isoamaranthin**(13)** in crude extracts at as verified and reported by HPLC analysis. Refrigeration significantly increased pigment retention of aqueous betaxanthin**(11)**was reported (Cai *et al.*, 2001).

The betalains of yellow, orange and red inflorescences of common cockscomb (*Celosia argentea* var. *cristata*) were compared and reported to be qualitatively identical to those of feathered amaranth (*C. argentea* var. *plumosa*). In addition to the known compounds were reported as amaranthin**(12)** and betalamic acid**(14)**, the structures of three yellow pigments were elucidated and reported to be immonium conjugates of betalamic acid**(14)** with dopamine**(15)**,3-methoxytyramine **(16)** and (S)-tryptophan**(17)** by various spectroscopic techniques and comparison to synthesized reference compounds; latter found and reported that two compounds are new to plants of *C. argentea*. Among the betacyanin**(18)** occurring in yellow inflorescences intrace amounts, the presence of 2-descarboxy-betanidin**(19)**, dopamine**(15)**derived betacyanin**(18)**, miraxanthinV **(20)**, betanidin **(21)**, 3-methoxytyramine-betaxanthine**(22)** and (S)-tryptophan-betaxanthine **(23)** have been ascertained through preparative HPLC analysis.

Table 5.1: Phytoconstituents Classification and its Chemical Nature of *Celosia* Genus

Phytoconstituents	*Number of Phytochemicals*
Alkaloids	*
Amino acids	7
Carbohydrate	5
Fats and fatty acids	7
Flavonoids	11
Glycosides	5
Gums	*
Mucilages	*
Phenolic compounds	1
Proteins	*
Reducing sugar	*
Saponins	*
Steroids	5
Tannins	*
Triterpenoids saponins	8
Vitamins	5
Bycyclic peptides	18
Colour pigments	13
Intermediate compounds	5
Total	**90**

Note: *Reported in preliminary phytochemical analysis only.

The detection of highdopamine **(15)** concentration may be of toxicological relevance in use of yellow inflorescences as a vegetable and in traditional Chinese medicine uses for the red inflorescences of common cockscomb was reported (Willibald *et al.*, 2001).

The phytochemicals, celogentins A-C **(24-26)** and moroidin **(8)** have been found and reported to inhibit the polymerization of tubulin as anti-mitotic activity in *C. argentea* (Morita *et al.*, 2000).The plant *C. argentea* has been reported its root powder was used and reported for asthma (Jagtap*et al.*, 2013).Three new bicyclic peptides, celogentins A **(24)**, B **(25)** and C **(26)**, have been isolated together with a known related peptide, moroidin **(8)**, from the seeds of *C. argentea* and their structures including absolute stereochemistry were determined and reported through using extensive NMR methods and chemical means. The celogentins A **(24)**, B **(25)** and C **(26)** inhibited the polymerization of tubulin and celogentin C **(26)** was four times more potent than moroidin **(8)** in the inhibitory activity and would be important for their biological activity was reported (Kobayashi *et al.*, 2001).

The plant C. *argentea* along with *Amaranthus cruentus* were analyzed for anti-oxidant activities. The anti-oxidant activities of total phenol and flavonoids concentrations of C. *argentea* were reported. The antioxidant activities of C. *argentea* were higher than that of *A. cruentus* was reported (Oloyede *et al.*, 2013).

Six new bicyclic peptides, celogentins D-H **(27-31)** and J **(32)** have been isolated and reported from the seeds of C. *argentea* and the structures including its absolute stereochemistry were determined by using extensive NMR methods and chemical means. The celogentins E-H**(29-31)** and J **(32)** were reported as potent inhibition of tubulin polymerization, while the inhibitory activity of celogentin D **(27)** was modest reported. Structure activity relationship study indicates that ring size of the bicyclic ring system including unusual β^s-leucine **(33)**, tryptophan **(34)** and histidine **(35)** residues were reported for their important biological activity.On the other hand, some derivatives of moroidin **(8)** such as the Glu^1 derivative **(36)**, the Glu^1derivative with bondcleavage between the $Gly^7$and His^8 residues **(37)**, the Glu^1derivative with bond cleavage between the $Arg^6$and Gly^7 residues **(38)**, and the derivative without the $PyroGlu^1$ residue **(39)**, the last of which was prepared by treatment of moroidin **(8)** were found to be less potent than(IC50,20mM for **(36)**; 80 mM for **(37)**; 100 mM for **(38)**; 100 mM for **(39)**, respectively) indicating that the presence of the $PyroGlu^1$residue is important for the activity as anti-mitotic activity (Hayato *et al.*, 2003).

The Oligogalacturonic acid**(40)**has been isolated and reported from the C. *argentea* seedlings and further biological effects were reported (Suzuki *et al.*, 2002).A new cyclic nonapeptide, celogenamide A**(41)**, has been isolated and reported from the seeds of C. *argentea* and the structure including absolute stereochemistry was determined by using extensive 2D NMR and MS-MS methods and chemical means (Morita *et al.*, 2004)

A novel cyclic peptide with a 3-hydroxyoxindole ring, celogentin K **(42)** has been isolated and reported from the seeds of C. *argentea* and the structure including its absolute stereochemistry was assigned by using extensive NMR, MS and CD spectra. The stereo structure of a known related bicyclic peptide, moroidin **(8)** was confirmed by a single crystal XRD analysis (Hayato *et al.*, 2004).

The plant C. *argentea* was reported the fingerprint analysis of HPLC for the different habitats. The study was reported on Hypersil ODS2 column with acetonitrile, glacial acetic acid as the mobile phase and elutes were detected by ELSD and similarity was verified through SPSS. The chromatographic profiles of the samples from different regions were very similar reported. The chemical constituents of C. *argentea* vary with different habitats so selection of material habitat is very important for quality control of C. *argentea*. The fingerprint with high individuality and specificity could be applied in the identification and quality control of the material of C. *argentea* was reported (Wang *et al.*, 2008).

The study has been reported that the chemical constituents of *C. argentea* seeds by solvent extraction and chromatography analysis. As a result,8 compounds were identified and reported through spectral analysis such as MS, UV,IR,^{1}H-NMR,^{13}C-NMR,MS,HMBC,HMQC,COSY and NOSEY. They were stigmasterol **(43)**, β-sitosterol **(44)**, celosin A **(45)**, celosin B**(46)**,β-daucosterol **(47)**, n-hexacosnic acid**(48)**, palmitic acid**(49)** and stearic acid**(50)**.The study investigated and reported the hepatoprotective effects of the crude ethanol extract of *C. argentea* seeds against the CCl_4-induced liver injury. The results showed and reported that the crude ethanol extract of *Celosia argentea* seeds remarkably lowered the serum AST and ALT compared to the CCl_4treated group significantly. This study lay and reported a foundation for the development and utilization of *C. argentea* and possessed significant hepatoprotective activity against the CCl_4 toxicity through above compounds (Sun *et al.*, 2009).

A new phenolic glycoside, stigmasterol **(43)**; β-sitosterol **(44)**;4-O-β-D-glucopyranosyl-2-hydroxy-6-methoxyace-tophenone **(51)**; 4-O-β-D-apifuranosyl-(1,2)-β-D-glucopyranosyl-2-hydroxy-6-methoxyacetophenone**(52)**; eugenyl-O-β-D-glucopyranoside**(53)**; sucrose**(54)**; quercetin-3-O-β-D-glucopyranoside**(55)**; isorhamnetin-3-O-β-D-glucopyranoside **(56)**; rhamnazin-3-O-β-D-glucopyranoside **(57)**; isorhamnetin-3-O-α-L-rhamnopyranosyl-(1,2)-β-D-glucopyranoside**(58)**; stigmasterol-3-O-β-D-glucopyranoside **(59)** andoleanolic acid **(60)** known compounds were isolated and reported from the methanolic extract of *C. argentea*. The structures of the compounds were elucidated on the basis of spectroscopic analysis and chemical methods. Among the isolated compounds, stigmasterol **(43)** showed and reported moderate inhibitory activities against SGC-7901 and BEL-7404 cell lines (Shuo *et al.*, 2010).

A review of the *C. argentea*, primary metabolites are reported as carbohydrates as sucrose **(54)** in seeds; lipids such as fatty acids, waxes, glycerides**(61)**, phospholipids, sterols **(62)**, vitamins A**(63)**, D**(64)**, E**(65)** and K**(66)**,stigmasterol**(43)**, β-sitosterol**(44)**, β-daucosterol**(47)**palmitic acid**(49)** and oleanolic acid **(60)** in whole plants of *C. argentea*. Amino acids and bicyclic peptides such as celogentins A **(24)**, B **(25)**, C **(26)**, D **(27)**, E **(28)**, F **(29)**, G **(30)**, H **(31)** and J**(32)**; moroidin**(8)**; cyclic peptide: celogenamide A**(41)**, celogentin K**(42)** were reported in seeds of *C. argentea* of betalain such as betaxanthin **(11)**, amaranthine **(12)**, isoamaranthine**(13)**, betalamic acid**(14)**, (S)-tryptophan**(17)**,betacyanins**(18)**, miraxanthin V**(20)**,3-methoxytyramine-betzaxanthin **(22)**, (S)-tryptophan-betaxanthin **(23)**and yellow betaxanthin**(11)** were reported in flowers of *C. argentea* (Ramesh *et al.*, 2013).

About the secondary metabolites are steroids such as cristatain**(67)**, saponin, celosin A**(45)**, B**(46)**, C**(68)**, D **(69)**, E**(70)**, F**(71)** andG**(72)** were reported in seeds of *C. argentea*. Diterpenes such as giberellic acid**(73)** was reported in seeds of *C. argentea*; flavonoids are isoflavones: 5-methoxy-6,7-methylenedioxy-2-hydroxyisoflavone**(74)** and 2'-methoxy derivative: tlatlancuayin**(75)** in aerial parts of *C. argentea*(Ramesh *et al.*, 2013).

Phenolic glycosides such as 4-O-β-D-apifuranosyl-(1→2)-β-D-glucopyranosyl-2-hydroxy-6-methoxyacetophenone**(52)**, eugenyl-O-β-D-glucopyranoside**(53)**, sucrose**(54)**, quercetin-3-O-β-D-glucopyranoside**(55)**, isorhamnetin-3-O-β-D-glucopyranoside **(56)**, rhamnatin-3-O-β-D-glucopyranoside **(57)**, isorhamnetin-3-O-β-L-rhamnopyranosyl-(1,2)-β-D-glucopyranoside **(58)**,stigmasterol **(43)**, β-sitosterol **(44)**and stigmasterol-3-O-β-D-glucopyranoside **(59)** were reported in whole plant of *C. argentea*. Phenols and phenolic acids such as 1-(4-O-β- glucopyranosyl-3 methoxyphenyl) propan-2-ene (citrusin C) **(3)**, 3-O- β-glucopyranosyl-1H-indole (indican) **(4)**, (3Z)-hexenyl-1-O-α-rhamnopyranosyl-β-glucopyranoside **(5)**, (3Z)-hexenyl-1-O-β-glucopyranoside **(6)**,(7E)-6,9-dihydromegastigma-7-ene-3-one-9-O-β-glucopyranoside **(7)**and trans-ferulic acid **(76)** were present and reported in leaves of *C. argentea* (Ramesh *et al.*, 2013).

The reported study was investigate the allelopathic effects of different concentrations of *C. argentea* weed on seed germination and seedling growth of surrounding crop plants *Sorghum bicolor, Phaseolous aureus, Arachis hypogaea* and *Vigna unguiculata*. The percentage seed germination and seedling growth of crop plant seeds decreased with concentration of *C. argentea* leaf extracts used and reported. The reduction of germination percentage may be due to the presence of allele chemicals of *C. argentea* which is reported to contain hyaluronic acid**(77)**,betanin **(78)**,celosianin**(79)**and isocelosianin**(80)**. Among the five crop plant seeds, *C. argentea* leaf extract showed more inhibitory effect on growth of seedlings of groundnut (Saritha *et al.*, 2013).

A new phenolic glycoside, 4-O-β-D-apifuranosyl-(1→2)-β-D-glucopyranosyl-2-hydroxy-6-methoxyacetophenone**(52)** and 11 known compounds were reported from the methanolic extract of plant *C. argentea*. The structures of the compounds were reported on the basis of spectroscopic and chemical means. Among the isolated compounds, stigmasterol **(43)**reported as moderate inhibitory activities against SGC-7901 and BEL-7404 cell lines (Shen *et al.*, 2010).

The plant *C. argentea* was reported phytochemicals presence and pharmacognostical standardization through WHO guidelines. Further, reported the quercetin **(81)** as flavonoids presence through preliminary and HPLC techniques (Rajni and Deokule, 2013).

Celosia cristata

The *Celosia cristata* AVPs have been reported to be containing very high neutral sugar **(82)** contents in the range of 26–28 per cent (Balasubrahmanyam *et al.*, 2000).

The two antiviral glycoproteins, active against mechanical transmission of two TMV and SRV and CRSV (ungrouped) were purified from the dried leaves of *C. cristata*. The proteins CCP-25 and CCP-27, have M_r 25 and 27 kDa, respectively. They were resistant to proteases in the native state, but were readily digested when denatured. Both of them imparted actinomycin D**(83)** sensitive resistance by inhibiting local lesions on *Nicotiana tabacum* by TMV. Their applications are individually reported in high resistance in systemic hosts to SRV, and CRSV, respectively through ammonium sulphate precipitation, IEC, CM-sepharose chromatography, GFC, protein characterization, estimation and systemic hosts methodology (Balasubrahmanyam *et al.*, 2000).

Celosianin **(79)** and Isocelosianin **(80)** were isolated and reported from the *C. cristata* through RP-HPLC method (Cai *et al.*, 2005).

The cochliophilin A (5-hydroxy-6,7-methylenedioxyflavone)**(84)** as a host-specific attractant towards the zoospores of *Aphanomyces cochlioides*, a cause of root rot and damping off diseases of Chenopodiaceae. The compound was found and reported in *C. cristata* that is susceptible to the pathogen. A new isoflavone, cristatain (5-hydroxy-6-hydroxymethyl-7,2'-dimethoxyisoflavone)**(67)** identified and reported (Wen *et al.*, 2006).

A new saponin, cristatain (**67**), together with four other saponin, celosin A **(45)**, B **(46)**, celosin C **(68)** and D **(69)** were isolated and reported from the seeds of *C. cristata*. Their structures were established and reported through chemical and spectroscopic analyses. The compound cristatain **(67)** exhibited significant hepatoprotective effect on CCl_4 and DMF induced hepatotoxicity in mice evidenced by significant decreases in the values of AST, ALT and ALPof serum and histopathological analysis was reported (Yan *et al.*, 2010).

The chemical constituents of *C. cristata* were reported through solvent extraction, SGCC, Sephadex and recrystallization were used for the isolation and purification of the constituents. The structures of the extracts were identified and reported on the basis of spectral analysis *viz.*, UV,IR,^{1}HNMR,^{13}CNMR and MS. The ethanolic extract of *C. cristata*, 6 compounds were isolated and reported as stigmasterol **(43)**, β-sitosterol**(44)**, 4-hydroxyphenethyl alcohol**(85)**, kaempferol**(86)**, quercetin **(80)** and 2-hydroxyoctadecanoic acid **(87)** (Chi *et al.*, 2010).

Two new oleanolic acid **(60)** saponins, namely celosin C **(68)** and celosin D **(69)** were isolated and reported from the ethanol extract of *S. celosiae*as *C. cristata*. Their structures were identified by spectroscopic and chemical analyses. The hepatoprotective activities were investigated and reported on CCl_4induced hepatotoxicity in mice and reported significant activity (Zhen *et al.*, 2010).

A new triterpenoid saponin, named semenoside A**(88)**, was isolated from *S.C. cristata*. Its structure was elucidated on the basis of 1D, 2D NMR, HR-FAB-MS and ESI-MS techniques and physico-chemical properties. The hepatoprotective activity of semenoside A**(88)** was investigated and reported significant activity on CCl_4 induced hepatotoxicity in mice(Sun *et al.*, 2011).

The chloroform, methanol and aqueous extracts of *C. cristata* leaves were reported for the phytochemical presence and anthelmintic activity. The plant *C. cristata* was possessed and reported as alkaloids, flavonoids, mucilages, phenolic compounds, tanninsand tripterpenoids,. Aqueous and methanolic extracts were reported as significant anthelmintic activity and not significant activity in chloroform extract of *C. cristata* due to the reason absence of tannins and triterpenoids (Danabalan, 2012).

Infrared spectra of *S. celosiae* and *S.C. cristata* were reported directly, quickly and accurately by FTIR with OMNI sampler. The difference of IR spectra between *S.celosia* and *S.C. cristata* was extruded greatly. One dimensional continuous wavelet transform was implemented for the IR spectra of *S. celosiae* and its confusable varieties. The results were reported that the study effective to apply continuous wavelet transform on FTIR analysis to identify the traditional Chinese medicinal materials, which are the same genus but different species (Zhang *et al.*, 2007).

Pharmacological and Therapeutical Applications of the *Celosia* Genus

Celosia argentea

In folklore practice, the seeds of *C. argentea* have been used traditionally for removing liver heat, improving eyesight, clearing wind heat, lowering the blood pressure and anti-inflammation according to Chinese Pharmacopoeia. Also, the plant *C. argentea* seeds are reported widely used in Indian folk medicine for the treatment of diabetes mellitus and used to dissolve urinary stones (Sonawane *et al.*, 2012). The *C. argentea* was traditionally used as an antidote for snake poison, anti-ulcer and in the treatment of eye diseases, glandular swellings, eczema and constipation (Priyanka *et al.*, 2012).

Table 5.2: Pharmacological and Therapeutical Applications of *Celosia* Genus in Folk Medicine, Research and Development

C. argentea	**Folk medicines**	Abnormal vaginal discharge
		Asthma
		Redness
		Swelling of the eyes
		Photophobia
		Lacrimation
		Headache
		Colic
		Gonorrhea
		Eczema
		Antipyretic
		Aphrodisiac
		Biliousness
		Dysentery
		Diarrhea
		Inflammation
		Sores
	Research and Development	Anti-bacterial activity
		Anti-cancer activity
		Anti-diarroheal activity
		Anti-inflammatory activity
		Anti-mitotic activity
		Anti-muscarinic activity
		Anti-oxidant activity
		Anti-lytic activity
		Anti-ulcer activity

Contd...

Table 5.2–*Contd...*

		Anti-urolithiatic activity
		Anti-viral activity
		Hepatoprotective activity
		Immunomodulating activity
		Inhibitory activity
C. cristata	**Folk medicine**	Liver heat
		Improving eyesight
		Clearing wind heat
		Lowering the blood pressure
		Fatigue
		Atherosclerosis
		Leucorrhoea
		Osteoporosis
		Antiinflammation
	Research and Development	Anthelmintic activity
		Anti-diabetic activity
		Anti-inflammatory activity
		Hepatoprotective activity
C. isertii	**Research and Development**	Anti-inflammatory activity
		Analgesic activity
C. laxa	**Folk medicine**	Anthelmintic activity
C. leptostachya	**Folk medicine**	Opthalmic disease
C. polygonoids	**Research and Development**	Anti-inflammatory activity

The celosian, an acidic polysaccharide from the seeds of *C. argentea* was found and reported a potent anti-hepatotoxic agent for chemical and immunological liver injury models in animals. The immunomodulating action of celosian was reported for the preventive mechanism of celosian on liver injuries. The celosian was induced and reported TNF-α production in mice and also induced the production of IL-1β and nitric oxide in macrophage cell line J774 in a concentration dependent manner. In addition, celosian enhanced and reported to gamma IFN-γ production activity of concanavalin A in mice spleen cells, though celosian alone did not significantly influence IFN-γ production. These results reported that celosian is an immunostimulating agent in addition to anti-hepatotoxic effects (Hase *et al.*, 1997).

The plant *C. argentea* seed extracts administration resulted in increased production of IL-2, IFN-γ and decreased production of a Th2 cytokine IL-4 from splenocytes. Thus, the anti-metastatic effect reported on *C. argentea* is based in its immunomodulating properties including induction of cytokines such as IL-12, IL-2 and IFN-γ leading to a Th1 dominant immune state and activating macrophages to the tumoricidal state. The study has been reported to possess the inhibition of cancer metastasis and having significant immunomodulating activity (Hayakawa *et al.*, 1998).

This study was reported of an alcoholic extract of *Celosia argentea* seeds reduce the increased blood glucose in Alloxan induced diabetic rats. Also, the extract prevented a decrease in body weight in Alloxan induced diabetic rats and further the study reported that significant anti-diabetic activity in Alloxan induced diabetic rats (Thangarasu *et al.*, 2002).

The anti-bacterial activity of crude alcoholic extract of *Datura alba* and *C. argentea* leaves were reported against pathogens isolated from infected burn patients. The disc diffusion method showed significant zone of lysis against the pathogens studied and the results are comparable to the conventional antibiotic cream namely SSD was reported. While compare with two extract of *D. alba* exhibited more than 50 per cent antibacterial activity when compared with *C. argentea* (Gnanamani *et al.*, 2003).

The study was reported that investigating the healing efficacy of *C. argentea* extract in an ointment using a rat wound burn model. Granulation tissue collected on every fifth day of healing showed an increase in collagen and hexosamine**(89)** content at a faster rate in the treated wounds. The two major cellular responses are cell proliferation and cell motility in two key cell lineages, fibroblasts and keratinocytes were reported on *C. argentea* treatment and they was nontoxic in fibroblasts and keratinocytes. The study reported a salutary action of the *C. argentea* extract on wound healing and reported that the action due to mitogenic and motogenic promotion of dermal fibroblasts (Priya *et al.*, 2004).

Anti-oxidant and cytoprotective activities of boiled, cold and methanolic extracts of nine edible vegetables in Southwest Nigeria were evaluated in the DPPH free radical assay and hemagglutination assay in bovine erythrocytes, respectively. *Crassocephalum rubens, Solanum americanum* and *Vernonia amygdalina, Solanum macrocarpon, Telfaria ccidentalis, Amaranthus hybridus, Jatropha tanjorensis, Talinum triangulare* were showed and reported anti-oxidant property along with *C. argentea.* The membrane stabilizing capacity of the extracts supports and reported the plants anti-oxidant activity (Iwalewa *et al.*, 2005).

The flavonoids fraction from ethanolic extract of the leaves *Celosia argentea* was investigated for anti-inflammatory activity in Wistar Albino rats model. Results of the study reported that, flavonoids possesses significant anti-inflammatory properties when investigated by employing carrageenan induced rat paw edema and cotton pellet induced chronic inflammatory models. The study showed significant dose dependent antiinflammatory activities in both the models and also confirmed that activity the presence of flavonoids in *C. argentea* (Santosh *et al.*, 2008).

Amongst, 34 plants treatment for the various diseases assessment, the whole plant *Celosia argentea* aqueous and ethanolic extracts were showed and reported for therapy in anti-pyretic, aphrodisiac, biliousness, dysentery, diarrhea, inflammation and sores. Further, reported as anti-bacterial activity against the *Staphylococcus* sp. (Jigna, 2008).

The leaves of *Hibiscus sabdariffa* and *C. argentea* were studied for their anti-viral activities against MSV as well as the effects of the extracts on Hep2 cells. The anti-viral activity of *C. argentea* could not be established at 10 and 15 mg/ml but did not show and reported any antiviral activity (Omilabu *et al.*, 2010).

The study was reported that the anti-diabetic hypoglycaemic properties of an ethanolic extract of the root of *C. argentea*. An ethanolic extract of *C. argentea* root was reported to possess lower blood glucose in basal conditions and after a heavy glucose load in normal Wistar Albino rats. Maximum reduction in serum glucose was observed and also body weight, but petroleum ether and chloroform extracts did not reduce the serum glucose. Ethanolic extract of *C. argentea* was also found to reduce the increase of blood sugar found in Streptozotocin induced diabetic rats. Chronic administration of the extract significantly reduced the blood sugar in Streptozotocin induced diabetic rats for several days. These results indicate and reported that *C. argentea* root extracts are able to ameliorate biochemical damages induced by Streptozotocin in diabetic rats (Santosh *et al.*, 2010).

The study was reported that the anti-diarrhoeal properties of alcoholic extract of leaves of *C. argentea* on diarrhoea by using different experimental models. Anti-diarrhoeal effect was evaluated by castor oil induced diarrhoea, charcoal meal test and PGE 2 induced diarrhoea. The reported study was compared and reported with Loperamide and Atropine as standard drugs. Results suggest and report that it may act centrally and may inhibit the PGE2 to give anti-diarrhoeal activiey. Also charcoal meal test also reported for anti-muscarinic activity (Praveen *et al.*, 2010).

The study has been suggested and reported that the high membrane stabilizing activity of the extract of *C. argentea* which has potential to protect the erythrocyte membrane from free radical damage. The membrane stabilizing activity of the extract reported due to the presence of flavonoids, alkaloids, tannins and/or saponins (Malomo *et al.*, 2011).

The triterpenoid saponins from the seeds of *C. argentea* and their anti-inflammatory and anti-tumor activities were reported. Three new triterpenoid saponins such as celosin E **(70)**, celosin F **(71)** and celosin G **(72)**, together with a knowncompound cristatain **(71)**, were isolated and reported from the seeds of *C. argentea* extensive NMR and MS experiments. The anti-inflammatory and anti-tumor activities of the four compounds were reported *in vitro* models (Wu *et al.*, 2011).

The plant *C. argentea* aqueous and ethanolic extracts were reported for calcium, phosphate solubilization, precipitation reaction and higher percentage inhibition for the significant urolithiasis activity (Bhavik *et al.*, 2011). The *C. argentea* also reported to possesses for the treatment of stone diseases in Bhopal district (Kumkum and Ranjana, 2012).

The *C. argentea* plant seeds have been reported to possess that anti-urolithiatic activity in ethylene glycol induced urolithiasis in rats. The results have been reported the physical parameters, promoters, inhibitors, renal function markers in urine and serum samples, anti-oxidant parameters and histopathology of kidney tissues were observed very normal and significant anti-urolithiatic activity while compare with the standard drug Cystone (Joshi *et al.*, 2012). The above same study has been carried and reported with methanolic extract of *C. argentea* roots having significant anti-urolithiatic activity (Kachchhi *et al.*, 2012)

Two new triterpenoid saponins, celosin A **(45)** and celosin B **(46)** were isolated and reported from the seeds of *C. argentea*. The structures of the two new compounds were elucidated and reported based on chemical analysis and spectral methods such as IR, 1D and 2D NMR, ESI-MS and HR-ESI-MS. They exhibited significant hepatoprotective effect on $CCl_4$and DMF induced hepatotoxicity in mice (Wu *et al.*, 2013).

Along with *Lactuca scariola, C. argentea* was found and reported to possess the anti-oxidant activity on DPPH, hydrogen peroxide and nitric oxide scavenging method scavenging methods. The aqueous extract of *L. serriola* showed higher free radical scavenging activity than *C. Argentea.* The methanolic extract of *C. argentea* was found to be very minimal anti-oxidant activity due to the presence of carbohydrates, glycosides, phenolic compounds, flavonoids, saponins and tannins; while compare with *L. scariola* (Urmila *et al.*, 2013).

The plant *C. argentea* acts as pan-PHA to cat red blood cells. Therefore the author reported agglutination inhibition test on *C. argentea* extract to cat red blood cells with human sera and the results are compared with known blood and serum types. One group inhibited and reported in agglutinating activities of *C. argentea* extracts to cat red blood cells. The agglutination inhibition phenomena of human sera on *C. argentea* extract to cat red blood cells were reported no constant relation with known blood and serum types such as ABO, H, Cl, ab.C, Cea, CCI and PVI respectively (Lim *et al.*, 1987).

Celosia cristata

An antiviral protein isolated and reported from leaves of *C. cristata* (CCP 25) was tested for depurination study on ribosomal RNA from yeast. Ribosomal RNA yielded 360 nucleotide base fragments after treatment with CCP 25 indicating that CCP 25 was a ribosome inactivating protein. CCP 25 also inhibited translation of brome mosaic virus BMV and PMV RNAs in rabbit reticulocyte translation system. The radioactive assay showed that incorporation of [35S]-methionine **(90)**was less in translation proteins of BMV nucleic acid when CCP 25 was added to translation system. The reported study suggested that anti-viral protein from *C. cristata* not only depurinated ribosomal RNA but also inhibited translation of viral RNA *in vitro* model (Baranwal *et al.*, 2002).

The anti-oxidant activity of *C. cristata* antiviral proteins (CCP-25 and CCP-27) using FRAP assay *in vitro* method reported that the proteins are strong anti-oxidants. The increase in activities of redox enzymes such as peroxidase, catalase and polyphenol oxidase on TMV inoculation of plants was inhibited when plants were treated with CCP-25 before TMV inoculation. The activity of phenylalanine ammonia lyase, involved in biosynthesis of anti-oxidative compounds was reported as inhibited. A possible correlation between anti-oxidant activity of CCPs and their anti-viral activity is speculated was reported (Gholizadeh, 2004).

Cystatins have been reported to use in plants as anti-viral strategy against viruses whose replication involves cysteine proteinase activity. Cystatins may confer resistance by inhibition of a virus induced cell death phenomenon in which cysteine proteinases are reported as active. A full length cDNA library was constructed from the pre-flowering stage of *C. cristata* leaves and a cDNA clone with cystatin domain was isolated using an oligonucleotide probe reported on the basis of the conserved peptide of plant cystatins. The activity of cysteine proteinases is not involved in the TMV replication process, that inhibition of the hypersensitive response by celostatain may be due to the inactivation of proteolysis involved in the plant cell death programme (Gholizadeh *et al.*, 2005).

The small cDNA fragment coding strong anti-viral peptide from *Celosia cristata* were transferred to *E. coli* for a good candidate to investigate further the location of the anti-viral determinants was reported (Gholizadeh *et al.*, 2005).

An antiviral protein CCP27 was purified and reported from the leaves of *C. cristata* at the post flowering stage byanion exchange, cation exchange, and GFC. It reported to exhibit resistance against SRVin its test host *Cyamopsis tetragonoloba* (Begam *et al.*, 2006).

A full length cDNA clone, encoding a RIP/AVP was isolated from the cDNA library of post flowering stage of *C. cristata* leaves. The coding region of the cDNA was amplified by PCR, cloned and expressed in *E. coli* as recombinant protein of 72 kDa. The expressed fusion product was confirmed and reported by Western analysis and purification by affinity chromatography. Both the recombinant protein (reCCP-27)and purified expressed protein (eCCP-27) inhibited translation in rabbit reticulocytes. The purified product also showed and reported N-glycosidase activity towards tobacco ribosomes and anti-viral activity towards TMV and SRV. The above study has been proved and reported through the SDS PAGE, cDNA clone, rRNA N-glycosidase activity, *In vitro* protein synthesis inhibition and anti-viral bioassay of the purified expressed protein (Mehbuba *et al.*, 2006).

The report was presented a full lengthDUF538cDNA (encoding 170 amino acid residues) clone, which was randomly isolated and reported from *C. cristata* leaf cDNA library constructed under normal growth conditions and consistently amplified from leaf cDNA populations prepared from no stressed and drought stressed leaves (Gholizadeh *et al.*, 2010).

The reported study was to evaluate the anti-oxidant compounds and anti-oxidant activities of the methanolic extracts and solvent fractions from *C. cristata* flowers. The contents of total polyphenol, flavonoid and tannin were reported by spectrophotometric methods. The above compounds total polyphenol, flavonoids and tannin contents of methanolic extracts of *C. cristata* were reported for anti-oxidative activities by DPPH, ABTS radical scavenging activities and reported as notable antioxidant activities in cockscome flowers are considered to have significant health benefits (Woo *et al.*, 2011).

A ClpS homologue from *C. cristata* was expressed as maltose binding fusion protein under the control of strong inducible tacpromoter of pMALc2X vector in TB1 strain of *E. coli*. SDS-PAGE analysis reported that fused ClpS is produced as about 63 kDa protein in recombinant bacteria.This prediction as reported to understanding the functional relationship between proteolytic andanti-oxidative systems in future biological investigations (Gholizadeh, 2011).

The report of tissue culture studies of *C. cristata* were established from various explants and the effects of various hormones on morphogenesis of this species were reported. *In vitro* plants produced lower mean cell areas but higher nuclear areas when compared to *in vivo* plants. Thus, no occurrence of somaclonal variation was detected and this was reported by morphological features of the *in vitro* plants of *C. cristata* (Taha *et al.*, 2012).

The traditional medicinal herb *C. cristata* used for treatment of fatigue, atherosclerosis, leucorrhoea and osteoporosis due to presence chemical constituent of this plant include mainly flavonoids was reported (Ramesh *et al.*, 2013).

Celosia isertii

The phytochemical studies carried out on aqueous and ethanol extract of *C. isertii* leaves showed and reported the presence of alkaloids, cardiac glycosides, flavonoids, phenols, phlobatannins, phytosterols, saponins, steroids, tannins and triterpenoids. The anti-inflammatory study carried out on the albino rats that had their paw injected with carrageenan injection to induce edema in the anti-inflammatory activity of the ethanol extract. While the analgesic study done on mice in which acetic acid was injected to induce pain was greater in aqueous extract but less than the ethanol extract of *C. isertii*. The results were obtained and reported that *C. isertii* has potent anti-inflammatory and analgesic properties (Anthony *et al.*, 2013).

Celosia globosa

In Zaire it is taken as a leaves and vegetable for food consumption and various ailments in the plant of *Celosia globosa* (Ramesh *et al.*, 2013).

Celosia laxa

Extracts of *C. laxa*, *Neocarya macrophylla* and *Zanthoxylum zanthoxyloides* leaves were screened for anthelminthic activities on *Ascaris lumbricoides*. *C. laxa* appeared to be more potent against *A. lumbricoides* with rapid recovery of weight while compare *Z. zanthoxyloides* has the lowest potency (Barnabas *et al.*, 2011).

Celosia leptostachya

In Nigeria the fruits and seeds are pounded up for topical application in cases of ophthalmiain the plant *C. leptostachya* (Ramesh *et al.*, 2013).

Celosia polygonoids

The study was reported to possess the anti-inflammatory activity of whole plant of *C. polygonoids*. The anti-inflammatory activity was reported by using carrageenan, cotton pellet, croton oil, formaldehyde, histamine and serotonin induced oedema of Wistar Albino rats model. The plant has various phytoconstituents such as flavonoids, reducing sugars, saponins, starch and steroids. Significantly one among those phytoconstituents lessens to oedema of the rats in dose dependent manner was reported (Rajesh *et al.*, 2009)

Celosia trigyna

The plant *C. trigyna* was reported and not possesses any cytotoxicity activity in dichloromethane, ethanol, water extracts in significant manner through KB and Vero cell lines assay. Also, the TLC method was reported for the presence of amino acids, flavonoids, proteins, saponins, tannins and terpenoids (Marie, 2011).

Various Activities of the *Celosia* Genus

Celosia argentea

The experimental result reported tha sulphur is a crucial and essential element for optimum *Celosia* production. The optimum Sulphur rate for *Celosia* leaf formation, plant height, leaf area, shoot production,Vitamin A **(63)**and crude protein syntheses was applied nutrient solution (David *et al.*, 1999).

The plant extract of *Senna alata* was chosen and reported to stimulation of growth and development of *C. argentea* (Agbagwa *et al.*, 2003).The 100 per cent of the recommended nitrogen dose showed and reported statistically same flower yield as that inoculation with *Azospirillum* in combination with application from recommended doses of nitrogen indicating the possibility of reducing input cost of N-fertilizers; biofertilization of *C. argentea* with *Azospirillum* substitutes of nitrogen to soil and reduce the input cost and sustains soil fertility (Rawia *et al.*, 2006).

Very little information is available on organic specialty cut flower production, especially fertilization requirements. In order to better understandorganic fertilization requirements of two specialty cut flower crops, *Limonium sinuatum* and *C. argentea*, the study reported the greenhouse experiment to study the effect of compost (organic) and conventional (inorganic) fertilization treatments on the growth and productivity of these crops. The study reported as organic fertilizer recommendations, in the form of animal manure composts should be based on phosphorus content of the compost rather than nitrogen content especially for soils high in initial phosphorous content (Sven and Louis,2007).

The great genetic diversity reported in *C. argentea* populations by significant geographic difference was confirmed by a high level of population genetics. The information was beneficial to future breeding and conservation for populations of *C. argentea* (Na Feng *et al.*, 2009).

The effects of *C. argentea* seed water absorption, temperature, light, darkness, giberellic acid**(73)** on seed germination of *C. argentea* were reported. The germination rate of the seeds of *C. argentea* was higher when treated with giberellic acid **(73)** (Xu *et al.*, 2009).

Celosia bonnivairii

The Zaire country has been reported to use as a vegetable food and used for fishing; also for various ailments from the plant *C. bonnivairii* (Ramesh *et al.*, 2013).

Celosia cristata

The reuse of saline wastewaters and ground waters becomes a viable option for the irrigation of salt tolerant floral crops as competition for high quality water increases as reported. A completely randomized design with three replications was used to assess the effects of two water ionic compositions and six salinity levels on the production and mineral accumulation of two cultivars of *C. argentea* var. *cristata* under greenhouse conditions. Ionic water compositions mimicked sea water and saline drainage waters of the Imperial and Coachella Valleys of California. Producing cut flowers in saline waters also reduces excessive stem length, which occurs when plants are grown in control treatments (Christy *et al.*, 2005).

In this study reported the effects of irrigation with municipal secondary treated effluents on growth of sunflower (*Helianthus annuus*) and *Celosia* (*C. argentea* var *cristata*) plants for cut flower production. The results reported that secondary treated municipal effluents can be used for production of sunflower and *Celosia* for cut flowers (Haya *et al.*, 2007).

Celosia palmeri

Flowering plant, that is native to the lower Rio Grande Valley of Texas in the United States as well as Northeastern Mexico and used for various ailments from the *C. palmeri* (Ramesh *et al.*, 2013).

Celosia trigyna

The leaves are boiled like cabbage and are known as torchata. It is also eaten as a vegetable in Africa was reported from the plant *C. trigyna* (Ramesh *et al.*, 2013).

Conclusion

In conclusion, the *Celosia* genus was found and reported in the review chapter about 90 compounds from amino acids, carbohydrates, fatty acids, flavonoids, glycosides, phenolic compounds, triterpenoids, saponins, vitamins, bycyclic peptides, colour pigments and intermediate compounds from different species as *Celosia argentea, C. bonnivairii, C. cristata, C. globosa, C. isertii, C. laxa, C. leptostachya, C. palmeri, C. polygonoids* and *C. trigyna*. The secondary metabolites such as β-sitosterol, citrusin C, colchicines, moroidin, isorhamnetin-3-O-β-D-glucopyranoside, indicant, quercetin-3-O-b-D-glucopyranoside and stigmasterol were reported from the above mentioned species. Also, the chapter reviewed the pharmacological and therapeutical applications of the reported compounds for various ailments. The chapter concluded that, reported *Celosia* genus consisting species having wide range of phytochemicals and is therapeutical applications will useful for the further research and development.

Abbreviations

1D and 2D NMR: 1 and 2 Dimensional nuclear magnetic resonance spectroscopy

^{13}C NMR: Carbon 13 nuclear magnetic resonance spectroscopy

^{1}H NMR: Hydrogen 1 nuclear magnetic resonance spectroscopy

ABTs: 2,2'-azino-bis(3-ethylbenzothiazoline-6-sulphonic acid)

ALP: Alkaline phosphatase

ALT: Alanine amino transferase

AST: Aspartate amino transferase

CCl_4: Carbon tetrachloride
cDNA: Complementary DNA
CD-spectra: Circular dichroism spectroscopy
CM-Sepharose: Carboxymethyl sepharose
COSY: Correlation spectroscopy
SRSV: Citrus Ring Spot Virus
DPPH: 2,2-diphenyl-1-picrylhydrazyl
ESI-MS: Electro spectro ionization - mass spectroscopy
ELSD: Evaporative light scattering detector
FRAP: Ferric reducing antioxidant power
FTIR: Fourier transform infrared spectroscopy
GFC: Gel filtration chromatography
HMBC: Heteronuclear multiple bond correlation
HMQC: Heteronuclear multiple quantum correlation
HPLC: High performance liquid chromatography
HR-ESI-MS: High resolution - electro spectro ionization - mass spectroscopy
HR-FAB-MS: High resolution -fast-atom bombardment- mass spectroscopy
IEC: Ion exchange chromatography
IFN-γ: Interferon-γ
IL-β: Interleukin-β
IL-1β: interleukin-1β
IR: Infrared spectroscopy
MSV: Measles Virus
MS: Mass spectroscopy
N,N DMF: N,N - dimethylformamide
NOSEY: Nuclear overhauser enhancement spectroscopy
PCR: Polymerase chain reaction
PGE2: Prostaglandin 2
PMV: Pokeweed mosaic virus
RIP/AVP: Ribosome inactivating/antiviral protein
RP-HPLC: Reversed phase – high performance liquid chromatography
SDS-PAGE: Sodium dodecyl sulphate - Poly acrylamide gel electrophoreses
SGCC: Silica gel column chromatography
SPSS: Statistical Package for the Social Sciences
SRV: Sunnhemp rosette virus
SSD: Silver sulphadiazine
TMV: Tobacco mosaic virus

TNF-α: Tumor necrosis factor-α
UV: Ultraviolet spectroscopy
WHO: World health organization
XRD: X-ray diffraction

References

Agbagwa, I.O., Onofeghara, F.A., and Mensah S.I. (2003). Stimulation of Growth and Development of *Celosia argentea* L. by crude extracts of *Senna alata* L. Roxb.*Journal Applied Science Environmental Management*, **7(1)**: 9-13.

Akiyoshi Sawabe., Takao Obata., Yoshinobu Nochika., Masanori Morita., Natsuki Yamashita.,Yoshiharu Matsubara and Tadashi Okamoto. (1999). Investigation of functional molecules in African *Celosia argentea* L. *Studies in Plant Science*,**6**: 290-296.

Alqasim Abdullahi Mustapha. (2013). Ethnomedicinal studies of medicinal plants with antifungal activities in Keffi local government, Nasarawa state, Nigeria. *Asian Journal of Plant Science and Research*,**3(4)**: 109-115.

Anthony E. Ojieh., Ese C. Adegor., and Ewhre O. Lawrence. (2013). Preliminary phytochemical screening, analgesic and antiinflammatory properties of *Celosia isertii*. *European Journal of Medicinal Plants*,**3(3)**: 369-380.

Balasubrahmanyam, A., Baranwal, V.K., Lodha, M.L., Varma, A., and Kapoor H.C. (2000). Purification and properties of growth stage dependent antiviral proteins from the leaves of *Celosia cristata*. *Plant Science*,**154**: 13-21.

Baranwal, V.K., Tumer, N.E., and Kapoor, H.C. (2002). Depurination of ribosomal RNA and inhibition of viral RNA translation by an antiviral protein of *Celosia cristata*. *Indian Journal Experimental Biology*,**40(10)**: 1195-7.

Barnabas, B.B., Mann A., Ogunrinola, T.S., and Anyanwu, P.E. (2011). Screening for anthelminthic activities from extracts of *Zanthoxylum zanthoxyloides*, *Neocarya macrophylla* and *Celosia laxa* against *Ascaris* infection in rabbits. *International Journal of Applied Research in Natural Products*, **3(4)**: 1-4.

Begam,M., NarwalS., Roy, S., Kumar, S., Lodha, M.L., and Kapoor,H.C. (2006). An antiviral protein having deoxyribonuclease and ribonuclease activity from leaves of the post flowering stage of *Celosia cristata*. *Biochemistry*, **71(1)**: S44-S48.

Bhavik Patel., Paresh Patel., and Rakesh Patel.(2011). Effect of different extracts from *Celosia argentea* on calcium and phosphate inhibition *in vitro*. **3(4)**: 337-339.

Cai, Y., Sun M., Schliemann, W., and Corke, H. (2001). Chemical stability and colorant properties of betaxanthin pigments from *Celosia argentea*. *Journal of Agricultural Food Chemistry*, **49(9)**: 4429-35.

Cai, Y., Sun, M., and Corke, H. (2005). HPLC characterization of betalains from plants in the Amaranthaceae. *Journal of Chromatographic Science*,**43**: 454-460.

Chi Xiang., Guo Mei Li., Song Hui and Chen Ya dan (2010). Study on chemical constituents of *Celosia cristata* seed. *Journal of Jilin Agricultural University*, **32(6)**: 657-660.

Christy, T.C., Catherine, M.G., James, A.P., and Donald, L.S. (2005). Production and ion uptake of *Celosia argentea* irrigated with saline wastewaters. *Scientia Horticulturae*, **106**: 381-394.

Danabalan Rubini., Dharmalingam Sudhahar and Kalusalingam Anandarajagpal. (2012). Phytochemical investigation and anthelmintic activity of *Celosia cristata* leaf extracts. *International Research Journal of Pharmacy*,**3(5)**: 335-337.

David O. Ojo., and Otunuya C. Oputa. (1999). Influence of sulphuron growth, yield and quality of *Celosia*. *Journal of Plant Nutrition*, **22(9)**: 1379-1388.

Ekka, R.N. and Dixit, V.K. (2007). Ethno-pharmacognostical studies of medicinal plants of Jaspur District, Chattisgarh. *International Journal of Green Pharmacy*,**1(1)**: 2-4.

Gautam Girendra Kumar., and Vidhyasagar Gali. (2011). Phytochemical screening of *Abutilon muticum* (Del.Ex.Dc.) and *Celosia argentea* Linn. *International Journal of Pharma and Biosciences*, **2(3)**: 463-467.

Gholizadeh, A., Santha, I.M., Kohnehrouz, B.B., Lodha, M.L., and Kapoor H.C. (2005). Cystatins may confer viral resistance in plants by inhibition of a virus-induced cell death phenomenon in which cysteine proteinases are active: cloning and molecular characterization of a cDNA encoding cysteine-proteinase inhibitor (celostatin) from *Celosia cristata* (crested cock's comb). *Biotechnology and Applied Biochemistry*, **42(3)**: 197-204.

Gholizadeh, A. (2011). Over expression, purification and functional characterization of *Celosia* ClpS as a fused protein in *Escherichia coli*. *Applied Biochemistry and Microbiology*, **47(4)**: 366–372.

Gholizadeh, A., Baghban Kohnehrouz, B., Santha, I.M., Lodha, M.L., and Kapoor H.C. (2005). Cloning and expression of small cDNA fragment encoding strong antiviral peptide from *Celosia cristata* in *Escherichia coli*. *Biochemistry*,**70(9)**: 1005-1010.

Gholizadeh, A., and Baghban Kohnehrouz, B. (2010).Identification of DUF538cDNA clone from *Celosia cristata* expressed sequences of non stressed and stressed leaves. *Russian Journal of Plant Physiology*, **572**: 247–252.

Gholizadeh Ashraf., Manoj Kumar., Balasubrahmanyam, A., Sushil Sharma.,Sneh Narwal., M.L., and Lodha, H.C. Kapoor. (2004). Antioxidant activity of antiviral proteins from *Celosia cristata*. *Journal of Plant Biochemistry and Biotechnology*, **13(1)**: 13-18.

Gnanamani, A., Shanmuga Priya, K., Radhakrishnan, N., and Mary Babu. (2003). Antibacterial activity of two plant extracts on eight burn pathogens. *Journal of Ethnopharmacology*, **86(1)**: 59-61.

Hase, K., Basnet, P., Kadota, S., and Namba, T. (1997). Immunostimulating activity of celosian, an antihepatotoxic polysaccharide isolated from *Celosia argentea. Planta Medica,* **63(3)**: 216-219.

Haya Friedman., Nirit Bernstein., Moshe Bruner., Ilona Rot., Zeev Ben-Noon., Atara Zuriel., Ron Zuriel., Sima Finkelstein., Nakdimon Umiel., and Amir Hagiladi. (2007). Application of secondary-treated effluents for cultivation of sunflower (*Helianthus annuus* L.) and celosia (*Celosia argentea* L.) as cut flowers. *Scientia Horticulturae,* **115**: 62–69.

Hayakawa, Y., Fujii, H., Hase, K., Ohnishi, Y., Sakukawa, R., Kadota, S., Namba, T., and Saiki, I. (1998). Anti-metastatic and immunomodulating properties of the water extract from *Celosia argentea* seeds. *Biological Pharma Bulletin,* **21(11)**: 1154-9.

Hayato Suzuki., Hiroshi Morita., Motoo Shiro., and Junichi Kobayashi (2004). Celogentin K, A new cyclic peptide from the seeds of *Celosia argentea* and X-ray structure of moroidin.*Tetrahedron,* **60**: 2489-2495.

Hayato Suzuki., Hiroshi Morita., Shigeo Iwasaki., and Junichi Kobayashi .(2003). New antimitotic bicyclic peptides, celogentins D-H, and J, from the seeds of *Celosia argentea. Tetrahedron,* **59**: 5307-5315.

Hiroshi Morita., Kazutaka Shimbo., Hideyuki Shigemori., and Junichi Kobayashi (2000). Antimitotic Activity of moroidin, a bicyclic peptide from the seeds of *Celosia argentea. Bioorganic and Medicinal Chemistry Letters,* **10**: 469-471.

Iwalewa, E.O., Adewunmi, C.O., Omisore, N.O.A., Adebanji, O.A., Azike, C.K., Adigun, A.O., Adesina, O.A., and Olowoyo, O.G. (2005). Pro- and antioxidant effects and cytoprotective potentials of nine edible vegetables in Southwest Nigeria. *Journal of Medicinal Food,* **8(4)**: 539-544.

Jadhao, A.B. (2013). An ethnobotanical and phytochemical screening some medicinal plants from Tahshil (Maharashtra), India. *International Journal of Pharmaceutical Science Invention,* **81(2)**: 19-21.

Jagtap, D.K., Patil, H.S. and Jakhi, P.S. (2013). Ethno-medicinal survey of some plants from villages of Khatav Tahashil (M.S.) India. *International Journal of Life Sciences,* **1(4)**: 264-269.

Jigna Parekh., and Sumitra V. Chanda (2008). Antibacterial activity of aqueous and alcoholic extracts of 34 Indian medicinal plants against some *Staphylococcus* species. *Turkish Journal of Biology,* **32**: 63-71.

Joshi Pranav, C., Patil Suhas, A., and Sambrekar, S.N. (2012). Evaluation of the antiurolithiatic activity of ethanolic extract of *Celosia argentea* (seeds) in rats. *Universal Journal of Pharmacy,* **1(1)**: 52-60.

Kachchhi, N.R., Parmar, R.K., Tirgar, P.R., Desai, T.R., and Bhalodia, P.N. (2012). Evaluation of the antiurolithiatic activity of methanolic extract of *Celosia argentea* roots in rats. *International Journal of Phytopharmacology,* **3(3)**: 249-255.

Kobayashi, J., Suzuki, H., Shimbo, K., Takeya, K., and Morita, H. (2001).Celogentins A-C, new antimitotic bicyclic peptides from the seeds of *Celosia argentea*. *Journal of Organic Chemistry*,**66(20)**: 6626-33.

Kumkum Agarwal., and Ranjana Varma. (2012). Some ethnomedicinal plants of Bhopal district used for treating stone diseases. *International Journal of Pharmacy and Life Sciences,* **3(1)**: 1356-1362.

Lim, S.G., and Moon G.J. (1987). New classification of human sera by phytoagglutinin *Celosia argentea* L. and cat red blood cells. *Korea University Medical Journal,* **24(3)**: 17-24.

Malomo, S.O., Ore, A., and Yakubo, M.T. (2011). *In vitro* and *in vivo* antioxidant activity of the aqueous extract of *Celosia argentea* leaves. *Indian Journal of Pharmacology,* **43(3)**: 278-285.

Marie Claire., Makambila Koubemba., Bernard Mbatchi., Denis Ardid., Agathe Gelot., Cindy Henrion., Robin Janisson., Ange A. Abena., and Jean Theophile Banzouzi. (2011). Pharmacological studies of Ten Medicinal plants used for Analgesic purposes in Congo Brazzaville. *International Journal of Pharmacology,* **7(5)**: 608-615.

Mehbuba Begama., Sushil Kumar., Sribash Roy., James J. Campanella., and H.C. Kapoor (2006). Molecular cloning and functional identification of a ribosomeinactivating/antiviral protein from leaves of post-flowering stage of *Celosia cristata* and its expression in *E. coli*. *Phytochemistry,* **67**: 2441-2449.

Morita, H., Shimbo, T., and Shigemori H. Kobayashi. (2000). *Journal of Bioorganic Medicinal Chemistry Letters*., **10**: 469-471.

Morita, H., Suzuki, H., and Kobayashi, J. (2004).Celogenamide A, a new cyclic peptide from the seeds of *Celosia argentea*. *Jounral of Natural Products,* **67(9)**: 1628-30.

Mossi, A.J., Mazutti Paroul, M., Corazza, N., Dariva, M.L., Cansian, C., and Oliveira, R.L. (2009) Chemical variation of tannins and triterpenes in Brazilian populations of *Maytenus ilicifolia* Mart. *Ex Reiss Brazilian Journal of Biology,* **69 (2)**: 339-345.

Na Feng., Qie Xue., Qinghua Guo., and Ru Zhao.(2009). Genetic diversity and population structure of *Celosia argentea* and related species revealed by SRAP. *Biochemical Genetics,* **47**: 521-532.

Oloyede, F.M., Oloyede, F.A. and Obuotor, E.M. (2013). Effect of plant maturity on the antioxidant profile of *Amaranthus cruentus* L. and *Celosia argentea* L. *Bulletin Environmental Pharmacological Life Science,* **2(2)**: 18-21.

Omilabu A. Sunday., A. Bankole Munir, O., Oyefolu Akeeb, A., Adesanya Bolanle., and Badaru, S.O. (2010). Antiviral effect of *Hibiscus sabdariffa* and *Celosia argentea* on measles virus. *African Journal of Microbiology Research,* **4(4)**: 293-296.

Palada, M.C., and Crossman, S.M.A. (1999). Evaluation of tropical leaf vegetables in the Virgin Islands. In: Janick, J. (Ed.), *Perspectives on New Crops and New Uses*. ASHS Press, Alexandria, pp. 388–393.

Patel, B.A., Patel, P.U., and Patel, R.K. (2010). Physico-chemical and phytochemical investigations of seeds of *Celosia argentea. International Journal of Pharmaceutical and Applied Sciences,***1(1)**: 124-126.

Prakash, D., Nath, P., and Pal, M. (1995). Composition and variation in Vitamin C, carotenoids, protein, nitrate and oxalate contents in *Celosia* leaves. *Plant Foods for Human Nutrition,* **47(3)**: 221-6.

Praveen Sharma., Gali Vidyasagar., Sunder Singh., Santosh Ghule., and Bimlesh Kumar. (2010). Antidiarrhoeal activity of leaf extract of *Celosia argentea* in experimentally induced diarrhoea in rats. *Journal of Advanced Pharmaceutical Technology and Research,* **1(1)**: 41-48.

Priya, K.S., Arumugam, G., Rathinam, B., Wells, A., and Babu, M. (2004). *Celosia argentea* L. leaf extract improves wound healing in a rat burn wound model. *Wound Repair Regeneration,* **12(6)**: 618-625.

Priyanka Ghorpade., Areej Siddiqui., Manohar J. Patil., and Rukhsana A. Rub. (2012). Pharmacognostic and phytochemical evaluation of *Celosia argentea. Pharmacognosy,* **4(33)**: 7-15.

Rajesh, P., Latha, S., Selvamani, P., Saraswathy, A., and Rajesh Kannan, V. (2009). Role of *Celosia polygonoids* Juss. in the experimental model of inflammation in Wistar rats. *Journal of Phytology,* **1(3)**: 185-192.

Rajni, B. Ranjan., and Deokule, S.S. (2013). Pharmacognostic and phytochemical investigation of *Celosia argentea. International Research Journal of Pharmacy,* **4(6)**: 138-144.

Ramesh B. Nidavani., Mahalakshmi, A.M., and Mallappa Shalawadi. (2013). Towards a better understanding of an updated ethnopharmacology of *Celosia argentea* L. *International Journal of Pharmacy and Pharmaceutical Sciences,* **5(3)**: 54-59.

Rawia, A. Eid., Abo Sedera,S.A., and Attia, M. (2006). Influence of nitrogen fixing bacteria incorporation with organic and/or inorganic nitrogen fertilizers on growth, flower yield and chemical composition of *Celosia argentea. World Journal of Agricultural Sciences,* **2(4)**: 450-458.

Santosh Ghule., Prakash, T., Kotresha, D., Roopa Karki., Surendra, V., and Divakar Goli (2010). Anti-diabetic activity of *Celosia argentea* root in streptozotocin-induced diabetic rats. *International Journal of Green Pharmacy,* p. 206-211.

Santosh S. Bhujbal., Sohan. S. Chitlange., Anupama. A. Suralkar., Devanand. B. Shinde., and Manohar J. Patil (2008). Anti-inflammatory activity of an isolated flavonoids fraction from *Celosia argentea* Linn. *Journal of Medicinal Plants Research,* **2(3)**: 52-54.

Saritha P.,and Sreeramulu, A. (2013). Allelopathic effects of *Celosia argentea* L. Leaf extracts on crop plant seed germination. *International Journal of science and Pharma Reviews,* **2(1)**: 56-64.

Shen, S., Ding, X., Ouyang, M.A., Wu, Z.J., and Xie, L.H. (2010). A new phenolic glycoside and cytotoxic constituents from *Celosia argentea*. *Journal of Asian Natural Products Research,* **12(9)**: 821-7.

Shuo Shen., Xiao Ding., Ming-An Ouyang., Zu-Jian Wu.,and Lian-Hui Xie.(2010). A new phenolic glycoside and cytotoxic constituents from *Celosia argentea*. *Journal of Asian Natural Products Research,* **12(9)**: 821-827.

Sonawane, V.B., Saler, R.S., Sonawane, M.D. and Kadam, V.B. (2012). Ethnobotanical studies of Mokhada, district Thane. *International Journal of Life Sciences and Pharma Research,* **2(2)**: 88-93.

Sun Zhen Liang (2009). Study on chemical constituents and hepatoprotective effect of *Celosia Argentea* L. seeds. *Masters thesis.*

Sun, Z.L., Gao, G.L., Xia, Y.F., Feng, J., and Qiao, Z.Y. (2011). A new hepoprotective saponin from Semen *Celosia cristatae*. *Fitoterapia,* **82(4)**: 591-4.

Suzuki, T., Tomita-Yokotani, K., Yoshida, S., Takase, Y., Kusakabe, and Hasegawa, K. (2002). Preparation and isolation of oligogalacturonic acids and their biological effects in cockscomb (*Celosia argentea* L.) seedlings. *Journal Plant Growth Regulation,* **21**: 209-215.

Sven Verlinden., and Louis McDonald (2007). Productivity and quality of statice (*Limonium sinuatum* cv. Soiree Mix) and cockscomb (*Celosia argentea* cv. Chief Mix) under organic and inorganic fertilization regiments. *Scientia Horticulturae,* **114**: 199-206.

Taha, R.M. and Wafa, S.N.(2012). Plant regeneration and cellular behaviour studies in *Celosia cristata* grown *in vivo* and *in vitro*. *Scientific World Journal,* p. 359-413.

Thangarasu Vetrichelvan., Maniappan Jegadeesan., and Bangaru Adigalar Uma Devi. (2002). Anti-diabetic activity of alcoholic extract of *Celosia argentea* L. seeds in rats. *Biology Pharmcology Bulletin,* **25(4)**: 526-528.

Urmila, G.H., Ganga Rao, B., and Satyanarayana, T. (2013). Physico-chemical and preliminary phytochemical screening for medicinal plants. *International Journal of Pharmaceutical and Chemical Sciences,* **2(4)**: 1738-1742.

Urmila, G.H., Ganga Rao, B., and Satyanarayana, T. (2013). Phytochemical and *in-vitro* antioxidant activity of methanolic extract of *Lactuca scariola* and *Celosia argentea* leaves. *Journal of Drug Delivery and Therapeutics,* **3(4)**: 114-117.

Vetrichelvan, T., Jegadeesan, M., and Devi, B.A. (2002). Anti-diabetic activity of alcoholic extract of *Celosia argentea* Linn. seeds in rats. *Biology Pharmacology Bulletin,* **25(4)**: 526-528.

Wang, Y, Guo, M.L., Wang, X.K., and Yin, J. (2008). HPLC fingerprints in seed of *Celosia argentea. Zhongguo Zhong Yao Za Zhi,* **33(1)**: 51-3.

Wen, Y., Islam, M.T., and Tahara, S. (2006). Phenolic constituents of *Celosia cristata* L. susceptible to spinach root rot pathogen *Aphanomyces cochlioides*. *Bioscience Biotechnology Biochemistry,* **70(10)**: 2567-70.

Willibald Schliemanna., Yizhong Cai., Thomas Degenkolb., Jurgen Schmidt., and Harold Corke. (2001). Betalains of *Celosia argentea. Phytochemistry*, **58**: 159-165.

Wong, K.Y. (1994). *Chinese Herbal Medicine*. Wokman Press, Hong Kong.

Woo, K., Ko, J., Song, S., Lee, J., Kang, J., Seo, M., Kwak, D., Oh, B., Nam, M., and Jeong, H. (2011). Antioxidant compounds and antioxidant activities of the methanolic extracts from cockscome (*Celosia cristata* L.) flowers. *Planta Med/*. p. 77-78.

Wu, Q., Wang, Y., and Guo, M. (2011).Triterpenoid saponins from the seeds of *Celosia argentea* and their antiinflammatory and antitumor activities. *Chemistry Pharma Bulletin*, **59(5)**: 666-671.

Wu, Q.B., Wang, Y., Liang, L., Jiang, Q., Guo, M.L. and Zhang, J.J. (2013). Novel triterpenoid saponins from the seeds of *Celosia argentea* L. *Natural Products Research*, **27(15)**: 1353-60.

Xu, G.J. (1996). Ji Guan Zi (No. 12768), Ji Guan Hua andJi Guan Miao (No. 12770). In: Yan, Y.Q., Yu, C.L., Huang, T.K.(Eds.), *Encyclopedia of Chinese Medicine, China Medicine*. Science Technology, Beijing,Vol. 2.**pp.** 509–511.

Xu LiangZheng., Luo LaiHui., Liu HuiNa., and Liao FuLin. (2009).A preliminary study on seed germination of wild medicinal and edible plant *Celosia argentea*. *Plant Physiology Communications*, **45(6)**: 583-585.

Yan Wang., Ziyang Lou., Qing-Bin Wu., and Mei-Li Guo. (2010).A novel hepatoprotective saponin from *Celosia cristata* L. **81(8)**: 1246-1252.

Zhang, C.J., Li, D.T., Liang, J.Z., and Cheng, C.G. (2007). Identification of *Semen celosiae* and *Semen celosiae cristatae* using continuous wavelet transform with FTIR. *Guang Pu Xue Yu Guang Pu Fen Xi*, **27(1):** 50-3.

ZhenLiang Sun., Ying Wang., Mei-Li Guo., and Yu-Xian Li. (2010). Two new hepaprotective saponins from *Semen celosiae. Fitoterpia*, **81(5):** 375-380.

Zingare, A.K. (2012). Ethnomedicinal uses of plants among the Halba tribe of Gondia District of Maharashtra, India. *Bionano Frontier*, p. 121-125.

Natural Products: Research Reviews Vol. 3 (2016) *Pages* **147–159**
Editor: **V.K. Gupta**
Published by: **DAYA PUBLISHING HOUSE, NEW DELHI**

6

Green-Synthesis and Antimicrobial Activities of Silver Nano Particles

Ali Parsaeimehr[1*], Yi-Feng Chen[1] and Amir Reza Jassbi[2]

ABSTRACT

Silver is known as an antimicrobial harmless element to the human's heath for a long time. Nowadays, by emerging multidrug-resistant microorganisms, application of silver in its nano-particle (SNPs) form has raised interests of many researchers in the field of medicinal chemistry. Currently, different strategies of usage and synthesis of SNPs are explored based on their effectiveness, cost and impacts on the ecosystem. Green synthesis is accepted as an alternate synthetic method of SNPs. This paper presents an overview on green synthesis of SNPs and their inhibitory role on the control of pathogenic microorganisms and parasites.

Keywords: Green-synthesis, Antimicrobial activities, SNPs.

Introduction

The rapid developments in the area of science and technology had tremendous impacts on different aspects of the human's life during past decades. Technological innovations have allowed us to manipulate miniature structures of atoms and molecules. Recently nanotechnology has been rising as a rapidly growing field with sensible applications in different branches of science and technology including, medicine, cosmetics, food product, energy, clothing and biology. The term nanotechnology was coined by Professor Norio Taniguchi from Tokyo Science

1 Institute of Biotechnology, Jiangsu Academy of Agricultural Sciences; Key Laboratory of Agrobiology of Jiangsu Province, Nanjing 210014, Jiangsu, China.

2 Medicinal and Natural Products Chemistry Research Center, Shiraz University of Medical Sciences, Shiraz, Iran.

* *Corresponding author*: E-mail: ali.parsaeimehr@gmail.com

University in 1974 describing this fact that the essence of materials transforms by changing one molecule or an atom (Rai and Aniket, 2009). In the sphere of nanotechnology, metallic nanoparticles (NPs) have attracted the consideration of scientists based on their interesting functions in the field of biomedical sciences.

Silver has been used since ancient time in Greece and Rome to cure infections and wounds. They used to keep silver coins in the jars of drinks to save them from spoilage. Silver in its metallic form is reported to be non-toxic to human and the most effective agent against microbes; even though in its ionic forms has shown interaction with biological fluids and long-term usage causes abnormality and interruption in wound healing process (Panácek *et al.*, 2006; Atiyeh *et al.*, 2007). Currently, SNPs (smaller than 100 nm) with different crystal forms spherical to triangular nano-plates is known to exhibit outstanding biological activities. SNPs have been broadly used in drugs, gene delivery, implantable materials, biosensors and to treat infections and wounds (Cuenca *et al.*, 2006; Kumar and Yadav, 2009). In point of fact, larger surface area plus the high surface energy of SNPs demonstrates more efficient means at a very low concentration for SNPs biological activities (Kittler *et al.*, 2010). SNPs have shown antimicrobial activities against broad spectrum of bacteria (*i.e., Pseudomonas aeruginosa, Staphylococcus* spp., *Escherichia coli, Klebsiella pneumoniae, Vibrio cholerae*), fungi (*i.e., Candida* spp., *Saccharomyces cerevisiae, Aspergillus niger* and *Rhizopus arrhizus*) and viruses (*i.e.*, influenza virus, *Hepatitis B virus*, Arenavirus and *HIV1).*

The common methods for the synthesis of SNPs involve the reduction of metal ions in solutions or in high temperature gas phase, these processes includes application of toxic chemicals, unsafe reaction conditions; and are expensive, while the green synthesized SNPs process are eco-friendlier, cheaper and their influence is comparable to other biomedical. The green synthesis of SNPs using plant's extracts, bacterial and fungal biotransformation resulted in the production of pure and well crystallized products (Chernousova *et al.*, 2013; Zhang *et al.*, 2013; Li *et al.*, 2013). Moreover, a better manipulation and control on nano-crystals formation and stabilization presents by slower kinetics of green synthesis compared to other methods. Mainly, green synthesis of SNPs involves in selection of appropriate solvent, reducing agent, and nontoxic materials for the SNPs stability (Singh *et al.*, 2010; Yan *et al.*, 2010).

Green Synthesis of Silver Nanoparticles (SNPs)

Unicellular and multicellular organisms are able to produce inorganic compounds either intra- or extra-cellular, this fact raised the idea of using them as reducing or capping agents for synthesis of SNPs. The SNPs can be biosynthesized by different methods using plant extracts, bacteria, fungi and biodegradable polymers. Different reaction parameters such as, time, pH, concentration of $AgNO_3$, temperature, reducing agents are responsible for the size control of SNPs.

Green Synthesis of SNPs Using Plant's Extracts and Products

White sugar and sodium hydroxide (NaOH) in the presence of sunlight converted the silver ions to SNPs in 15-30 nm diameters and the presence of gluconic acid as capping agent promotes the stability of SNPs in the colloids (Meshram *et al.*, 2013).

Jagtap and Bapta (2013) presented another approach toward green synthesis of SNPs using jackfruit *(Artocarpus heterophyllus* Lam.) seed powder extract as a reducing agent. They mentioned that the seeds of this plant contain a lectin named Jacalin which can be used as a bonding agent for SNPs. The synthesized SNPs were reported to be totally efficient against *Bacillus cereus, B. subtilis* and *S. aureus* and *P. aeruginosa.* It is confirmed that, SNPs are able to bond to proteins via free amine groups or cystein residues of proteins which helps to enhance their stabilization (Gole *et al.*, 2011).

Myrrh is an aromatic resin from *Commiphora myrrha* tree which was used as green-reducing and stabilizer agent for SNPs synthesis. The produced SNPs showed significant antibacterial activity against *Bacillus thuringiensis* and *P. aeruginosa*. SNPs were quickly synthesized using silver ions in the presence of paprika (*Capsicum annuum* L.) extract, *C. annuum* proteins showed a reducing and controlling role during the formation of SNPs and secondary structure of the proteins alters after reaction with silver ions (El-Sherbiny *et al.*, 2013).

The latex of *Jatropha curcas* was used as reducing and capping agent to synthesize SNPs with a centered cubic shape, the cavity radius of cyclic peptides existed within the latex exposed that particles with radius 10–20 nm are typically stabilized by the cyclic peptides (Bar *et al.*, 2009). The leaf extract of Indian acalypha (*Acalypha indica)* were used for the formation of SNPs with the size of 20–30 nm. Interestingly the synthesized SNPs showed antimicrobial activity against two water borne bacterial pathogens *E. coli* and *Vibrio cholerae*. Alteration in the membrane permeability and respiration of the SNPs treated bacterial cells were the main reasons for their antimicrobial activities (Krishnaraj *et al.*, 2013). Madagascar rosy periwinkle (Catharanthus roseus) has proven to be a good source for synthesis of 35-55 nm size SNPs with significant antiplasmodial activity against *Plasmodium falciparum* a protozoan parasite (Ponarulselvam *et al.*, 2012).

Application of different concentration of the plant extracts and silver ions as the reducing agents and precursors of the SNPs showed a linear correlation with the size and the reaction rate of the SNPs formation (Khan *et al.*, 2013). The SNPs were synthesized from the reaction of the silver ions in leaf extract of *Spanish cherry* (*Mimusops elengi* L.) at room temperature. The formation of stable SNPs at 1 mM concentrations of silver nitrate yielded spherical shaped particles with a diameter range of 55-83 nm (Prakash *et al.*, 2013). Interestingly, the *SNPs* in the environment can be accumulated in plants and can be used for medicinal purposes. Using this approach, Jorge *et al.* (2002) reported Lucerne (*Alfalfa sprouts*) as a plant that can form SNPs by up taking silver metals from a silver-rich solid medium. Mustard greens *(Brassica juncea)* is also able to up take $AgNO_3$, $Na_3Ag(S_2O_3)_2$, and $Ag(NH_3)_2NO_3$ solutions under the hydroponically growth condition. And reduction of ionic silver to its metallic form of SNPs in a range size of 2 to 35 nm in 0.35 wt per cent Ag on a dry plant basis (Gardea-Torresdey *et al.*, 2003).

Aqueous cortexes extract of *B. javanica* were reported to produce spherical form of SNPs at the average size about 38±14 nm at the room temperature as a medium/reducing system for the formation of SNPs from $AgNO_3$ (Yudha, 2013). Song and Kim (2009) reported the syntheses of 15 to 500 nm SNPs using the leaves extracts of pine, persimmon, ginkgo, magnolia and platanus. The size of SNPs can be adjusted by changing the concentrations of $AgNO_3$, leaf extract and the reaction temperature. Authors have also reported that among the chosen plants Magnolia leaf extracts was the best choice based on SNPs synthesis rate.

Green Synthesis of SNPs Using Fungal Biotransformation

Puccinia graminis is a plant-symbiotic fungi belonging to the family of *Pucciniaceae and reported as* a reducing and capping agent for production of SNPs in a range from 30 to120 nm (Kirthi *et al.*, 2012). *Penicillium diversum* was also used for synthesis of SNPs in a size of 5 to 45 nm. Apparently the existed reductase enzyme in the fungal extract are responsible for the reduction and stabilization of the SNPs, and the produced SNPs showed significant antimicrobial activity against *E. coli, Salmonella typhi, V. cholerae*, and *Paratyphia* spp. (Ganachari *et al.*, 2012). Furthermore, a stable silver hydrosol was shaped when the aqueous silver ions were treated with *Fusarium oxysporum*. The sizes of the silver particles were in a range of 5–50 nm and they were stabilized in the solution by the fungal-secreted proteins (Ahmad *et al.*, 2003). Literature surveys indicated that, the existed water soluble organic compounds in the bark extracts of Dalchini (*Cinnamon zeylanicum*) were the major agents for the reduction of silver ions to nanostructure. The biosynthesized SNPs using bark extracts showed inhibitory influence on *Escherichia coli* BL-21 strain (Sathishkumar *et al.*, 2009).

Green Synthesis of SNPs Using Bacterial Biotransformation

The SNPs can be biotransformed from the silver ions using bacterial suspension of *Streptococcus*, *Proteus* and *Pseudomonas* species, and the products had different antimicrobial activity, for instance the synthesized SNPs from *Streptococcus* spp., demonstrated more antimicrobial activity compared to the other bacteria against *S. typhi* (Nanda *et al.*, 2009). Bacterial exopolysaccharide as both a reducing and stabilizing agent were utilized to synthesis SNPs with a size from 2 to 15 nm with antimicrobial activities against multidrug resistant *P. aeruginosa, K. pneumoniae, Listeria monocytogenes, Aspergillus* spp. and *Penicillum* spp. (Kanmani and Lim, 2013). SNPs were extracellular biosynthesized using *Epicoccum nigrum* isolated from *Phellodendron amurense*, and it was shown that the reaction rate of the SNPs formation was increased with raising the pH and the temperature (Qian *et al.*, 2013).

Biological Activities of SNPs

Antibacterial Activity of SNPs

Antibacterial are those agents that kill bacteria or retard their growth; SNPs are used efficiently to control and prevent dangerous infections and have demonstrated synergistic effect when combined with the most common antibiotics such as, ceftazidime, amoxicillin, erythromycin, clindamycin, and vancomycin, streptomycin, kanamycin, ampiclox, polymyxin B. In some cases antagonistic activity of SNPs with antibiotics like, chloramphenicol are also reported (Fayaz *et al.*, 2010; Shahverdi *et al.*, 2007).

In fact, the antibacterial activity of SNPs has a dual action mechanism: first the bactericidal effect of Ag+ and the second is its membrane-disrupting effect (Sharma,

2009). Different bioassays showed that SNPs inhibited the growth of drug resistant bacteria such as: multidrug-resistant *P. aeruginosa*, ampicillin-resistant *E. coli* and erythromycin-resistant *Streptococcus pyogenes*. Li *et al.* (2013) reported the antibiotic activity of 21 different nano particles on multidrug resistant *Neisseria gonorrhoeae*, their finding indicated that SNPs (approximate size: 120 nm) had a great potential to inhibit *N. gonorrhoeae* colony formation while it showed no cytotoxicity in human fibroblasts or epithelial cells and had a synergistic effect with cefmetazole. A novel strategy to synthesize β- lactamase resistant penicillin G molecules is suggested in the presence of SNPs (Ahmed *et al.*, 2010). In this procedure the free amino groups were attached on the surface of SNPs by coating a uniform layer of polyaniline and attaching to the C3 carboxyl group of the β lactam ring of the penicillin G molecule in the presence of N-(3-dimethylaminopropyl)-N2-ethylcarbodiimide hydrochloride (EDAC).The synthesized penicillin was totally effective against β lactam resistant- *E. coli* and *Staphylococcus aureus*. Also, the biosynthesized SNPs from *Cryphonectria* spp. showed reliable antimicrobial activity against multidrug resistant bacteria including, *S. aureus*, *E. coli* and *S. typhi* (Dar *et al.*, 2013).Brucellosis (Mediterranean fever) disease caused by unsterilized milk or meat from infected animals by rod shaped bacteria recognized as *Brucella* spp. It is known that intramacrophage survival is the main factor of its virulence; SNPs showed they are able to neutralize the *Brucella* spp. in concentrations from 4 to 6 ppm within 40 min (Alizadeh *et al.*, 2013). Evidently, the isolated *Pencillium* spp. fungus from leaves of *C. longa* are capable to generate spherical SNPs with the size of 25 nm and they can be used against *P. aeruginosa*, *K. pneumonia* (Singh *et al.*, 2013). The extracellular enzyme from *Exiguobacterium* sp. KNU1 showed that it can be used for the reduction of $AgNO_3$ solutions to spherical-shaped SNPs in a rage of 5-50 nm, with significant antimicrobial activity against *S. typhimurium*, *P. aeruginosa*, *E. coli* and *S. aureus*. It is shown that *E. coli* displayed DNA fragmentation after being treated with the SNPs (Tamboli *et al.*, 2013). It is known that, Ag^+ is able to reduce phosphate uptake and exchange in *E. coli* and cause efflux of accumulated phosphate may relocate metal cations from their common binding sites in enzymes or hit the respiratory chain and cell division. Apparently, SNPs can penetrate barrier of outer membrane permeability, peptidoglycan and periplasm and are able to enhance the permeability of the bacterial membrane for reducing sugars and proteins which causes remarkable breaks in the cell membrane (Schreurs and Rosenberg, 1982; Li *et al.*, 2010).

Antifungal Activity of SNPs

Only small numbers of fungi, around 100 species, are linked to human diseases, but fungal pathogens have caused a wide range of diseases including, allergies, skin infections, and invasive mycoses (Huffnagle and Noverr, 2013). In addition to different types of effective fungicide such as ketoconazole, imidazoles, allylamines and echinocandins, SNPs are other alternative antifungal agents.

SNPs inhibited hyphal growth of fungal sclerotium-forming species such as, *Rhizoctonia solani*, *Sclerotinia sclerotiorum* and *S. minor*, with a mechanism by which the layers of hyphal wall are broken up and sclerotial germination are blocked (Min *et al.*, 2009). Chitosan is a linear polysaccharide which is used to control fungal contamination in food industries. SNPs are as a complex with chitosan to control fungi such as, *Ophiostoma flexuosum, O. tetropii, O. polonicum* and *O. ips*. The SNPs inhibited the growth of *Candida* spp. at concentrations as low as 0.21 mg/L of Ag. The inhibitory effect of SNPs was enhanced through their stabilization (Velmurugan *et al.*, 2009). Moreover, SNPs have also revealed remarkable antifungal activity against *Trichophyton mentagrophytes* and *Candida* species. The activity of SNPs was similar to amphotericin B, but greater than fluconazole antifungal drugs; moreover, the SNPs exposed significant influence on mycelium (Kim *et al.*, 2008). SNPs have also shown outstanding antifungal effect with reduction of the colonies of two plant-pathogenic fungi: *Bipolaris sorokiniana* and *Magnaporthe grisea*, (Jo *et al.*, 2009). The biosynthesized SNPs from fungi *Alternaria alternate* showed antifungal activities against *Phoma glomerata, P. herbarum, Fusarium semitectum, Trichoderma* spp., and *Candida albicans*. Fluconazole antifungal activity was improved by its combination with SNPs as a complex. The combination of SNPs with nystatin or chlorhexidine digluconate had synergistic effects and it demonstrated remarkable antifungal activity against biofilms of *Candida albicans* and *Candida glabrata* (Gajbhiye *et al.*, 2009; Monteiro *et al.*, 2013).

Antifungal potential of SNPs, biosynthesized by a polyphenol named "Bayberry tannin" and attached collagen fiber as supporter were determined by using *Saccharomyces cerevisiae*, viability of the cells of *S. cerevisia* suspension were severely diminished to zero when they were exposed to the complex of SNPs and "Bayberry tannin" (10^5 CFU/m) (He *et al.*, 2012). The strain GP-23 of the bacteria *Bacillus* cell free culture supernatant was used to synthesize SNPs. The resulting SNPs demonstrated substantial antifungal activity against *Fusarium oxysporum* at the range of 8 µg/ml. In another experiment the SNPs were synthesized by biological reduction of silver nitrate in a filtrated culture of a marine bacterium, *Streptomyces* spp. The SNPs inhibited the growth of mycelia of *Aspergillus niger, A. flavus* and *A. fumigates* within the range from 62 to 75 per cent antifungal index (Gopinath *et al.*, 2013; Thenmozhi *et al.*, 2013).

Antiviral Activity of SNPs

The study of Lara and co workers (2010) proved the antiviral activity of SNPs. They showed that SNPs are capable to bond the envelope glycoprotein GP120 and inhibit CD4-dependent virion binding. This leads to an inhibition in post-entry phase of the HIV-1 virus at the early stage of viral replication. The SNPs are also able to inhibit the formation of intercellular Hepatitis B Virus (HBV) RNA and extracellular virions. In addition to the above mechanisms, SNPs also can bind to HBV DNA with a binding constant of $(8.8 \pm 1.0) \times 10^5$ dm^3/mol (Lu *et al.*, 2008). It is documented that, SNPs unified into polysulfone ultrafiltration membranes show evidence of antiviral properties on bacteriophage MS2, a single-stranded RNA virus (Zodrow *et al.*, 2009).

Mehrbod *et al.* (2009) suggested three mechanisms for interaction of SNPs and the viruses, in the first SNPs are able to interact with viral membrane fusion by inhibition of penetration, second they induce inhibition during adsorption/ penetration process, and third, is the reduction of the interactions between glycoprotein knobs and antibodies. Considerably, SNPs with mean particle diameters of 10 nm are capable to demonstrate antiviral activity against H1N1 influenza A virus (Xiang *et al.*, 2011). The SNPs are capable to inhibit prototype arenavirus at non-toxic concentrations and efficiently reduce arenavirus replication when administered prior to viral infection or early after initial virus exposure. It is also reported that, SNPs have antiviral activity against Tacaribe virus at the range more than 50 µg/ml; moreover, by using transmission electron microscopy techniques it is demonstrated that, Tacaribe virus and SNPs taken up into the same cell, since the vRNA appears to be inhibited in SNPs-treated Tacaribe virus infections. The above mentioned observations may suggest that the SNPs could be interfering with the Tacaribe virus RNA-dependent RNA polymerase (L protein). Two arenavirus proteins, the nucleoprotein and the small Z protein, have been shown to possess Zn-binding motifs. The viral neutralization capabilities of SNPs may be related to their inferring with the zinc (Zn)-finger motif of the virus. (Speshock *et al.*, 2010).

Antiparasitic Activity of SNPs

Antiparasitics are a class of drugs that are able to treat parasitic diseases including nematodes, trematodes and amoebas. Recently, SNPs have demonstrated antiparasitic activities in addition to the classical anti parasitic agents such as, pyrantel, niclosamide, benznidazole, pyrimethamine, chloroquine, etc. (Demir *et al.*, 2011; Said *et al.*, 2012).

It is claimed that, the green synthesized SNPs using aqueous leaf extract of Sensitive plant (*Mimosa pudica)* showed antiparasitic activities on the larvae of *Anopheles subpictus* (primary vector of malaria), *Culex quinquefasciatus* (vector of west nile virus, filariasis, and Japanese encephalitis and avian malaria) and *Rhipicephalus microplus* (vector of bovine babesiosis and anaplasmosis). They also mention that, formations of nanoparticles were observed after 6 hours using leaf extract of *M. pudica* (Marimuthu *et al.*, 2011). Leishmania is a genus of trypanosomatid and responsible for the disease leishmaniasis; Trypanothione/trypanothione reductase (TR) system in Leishmania is known to play an essential defensive action against mammalian host defense systems, trypanothione recycles trypanothione disulphide by the enzyme trypanothione reductase. The SNPs inhibited TR system and suppress the Leishmanial life cycle. This process is induced by the bonding of SNPs to the catalytic triads, Cys52, Cys57, and His 461.

Also, SNPs are known to produce reactive oxygen species and disrupt the metabolic activity of Leishmania parasites; moreover they significantly inhibit the survival of amastigotes in the host cells (Baiocco *et al.*, 2010). The SNPs were prepared by aqueous extract of the stems of Asthisamharaka (*Cissus quadrangularis*) and showed significant anti-adulticidal and larvicidal activity against *Hippobosca maculate* and *Rhipicephalus microplus* (Santhoshkumar *et al.*, 2012). The synthesized SNPs from the leaf extracts of *prostrate sandmat* (*Euphorbia prostrate)* were used as antiparasitic agent against adult cattle tick *Haemaphysalis bispinosa* Neumann and the haematophagous fly *Hippobosca maculate* Leach, considerably SNPs showed 100 per cent effectiveness at a concentration of 10 mg l^{-1} (Zahir and Abdul Rahuman, 2012).

Mechanism of Antiinfection of SNPs

Even though that most aspects of the antimicrobial activity of SNPs remained in the shadow, but evidences point to this fact that, SNPs inhibitory activity is significantly linked to their sizes, distribution, surface charge and stability in the medium. As a matter of fact based on equivalent silver mass content, the smaller particles provide superior activities. SNPs with high electrolyte content cause diminishes in antibacterial activities, smaller SNPs can penetrate inside the cells easier compared to the biggest particles and can generate reactive oxygen species, induce oxidative stress, affect the ribosomal subunit protein and some enzymes important for the bacterial cells, they can significantly inhibit synthesis of nucleic acid and the protein synthesis mediated by the 30s ribosomal subunit, DNA may loses its replication ability under the influence of SNPs. Also, SNPs are able to interact with disulfide bonds of proteins and alter three-dimensional arrangement of proteins. Furthermore, they are able to disturb membrane permeability, inhibit the cell wall synthesis, and as a great deal release of silver ions from particles can inhibit the microbial cellular respiration (Morones *et al.*, 2005; Li *et al.*, 2010; Lara *et al.*, 2010).

Conclusion

The rapid progresses in the area of nanotechnology are not limited to physics and chemistry and it has expended its borders to the biological sciences. Low cost, eco-friendly and effectiveness of synthesized SNPs by green synthesis methods can be used efficiently against microorganisms and parasites. Furthermore, the usage of SNPs demonstrates a rational approach against multidrug resistant microorganism based on their effectiveness and synergism with most of antibiotics. Undoubtedly biological sciences are capable to be used as a tool in the hand of scientists in the area of nanotechnology for biosynthesis of efficient SNPs using green methods.

References

Ahmad, A., Mukherjee, P., Senapati. S., Mandal, D., Khan, M. I., Kumar, R., and Sastry, M. (2003). Extracellular biosynthesis of silver nanoparticles using the fungus *Fusarium oxysporum*. *Colloids and Surfaces B: Biointerfaces*, **28**: 313-318

Ahmed, V., Kumar, J., and Kumar, M. (2013).Synthesis, characterization of penicillin G capped silver nanoconjugates to combat β-lactamase resistance in infectious microorganism. *Journal of Biotechnology*, **163**: 419-424.

Albers, C. E., Hofstetter, W., Siebenrock, K. A., Landmann, R., and Klenke, F. M. (2013). *In vitro* cytotoxicity of silver nanoparticles on osteoblasts and osteoclasts at antibacterial concentrations. *Nanotoxicology*, **7**: 30-36.

Alizadeh, H., Salouti, M., and Shapouri, R. (2013). Intramacrophage antimicrobial effect of silver nanoparticles against *Brucella melitensis* 16M. *Scientia Iranica*, **20**: 1035-1038

Atiyeh, B. S., Costagliola, M., Hayek, S. N., and Dibo, S. A. (2007). Effect of silver on burn wound infection control and healing: review of the literature. *Burns*, **33**: 139-148.

Baiocco, Paola, Andrea Ilari, Pierpaolo Ceci, Stefania Orsini, Marina Gramiccia, Trentina Di Muccio, and Gianni Colotti. (2010).Inhibitory effect of silver nanoparticles on trypanothione reductase activity and *Leishmania infantum* proliferation. *ACS Medicinal Chemistry Letters*, **2**: 230-233.

Bar, H., Bhui, D. K., Sahoo, G. P., Sarkar, P., Pyne, S., and Misra, A. (2009). Green synthesis of silver nanoparticles using seed extract of *Jatropha curcas*. *Colloids and Surfaces A: Physico-chemical and Engineering Aspects*, **348**: 212-216.

Chernousova, S., and Epple, M. (2013). Silver as antibacterial agent: ion, nanoparticle, and metal. *Angewandte Chemie International Edition*, **52**: 1636-1653.

Cuenca, A. G., Jiang, H., Hochwald, S. N., Delano, M., Cance, W. G., and Grobmyer, S. R. (2006). Emerging implications of nanotechnology on cancer diagnostics and therapeutics. *Cancer*, **107**: 459-466.

Dar Mudasir, A., Avinash Ingle, and Mahendra Rai (2013). Enhanced antimicrobial activity of silver nanoparticles synthesized by *Cryphonectria* sp. evaluated singly and in combination with antibiotics. *Nanomedicine: Nanotechnology, Biology and Medicine*, **9**: 105-110.

Demir, E., Vales, G., Kaya, B., Creus, A., and Marcos, R. (2011). Genotoxic analysis of silver nanoparticles in *Drosophila*. *Nanotoxicology*, **5**: 417-424.

El-Sherbiny, I. M., Salih, E., and Reicha, F. M. (2013). Green synthesis of densely dispersed and stable silver nanoparticles using myrrh extract and evaluation of their antibacterial activity. *Journal of Nanostructure in Chemistry*, **3**: 1-7.

Fayaz, A. M., Balaji, K., Girilal, M., Yadav, R., Kalaichelvan, P. T., and Venketesan, R. (2010). Biogenic synthesis of silver nanoparticles and their synergistic effect with antibiotics: a study against gram-positive and gram-negative bacteria. *Nanomedicine: Nanotechnology, Biology and Medicine*, **6**: 103-109.

Gajbhiye, M., Kesharwani, J., Ingle, A., Gade, A., and Rai, M. (2009). Fungus-mediated synthesis of silver nanoparticles and their activity against pathogenic fungi in combination with fluconazole. *Nanomedicine: Nanotechnology, Biology and Medicine*, **5**: 382-386

Ganachari, S. V., Bhat, R., Deshpande, R., and Venkataraman, A. (2012). Extracellular biosynthesis of silver nanoparticles using fungi *Penicillium diversum* and their antimicrobial activity studies. *BioNanoScience*, **2**: 316-321.

Gardea-Torresdey Jorge, L., Eduardo Gomez Jose, R., Peralta-Videa Jason, G., Parsons Horacio Troiani, and Miguel Jose-Yacaman (2003). Alfalfa sprouts: a natural source for the synthesis of silver nanoparticles. *Langmuir*, **19**: 1357-1361.

Gole, A., Dash, C., Ramakrishnanm, V., Sainkar, S. R., Mandale, A. B., Rao, M., and Sastry, M. (2001). Pepsin-gold colloid conjugates: preparation, characterization, and enzymatic activity. *Langmuir*, **17**: 1674-1679.

Gopinath, V., and Velusamy, P. (2013).Extracellular biosynthesis of silver nanoparticles using *Bacillus* sp. GP-23 and evaluation of their antifungal activity towards *Fusarium oxysporum*. *Spectrochimica Acta Part A: Molecular and Biomolecular Spectroscopy*. **106**: 170-174.

He Li, Hao Wu, Siying Gao, Xuepin Liao, Qiang He, and Bi Shi. (2012).Silver nanoparticles stabilized by tannin grafted collagen fiber: synthesis, characterization and antifungal activity. *Annals of microbiology*, **62**: 319-327.

Huffnagle, G. B., and Noverr, M. C. (2013). The emerging world of the fungal microbiome. *Trends in microbiology*, **21**: 334-341.

Jagtap, U. B., and Bapat, V. A. (2013). Green synthesis of silver nanoparticles using *Artocarpus heterophyllus* Lam. seed extract and its antibacterial activity. *Industrial Crops and Products*, **46**: 132-137.

Jo, Young-Ki, Byung. H. Kim, and Geunhwa Jung (2009). Antifungal activity of silver ions and nanoparticles on phytopathogenic fungi. *Plant Disease*, **10**: 1037-1043.

Kanmani, Paulraj, and Seung Taik Lim. (2013).Synthesis and structural characterization of silver nanoparticles using bacterial exopolysaccharide and its antimicrobial activity against food and multidrug resistant pathogens. *Process Biochemistry*, **48**: 1099-1106.

Khan, M., Khan, M., and Adil, S. F. (2013). Green synthesis of silver nanoparticles mediated by *Pulicaria glutinosa extract[J]. International journal of nanomedicine*, **8**: 1507.

Kim, K. J., Sungm W. S., Moon, S. K., Choi, J. S., Kim, J. G., and Lee, D. G. (2008). Antifungal effect of silver nanoparticles on dermatophytes. *Journal of microbiology and biotechnology*, **18**: 1482-4.

Kirthi, Arivarasan Vishnu, Abdul Abdul Rahuman, Chidambaram Jayaseelan, Loganathan Karthik, Sampath Marimuthu, Thirunavukkarasu Santhoshkumar and Jayachandran Venkatesan (2012). Novel approach to synthesis silver nanoparticles using plant pathogenic fungi, *Puccinia graminis*. *Materials Letters*, **81**: 69-72.

Kittler, S., Greulich, C., Diendorf, J., Koller, M., and Epple, M. (2010). Toxicity of silver nanoparticles increases during storage because of slow dissolution under release of silver ions. *Chemistry of Materials*, **22**: 4548-4554.

Krishnaraj, C., Jagan, E. G., Rajasekar, S., Selvakumar, P., Kalaichelvan, P. T., and Mohan, N. (2010). Synthesis of silver nanoparticles using Acalypha indica leaf extracts and its antibacterial activity against water borne pathogens. *Colloids and Surfaces B: Biointerfaces*, **76**: 50-56.

Kumar, V., and Yadav, S. K. (2009). Plantmediated synthesis of silver and gold nanoparticles and their applications. *Journal of Chemical Technology and Biotechnology*, **84**: 151-157.

Lara, H. H., Ayala-Nuñez, N. V., Ixtepan-Turrent, L., and Rodriguez-Padilla, C. (2010). Mode of antiviral action of silver nanoparticles against HIV-1. *J. Nanobiotechnology*, **8**: 1-8.

Li, Wen-Ru., Xiao-Bao, Xie., Qing-Shan. Shi., Hai-Yan, Zeng., OU-Yang, You-Sheng., and Yi-Ben, Chen. (2010).Antibacterial activity and mechanism of silver nanoparticles on *Escherichia coli*. *Applied microbiology and biotechnology*, **85**: 1115-1122.

Li, Wen-Ru., Xiao-Bao Xie., Qing-Shan Shi., Hai-Yan Zeng., OU-Yang You-Sheng., and Yi-Ben Chen. (2010).Antibacterial activity and mechanism of silver nanoparticles on *Escherichia coli*. *Applied microbiology and biotechnology*, **85**: 1115-1122.

Li, C., Wang, X., Chen, F., Zhang, C., Zhi, X., Wang, K., and Cui, D. (2013). The antifungal activity of graphene oxide–silver nanocomposites. *Biomaterials*, **34**: 3882-3890.

Lu Lei, R. W., Sun, Rong Chen,, Chee-Kin Hui,, Chi-Ming Ho,, John M. Luk,, G. K. Lau, and Chi-Ming Che. (2008). Silver nanoparticles inhibit hepatitis B virus replication. *Antiviral therapy*, **13**: 253

Marimuthu, Sampath, Abdul Abdul Rahuman, Govindasamy Rajakumar, Thirunavukkarasu Santhoshkumar, Arivarasan Vishnu Kirthi, Chidambaram Jayaseelan, Asokan Bagavan, Abdul Abduz Zahir, Gandhi Elango, and Chinnaperumal Kamaraj. (2011). Evaluation of green synthesized silver nanoparticles against parasites. *Parasitology research*, **108**: 1541-1549.

Mehrbod, P. N., Motamed, M., Tabatabaian, R., Soleimani Estyar, E., Amini, M., Shahidi, and M. T. Kheiri (2013). *In vitro* antiviral effect of "nanosilver" on influenza virus. *DARU Journal of Pharmaceutical Sciences*, **17**: 88-93.

Meshram, S. M., Bonde,S. R., Gupta, I. R., Gade, A. K., and Rai, M. K. (2013). Green synthesis of silver nanoparticles using white sugar. *Nanobiotechnology, IET*, **7**: 28-32.

Min, Ji Seon, Kyoung Su Kim, Sang Woo Kim, Jin Hee Jung, Kabir Lamsal, Seung Bin Kim, Mooyoung Jung, and Youn Su Lee. (2009).Effects of colloidal silver nanoparticles on sclerotium-forming phytopathogenic fungi. *Plant Pathol J.*, **25**: 376-38

Monteiro Douglas, R., Sónia Silva, Melyssa Negri, Luiz. F., Gorup Emerson, R., Camargo, Rosário Oliveira, Debora B., Barbosa, and Mariana Henriques.(2013). Antifungal activity of silver nanoparticles in combination with nystatin and *chlorhexidine digluconate* against *Candida albicans* and *Candida glabrata* biofilms. *Mycoses*, 56: 672-680.

Morones, J. R., Elechiguerra, J. L., Camacho, A., Holt, K., Kouri, J. B., Ramírez, J. T., and Yacaman, M. J. (2005). The bactericidal effect of silver nanoparticles. *Nanotechnology*, **16**: 2346.

Nanda, Anima, and M. Saravanan (2009). Biosynthesis of silver nanoparticles from *Staphylococcus aureus* and its antimicrobial activity against MRSA and MRSE. *Nanomedicine: Nanotechnology, Biology and Medicine* **5**: 452-456.

Panácek, A., Kvitek, L., Prucek, R., Kolar, M., Vecerova, R., Pizurova, N., and Zboril, R. (2006). Silver colloid nanoparticles: synthesis, characterization, and their antibacterial activity. *The Journal of Physical Chemistry B*, **110**: 16248-16253.

Ponarulselvam, S., Panneerselvam, C., Murugan, K., Aarthi, N., Kalimuthu, K., and Thangamani, S. (2012). Synthesis of silver nanoparticles using leaves of

Catharanthus roseus Linn. G. Don and their antiplasmodial activities. *Asian Pacific journal of tropical biomedicine*, **2**: 574-580

Prakash, P. P., Gnanaprakasam, R., Emmanuel, S., Arokiyaraj, and M. Saravanan. (2013). Green synthesis of silver nanoparticles from leaf extracts of *Mimusops elengi* Linn. for enhanced antibacterial activity against multi drug resistant clinical isolates. *Colloids and Surfaces B: Biointerfaces,* **108**: 255-259.

Qian, Yongqing, Huimei Yu, Dan He, Hui Yang, Wanting Wang, Xue Wan, and Li Wang. (2013). Biosynthesis of silver nanoparticles by the endophytic fungus Epicoccum nigrum and their activity against pathogenic fungi. *Bioprocess and biosystems engineering,* **36**: 1613-1619.

Rai, M., Alka, Y., and Aniket, G. (2009). Silver nanoparticles as a new generation of antimicrobials. *Biotechnology advances,* **27**: 76-83.

Said, D. E., ElSamad, L. M., and Gohar, Y. M. (2012). Validity of silver, chitosan, and curcumin nanoparticles as anti-giardia agents. *Parasitology research,* **111**: 545-554.

Santhoshkumar, Thirunavukkarasu, Abdul Abdul Rahuman, Asokan Bagavan, Sampath Marimuthu, Chidambaram Jayaseelan, Arivarasan Vishnu Kirthi and Chinnaperumal Kamaraj (2012). Evaluation of stem aqueous extract and synthesized silver nanoparticles using *Cissus quadrangularis* against *Hippobosca maculata* and *Rhipicephalus boophilus* microplus. *Experimental Parasitology,* **132**: 156-165.

Sathishkumar, M., Sneha, K., and Won, S. W(2009). Cinnamon zeylanicum bark extract and powder mediated green synthesis of nano-crystalline silver particles and its bactericidal activity. *Colloids and Surfaces B: Biointerfaces,* **73**: 332-338.

Schreurs, W. J., and Rosenberg, H. (1982). Effect of silver ions on transport and retention of phosphate by *Escherichia coli. Journal of bacteriology,* **152**: 7-13.

Shahverdi Ahmad, R., Ali Fakhimi Hamid, R., Shahverdi, and Sara Minaian. (2007). Synthesis and effect of silver nanoparticles on the antibacterial activity of different antibiotics against *Staphylococcus aureus* and *Escherichia coli. Nanomedicine: Nanotechnology, Biology and Medicine,* **2**: 168-171.

Sharma, Virender, K., Ria, A., Yngard, and Yekaterina, Lin. (2009). Silver nanoparticles: green synthesis and their antimicrobial activities. *Advances in Colloid and Interface Science,* **145**: 83-96.

Singh, A., Jain, D., Upadhyay, M. K., Khandelwal, N., and Verma, H. N. (2010). Green synthesis of silver nanoparticles using *Argemone mexicana* leaf extract and evaluation of their antimicrobial activities. *Dig J Nanomater Bios.,* **5**: 483-489.

Singh, D., Rathod, V., Ninganagouda, S., Herimath, J., and Kulkarni, P. (2013). Biosynthesis of silver nanoparticle by endophytic fungi *Pencillium* sp. isolated from *Curcuma longa* (turmeric) and its antibacterial activity against pathogenic gram negative bacteria. *Journal of Pharmacy Research,* **7**: 448-453.

Song, Jae Yong, and Beom Soo Kim.(2009). Rapid biological synthesis of silver nanoparticles using plant leaf extracts. *Bioprocess and biosystems engineering*, **32**: 79-84.

Speshock, J. L., Murdock, R. C., Braydich-Stolle, L. K., Schrand, A. M., and Hussain, S. M. (2010). Interaction of silver nanoparticles with Tacaribe virus. *Journal of nanobiotechnology*, **8**: 19-27.

Tamboli Dhawal, P., and Dae Sung Lee. (2013). Mechanistic antimicrobial approach of extracellularly-synthesized silver nanoparticles against Gram positive and Gram negative bacteria. *Journal of Hazardous Materials*, **260**: 878-884.

Thenmozhi, M., Kannabiran, K., Kumar, R., and Gopiesh Khanna. V. (2013). Antifungal activity of *Streptomyces* sp. VITSTK7 and its synthesized Ag 2 O/Ag nanoparticles against medically important Aspergillus pathogens. *Journal de Mycologie Médicale/Journal of Medical Mycology*, **23**: 97-103.

Velmurugan Natarajan, G., Gnana Kumar, Sang Sub Han, S. K. Nahm, and S. Y. Lee. (2009). Synthesis and characterization of potential fungicidal silver nano-sized particles and chitosan membrane containing silver particles. *Iranian polymer Journal*, **18**: 383-392.

Xiang, Dong-xi, Qian Chen, Lin Pang, and Cong-long Zheng. (2011). Inhibitory effects of silver nanoparticles on H1N1 influenza A virus *in vitro*. *Journal of virological methods*, **178**: 137-142.

Yan, Ning, Chaoxian Xiao, and Yuan Kou. (2010).Transition metal nanoparticle catalysis in green solvents. *Coordination Chemistry Reviews*, **254**: 1179-1218.

Yan, J. K., Cai, P. F., Cao, X. Q., Ma, H. L., Zhang, Q., Hu, N. Z., and Zhao, Y. Z. (2013). Green synthesis of silver nanoparticles using 4-acetamido-TEMPO-oxidized curdlan. *Carbohydrate Polymers*.

Yudha, S. S., Notriawan, D., and Angasa, E. (2013). Green Synthesis of Silver Nanoparticles Using Aqueous Rinds Extract of *Brucea javanica* (L.) Merr at Ambient Temperature. *Materials Letters*, **97**: 181-183.

Zahir, A. Abduz, and A. Abdul Rahuman. (2012). Evaluation of different extracts and synthesised silver nanoparticles from leaves of *Euphorbia* prostrata against *Haemaphysalis bispinosa* and *Hippobosca maculata. Veterinary Parasitology*, **187**: 511-520.

Zhang, Hongyin, and Vinka Oyanedel-Craver. (2013). Comparison of the bacterial removal performance of silver nanoparticles and a polymer based quaternary amine functiaonalized silsesquioxane coated point-of-use ceramic water filters. *Journal of Hazardous Materials*, **260**: 272-277.

Zodrow, Katherine, Léna Brunet, Shaily Mahendra, Dong Li, Anna Zhang, Qilin Li, and Pedro JJ Alvarez. (2009). Polysulfone ultrafiltration membranes impregnated with silver nanoparticles show improved biofouling resistance and virus removal. *Water research*, **43**: 715-723.

Natural Products: Research Reviews Vol. 3 (2016) *Pages* **161–176**
Editor: **V.K. Gupta**
Published by: **DAYA PUBLISHING HOUSE, NEW DELHI**

7

Botanical Based Complementary and Alternative Treatments for Attention Deficit Hyperactivity Disorder (ADHD): A Retrospective

Abhijit Dey[1], Anuradha Mukherjee[2] and (Late) Jitendranath De[3*]

ABSTRACT

Attention deficit hyperactivity disorder (ADHD) is a common disorder of 3 per cent to 6 per cent of school-going children. Conventional treatment includes stimulant medications for symptomatic relief but these are associated with non-responsiveness and side effects. Complementary or alternative medications (CAM) are often preferred over conventional treatment and herbal medicine serves as a popular CAM. Extracts of Panax quinquefolius, Ginkgo biloba, Passiflora incarnata and omega-3 fatty acid supplementation etc. have been investigated against ADHD. However, further research is needed to measure the effectiveness and reliability of these alternatives. In the present review, the authors have tried to explore the possible role of herbal extracts, singly or in combination and the effects of bioactive molecules against ADHD in children as well as in adults. We have also added a note on possible use of other neuroprotective extracts and phytochemicals against ADHD which are otherwise used

1 Assistant Professor, UG and PG Department of Biological Sciences, Presidency University (Erstwhile Presidency College), 86/1 College Street, Kolkata – 700 073, India.

2 Assistant Teacher, Biology, NMHS, South 24 Paraganas, West Bengal, India.

3 Retired Reader and HOD, Department of Botany, Charuchandra College, 22, Lake Road, Kolkata – 700 029, West Bengal, India.
Present address: Flat no. C-103, Sunrise Estate, 14C Radhanath Chaudhuri Road, Kolkata – 700 015, West Bengal, India.

* *Corresponding author*: E-mail: jnde@rocketmail.com

against a number of neurological and psychological disorders such as Alzheimer's disease, Parkinson's disease, Huntington's disease, schizophrenia, borderline personality disorder etc.

Keywords: Attention deficit hyperactivity disorder (ADHD), Botanicals, Complementary or alternative medications (CAM), Parkinson's disease.

Introduction

Attention-Deficit/Hyperactivity Disorder (ADHD) is a commonly encountered childhood brain or psychiatric disorder of the neurodevelopmental type which may continue through adolescence and adulthood. The symptoms include lack of staying focused and paying attention, emotional self-control and motivation, hyperactivity, or acting impulsively (Ramos-Quiroga *et al.*, 2014). The symptoms begin at the age range of six to twelve and inattention and inability to focus in school going children suffering from ADHD results in poor academic performance. Upto 1 in 20 children in the United States is affected by ADHD (Pellow *et al.*, 2011). In about 60 per cent of children with ADHD, the symptoms continue into adulthood. The occurance of adult ADHD in the general population has been reported to be between 2-5 per cent (Ramos-Quiroga *et al.*, 2014). ADHD has been correlated with risky sexual behavior (RSB) (Sarver *et al.*, 2014), psychoactive substance use disorder, nicotine dependence in adolescence and comorbid conduct disorder (CD) (Groenman *et al.*, 2013) in adolescents and adults. Adults with ADHD have also exhibited increased subjective daytime sleepiness (Sobanski *et al.*, 2014). Further studies were indicated to elucidate the remaining queries related to adult ADHD (Ramos-Quiroga *et al.*, 2014).

Pathophysiology

The underlying pathogenesis for ADHD is considered as diverse which include neurological, biochemical and genetic factors. Risk factors include nutritional deficiencies (such as omega-3 fatty acid deficiency, amino acid deficiency etc.), exposure to environmental toxic substances (such as heavy metals, pesticides etc.), electronic media (such as mobile phones and television) and culture (Pellow *et al.*, 2011; Millichap and Yee, 2012; Byun *et al.*, 2013; Kim *et al.*, 2013; Marinelli *et al.*, 2014). Genetic variations in fat metabolism influenced by culture and practice, (Field, 2014), sex-specific association of brain-derived neurotrophic factor (BDNF) (Li *et al.*, 2014), monoamine oxidase A (MAOA) gene variants' expression (Karmakar *et al.*, 2014), role of dopaminergic gene variants (Maitra *et al.*, 2014) etc. are considered among other factors affecting disease etiology. The pathophysiology of co-occurrence of autistic spectrum disorder (ASD) and ADHD might also open some new highlights into the therapeutic interventions of the diseases (Leitner, 2014).

Conventional Treatment

Behavioral counseling, medications, individual and family counseling are the different conventional treatments for ADHD. Drugs used to treat ADHD are mostly the stimulus to increase the concentration of neurotransmitters and to reduce hyperactivity and impulsiveness. Stimulant drugs include methylphenidate and dextroamphetamine/amphetamine. Non-stimulant drugs include atomoxetine which is a selective norepinephrine reuptake inhibitor. A third class of drugs known as antidepressants affecting neurotransmitters. All of these are with their own adverse effects which include loss of appetite, nausea, dry mouth, headaches, insomnia, depression, mood swings, anxiety, irritability, fatigue, cough, jaundice, stomachaches, increased blood pressure, erectile dysfunction etc.

Alternative Treatment

Complementary or alternative medications (CAM) are often preferred over conventional treatment due to their lesser side effects and cheap price and herbal medicine serves as a popular CAM. Its holistic approach to treat patients individually based on the pathogenesis has popularized CAM even in the treatment of ADHD (Chang *et al.*, 2007; Pellow *et al.*, 2011). Diet, elimination diets, nutritional supplements (especially essential fatty acid supplementation), exercise, yoga, massage, homeopathy and botanicals have been tested against ADHD (Rojas and Chan, 2005; Sawni, 2008; Pellow *et al.*, 2011). Both single and multi-ingredient formulas have been reported effective in a number of studies and combination therapies were found to be useful against cognitive and behavioral problems (Rucklidge *et al.*, 2009). However, the safety and efficacy of CAM treatments especially for children need to be further investigated (Weber and Newmark, 2007). Effectiveness of CAM as single therapies needs to be elucidated to eliminate the controversies of using CAM against ADHD (Rojas and Chan, 2005). Moreover, conventional treatments against ADHD were suggested preferable than using unsuccessful CAM treatments (Brue and Oakland, 2002). However, CAM treatments were justified with some scientific evidences (Swani, 2008) and further investigations in this regard might be needed to establish its efficacy also considering its safety of using in children.

Herbal Treatment

Plants are popularly used in different parts of the globe as an alternative treatment to alleviate diseases (Dey and De, 2012a, b). Herbs and herb derived compounds were found to be effective in numerous *in vitro, in vivo* and clinical investigations. Medicinal plants were also reported effective in a number of psychological and neurological disorders with their own and unique underlying mechanism of pathogenesis (Dey and De, 2015). Scientific rationale behind using such herbal therapy from age old traditional medicine to the modern day botanical based alternative medicines has been elucidated in various scientific evidences. In the following section, a number of plants are tabulated with their scientific names, author citations with their mode of action as found in different clinical trials conducted among children, adolescents and adults suffering from ADHD (Table 6.1). The nomenclature was verified from the website http://www.tropicos.org/and the data was extracted from the leading scientific databases *viz.* Pubmed, Google Scholar, Sciencedirect, Scopus, Springer, Jstor etc. Figure 6.1 represents the chemical structures (www.chem-spider.com) of some neuroprotective molecules isolated from the botanicals discussed in the text.

Table 7.1: A List of Anti-ADHD Medicinal Plants with their Mode of Action

Plant	*Family*	*Sample*	*Mode of Action*	*Reference*
Hypericum perforatum L.	Hypericaceae	14-16-year-old 3 male psychiatric outpatients with ADHD	↑Score for Conners' hyperactivity, !inattention, !immaturity factors	Niederhofer *et al.*, 2010
		54 children aged 6 to 17 years with ADHD	No improvement of symptoms	Weber *et al.*, 2008
Matricaria chamomilla L.	Asteraceae	14-16-year-old 3 male psychiatric outpatients with ADHD	↑score for Conners' hyperactivity, ↑inattention, ↑immaturity factors	Niederhofer *et al.*, 2009
Passiflora incarnata L.	Passifloraceae	34 children with ADHD	↑Significant clinical benefit	Akhondzadeh *et al.*, 2005
Bacopa monnieri (L.) Pennell	Plantaginaceae	Childrenwith ADHD	↑Memory and learning	Negi *et al.*, 2000
		31 children with 6-12 years of age with ADHD	↑Restlessness, ↑self-control, ↑attention-deficit symptoms, ↑learning problems, ↑impulsivity, ↑psychiatric problems	Dave *et al.*, 2014
Pinus pinaster Aiton bark extract (Pycnogenol®)	Pinaceae	61 children with ADHD	↑Hyperactivity, ↑attention, ↑visual-motoric coordination, ↑concentration	Trebatická *et al.*, 2006
		Children with ADHD	↑DNA damage, ↑total antioxidant status (TAS), ↑attention	Chovanová *et al.*, 2006
		Children with ADHD	↓Oxidized (GSSG) glutathione, ↑reduced (GSH) glutathione, ↑TAS	Dvoráková *et al.*, 2006
		24 adults (24 to 53 years old) with ADHD	↑ADHD symptoms	Tenenbaum *et al.*, 2002
		ADHD children	Normalize catecholamine concentrations, ↑hyperactivity, ↓oxidative stress	Dvoráková *et al.*, 2007

Contd...

Table 7.1–*Contd...*

Plant	*Family*	*Sample*	*Mode of Action*	*Reference*
Ginkgo biloba L.	Ginkgoaceae	Children and adolescents with ADHD	↑Parent and Teacher Rating Scale scores	Salehi *et al.*, 2010
		6 psychiatric outpatients diagnosed with ADD	↑Hyperactivity, ↑inattention, ↑immaturity factors	Niederhofer, 2010
Panax ginseng C.A. Mey.	Araliaceae	14-17-year-old 3 male psychiatric outpatients with ADHD	↑Score for Conners' ↑hyperactivity, ↑inattention, ↑immaturity factors	Niederhofer, 2009
		18 subjects aged between 6 and 14 diagnosed with ADHD	↑inattentiveness	Lee *et al.*, 2011
Panax quinquefolius L. and *G. biloba* L.	Araliaceae and Ginkgoaceae	36 children from 3 to 17 years with ADHD	↑Anxious-shy attribute, ↑psychosomatic attribute, ↑social problems attribute, ↑Conners' ADHD index, ↑DSM-IV hyperactive-impulsive attribute	Lyon *et al.*, 2001

↑:improved; ↓: Reduced

Hypericum perforatum

H. perforatum (St John's wort) are popularly used in the treatment of depression. By using forced-swimming and tail-suspension methods in mice, the antidepressant activity of the plant was noted (Bach-Rojecky *et al.*, 2004). Synergistic antidepressant effect in mice was also noted when quercetin and *H. perforatum* were applied together in a particular ratio (Liu *et al.*, 2013). However, in a double-blind, randomized, placebo-controlled trial among adult outpatients (n = 340), the plant was found to be inefficacious against moderately severe major depression (Hypericum Depression Trial Study Group, 2002). In another study conducted among 54 children with ADHD aged 6 to 17 years, no improvement of symptoms were noted following *H. perforatum* treatment (Weber *et al.*, 2008). However, contradictory results were reported with improved score for Conners' hyperactivity, inattention and immaturity factors in the tested samples of 14-16-year-old three male psychiatric outpatients with ADHD (Niederhofer *et al.*, 2010).

Matricaria chamomilla

Another species, *M. recutita* showed antidepressant activity (Amsterdam *et al.*, 2012) and anxiolytic properties (Amsterdam *et al.*, 2009). In 14-16-year-old three male psychiatric outpatients with ADHD, use of the *M. chamomilla* improved score for Conners' hyperactivity, inattention and immaturity factors (Niederhofer *et al.*, 2009).

Passiflora incarnata

Aerial parts from *P. incarnata* are popularly used as herbal sedatives (Krenn, 2002). In a double-blind, placebo-controlled trial, administration of preoperative oral *P. incarnata* was found to decrease anxiety in ambulatory surgery patients (Movafegh *et al.*, 2008). Anti-anxiety effect of the plant was also depicted earlier where the bioactivity was attributed to some possible compound with benzoflavone nucleus (Dhawan *et al.*, 2001). *P. incarnata* herbal tea was also found to provide short-term subjective sleep benefit in human subjects (Ngan and Conduit, 2011). In a clinical trial among 34 children with ADHD, significant clinical benefit was noted after applying the herb (Akhondzadeh *et al.*, 2005).

Bacopa monniera

Clinical and experimental studies have shown that the plant is efficacious in cognition and memory enhancement (Stough *et al.*, 2001; Roodenrys *et al.*, 2002). In a recent study, standardized extract of *B. monnieri* was found to ameliorate ADHD symptoms in thirty-one children of 6-12 years of age and the medication was well tolerated by the patients (Dave *et al.*, 2014). Earlier, *B monnieri* was found to improve memory and learning in ADHD affected children (Negi *et al.*, 2000).

Pinus pinaster

Pycnogenol® (extract of French maritime pine *Pinus pinaster* bark) contains 70 per cent procyanidin which is a strong antioxidant (Schoonees *et al.*, 2012). It also contains bioflavonoids, catechins, and phenolic acids and acts as a chelating agent and stimulators of a number of enzymes (Dvoráková *et al.*, 2006). In a study conducted among 61 children with ADHD, the extract improved hyperactivity, attention, visual-motoric coordination and concentration in affected children (Trebatická *et al.*, 2006). In ADHD children, it also improved DNA damage and total antioxidant status (TAS) (Chovanová *et al.*, 2006) and decreased oxidized (GSSG) glutathione and increased reduced (GSH) glutathione (Dvoráková *et al.*, 2006). In another double-blind, placebo-controlled, crossover study among 24 adults (24 to 53 years old) with ADHD, the anti-ADHA effect of Pycnogenol® was compared with the effect of methylphenidate and the former was found to improve ADHD symptoms in adults (Tenenbaum *et al.*, 2002). It normalized catecholamine concentrations, improved hyperactivity and reduced oxidative stress in ADHD children were detected by measuring urine catecholamine concentrations (Dvoráková *et al.*, 2007).

Ginkgo biloba

G. biloba was reported to improve memory (Sakatani *et al.*, 2014) and against ischemia, epilepsy, and peripheral nerve damage due to its antioxidant and vasoactive properties (Brondino *et al.*, 2013). The plant was also found to provide cognitive benefits in Alzheimer's disease patients under cholinesterase inhibitors treatment (Canevelli *et al.*, 2014). *G. biloba* improved Parent and Teacher Rating Scale scores in children and adolescents with ADHD (Salehi *et al.*, 2009). In six psychiatric outpatients diagnosed with attention-deficit disorder (ADD), use of the plant improved hyperactivity, inattention, and immaturity factors (Niederhofer, 2010).

Panax ginseng and *Panax quinquefolius*

Ginseng and ginsenosides are known herbal medicines against a number of CNS (central nervous system) disorders such as Alzheimer's disease (AD), Parkinson's disease (PD), cerebral ischemia, depression etc. (Kim *et al.*, 2013). Ginsenoside Rg1 exhibited antidepressant properties in mice which were attributed to its ability to induce BDNF (brain-derived neurotrophic factor) signaling and hippocampal neurogenesis regulation (Jiang *et al.*, 2012). *Panax quinquefolium* saponins showed protective effects against ischemic injury in rats possibly via antioxidation and anti-apoptosis (Xu *et al.*, 2014). In three 14-17-year-old male psychiatric outpatients with ADHD, the plant improved score for Conners' hyperactivity, inattention and immaturity factors (Niederhofer, 2009). Moreover, in eighteen subjects aged between 6 and 14 diagnosed with ADHD, it improved inattentiveness (Lee *et al.*, 2011). *P. quinquefolium* and *G. biloba* improved anxious-shy attribute, psychosomatic attribute, social problems attribute, Conners' ADHD index and the DSM-IV hyperactive-impulsive attribute in 36 children ranging in age from 3 to 17 years with ADHD (Lyon *et al.*, 2001).

Discussion

The aforesaid plant extracts or preparations either singly or in combination improved the ADHD symptoms in children as well as adults with a few contradictory results. Overall, the botanicals were found to improve score for Conners' hyperactivity, inattention, immaturity factors, memory and learning, restlessness, self-control, attention-deficit symptoms, impulsivity, psychiatric problems, visual-motoric coordination, concentration, DNA damage, total antioxidant status (TAS), reduced (GSH) glutathione, Parent and Teacher Rating Scale scores, anxious-shy attribute,

Figure 7.1(a-f): Neuroprotective Phytochemicals from the Anti-ADHD Botanicals.

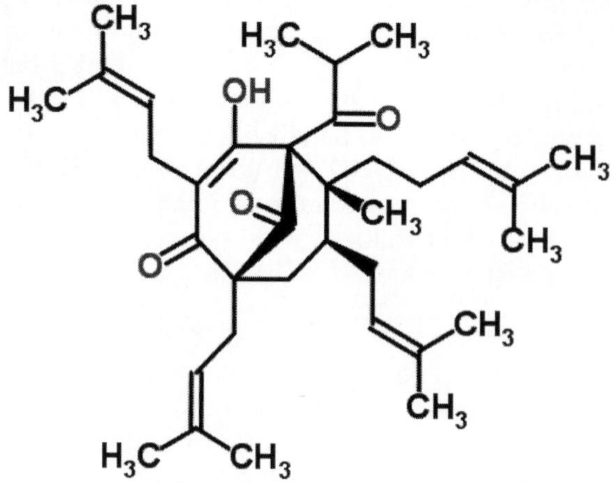

a. Hyperforin

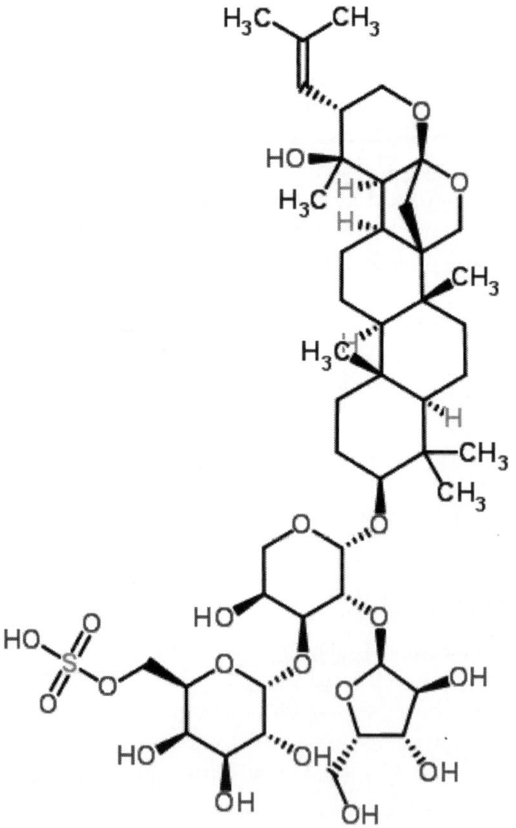

b. Bacopaside I

Contd...

Figure 7.1–*Contd...*

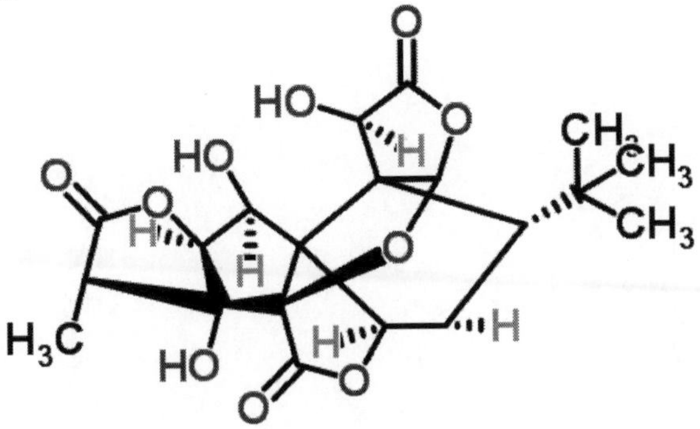

c. Ginkgolide B

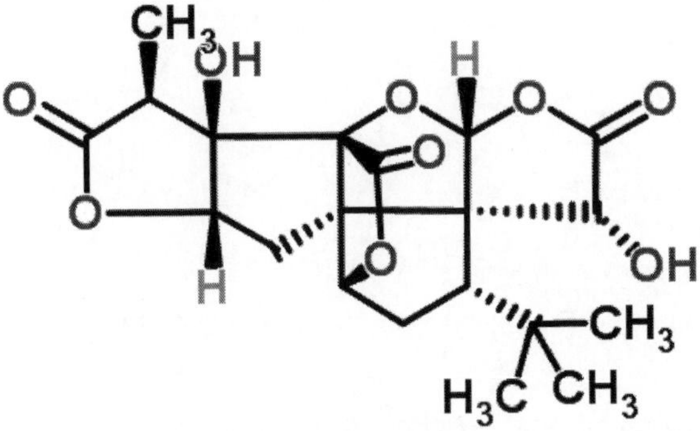

d. Ginkgolide A

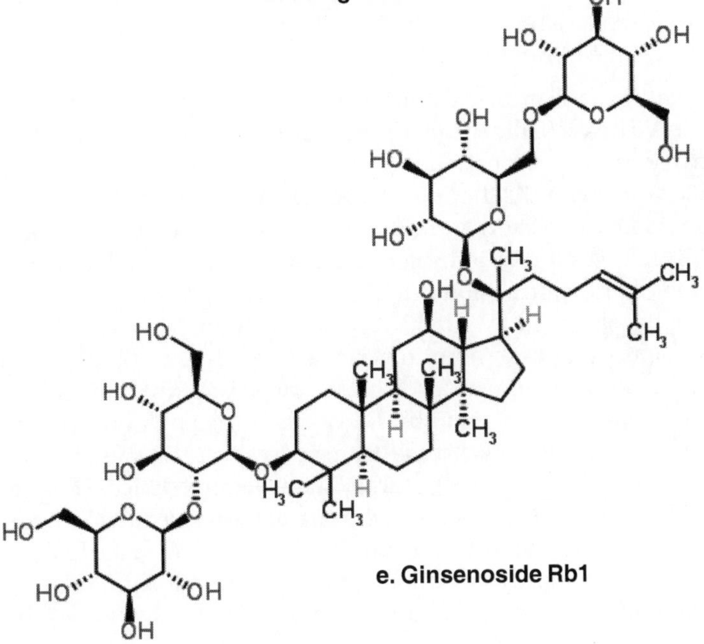

e. Ginsenoside Rb1

Contd...

Figure 7.1–*Contd...*

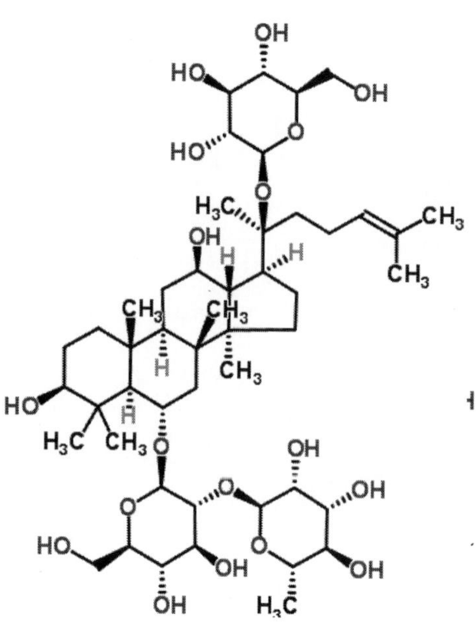

f. Ginsenoside Re

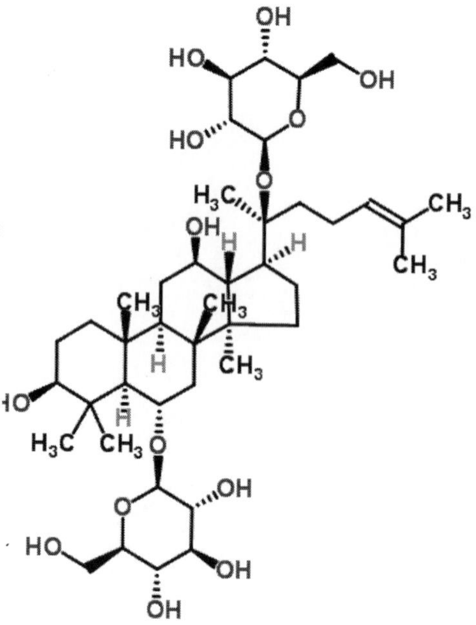

f. Ginsenoside Rg1

psychosomatic attribute, social problems attribute, Conners' ADHD index, DSM-IV hyperactive-impulsive attribute etc. Some of the herbal medications were found to decrease oxidized (GSSG) glutathione and oxidative stress and normalize catecholamine concentrations.

H. perforatum standardized extracts exhibited neuroprotection in rotenone model of Parkinson's disease (Gómez del Rio *et al.*, 2013). An extract of the plant with final hyperforin (Figure 7.1a) concentration 5 per cent decreased beta-amyloid accumulation in a double transgenic Alzheimer's disease mouse model (Brenn *et al.*, 2014). Hyperforin is known to exert antidepressive properties on brain cells (Bouron and Lorrain, 2014). It was found to protect against brain damage induced by transient middle cerebral artery occlusion (MCAO) in rats (Lin *et al.*, 2013). Antidepressant-like effect of bacopaside I (Figure 7.1b) isolated from *B. monnieri* was tentatively attributed to antioxidant and noradrenergic activation (Liu *et al.*, 2013a). Bacopaside I has been reported to offer neuroprotection against ischemic brain injury in rats (Liu *et al.*, 2013b). Effects of *G. biloba* have been investigated against a number of neurological disorders such as Alzheimer's disease (Canevelli *et al.*, 2014) and Parkinson's disease (Dey and De, 2015) and also acts on learning and memory deficits (Raghu *et al.*, 2013). Ginkgolide B (Figure 7.1c), isolated from the plant has exhibited neuroprotective properties on bupivacaine-induced apoptosis in SH-SY5Y cells (Li *et al.*, 2013). The compound was also found to ameliorate migraine aura in human samples with no serious side effects (Allais *et al.*, 2013). Ginkgolide A (Figure 7.1d) has shown anxiolytic-like effect in mice (Kuribara *et al.*, 2003). Ginsenosides (Figures 7.1e,f) isolated from various *Panax* sp. are known to possess beneficial effects on the central nervous system (CNS) (Radad *et al.*, 2005; 2011). Ginsenoside Rb1, Rg1, Re, Rh1 and Rg2 have been tested in different PD models with positive results (Dey and De, 2015). Ginsenosides Rb1 and Rg1 have exhibited *in vitro* neuroprotection on spinal cord neurons (Liao *et al.*, 2002) and exerted neurotrophic and neuroprotective roles on dopaminergic neurons (Radad *et al.*, 2004). Ginsenoside Rd showed neuroprotection for acute ischemic stroke (Ye *et al.*, 2013).

Concluding Remarks

Although a number of herbal medicines are used worldwide in the treatment of various neurological disorders including ADHD, one of the major constraints in anti-ADHD CAM research is use of children and corresponding efficacy and safety issues. Moreover, the present review discusses the role of few phytochemicals as neuroprotective agents which have not been tested on ADHD patients. Unlike AD, PD and Huntington's disease (HD), no *in vitro* and/or *in vivo* model and neurotoxin have been designed which can mimic at least some of the pathophysiological features of ADHD. As anti-ADHD treatment mostly rely on human sample especially children, rigorous preclinical and clinical trials are needed in anti-ADHD drug discovery program.

References

Akhondzadeh, S., Mohammadi, M. R., and Momeni, F. (2005). *Passiflora incarnata* in the treatment of attention-deficit hyperactivity disorder in children and adolescents. *Therapy*, **2(4)**: 609-614.

Allais, G., D'Andrea, G., Maggio, M., and Benedetto, C. (2013). The efficacy of ginkgolide B in the acute treatment of migraine aura: an open preliminary trial. *Neurological Sciences*, **34(1)**: 161-163.

Amsterdam, J. D., Li, Y., Soeller, I., Rockwell, K., Mao, J. J., and Shults, J. (2009). A randomized, double-blind, placebo-controlled trial of oral *Matricaria recutita* (chamomile) extract therapy of generalized anxiety disorder. *Journal of clinical psychopharmacology*, **29(4)**: 378-382.

Amsterdam, J. D., Shults, J., Soeller, I., Mao, J. J., Rockwell, K., and Newberg, A. B. (2012). Chamomile (*Matricaria recutita*) may have antidepressant activity in anxious depressed humans-An exploratory study. *Alternative therapies in health and medicine*, **18(5)**: 44.

Bach-Rojecky, L., Kalodjera, Z., and Samarzija, I. (2004). The antidepressant activity of *Hypericum perforatum* L. measured by two experimental methods on mice. *Acta Pharmaceutica-Zagreb.*, **54(2)**: 157-162.

Bouron, A., and Lorrain, E. (2013). Cellular and molecular effects of the antidepressant hyperforin on brain cells: Review of the literature. *L'Encephale.*, **40(2)**: 108-113.

Brenn, A., Grube, M., Jedlitschky, G., Fischer, A., Strohmeier, B., Eiden, M., and Vogelgesang, S. (2014). St. John's Wort reduces betaamyloid accumulation in a double transgenic Alzheimer's disease mouse model–Role of PGlycoprotein. *Brain Pathology*, **24(1)**: 18-24.

Brue, A. W., and Oakland, T. D. (2001). Alternative treatments for attention-deficit/ hyperactivity disorder: does evidence support their use? *Alternative Therapies in Health and Medicine*, **8(1)**: 68-70.

Byun, Y. H., Ha, M., Kwon, H. J., Hong, Y. C., Leem, J. H., Sakong, J., and Kim, N. (2013). Mobile phone use, blood lead levels, and attention deficit hyperactivity symptoms in children: a longitudinal study. *PloS one*, **8(3)**: e59742.

Canevelli, M., Adali, N., Kelaiditi, E., Cantet, C., Ousset, P. J., and Cesari, M. (2014). Effects of *Gingko biloba* supplementation in Alzheimer's disease patients receiving cholinesterase inhibitors: Data from the ICTUS study. *Phytomedicine*, **21(6)**: 888-892.

Chang, H. H., Chang, C. S., and Shih, Y. L. (2007). The process of assisting behavior modification in a child with attention-deficit hyperactivity disorder. *Journal of Nursing Research*, **15(2)**: 147-155.

Chovanová, Z., Muchová, J., Sivonova, M., Dvorakova, M., Zitnanova, I., Waczulíková, I. Trebatická, J., Skodácek,I., and Ïuracková, Z. (2006). Effect of polyphenolic extract, Pycnogenol®, on the level of 8-oxoguanine in children suffering from attention deficit/hyperactivity disorder. *Free radical research*, **40(9)**: 1003-1010.

Dave, U. P., Dingankar, S. R., Saxena, V. S., Joseph, J. A., Bethapudi, B., Agarwal, A., and Kudiganti, V. (2013). An open-label study to elucidate the effects of standardized *Bacopa monnieri* extract in the management of symptoms of attention-deficit hyperactivity disorder in children. *Advances in mind-body medicine*, **28(2)**: 10-15.

Dey, A., and De, J. N. (2015). Possible anti-Parkinson's disease therapeutics from nature-A review. *In*: Studies in Natural Product Chemistry, Vol. 44, Ed. By Atta-ur-Rahman, Elsevier Science, Amsterdam, The Netherlands, pp. 447-520.

Dey, A. and De, J. N. (2012). Anti snake venom botanicals used by the ethnic groups of Purulia District, West Bengal, India. *Journal of Herbs, Spices and Medicinal Plants*, **18(2)**: 152-165.

Dey, A. and De, J. N. (2012). Ethnomedicinal plants used by the tribals of Purulia district, West Bengal, India against gastrointestinal disorders. *Journal of Ethnopharmacology*, **143(1)**: 68-80.

Dhawan, K., Kumar, S., and Sharma, A. (2001). Anti-anxiety studies on extracts of *Passiflora incarnata* Linneaus. *Journal of Ethnopharmacology*, **78(2)**: 165-170.

Dvoøáková, M., Ježová, D., Blažíèek, P., Trebatická, J., Škodáèek, I., Šuba, J., and Ïuraèková, Z. (2007). Urinary catecholamines in children with attention deficit hyperactivity disorder (ADHD): modulation by a polyphenolic extract from pine bark (Pycnogenol®). *Nutritional neuroscience*, **10(3-4)**: 151-157.

Dvoøáková, M., Sivoòová, M., Trebatická, J., Škodáèek, I., Waczuliková, I., Muchová, J., and Ïuraèková, Z. (2006). The effect of polyphenolic extract from pine bark,

Pycnogenol® on the level of glutathione in children suffering from attention deficit hyperactivity disorder (ADHD). *Redox Report*, **11(4)**: 163-172.

Field, S. S. (2014). Interaction of genes and nutritional factors in the etiology of autism and attention deficit/hyperactivity disorders: A case control study. *Medical hypotheses*, **82(6)**: 654-661.

Gomez del Rio, M. A., Sanchez-Reus, M. I., Iglesias, I., Pozo, M. A., Garcia-Arencibia, M., Fernandez-Ruiz, J., and Benedi, J. (2013). Neuroprotective properties of standardized extracts of *Hypericum perforatum* on rotenone model of Parkinsons disease. *CNS and Neurological Disorders-Drug Targets (Formerly Current Drug Targets-CNS and Neurological Disorders)*, **12(5)**: 665-679.

Groenman, A. P., Oosterlaan, J., Rommelse, N., Franke, B., Roeyers, H., Oades, R. D., Sergeant, J. A., Buitelaar, J. K., and Faraone, S. V. (2013). Substance use disorders in adolescents with attention deficit hyperactivity disorder: a 4year followup study. *Addiction*, **108(8)**: 1503-1511.

Hypericum Depression Trial Study Group. (2002). Effect of *Hypericum perforatum* (St John's wort) in major depressive disorder: a randomized controlled trial. *Jama*, **287(14)**: 1807-1814.

Jiang, B., Xiong, Z., Yang, J., Wang, W., Wang, Y., Hu, Z. L., and Chen, J. G. (2012). Antidepressantlike effects of ginsenoside Rg1 are due to activation of the BDNF signalling pathway and neurogenesis in the hippocampus. *British journal of pharmacology*, **166(6)**: 1872-1887.

Karmakar, A., Maitra, S., Verma, D., Chakraborti, B., Goswami, R., Ghosh, P., and Mukhopadhyay, K. (2014). Potential contribution of monoamine oxidase a gene variants in ADHD and behavioral co-morbidities: scenario in eastern Indian probands. *Neurochemical research*, **39(5)**: 843-852.

Kim, H. J., Kim, P., and Shin, C. Y. (2013). A comprehensive review of the therapeutic and pharmacological effects of ginseng and ginsenosides in central nervous system. *Journal of ginseng research*, **37(1)**: 8-29.

Kim, S., Arora, M., Fernandez, C., Landero, J., Caruso, J., and Chen, A. (2013). Lead, mercury, and cadmium exposure and attention deficit hyperactivity disorder in children. *Environmental research*, **126**: 105-110.

Krenn, L. (2001). Passion Flower (*Passiflora incarnata* L.)–a reliable herbal sedative. *Wiener medizinische Wochenschrift (1946)*, **152(15-16)**: 404-406.

Kuribara, H., Weintraub, S. T., Yoshihama, T., and Maruyama, Y. (2003). An anxiolytic-like effect of *Ginkgo biloba* extract and its constituent, ginkgolide-A, in mice. *Journal of natural products*, **66(10)**: 1333-1337.

Lee, S. H., Park, W. S., and Lim, M. H. (2011). Clinical effects of Korean red ginseng on attention deficit hyperactivity disorder in children: An observational study. *Journal of ginseng research*, **35(2)**: 226.

Leitner, Y. (2014). The co-occurrence of autism and attention deficit hyperactivity disorder in children–what do we know?. *Frontiers in human neuroscience*, **8**: 268.

Li, H., Liu, L., Tang, Y., Ji, N., Yang, L., Qian, Q., and Wang, Y. (2014). Sex-specific association of brain-derived neurotrophic factor (BDNF) Val66Met polymorphism and plasma BDNF with attention-deficit/hyperactivity disorder in a drug-naïve Han Chinese sample. *Psychiatry research*, **217(3)**: 191-197.

Li, L., Zhang, Q. G., Lai, L. Y., Wen, X. J., Zheng, T., Cheung, C. W., and Xu, S. Y. (2013). Neuroprotective effect of ginkgolide B on bupivacaine-induced apoptosis in SH-SY5Y cells. *Oxidative medicine and cellular longevity*, **2013**: 159864.

Liao, B., Newmark, H., and Zhou, R. (2002). Neuroprotective effects of ginseng total saponin and ginsenosides Rb1 and Rg1 on spinal cord neurons *in vitro*. *Experimental neurology*, **173(2)**: 224-234.

Lin, Y., Zhang, J. C., Fu, J., Chen, F., Wang, J., Wu, Z. L., and Yuan, S. Y. (2013). Hyperforin attenuates brain damage induced by transient middle cerebral artery occlusion (MCAO) in rats via inhibition of TRPC6 channels degradation. *Journal of Cerebral Blood Flow and Metabolism*, **33(2)**: 253-262.

Liu, J. X., Fang, Y. Q., Wei, Z. X., Yang, X. Q., and Zeng, L. H. (2013). Synergic antidepressive effect of quercetin and *Hypericum perforatum* extract in mice.*Zhejiang da xue xue bao. Yi xue ban= Journal of Zhejiang University. Medical sciences*, **42(6)**: 615-619.

Liu, X., Liu, F., Yue, R., Li, Y., Zhang, J., Wang, S., and Zhang, W. (2013). The antidepressant-like effect of bacopaside I: Possible involvement of the oxidative stress system and the noradrenergic system. *Pharmacology Biochemistry and Behavior*, **110**: 224-230.

Liu, X., Yue, R., Zhang, J., Shan, L., Wang, R., and Zhang, W. (2013). Neuroprotective effects of bacopaside I in ischemic brain injury. *Restorative neurology and neuroscience*, **31(2)**: 109-123.

Lyon, M. R., Cline, J. C., de Zepetnek, J. T., Shan, J. J., Pang, P., and Benishin, C. (2001). Effect of the herbal extract combination *Panax quinquefolium* and *Ginkgo biloba* on attention-deficit hyperactivity disorder: a pilot study. *Journal of Psychiatry and Neuroscience*, **26(3)**: 221.

Maitra, S., Sarkar, K., Ghosh, P., Karmakar, A., Bhattacharjee, A., Sinha, S., and Mukhopadhyay, K. (2014). Potential contribution of dopaminergic gene variants in ADHD core traits and co-morbidity: A study on eastern Indian probands. *Cellular and molecular neurobiology*, **34(4)**: 549-564.

Marinelli, M., Sunyer, J., Alvarez-Pedrerol, M., Iñiguez, C., Torrent, M., Vioque, J., and Julvez, J. (2014). Hours of television viewing and sleep duration in children: A multicenter birth cohort study. *JAMA pediatrics*, **168(5)**: 458-464.

Millichap, J. G., and Yee, M. M. (2012). The diet factor in attention-deficit/hyperactivity disorder. *Pediatrics*, **129(2)**: 330-337.

Movafegh, A., Alizadeh, R., Hajimohamadi, F., Esfehani, F., and Nejatfar, M. (2008). Preoperative oral *Passiflora incarnata* reduces anxiety in ambulatory surgery patients: A double-blind, placebo-controlled study. *Anesthesia and Analgesia*, **106(6)**: 1728-1732.

Negi, K. S., Singh, Y. D. and Kushwaha, K.P. (2002). Clinical evaluation of memory enhancing properties of Memory Plus in children with attention deficit hyperactivity disorder. *Indian Journal of Psychiatry*, **42(2)**.

Ngan, A., and Conduit, R. (2011). A doubleblind, placebocontrolled investigation of the effects of *Passiflora incarnata* (Passionflower) herbal tea on subjective sleep quality. *Phytotherapy Research*, **25(8)**: 1153-1159.

Niederhofer, H. (2009). Observational study: *Matricaria chamomilla* may improve some symptoms of attention-deficit hyperactivity disorder. *Phytomedicine*, **16(4)**: 284-286.

Niederhofer, H. (2009). *Panax ginseng* may improve some symptoms of attention-deficit hyperactivity disorder. *Journal of dietary supplements*, **6(1)**: 22-27.

Niederhofer, H. (2010). St. John's wort may improve some symptoms of attention-deficit hyperactivity disorder. *Natural product research*, **24(3)**: 203-205.

Pellow, J., Solomon, E. M., and Barnard, C. N. (2011). Complementary and alternative medical therapies for children with attention-deficit/hyperactivity disorder (ADHD). *Alternative Medicine Review*, **16(4)**: 323-337.

Radad, K., Gille, G., Liu, L., and Rausch, W. (2006). Use of ginseng in medicine with emphasis on neurodegenerative disorders. *Journal of pharmacological sciences*, **100(3)**: 175-186.

Radad, K., Gille, G., Moldzio, R., Saito, H., and Rausch, W. D. (2004). Ginsenosides Rb_1 and Rg_1 effects on mesencephalic dopaminergic cells stressed with glutamate. *Brain research*, **1021(1)**: 41-53.

Radad, K., Moldzio, R., and Rausch, W. D. (2011). Ginsenosides and their CNS targets. *CNS neuroscience and therapeutics*, **17(6)**: 761-768.

Raghu, J., Raghuveer, V. C., Rao, M. C., Somayaji, N. S., and Babu, P. B. (2013). The ameliorative effect of ascorbic acid and *Ginkgo biloba* on learning and memory deficits associated with fluoride exposure. *Interdisciplinary toxicology*, **6(4)**: 217-221.

Ramos-Quiroga, J. A., Nasillo, V., Fernández-Arana, F., and Casas, M. (2014). Addressing the lack of studies in attention-deficit/hyperactivity disorder in adults. *Expert review of neurotherapeutics*, **14(5)**: 553-67.

Rojas, N. L., and Chan, E. (2005). Old and new controversies in the alternative treatment of attentiondeficit hyperactivity disorder. *Mental Retardation and Developmental Disabilities Research Reviews*, **11(2)**: 116-130.

Roodenrys, S., Booth, D., Bulzomi, S., Phipps, A., Micallef, C., and Smoker, J. (2002). Chronic effects of Brahmi (*Bacopa monnieri*) on human memory. *Neuropsychopharmacology*, **27(2)**: 279-281.

Rucklidge, J. J., Johnstone, J., and Kaplan, B. J. (2009). Nutrient supplementation approaches in the treatment of ADHD. *Expert Rev Neurother.*, **9(4)**: 461-476.

Sakatani, K., Tanida, M., Hirao, N., and Takemura, N. (2014). *Ginko biloba* extract improves working memory performance in middle-aged women: role of asymmetry of prefrontal cortex activity during a working memory task. *Advances in experimental medicine and biology*, **812**: 295-301.

Salehi, B., Imani, R., Mohammadi, M. R., Fallah, J., Mohammadi, M., Ghanizadeh, A., Tasviechi, A. A., Vossoughi, A., Rezazadeh, S. A., and Akhondzadeh, S. (2010). *Ginkgo biloba* for Attention-Deficit/Hyperactivity Disorder in children and adolescents: A double blind, randomized controlled trial. *Progress in Neuro-Psychopharmacology and Biological Psychiatry*, **34(1)**: 76-80.

Sarver, D. E., McCart, M. R., Sheidow, A. J., and Letourneau, E. J. (2014). ADHD and risky sexual behavior in adolescents: Conduct problems and substance use as mediators of risk. *Journal of Child Psychology and Psychiatry*. **55(12)**: 1345-1353.

Schoonees, A., Visser, J., Musekiwa, A., and Volmink, J. (2012). Pycnogenol®(extract of French maritime pine bark) for the treatment of chronic disorders. *Cochrane Database Syst Rev.*, **4**: CD008294.

Sobanski, E., Alm, B., Hennig, O., Riemann, D., Feige, B., and Schredl, M. (2014). Daytime sleepiness in adults with ADHD a pilot trial with a multiple sleep latency test. *Journal of attention disorders*, online first.

Stough, C., Lloyd, J., Clarke, J., Downey, L., Hutchison, C., Rodgers, T., and Nathan, P. (2001). The chronic effects of an extract of *Bacopa monniera* (Brahmi) on cognitive function in healthy human subjects. *Psychopharmacology*,**156(4)**: 481-484.

Tenenbaum, S., Paull, J. C., Sparrow, E. P., Dodd, D. K., and Green, L. (2002). An experimental comparison of Pycnogenol® and methylphenidate in adults with Attention-Deficit/Hyperactivity Disorder (ADHD). *Journal of Attention Disorders*, **6(2)**: 49-60.

Trebatická, J., Kopasová, S., Hradeèná, Z., Èinovský, K., Škodáèek, I., Šuba, J., and Ïuraèková, Z. (2006). Treatment of ADHD with French maritime pine bark extract, Pycnogenol®. *European child and adolescent psychiatry*, **15(6)**: 329-335.

Weber, W., and Newmark, S. (2007). Complementary and alternative medical therapies for attention-deficit/hyperactivity disorder and autism. *Pediatric Clinics of North America*, **54(6)**: 983-1006.

Weber, W., Vander Stoep, A., McCarty, R. L., Weiss, N. S., Biederman, J., and McClellan, J. (2008). *Hypericum perforatum* (St John's wort) for attention-deficit/hyperactivity disorder in children and adolescents: a randomized controlled trial. *Jama*, **299(22)**: 2633-2641.

Xu, H., Yu, X., Qu, S., Chen, Y., Wang, Z., and Sui, D. (2014). Protective effect of *Panax quinquefolium* 20 (S)protopanaxadiol saponins, isolated from *Pana quinquefolium*, on permanent focal cerebral ischemic injury in rats. *Experimental and therapeutic medicine*, **7(1)**: 165-170.

Ye, R., Zhao, G., and Liu, X. (2013). Ginsenoside Rd for acute ischemic stroke: translating from bench to bedside. *Expert review of neurotherapeutics*, **13(6)**: 603-613.

Natural Products: Research Reviews Vol. 3 (2016) *Pages* **177–193**
Editor: **V.K. Gupta**
Published by: **DAYA PUBLISHING HOUSE, NEW DELHI**

8

Effect of Different Extraction Methods of Indian Medicinal Plants to Improve Multidrug Resistance in Gram Positive Bacteria

Harsha Patel[1], Yogeshkumar Vaghasiya[2], and Sumitra Chanda[2*]

ABSTRACT

Gram-positive bacteria cause a broad range of disease in community-acquired and hospital-acquired infections. Methicillin-resistant Staphylococcus aureus (MRSA), vancomycin-resistant enterococci (VRE) and multidrug-resistant pneumococci are an important cause of morbidity and mortality in patients. The increasing incidence of antibiotic resistance among these pathogens necessitates plant extracts/phytochemicals as an alternate therapy in restricting the resistant infectious organisms. The most established herbal therapeutic systems are Ayurveda, Unani and Siddha of Indian origin. Plant extracts contain various bioactive components having antibacterial action. Various extraction methods with different solvents are used to check antibacterial activity. The basis of extract preparation from plants is under ideal conditions and methods. This review aims to provide a better understanding of the most suitable extraction method from Indian medicinal plants to combat multidrug resistant Gram positive pathogens.

Keywords: Antibacterial, Extraction methods, Indian medicinal plants, Multi-drug resistant Gram positive bacteria.

1 Gujarat Agricultural Biotechnology Institute, Navsari Agricultural University, Surat, Gujarat, India.

2 Phytochemical, Pharmacological and Microbiological Laboratory, Department of Biosciences, Saurashtra University, Rajkot – 360 005, Gujarat, India.

* *Corresponding author*: E-mail: svchanda@gmail.com

Introduction

Bacteria, the oldest forms of life on earth, are remarkably diverse and exist in surprising numbers. Diseases caused by bacteria include some of the most common infections in the world, as well as some of the most important human nuisance in past, present, and probably future (Relman, 2002). The aerobic and facultatively anaerobic Gram-positive cocci were first divided into two families, *Micrococcaceae* and *Streptococcaceae*, on the basis of catalase activity, cell morphology and aggregation. Multidrug resistance in Gram-positive pathogens are important causes of nosocomial infection especially in hospitals where patients are vulnerable to infection and community-acquired infections, and these pathogens have revealed an extraordinary ability to develop resistance to antimicrobial agents in the last two decades (Rivera and Boucher, 2011; Patel *et al.*, 2012; Tomayko *et al.*, 2014). The worldwide spread of methicillin-resistant *Staphylococcus aureus* (MRSA), glycopeptide-resistant enterococci (GRE) and pneumococci resistant to penicillin and macrolides, and the more recent emergence of vancomycin-intermediate *S. aureus* (VISA) and vancomycin-resistant *S. aureus* (VRSA) become a major public health problem on a global scale (Nseir *et al.*, 2011; Wyllie *et al.*, 2011; Ahmed *et al.*, 2014).

Drug-Resistant Gram-Positive Bacteria: Growing Global Public Health Threat

Bacterial resistance to important antibiotic classes has become a major public health problem. The tendency to use antibiotics that promote the emergence of resistant pathogens is called antibiotic pressure, and there are many reports of resistance rising during increased antibiotic use and falling after a reduction in use (McGowan 1983). For example serious infections due to staphylococci were curable with the discovery of penicillin, but after decades of misuse and over prescription of antibiotics created resistance to penicillin. *S. aureus* was a consequence of the acquisition of a plasmid, coding for penicillinase, a penicillin-hydrolyzing enzyme, which is able to cleave the β-lactam ring and thus inactivate antibiotic molecule (Turlej *et al.*, 2011). After that methicillin was used to treat infections caused by *S. aureus*, which has also become resistant to methicillin after some time (Boyce *et al.*, 1997). The worldwide increase of methicillin-resistant *Staphylococcus aureus* and the more recent emergence of vancomycin-intermediate *S. aureus* and vancomycin-resistant *S. aureus* strains have a major impact on antibiotic policies and prompted the discovery and development of new antibiotics to combat difficult-to treat infections caused by such pathogens (Cornaglia and Rossolini, 2009). Second most common vancomycin- resistant enterococci (VRE) as a ubiquitous low-GC percent Gram-positive pathogen is a striking example of this new danger to vulnerable patients and causes of endocarditis, bacteraemia and urinary tract infections (Peel *et al.*, 2012). Most common enterococcal species *Enterococcus faecalis* and *Enterococcus faecium* are natural members of the digestive microbiota in humans. Enterococci with high level resistance to aminoglycosides, β-lactamase production and glycopeptide resistance including vancomycin resistance are posing a therapeutic challenge not only for clinicians but also for healthcare institutions. Risk factors like indiscriminate use of antibiotics, prolonged hospital stay, severity of illness and immunosuppression are responsible for nosocomial acquisition of drug resistant enterococci ultimately leading to environmental contamination and cross infections (Drees *et al.*, 2012). *Streptococcus pneumoniae* is a significant pathogen of respiratory tract infections such as pneumonia, acute otitis media, bacteremia and meningitis. Spores of *Bacillus* and *Clostridium* species are extremely resistant to a variety of agents, including high temperatures, radiation, and many chemicals (Setlow, 2006). This extreme spore resistance has major applied implications, because spores of *Bacillus* and *Clostridium* species are vectors for food spoilage and a number of food born diseases (Sanchez-Salas *et al.*, 2011). Rising incidences of morbidity and mortality worldwide have led to research into and development of advanced antibacterial with improved activity against Gram-positive strains (Bassetti *et al.*, 2011; Shibl *et al.*, 2014).

The Impact of Antimicrobial Resistance

Infectious diseases are caused by bacteria, viruses, parasites and fungi, and it is due to a complex interaction between the pathogen, host and the environment. The discovery of antibiotics is the most significant achievement of scientists because it decreases the spread and severity of microbial infections. Antimicrobial resistance has a significant negative impact on the outcome of therapy and increases the risk of cross-infection in hospitals. Resistance leads to improper empirical therapy, poor hygiene practice and the use of less effective, more toxic and more expensive drugs (Kollef, 2000) and this produced serious problems by the early 1990s prompting several authorities to publish serious warnings of impending disaster in major international journals, with titles such as: 'The crisis in antibiotic resistance' (Neu, 1992), 'Resistance to antimicrobial drugs – a worldwide calamity' (Kunin, 1993). The lower effectiveness of antibiotics causes thousands of deaths worldwide (Allahverdiyev *et al.*, 2011). The increase in isolates of Gram positive bacteria with resistance to β-lactum antibiotics and decreased susceptibility to vancomycin has created an urgent need for the development of new drug against Gram positive pathogens (Saravolatz *et al.*, 2010) that kill the resistant mutants. So, the identification of an alternative and safer drug for the control of multi-drug resistant Gram positive bacteria is necessary to combat this worldwide problem. Nature is the only source to provide a variety of chemical compounds that can be used for new drug discovery especially from mineral, plant and animal products.

Medicinal Plants: An Alternative Source as Antibacterial

Natural products play a major role from ancient civilizations to the current 21th century and more than half of the drugs in the market are natural products or their derivatives. Nature has served as a rich repository of medicinal plants for thousands of years and an impressive number of modern drugs have been isolated from natural sources, notably of plant origin (Cowan, 1999). Plants can be the major source for new drug development against pathogens (Brighenti *et al.*, 2014). The control of major diseases by synthetic products is decreasing there is increased interest in the revival of herbal medicines leading to the current widespread belief that "green medicine" is safe, more accessible and more affordable (Aref *et al.*, 2010; Wu *et al.*, 2008) than costly synthetic drugs, many of which have adverse side effects. The use of the plants in medical practice is very old and common to all societies especially in the developing nations. Plants produce a higher number of naturally occurring secondary metabolites, many of them with unique pharmacologic activities. These metabolites include flavonoids, phenols and phenolic glycosides, saponins, cyanogenic glycosides, unsaturated lactones and glucosinolates (Kappel *et al.*, 2008). Recently, pharmaceutical industries have started taking more interest in plant products to isolate active constituents from different plant parts and use them directly as drug or design them as pharmacologically active compounds with or without addition of synthetic ones (Tiwari *et al.*, 2005). Fifty percent of currently available drugs in the market are natural products or derivatives of natural products. Recent studies have shown that the plants either as pure compounds or as standardized extracts demonstrated unlimited opportunities for new drug discovery with different chemical nature. These types of studies should continue and a number of pivotal quality standards need to be set at the level of extract processing and primary evaluation in pharmacological screening models. Many researchers have conducted antimicrobial screening of plant extracts or isolated compounds (Dadgar *et al.*, 2006; Quave *et al.*, 2008; Anthony *et al.*, 2011; Vaghasiya *et al.*, 2011). The screening of plant extracts for antimicrobial activity is by itself a research area of major significance with resistance modifying action of interest.

Indian Medicinal Plants and Different Extraction Methods

India is one of the world's 12 hotspots, having the largest plant biodiversity, and it has almost 45,000 plant species of which 15,000 to 20,000 are used for medicinal purposes (Krishna Kumar, 1996). India is a varietal emporium of medicinal plants with regard to genetic resources, because it has a vast area with wide variation in climate, soil, altitude and latitude (Martins *et al.*, 2011). In India, the traditional folklore healthcare system has a long history and is very deeply rooted in rural and tribal populations. The art of herbal healing is deeply rooted in Indian culture and folklore (Katewa, 2009). Ethnomedicinal practices are the mother of all other traditional systems such as Ayurveda, Siddha, Unani and even today's most popular modern therapeutic system. About 60 per cent of the total global population remains dependent on traditional medicines for their health care system; whereas about 80 per cent of the population of India is partially or fully dependent upon herbal drugs for primary healthcare, and higher plants are known to be the main source of drug therapy in traditional medicine (Shrestha and Dhillion, 2003; Ramawat and Goyal, 2008). Herbal medicine is not just a poor man's substitute for conventional medicine but a valuable form of treatment in its own right.

When referring to a plant product in a scientific report, details such as the scientific name of the plant, plant part, preparation, formulation, and dose must be included as the basis of any discussion of therapeutics. Without a full description of the test material, there can be no assurance of a reproducible effect. Extraction is an important step involved in the discovery of bioactive components of plant material (Annegowda *et al.*, 2012; Redfern *et al.*, 2014). Plant extracts have a very special characteristic that distinguishes them from chemical drugs: a single plant may contain a great number of bioactive phytocompounds and a combination of plants even more. This complexity is one of the most important challenges to researchers attempting to identify a single bioactive compound or chemical group in the enormous universe that comprises a single crude extract. Successful prediction of botanical compounds from plant material is largely dependent on the type of solvent used in the extraction procedure. Traditional healers use normally water as a solvent for crude extraction. Known amount of dry plant powder is added in water and boiled it for a period of time. Cold extraction method is also used specially for antimicrobial studies (Rajan *et al.*, 2011). Decoctions are often made of roots, bark, and berries, which may require more forceful treatment than more fragile plant parts such as leaves and flowers. Amount of water and time required for extraction is dependent on plant material. Antibacterial potentiality of aqueous extract has been studied by many researchers and they also found remarkable results (Cutler and Wilson, 2004; Chauhan *et al.*, 2007; Assam *et al.*, 2010). In contrast, many studies showed that aqueous extracts are less effective than organic solvent extracts against bacterial strains (Nair and Chanda, 2008; Vaghasiya *et al.*, 2008; Hammuel *et al.*, 2011). These observations can be rationalized in terms of the polarity of the compounds being extracted by each solvent and, in addition to their intrinsic bioactivity, by their ability to dissolve or diffuse in the different media used in the assay. Many organic solvents are used for extraction purpose by number of researches for antimicrobial activity *i.e.* ethanol, methanol, ethyl acetate, acetone, chloroform, hexane, petroleum ether etc. For organic solvent extractions, known amount of dried plant powder is soaked in a solvent for a definite period of time. Then the slurry is filtered, the filtrate is centrifuged, supernatant is collected, and solvent is evaporated. The dried extract of the solvent is stored at 4°C in airtight bottles until use. Several other extraction methods are also used by researchers for antimicrobial studies *i.e.* soxhlet extraction, ultrasonic extraction, accelerated solvent extraction, successive extraction, microwave extraction, extraction by reflux, etc. After testing the crude extracts, fractions or pure compounds can be isolated by a variety of methods. Column chromatography is normally used for isolation of fractions or pure compounds. Further isolation and characterization of pure compounds can be done with High Performance Liquid Chromatography (HPLC) and Liquid Chromatography - Mass Spectroscopy (LC-MS) equipments. Some plant extracts or isolated pure compounds have been reported to have resistance modifying activities *in vitro* (Table 8.1). This has prompted the search for such compounds from a variety of medicinal plants.

Synergistic Activity of Plant Extracts with Antibiotics

When two different compounds combine and enhance their individual activity, it is called synergism between these compounds. If this combination decreases the activity, it is called antagonism (Chung *et al.*, 2011). When antibacterial effect of single antibiotic is reduced, combination therapy is used with other antibiotics. There are many examples which showed the effective antibacterial therapy with combination of different antibiotics (Liu *et al.*, 2007; Alexopoulos *et al.*, 2011; Toroglu *et al.*, 2011). But some bacterial strains are resistant to these combinations of antibiotics also; for example methicillin-resistant *S. aureus*, vancomycin resistant enterococci, multidrug resistant Gram negative bacteria etc. Inhibitory activity of different antibiotics on various pathogens is either by killing them outright (microbicidal action) or by arresting the growth of the causal organism (microbistatic action). The impact of antibiotic on a pathogen is specific and differs from pathogen to pathogen and vice-versa. Sometimes

Table 8.1: Details of some Indian Plants Active against Gram Positive Bacteria

Name of Indian Plants (Family)	*Extraction Methods*	*Solvents Used*	*Potential Extracts/ Compounds*	*Susceptible Gram Positive Strains*	*References*
Abelmoschus moschatus Medik. (Malvaceae)	CRUC	DW, ET	DW and ET extracts	*S. aur, B. sub*	Gul *et al.*, 2011
Acacia catechu (Linn. f.)Willd. (Mimosaceae)	CRUC, CRUH	DW, MET, HEX, ACE	MET extract	*S. aur*	Negi and Dave, 2010
Albizia odoratissima Benth. (Mimosaceae)	SEQ	MET	MET extract	*S. aur, S. mut*	Kumar *et al.*, 2011
Allium sativum L. (Liliaceae)	CRUC	DW, MET, HEX, EA, CHL	DW and MET extracts	*S. aur, B. sub*	Meriga *et al.*, 2012
Callistemon rigidus R.Br. (Myrtaceae)	SOXS	MET	CHL:MET (1:1) fraction	*S. aur*	Gomber and Saxena, 2007
Calotropis gigantea L. (Asclepiadaceae)	CRUC, SOXSUC	DW, CHL, EA, DCHM	EA and DCHM extracts	*B. sub, S. aur, M. lut*	Bharathi *et al.*, 2011
Curcuma amada Roxb. (Zingiberaceae)	SEQ	HEX, CHL, EA, ACE and MET	Difurocumenonol	*M. lut, S. aur, E. fae, B. cer, B. sub, L. mon*	Policegoudra *et al.*, 2007
Curcuma longa L. (Zingiberaceae)	SD	–	Essential oil	*S. mut*	Lee *et al.*, 2011
Curcuma longa L. (Zingiberaceae)	SOXS	ET	ET extract	*B. sub*	Naz *et al.*, 2010
Ixora coccinea L. (Rubiaceae)	CRUC	ET, DW, PE, BEN, CHL and EA	ET extract	*S. aur, B. pum, E. fae*	Selvaraj *et al.*, 2011
Launaea procumbens Roxb. (Labiateae)	CRUC, CRUH	DW and ET	ET	*B. cer, B. sub, S. aur, S. epi*	
Mangifera indica L. (Anacardiaceae)	CRUC	MET	MET extract	*Ent, S. alb, S. aur, Staph*	Vaghasiya *et al.*, 2011b

Contd...

Table 8.1–*Contd...*

Name of Indian Plants (Family)	*Extraction Methods*	*Solvents Used*	*Potential Extracts/ Compounds*	*Susceptible Gram Positive Strains*	*References*
Marrubium vulgare L. (Lamiaceae)	SD	–	Essential oil	*S. epi, S. aur, M. lut, B. sub*	Zarain *et al.*, 2011
Myrtus communis (Myrtaceae)	PARCH	MET : DW and HEX, HEX : ACE fractions	Eucalyptol, linalool, and myrtenol	*S. aur, E. fae, M. lut*	Sacchetti *et al.*, 2007
Ocimum sanctum L. (Labiatae)	OIL, CRUC	DW, CHL, MET	CHL extract and oil	*S. aur*	Mishra and Mishra 2011
Peltophorum vogelianum Benth. (Caesalpiniaceae)	Refluxing	CHL, MET	Peltophorumyl-β-D-glucuropyranoside	*S. aur*	Parveen *et al.*, 2010
Physalis minima L. (Solanaceae)	CRUH, REF	MET, CHL	MET and CHL extracts	*B. cer, B. sub, B. bro, E. fae, S. aur, S. epi*	Patel *et al.*, 2011
Polyalthia longifolia (Sonn.) Thw. (Annonaceae)	CRUC	MET, ACE, DMF	MET, ACE, DMF	*S. aur, S. epi, B. cer, B. sub*	Chanda and Nair, 2010
Punica granatum L. (Punicaceae)	COLCH	Aceticacid: hydrochloric acid: DW (30:3:10), butanol:acetic acid:DW (4:1:5)	Fresh juice, Pelargonidin-3-galactose, Cyanidin-3-, glucose Quercetin, Myrecetin, Gallic acid	*B. sub, C. dip, M. lys, S. aur, S. epi, S. sap, E. fae, S. pne, S. pyo*	NAZ *et al.*, 2007
Terminalia catappa L. (Combretaceae)	CRUC	MET, ACE, DMF	MET, ACE, DMF	*S. aur, S. sub, M. fla*	Chanda *et al.*, 2011
Sapindus emarginatus Vahl. (Sapindaceae)	SOXS, SOXSUC	MET and DW	DW extract	*S. aur*	Deepa *et al.*, 2012
Tamarindus indica L. (Fabaceae)	CRUC	DW, ACE, ET	DW, ACE and ET extracts	*S. aur, B. sub, S. pyo*	Doughari, 2006

Contd...

Table 8.1–*Contd...*

Name of Indian Plants (Family)	*Extraction Methods*	*Solvents Used*	*Potential Extracts/ Compounds*	*Susceptible Gram Positive Strains*	*References*
Toddalia asiatica (L.) Lam. (Rutaceae)	CRUC, COLCH	HEX, CHL, EA, MET and DW	Ulopterol isolated from EA extract	*S. epi*	Karunai Raj *et al.*, 2012
Woodfordia fruticosa Kurz. (Lythraceae)	CRUC	MET	MET extract	*B. cer, S. aur, S. sub*	Parekh and Chanda, 2007
Kyllinga nemoralis Hutch and Dalz (Cyperaceae)	CRUC	MET	MET extract	*S. aur, E. fae, S. mut, P. pne, S. sap*	Sindhu *et al.*, 2014
Alchornea floribunda Mull. Arg. (Euphorbiaceae)	CRUC	MET, ET, EA, CHL	MET, ET, EA and CHL extracts	*S. aur, K. pne, S. sap*	Siwe Noundou *et al.*, 2014

S. aur: Staphylococcus aureus; B. sub: Bacillus subtilis; C. dip: Corynebacterium diphtheria; M. lys: Micrococcus lysodiecticus; S. epi: Staphylococcus epidermidis; S. sap: Staphylococcus saprophyticus; E. fae: Enterococcus faecalis; S. pne: Streptococcus pneumoni; S. pyo: Streptococcus pyogenes; S. sub: Staphylococcus subflava; B. bro: Bordetella bronchiseptia; B. cer: Bacillus cereus; S. alb: Staphylococcus albus; B. pum: Bacillus pumilius; M. lut: Micrococcus luteus; S. mut: Streptococcus mutans; L. mon: Listeria mon; Ent: Enterococci species; Staph: *Staphylococcus* species; *P. pne*: *Ptococcus pnemonia*; CRUC: Crude extraction cold method; CRUH: Crude extraction hot method; SOXS: Soxhlet extraction separately; SOXSUC: Soxhlet extraction successively; SEQ: Sequential extraction; SD: Steam distillation; REF: refluxing; PARCH: Partition chromatography; OIL: Olil extraction; COLCH: Column chromatography; DW: Distilled water; ET: Ethanol; MET: Methanol; HEX: Hexane; CHL: Chloroform; EA: Ethyl acetate; DCHM: Dichloromethane; PE: Petrolium ether; DMF: N; N: dimethylformamide; BEN: benzene.

Table 8.2: Combination of Plant Extract/Compound and Antibiotics for Antibacterial Studies

Name of Plants	*Combinations*	*Gram Positive Strains*	*References*
Afzelia africana (Smith.)	MET extract + various antibiotics	*E. fae, S. aur, B. pum, M. lut, B. sub, S. epi*	Aiyegoro *et al.*, 2011
Alpinia officinarum Hance.	Galangin, quercetin and baicalein + ceftazidime	*S. aur*	Eumkeb *et al.*, 2010
Camellia sinensis (L.) Kuntze	Green tea extract + Imipenem	*S. aur*	Aboulmagd *et al.*, 2011
Ecballium elaterium (L.) A.Rich.	ET extract + penicillin	*S. aur*	Adwan *et al.*, 2011
Garcinia kola Heckel	ACE extract + Amoxycillin	*S. aur. E. fae*	Sibanda and Okoh, 2008
Laurus nobilis L.	· Kaempferol- 3-*O*-α -L- (2",4"-di-*E*-*p*-coumaroyl)- hamnoside + norfloxacin/ciprofloxacin	*S. aur* Ent	Liu *et al.*, 2009
	· Kaempferol-3-*O*-α -L- (2"-*E*-*p*-coumaroyl-4"-*Z*-*p*-coumaroyl)- rhamnoside + norfloxacin/ciprofloxacin		
Salvia officinalis L.	· Amoxicillin + ACE/EA extract	*S. aur*	Stefanovic *et al.*, 2012
	· Chloramphenicol + EA extract	*B. sub*	
Selaginella tamariscina (Beauv.)	Isocryptomerin + cefotaxime	*S. aur*	Lee *et al.*, 2009
Smallanthus sonchifolius (Poepp. and Endl.)	MET extract + ampicillin or oxacillin	*S. aur*	Joung *et al.*, 2010
Turnera ulmifolia L.	ET extract and aminoglycosides	*S. aur*	Coutinho *et al.*, 2009
Vangueria spinosa Roxb.	ET extract +doxycycline/ofloxacin	*S. aur*	Chatterjee *et al.*, 2009
Varthemia iphionoides Boiss et Blanche.	ET extract + Cefotaxime	*S. aur, B. sub, S. epi*	Abu Hijleh *et al.*, 2009
Libanotis montana Crantz (Apiaceae)	Essential oil + antibiotics	*S. aur, K. pne, B. cer*	Miladinovic *et al.*, 2014

S. aur: Staphylococcus aureus: B. sub: Bacillus subtilis: CS. epi: Staphylococcus epidermidis: B. cer: Bacillus cereus: S. alb: Staphylococcus albus: M. lut: Micrococcus luteus: L. mon: Listeria mon: Ent: Enterococci species: ET: Ethanol: MET: Methanol: HEX: Hexane: CHL: Chloroform: EA: Ethyl acetate.

the use of a single antibiotic does not produce the desired or effective inhibitory effects and to overcome this, combination of drugs often exercises their synergistic effect which surpasses their individual performance (Cai *et al.*, 2009). Plants are the most effective therapeutic agents against pathogens because they contain a variety of phytochemicals. Combinations of antimicrobials with plant extracts demonstrate an *in vitro* synergism against infecting strains and are more likely to result in successful therapy (Aqil *et al.*, 2006; Fernandes *et al.*, 2012). When bacterial species are resistant to a drug, that drug can be reused after combination with plant extracts or compounds, which may result in resistant strains being converted to susceptible strains. Isolation of compounds in their pure form and evaluating them for their biological properties leading to drug discovery is a long, tedious, time-consuming and expensive path. Combination therapy may be useful to save time and money. Some examples of successful *in vitro* antibacterial studies of combination of plant extracts/compounds and antibiotics against Gram positive bacteria are given in Table 8.2.

Future Directions

The increasing phenomenon of acquisition of resistance among Gram positive strains against higher antibacterial drugs is attributed to their indiscriminate and improper use of antibiotics and the genetic plasticity of microorganisms converting themselves into drug resistant super infectious agents. These are difficult to treat and are responsible for a variety of chronic infections. Therefore, actions must be taken to control the use of antibiotics, intensify research to better understand the genetic mechanisms of resistance and to continue studies to develop new drugs. It has focused on the isolation and characterization of new antimicrobial compounds from a variety of sources including medicinal plants. The large varieties of compounds produced by plants have proved to have therapeutic potentials as antimicrobials and as resistance modifiers. However, there are some phytochemicals which enhance the bioactivity of other compounds, referred to as biopotentiators or bioenhancers. These bioenhancers also work as synergists by reversing the antimicrobial drug resistance. The mechanisms of drug resistance reversal are being looked at for development of newer and effective drug formulations of bioenhancers to overcome antimicrobial drug resistance. India, endowed with number of medicinal plants, promises to be a potential source of therapeutically useful compounds, especially from the perspective of their potentials in combination with antimicrobial drugs which should form the subject of further extensive study.

References

Aboulmagd, E., Al-Mohammed, H.I., and Al-Badry, S. (2011). Synergism and post antibiotic effect of green tea extract and imipenem against methicillin-resistant *Staphylococcus aureus*. *Microbiology Journal*, **1**: 89-96.

Abu-Hijleh, A., Jarrar, N., and Adwan, K. (2009). Antibacterial Activity of common varthemia, *Varthemia iphionoides* ethanol extract alone and in combination with cefotaxime. *Advances in Biological Research*, **3**: 144-147.

Adwan, G., Salameh, Y., and Adwan, K. (2011). Effect of ethanolic extract of *Ecballium elaterium* against *Staphylococcus aureus* and *Candida albicans*. *Asian Pacific Journal of Tropical Biomedicine*, **1**: 456-460.

Ahmed, M.H., Ibrahim, M.A., Zhang, J., Melek, F.R., El-Hawary, S.S., Jacob, M.R., and Muhammad I (2014) Methicillin-resistant *Staphylococcus aureus*, vancomycin-resistant *Enterococcus faecalis* and *Enterococcus faecium* active dimeric isobutyrylphloroglucinol from *Ivesia gordonii*. *Natural Products Communications*, **9**: 221-224.

Aiyegoro, O., Adewusi, A., Oyedemi, S., Akinpelu, D., and Okoh, A. (2011). Interactions of antibiotics and methanolic crude extracts of *Afzelia africana* (Smith.) against drug resistance bacterial isolates. *International Journal of Molecular Sciences*, **12**: 4477-5503.

Alexopoulos, A., Kimbaris, A. C., Plessas, S., Mantzourani, I., Theodoridou, I., Stavropoulou, E., Polissiou, M. G., and Bezirtzoglou, E. (2011). Antibacterial activities of essential oils from eight Greek aromatic plants against clinical isolates of *Staphylococcus aureus*. *Anaerobe*, **17**: 399-402

Allahverdiyev, A. M., Kon, K. V., Abamor, E. S., Bagirova, M., and Rafailovich, M. (2011). Coping with antibiotic resistance: combining nanoparticles with antibiotics and other antimicrobial agents. *Expert Review of Anti-Infective Therapy*, **9**: 1035-1052.

Annegowda, H., Mordi, M., Ramanathan, S., Hamdan, M., and Mansor, S. (2012). Effect of extraction techniques on phenolic content, antioxidant and antimicrobial activity of *Bauhinia purpurea*: HPTLC determination of antioxidants. *Food Analytical Methods*, **5**: 226-233.

Anthony, K. G., Strych, U., Yeung, K. R., Shoen, C. S., Perez, O., Krause, K. L., Cynamon, M. H., Aristoff, P. A., and Koski, R. A. (2011). New classes of alanine racemase inhibitors identified by high-throughput screening show antimicrobial activity against *Mycobacterium tuberculosis*. *PLoS One*, **6**: e20374.

Aqil, F., Ahmad, I., and Owais, M. (2006). Evaluation of anti-methicillin-resistant *Staphylococcus aureus* (MRSA) activity and synergy of some bioactive plant extracts. *Biotechnology Journal*, **1**: 1093-1102.

Aref, H. L., Salah, K. B., Chaumont, J. P., Fekih, A., Aouni, M., and Said, K. (2010). *In vitro* antimicrobial activity of four *Ficus carica* latex fractions against resistant human pathogens. *Pakistan Journal of Pharmaceutical Sciences*, **23**: 53-58.

Assam, A. J., Dzoyem, J. P., Pieme, C. A., and Penlap, V. B. (2010). *In vitro* antibacterial activity and acute toxicity studies of aqueous-methanol extract of *Sida rhombifolia* Linn. (Malvaceae). *BMC Complementry and Alternative Medicine*, **10**: 40.

Bassetti, M., Ginocchio, F., and Giacobbe, D. R. (2011). New approaches for empiric therapy in Gram-positive sepsis. *Minerva Anestesiologica*, **77**: 821-827.

Bharathi, P., Alex, T., Ansa T., Krishnan, S., and Ravi, T.K. (2011). Antibacterial activity of leaf extracts of *Calotropis gigantea* Linn. against certain Gram negative and Gram positive bacteria. *International Journal of Chemical Sciences*, **9**: 919-923.

Boyce, J. M., Potter-Bynoe, G., Chenevert, C., and King, T. (1997). Environmental contamination due to methicillin-resistant *Staphylococcus aureus*: possible infection control implications. *Infection Control and Hospital Epidemiology*, **18**: 622-627.

Brighenti, FL., Salvador, M.J., Delbem, A.C., Oliveira, M.A., Soares, C.P., Freitas, L.S., Koga-Ito, C.Y. (2014). Systematic screening of plant extracts from the Brazilian pantanal with antimicrobial activity against bacteria with cariogenic relevance. *Caries Research*, **48**: 353-360.

Cai, Y., Fan, Y., Wang, R., An, M. M., and Liang, B. B. (2009). Synergistic effects of aminoglycosides and fosfomycin on *Pseudomonas aeruginosa in vitro* and biofilm infections in a rat model. *Journal of Antimicrobial Chemotherapy*, **64**: 563-566.

Chanda, S., and Nair, R. (2010). Antimicrobial activity of *Polyalthia longifolia* (Sonn.) Thw. var. Pendula leaf extracts against 91 clinically important pathogenic microbial strains. *Chinese Medicine*, **1**: 31-38

Chanda, S., Rakholia, K., and Nair, R. (2011). Antimicrobial activity of *Terminalia catappa* L. leaf extracts against some clinically important pathogenic microbial strains. *Chinese Medicine*, **2**: 171-177.

Chatterjee, S. K., Bhattacharjee, I., and Chandra, G. (2009). In vitro synergistic effect of doxycycline and ofloxacin in combination with ethanolic leaf extract of *Vangueria spinosa* against four pathogenic bacteria. *Indian Journal of Medical Research*, **130**: 475-478.

Chauhan, A. S., Negi, P. S., and Ramteke, R. S. (2007). Antioxidant and antibacterial activities of aqueous extract of Seabuckthorn (*Hippophae rhamnoides*) seeds. *Fitoterapia*, **78**: 590-592.

Chung, P. Y., Navaratnam, P., and Chung, L. Y. (2011). Synergistic antimicrobial activity between pentacyclic triterpenoids and antibiotics against *Staphylococcus aureus* strains. *Annals of Clinical Microbiology and Antimicrobials*, **10**: 25.

Cornaglia, G., and Rossolini, G. M. (2009). Forthcoming therapeutic perspectives for infections due to multidrug-resistant Gram-positive pathogens. *Clinical Microbiology Infection*, **15**: 218-223.

Coutinho, H. D., Costa, J. G., Lima, E. O., Falcao-Silva, V. S., and Siqueira, J. P., Jr. (2009). Herbal therapy associated with antibiotic therapy: potentiation of the antibiotic activity against methicillin–resistant *Staphylococcus aureus* by *Turnera ulmifolia* L. *BMC Complementary Alternative Medicine*, **9**: 13.

Cowan, M. M. (1999). Plant products as antimicrobial agents. *Clinical Microbiology Reviews* 12, 564-82.

Cutler, R. R., and Wilson, P. (2004). Antibacterial activity of a new, stable, aqueous extract of allicin against methicillin-resistant *Staphylococcus aureus*. *British Journal of Biomedical Science*, **61**: 71-74.

Dadgar, T., Asmar, M., Saifi, A., Mazandarani, M., Bayat, H., Moradi, A., Bazueri, M., and Ghaemi, E. (2006). Antibacterial activity of certain Iranian medicinal plants against methicillin-resistant and sensitive *Staphylococcus aureus*. *Asian Journal of Plant Sciences*, **5**: 861-866.

Deepa, T., Elamathi, R., Kavitha, R., Kamalakannan., Sridhar, S., Suresh Kumar, J. (2012). Screening for physical, phytochemical and antimicrobial activities of leaf extracts of *Sapindus emarginatus* Vahl. *International Journal of PharmTech Research*, **4**: 392-397.

Doughari, J.H. (2006). Antimicrobial Activity of *Tamarindus indica* Linn. *Tropical Journal of Pharmaceutical Research*, **5**: 597-603.

Drees, M., Snydman, D. R., Schmid, C. H., Barefoot, L., Hansjosten, K., Vue, P. M., Cronin, M., Nasraway, S. A., and Golan, Y. (2008). Antibiotic exposure and room contamination among patients colonized with vancomycin-resistant enterococci. *Infection Control and Hospital Epidemiology*, **29**: 709-715.

Eumkeb, G., Sakdarat, S., and Siriwong, S. (2010). Reversing beta-lactam antibiotic resistance of *Staphylococcus aureus* with galangin from *Alpinia officinarum* Hance and synergism with ceftazidime. *Phytomedicine*, **18**: 40-45.

Fernandes, T.G., de Mesquita, A.R., Randau, K.P., Franchitti, A.A., and Ximenes,. E.A. (2012) *In vitro* synergistic effect of *Psidium guineense* (Swartz) in combination with antimicrobial agents against methicillin-resistant *Staphylococcus aureus* strains. *Scientific World Journal*, **2012**: 158237.

Gomber, C., and Saxena, S. (2007). Anti-staphylococcal potential of *Callistemon rigidus*. Central *European Journal of Medicine*, **2**: 79-88.

Gul, M. Z., Bhakshu, L. M., Ahmad, F., Kondapi, A. K., Qureshi, I. A., and Ghazi, I. A. (2011). Evaluation of *Abelmoschus moschatus* extracts for antioxidant, free radical scavenging, antimicrobial and anti-proliferative activities using *in vitro* assays. *BMC Complementary Alternative Medicine*, **11**: 64.

Hammuel, C., Yebpella, G. G., Shallangwa, G. A., Magomya, A. M., and Agbajp, A. S. (2011). Phytochemical and antimicrobial screening of methanol and aqueous extracts of *Agave sisalana*. *Acta Poloniae Pharmaceutica*, **68**: 535-539.

Joung, H., Kwon, D. Y., Choi, J. G., Shin, D. Y., Chun, S. S., Yu, Y. B., and Shin, D. W. (2010). Antibacterial and synergistic effects of *Smallanthus sonchifolius* leaf extracts against methicillin-resistant *Staphylococcus aureus* under light intensity. *Journal of Natural Medicine*, **64**: 212-215.

Kappel, V. D., Costa, G. M., Scola, G., Silva, F. A., Landell, M. F., Valente, P., Souza, D. G., Vanz, D. C., Reginatto, F. H., and Moreira, J. C. (2008). Phenolic content and antioxidant and antimicrobial properties of fruits of *Capsicum baccatum* L. var. *pendulum* at different maturity stages. *Journal of Medicinal Food*, **11**: 267-274.

Karunai Raj, M., Balachandran, C., Duraipandiyan, V., Agastian, P., and Ignacimuthu, S. (2012). Antimicrobial activity of Ulopterol isolated from *Toddalia asiatica* (L.) Lam.: a traditional medicinal plant. *Journal of Ethnopharmacology*, **140**: 161-165.

Katewa, S.S. (2009). Indigenous People and Forests: Perspectives of an Ethnobotanical Study from Rajasthan (India). Ramawat K.G. (ed.), Herbal Drugs: Ethnomedicine to Modern Medicine. Springer-Verlag Berlin Heidelberg.

Kollef, M. H. (2000). Inadequate antimicrobial treatment: an important determinant of outcome for hospitalized patients. *Clinical Infectious Diseases*, **31**: S131-S138.

Krishna Kumar, P.R. (1996). Indian medicine industry under the emerging patent regimes. *Ancient Sciences Life*, **15**: 161

Kumar, D., Kumar, S., Kohli, S., Arya, R., and Gupta, J. (2011). Antidiabetic activity of methanolic bark extract of *Albizia odoratissima* Benth. in alloxan induced diabetic albino mice. *Asian Pacific Journal of Tropical Medicine*, **4**: 900-903.

Kunin, C. M. (1993). Resistance to antimicrobial drugs-a worldwide calamity. *Annals of Internal Medicine*, **118**: 557-661.

Lee, J., Choi, Y., Woo, E. R., and Lee, D. G. (2009). Antibacterial and synergistic activity of isocryptomerin isolated from *Selaginella tamariscina*. *Journal of Microbiology Biotechnology*, **19**: 204-207.

Lee, K. H., Kim, B. S., Keum, K. S., Yu, H. H., Kim, Y. H., Chang, B. S., Ra, J. Y., Moon, H. D., Seo, B. R., Choi, N. Y., and You, Y. O. (2011). Essential oil of *Curcuma longa* inhibits *Streptococcus mutans* biofilm formation. *Journal of Food Sciences*, **76**: 226-230.

Liu, C. S., Cham, T. M., Yang, C. H., Chang, H. W., Chen, C. H., and Chuang, L. Y. (2007). Antibacterial properties of Chinese herbal medicines against nosocomial antibiotic resistant strains of *Pseudomonas aeruginosa* in Taiwan. *The American Journal of Chinese Medicine*, **35**: 1047-1060.

Liu, M. H., Otsuka, N., Noyori, K., Shiota, S., Ogawa, W., Kuroda, T., Hatano, T., and Tsuchiya, T. (2009). Synergistic effect of kaempferol glycosides purified from *Laurus nobilis* and fluoroquinolones on methicillin-resistant *Staphylococcus aureus*. *Biological and Pharmaceutical Bulletin*, **32**: 489-392.

Martins, A.P., Salgueiro, L., Goncalves, M.J., da Cunha, A.P., Vila, R., Canigueral, S., Mazzoni, V., Tomi, F., and Casanova J. (2001). Essential oil composition and antimicrobial activity of three Zingiberaceae from *S. Tome* e Principe. *Planta Medica*, **67**: 580-584.

McGowan, J. E., Jr. (1983). Antimicrobial resistance in hospital organisms and its relation to antibiotic use. *Reviews of Infectious Diseases*, **5**: 1033-1048.

Meriga, B., Mopuri, R., and Muralikrishna, T. (2012). Insecticidal, antimicrobial and antioxidant activities of bulb extracts of *Allium sativum*. *Asian Pacific Journal of Tropical Medicine*, **5**: 391-395.

Miladinovic, D.L., Ilic, B.S., Mihajilov-Krstev, T.M., Jovic, J.L., and Markovic, M.S. (2014). *In vitro* antibacterial activity of *Libanotis montana* essential oil in combination with conventional antibiotics. *Natural Product Communications*, **9**: 281-286.

Mishra, P., and Mishra S. (2011). Study of antibacterial activity of *Ocimum sanctum* extract against Gram positive and Gram negative bacteria. *American Journal of Food Technology*, **6**: 336-341.

Nair, R., and Chanda, S. (2008). Antimicrobial activity of *Terminalia catappa*, *Manilkara zapota* and *Piper betel* leaf extract. *Indian Journal of Pharmaceutical Sciences*, **70**: 390-393.

Naz, S., Jabeen, S., Ilyas, S., Manzoor, F., Aslam, F., and Ali, A. (2010). Antibacterial activity of *Curcuma longa* varieties against different strains of bacteria. *Pakistan Journal of Botany*, **42**: 455-462.

Naz, S., Siddiqi, R., Ahmad, S., Rasool, S. A., and Sayeed, S. A. (2007). Antibacterial activity directed isolation of compounds from *Punica granatum*. *Journal Food Sciences*, **72**: 341-345.

Negi, B. S., and Dave, B. P. (2010). *In vitro* antimicrobial activity of *Acacia catechu* and its phytochemical analysis. *Indian Journal Microbiology*, **50**: 369-374.

Neu, H. C. (1992). The crisis in antibiotic resistance. *Science*, **257**: 1064-1073.

Nseir, S., Blazejewski, C., Lubret, R., Wallet, F., Courcol, R., and Durocher, A. (2011). Risk of acquiring multidrug-resistant Gram-negative bacilli from prior room occupants in the intensive care unit. *Clinical Microbiology and Infection*, **17**: 1201-1208.

Parekh, J., and Chanda, S. (2007). *In vitro* antibacterial activity of the crude methanol extract of *Woodfordia fruticosa* Kurz. flower (Lythraceae). *Brazilian Journal of Microbiology*, **38**: 204-207.

Parekh, J., and Chanda, S. (2009). *In vitro* antimicrobial activities of extracts of *Launaea procumbens* roxb.(Labiateae), *Vitis vinifera* L. (Vitaceae) and *Cyperus rotundus* L. (Cyperaceae). *African Journal of Biomedical Research*, **9**: 89-93.

Parveen, M., Ghalib, R. M., Khanam, Z., Mehdi, S. H., and Ali, M. (2010). A novel antimicrobial agent from the leaves of *Peltophorum vogelianum* (Benth.). *Natural Product Research*, **24**: 1268-1273.

Patel, H., Vaghasiya, Y., Vyas, B., and Chanda, S. (2012). Emergence of methicillin-resistant *Staphylococcus aureus* (MRSA) as a public-health threat and future directions of antibiotic therapy for MRSA infections. *Anti-Infective Agents*, **10**: 149-157.

Patel, T., Shah, K., Jiwan, K., and Shrivastava, N. (2011). Study on the antibacterial potential of *Physalis minima* Linn. *Indian Journal Pharmaceutical Science*, **73**: 111-115.

Peel, T., Cheng, A. C., Spelman, T., Huysmans, M., and Spelman, D. (2012). Differing risk factors for vancomycin-resistant and vancomycin-sensitive enterococcal bacteraemia. *Clinical Microbiology and Infection*, **18**: 388-394.

Policegoudra, R. S., Abiraj, K., Channe Gowda, D., and Aradhya, S. M. (2007). Isolation and characterization of antioxidant and antibacterial compound from mango ginger (*Curcuma amada* Roxb.) rhizome. *Journal of Chromatography. B, Analytical Technologies in the Biomedical and Life Sciences*, **852**: 40-48.

Quave, C. L., Plano, L. R., Pantuso, T., and Bennett, B. C. (2008). Effects of extracts from Italian medicinal plants on planktonic growth, biofilm formation and adherence of methicillin-resistant *Staphylococcus aureus*. *Journal of Ethnopharmacology*, **118**: 418-428.

Rajan, S., Thirunalasundari, T., and Jeeva, S. (2011). Anti-enteric bacterial activity and phytochemical analysis of the seed kernel extract of *Mangifera indica* Linnaeus against *Shigella dysenteriae* (Shiga, corrig.) Castellani and Chalmers. *Asian Pacific Journal of Tropical Medicine*, **4**: 294-300.

Ramawat, K. G., and Goyal, S. (2008). Indian herbal drugs scenario in global perspectives. In: Ramawat KG, Merillon JM (eds) Biotechnology: Bioactive Molecules. Springer, Berlin Heidelberg New York, p 323.

Redfern, J., Kinninmonth, M., Burdass, D., and Verran, J. (2014) Using soxhlet ethanol extraction to produce and test plant material (essential oils) for their antimicrobial properties. *Journal of Microbiology and Biology Education,* **15**: 45-46.

Relman, D. A. (2002). The human body as microbial observatory. *Nature Genetics,* **30**: 131-133.

Rivera, A. M., and Boucher, H. W. (2011). Current concepts in antimicrobial therapy against select Gram-positive organisms: methicillin-resistant *Staphylococcus aureus*, penicillin-resistant pneumococci, and vancomycin-resistant enterococci. *Mayo Clinic Proceedings,* **86**: 1230-1243.

Sacchetti, G., Muzzoli, M., Statti, G. A., Conforti, F., Bianchi, A., Agrimonti, C., Ballero, M., and Poli, F. (2007). Intra-specific biodiversity of Italian myrtle (*Myrtus communis*) through chemical markers profile and biological activities of leaf methanolic extracts. *Natural Product Research*, **21**: 167-179.

Sanchez-Salas, J.-L., Setlow, B., Zhang, P., Li, Y.-q., and Setlow, P. (2011). Maturation of released spores is necessary for acquisition of full spore heat resistance during *Bacillus subtilis* sporulation. *Applied and Environmental Microbiology*, **77**: 6746–6754.

Saravolatz, L. D., Pawlak, J., and Johnson, L. B. (2010). *In vitro* activity of oritavancin against community-associated methicillin-resistant *Staphylococcus aureus* (CA-MRSA), vancomycin-intermediate *S. aureus* (VISA), vancomycin-resistant *S. aureus* (VRSA) and daptomycin-non-susceptible *S. aureus* (DNSSA). *International Journal of Antimicrobial Agents*, **36**: 69-72.

Selvaraj, N., Lakshmanan, B., Mazumder, P. M., Karuppasamy, M., Jena, S. S., and Pattnaik, A. K. (2011). Evaluation of wound healing and antimicrobial potentials of *Ixora coccinea* root extract. *Asian Pacific Journal Tropical Medicine,* **4**: 959-4963.

Setlow, P. (2006). Spores of *Bacillus subtilis*: their resistance to and killing by radiation, heat and chemicals. *Journal of Applied Microbiology,* **101**: 514-525.

Shibl, A.M., Memish, Z.A., Kambal, A.M., Ohaly, Y.A., Ishaq, A., Senok, A.C., and Livermore, D.M. (2014) National surveillance of antimicrobial resistance among Gram-positive bacteria in Saudi Arabia. *Journal Chemotherapy*, **26**: 13-18.

Shrestha, P. M., and Dhillion, S. S. (2003). Medicinal plant diversity and use in the highlands of Dolakha district, Nepal. *Journal of Ethnopharmacology*, **86**: 81-96.

Sibanda, T., and Okoh, A.I. (2008). *In vitro* evaluation of the interactions between acetone extracts of *Garcinia Kola* seeds and some antibiotics. *African Journal of Biotechnology*, **7**: 1672-1678.

Sindhu, T., Rajamanikandan, S., and Srinivasan, P. (2014) *In vitro* antioxidant and antibacterial activities of methanol extract of *Kyllinga nemoralis*. *Indian Journal of Pharmaceutical Science*, **76**: 170-174.

Siwe Noundou, X., Krause, R.W., van Vuuren, S.F., Tantoh Ndinteh, D., and Olivier, D.K. (2014) Antibacterial activity of the roots, stems and leaves of *Alchornea floribunda*. *Journal of Ethnopharmacology*, **151**: 1023-1027.

Stefanovic, O. D., Stanojevic, D. D., and Comic, L. R. (2012). Synergistic antibacterial activity of *Salvia officinalis* and *Cichorium intybus* extracts and antibiotics. *Acta Poloniae Pharmaceutica*, **69**: 457-663.

Tiwari, T. P., Bharti, S. K., Kaur, H. D., Dikshit, R. P., and Hoondal, G. S. (2005). Synergistic antimicrobial activity of tea and antibiotics. *Indian Journal of Medical Research*, **122**: 80-84.

Tomayko, J.F., Rex, J.Ḥ., Tenero, D.M., Goldberger, M., and Eisenstein, B.I. (2014) The challenge of antimicrobial resistance: New regulatory tools to support product development. *Clinical Pharmacology and Therapeutics*, **96**: 166-168.

Toroglu, S. (2011). *In-vitro* antimicrobial activity and synergistic/antagonistic effect of interactions between antibiotics and some spice essential oils. *Journal of Environmental Biology*, **32**: 23-29.

Turlej, A., Hryniewicz, W., and Empel, J. (2011). Staphylococcal cassette chromosome mec (Sccmec) classification and typing methods: an overview. *Polish Journal of Microbiology*, **60**: 95-103.

Vaghasiya, Y., Nair, R., and Chanda, S. (2008). Antibacterial and preliminary phytochemical and physico-chemical analysis of *Eucalyptus citriodora* Hk leaf. *Natural Product Research*, **22**: 754-762.

Vaghasiya, Y., Parekh, J., Shukla, V., and Chanda, S. (2011). Antimicrobial and anti-inflammatory screening of four Indian medicinal plants. *Latin American Journal of Pharmacy*, **30**: 661-662.

Vaghasiya, Y., Patel, H., and Chanda, S. (2011). Antibacterial activity of methanol extract of *Mangifera indica* against some human pathogens and its phytochemical study. *African Journal of Biotechnology*, **10**: 15788-15794.

Wu, K. M., Ghantous, H., and Birnkrant, D. B. (2008). Current regulatory toxicology perspectives on the development of herbal medicines to prescription drug products in the United States. *Food and Chemical Toxicology*, **46**: 2606-2610.

Wyllie, D., Paul, J., and Crook, D. (2011). Waves of trouble: MRSA strain dynamics and assessment of the impact of infection control. *Journal of Antimicrobial and Chemotherapy*, **66**: 2685-2688.

Zarai, Z., Kadri, A., Ben Chobba, I., Ben Mansour, R., Bekir, A., Mejdoub, H., and Gharsallah, N. (2011). The *in-vitro* evaluation of antibacterial, antifungal and cytotoxic properties of *Marrubium vulgare* L. essential oil grown in Tunisia. *Lipids Health Diseases*, **10**: 161.

Natural Products: Research Reviews Vol. 3 (2016) *Pages* **195–303**
Editor: **V.K. Gupta**
Published by: **DAYA PUBLISHING HOUSE, NEW DELHI**

9

Antioxidant Properties of African Herbal Remedies

M. Gulumian[1,2*], S. Adebayo[3,4] and V. Steenkamp[5]

ABSTRACT

The use of traditional herbal remedies as alternative medicine plays a significant role in Africa since it features significantly in primary health care. Investigations into the phytochemistry and biological activity of herbal remedies provide an understanding of some of the mechanisms involved in their therapeutic effects. However, of the large amount of plants used as remedies in Africa, only a few have been thoroughly studied for all aspects of their potential therapeutic value. This paper reviews African herbal remedies with antioxidant activity where, firstly the compounds of herbals known to contain activity have been identified and secondly where the herbal has proven antioxidant activity but the compound responsible for the activity has as yet not been identified. In the latter case we have attempted to ascribe the activity to a compound known to be present in the plant family and/or species, where related

---

1 National Institute for Occupational Health, P O Box 4788, Johannesburg 2000, South Africa.

2 Haematology and Molecular Medicine, University of the Witwatersrand, Johannesburg, South Africa.

3 Phytomedicine Programme, University of Pretoria, Ondersterpoort, Pretoria 0110, South Africa.

4 Department of Biomedical Sciences, Tshwane University of Technology, Pretoria 0001, South Africa.

5 Department of Pharmacology, University of Pretoria, Private Bag X323, Arcadia 0007, South Africa.

* *Corresponding author*: E-mail: gulumian@nioh.nhls.ac.za

activity has previously been documented for another genus of the species or the compound has proven antioxidant activity. A growing body of evidence suggests that at least part of the therapeutic value of herbals may be explained by their antioxidant activity.

Keywords: Antioxidant properties, African herbal remedies, Phytochemistry, Biological activity, Therapeutic values.

Introduction

Human cells are continuously exposed to exogenous oxidants as well as to those produced endogenously during normal physiological processes. Antioxidants form part of protective mechanisms that exist in human cells to scavenge and neutralize these oxidants. Oxidants such as the reactive oxygen species (ROS) and in particular oxygen centred free radicals, are involved in several diseases (Karabina *et al.*, 2005; MacNee, 2005; Mariani *et al.*, 2005). Antioxidant defences are defective in these diseases and therefore it is possible to limit oxidative damage and ameliorate disease progression with antioxidant supplementation (Rennard, 2004; Violi and Cangemi, 2005). Naturally occurring antioxidants are favoured over the synthetic antioxidants as the latter are suspected to cause or promote negative health effects (Calviello *et al.*, 2005). Subsequently, the use of synthetic antioxidants as food additives is restricted in several countries due to these undesirable health effects (Nakatani, 1997).

With the aim of listing African medicinal plants with antioxidant activities, a comprehensive literature search was conducted. Reviews have listed a number of these African medicinal plants with antioxidant properties (Atawodi, 2005; Iwalewa *et al.*, 2007; Mahomoodally, 2013). The present work has therefore aimed to expand the list to include medicinal plant species with antioxidant properties that are used in different African countries including those from Madagascar and Mauritius. For the sake of inclusivity, plants that have been shown to contain compounds that hold the potential of being novel antioxidants are also considered. In addition, those with anti-inflammatory properties were also included due to an earlier observation that the anti-inflammatory activities of the same extracts could be explained, at least in part, by their antioxidant properties (Schinella *et al.*, 2002). Additional efforts were also made to include information, if they were available, on their vernacular names, their regional distribution, and medicinal use and plant parts used for these preparations or for the isolation of the antioxidant ingredient(s). Table 9.1 lists medicinal plants that are investigated and are shown to have confirmed antioxidant and/or antinflammatory activity and are also shown to contain compounds that are known to have such activitites. Table 9.2 on the other hand, lists medicinal plants that are either investigated to have confirmed antioxidant activity. Compounds responsible for the antioxidant activity have not yet been identified or plants containing compounds with antioxidant activity but these compounds have not been confirmed to be responsible for the activity.

A number of edible and culinary herbs and condiments were also included in these two tables as they are used in certain instances as medicinal herbs to treat

Table 9.1: Medicinal Plants with Confirmed Antioxidant Activity and also Shown to Contain Compounds that are Known to have Such Activity

Family and Genus	*Vernacular Name*	*Plant Part*	*Country/ Area*	*Medicinal Use and/or Experimental Validation*	*Compounds Isolated*	*Reference*
Aloaceae						
Aloe barbadensis Mill.	Burn plant, siber, sbar/ essouktouri/ mar, sbar	Leaf exudate	Algeria, Moroco, Tunisia	Antioxidant activity. Used as laxative, purgative, diuretic, asthma, calvicie, cuts, bounds, skin rash.	Flavonoids, two dihydro-coumarin derivatives and two flavone glyxosides	(Hu *et al.*, 2003; Moniruzzaman *et al.*, 2012; Rajasekaran *et al.*, 2005a; Rajasekaran *et al.*, 2005b; Zhang *et al.*, 2006;)
Aloe claviflora Burch.	Kraal aloe	Leaf exudate	South Africa	Radical scavenging activity	Chromone glycoside and moderate activity in the lipid peroxidation assay	(Amoo *et al.*, 2014; Lindsey *et al.*, 2003)
A. saponaria (Ait.) Haw.	Mpelu Mnemvu Soap aloe, African aloe	Leaf exudate	South Africa	Radical scavenging activity and moderate activity in the lipid peroxidation assay	Chromone glycoside	(Lindsey *et al.*, 2003; Zapata *et al.*, 2013)
A. thraskii Baker	Dune aloe, ikhala, umhlaba	Leave exudate	South Africa	Radical scavenging activity and moderate activity in the lipid peroxidation assay	Chromone glycoside	(Lindsey *et al.*, 2003; Amoo *et al.*, 2014)
Amaranthaceae						
Amaranthus caudatus L.	Tassel flower	Seed; Young shoots	Ethopia	Antioxidant properties	Tocopherols, phenolic acids	(Bruni *et al.*, 2002; Conforti *et al.*, 2005; Klimczak *et al.*, 2002; Veeru *et al.*, 2009)
Anacardiaceae						
Lannea edulis Engl.	Wild Grape	Root-bark	Zimbabwe	Semipolar extracts high activity both as radical scavengers and lipoxygenase inhibitors. Lipophilic extracts inhibitor of 15-lipoxygenase. Used for painful menstruation, urogenital infection, STD.	Two Alkylphenols (cardonol 7 and cordonol 13) and three dihydroalkylhexenones	(Maroyi, 2013; Munodawafa *et al.*, 2013; Queiroz *et al.*, 2003)

Contd...

Table 9.1–*Contd...*

Family and Genus	*Vernacular Name*	*Plant Part*	*Country/ Area*	*Medicinal Use and/or Experimental Validation*	*Compounds Isolated*	*Reference*
Lannea velutina A. Rich	Bemmbeyi Raisinier velu, Lannéa velouté	Leaves, bark, root	Mali	Antioxidant properties	Proanthocynaidins	(Diallo *et al.*, 2001; Maiga *et al.*, 2006; Maiga *et al.*, 2007; Quattara *et al.*, 2011)
Mangifera indica L.	Mango Mangoro	Leaves, seeds, stem-bark	Benin Burkina Faso	Antiinflammatory, analgesic, and hypoglycemic effects. Used to treat urogenital infection, tonic, diarrhoea, tooth ache, gingivitis, liver disease, diabetes.	Polyphenolics, flavonoids	(Igoli *et al.*, 2005; Ojewole, 2005; Kim *et al.*, 2010; Ling *et al.*, 2009)
Apiaceae						
Centella asiatica (L.) Urb.	Gotu kola	Leaves	South Africa	Antioxidant and anti-inflammatory activities. Used for wound healing. Protection against radiation-induced injury. Cardio protective effect. Oral treatment increased antioxidant enzymes.	Quercetin and tetrandrine	(Arora *et al.*, 2005; Chen *et al.*, 1999; Jayashree *et al.*, 2003; MacKay and Miller, 2003; Pittella *et al.*, 2009; Sharma and Sharma, 2002; Sharma and Sharma, 2005; Shukla *et al.*, 1999a, 1999b)
Apocynaceae						
Catharanthus roseus (L.) G. Don	Madagascar Periwinkle kaka poul, karaktè dezosm blan, zèb sosyé	Whole plant	Madag-ascar	Antioxidant activity and ability to increase antioxidant enzymes. Used for conjunctivitis.	Phenols	(Zheng and Wang, 2001)
Asclepiadaceae						
Secamone afzelii Rhoem.	Ahaban Kroratima	Stem	Central Africa	Antioxidant and anti-inflammatory properties. Used for wound healing.	Flavonoids, caffeic acid derivatives and-tocopherol.	(Houghton *et al.*, 2005; Mensah *et al.*, 2004; Zabri *et al.*, 2008)

Contd...

Table 9.1–*Contd...*

Family and Genus	*Vernacular Name*	*Plant Part*	*Country/ Area*	*Medicinal Use and/or Experimental Validation*	*Compounds Isolated*	*Reference*
Asphodelaceae						
Bulbine capitata Poelln.	Scented grass bulbine	Roots Aerial parts	South Africa	Anti-inflammatory and weak antioxidant and free radical scavenging and lipid peroxidation inhibition activities. Knipholone as a selective inhibitor of leukotriene metabolism. Used for mild purgative and to cure gonorrhoeal infections.	Anthroquinone Knipholone 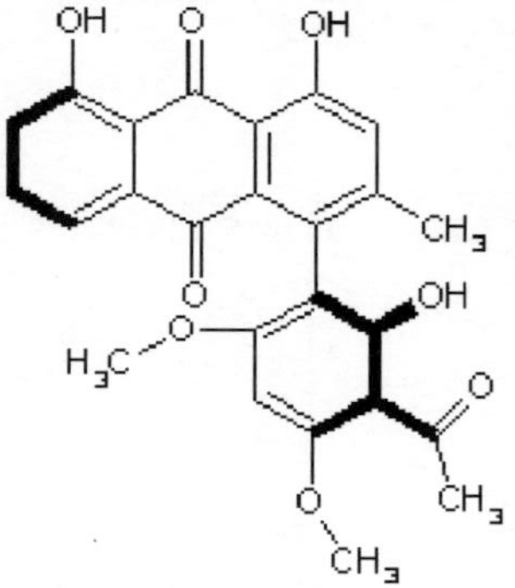	(Abegaz, 2002; Abegaz *et al.,* 2002a; Bezabih and Abegaz, 1998; Bezabih *et al.,* 2001; Bezabih *et al.*, 1997; Bosch, 2008; Majinda *et al.,* 2001; Qhotsokoane-Lusunzi and Karuso, 2001; Wube *et al.,* 2006)
Bulbine frutescens Willd.	Snake flower, cat's tail, burn jelly plant	Leaf juice Roots	South Africa	Anti-inflammatory and weak antioxidant and free radical scavenging and lipid peroxidation inhibition activities. Knipholone is a selective inhibitor of leukotriene metabolism. Used for burns, rashes, blisters, insect bites, cracked lips, acne, cold sores, mouth ulcers and areas of cracked skin.	Phenylanthraquinones, Isofuranonaphthoquinones, Gaboroquinones A and B and 4'-O-demethylknipholone-4'-O-beta-D-glucopyranoside, and Knipholone (anthroquinone)	(Abegaz *et al.,* 2002a; Majinda *et al.,* 2001; Menche, 2002; Mutanyatta *et al.,* 2005; Tambama *et al.,* 2014; Wube *et al.,* 2006)

Contd...

Table 9.1–*Contd...*

Family and Genus	*Vernacular Name*	*Plant Part*	*Country/ Area*	*Medicinal Use and/or Experimental Validation*	*Compounds Isolated*	*Reference*
Kniphofia foliosa Hochst.	Red-not-peker		Kenya	Anti-inflammatory and weak antioxidant and free radical scavenging and lipid peroxidation inhibition activities. Knipholone as a selective inhibitor of leukotriene metabolism. Used for abdominal cramps, wound healing	Anthroquinone: Knipholone	(Adams and Baeur, 2008; Dagne and Steglich, 1984; Dagne and Yenesew, 1994; Habtemariam, 2010; Wube *et al.,* 2006)
Asteraceae						
Artemisia abyssinica Sch.Bip.	Chikugn (Amharic) *Arrtta bera* (*Or*)	Whole plant	Ethopia	Radical scavenging and antioxidant activities. Used for stomach pain and wound healing.	Essential oils: and flavonoids	(Burits *et al.,* 2001; Emami *et al.,* 2007; Esteban *et al.*, 1986; Naidoo *et al.,* 2008)
				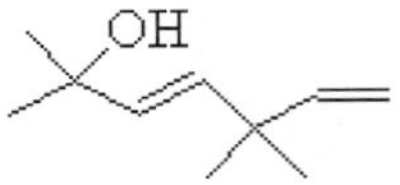 OH 1	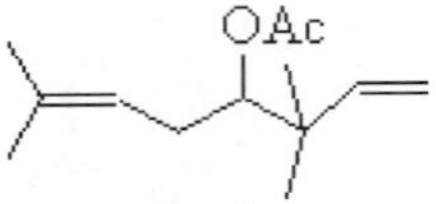 OAc 6	
A. afra Jacq.ex Willd.	African wormwood Wild wormwood	Roots, stems and leaves	Ethopia South Africa	Radical scavenging and antioxidant activities. Used for stomach pain, coughs, colds, fever, loss of appetite, colic, headache, earache, intestinal worms to malaria.	Essential oils and flavonoids	(Burits *et al.,* 2001; Buwa and Afolayan, 2009; Naidoo *et al.,* 2008; Van Wyk *et al.*, 1997; Watt and Breyer-Brandwijk, 1962)
A. arvensis L.	Mugwort Wormwood	Whole plant	Algeria	Radical scavenging and antioxidant activities.	Phenolic compounds and flavonoids.	(Djeridane *et al.,* 2006; 2007)

Contd...

Table 9.1–*Contd...*

Family and Genus	Vernacular Name	Plant Part	Country/ Area	Medicinal Use and/or Experimental Validation	Compounds Isolated	Reference
A. campestris L.	Field sagewort Field worm-wood	Whole plant	Algeria	Radical scavenging and antioxidant activities. Used to treat insomnia	Phenolic compounds and flavonoids.	(Akrout *et al.*, 2011; Djeridane *et al.*, 2006; Naili *et al.*, 2010)
Bidens pilosa L.	Black jack	Leaves Roots	South Africa	Antioxidant and anti-inflammatory, antibacterial, antihypertensive activities. Used to treat diabetes and backache.	Phenolic compounds: quercetin 3-O-rabinobioside, quercetin 3-O-rutinoside. Two novel methoxylated flavone glycosides : quercetin 3,3'-dimethyl ether 7-O-c′-L-rhamnopyranosyl-(1 ~ 6)-fl-D-glucopyranoside and the known quercetin 3,3'-dimethyl ether 7-O-fl-D-glucopyranoside	(Brandao *et al.*, 1998; Chiang *et al.*, 2004; Lindsey *et al.*, 2002; Yuan *et al.*, 2008; Deba *et al.*, 2008)
Cynara scolymus L.	Globeartichoke	Leaves	Ethiopia	Antioxidative and lipid-lowering properties and eNOS up-regulating ability. Used to treat chronic liver and gall bladder diseases, jaundice, hepatitis and aterosclerosis.	Polyphenolic flavonoid compounds	(Fratianni *et al.*, 2007; Gebhardt, 1997; Li *et al.*, 2004; Speroni *et al.*, 2003)
Helichrysum dasyanthum Sweet	Afrikaans common name of kooigoed (bedding material)	Leaves	South Africa	Antioxidant, radical scavenging and anti-inflammatory activities. Used for wounds, infections, respiratory conditions.	Essential oils	(Albayrak *et al.*, 2010; Lourens *et al.*, 2004; 2008)

Contd...

Table 9.1–*Contd...*

Family and Genus	*Vernacular Name*	*Plant Part*	*Country/ Area*	*Medicinal Use and/or Experimental Validation*	*Compounds Isolated*	*Reference*
H. petiolare Hilliard and B.L. Burtt.	Everlasting, Imphepho	Leaves	South Africa	Antioxidant, radical scavenging and anti-inflammatory activities. Used for wounds, infections, respiratory conditions, asthma, chest problems and high blood pressure	Essential oils	(Albayrak *et al.*, 2010; Lourens *et al.*, 2004; 2008)
Tagetes minuta L.	Khaki bush stinking roger muster John Henry, wild marigold	Leaves	Madagascar	Antimicrobial and antioxidant activity. Anthelmintic, anti-spasmodic, purgative and for the treatment of gastritis, indigestion and internal worms.	Essential oils	(Juliani *et al.*, 2004; Ranilla *et al.*, 2010)
Balanophoraceae						
Thonningia sanguinea Vahl.	Nkomango	Roots	Ghana	Antioxidative and radical scavenging activities and lipid peroxidation inhibitory activity. Used for bronchial asthma, rheumatoid arthritis, artherosclerosis and diabetes.	Ellagitannins: Thonningianin A and B	(Gyamfi *et al.*, 2000; Gyamfi *et al.*, 2004a; Gyamfi *et al.*, 2004b; Gyamfi *et al.*, 1999; Ohtani *et al.*, 2000; N'Guessan *et al.*, 2007)
Balanitaceae						
Balanites aegyptiaca (L.) Delile	Hausa: aduwa Desert date	Bark and date	East Africa	Antioxidant properties *in vitro* confirmed. The bark and roots are used as laxatives, and for colic. The bark is used for sore throats, and as a remedy for sterility, mental diseases, epilepsy,	Cumarins, flavonoids, saponins (Balanin 1 (3β, 12β,14β,16β) cholest-5-ene-3,16-diyl bis (β-d -glucopyranoside)- 12-sulphate, a new sterol-sulfonated and Balanin 2	(Chothani and Vaghasiya, 2011; Cook *et al.*, 1998; El Tahir *et al.*, 1998; Speroni *et al.*, 2005)

Contd...

Table 9.1–*Contd...*

Family and Genus	*Vernacular Name*	*Plant Part*	*Country/ Area*	*Medicinal Use and/or Experimental Validation*	*Compounds Isolated*	*Reference*
				yellow fever, syphilis, and tooth aches.	(3β,*20S,22R,25R*)-26-hydroxy-22-acetoxyfurost-5-en-3- yl-rhamnopyranosyl-(1 →2)-glucopyranoside, a novel furostanol saponin)	
Bignoniaceae						
Spathodea campanulata P.Beauv.	African tulip	Stem-bark	Nigeria, Ghana, Cameroon (Yaounde region)	Anti-inflammatory, anti-oxidant, hypoglycemic, anticomplement and anti-HIV activities. Used to treat itching, arthritis, and diabetes.	Flavonoids and caffeic acid derivatives	(Houghton *et al.*, 2005; Ofori-Kwakye *et al.*, 2009)
Capparaceae						
Cleome arabica L.	Cleome efeina	Leaves	Egypt	Antioxidant activity, inhibited lipoxygenase activity and calcium ionophore-stimulated LTB4 synthesis in human neutrophils. Used to treat wounds and prevent inflammation	Rutin and quercetin.	(Akula and Odhav, 2008; Djeridane *et al.*, 2010; Selloum *et al.*, 1997)
Clusiaceae						
Garcinia kola Heckel	Bitter cola/ aku ilu, agbu ilu. Nigeria Hausa: Góórò *pl.* gwârráá *or* gòòràrràkáí	Seeds	Nigeria	Inhibit lipid peroxidation and protective against H_2O_2-induced DNA strand breaks and oxidized bases. Used for laryngitis, coughs, liver disease, bronchitis and throat infections. Inhibits Aflatoxin B1 induced genotoxicity.	Biflavonoid: kolaviron	(Adaramoye and Akinloye, 2000; Adaramoye *et al.*, 2005; Adegoke *et al.*, 1998; Farombi *et al.*, 2000; 2004; 2009; Nwankwo *et al.*, 2000; Olaleye *et al.*, 2000)

Contd...

Table 9.1–*Contd...*

Family and Genus	*Vernacular Name*	*Plant Part*	*Country/ Area*	*Medicinal Use and/or Experimental Validation*	*Compounds Isolated*	*Reference*
Harungana madagascariensis Poir.	Otori	Stem-bark	Eastern Nigeria	Significant antioxidant activity. Used to treat skin diseases.	**Anthronoids**: Harunganol B and harungin anthrone. Prenylated anthronoids: harunmadagascarins A and B	(Biapa *et al.*, 2007; Iwalewa *et al.*, 2008; Kouam *et al.*, 2005)
Hypericum carinatum Griseb.	Not signalized	Leaves	Egypt	Antioxidant and radical scavenging activities.	**Benzophenones**: Cariphenone A (6-benzoyl-5,7-dihydroxy-2,2,8-trimethyl-2H-chromene) (1) and cariphenone B (8-benzoyl-5,7-dihydroxy-2,2,6-trimethyl-2H-chromene) (2).	(Bernardi *et al.*, 2005; Nunes *et al.*, 2010)
H. perforatum L.	Common St.-John' wort	Whole plant	Egypt	Anti-inflammatory and anti-oxidant activities. Free radical scavenging, metal-chelation, and reactive oxygen quenching activities. Protective against scopolamine-induced altered brain oxidative stress status and amnesia in rats. Ability to suppress the activities of 5-lipoxygenase (5-LO) and cyclooxygenase-2 (COX-2), key enzymes in the formation of proinflammatory eicosanoids from arachidonic acid (AA). Analgesic, antiseptic, antispasmodic, digestive, diuretic and sedative.	**Flavonoids**: Rutin, hyperoside, isoquercitrin, avicularin, quercitrin, and quercetin.	(Bernardi *et al.*, 2005; El-Sherbiny *et al.*, 2003; Herold *et al.*, 2003a; Herold *et al.*, 2003b; Hosseinzadeh *et al.*, 2005; Orèiæ *et al.*, 2011; Zou *et al.*, 2004)

Contd...

Table 9.1–*Contd...*

Family and Genus	*Vernacular Name*	*Plant Part*	*Country/ Area*	*Medicinal Use and/or Experimental Validation*	*Compounds Isolated*	*Reference*
Cochlospermaceae						
Cochlospermum tinctorium A.Rich.	N´tiribara	Roots	Sudan, Uganda West Africa	Antioxidant activity. Used for malaria, jaundice.	**Polyphenols**: Gallotanins and ferulic acids	(Nergard *et al.,* 2005)
Combretaceae						
Combretum woodii Drum.	Large-leaved forest bush willow	Leaf	South Africa	Antioxidant and anti-bacterial activities. Also tannins showed inhibitory effect on Fe^{2+}-induced lipid peroxidation and radical scavenger activity. Used for pneumonia, syphilis, abdominal pain and conjuctivitis.	**Polyphenols:** Combretastatin B5 (2',3'4-trihydroxyl, 3,5,4'-trimethoxybibenzyl). Tannins.	(Aderogba *et al.,* 2012; Coombes and Rogers, 2002; Eloff *et al.,* 2005a; Eloff *et al.,* 2005b; Masoko and Eloff, 2008; Zishiri, 2005)
Combretum imberbe	Not signified		South Africa	Combretum species are widely used for treating abdominal disorders (*e.g.* abdominal pains, diarrhea) backache, bilharziasis, chest coughs, colds, conjunctivitis, dysmenorrhoea, earache, fattening babies, fever, headache	1α,3β-dihydroxy-12-oleanen-29-oic (1), 1-hydroxy-12-olean-30-oic acid (2), 3,30-dihydroxyl-12-oleanen-22-one (3), and 1,3,24-trihydroxyl-12-olean-29-oic acid (4), a new pentacyclic triterpenoid (1α, 23-dihydroxy-12-oleanen-29-oic acid-3β-O-2,4-di-acetyl-l-rhamnopyranoside) (5)	(Angeh *et al.,* 2007)

Contd...

Table 9.1–*Contd...*

Family and Genus	*Vernacular Name*	*Plant Part*	*Country/ Area*	*Medicinal Use and/or Experimental Validation*	*Compounds Isolated*	*Reference*
Guiera senegalensis J.F.Gmel.	N´kundjè	Leaf	Western Africa	Antioxidant and radical scavenging activities. Used against dysentery, diarrhoea, gastro-intestinal pains and disorders, rheumatism, antidiabetics and fever.	Flavonol aglycones, flavonol glycosides and flavonoids (catechin, myricitrin, rutin and quercetin) as well as tannins (galloylquinic acids (hydrolysable tannins).	(Amos *et al.*, 2001; Bouchet *et al.*, 1998; Ficarra *et al.*, 1997; Males *et al.*, 1998; Sombie *et al.*, 2011)
Terminalia sericea Burch. ex DC.	Silver cluster-leaf	Bark	South Africa	Radical scavenging and antioxidant activities. Used to treat diabetes and pneumonia and to relieve colic	Pentacyclic triterpenoids Anolignan B	(Eldeen *et al.*, 2006; Masoko *et al.*, 2005; Masoko and Eloff, 2008; Steenkamp *et al.*, 2004)
Commelinaceae						
Commelina diffusa Burm.f.	Wandering Jew Climbing day flower	Leaves	Ghanna	Anti-inflammatory and antioxidant properties. Used to treat fever and is diuretic	Flavonoids	(Ezea *et al.*, 2014; Houghton *et al.*, 2005)
Palisota hirsuta K.Schum.,	Not signified	Aqueous leaf extracts	Nigeria	Anti-inflammatory effects against carrageenan induced hind paw oedema	Not identified	(Akah and Nwambie, 1994; Boakye-Gyasi *et al.*, 2011)
Crassulaceae						
Bryophyllum pinnatum (Lam.) Oken Synonyme: *Kalanchoe pinnata* (Lam.) Pers.	Ufu ivo	Leaves	Nigeria, South Africa	Anti-inflammatory properties. Used for earache.	Flavonoids, polyphenols, triterpenoids	(Igoli *et al.*, 2005; Ojewole, 2005b; Tatsimo *et al.*, 2012)
Cupressaceae						
Juniperus procera Hochst ex. Endl.	African Juniper	Young twigs and buds	Ethiopia	Antioxidant and free radical scavenging activities. Used to relieve stomach pain.	Essental oils	(Alqasoumi and Abdel-Kader, 2012; Burits *et al.*, 2001; Orhan *et al.*, 2011)

Contd...

Table 9.1–*Contd...*

Family and Genus	*Vernacular Name*	*Plant Part*	*Country/ Area*	*Medicinal Use and/or Experimental Validation*	*Compounds Isolated*	*Reference*
Dioscoreaceae						
Dioscorea dumetorum (Kunth) pax	Yam	Tubers	Nigeria	Antioxidant activity to modify serum lipid and anti-inflammatory activity. Used to treat diabetes.	Dioscorea and Dioscoretine	(Araghiniknam *et al.*, 1996; Iwu *et al.*, 1990; Sonibare and Abegunde, 2012)
Drosera madagascariensis (DC.) *D. ramentacea* Burchell	Sundew	Roots and flowers	Madagascar	Anti-inflammatory effects. Used to treat coughs and asthma	Flavonoids: hyperoside, quercetin and isoquercitrin	(Egan and van der Kooy, 2013; Paper *et al.*, 2005)
Drosera rotundifolia L.	Round-leaf Sundew	Roots and flowers	Madagascar	Anti-inflammatory effects. Used to treat coughs and asthma	Flavonoids: hyperoside, quercetin and isoquercitrin	(Hohtola, 2010; Paper *et al.*, 2005)
Euphorbiaceae						
Alchornea laxiflora (Benth) Pax and K. Hoffm.	Wild banana	Leaf and root	Nigeria	Anti-oxidant and anti-microbial activity. Used to treat jaundice and liver disorders. Also used in food preservation.	Quercetin-7,4'-disulphate (1) quercetin (2) quercetin-3',4'-disulphate (3) quercetin-3,4'-diacetate (4) rutin (5) and quercetrin (6)	(Farombi *et al.*, 2003; Ogundipe *et al.*, 2001; Okigbo *et al.*, I, 2009; Oloyede *et al.*, 2010)
Bridelia ferruginea Benth.	Ora	Leaves, stem and bark	West Africa Democratic republic of Congo, Nigeria	Anti-inflammatory. To treat diarrhea, dysentery, gastro-intestinal disorders, gynecological disorders (including sterility), and rheumatic pains.	A bioflavonoid: Gallocatechin-(4'→O →7)-Epigallocatechin.	(Adetutu *et al.*, 2011; Bakoma *et al.*, 2012; De Bruyne *et al.*, 1997; Cimanga *et al.*, 1999; Fabiyi *et al.*, 2012; Igoli *et al.*, 2003; Igoli *et al.*, 2005; Olajide *et al.*, 2000b)
Mallotus oppositifolius (Geiseler) Muell. Arg.	Jororo Káfàr mútúwàà Senampendi Mvundza jembe	Leaves, roots	West Africa Nigeria	Antioxidant, anti-inflammatory and antimicrobial activities. Used for abortion.	Flavonoids: quercetin and quercitrin.	(Barku *et al.*, 2013; Chukwujekwu *et al.*, 2005; Farombi, 2003; Farombi e *et al.*, 2001; Kamgang *et al.*, 2006; Nwaehujor *et al.*, 2012)

Contd...

Table 9.1–*Contd...*

Family and Genus	*Vernacular Name*	*Plant Part*	*Country/ Area*	*Medicinal Use and/or Experimental Validation*	*Compounds Isolated*	*Reference*
Fabaceae						
Aspalathus linearis (Brum. F.) R. Dahlgr.	Rooibos	Leaves	South Africa	Radical Scavenging Capacity Used to treat stomach cramps, insomnia, and to reduce stress.	Phenolic Fractions, Tannins and monomeric flavonoids aspalathin, nothofagin, quercetin, rutin, isoquercitrin, orientin, isoorientin, luteolin, vitexin, isovitexin, and chrysoeriol.	(Bramati *et al.*, 2003; Bramati *et al.*, 2002; Inanami *et al.*, 1995; Koeppen and Roux, 1966; Koeppen *et al.*, 1962; Lindsey *et al.*, 2002; Rabe *et al.*, 1994; Snijman *et al.*, 2009; Steenkamp *et al.*, 2004; Ulicna *et al.*, 2005)
Burkea africana Hook	Wild Syringa	Bark	Mali and Sub-Saharan Africa	Antioxidant and radical scavenging activity. Used to treat coughs, colds, stomach obstruction, infusions against gonorrhoea and syphilis.	Proanthocyanidins; fisetinidol-(4alpha- >8)-catechin 3-gallate and bis-fisetinidol-(4alpha- >6, 4alpha- >8)-catechin 3-gallate, with smaller amounts of flavan-3-ols (catechin, epicatechin and fisetinidol)	(Chanda and Dave, 2009; Diallo *et al.*, 2001; Mathisen *et al.*, 2002)
Crotalaria podocarpa DC.	Crotalaria	Roots	South Africa	Anti-inflammatory activity. Used for the treatment of sore-eyes and boils. Expectorant.	Flavonoids	(Majinda *et al.*, 2001; Stodard, 2013)
Cyclopia intermedia E. Mey. and *C. subternata* Vog.	Honeybush	Leaves and stem	South Africa	Antioxidant activity. Used as tonic for colds, catarrh and tuberculosis.	Pinitol, shikimic acid, p-coumaric acid, 4-glucosyltyrosol, epigallocatechin gallate, the isoflavone orobol, the flavanones hesperedin, narirutin and eriocitrin, a glycosylated flavan, the flavones luteolin, 5-deoxy-	(Joubert *et al.*, 2008; Kamara *et al.*, 2004; Kamara *et al.*, 2003; Lindsey *et al.*, 2002; McKay and Bloomberg, 2007; Steenkamp *et al.*, 2004)

Contd...

Table 9.1–*Contd...*

Family and Genus	*Vernacular Name*	*Plant Part*	*Country/ Area*	*Medicinal Use and/or Experimental Validation*	*Compounds Isolated*	*Reference*
					luteolin and scolymoside, the xanthone mangiferin and the flavonol C-6-glucosylkaempferol. Phenolic content: tyrosol and a methoxy analogue, 2-[4-[O-alpha-apiofuranosyl-(1' '–>6')-beta-d-glucopyranosyloxy] phenyl] ethanol, 4-[O-alpha-apiofuranosyl-(1' '–>2')-beta-d-glucopyranosyloxy]benzaldehyde, five glycosylated flavonols, two isoflavones, four flavanones, two isoflavones, and two flavones	
Erythrina latissima E. Mey.	Broad-leaved coral tree	Stem Wood Root wood Seeds	South Africa Botswana	Antimicrobial activity and weak radical scavenging properties. Purgative.	Flavonoids and isoflavonoids. Isoflavones: erylatissin A and B. Flavanone: erylatissin C and flavonoids and Isoflavone glycosides: 4'-hydroxyisoflavone-7-O-beta-D-glucopyranoside (compound 1); 4'-hydroxyisoflavone-7-O-alpha-L-rhamnosyl (1–>6)-beta-D-glucopyranoside (compound 2); and a new compound 4', 8-dimethoxy isoflavone-7-O-alpha-L-rhamnosyl (1–>6) glucopyranoside (8-O-methylretusin-	(Chacha *et al.*, 2005; Majinda *et al.*, 2001; Wanjala *et al.*, 2002; Wanjala and Majinda, 2001)

Contd...

Table 9.1–*Contd...*

Family and Genus	*Vernacular Name*	*Plant Part*	*Country/ Area*	*Medicinal Use and/or Experimental Validation*	*Compounds Isolated*	*Reference*
					7-O-alpha-L-rhamnosyl (1-6)-beta-D-glucopyranoside) (compound 3) Isoflavonoids: 5,7-dihydroxy-2',4',5'-trimethoxyisoflavanone.	
E. lysistemon Hutch.	Common coral tree; lucky bean tree	Bark	South Africa	Mild antioxidant activity. Used to treat sores, wounds, abscesses and arthritis.	Three prenylated flavonoid derivatives; 5,7,4'-trihydroxy-8-(3"'-methylbut-2"'-enyl)-6-(2"-hydroxy-3"-methylbut-3" enyl) isoflavone (isoerysenegalensein E), 5,7,2'-trihydroxy-4'-methoxy-5'-(3"-methylbut-2"-enyl) isoflavanone (lysisteisoflavanone), 5, 4'-dihydroxy-6-(3"'-methylbut-2"'-enyl)-2"-hydroxyisopropyl dihydrofurano [4",5": 8,7] isoflavone (isosenegalensin), together with the four known flavonoids abyssinone V-4'-methylether, alpinumisoflavone, wighteone and burttinone	(el-Masry *et al.,* 2002; Juma and Majinda, 2004; Mabona and van Vuuren, 2013; Steenkamp *et al.,* 2005b)
Melilotus elegans Salzm. ex Ser. (syn. *M. abyssinica* Baker)	Egug, Gugi, Yemen berri Elegant sweet clover	Leaves	Ethiopia	Anti-inflammatory properties. Used for asthma, haemorrhoid, wounds, excavated sore, piles, ulcers mouth infection, lacerated wounds, haemorrhoids, bronchial asthma (personal communication)	Flavonoids: kaempferol	(Asres *et al.,* 2005; Chorepsima *et al.,* 2013; Gebre-Mariam *et al.,* 2005; Gebre-Mariam *et al.,* 2006)

Contd...

Table 9.1–*Contd...*

Family and Genus	*Vernacular Name*	*Plant Part*	*Country/ Area*	*Medicinal Use and/or Experimental Validation*	*Compounds Isolated*	*Reference*
Millettia griffoniana	Not signalized	Root-bark and seeds	Cameroon	Anti-inflammatory activity. Used as antimalarial.	Coumarin: 4-hydroxy-3-(3,4-methylenedioxyphenyl)-5,6,7-trimethoxycoumarin, durmillone, odorantin, 7-methoxyebenosin, calopogonium isoflavone B and 7,2'-dimethoxy-4',5'-methylenedioxy isoflavone maximaisoflavone G (5) and 7-hydroxy-6-meth oxy-3',4'-methylenedioxy isoflavone and new prenylated isoflavonoids griffonianones A, B, C, D and E. Griffonianone D ((7E)-(6",7"-dihydroxy-3", 7"-dimethyloct-2"-enyl) oxy-4'-methoxyisoflavone), an isoflavone. 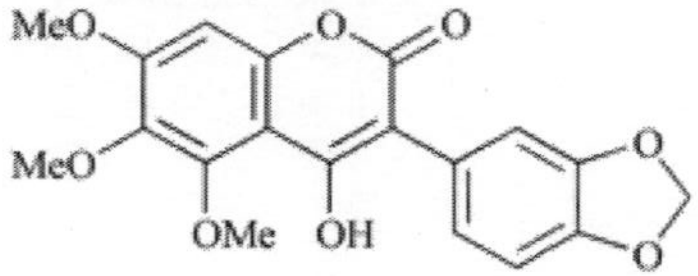 MeO MeO OMe OH O O O O	(Combes *et al.*, 2002; Baill. Ngamga *et al.*, 2005a; Ngamga *et al.*, 2005b; Yankep *et al.*, 1998; Yankep *et al.*, 1997; Yankep *et al.*, 2001, 2003; Zingue *et al.*, 2013)
Parkia biglobosa (Jacq.) Benth	African Locust Bean Nèrè Ojinyi	Bark Seeds	Mali Sudan Ivory Coast	Antiinflammatory activity. Used as antiseptic and to treat coughs, chest pain, and wound healing	Tocopherol, ascorbic acid (Seeds)	(Alabi *et al.*, 2005; Diallo *et al.*, 2001; Igoli *et al.*, 2005; Kouadio *et al.*, 2000; Lamien-Meda *et al.*, 2008)

Contd...

Table 9.1–*Contd...*

Family and Genus	*Vernacular Name*	*Plant Part*	*Country/ Area*	*Medicinal Use and/or Experimental Validation*	*Compounds Isolated*	*Reference*
Peltophorum africanum Sond.	Weeping Wattle	Root and bark	South Africa	Antioxidant and anti-bacterial activities. Used to treat diarrhoea, dysentery, sore throat, wounds, back and joint pains, HIV-AIDS, venereal diseases and infertility.	Flavonol glycosides and flavonol glucoside gallates	(Bizimenyera *et al.,* 2005; El Sherbeiny *et al.*, 1977; Mongalo, 2013; Shai *et al.,* 2011)
Piliostigma thonningii (Schum.) Milne-Redh	Camel's foot tree, Monkey Bread Niama (Mali).Abefe Kalgo Okpoatu Omepa	Root, bark, pods, leaves	Nigeria, Ethopia Botswana, Kenya, Namibia, Senegal, South Africa, Sudan, Tanzania, Uganda, Zambia	Anti-oxidant and anti-inflamatory properties. Used to treat wounds, chronic ulcers, cough, respiratory disorders and toothache, gum inflammation, arthritis, headache, bachache, and antioxidant supplement.	Proanthocyanidins epicatechin, catechin trimers and oligomers, flavonoids, polyphenolics, C-methylflavonols (in the leaf extract)	(Aderogba *et al.,* 2004; Ajiboye, 2010; Bombardelli *et al.*, 1994; Fakae *et al.,* 2000; Ibewuike *et al.*, 1996; Ibewuike *et al.*, 1997; Ighodaro *et al.,* 2012; Igoli *et al.,* 2005; Jimoh and Oladiji, 2005)
Sutherlandia frutescens R.Br.	Cancerbush Phetola	Leaves	South Africa	Superoxide and hydrogen peroxide scavenging activities. Used as tonic to boost the immune system.	Canavanine, pinitol	(Fernandes *et al.,* 2004; Tai *et al.,* 2004; van Wyk and Albrecht, 2008)
Trigonella foenu-mgraecum L.	Fenugreek	Seeds	Ethiopia, Moroco	Protective effect against Oxidative stress during ischemia-reperfusion. It is hypolipidemic, and is also used to treat boils and to improve appetite.	Free phenolics and Vit C.	(Bordia *et al.*, 1997; Dilsiz *et al.,* 2006; Kaviarasan *et al.,* 2007; Randhir *et al.,* 2004; Saleh *et al.*, 1977; Srinivasan *et al.,* 2004; Wojdylo *et al.,* 2007)

Contd...

Table 9.1–*Contd...*

Family and Genus	*Vernacular Name*	*Plant Part*	*Country/ Area*	*Medicinal Use and/or Experimental Validation*	*Compounds Isolated*	*Reference*
Humiriaceae						
Sacoglottis gabonensis Urb.	Cherry tree, ozouga	Stem-bark	West Africa	Antioxidant activity.	Bergenin Structure of bergenin	(Akintonwa and Maduka, 2005; Maduka and Okoye, 2002; Maduka *et al.*, 2002; Maduka and Okoye, 2001; Patel *et al.*, 2012)
Hypoxidaceae						
Hypoxis hemerocallidea Fisch. and C.A. Mey.	African potato	Corms	South Africa	Antioxidant activity. Used to treat tuberculosis, cancer, bladder disorders, benign prostatic hyperplasia.	Rooperol	(Nair *et al.*, 2007; Owira and Ojewole, 2009; Steenkamp *et al.*, 2005a; Van der Merwe *et al.*, 1993)
Lamiaceae						
Ocimum basilicum L.	Mükandu	Leaves	Burkina Faso	Intermediate antioxidant activity and high anti-bacterial activity.	Linalool basil oil	(Gulcin *et al.*, 2007; Juliani *et al.*, 2004)
	Basil		Ethopia	Used in Ethopia to treat Conjunctivitis and in Kenya to treat colds and stomache-ache.	Methyl chavicol, eugenol, (E)-methyl cinnamate, thymol, linalool etc. (Jansen, 1981).	

Contd...

Table 9.1–*Contd...*

Family and Genus	*Vernacular Name*	*Plant Part*	*Country/ Area*	*Medicinal Use and/or Experimental Validation*	*Compounds Isolated*	*Reference*
Ocimum gratissimum L.	The tea bush, Scent leaf/ Nchuanwu. Ujuju okpevu Basil	Leaves	Popular republic of Congo (ex Brazaville Congo) Eastern Nigeria	Antioxidant activity Popular republic of Congo it is used as a laxative, purgative, and to treat snakebite, diabetes, tooth ache, gingivitis.	Xanthomicrol, cirsimaritin, rutin, kaempferol 3-*O*-rutinoside and vicenin-2 were identified as the major flavonoids, whereas luteolin 5-*O*-glucoside, luteolin 7-*O*-glucoside, apigenin 7-*O*-glucoside, vitexin, isovitexin, quercetin 3-*O*-glucoside and isothymusin were detected as minor constituents.	(Akinmoladun *et al.*, 2007; Fakae *et al.*, 2000; Grayer *et al.*, 2000; Igoli *et al.*, 2005; Odukoya *et al.*, 2005; Prakash *et al.*, 2011; Vieira *et al.*, 2001)
Lauraceae						
Cinnamomum zeylanicum Breyne	Cinnamon leaf	Leaves	Madaga-scar Ethiopia	Very high antioxidant and high antimicrobial activities. Used to treat diarrhoea, rheumatism, colds and hypertension	Cinnamaldehyde, eugenol and eugenyl acetate to be the main constituents of cinnamon oil.	(Baratta *et al.*, 1998; Jayaprakasha *et al.*, 2007; Juliani *et al.*, 2004; Mancini-Filho *et al.*, 1998)
Ocotea bullata (Burch.) Baill.	Black stinkwood Unukane (Zulu)	Bark	South Africa	Anti-inflamatory, cyclooxygenase inhibitory activity. Urinary disorders, headaches.	Monoterpenoids	(Jäger and van Staden, 2005; Steenkamp *et al.*, 2005b)
Ravensara aromatica Sonn.	Nutmeghavozo	Bark Leaf	Madaga scar	Low antioxidant and antimicrobial activity. Useful for chronic respiratory conditions, and sometimes helpful in cases of asthma.	Essential oils, principally composed of the monoterpene hydrocarbons a-pinene, sabinene, myrcene, limonene, and the azulene: iso-ledene. In barks, estragole (methyl chavicol) but leaves contain b-myrcene, 1,8-cineole, linalool, and carotol.	(Juliani *et al.*, 2004; Möllenbeck *et al.*, 1997; Zabka *et al.*, 2009)

Contd...

Table 9.1–*Contd...*

Family and Genus	*Vernacular Name*	*Plant Part*	*Country/ Area*	*Medicinal Use and/or Experimental Validation*	*Compounds Isolated*	*Reference*
Malvaceae						
Hibiscus sabdariffa L.	Red tea, sorelle Rosella	Flowers	Nigeria South Africa	Antimutagenic activity and free radical scavenging effects on active oxygen species. Used against insomnia, colic.	Flavonol glucoside hibiscitrin Anthocyanins. Such as cyanidin 3-*O*-β-D-glucopyranoside, cyanidin 3-*O*-(2-*O*-β-D-xylopyranosyl)-β-D-glucopyranoside, delphinidin 3-*O*-β-D-glucopyranoside and delphinidin 3-*O*-(2-*O*-β-D-xylopyranosyl)-β-D-glucopyranoside.	(Farombi and Fakoya, 2005; Lindsey *et al.*, 2002; Palé *et al.*, 2004; Prenesti *et al.*, 2007; Steenkamp *et al.*, 2004; Wang *et al.*, 2000)
Meliaceae						
Trichilia roka Chiov.	Soulafinzan	Root	Tropical Africa Mali	Significantly protective against CCl_4-induced liver damage and prevented perisinusoidal fibrosis. Used to treat malaria, abdominal pain and dermatitis.	Polyphenols	(Germano *et al.*, 2001; Nana *et al.*, 2013)
Menispermaceae						
Sphenocentrum jollyanum Pierre	Akerejupon ajo	Fruit Root	West Africa	Anti-inflammatory activity. Used to treat inflammatory-based diseases	Furanoditerpenes: columbin, isocolumbin. Flavonoids-rich fraction.	(Moody *et al.*, 2005; Olorunisola *et al.*, 2011; Olorunisola and Afolayan, 2013)
Tinospora bakis		Whole plant	Sudan	Anti-inflammatory activity. To treat headache and rheumatism	A diterpenoid furanolactone, columbin	(Abdelwahab *et al.*, 2012)

Contd...

Table 9.1–*Contd...*

Family and Genus	*Vernacular Name*	*Plant Part*	*Country/ Area*	*Medicinal Use and/or Experimental Validation*	*Compounds Isolated*	*Reference*
Moraceae						
Dorstenia barteri var. *subtriangularis* (Engler) M.E.E. Hijman and C.C. Berg	Contrayerva	Twigs/ leaves	Cameroon	Antioxidant properties account for the anti-inflammatory action of these extracts. Used to treat arthritis, rheumatism, gout, headache and other forms of body pains.	Prenylated flavonoids: Three **diprenylated chalcones**: **bartericins A** (-)-3-(3,3-dimethylallyl)-5'-(2-hydroxy-3-methylbut-3-enyl)-4,2',4'-trihydroxychalcone, **bartericins B** (+)-3-(3,3-dimethylallyl)-4',5'-[2'''-(1-hydroxy-1-methylethyl)-dihydrofurano]-4,2'-dihydroxychalcone and **bartericins C** 3,4-(6",6"-dimethyldihydropyrano)-4',5'-[2''',-(1-hydroxy-1-methylethyl)-dihydrofurano]-2'-hydroxychalcone and also two novel **diprenylated chalcones**: 3,5'-di-(2-hydroxy-3-methylbut-3-enyl)-4,2',4'-trihydroxychalcone, 3, 4-(2,2-dimethylpyrano)-3'-(2-hydroxy-3-methylbut-3-enyl)-2',4'-dihydroxychalcone, 4,2', 4'-trihydoxy-3'-prenylchalcone and 4,2',4'-trihydoxy-3, 3'-diprenylchalcone; and 5,7,4'-trihydoxy-8-prenylflavone. Also known compounds such as stipulin, 4-hydroxylonchocarpin, kanzonol B, 3'-(2-	(Kuete *et al.*, 2010; Majinda *et al.*, 2001; Ngadjui *et al.*, 2005; Ngameni *et al.*, 2004; Omisore *et al.*, 2005; Omisore *et al.*, 2004; Tsopmo *et al.*, 1999)

Contd...

Table 9.1–*Contd...*

Family and Genus	*Vernacular Name*	*Plant Part*	*Country/ Area*	*Medicinal Use and/or Experimental Validation*	*Compounds Isolated*	*Reference*
					Hydroxy-3-methylbut-3-enyl)-5'-(3,3-dimethylallyl)-4,2',4'-trihydroxychalcone, and dorstenone.	
D. ciliata Engl.	Contrayerva	Aerial parts	Cameroon Central Africa	Antiradical and anti-oxidant activities. Used as food additive.	Phenolic compound (6-prenyl-apigenin) Flavones: (ciliatin A) 5,4'-Dihydroxy-5"-isopropenyl-dihydrfuranol[2",3":7,6]flavone 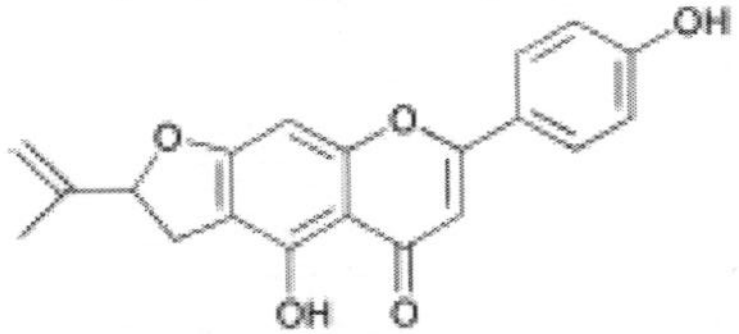 (ciliatin B) 7,4'-Dihydroxy-3'-methoxy-6",6"-dimethyl-dihydropyranol[2",3":5,6]. 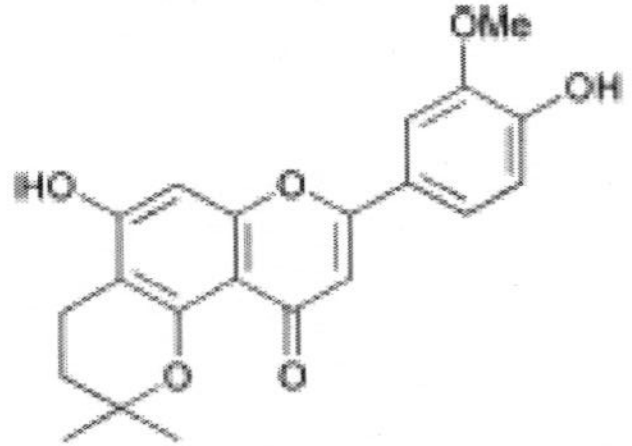	(Al-Jaber *et al.*, 2011; Kansci *et al.*, 2003; Ngadjui *et al.*, 2002b)

Contd...

Table 9.1–*Contd...*

Family and Genus	*Vernacular Name*	*Plant Part*	*Country/ Area*	*Medicinal Use and/or Experimental Validation*	*Compounds Isolated*	*Reference*
D. convexa De Wild.	Contrayerva	Twigs and leaves	Democratic Republic of the Congo	Antioxidant properties account for the anti-inflammatory action of these extracts. Used to treat	Prenylated flavonoids	(Majinda *et al.*, 2001; Omisore *et al.*, 2005; Omisore *et al.*, 2004)
D. mannii Hook.f.	Contrayerva	Twigs/ leaves. Aerial parts	Central Africa	Antioxidant action against copper-induced LDL oxidation, this activity is similar to the non-prenylated flavonoid quercetin. Also inhibition of platelet aggregation and influence of cyclooxygenase and lipoxygenase activity. Used to treat rheumatism, stomach disorders. Anti-trichomonal activity.	Grenylated and prenylated flavonoids and flavonones: Flavonones: 6,8-diprenyl-5, 7,3'4'-tetrahydroxyflavanone, 4-hydroxylonchocarpin, 4-methoxylonchocarpin, 6-prenylchrysoeriol, 6,8-diprenyleriodictyol, gancaonin P and Prenylated flavonoids: 6,8-diprenyleriodictyol, dorsmanin C 7,8-(2,2-Dimethylchromeno)-6-geranyl-3,5,3', 4'-tetrahydroxyflavone and dorsmanin D 6,8-Diprenyl-3,5,7,4'-tetrahydroxy-3'-methoxyflavone, dorsmanins 1, J and 2"-epimers of dorsmanins F (6,7-(2,2-dimethylpyrano)-8-prenyl-5, 3',4'-trihydroxyflavanone, G (6,7-(2,2-dimethyldihydropyrano)-8-prenyl-5,3',4'-trihydroxflavanone). Also dorsmanins F and G. Four new prenylated **flavanones**, named dorsa-	(Dufall *et al.*, 2003; Majinda *et al.*, 2001 Mbaveng *et al.*, 2012; Ngadjui *et al.*, 1999a; Ngadjui *et al.*, 1998a; Ngadjui *et al.*, 2000)

Contd...

Table 9.1–*Contd...*

Family and Genus	*Vernacular Name*	*Plant Part*	*Country/ Area*	*Medicinal Use and/or Experimental Validation*	*Compounds Isolated*	*Reference*
					mine F (7,8-[2 "-(1-hydroxy-1-methylethyl)-dihydrofurano]-6-prenyl-5,3',4'-trihydroxy-flavanone), dorsmaine G (6,7-[2 "-(1-hydroxy-1-methylethyl) dihydrofurano]-8-prenyl-5,3',4'-trihydroxyflavanone) and dorsamine H (6-prenyl-8-(2-hydroxy-3-methylbut-3-enyl)-5,7,3',4'-tetrahydroxy-flavanone). **Dorsmanin I** 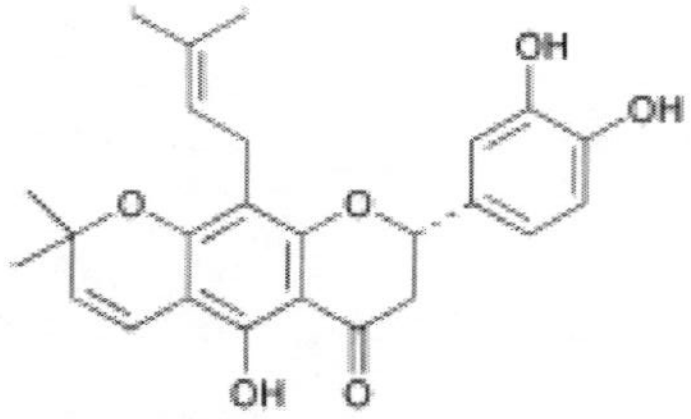 OH OH O OH O **Dorsmanin J** OH OH O OH O 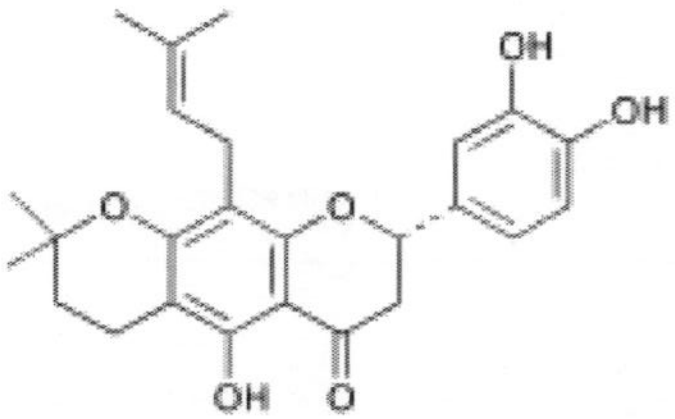	

Contd...

Table 9.1–*Contd...*

Family and Genus	*Vernacular Name*	*Plant Part*	*Country/ Area*	*Medicinal Use and/or Experimental Validation*	*Compounds Isolated*	*Reference*
					Epi-dorsmanin F	
					Epi-dorsmanin G	
D. poinsettifolia var. *angusta* Engl.	Dingetenga	Whole plant	Cameroon	Antiradical and anti-oxidant activities. Used to treat infected wounds.	Grenylated and prenylated flavonoids. The unusual 4-phenyl-substituted dihydro-coumarin, and the rare geranyl-and prenyl-substituted Chalcone.	(Kuete and Sandjo, 2012; Ngadjui *et al.*, 1999b; Tsopmo *et al.*, 1998)

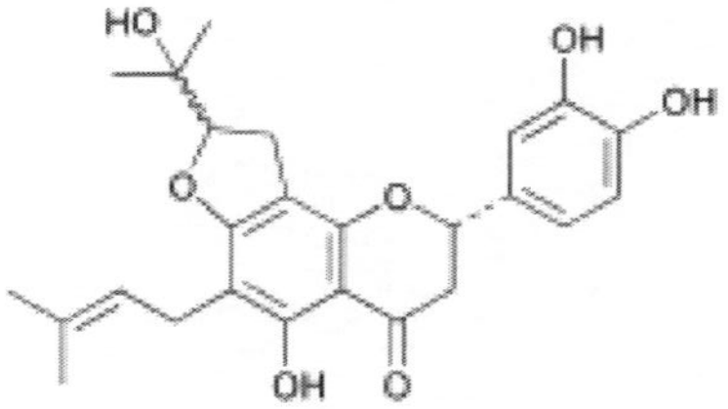

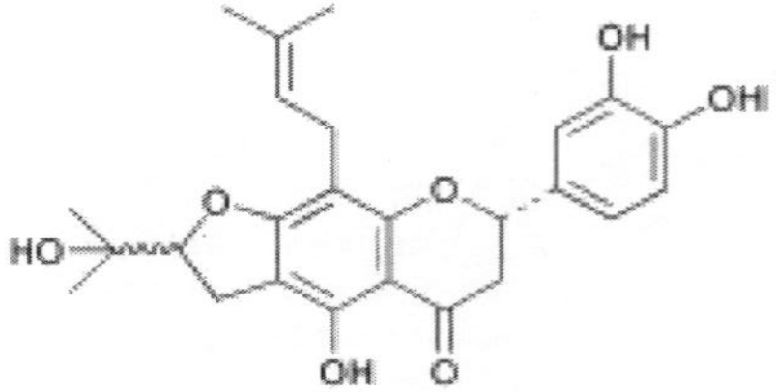

Contd...

Table 9.1–*Contd...*

Family and Genus	*Vernacular Name*	*Plant Part*	*Country/ Area*	*Medicinal Use and/or Experimental Validation*	*Compounds Isolated*	*Reference*
D. psilurus Welw.	Dingetenga	Roots	Cameroon Central Africa	Antiradical and anti-oxidant activities. Used against snakebite and to treat rheumatism, headache and stomach disorders.	Grenylated and prenylated flavonoids.Three **phenolic compounds**: 6,8-diprenyl-3'[O],4'-(2,2-dimethylpyrano)-3,5,7-trihydroxyflavone, 3,6-diprenyl-8-(2-hydroxy-3-methylbut-3-enyl)-5,7,2',4'-tetrahydroxyflavone and an unusualB/C ring modified flavonoid derivative for which the names dorsilurins C, D and E, respectively, are proposed. Two new **flavones**, dorsilurins A and B, and a new benzofuran derivative have been isolated from Dorstenia psilurus, together with three known phenyl-propanoid derivatives, stearyl-p-coumarate [octadecanyl 3-(4-hydroxyphenyl)prop-2-enoate], stearyl ferulate [octa-decanyl 3-(4-hydroxy-3-meth-oxyphenyl)prop-2-enoate] and psoralen.	(Abegaz *et al.*, 1999; Etoundi *et al.,* 2010; Kansci *et al.,* 2003; Ngadjui *et al.*, 1998b; Ngadjui *et al.*, 1999c)
Myrtaceae						
Eugenia elliptica Sm. *Syzygium smithii* (Poir.) Nied.	Lilly Pilly	Leaves	Mauritius	Modulate the expression of the antioxidant enzyme genes.	Quercetin-3-O-galactoside (hyperoside), kaempferol-3-glucoside (astragalin), quercetin-3-O-glucoside (isoquercitrin), (+)-catech	(Angaji *et al.,* 2012; Toyokuni *et al.,* 2003)

Contd...

Table 9.1–*Contd...*

Family and Genus	*Vernacular Name*	*Plant Part*	*Country/ Area*	*Medicinal Use and/or Experimental Validation*	*Compounds Isolated*	*Reference*
E. fasciculata Wall.	Not signalized	Leaves	Mauritius	Modulate the expression of the antioxidant enzyme genes.	Quercetin-3-O-galactoside (hyperoside), kaempferol-3-glucoside (astragalin), quercetin-3-O-glucoside (isoquercitrin), (+)-catech. procyanidin B2 dimer and (-)-epicatechin	(Toyokuni *et al.*, 2003)
E. orbiculata Lam.	Not signalized	Leaves	Mauritius	Modulate the expression of the antioxidant enzyme genes.	Quercetin-3-O-galactoside (hyperoside), kaempferol-3-glucoside (astragalin), quercetin-3-O-glucoside (isoquercitrin), (+)-catech. quercetin-3-O-rutinoside (rutin)	(Neergheen *et al.*, 2007; Toyokuni *et al.*, 2003)
E. pollicina J.Gueho and A.J.Scott	Not signalized	Leaves	Mauritius	Modulate the expression of the antioxidant enzyme genes.	Quercetin-3-O-galactoside (hyperoside), kaempferol-3-glucoside (astragalin), quercetin-3-O-glucoside (isoquercitrin), (+)-catech. (-)-epicatechin gallate	(Toyokuni *et al.*, 2003; Ramful *et al.*, 2011)
Monimiastrum acutisepalum J. Gueho and A.J. Scott.	Not signalized	Leaves	Mauritius	Modulate the expression of the antioxidant enzyme genes.	Quercetin-3-O-galactoside (hyperoside), kaempferol-3-glucoside (astragalin), and quercetin-3-O-glucoside (isoquercitrin). (+)-catechin	(Anganji *et al.*, 2012; Neergheen *et al.*, 2007; Toyokuni *et al.*, 2003)
M. globosum J. Gueho and A.J. Scott	Not signalized	Leaves	Mauritius	Modulate the expression of the antioxidant enzyme genes.	Quercetin-3-O-galactoside (hyperoside), kaempferol-3-glucoside (astragalin), and quercetin-3-O-glucoside (isoquercitrin). (-)-epicatechin gallate	(Toyokuni *et al.*, 2003)

Contd...

Table 9.1–*Contd...*

Family and Genus	*Vernacular Name*	*Plant Part*	*Country/ Area*	*Medicinal Use and/or Experimental Validation*	*Compounds Isolated*	*Reference*
Syzygium aromaticum (L.) Merr. and L.M. Perry	Clove bud	Dried flowers Buds	Mada-gascar Sudan	Antioxidant and anti-microbial activities. Used to treat tooth ache and throat inflammation.	Eugenol Methyleugenol	(Barakat, 2014; Juliani *et al.*, 2004; Nassar *et al.*, 2007)
S. coriaceum J.Bosser and J.Guého	Bois de pomme		Mauritius	Abilities to modulate the expression of the antioxidant enzyme genes.	**Phenols and flavonoids**: Quercetin-3-*O*-rutinoside, kaempferol-3-glucoside (astragalin) and quercetin-3-O-glucoside (isoquercitrin), (+)-catechin, procyanidin B1 dimer, (-)-epicatechin gallate	(Toyokuni *et al.*, 2003)
S. glomeratum DC. pomme	Bois de	Leaves	Mauritius	Abilities to modulate the expression of the anti-oxidant enzyme genes. Used to treat boils, abscesses, fever and wounds and as expectorant.	**Phenols and flavonoids**: kaempferol-3-glucoside (astragalin) and quercetin-3-O-glucoside (isoquercitrin), procyanidin B1 dimer, (-)-epicatechin gallate, chloro-genic acid, (-)-epicatechin	(Toyokuni *et al.*, 2003)
S. guehoii	Not signalized		Mauritius	Abilities to modulate the expression of the antioxidant enzyme genes.	**Phenols and flavonoids**: quercetin-3-O-rutinoside (rutin), kaempferol-3-glucoside (astragalin) and quercetin-3-O-glucoside (isoquercitrin), (+)-catechin, chlorogenic acid, procyanidin B2 dimer	(Toyokuni *et al.*, 2003)

Contd...

Table 9.1–*Contd...*

Family and Genus	*Vernacular Name*	*Plant Part*	*Country/ Area*	*Medicinal Use and/or Experimental Validation*	*Compounds Isolated*	*Reference*
S. mauritianum J.Gueho and A.J.Scott	Not signalized	Leaves	Mauritius	Abilities to modulate the expression of the anti-oxidant enzyme genes.	**Phenols and flavonoids**: quercetin-3-*O*-rutinoside (rutin), kaempferol-3-glucoside (astragalin) and quercetin-3-*O*-glucoside (isoquercitrin), (+)-catechin, chlorogenic acid	(Toyokuni *et al.*, 2003)
S. petrinense J.Bosser and J.Guého	Not signalized		Mauritius	Abilities to modulate the expression of the anti-oxidant enzyme genes.	**Phenols and flavonoids**: quercetin-3-*O*-rutinoside (rutin), kaempferol-3-glucoside (astragalin) and quercetin-3-*O*-glucoside (isoquercitrin), procyanidin B1 dimer, chlorogenic acid	(Toyokuni *et al.*, 2003)
S. venosum (Lam.) J.Gueho and A.J.Scott	Not signalized		Mauritius	Abilities to modulate the expression of the anti-oxidant enzyme genes.	**Phenols and flavonoids**: quercetin-3-*O*-rutinoside (rutin), kaempferol-3-glucoside (astragalin) and quercetin-3-*O*-glucoside (isoquercitrin), (+)-catechin, procyanidin B2 dimer	(Ajaib *et al.*, 2011; Neergheen *et al.*, 2007; Toyokuni *et al.*, 2003)
Oleaceae						
Olea europaea subsp africana (Mill.)P.S. Green	African wild olive	Leaves	South Africa	Potent antioxidant activity. Used as eye lotions and tonics, lower blood pressure, improve kidney function and deal with sore throats. The early Cape settlers used the fruits to treat diarrhoea	Oleuafricein (mixture of oleanolic acid and ursolic acids), Triterpenoids and oleoropein.	(Lee and Lee, 2010; Somova *et al.*, 2003; Van Wyk *et al.*, 1997)

Contd...

Table 9.1–*Contd...*

Family and Genus	*Vernacular Name*	*Plant Part*	*Country/ Area*	*Medicinal Use and/or Experimental Validation*	*Compounds Isolated*	*Reference*
Pedaliaceae						
Harpagophytum procumbens DC. ex Meissner	Devil's claw	Root	South Africa Native to the Kalahari Desert of southern Africa, Namibia and Botswana.	Antiinflammatory and ability to inhibit the expression of cyclooxygenase-2 and inducible nitric oxide by suppression of NF-kappaB activation. Used for pain, muscular tension, osteo-arthritis, degenerative rheumatism or painful arthrosis and tendonitis as well as tonic for loss of appetite and dyspeptic complaints. Can also be genotoxic.	Roots contain iridoid glycosides mainly harpagoside. Other constituents are flavones and flavonols kaempferol, and luteolin.	(Betancor-Fernandez *et al.*, 2003; Gagnier *et al.*, 2004; Gobel *et al.*, 2001; Grant *et al.*, 2007; Huang *et al.*, 2006; Kaszkin *et al.*, 2004; Mahomed *et al.*, 2005; Mahomed and Ojewole, 2004; Mahomed and Ojewole, 2006a, 2006b; McGregor *et al.*, 2005)
Piperaceae						
Piper guineense Schum. and Thonn.	West African black pepper Bush pepper Ikom, Amana kakwale iyeyeh ashoesie taquale Meshoro	Fruit, seed and leaf	Ghana, West Africa Nigeria Cameroon	Antioxidant activity.	Volatile oil components-monoterpenes, sesquiter penes, terpenoids, lignans and sterols.	(Abila *et al.*, 1993; Agbor *et al.*, 2005; Agbor *et al.*, 2007; Isong and Essien, 1996; Natarajan *et al.*, 2006)
Podocarpaceae						
Podocarpus species *Podocarpus elongates Podocarpus falcatus,*		Leaves and young stems	Eastern and Southern Africa	These species are used to treat fevers, asthma, coughs, cholera, chest complaints, arthritis, rheumatism, painful joints and venereal diseases	Diterpenoids, bioflavonoids and	(Abdillahi *et al.*, 2011)

Contd...

Table 9.1–*Contd...*

Family and Genus	*Vernacular Name*	*Plant Part*	*Country/ Area*	*Medicinal Use and/or Experimental Validation*	*Compounds Isolated*	*Reference*
Podocarpus henkelii and *Podocarpus latifolius*						
Ranunculaceae						
Nigella sativa L.	Black cumin	Seed	African countries in the Mediterranean region	Antioxidant potentials through scavenging ability of different free radicals including the superoxide anion radical, inhibition of lipid peroxidation, and protection of liver against carbon tetrachloride (CCl_4)-induced liver fibrosis in rabbits. Used to treat diarrhoea, asthma, and as gastroprotective agent.	Oil: Thymoquinone	(Badary *et al.*, 2003; Erkan *et al.*, 2008; Farah *et al.*, 2005; Meral *et al.*, 2001; Ramadan *et al.*, 2003)
Rosaceae						
Crataegus monogyna Jacq.	Hawthorn, May Blossom, May Day Flower, White Thorn.	Fresh vegetative and reproductive organs.	Mauritius, Northern Africa	Antioxidant activities. Used for its neuro- and cardiosedative actions.	**Polyphenols**: (proanthocyanidin, flavonoid, antho cyanin, (-)-epicatechin, procyanidin B2, chlorogenic acid). **Flavonoids**, quercetin and quercetrin, glycosides, proanthocyanidins, anthocynaidins, saponins, tannins, and cratetegin (*most prevalent in the flowers, then leaves, then berries*). Also, Vitamin C.	(Bahorun *et al.*, 2003; Bahorun *et al.*, 1996; Bahorun *et al.*, 1994; Bernatonienë *et al.*, 2008)

Contd...

Table 9.1–*Contd...*

Family and Genus	*Vernacular Name*	*Plant Part*	*Country/ Area*	*Medicinal Use and/or Experimental Validation*	*Compounds Isolated*	*Reference*
Leucosidea sericea		Leaf, bark and roots	Southern Africa	Antimicrobial and anti-inflammatory properties	Phenolics, alkaloids and saponins	(Aremu *et al.,* 2010)
Pygeum africanum Hook. f.	African plum tree Red Stinkwood	Bark	South Africa	Anti-inflammatory. Used to treat against benign prostatic hyperplasia, prostatitis	14 per cent triterpenes (urolic acids, oleanolic acid, cratae-golic acid), 0.5 per cent n-docosanol Phytosterol (β-sitosterol, β-sitosterone, Campesterol	(Breza *et al.*, 1998; Ishani *et al.,* 2000; Paubert-Braquet *et al.*, 1994; Steenkamp *et al.,* 2005a; Wang *et al.,* 2010)
Rubiaceae						
Crossopteryx febrifuga Benth.	Roger Blench ”rima jogoo-hi/je”	Seeds Leaf and roots	Mali Nigeria	Radical scavenging and lipoxygenase inhibition activities. Used to treat fever and various respiratory diseases	Flavonoids	(Maiga *et al.,* 2006; Salawu *et al.,* 2011; Occhiuto *et al.*, 1999; Tomas-Barberan and Hostettmann, 1988)
Rutaceae						
Agathosma betulina (Berg.) Pillans.	Round-leaf buchu	Leaves, stems	South Africa	Hydroxyl radical ion scavenging ability. Used for stomach problems, kidney and urinary track diseases.	Essential oils and flavonoids	(Lis-Balchin, Hart and Simpson, 2001; Steenkamp *et al.,* 2005a, 2005b)
A. crenulata (L.) Pillans	Oval-leaf buchu	Leaves, stems	South Africa	Anti-inflammatory activity. Used to treat benign prostatic hyperplasia, prostatitis, diabetes, inflammation of the colon, gums, and mucous membranes. Leaves chewed to relieve stomach complaints.	Essential oils and flavonoids	(Lis-Balchin *et al.,* 2001; Steenkamp *et al.,* 2005a; Steenkamp *et al.,* 2005b; Van Wyk *et al.*, 1997)

Contd...

Table 9.1–*Contd...*

Family and Genus	*Vernacular Name*	*Plant Part*	*Country/ Area*	*Medicinal Use and/or Experimental Validation*	*Compounds Isolated*	*Reference*
Fagara zanthoxyloides Lam.	xeti, xe Wô	Roots, root-bark.	Cameroon, Unganda	Antioxidant activity. Used to treat gingivitis, toothache, urinary and venereal diseases, rheumatism and lumbago, malaria and other infections.	Phenylethanoid derivative, lignans and fagaronine	(Chaaib *et al.*, 2003; Imaga *et al.*, 2011; Messmer *et al.*, 1972; Ogwal-Okeng *et al.*, 2003; Rotimi *et al.*, 1988)
Sapindaceae						
Dodonaea viscosa Jacq. Synonyms: *Dodonaea angustifolia* L. f.; *Ptelea viscosa* L.	Umusasa	Leaves	Rwanda	Anti-inflammatory activity by inhibiting the synthesis of prostaglandin (PG) E(2). Used to treat rheumatism, skin infections, diarrhea, stomachaches, pains of hepatic and splenic origin, uterine colic. It is also used as an antipuritic in skin rashes and for the treatment of some throat, dermatitis and hemmorroids.	Quercetin, isorhamnetin	(Getie *et al.*, 2003; Getie *et al.*, 2002; Mothana *et al.*, 2010; Rojas *et al.*, 1996; Teffo *et al.*, 2010)
Xanthorrhoeaceae						
Aloe ferox Mill.	Bitter aloe or Cape aloe	Leaves	South Africa, Lesotho	*A. ferox* gel contains at least 130 medicinal agents with anti-inflammatory, analgesic, calming, antiseptic, antiviral, antiparasitic and anticancer effects	Chromones, anthraquinones, anthrone, anthrone-C-glycosides, and other phenolic compounds Barbaloin	(Mahomoodally, 2013)

Contd...

Table 9.1–*Contd...*

Family and Genus	*Vernacular Name*	*Plant Part*	*Country/ Area*	*Medicinal Use and/or Experimental Validation*	*Compounds Isolated*	*Reference*
Zingiberaceae						
Siphonochilus aethiopicus (Schweinf.) B.L. Burtt.	Wild ginger Natal ginger African Ginger	Rhizome	South Africa	Anti-inflamatory activity through cyclooxygenase inhibitory (prostaglandin-synthetase inhibition), activity. Used to treat Coughs, colds, asthma.	Sesquiterpenoid	(Jäger and van Staden, 2005; Opoku *et al.*, 2011; Steenkamp *et al.*, 2005b)

Table 9.2: Medicinal Plants with Confirmed Antioxidant Activity where the Compounds Responsible for the Activity have not yet been Identified and Medicinal Plants that Contain Compounds that are known to have Antioxidant Activity but these Compounds have not been Confirmed to be Responsible for the Activity.

Family Plant Name	*Vernacular Name*	*Plant Part*	*Country/ Area*	*Medicinal Use and/or Experimental Validation*	*Compounds Isolated*	*Reference*
Acanthaceae						
Barleria species *B. albostellata, B. greenii, B. prionitis*		Leaves, twigs and roots	South Africa	Anti-inflammatory and antioxidant activities	**Not identified**	(Amoo *et al.,* 2009; Amoo *et al.,* 2011)
Hypoestes rosea Decne.	Not signalized	Leaf extract	Nigeria	Anti-inflammatory activity due in part to its ability to inhibit NF-kappaB activation through direct inhibition of IkappaB kinase (IKK).	**Diterpene**: Hypoestoxide (a bicyclo [9,3,1] pentadecane)	(Ojo-Amaize *et al.,* 2001; Ojo-Amaize *et al.,* 2002)
Aizoaceae						
Glinus lotoides L.	"Mettere" Hairy carpet-weed	Seeds	Cameroon Ethiopia, Sudan, Uganda, Egypt	Used to treat cardiovascular and gastrointestinal system.	Three **flavonoids**: apigenin-7-O-glucoside, isovitexin, and luteolin-7-O-glucoside Three **isoflavonoids**: 5,7, 2',4'-tetrahydroxy-6-(3,3-dimethylallyl)isoflavone, 5,7,4'-trihydroxy -6,3'-di-(3,3-dimethylallyl)isoflavone, and 5,7,2',4'-tetrahydroxy-6,3'-di-(3,3-dimethylallyl) isoflavone.	(Abegaz *et al.,* 1999; Chernishov, Arragie and Etana, 1978; Endale, 2005; Endale *et al.,* 2005; Endale, *et al.,* 2004; Mohamed, 1998)
G. oppositifolius (L.) Aug. DC.	Balasa	Whole plant	Mali	Antioxidant and radical scavenging abilities.	Not identified	(Diallo *et al.,* 1999; Diallo *et al.,* 2001)

Contd...

Table 9.2–*Contd...*

Family Plant Name	*Vernacular Name*	*Plant Part*	*Country/ Area*	*Medicinal Use and/or Experimental Validation*	*Compounds Isolated*	*Reference*
Aloaceae						
Aloe claviflora Burch.	Kraal aloe		South Africa	Free radical scavenging and moderate inhibition in lipid peroxidation. Used as a purgative.	Not identified	(Lindsey *et al.,* 2003)
A. maculata Forssk. (=A. saponaria)	'Yellow Form' Tiger Aloe, Soap Aloe		South Africa	Free radical scavenging and moderate inhibition in lipid peroxidation. Used as a purgative.	Not identified	(Lindsey *et al.,* 2003)
A. thraskii Baker	Dune aloe		South Africa	Free radical scavenging and moderate inhibition in lipid peroxidation. Used as a purgative.	Not identified	(Lindsey *et al.,* 2003)
Anacardiaceae						
Anacardium occidentale L.	Not signalized	Stem bark	Nigeria	Anti-inflammatory properties.	Not identified	(Olajide *et al.,* 2004)
Sclerocarya birrea (A.Rich.) Hochst	Marula	Stem-bark		Anti-inflammatory activity. Used to treat diabetes, tonsillitis, snake bite and also diarrhoea.	Not identified	(Ojewole, 2004)
Annonaceae						
Enantia chlorantha Oliver	Erenbavbogo, Mföl Muamba	Root, stem-bark	Nigeria	Antiinflammatory activity. Used to treat ulcers and leprous spots wounds. Bark sap is taken as decoction against diarrhoea.	Not identified	(Agbaje *et al.,* 2003; Atata *et al.,* 2003; Moody *et al.,* 1995; Tan *et al.,* 2002)

Contd...

Table 9.2–*Contd...*

Family Plant Name	*Vernacular Name*	*Plant Part*	*Country/ Area*	*Medicinal Use and/or Experimental Validation*	*Compounds Isolated*	*Reference*
Uvaria afzelii Sc. Elliot	Pareho-houon, Bahie oulin	Leaves, roots and stem-bark	Ivory Coast	Used as for its antiparasitic activity	Anthocyanins and other flavonoids	(Koffi *et al.*, 2004; *et al.*, 2006; Okwu and Iroabuchi, 2009)
U. chamae P.Beauv.	Okandii Anweda tsoGa	Stem, bark Leaves, root	Ivory Coast Nigeria	Used as for its antiplasmodial activity.	Polyphenols	(Fall *et al.*, 2002; Igoli *et al.*, 2005; Koffi *et al.*, 2004; Philipov *et al.*, 2000)
Apocynaceae						
Alstonia boonei De Wild.	Awun, Egbu	Stem-bark Root-bark	Nigeria Ghana	Anti-inflammatory activity. Used for its analgesic and anti-inflammatory properties.	Not identified	(Fakae *et al.*, 2000; Kweifio-Okai, 1991a; Kweifio-Okai, 1991b; Olajide *et al.*, 2000a)
Picralima nitida Th. and H. Dur.	Ghana: Kpetepetetso, Kanwini, Kanwinu Cameroon: *motoko-toko*	Seeds Stem-bark	Ghana	Anti-inflammatory activity. Used for its analgesic and anti-inflammatory properties.	Not identified	(Betti, 2004; Chukwujekwu *et al.*, 2005; Duwiejua *et al.*, 2002; Ezeamuzie *et al.*, 1994; Fakeye *et al.*, 2000)
Rauvolfia vomitoria Afzel.	Asofeyeje, adapopo Mwanje	Root bark	Ghana	Anti-inflammatory activity. Used for its analgesic, antipyretic and anti-inflammatory activities. Also to treat scabies, high blood pressure, fever and snakebites.	Not identified	(Kweifio-Okai, 1991a; Kweifio-Okai, 1991b)
Araliaceae						
Cussonia barteri Seem.	Cabbage tree	Leaves Roots	Nigeria, Mali	Antioxidant and radical scavenging abilities. Inhibitory activity on 5-lipoxygenase and cyclooxygenase-1.	Not identified	(Diallo *et al.*, 2001; Dubois *et al.*, 1986; Papajewski *et al.*, 2001)

Contd...

Table 9.2–*Contd...*

Family *Plant Name*	*Vernacular Name*	*Plant Part*	*Country/ Area*	*Medicinal Use and/or Experimental Validation*	*Compounds Isolated*	*Reference*
Arecaceae						
Elaeis guineensis Jacq.	Ori	Nuts	Ghana Nigeria	Anti-inflammatory activity. Used to treat rhuematoid arthritis.	Not identified	(Igoli *et al.*, 2005; Kweifio-Okai, 1991a; Kweifio-Okai, 1991b)
Hyphaene thebaica Mart.	Not signalized	Shell	Niger	Antioxidant activity	Not identified	(Cook *et al.*, 1998)
Asclepiadaceae						
*Calotropis procera (*Aiton) W.T.Aiton	African milk weed Sodom apple/ Giant milk-weed/ Swallow-wort/ Auricula tree	Latex	Ethiopia Sudan	Anti-inflammatory and antioxidant activities. Used to control dermal fungal infections and for pain relief. Latex used against scorpion stings and roots for jaundice.	Not identified	(Roy *et al.*, 2005)
Gongronema latifolium Benth.	Not signalized	Leaves	Nigeria	Antioxidant activity	Not identified	(Ugochukwu and Babady, 2002, 2003; Ugochukwu *et al.*, 2003)
Leptadenia hastata Decne.	Not signalized	Leaves	Niger	Antioxidant activity	Not identified	(Cook *et al.*, 1998)
Pachycarpus rigidus E. Mey.	Not signalized	Bark	South Africa	Antioxidant activity. Used to treat pain in the joints	Not identified	(Steenkamp *et al.*, 2005b)
Asparagaceae						
Asparagus virgatus Baker Refug. Bot. (Saunders)	Broom asparagus	Bark	South Africa	Antioxidant activity. Used to treat syphilis, anthelmintic	Not identified	(Steenkamp *et al.*, 2005b)

Contd...

Table 9.2–*Contd...*

Family Plant Name	*Vernacular Name*	*Plant Part*	*Country/ Area*	*Medicinal Use and/or Experimental Validation*	*Compounds Isolated*	*Reference*
Asteraceae						
Ageratum conyzoides L.	Inkuruba Herbe à bouc	Whole plant	Central Africa, Rwanda Ethiopia	Antioxidant and anti-inflammatory properties. Used to treat mastitis and urogenital infections and to dress wounds. Also gastroprotective.	Not identified	(Igoli *et al.*, 2005; Nébié *et al.*, 2004; Shirwaikar *et al.*, 2003)
Artemisia herba-alba	Desert wormwood, shih	Aerial parts	Algeria, Tunisia, Israel, Morocco	Herbal tea from *A. herba-alba* has been used as analgesic, antibacterial, antispasmodic, and hemostatic agents in folk medicines	Camphor (17–33 per cent), 5-thujone (7–28 per cent), and chrysanthenone (4–19 per cent)	(Mahomoodally, 2013)
Artemisia judaica L. disorders	Wormwood	Leaves	Egypt	Used for gastrointestinal	Flavonoids with antioxidant activitities.	(Liu *et al.*, 2004)
Callilepis laureola DC.	Ox-eye daisy, Impila	Tuber	South Africa	Antioxidant and radical scavenging activities. Used to induce fertility, impotence, tapeworm infestations but induces hepatic and renal tubular necrosis.	Not identified	(Popat *et al.*, 2001; Steenkamp *et al.*, 2005b; Steenkamp *et al.*, 1999; Watson *et al.*, 1979)
Psiadia punctulata (DC) Vatke	Mwendathigo	Leaf exudate	Kenya, East Africa	Used to treat colds, fevers and abdominals pains.	Flavones: 5,7-dihydroxy-2',3',4',5'-tetramethoxy-flavone, 5,4'-dihydroxy-7,2',3',5'-tetramethoxyflavone, 5,7,4'-trihydroxy-2',3',5'-trimethoxyflavone, 5-hydroxy-7,2',3',4',5'-pentamethoxy-flavone and 5,7,3'-trihydroxy-2',4',5'-trimethoxyflavone.	(Juma *et al.*, 2001; Midiwo *et al.*, 2002)

Contd...

Table 9.2–*Contd...*

Family Plant Name	*Vernacular Name*	*Plant Part*	*Country/ Area*	*Medicinal Use and/or Experimental Validation*	*Compounds Isolated*	*Reference*
Vernonia kotschyana Sch. Bip. ex Walp.	Buaye	Leaves, roots	Mali	Anti-inflammatory activity. Used to treat gastritis, gastro duodenal ulcers, as an aid to ameliorate digestion and as a wound healing remedy. Immunomodulating activities.	Not identified	(Diallo *et al.*, 2001; Nergard *et al.*, 2004).
Bignoniaceae						
Jacaranda mimosaefolia D.Don.	Sharpleaf Jacaranda	Leaves Stem-bark	Nigeria	Shown to have antimicrobial activity and used to treat infections	Tannins, flavonoids, alkaloids, quinones and traces of saponins.	(Binutu and Lajubutu, 1994; Rojas *et al.*, 2006)
Kigelia pinnata DC.	Suasage tree, Cucumber tree	Root fruit	Egypt	Used as dressing for ulcers and used to treat rheumatism Anti-inflammatory activity	**Naphthoquinones**: kigelinone (1), isopinnatal (2), dehydro-alpha-lapachone (3), and lapachol (4) and the **phenylpropanoids:** p-coumaric acid (5) and ferulic acid (6) (root) and kigelinone (1) and caffeic acid (7) (fruits).	(Carey *et al.*, 2010; Gouda *et al.*, 2003)
Tecoma stans (L.) H.B. and K.	Yellow trumpet bush	Leaves Stem-bark	Nigeria	Anti-diabetic activity is shawn.	Tannins, flavonoids, alkaloids, quinones and traces of saponins.	(Binutu and Lajubutu, 1994)
Tabebuia rosea (Bertol.) DC.	Pink tecoma Pink trumpet tree	Leaves Stem-bark	Nigeria	Used to treat arthritis.	Tannins, flavonoids, alkaloids, quinones and traces of saponins	(Binutu and Lajubutu, 1994)
Crescentia cujete L.	Calabash Gourd tree	Leaves Stem-bark	Nigeria	Used as purgative and to treat coughs.	Tannins, flavonoids, alkaloids, quinones and traces of saponins	(Binutu and Lajubutu, 1994)

Contd...

Table 9.2–*Contd...*

Family Plant Name	*Vernacular Name*	*Plant Part*	*Country/ Area*	*Medicinal Use and/or Experimental Validation*	*Compounds Isolated*	*Reference*
Bombacaceae						
Bombax costatum Pellegrin and Vuillet	Not signalized	Fruit	Niger	Antioxidant activity	Not identified	(Cook *et al.*, 1998)
Boraginaceae						
Heliotropium indicum L.	Nonsikou	Leaves	Mali	Moderate antioxidant activity. Used for wound healing and for ocular infection.	Not identified	(Reddy *et al.*, 2002; Souza *et al.*, 2005; Togola *et al.*, 2005)
Buddlejaceae						
Buddleja madagascariensis Lam.	Butterfly-bush	Leaves	Egypt	Used to treat coughs, asthma, and bronvhitis.	Flavonoids triglycosides: hesperetin and diosmetin 7-O (2",6"- di-O-alpha-L-rhamnopyranosyl)-beta-D-glucopyranosides	(Emam *et al.*, 1998)
Caesalpiniaceae						
Cassia fistula L.	Golden shower tree	Fruit	Mauritius	Laxative	Phenolics and flavonoids	(Luximon-Ramma *et al.*, 2002)
Canellaceae						
Warburgia salutaris (Bertol. F.) Chiov	Pepper-bark tree Isibaha	Bark	South Africa	Antioxidant and radical scavenging activities. Used to treat coughs, stomach ulcers, malaria, rheumatism, liver and veneral dmiseases	Not identified	(Steenkamp *et al.*, 2005b)

Contd...

Table 9.2–*Contd...*

Family Plant Name	*Vernacular Name*	*Plant Part*	*Country/ Area*	*Medicinal Use and/or Experimental Validation*	*Compounds Isolated*	*Reference*
W. ugandensis Sprague	Fever tree	Stem-bark Leaves	Kenya Ethopia	Used to treat stonachache, chest pains, malaria, toothache and coughs.	Flvonol glycoside Kaempferol, kaempferol 3-rhamnoside, kaempferol3-Rhamnoside-7,4'-digalactoside and Quercetin: 3-Rhamnosyl (1→6[glucosyl(1→2)glucoside]-7-rhamnoside, kaempferide 3-O-beta-xylosyl (1→2)-beta-glucoside, kaempferol 3-O-alpha-rhamnoside-7,4'-di-O-beta-galactoside, kaempferol 3,7,4'-tri-O-beta-glucoside, kaempferol 3-rutinoside, myricetin, quercetin 3-rhamnoside, kaempferol 3-arabinoside, quercetin 3-glucoside, quercetin, kaempferol 3-rhamnoside-4'-galactoside, myricetin 3-galactoside and kaempferol 3-glucoside.	(Manguro *et al.*, 2003a, 2003b; Wube *et al.*, 2005a, 2005b)
Capparaceae						
Boscia senegalensis (Pers.) Lam. ex Poiret	Senegal Boscia	Fruit hull Roots and leaf	Mali Niger	Antioxidant activity. Used to treat diarrhoea, cholera, tachycardia, pectoral pain.	Not identified	(Cook *et al.*, 1998; Diallo *et al.*, 2001)
Gynandropsis gynandra Merr.	Not signalized	Leaves	Niger	Antioxidant activity	Not identified	(Cook *et al.*, 1998)

Contd...

Table 9.2–*Contd...*

Family Plant Name	*Vernacular Name*	*Plant Part*	*Country/ Area*	*Medicinal Use and/or Experimental Validation*	*Compounds Isolated*	*Reference*
Celastraceae						
Salacia leptoclada Tul.	Lemon rope	Root	South Africa	Antioxidant activity. Used as an aphrodisiac	Not identified	(Steenkamp *et al.*, 2005b)
Chenopodiaceae						
Salsola somalensis N.E.Br.	Dingetegna	Roots	Ethopia	Used as taenicide.	Nine new isoflavones, 5,3'-dihydroxy-6,7,2'-trimethoxy isoflavone, 5,8,3'-trihydroxy-7,2'-dimethoxyisoflavone, 8,3'-dihydroxy-5,7,2'-trimethoxy-isoflavone, 5,6,3'-trihydroxy-7,2'-dimethoxyisoflavone, 6,7,3' -trihydroxy-5,2'-dimethoxyisoflavone, 5,8,3'-trihydroxy -2'-methoxy-6,7-methylenedioxyiso-flavone, or 5,6,3'-trihydroxy-2'-methoxy-7,8-methyl-enedioxyisoflavone, 3'-hydroxy-5,6,7,2'-tetrameth-oxyisoflavone, 7,3'-dihydroxy-5,6,2'-trimethoxyisoflavone and 6,3'-dihydroxy-5,7,2'-trimethoxyisoflavone.	(Abegaz and Woldu, 1991)
Clusiaceae						
Harungana madagascariensis Lam. ex Poir.	Mtunu *Muithathu*		Cameroon	Antioxidant activity. Used as laxative and to treat urogenital infection, diabetes, liver disease.	Prenylated Anthronoids	(Kouam *et al.*, 2005)

Contd...

Table 9.2–*Contd...*

Family Plant Name	*Vernacular Name*	*Plant Part*	*Country/ Area*	*Medicinal Use and/or Experimental Validation*	*Compounds Isolated*	*Reference*
Psorospermum guineense Hochr.	Karidjakouma	Leaves	Mali	Antioxidant activity. Used as diuretic and febrifuge.	Not identified	(Diallo *et al.*, 2001)
Combretaceae						
Pteleopsis suberosa Engl. and Diels.	Girga	Stem-bark	Mali	Antioxidant properties. Used to treat gastric and duodenal ulcers.	Not identified	(Diallo *et al.*, 2001; Germano *et al.*, 1998; Occhiuto *et al.*, 1999)
Dioscoreaceae						
Dioscorea dumetorum Th.Dur.et Schinz	Cluster yam African bitter yam Trifoliate yam	Tubers	Nigeria Tropical West Africa	Antioxidant and hypo-lipidemic activities. Used to treat diabetes.	Not identified	(Araghiniknam *et al.*, 1996; Iwu *et al.*, 1990; Nimenibo-Uadia, 2003)
Ebenaceae						
Diospyros abyssinica (Hiern) F. White	Giant diospyros	Leaves, roots Root-bark	Mali	Radical scavengers and lipoxygenase inhibitors.	Not identified	(Diallo *et al.*, 2001; Maiga *et al.*, 2006)
Euclea divinorum Hiern	Diamond-leaved euclea Magic guarri	Roots	Ethopia	Used to treat venereal diseases, chest pains, pneumonia, internal body pains, stomach-ache and diarrhea. Chewed roots ease toothache.	Flavonoids	(Dagne *et al.*, 1993)
Euphorbiaceae						
Acalypha hispida Burm. f.	Chenille plant Red-hot cattail	Leaves Flowers	Nigeria	Used as anti-bacterial agent.	Gallic acid and Quercetin 3-O-rutinoside and kaempferol 3-O-rutinoside. The main anthocyanin is the known cyanidind 3-*O*-	(Adesina *et al.*, 2000b; Reiersen *et al.*, 2003)

Contd...

Table 9.2–*Contd...*

Family Plant Name	*Vernacular Name*	*Plant Part*	*Country/ Area*	*Medicinal Use and/or Experimental Validation*	*Compounds Isolated*	*Reference*
					(2-*O*-galloylgalactose, but a minor pigment (5 per cent) is the new cyanidin Cy 3-*O*-(2-*O*-galloyl-6-*O*-rhamnosylgalactoside 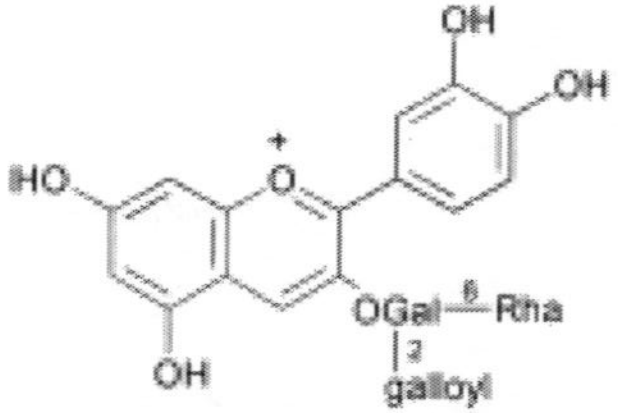	
A. wilkesiana Müll. Arg.	Copper leaf	Leaves	Nigeria	Used to treat ailments of microbial origin	Gallic acid and Quercetin 3-O-rutinoside and kaempferol 3-O-rutinoside	(Adesina *et al.,* 2000a; Akinyemi *et al.,* 2005)
Croton gratissimus Burch.	Lavender fever-berry	Bark	South Africa	Used as purgative for abdominal disorders, fever. The charred and powdered bark is used to treat bleeding gums	Flavonoids.	(Steenkamp *et al.,* 2005b)
Euphorbia hirta L.	Kasandasanda Ufu idire	Whole plant Leaves	Ethiopia	Used to treat diarrhoea and asthma.	**Flavonoid**: quercitrin Flavonol: Euphorbianin (3-(6'''-Acetylglucosyl) (1→3)galactoside)	(Galvez *et al.*, 1993; Harborne and Williams, 2001; Igoli *et al.*, 2005)

Contd...

Table 9.2–*Contd...*

Family Plant Name	*Vernacular Name*	*Plant Part*	*Country/ Area*	*Medicinal Use and/or Experimental Validation*	*Compounds Isolated*	*Reference*
Fabaceae						
Acacia caffra (Thunb.) Wild.	Hook-thorn Cat-thorn	Bark	South Africa	Used to treat diarrhoea and as emetics.	Proanthocyanidins : oritin-(4alpha→5)-epioritin-4beta-ol, ent-epioritin-(4alpha→5)-epioritin-4beta-ol and epioritin-(4beta→5)-epioritin-4alpha-ol and ent-oritin-(4beta→5)-epioritin-4alpha-ol.	(Bennie *et al.,* 2002a; Bennie *et al.,* 2002b; Bennie *et al.,* 2000)
A. galpinii Burtt Davy.	Monkey-thorn	Bark	South Africa	Used to treat diarrhoea.	Proanthocyanidins : oritin-(4alpha→5)-epioritin-4beta-ol, ent-epioritin-(4alpha→5)-epioritin-4beta-ol and epioritin-(4beta→5)-epioritin-4alpha-ol and ent-oritin-(4beta→5)-epioritin-4alpha-ol.	(Bennie *et al.,* 2002a, 2002b)
Afzelia bella Harms	Pretty Afzelia	Stem-bark	Ivory Coast	Used to treat skin diseases and cough.	An acylated dihydroflavonol glycoside identified as 2R,3R-trans-aromadendrin-7-O-beta-D-glucopyranoside-6"-(4"-hydroxy-2"-m ethylene butanoate), along with five known flavonoids and the lignan glycoside (+)-isolariciresinol 9-O-xyloside.	(Binutu and Cordell, 2001)
Bolusanthus speciosus Harms	Tree Wisteria	Root Stem-bark	South Africa, Botswana, Mozam-	Used to treat abdominal pains, emetism and tuberculosis.	Three new **flavonoids from the root:** 5,7,4'-trihydroxy-6-[1-hydroxy-2-methylbuten-2-yl]isoflavone (isogancaonin	(Bojase *et al.,* 2001; Erasto *et al.,* 2004; Majinda *et al.,* 2001)

Contd...

Table 9.2–*Contd...*

Family Plant Name	*Vernacular Name*	*Plant Part*	*Country/ Area*	*Medicinal Use and/or Experimental Validation*	*Compounds Isolated*	*Reference*
			bique, Zimbabwe, Zambia.		C),7,2'-dihydroxy-4'-methoxy-isoflav-3-ene (bolusanthin III), 6,6'-dihydroxy-4'-methoxy-2-arylbenzofuran (bolusanthin IV) in addition to eight known **flavonoids**: derrone, medicarpan, genistein, wighteone, lupiwighteone, gancaonin C, 7-hydroxy-4'-methoxyisoflavone and 7,3'-dihydroxy-4'-methoxyisoflavone flavonoids 2*R*,3*R*-Aromadendrin 7-(6-[4-hydroxy-2-methylenebutanoyl]glucoside). 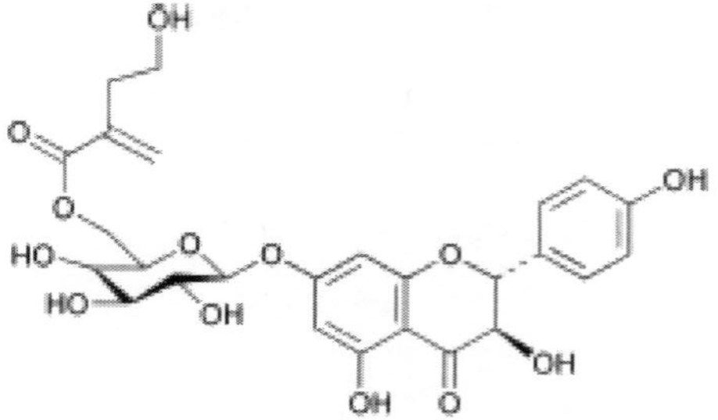 Two new **isoflavonoids** from the combined ethyl acetate/ methanolic extracts of the stem bark of Bolusanthus speciosus have been	

Contd...

Table 9.2–*Contd...*

Family Plant Name	*Vernacular Name*	*Plant Part*	*Country/ Area*	*Medicinal Use and/or Experimental Validation*	*Compounds Isolated*	*Reference*
					established as 4,7,2'-trihydroxy-4'methoxyisoflavanol (**1**) and 5,7,3',4'-tetrahydroxy-5'-(2-epoxy-3-methylbutyl) isoflavanone (**2**). Five other known isoflavonoids, 5,7,3'-trihydroxy-4'-methoxy-5'-γ,γ-dimethylallyisoflavanone, 5,7,2'trihydroxy-4'-methoxy-6,5'-di(γ,γ-dimethyla)isoflavanone, 5,7,2',4'-tetrahydroxy-8,5'-di(γ,γdimethylallyl) isoflavanone, 5,7,2',4'-tetrahydroxy-8,3'-di(γ,γ-dimethylallyl)-isoflavanone, and derrone.	
Crotalaria lanceolata E. Mey.	Lanceleaf rattlebox	Root	South Africa	Antioxidant activity. Used to treat coughs.	Not identified	(Steenkamp *et al.,* 2005b)
Derris trifoliata Lour.	Common derris	Root-bark Stem-bark Seeds.	Kenya	Used for prevention of cancer. Entire plant is used as stimulant, anti-spasmodic. Bark is used as an alternative in rheumatism.	An isoflavonoid derivative, named 7a-O-methyldeguelol, a modified rotenoid with an open ring-C, representing a new sub-class of isoflavonoids (the sub-class is here named as rotenoloid). In addition, the known rotenoids, rotenone, deguelin and alpha-toxicarol. In addition, two unusual rotenoid derivatives, a rotenoloid (named 7a-O-	(Ito *et al.,* 2004a, 2004b; Yenesew *et al.,* 2006; Yenesew *et al.,* 2005b)

Contd...

Table 9.2–*Contd...*

Family Plant Name	*Vernacular Name*	*Plant Part*	*Country/ Area*	*Medicinal Use and/or Experimental Validation*	*Compounds Isolated*	*Reference*
					methyl-12a-hydroxydeguelol) and a spirohomooxarotenoid (named spiro-13-homo-13-oxaelliptone). In addition a rare natural chromanone (6,7-dimethoxy-4-chromanone) and the known rotenoids rotenone, tephrosin and dehydrodeguelin were identified. Also one new rotenoid, 6-alpha, 12-alpha-12a-hydroxyelliptone.	
Entada africana Guill. and Perr.	Samanere	Leaves	Mali Niger	Antioxidant properties. Protective against carbon tetrachloride-induced liver damage. Used to treat fever and various respiratory diseases	Not identified	(Diallo *et al.,* 2001; Occhiuto *et al.*, 1999; Ragusa *et al.,* 2001; Sanogo *et al.*, 1998)
Eriosema robustum		Twigs	Burundi, Ethiopia, Kenya, Rwanda, Tanzania, Uganda, Democratic Republic of Congo and Cameroon	It is used traditionally for the treatment of coughs in East Africa and skin diseases in Central Africa	2',3',5',5,7-pentahydroxy-3,40-dimethoxyflavone (1) and 2',3,5',5,7-pentahydroxy-4'-methoxyflavone (2), along with five known compounds: 6-prenylpinocembrin (3), 1-O-heptatriacontanoyl glycerol (4), b-sitosterol (5), stigmasterol (6) and 3-O-b-D-glucopyranoside of sitosterol (7)	(Awouafack *et al.,* 2013; Kokwaro, 2009)

Contd...

Table 9.2–*Contd...*

Family Plant Name	*Vernacular Name*	*Plant Part*	*Country/ Area*	*Medicinal Use and/or Experimental Validation*	*Compounds Isolated*	*Reference*
Erythrina abyssinica Lam.	Red hot poker tree	Stem bark Root bark	Kenya	Used to treat malaria.	New isoflav-3-ene [7,4'-dihydroxy-2',5'-dimethoxy-isoflav-3-ene] in addition to the known compounds erycristagallin, licoagro-chalcone A, octacosyl ferulate and triacontyl 4-hydroxycinnamate were identified. A new chalcone, 2',3,4,4'-tetrahydroxy-5-prenylchalcone (trivial name 5-prenylbutein) and a new flavanone, 4',7-dihydroxy-3'-methoxy-5'-prenylflava-none (trivial name, 5-deoxy-abyssinin II) along with known flavonoids	(Yenesew *et al.*, 2003a; Yenesew *et al.*, 2004)
E. burttii Baker f.	Not signalized	Stem-bark Root-bark	Kenya	Used as antifungal and antibacterial agent.	Two new flavanones: 5,7-dihydroxy-4'-methoxy-3',5'-di-(3-methylbut-2-enyl) flavanone (trivial name, abyssinone V-4'-methyl ether) and 5,7-dihydroxy-4'-methoxy-3'-(3-hydroxy-3-methylbut-1-enyl)-5'-(3-methylbut-2-enyl) favanone (trivial name, burttinone). A new isoflavone, 5,2',4'-trihydroxy-7-methoxy-6-(3-methylbut-2-enyl)isoflavone (trivial name, 7-O-methyllu-	(Yenesew *et al.*, 2005a; Yenesew *et al.*, 2003b; Yenesew *et al.*, 2002; Yenesew *et al.*, 1998b)

Contd...

Table 9.2–*Contd...*

Family Plant Name	*Vernacular Name*	*Plant Part*	*Country/ Area*	*Medicinal Use and/or Experimental Validation*	*Compounds Isolated*	*Reference*
					teone) and a new flavanone, 5,7-dihydroxy-4'-methoxy-3'-(3-methylbutadienyl)-5'-(3-methylbut-2-enyl)flavanone, 3 isoflavonoids (8-prenylluteone, 3-O-methylcalopocarpin and genistein) Three isoflav-3-enes, 7,4'-dihydroxy-2'-methoxy-6-(1",1"-dimethylallyl)isoflav-3-ene (trivial name, burttinol-A), 4'-hydroxy-2'-methoxy-2",2"-dimethylpyrano[5",6":8,7]isoflav-3-ene (trivial name, burttinol-B), 7,4'-dihydroxy-2'-methoxy-8-(3",3"-dimethylallyl)isoflav-3-ene (trivial name, burttinol-C), and 2-arylbenzofuran, 6,4'-dihydroxy-2'-methoxy-5-(1",1"-dimethylallyl)-2-arylbenzofuran (trivial name, burttinol-D).	
E. eriotricha Harms.	Not signalized	Root-bark	Cameroon	Anti-microbial activity	A novel isoflavanone, named eriotrichin B, one new prenylated flavanone, named sigmoidin L, one flavanone (sigmoidin A), four isoflavones (scandenone, 6,8-diprenylgenistein), flemiphilippinin B and 8-prenyldaidzein	(Nkengfack *et al*., 1995; Nkengfack *et al*., 1997)

Contd...

Table 9.2–*Contd...*

Family Plant Name	*Vernacular Name*	*Plant Part*	*Country/ Area*	*Medicinal Use and/or Experimental Validation*	*Compounds Isolated*	*Reference*
E. sacleuxii Hua	Kinyarwanda	Bark	Kenya	Used to treat fever, malaria and leprosy.	Two new isoflavanones, (R)-5,7-dihydroxy-2',4',5'-trimethoxyisoflavanone (trivial name, (R)-2,3-dihydro-7-demethylrobustigenin) and (R)-5-hydroxy-2',4',5'-trimethoxy-2",2"-dimethylpyrano[5",6":6,7] isoflavan one (trivial name, (R)-saclenone)	(Andayi *et al.*, 2006; Yenesew *et al.*, 1998a; Yenesew *et al.*, 2000)
Millettia ferruginea (Hochst.) Baker	Birbira Sotallo Sari	Bark	Ethopia	Used for skin disorders.	O-Geranylated and O-prenylated flavonoids, C-prenylated isoflavones Geranylated and prenylated flavonoids	(Dagne and Bekele, 1990; Dagne *et al.*, 1990)
M. dura Dunn.	Runyankore Uumuyogoro	Stem-bark	Rwanda Uganda	Used for blood parasitism	Flavonoids: A new isoflavone (7,3'-dimethoxy-4',5'-methylenedioxyisoflavone) and three known isoflavones [isoerythrinin A 4'-(3-methylbut-2-enyl) ether, isojamaicin and nordurlettone].	(Dagne *et al.*, 1991; Derese *et al.*, 2003)
Ostryoderris stuhlmannii (Taub.) Dunn ex Harms	Mnyinga	Leaves	Mali	Antioxidant activity. Used to treat painful menstruation, peritonitis, gastritis, colitis and gingivitis.	Not identified	(Diallo *et al.*, 2001)
Piliostigma reticulatum (DC.) Hochst	Kalga	Leaves Bark	Nigeria	High antioxidant activity. Used to treat wounds, bronchitis, malaria, sterility (leaves) and diarrhoea and dysentery (bark).	Not identified	(Aderogba *et al.*, 2004)

Contd...

Table 9.2–*Contd...*

Family Plant Name	*Vernacular Name*	*Plant Part*	*Country/ Area*	*Medicinal Use and/or Experimental Validation*	*Compounds Isolated*	*Reference*
Sesbania pachycarpa DC.	Not signalized	Leaves	Niger	Antioxidant activity	Not identified	(Cook *et al.*, 1998)
Tephrosia polyphylla (Chiov.) J.B. Gillett	Hoary pea	Aerial part	Kenya		Flavonoids	(Dagne *et al.*, 1992)
T. deflexa Baker	Hoary pea	Aerial part	Senegal		Flavonoids: Rutin 1 – quercetine 3-O-a-L-rhamno-pyrannosyl (1-6) gluco-pyrannose – and morin 2 – 3, 5,7,2',4'-pentahydroxyflavone.	(Niassy *et al.*, 2004)
T. albifoliolis A.Nongonierma and T.Sarr	Hoary pea	Aerial part	Senegal		Flavonoids: Rutin 1 – quercetine 3-O-a-L-rhamno-pyrannosyl (1-6) glucopyran-nose – and morin 2 – 3,5,7, 2',4'-pentahydroxyflavone.	(Niassy *et al.*, 2004)
Taverniera abys-sinica A. Rich.	Dingetegna	Root	Ethiopia	Used to treat fever, discomfort and pain, stomach ache.	Four isoflavonoids	(Abegaz *et al.*, 1999; Duddeck *et al.*, 1987; Noamesi *et al.*, 1990)
Flacourtiaceae						
Flacourtia flavescens Willd.	Not signalized	Leaves	Mali	Antioxidant activity.	Not identified	(Diallo *et al.*, 2001)
Geraniaceae						
Pelargonium reniforme Spreng.	Xhosa (Umckaloabo)	Root	Southern Africa	Used to treat liver disorders, laxative, purgative, cancer, and pulmonary disorders	Polyphenols: catechol (3'4'-dihydroxy) element in the B-ring, which possesses higher antioxidant activity than ascorbic acid.	(Latté, 1999; Latte and Kolodziej, 2004; Wagner and Bladt, 1975)

Contd...

Table 9.2–*Contd...*

Family Plant Name	*Vernacular Name*	*Plant Part*	*Country/ Area*	*Medicinal Use and/or Experimental Validation*	*Compounds Isolated*	*Reference*
Gunneraceae						
Gunnera perpensa L.	River pumpkin Ugobho	Root Leaves and stem	South Africa	Decreased luciginin enhanced chemiluminescence. Used to treat wounds and psoriasis.	Not identified	(Drewes *et al.*, 2005; Steenkamp *et al.*, 2004)
Irvingiaceae						
Irvingia gabonensis (Aubry-Lecomte ex O'Rorke) Baill.	Bush mango Ono	Seeds	Nigeria Cameroon	Antioxidant activity. Used as laxative and for stomach and kidney pain. Shown to lower total cholesterol.	Not identified	(Agbor *et al.*, 2005; Igoli *et al.*, 2005; Ngondi *et al.*, 2005)
Lamiaceae						
Leonotis leonurus (L.) R.Br.	Wild dagga	Leaves	South Africa	Anti-inflammatory properties. Used to treat headaches, dysentery, coughs and colds.	Not identified	(Ojewole, 2005c)
Salvia stenophylla Burch. ex Benth.	Sage	Leaves	South Africa	Solvent extracts: antioxidant activity but poor anti-inflammatory properties. Essential oils: anti-inflammatory activity but poor anti-oxidant activity. Used against fever and digestive disorders.	Not identified	(Kamatou *et al.*, 2005)
S. repens Burch. ex Benth.	Not signalized	Leaves	South Africa	Solvent extracts: antioxidant activity but poor anti-inflammatory properties. Essential oils: anti-inflammatory activity but poor anti-oxidant activity. Used for fevers and digestive disorders.	Not identified	(Kamatou *et al.*, 2005)

Contd...

Table 9.2–*Contd...*

Family Plant Name	*Vernacular Name*	*Plant Part*	*Country/ Area*	*Medicinal Use and/or Experimental Validation*	*Compounds Isolated*	*Reference*
S. runcinata L.f.	Not signalized	Leaves	South Africa	Solvent extracts: antioxidant activity but poor anti-inflammatory properties. Essential oils: anti-inflammatory activity but poor anti-oxidant activity. Used against fever and digestive disorders.	Not identified	(Kamatou *et al.*, 2005)
Loranthaceae						
Tapinanthus globiferus Tiegh.	Not signalized	Leaves	Niger	Antioxidant activity	Not identified	(Cook *et al.*, 1998)
Malvacea						
Adansonia digitata (L.)	English: baobab, Afrikaans: kremetart, Hausa: kuka, Sotho: seboi, Tswana: mowana, Tsonga: shimuwu, Venda: muvhuyu, Arabic: tabladi	Leaves, root, bark and fruits	All over Africa, but limited trees in Central Africa	Antioxidant, analgesic and anti-inflammatory properties of extracts	L-ascorbic acid	(Sugandha *et al.*, 2013; Vertuani *et al.*, 2002)

Contd...

Table 9.2–*Contd...*

Family Plant Name	*Vernacular Name*	*Plant Part*	*Country/ Area*	*Medicinal Use and/or Experimental Validation*	*Compounds Isolated*	*Reference*
Mimosaceae						
Albizia lebbeck (L.) Benth.	East Indian walnut, frywood, koko, lebbek, lebbek tree, rain tree, raom tree, silver raintree, siris rain tree, siris tree, soros-tree, woman's tongue.	Leaves and bark	Egypt	Used to treat asthma and skin disorders (bark) and eye diseases and dysentery (leaves)	Two new tri-O-glycoside flavonols: kaempferol and quercetin 3-O-alpha-rhamno-pyranosyl(1→6)-beta-gluco-pyranosyl(1→6)-beta-galactopyranosides	(el-Mousallamy, 1998)
Moraceae						
Dorstenia angusticornis Engl.	Not signalized	Twigs	Cameroon	Used for snakebite and to treat infection, rheumatism, headache, cough and stomach pain.	Two novel **diprenylated chalcones**: 3,5'-di-(2-hydroxy-3-methylbut-3-enyl)-4,2',4'-trihydroxy-chalcone, 3, 4-(2,2-dimethyl-pyrano)-3'-(2-hydroxy-3-methylbut-3-enyl)-2',4'-dihydroxych alcone and the known stipulin. 3-(2-Hydroxy-3-methylbut-3-enyl)-5'-(3,3-dimethylallyl)-4,2',4'-trihydroxy chalcone and the known compounds: gancaonin Q, paratocarpins C, F, and lupeol.	(Majinda *et al.,* 2001; Ngadjui *et al.,* 2005)

Contd...

Table 9.2–*Contd...*

Family Plant Name	*Vernacular Name*	*Plant Part*	*Country/ Area*	*Medicinal Use and/or Experimental Validation*	*Compounds Isolated*	*Reference*
D. dinklagei Engl.	Not signalized	Twigs	Cameroon	Used for snakebite and to treat infection, rheumatism, headache, cough and stomach pain.	Three prenylated flavonoids, dinklagins A, B and C identified, respectively, as (dinklagin B): (+)-5,4',5"ξ-Trihydroxy-6",6"-dimethyldi-hydropyranol[2",3":7,6]flavone. 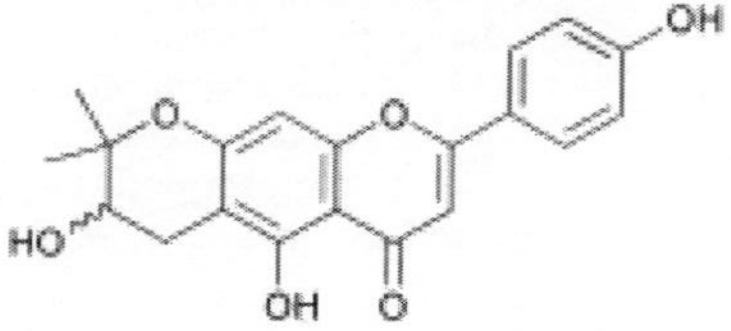 (dinklagin C): (+)-6-(2x-Hydroxy-3-methyl-3-butenyl)-5,7,4'-trihydroxyflavone 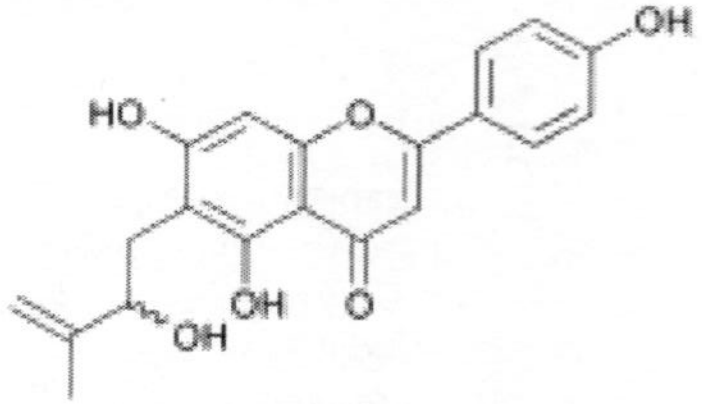 (-)-6-(3,3-dimethylallyl)-7-hydroxy-6''', 6'''-dimethyl-chromeno-(4',3',2''',3''')-flavanone, (+)-5,4',5"ξ-trihydroxy-6'',6''-dimethyl-	(Majinda *et al.,* 2001; Ngadjui *et al.,* 2002a)

Contd...

Table 9.2–*Contd...*

Family Plant Name	*Vernacular Name*	*Plant Part*	*Country/ Area*	*Medicinal Use and/or Experimental Validation*	*Compounds Isolated*	*Reference*
					chromano-(7,6,2",3")-flavone and (+)6-(2ξ-hydroxy-3-methyl-3-butenyl)-5,7,4'-trihydroxyflavone. 6-prenylapigenin, 4-hydroxyl-onchocarpin, stipulin and 5,4'-dihydroxy-6",6"-dimethylchromano-(7,6,2", 3")-flavone.	
D. elliptica Bur.	Not signalized	Twigs	Botswana	Used to treat eye infection.	Monoprenylated flavan	(Abegaz *et al.*, 2004)
D. Kameruniana. Engl.	Not signalized	Leaves	Botswana	Used for snakebite and to treat infection, rheumatism, headache, cough and stomach pain.	Two novel favonoids: 6,7-(2,2-dimethylchromano)-5,4'-dihydroxyfavone and 3,4-,4',5'-*bis*-(2,2-dimethyl-chromano)-2'-hydroxy-chalcone together with the known 6-(3-methylbut-2-enyl)apigenin and twochal-cones (*E*)-1-[2,4-dihydroxy-3-[3-methylbut-2-enyl] phenyl]-3-[4-hydroxyphenyl]-prop-2-en-1-one and (*E*)1-[2,4-dihydroxy-5-[3-methyl-but-2-enyl]phenyl]-3-[4-hydroxy-3-[3-methylbut-2-enyl]phenyl]-prop-2-en-1-one.	(Abegaz *et al.*, 1998)

Contd...

Table 9.2–*Contd...*

Family Plant Name	*Vernacular Name*	*Plant Part*	*Country/ Area*	*Medicinal Use and/or Experimental Validation*	*Compounds Isolated*	*Reference*
D. prorepens Engl.	Not signalized	Twigs	Botswana	Used for snakebite and to treat infection, rheumatism, headache, cough and stomach pain.	Digeranylated chalcone, 5,3'-(3,7-dimethyl-2,6-octadienyl)-3,4, 2',4'-tetrahydroxychalcone. 4-Hydroxylonchocarpin Chalcone: 3,4,2',4'-Tetrahy droxy-5,3'-digeranylchalcone 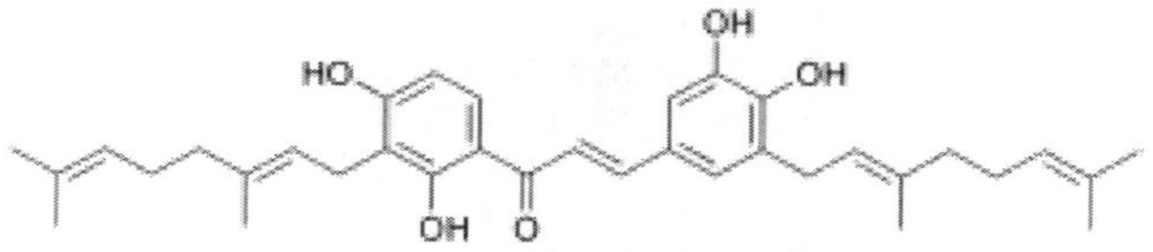	(Abegaz *et al.*, 2002b; Majinda *et al.*, 2001)
D. poinsettiifolia Engl.	Not signalized	Twigs	Botswana	Used for snakebite and to treat infection, rheumatism, headache, cough and stomach pain.	Grenylated and prenylated flavonoids. In addition, the flavone 5,7,4-trihydroxy-8-prenylflavone (licoflavone C), the chalcones 4,2',4'-trihydroxy-3'-prenylchalcone (isobavachalcone) and isobavachromene, the triterpene butyrospermol, and the carotenoid lutein.	(Majinda *et al.*, 2001; Ngadjui *et al.*, 1999b; Tsopmo *et al.*, 1998)
D. zenkeri Engl.	Not signalized	Twigs	Botswana	Used for snakebite and to treat infection, rheumatism, headache, cough and stomach pain.	3',4'-(3-hydroxy-2,2-dimethyl-dihydropyrano)-4,2'-dihydroxychalcone and a bichalcone. 4-Hydroxylonchocarpin. p-hydroxybenzaldehyde, dorsmanin A, 4,2',4'-trihydroxychalcone and 4,2',4'-trihydroxy-3'-prenylchalcone	(Abegaz *et al.*, 2002b; Majinda *et al.*, 2001)

Contd...

Table 9.2–*Contd...*

Family Plant Name	*Vernacular Name*	*Plant Part*	*Country/ Area*	*Medicinal Use and/or Experimental Validation*	*Compounds Isolated*	*Reference*
					Chalcones: 4,2',5"-Trihydroxy-6",6"-dimethyldihydropyranol [2",3":4',3']chalcone 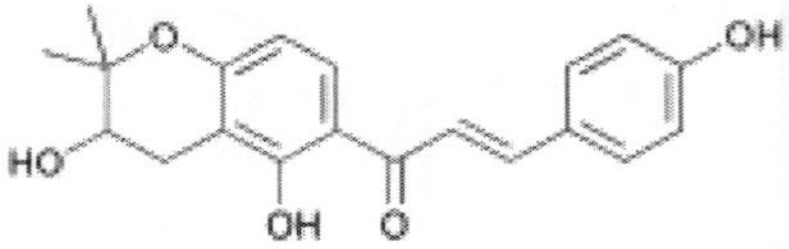	
Moringaceae						
Moringa oleifera Lam.	Horse-radish tree Drumstick Moringo Zakalanda	Root	West Africa Zimbawe	Anti inflammatory activity. Used as aphrodisiac and to treat asthma, gout and rheumatism.	Not identified	(Ezeamuzie *et al.*, 1996)
Myrtaceae						
Eucalyptus camaldulensis Dehnh.	Not signalized	Leaves	Egypt	Antioxidant activity	Not identified	(El-Ghorab et al., 2003)
Polygonaceae						
Polygonum senegalense Meisn.	Fotsimbarin' akoholahy	Leaves	Madagascar		Flavonoids: quercetin, kaempferol and luteolin and their glycosides such as dihydrochalcone glucoside and quercetin glycosides.	(Midiwo *et al.*, 1994; Midiwo *et al.*, 2002; Midiwo *et al.*, 1992)
Rumex abyssinicus Jacq.	Mekmeko	Leaves	N. Africa Ethiopia	Anti-inflammatory properties. Used to treat itching, skin eczema and leprosy.	Flavonoids.	(Getie *et al.*, 2003; Sachdev and Kulshreshtha, 1983)

Contd...

Table 9.2–*Contd...*

Family Plant Name	*Vernacular Name*	*Plant Part*	*Country/ Area*	*Medicinal Use and/or Experimental Validation*	*Compounds Isolated*	*Reference*
R. nervosus Vahl.	Alcgango Dengogo	Leaves	Ethopia	Anti-inflammatory properties. Used to treat achne, wounds, eczema, typhus and as an ophthalmic antiseptic.	Not identified	(Getie *et al.*, 2003)
Rubiaceae						
Nauclea latifolia Smith	Pin Cushion Tree Ìgíyàà	Leaves and root	Nigeria	Used as anthelmintic and to treat malaria, fever, stomach-ache and liver diseases.	Proanthocyanidins.	(Akpanabiatu *et al.*, 2005; Fakae *et al.*, 2000; Gidado *et al.*, 2005; Igoli *et al.*, 2005; Onyeyili *et al.*, 2001)
Solanaceae						
Datura stramonium L.	Thorn-apple rwiziringa	Seeds	South Africa	Antioxidant activity. Used to treat asthma, headaches and wounds.	Not identified	(Steenkamp *et al.*, 2005b)
Tiliaceae						
Grewia occidentalis L.	Cross-berry Four-corner	Bark	South Africa	Antioxidant activity. Used to treat bladder ailments, wounds, impotence and sterility, and to help in childbirth.	Not identified	(Steenkamp *et al.*, 2005b)
Vahliaceae						
Vahlia capensis (L.f) Thunb.	Vahlia of the Cape	Zimbabwe		Used to treat bacterial infections.	Kaempferol, quercetin, afzelin, astragalin, quercitrin, isoquercitrin, rutin, gallic acid, chiro-inositol, dulcitol, and a novel biflavonoid, VC-15B (vahlia biflavone)	(Majinda *et al.*, 1997)

Contd...

Table 9.2–*Contd...*

Family Plant Name	*Vernacular Name*	*Plant Part*	*Country/ Area*	*Medicinal Use and/or Experimental Validation*	*Compounds Isolated*	*Reference*
Vitaceae						
Cyphostemma natalitium (Szyszl.) J.v. d. Merwe	Tick-berry bush	Root	South Africa	Anti-inflammatory and anti-microbial agents with significant inhibition of cyclo-oxygenase (COX-1)	Not identified	(Lin *et al.*, 1999)
Rhoicissus digitata Gilg. and Brandt	Wilde patatat	Roots, stems and leaves	South Africa	At high concentrations possessed some pro-oxidative properties. Anti-inflammatory and anti-microbial agents with significant inhibition of cyclo-oxygenase (COX-1). Used to facilitate delivery.	Not identified	(Lin *et al.*, 1999; Opoku *et al.*, 2002)
R. rhomboidea (E. Meyer ex Harvey) Planchon	Glossy forest grape	Roots, stems and leaves	South Africa Mozambique	Radical scavenging activity, inhibitory effect on xanthine oxidase activity, prevention of lipid peroxidation and damage to DNA and ability to chelate iron. Anti-inflammatory through inhibition of cyclo-oxygenase (COX-1).	Not identified	(Lin *et al.*, 1999; Opoku *et al.*, 2002)
R. tomentosa (Lam.) Wild and R.B. Drum.	Wild grape Forest Grape, Monkey rope	Roots, stems and leaves	South Africa	Antioxidant and anti-inflammatory activities. Anti-inflammatory through inhibition of cyclo-oxygenase (COX-1). Used to facilitate delivery.	Not identified	(Lin *et al.*, 1999; Opoku *et al.*, 2002)

Contd...

Table 9.2–*Contd...*

Family Plant Name	*Vernacular Name*	*Plant Part*	*Country/ Area*	*Medicinal Use and/or Experimental Validation*	*Compounds Isolated*	*Reference*
R. tridentata (L.f.) Wild and Drum.	Bitter grape Bushman's grape Isinwazi	Roots, stems and leaves	South Africa: Venda	Radical scavenging activity, inhibitory effect on xanthine oxidase activity, prevention of lipid peroxidation and damage to DNA and ability to chelate iron. Anti-inflammatory through inhibition of cyclo-oxygenase (COX-1).Used to treat colds, infertility and stomach ailments.	Not identified	(Arnold and Gulumian, 1984; Katsoulis *et al.*, 2000; Lin *et al.*, 1999; Opoku *et al.*, 2002)

diseases. These include fruits and seeds of *Balanites aegyptiaca*, leaves of *Boscia senegalensis*, leaves of *Entada africana* and seeds of *Parkia biglobosa*, from Niger (Cook *et al*., 1998) and also leaves, seeds and stem-bark of *Mangifera indica* from Benin and Burkina Faso (Igoli *et al.*, 2005; Ojewole, 2005a), leaves of *Cynara scolymus* from Ethopia (Gebhardt, 1997; Li *et al.*, 2004) leaves of *Aspalathus linearis* from South Africa (Bramati *et al.*, 2002,2003; Inanami *et al.*, 1995; Koeppen and Roux, 1966; Koeppen *et al.*, 1962; Lindsey *et al.*, 2002; Rabe *et al.*, 1994; Steenkamp *et al.*, 2004; Ulicna *et al.*, 2005), leaves of *Cinnamomum zeylanicum* from Madagascar and Ethiopia (Baratta *et al.*, 1998; Juliani *et al.*, 2004; Mancini-Filho *et al.*, 1998), essential oils from the bark and leaves of *Ravensara aromatica* from Madagascar (Juliani *et al.*, 2004; Möllenbeck *et al.*, 1997), buds of *Syzygium aromaticum* from Madagascar (Juliani *et al.*, 2004; Barakat 2014), seeds of *Trigonella foenumgraecum* from Ethopia and Maroco (Bordia *et al.*, 1997; Dilsiz *et al.*, 2006; Randhir *et al.*, 2004; Saleh *et al.*, 1977; Srinivasan *et al.*, 2004) and oils in seeds of *Nigella sativa* from African countries of the Mediterranean region (Badary *et al.*, 2003; Farah *et al.*, 2005; Ramadan *et al.*, 2003).

Tests Used to Assess Antioxidant Activities of African Medicinal Plant Extracts

A variety of test systems were employed to assess the antioxidant properties of the medicinal plant extracts listed in Tables 9.1 and 9.2. Most of these tests were conducted *in vitro* but *in vivo* studies were also reported. For the determination of antioxidant activity, water as well as organic extracts were tested.

Tests used to determine the antioxidant activity included the assessment of the ability to scavenge free radicals such as 2,2'-diphenyl-p-picrylhydrazyl radical (DPPH), 2,2'-azino-bis(3-ethylbenzthiazoline-6-sulfonic acid) ($ABTS^{+}$) (Benedi *et al.*, 2004; Betancor-Fernandez *et al.*, 2003; Bouchet *et al.*, 1998; Bramati *et al.*, 2003; Chacha *et al.*, 2005; Diallo *et al.*, 2001; Djeridane *et al.*, 2006; Dufall *et al.*, 2003; Erasto *et al.*, 2004; Farombi *et al.*, 2003; Gyamfi *et al.*, 1999; Juma and Majinda, 2004; Kamatou *et al.*, 2005; Kansci *et al.*, 2003; Katerere and Eloff, 2004; Latte and Kolodziej, 2004; Lindsey *et al.*, 2002; Lindsey *et al.*, 2003; Lourens *et al.*, 2004; Maiga *et al.*, 2006; Mathisen *et al.*, 2002; Mensah *et al.*, 2004; Nergard *et al.*, 2005; Ohtani *et al.*, 2000; Opoku *et al.*, 2002; Palé, Kouda-Bonafos and Nacro, 2004), or the hydroxyl radicals (Burits *et al.*, 2001; Burits and Bucar, 2000; Farombi and Fakoya, 2005; Gyamfi and Aniya, 2002; Steenkamp *et al.*, 2005b) as well as the hydroperoxyl radicals by the Briggs-Rauscher reaction (Speroni *et al.*, 2005). The ability of the extracts to chelate metal ions was also determined as further indication of their ability to contribute in the reduction of free radicals such as the hydroxyl radical (Farombi *et al.*, 2004). In addition, assessment of the ability of these medicinal plant extracts to protect against lipid peroxidation was also included, which in turn was measured by the malondialdehyde-thiobarbituric acid (MDA) test (Bahorun *et al.*, 1994; Farombi and Nwaokeafor, 2005), the modified thiobarbituric acid reactive species (TBARS) assay (Baratta *et al.*, 1998; Inanami *et al.*, 1995) or conjugated diene formation (Farombi and Nwaokeafor, 2005). Moreover, lipid peroxidation was assessed using the fluorescent probe diphenyl-1-pyrenylphosphine (DPPP) (Steenkamp *et al.*, 2005b) or using the inhibition of Cu(2+)-mediated oxidation of human low density lipoprotein (LDL) (Dufall *et al.*, 2003;

Farombi and Nwaokeafor, 2005). The ability of extracts to protect against damage to DNA using the Comet assay was also employed (Farombi *et al.*, 2004; Steenkamp *et al.*, 2005b).

The antioxidant capacity of medicinal plant extracts was determined using either the Trolox equivalent antioxidant capacity (TEAC) assay or the ferric reducing antioxidant power (FRAP) (Agbor *et al.*, 2005; Bahorun *et al.*, 2003; Betancor-Fernandez *et al.*, 2003; Cook *et al.*, 1998; Djeridane *et al.*, 2006; Luximon-Ramma *et al.*, 2002). The ability of extracts to modulate the gene expression of the antioxidant enzymes, such as Cu, Zn-superoxide dismutase (Cu,Zn-SOD), Mn-superoxide dismutase (Mn-SOD), catalase, and glutathione dismutase (GPx), was also used as a measure of their antioxidant property (Toyokuni *et al.*, 2003). The Photochemilumiescence (PLC) assay method is a more recent antioxidant capacity assessment method used for plant extracts, and was used for evaluation of antioxidant capacity of baobab fruit pulp extracts (Sugandha *et al.*, 2013).

Anti-inflammatory properties of these extracts were assessed by their ability to inhibit 5-lipoxygenases (Bucar *et al.*, 1998; Bucar *et al.*, 2004; Lourens *et al.*, 2004; Maiga *et al.*, 2006) or cyclooxygenase (COX-1 and COX-2) activities (Chukwujekwu, van Staden and Smith, 2005; Papajewski *et al.*, 2001; Schneider and Bucar, 2005; Wube *et al.*, 2006; Wube *et al.*, 2005c; Abdillahi *et al.*, 2011; Abdelwahab *et al.*, 2012). Using the former (Lin *et al.*, 1999) and the latter (Jäger and van Staden, 2005; Steenkamp *et al.*, 2005a) methodologies, respectively, a great number of South African medicinal plant extracts were screened for their anti-inflammatory properties. The effect of medicinal extracts on the biosynthesis of different prostaglandins was assessed as a measure of their anti-inflammatory effect (Getie *et al.*, 2003; Hiermann and Bucar, 1994; Ibewuike *et al.*, 1997). Extracts of *Podocarpus* species were shown to inhibit the activites of cyclo-oxygenase enzymes (Abdillahi *et al.*, 2011). Once again, using this test, the anti-inflammatory property of the aqueous and ethanolic extracts of 39 plants used in traditional Zulu medicine were screened (Jager *et al.*, 1996). The HET-CAM-Assay (Hen´s Egg Test – Chorioallantoic Membrane) utilizing the CAM´s capillary system of breeded hen eggs was also used to assess the anti-inflammatory effects through anti-angiogenic effects of the ethanol and aqueous extracts from *Drosera rotundifolia* and *D. madagascariensis* (Paper *et al.*, 2005).

The antioxidant and anti-inflammatory abilities of the herbal extracts were further assessed by evaluating their ability to control the production of reactive oxygen species produced by oxidative burst in neutrophils stimulated with L-formyl-L-methionyl-L-leucyl-L-phenylalanine (FMLP) (Fernandes *et al.*, 2004; Steenkamp *et al.*, 2004). The inhibition of neutrophil elastase was used as a measure of anti-inflammatory activity and it was proposed that the presence of flavonoids such as hyperoside, quercetin and isoquercitrin in *D. rotundifolia* (Krenn *et al.*, 2004) and five flavonoid compounds in two *Polypodium* species (*P.decumanum* and *P.triseriale*) (Vasange *et al.*, 1997) were thought to contribute to this anti-inflammatory activity. These and other *in vitro* tests were used to assess the antioxidant properties of three Ghanaian species: *Spathodea campanulata*, *Commelina diffusa* and *Secamone afzelii* (Houghton *et al.*, 2005).

Inflammation is a complex mechanism with many pathways. Several extracts derived from medicinal plants have been shown to modulate or inhibit the activities of mediators of inflammation. For instance, Kolaviron, a bioflavonoid compound isolated from the seeds of *Garcinia kola* that has been reported to possess anti-inflammatory and antioxidant activities via its effects on COX-2 and inducible nitric oxide synthase (iNOS)s by inhibiting the exoressions of NF-κB (nuclear factor kappa B) (Farombi *et al.*, 2009). Quercetin is a flavonoid molecule ubiquitous in nature and functions as an anti-oxidant and anti-inflammatory agent. Dose- and time-dependent effect of quercetin has been investigated on proinflammatory cytokine expression and iNOS, focusing on its effects on NF-κB signal transduction pathways in lipopolysaccharide-stimulated RAW 264.7 cells by using real time polymerase chain reaction (RT-PCR) and immunoblotting. Curcumin, a yellow pigment of turmeric, has been shown to exhibit anti-inflammatory activities. Curcumin are effective for the treatment or control of chronic inflammatory conditions such as rheumatism, atherosclerosis, type II diabetes, and cancer (Sharma *et al.*, 2005). Calixto *et al.* (2003) reported that the anti-inflammatory action of active spice-derived components results from the disruption of the production of various inflammatory proteins (*e.g.*, cytokines such as tumour necrosis factor-alpha (TNF-α), iNOS, and COX-2.

Animal studies were also conducted to assess the antioxidant properties of a number of medicinal extracts. The antioxidant potential of *Hypericum perforatum,* containing many polyphenolic compounds, was evaluated on splanchnic artery occlusion (SAO) shock-mediated injury (De Paola *et al.*, 2005) and also against elevated brain oxidative status induced by amnestic dose of scopolamine in rats (El-Sherbiny *et al.*, 2003). Some medicinal plant extracts were tested for their ability to protect against carbon tetrachloride-, 2-acetylaminofluorene (2-AAF)- and galactosamine-induced liver and also aflatoxin B1 (AFB1)-induced genotoxicity. Using these tests, it was found that an extract of *Garcinia kola* seeds (Farombi, 2000; Farombi *et al.*, 2005; Farombi *et al.*, 2000), a decoction of *Trichilia roka* root (Germano *et al.*, 2001), extracts of *Entada africana* (Sanogo *et al.*, 1998) and *Thonningia sanguinea* (Gyamfi and Aniya, 1998; Gyamfi *et al.*, 1999) possessed protective abilities. The antioxidant properties of plant extracts against potassium bromate (KBrO(3))-induced kidney damage showed the ability of *G. kola* seed extract to protect the kidneys (Farombi *et al.*, 2002).

Animal studies were also used to determine the anti-inflammatory ability of medicinal plant extracts using the carrageenan-induced rat paw oedema model. Plants investigated include: seed extracts of *Picralima nitida* (Duwiejua *et al.*, 2002), crude methanol extract of the root of *Moringa oleifera* (Ezeamuzie *et al.*, 1996), powdered leaves and root of *Mallotus oppositifolium* (Farombi *et al.*, 2001), methanolic extract of *Picralima nitida* fruit (Ezeamuzie *et al.*, 1994), hot water extract of *Alstonia boonei* root-bark, *Rauvolfia vomitoria* root-bark, *Elaeis guineensis* nuts (Kweifio-Okai, 1991a), secondary root aqueous extract of *Harpagophytum procumbens* (Mahomed and Ojewole, 2004), crude extracts of *Sphenocentrum jollyanum* (Moody *et al.*, 2005), aqueous and methanolic extracts of *Hypoxis hemerocallidea* corm (Ojewole, 2006), aqueous and methanolic extracts of *Sclerocarya birrea* stem-bark (Ojewole, 2003), aqueous extract of *Mangifera indica* stem-bark (Ojewole, 2005a), aqueous extract of *Leonotis leonurus* leaves (Ojewole, 2005c), leaf extract of *Bryophyllum pinnatum* (Ojewole, 2005b), methanol extract of the stem-bark of *Alstonia boonei* (Olajide *et al.*, 2000a), aerial parts of *Amaranthus caudatus* (Kabiri *et al.*, 2010), methanolic extracts of *Kigelia pinnata* DC flowers (Carey *et al.*, 2010) and leaf and twig extracts of *Dorstenia barteri* (Omisore *et al.*, 2004). In all of these studies, the anti-inflammatory effect against carrageenan-induced rat paw oedema was attributed to flavonoids and other polyphenolic compounds. Animal tests were also employed to assess the anti-inflammatory effects of the medicinal plant extracts including inflammatory cell response such as neutrophil chemotaxis and degranulation (Selloum *et al.*, 1995, 1997), anti-atherosclerosis effects (Kabiri *et al.*, 2010) and pain assessment in experimental animals (Olaleye *et al.*, 2000).

The effect of medicinal plants on the induction or inhibition of drug metabolizing enzymes was also studied in animals. The effect of the aqueous extract of *Thonningia sanguinea* on 7-ethoxyresorufin O-deethylase (EROD, CYP1A1), 7-pentoxyresorufin O-dealkylase (PROD, CYP2B1/2), 7-methoxyresorufin O-demethylase (MROD, CYP1A2), aniline hydroxylase (aniline, CYP2E1), p-nitrophenol hydroxylase (PNPH, CYP2E1) and erythromycin N-demethylase (ERDM, CYP3A1) in rat liver was found to be a selective modulation of CYP isoenzymes (Gyamfi *et al.*, 2000) suppressing CYP3A2 and CYP1A2 gene expression (Gyamfi *et al.*, 2004b).

Compounds Isolated from African Medicinal Plant Extracts with Confirmed Antioxidant Activities

Several medicinal plant extracts were studied at research centres in African countries for their antioxidant properties. The major findings of these investigations have indicated that in addition to known antioxidant compounds such as ascorbic acid in the seeds of *Parkia biglobosa* (Alabi *et al.*, 2005) and fruits pulp of *Adansonia digitata* (Sugandha *et al.*, 2013), alpha-tocopherol in the methanol extract of the stems of *Secamone afzelii* (Mensah *et al.*, 2004) or from the seeds (Bruni *et al.*, 2002) and methanolic extract of leaves of *Amaranthus caudatus* (Veeru *et al.*, 2009), and apigenin and luteolin in aerial parts of *Bulbine capitata* (Bezabih and Abegaz, 1998), there are several novel compounds that were identified with antioxidant properties. The identified polyphenolic compounds included mainly flavonoids such as flavones and flavonols, flavone and flavonol glycosides, chalcones and dihydrochalchalcones and flavonones, although some anthocyanins, proanthocyanidins and anthrones were also isolated with antioxidant properties.

A wide range of plant extracts investigated have been shown to contain flavonoids. *Dorstenia* species are rich in flavonoids some of which are unique to this genus (Abegaz *et al.*, 2000; Majinda *et al.*, 2001) namely prenylated flavonoids as found in *Dorstenia kameruniana* and twigs of *D. mannii* (Ngadjui *et al.*, 1998b; Ngadjui *et al.*, 1999b). Earlier studies have shown that prenylated flavonoids passess antioxidant properties by protecting human LDL from oxidation (Miranda *et al.*, 2000). Flavonoids isolated from African medicinal plant extracts were also tested and their antioxidant properties were confirmed. The antioxidant activities of three prenylated flavonoids from *D. mannii* (6,8-diprenyleriodictyol, dorsmanin C, 7,8-(2,2-dimethylchromeno)-6-geranyl-3,5,32 ,42-tetrahydroxyflavonol and dorsmanin F, 7,8-[2 ″-(1-hydroxy-1-methylethyl)-dihydrofurano]-6-prenyl-5,3',4'-trihydroxyflavanone) against LDL oxidation and also their free radical scavenging activity was indicated (Dufall *et al.*, 2003). Similarly, the diprenylated chalcone, Bartericin A, present in *D. barteri* leaf and twig extracts was shown to have potent antioxidant properties. It was found that this and other prenylated and geranylated chalcones were as active as the prenylated flavones, possibly accounting for the anti-inflammatory action of these extracts (Omisore *et al.*, 2005). Free radical scavenging activity was also confirmed for prenylated anthronoids isolated from the stem-bark of *Harungana madagascariensis* (Kouam *et al.*, 2005) and for proanthocyanidins isolated from the bark of the tree *Burkea africana* (Mathisen *et al.*, 2002). Kolaviron, a biflavonoid isolated from a *Garcinia kola* seed extract has been reported to scavenge free radicals, protect against lipid peroxidation and H_2O_2-induced DNA strand breaks and oxidized bases, (Adaramoye and Akinloye, 2000; Adaramoye *et al.*, 2005; Adegoke *et al.*, 1998; Farombi *et al.*, 2004; Farombi *et al.*, 2000; Olaleye *et al.*, 2000). In addition, the ability of Thonningianin A and B, ellagitannins, isolated from *Thonningia sanguinea* to scavenge free radicals and inhibit lipid peroxidation has been reported (Gyamfi and Aniya, 2002; Ohtani *et al.*, 2000). The anti-inflammatory ability of Griffonianone D ((7E)-(6″,7''-dihydroxy-3'',7''-dimethyloct-2''-enyl)oxy-4'-methoxyisoflavone), an isoflavone present in *Millettia griffoniana* was established (Yankep *et al.*, 2003). Prenylated anthronoids, harunmadagascarins A (8,9-dihydroxy-4,4-bis-(3,3-dimethylallyl)-6-methyl-2,3-(2,2-dimethylpyrano)anthrone and B (8,9-dihydroxy-4,4,5-tris-(3,3-dimethylallyl)-6-methyl-2,3-(2,2-dimethylpyrano)anthrone), harunganol B and harungin anthrone from the stem-bark of *Harungana madagascariensis* have exhibited significant antioxidant activity (Kouam *et al.*, 2005). Saponins and isofuranonaphthoquinones isolated from different medicinal plant extracts with antioxidant properties include the saponin, Balanin 1 (3β,12β,14β,16β) cholest-5-ene-3,16-diyl bis (β-d -glucopyranoside)- 12-sulphate, sterol sulfonated, Balanin 2 (3β,20*S*,22*R*,25*R*)-26-hydroxy-22-acetoxyfurost-5-en-3- yl-rhamnopyranosyl-(1 →2)-glucopyranoside and a furostanol saponin isolated from *Balanites aegyptiaca* (Speroni *et al.*, 2005). Isofuranonaphthoquinones isolated from the roots of *Bulbine capitata*, 5,8-dihydroxy-1-tigloylmethylnaphtho[2,3-c]furan-4,9-dione, 1-acetoxymethyl-8-hydroxynaphtho[2,3-c]furan-4,9-dione, and 1-acetoxymethyl-5,8-dihydroxynaphtho[2,3-c]furan-4,9-dione possess antioxidant activities (Bezabih *et al.*, 2001).

Crude Extracts of African Medicinal Plants with Confirmed Antioxidant Activities

The antioxidant properties of a large proportion of African medicinal plants listed in Tables 9.1 and 9.2 were tested using either aqueous or organic plant extracts. After confirming antioxidant properties, a correlation was proposed between this property and the general groups of antioxidant compounds that are present in these extracts. No further attempts were made to isolate the specific compounds that may have contributed towards such activity. Flavonoids in *Aloe barbadensis* (Hu *et al.*, 2003; Rajasekaran *et al.*, 2005a), chromone glycoside in *A. claviflora* (Lindsey *et al.*, 2003), essential oils in *Artemisia abyssinica* and *Juniperus procera* (Burits *et al.*, 2001) as well as *Helichrysum dasyanthum*, *H. felinum*, *H. excisum* and *H. petiolare* (Lourens *et al.*, 2004), peoanthocyanidins in *Burkea africana* bark (Mathisen *et al.*, 2002), polyphenols in extracts of *Crataegus monogyna* (Bahorun *et al.*, 2003), saponins and alkaloids in extracts of *Leucosidae sericea* (Aremu *et al.*, 2010; Aremu *et al.*, 2011) are all considered as major ingredients that have contributed to the antioxidant properties of these plants. Reports on a number of *Barleria* species, which includes *B. albostellata*, *B. greenii* and *B. prionitis* has indicated their good anti-inflammatory (Amoo *et al.*, 2009) and antioxidant capacities (Amoo *et al.*, 2011).

Studies have focused on screening a vast number of plants, used in a specific region, so as to determine their antioxidant properties: Mali (Diallo *et al.*, 2001; Maiga *et al.*, 2006), South Africa (Lindsey *et al.*, 2002; Opoku *et al.*, 2002; Steenkamp *et al.*, 2005b; Steenkamp *et al.*, 2004), Cameroon (Agbor *et al.*, 2005; Awouafack *et al.*, 2013), Algeria (Djeridane *et al.*, 2006), Ghana (Gyamfi *et al.*, 1999), Burkina Faso (Palé *et al.*, 2004), Madagascar (Juliani *et al.*, 2004) and Mauritius (Toyokuni *et al.*, 2003) and anti-inflammatory properties: South Africa (Chukwujekwu *et al.*, 2005; Jager *et al.*, 1996; Jäger and van Staden, 2005; Lin *et al.*, 1999), and West Africa (Ezeamuzie *et al.*, 1994).

Discussion and Conclusion

The use of traditional herbal remedies as alternative medicine plays a significant role in Africa since it features extensively in primary health care. The search for natural antioxidants, especially from plant sources, as a potential preventive intervention for free-radical mediated diseases has turned into an important research field and an essential social issue for improvement of the quality of human life, especially for those in developing countries. This is the case as many polyphenols, including phenolic acids, flavonoids (anthocyanins and anthoxanthins), tannins and lignans act as antioxidants and protect against various pathological conditions such as coronary heart disease, in addition to their anti-inflammatory, antimicrobial and anti-cancer activities (Barron *et al.*, 2002; Cody *et al.*, 1988; Hertog *et al.*, 1993; Middleton *et al.*, 2000).

Flavonoids are a large group of compounds containing a number of hydroxyl groups on their ring structures with several medicinal properties. They include isoflavonoids and isoflavonoid glycosides, flavones and flavone glycosides, flavonols and flavonol glycosides, anthocyanins, chalcones and dihydrochalcones, aurones, flavonones and dihydroflavonols, and flavans and biflavonyls. To date, approximately 9000 different flavonoids have been identified in plant sources (Williams and Grayer, 2004). Great interest has been dedicated to the antioxidant properties of flavonoids that may function as potent free radical scavengers, reducing agents, and protectors against peroxidation of lipids (Hanasaki *et al.*, 1994; Miranda *et al.*, 2000). Reviews have been published on the antioxidant efficacy of flavonoids and phenolic compounds as well as on the relationship between their antioxidant activities, as hydrogen donating free radical scavengers, in relation to their chemical structures. Unsaturation in the C ring of quercetin (compared to catechin) results in the increased antioxidant activity. Also the position and number of hydroxyl groups on the phenolic rings results in increasing or decreasing the antioxidant properties of these compounds (Cao *et al.*, 1997; Chen *et al.*, 1996; Cotelle *et al.*, 1996; Middleton *et al.*, 2000; Pietta, 2000; Rice-Evans *et al.*, 1996).

Although numerous flavonoids have been isolated from different African medicinal plant extracts, the structure activity relationship of these compounds has as yet not been investigated. Recent studies have also shown that some flavonoids are modulators of pro-inflammatory gene expression, thus leading to the attenuation of the inflammatory response (Garcý´a-Lafuente *et al.*, 2009). Examples of these include the lipophilic flavones and flavonols 5,7-dihydroxy-2',3',4',5'-tetramethoxyflavone, 5,4'-dihydroxy-7,2',3',5'-tetramethoxyflavone, and 5,7,4'-trihydroxy-2',3',5'-trimethoxyflavone isolated from *Psiadia punctulata* (Juma *et al.*, 2001) and Dinklagin B and C isolated from *Dorstenia dinklagei* (Ngadjui *et al.*, 2002). Isolated flavone and flavonol glycosides include kaempferide 3-O-beta-xylosyl (1–>2)-beta-glucoside, kaempferol 3-O-alpha-rhamnoside-7,4'-di-O-beta-galactoside, kaempferol 3,7,4'-tri-O-beta-glucoside and quercetin 3-O-[alpha-rhamnosyl (1–>6)] [beta-glucosyl (1–>2)]-beta-glucoside-7-O-alpha-rhamnoside from *Warburgia ugandensis* and quercetin-7,4'-disulphate from *Alchornea laxiflora* (Ogundipe *et al.*, 2001). Flavanones and dihydroflavonols include Dorsmanin I and J and epi-dorsmanin F and G isolated from *Dorstenia mannii* (Ngadjui *et al.*, 1999a) and Dinklagins A, isolated from the twigs of *Dorstenia dinklagei* (Ngadjui *et al.*, 2002a) and two flavones isolated from the twigs of *Eriosema robustum* (Awouafack *et al.*, 2013) and 1α,3β-dihydroxy-12-oleanen-29-oic (1), 1-hydroxy-12-olean-30-oic acid (2), 3,30-dihydroxyl-12-oleanen-22-one (3), and 1,3,24-trihydroxyl-12-olean-29-oic acid (4), a new pentacyclic triterpenoid (1 α, 23-dihydroxy-12-oleanen-29-oic acid-3β-O-2,4-di-acetyl-l-rhamnopyranoside) (5) from *Combretum imberbe* (Angeh *et al.*, 2007). Anthocyanins isolated include; cyanidind 3-*O*-(2-*O*-galloylgalactose and cyanidin Cy 3-*O*-(2-*O*-galloyl-6-*O*-rhamnosyl-galactoside from *Acalypha hispida* (Reiersen *et al.*, 2003) and cyanidin 3-*O*-β-D-glucopyranoside, cyanidin 3-*O*-(2-*O*-β-D-xylopyranosyl)-β-D-glucopyranoside from *Hibiscus sabdariffa* (Palé *et al.*, 2004). Structure-activity studies on the isolated compounds from African medicinal extracts will therefore be of great interest.

Antioxidants may exert their protective effects via different mechanisms at different stages of the oxidation process. There are those that are able to inhibit the production of free radicals via their ability to chelate transition metal ions and those that are able to quench and stabilise free radicals (Gordon, 1990; Pinchuk and Lichtenberg, 2002). Additionally, they are subdivided further into categories according to their functions (Noguchi *et al.*, 2000). Such classification of the newly isolated antioxidant compounds from African medicinal plant extracts is warranted to better understand their antioxidant properties.

In conclusion, attempts should be made to identify the compounds responsible for the proven antioxidant properties and in cases where they have been isolated, their antioxidant properties should be investigated. Finally, in a country where traditional remedies form part of primary health care, the remedies containing antioxidant activity may be considered as easily accessible and affordable sources of natural antioxidants and therefore may provide an additional health benefit.

References

Abdelwahab S., I, Koko W., S.,, Mohan M., M., E., T., S., Achoui M., Abdulla M., A., Mustafa M., R., Ahmad S., Noordin M., I., Yong C., L., Sulaiman M., R., Othman R., and Hassan, A., A. (2012). *In vitro* and *in vivo* anti-inflammatory activities of columbin through the inhibition of cycloxygenase-2 and nitric oxide but not the suppression of NF-κB translocation. *European Journal of Pharmacology*, **678**: 61–70.

Abdillahi, H., S., Finnie, J., F., and van Staden J. (2011). Anti-inflammatory, antioxidant, anti-tyrosinase and phenolic contents of four *Podocarpus* species used in traditional medicine in South Africa. *J Ethnopharmacology*, **136**: 496– 503.

Abegaz, B.M. (2002). Novel phenylanthraquinones, isofuranonaphthoquinones, homoisoflavonoids, and biflavonoids from African plants in the genera *Bulbine, Scilla, Ledebouria,* and *Rhus*. *Phytochemistry Reviews*, **1**: 299 – 310.

Abegaz, B.M., Bezabih, M., Msuta, T., Brun, R., Menche, D., Muhlbacher, J., and Bringmann, G. (2002a). Gaboroquinones A and B and 4'-O-demethylknipholone-4'-O-beta-D-glucopyranoside, phenylanthraquinones from the roots of *Bulbine frutescens*. *J Nat Prod.*, **65**: 1117-21.

Abegaz, B.M., Ngadjui, B.T., Bezabih, M., and Mdee, L.K. (1999). Novel natural products from marketed plants of eastern and southern Africa. *Pure Appl. Chem.*, **71**: 919-926.

Abegaz, B.M., Ngadjui, B.T., Dongo, E., and Bezabih, M.T. (2000). Chemistry of the genus *Dorstenia*. *Current Organic Chemistry*, **4**: 1079-1090.

Abegaz, B.M., Ngadjui, B.T., Dongo, E., Ngameni, B., Nindi, M.N., and Bezabih, M. (2002b). Chalcones and other constituents of *Dorstenia prorepens* and *Dorstenia zenkeri*. *Phytochemistry*, **59**: 877-83.

Abegaz, B.M., Ngadjui, B.T., Dongo, E., and Tamboue, E. (1998). Prenylated chalcones and flavones from the leaves of *Dorstenia kameruniana*. *Phytochemistry*, **49**: 1147-1150.

Abegaz, B.M., Ngadjui, B.T., Folefoc, G.N., Fotso, S., Ambassa, P., Bezabih, M., Dongo, E., Rise, F., and Petersen, D. (2004). Prenylated flavonoids, monoterpenoid furanocoumarins and other constituents from the twigs of *Dorstenia elliptica* (Moraceae). *Phytochemistry*, **65**: 221-6.

Abegaz, B.M., and Woldu, Y. (1991). Isoflavonoids from the roots of *Salsola somalensis*. *Phytochemistry*, **30**: 1281-1284.

Abila, B., Richens, A., and Davies, J. A. (1993). Anticonvulsant effects of extracts of the west African black pepper, *Piper guineense*. *J Ethnopharmacol*, **39**: 113-7.

Adams, M., and Bauer, R. (2008). Inhibition of leukotriene biosynthesis by secondary plant metabolites. *Current Organic Chemistry*, **12**: 602-618.

Adaramoye, O.A., and Akinloye, O. (2000). Possible protective effect of kolaviron on CCl4-induced erythrocyte damage in rats. *Biosci Rep.*, **20**: 259-64.

Adaramoye, O.A., Nwaneri, V.O., Anyanwu, K.C., Farombi, E.O., and Emerole, G.O. (2005). Possible anti-atherogenic effect of kolaviron (a Garcinia kola seed extract) in hypercholesterolaemic rats. *Clin Exp Pharmacol Physiol.,* **32**: 40-6.

Adegoke, G.O., Kumar, M.V., Sambaiah, K., and Lokesh, B.R. (1998). Inhibitory effect of *Garcinia kola* on lipid peroxidation in rat liver homogenate. *Indian J Exp Biōl.,* **36**: 907-10.

Aderogba, M.A., Okoh, E.K., Adelanwa, T.A., and Obuotor, E.M. (2004). Antioxidant Properties of the *Nigerian Piliostigma* Species. *Journal of Biological Sciences,* **4**: 501-503.

Aderogba M., A., Kgatle, D., T., McGaw, L., J., and Eloff, J., N. (2012). Isolation of antioxidant constituents from *Combretum apiculatum* sub sp. *apiculatum*. *South African Journal of Botany*, **79**: 125–131.

Adesina, S.K. (2005). The Nigerian *Zanthoxylum*: chemical and biological values. *Afr. J. Trad. Comp. Alt. Med.,* **2**: 282 – 301.

Adesina, S.K., Idowu, O., Ogundaini, A.O., Oladimeji, H., Olugbade, T.A., Onawunmi, G.O., and Pais, M. (2000a). Antimicrobial constituents of the leaves of *Acalypha wilkesiana* and *Aacalypha hispida*. *Phytother Res.,* **14**: 371-374.

Adesina, S.K., Idowu, O., Ogundaini, A.O., Oladimeji, H., Olugbade, T.A., Onawunmi, G.O., and Pais, M. (2000b). Antimicrobial constituents of the leaves of *Acalypha wilkesiana* and *Acalypha hispida*. *Phytother Res.,* **14**: 371-374.

Adetutu, A, Morgana, W., and Corcorana O. (2011). Antibacterial, antioxidant and fibroblast growth stimulation activity of crude extracts of *Bridelia ferruginea* leaf, a wound-healing plant of Nigeria antibacterial, antioxidant and fibroblast growth stimulation activity of crude extracts of *Bridelia ferruginea* leaf, a wound-healing plant of Nigeria. *J Ethnopharmacology*, 133: 116–119.

Agbaje, E.O., Tijani, A.Y., and Braimoh, O.O. (2003). Effects of *Enantia chlorantha* extracts in laboratory-induced convulsion and inflammation. *Orient Journal of Medicine,* **15**: 68-71.

Agbor, G.A., Vinson J., A, Oben J., E., A., and Ngogang, J., Y. (2007). *In vitro* antioxidant activity of three *Piper* species. *Journal of Herbal Pharmacotherapy*, **7**(2) 49-64.

Agbor, G.A., Oben, J.E., Ngogang, J.Y., Xinxing, C., and Vinson, J.A. (2005). Antioxidant capacity of some herbs/spices from Cameroon: A comparative study of two methods. *J Agric Food Chem.,* **53**: 6819-24.

Ajaib, M, Khan Z., Khan, N., Abbasi, M., A, Shahwar, D, Wahab, M., and Saddiqui, M, F. (2011). Antibacterial and antioxidant activities of an ethnobotanically important plant *Sauromatum venosum* (ait.) Schott. of district Kotli, Azad Jammu and Kashmir. *Pak. J. Bot.,* **43(1)**: 579-585.

Ajiboye T.O. (2010). *In vivo* antioxidant potentials of *Piliostigma thonningi*i (Schum) leaves: Studies on hepatic marker enzyme, antioxidant system, drug detoxifying enzyme and lipid peroxidation. *Hum Exp Toxicol.,* doi: 10.1177/0960327110366785.

Akah, P. A., and Nwambie, A. I. (1994). Evaluation of Nigerian traditional medicines: 1. Plants used for rheumatic (inflammatory) disorders. *J Ethnopharmacol*., **42(3)**: 179-82.

Akinmoladun A., C., Ibukun E., O., Afor E., Obuotor E., M., and Farombi E., O. (2007). Phytochemical constituent and antioxidant activity of extract from the leaves of *Ocimum gratissimum*. *Scientific Research and Essay*, **2 (5)**: 163-166.

Akintonwa, D.A.A., and Maduka, H.C.C. (2005). The use of theoretical mechanistic biochemistry (TMB) in the rationalization of the experimental data on sacoglottis gabonensis on the natural antioxidant defense during 2,4-dinitrophenyl hydrazine administration *in-vivo*. *The Internet Journal of Gastroenterology*, http: / / www.ispub.com / ostia / index.php?xmlFilePath=journals / ijge / vol4n1 / tmb.xml.

Akinyemi, K.O., Oladapo, O., Okwara, C.E., Ibe, C.C., and Fasure, K.A. (2005). Screening of crude extracts of six medicinal plants used in South-West Nigerian unorthodox medicine for anti-methicillin resistant *Staphylococcus aureus* activity. *BMC Complement Altern Med.*, **5**: 6.

Akpanabiatu, M.I., Umoh, I.B., Udosen, E.O., Udoh, A.E., and Edet, E.E. (2005). Rat serum electrolytes, lipid profile and cardiovascular activity on *Nauclea Latifolia* leaf extract administration. *Indian Journal of Clinical Biochemistry*, **20**: 29-34.

Akrout, A., Gonzalez, L., El Jani, H., and Madrid P.C. (2011). Antioxidant and antitumor activities of *Artemisia campestris* and *Thymelaea hirsuta* from southern Tunisia. *Food and Chemical Toxicology*, **49**: 342–347.

Akula U., S., and Odhav B. (2008). *In vitro* 5-Lipoxygenase inhibition of polyphenolic antioxidants from undomesticated plants of South Africa. *Journal of Medicinal Plants Research*, **2(9)**: 207-212.

Alabi, D.A., Akinsulire, O.R., and Sanyaolu, M.A. (2005). Qualitative determination of chemical and nutritional composition of *Parkia biglobosa* (Jacq.) Benth. *African Journal of Biotechnology*, **4**: 812-815.

Albayrak, S., Aksoy, A., Sagdic, O., and Hamzaoglu, E. (2010.) Compositions, antioxidant and antimicrobial activities of *Helichrysum* (Asteraceae) species collected from Turkey. *Food Chemistry*, **119**: 114–122.

Al-Jaber, N. A., Awaad, A. S., and Moses J. E. (2011). Review on some antioxidant plants growing in Arab world. *Journal of Saudi Chemical Society*, **15**: 293–307.

Alqasoumi, S. I., and Abdel-Kader, M. S. (2012). Terpenoids from *Juniperus procera* with hepatoprotective activity. *Pak. J. Pharm. Sci.*, **25(2)**: 315-322.

Amoo, S.O., Finnie, J.F., A. I. van Staden, J. (2009). *In vitro* pharmacological evaluation of three *Barleria* species. *Journal of Ethnopharmacology*, **121**: 274–277.

Amoo, S.O., Ndhlala, A.R., Finnie, J. F., and van Staden J. (2011). Antifungal, acetylcholineserase inhibition, antioxidant and phytochemical properties of three *Barleria* species. *South Afr J Botany*, **77**: 435-445.

Amoo, S.O., Aremu, A.O., and van Staden J. (2014). Unraveling the medicinal potential of South African *Aloe* species. *J Ethnopharmacology,* **153**: 19–41.

Amos, S., Kolawole, E., Akah, P., Wambebe, C., A. I. and Gamaniel, K. (2001). Behavioral effects of the aqueous extract of *Guiera senegalensis* in mice and rats. *Phytomedicine,* **8**: 356-61.

Andayi, A.W., Yenesew, A., Derese, S., Midiwo, J.O., Gitu, P.M., Jondiko, O.J., Akala, H., Liyala, P., Wangui, J., Waters, N.C., Heydenreich, M., and Peter, M.G. (2006). Antiplasmodial flavonoids from *Erythrina sacleuxii*. *Planta Med.,* **72**: 187-189.

Anganji, S., A., Mousavi S., F., and Babapour E. (2012). Antioxidants: A few key points. *Annals of Biological Research,* **3(8)**: 3968-3977.

Angeh,J. E., Huang, X., Sattler I., Swan, G. E., Dahse, H., Hartl, A., and Eloff, J. N. (2007). Antimicrobial and anti-inflammatory activity of four known and one new triterpenoid from *Combretum imberbe* (Combretaceae). *J Ethnopharmacology,* **110**: 56–60.

Araghiniknam, M., Chung, S., Nelson-White, T., Eskelson, C., and Watson, R.R. (1996). Antioxidant activity of dioscorea and dehydroepiandrosterone (DHEA) in older humans. *Life Sci.,* **59**: 147-57.

Aremu, A. O., Amoo S., O., Ndhlala, A., R., Finnie, J., F., and van Staden, J. (2011). Antioxidant activity, acetylcholinesterase inhibition, iridoid content and mutagenic evaluation of *Leucosidea sericea. Food and Chemical Toxicology,* **49**: 1122–1128.

Aremu, A.O., Fawole, O.A., Chukwujekwu, J.C., Light, M.E., Finnie, J.F., and van Staden, J. (2010). *In vitro* antimicrobial, anthelmintic and cyclooxygenase-inhibitoryactivities and phytochemical analysis of *Leucosidea sericea. J Ethnopharmacol,* **131**: 22–27.

Arnold, H.J., and Gulumian, M. (1984). Pharmacopoeia of traditional medicine in Venda. *J Ethnopharmacol.,* 1**2**: 35-74.

Arora, R., Gupta, D., Chawla, R., Sagar, R., Sharma, A., Kumar, R., Prasad, J., Singh, S., Samanta, N., and Sharma, R.K. (2005). Radioprotection by plant products: present status and future prospects. *Phytother Res.,* **19**: 1-22.

Asres, K., Gibbons, S., Hana, E., and Bucar, F. (2005). Anti-inflammatory activity of extracts and a saponin isolated from *Melilotus elegans. Pharmazie,* **60**: 310-2.

Atata, R.F., Sani, A., and Ajewole, S.M. (2003). Effect of stem bark extracts of *Enantia chloranta* on some clinical isolates. *Biokemistri.,* **15**: 84-92.

Atawodi, S.E. (2005). Antioxidant potential of African medicinal plants. *African Journal of Biotechnology,* **4**: 128-133.

Awouafack M., D., Pierre Tane P., and Eloff J., N. (2013). Two new antioxidant flavones from the twigs of *Eriosema robustum* (Fabaceae). *Phytochemistry Letters,* **6**: 62-66.

Badary, O.A., Taha, R.A., Gamal el-Din, A.M., and Abdel-Wahab, M.H. (2003). Thymoquinone is a potent superoxide anion scavenger. *Drug Chem Toxicol.*, **26**: 87-98.

Bahorun, T., Aumjaud, E., Ramphul, H., Rycha, M., Luximon-Ramma, A., Trotin, F., and Aruoma, O.I. (2003). Phenolic constituents and antioxidant capacities of *Crataegus monogyna* (Hawthorn) callus extracts. *Nahrung.*, **47**: 191-8.

Bahorun, T., Gressier, B., Trotin, F., Brunet, C., Dine, T., Luyckx, M., Vasseur, J., Cazin, M., Cazin, J.C., and Pinkas, M. (1996). Oxygen species scavenging activity of phenolic extracts from hawthorn fresh plant organs and pharmaceutical preparations. *Arzneimittelforschung*, **46**: 1086-9.

Bahorun, T., Trotin, F., Pommery, J., Vasseur, J., and Pinkas, M. (1994). Antioxidant activities of *Crataegus monogyna* extracts. *Planta Med.*, **60**: 323-8.

Bakoma B., Bénédicte B., Kwashie E-G., Diallo A., Agbonon A., Aklikokou K., Gbeassor M., and Moore N. (2012). Total phenolic content, antioxidant activity, and *in vitro* inhibitory potential against key enzymes relevant for hyperglyceamia of *Bridelia ferruginea* extracts. *Research Journal of Phytochemistry*, **6**(4): 120-126.

Barakat H. (2014). Composition, antioxidant, antibacterial activities and mode of action of Clove (*Syzygium aromaticum* L.) buds essential oil. *British Journal of Applied Science and Technology*, **4**(13): 1934-1951.

Baratta, M.T., Dorman, H.J.D., Deans, S.G., Figueiredo, A.C., Barroso, J.G., and Ruberto, G. (1998). Antimicrobial and antioxidant properties of some commercial essential oils. *Flavour and Fragrance Journal*, **13**: 235-244.

Barku, V. Y. A., Opoku-Boahen, Y., Owusu-Ansah, E., Dayie, N., T. K., D., and Mensah,F. E. (2013). *In-vitro* assessment of antioxidant and antimicrobial activities of methanol extracts of six wound healing medicinal plants. *Journal of Natural Sciences Research*, **3(1)**: 74-80.

Barron, D., Di Pietro, A., Dumontet, C., and McIntosh, D.B. (2002). Isoprenoid flavonoids are new leads in the modulation of chemoresistance. *Phytochemistry Reviews*, **1**: 325-332.

Benedi, J., Arroyo, R., Romero, C., Martin-Aragon, S., and Villar, A.M. (2004). Antioxidant properties and protective effects of a standardized extract of *Hypericum perforatum* on hydrogen peroxide-induced oxidative damage in PC12 cells. *Life Sci.*, **75**: 1263-76.

Bennie, L., Coetzee, J., Malan, E., and Ferreira, D. (2002a). (4–>6)-Coupled proteracacinidins and promelacacinidins from *Acacia galpinii* and *Acacia caffra*. *Phytochemistry*, **60**: 521-32.

Bennie, L., Coetzee, J., Malan, E., and Ferreira, D. (2002b). Structure and stereochemistry of dimeric proteracacinidins possessing the rare C-4(C) –> C-5(D) interflavanyl linkage. *Phytochemistry*, **59**: 673-8.

Bennie, L., Malan, E., Coetzee, J., and Ferreira, D. (2000). Structure and synthesis of ether-linked proteracacinidin and promelacacinidin proanthocyanidins from *Acacia caffra*. *Phytochemistry*, **53**: 785-93.

Bernardi, A.P., Ferraz, A.B., Albring, D.V., Bordignon, S.A., Schripsema, J., Bridi, R., Dutra-Filho, C.S., Henriques, A.T., and von Poser, G.L. (2005). Benzophenones from *Hypericum carinatum*. *J Nat Prod.*, **68**: 784-6.

Bernatonienė, J., Masteikova, R, Majienė, D., Savickas, A., Kėvelaitis, E., Bernatonienė, R., Dvoráèková, K., Civinskienė, G., Lekas, R., Vitkevièius, K., and Peèiûra, R. (2008). Free radical-scavenging activities of *Crataegus monogyna* extracts. *Medicina (Kaunas)*, **44(9)**: 706-712.

Betancor-Fernandez, A., Perez-Galvez, A., Sies, H., and Stahl, W. (2003). Screening pharmaceutical preparations containing extracts of turmeric rhizome, artichoke leaf, devil's claw root and garlic or salmon oil for antioxidant capacity. *J Pharm Pharmacol.*, **55**: 981-6.

Betti, J.L. (2004). An ethnobotanical study of medicinal plants among the Baka pygmies in the DJA Biosphere Reserve, Cmeroon. *African Study Monographs*, **25**: 1-27.

Bezabih, M., and Abegaz, B.M. (1998). 4'-Demethylknipholone from aerial parts of *Bulbine capitata*. *Phytochemistry*, **48**: 1071-1073.

Bezabih, M., Abegaz, B.M., Dufall, K., Croft, K., Skinner-Adams, T., and Davis, T.M. (2001). Antiplasmodial and antioxidant isofuranonaphthoquinones from the roots of *Bulbine capitata*. *Planta Med.*, **67**: 340-4.

Bezabih, M., Motlhagodi, S., and Abegaz, B.M. (1997). Isofuranonaphthoquinones and phenolic and knipholone derivatives from the roots of *Bulbine capitata*. *Phytochemistry*, **46**: 1063-1067.

Biapa, P-C., N, Agbor G., A, Oben J., E., and Ngogang, J., Y. (2007). Phytochemical studies and antioxidant properties of four medicinal plants used in Cameroon. *Afr. J. Trad. CAM.*, **4** (4): 495 – 500.

Binutu, O.A., and Cordell, G.A. (2001). Constituents of *Afzelia bella* stem bark. *Phytochemistry*, **56**: 827-30.

Binutu, O.A., and Lajubutu, B.A. (1994). Antimicrobial potentials of some plant species of the *Bignoniaceae* family. *Afr J Med Med Sci.*, **23**: 269-73.

Bizimenyera, S.E., Swan, G.E., Chikoto, H., and Eloff, J.N. (2005). Rationale for using *Peltophorum africanum* (Fabaceae) extracts in veterinary medicine. *J S Afr Vet Assoc.*, **76**: 54-8.

Boakye-Gyasi, E., Ainooson, G. K., and Abotsi, W.K.M. (2011). Anti-inflammatory, antipyretic and antioxidant properties of a hydroalcoholic leaf extract of *Palisota hirsuta* K. Schum. (Commelinaceae). *West African Journal of Pharmacy*, **22 (1)**: 10 – 18.

Bojase, G., Wanjala, C.C., and Majinda, R.R. (2001). Flavonoids from the stem bark of *Bolusanthus speciosus*. *Phytochemistry*, **56**: 837-41.

Bombardelli, E., Cristoni, A., Lolla, A., Morazzoni, P., Mustich, G., Pace, R., and Piretti, M.V. (1994). Chemical and biological characterisation of *Piliostigma thonningii* polyphenols. *Fitoterapia*, **65**: 493-501.

Bordia, A., Verma, S.K., and Srivastava, K.C. (1997). Effect of ginger (*Zingiber officinale* Rosc.) and fenugreek (*Trigonella foenumgraecum* L.) on blood lipids, blood sugar and platelet aggregation in patients with coronary artery disease. *Prostaglandins Leukot Essent Fatty Acids*, **56**: 379-84.

Bosch C. H. (2008). Plant Resources of Tropical Africa: Medicinal Plants. PROTA Foundation, Wageningen, Netherlands, pg 128.

Bouchet, N., Barrier, L., and Fauconneau, B. (1998). Radical scavenging activity and antioxidant properties of tannins from *Guiera senegalensis* (Combretaceae). *Phytotherapy Research*, **12**: 159 – 162.

Bramati, L., Aquilano, F., and Pietta, P. (2003). Unfermented rooibos tea: quantitative characterization of flavonoids by HPLC-UV and determination of the total antioxidant activity. *J Agric Food Chem.*, **51**: 7472-4.

Bramati, L., Minoggio, M., Gardana, C., Simonetti, P., Mauri, P., and Pietta, P. (2002). Quantitative characterization of flavonoid compounds in Rooibos tea (*Aspalathus linearis*) by LC-UV/DAD. *J Agric Food Chem.*, **50**: 5513-9.

Brandao, M.G.L., Nery, C.G.C., Mamao, M.A.S., and Krettli, A.U. (1998). Two methoxylated flavone glycosides from *Bidens pilosa*. *Phytochemistry*, **48**: 397-399.

Breza, J., Dzurny, O., Borowka, A., Hanus, T., Petrik, R., Blane, G., and Chadha-Boreham, H. (1998). Efficacy and acceptability of tadenan (*Pygeum africanum* extract) in the treatment of benign prostatic hyperplasia (BPH): a multicentre trial in central Europe. *Curr Med Res Opin.*, **14**: 127-39.

Bruni, R., Guerrini, A., Scalia, S., Romagnoli, C., and Sacchetti, G. (2002). Rapid techniques for the extraction of vitamin E isomers from *Amaranthus caudatus* seeds: ultrasonic and supercritical fluid extraction. *Phytochem Anal.*, **13**: 257-61.

Bucar, F., Resch, M., Bauer, R., Burits, M., Knauder, E., and Schubert-Zsilavecz, M. (1998). 5-Methylflavasperone and rhamnetin from *Guiera senegalensis* and their antioxidative and 5-lipoxygenase inhibitory activity. *Pharmazie*, **53**: 875-8.

Bucar, F., Schneider, I., Ogmundsdottir, H., and Ingolfsdottir, K. (2004). Anti-proliferative lichen compounds with inhibitory activity on 12(S)-HETE production in human platelets. *Phytomedicine*, **11**: 602-6.

Burits, M., Asres, K., and Bucar, F. (2001). The antioxidant activity of the essential oils of *Artemisia afra*, *Artemisia abyssinica* and *Juniperus procera*. *Phytother Res.*, **15**: 103-8.

Burits, M., and Bucar, F. (2000). Antioxidant activity of *Nigella sativa* essential oil. *Phytother Res.*, **14**: 323-8.

Buwa, L., V., and Afolayan, A., J. (2009). Antimicrobial activity of some medicinal plants used for the treatment of tuberculosis in the Eastern Cape Province, *South Africa. African Journal of Biotechnology*, **8 (23)**: 6683-6687.

Calixto J. B., Otuki M. F., and Santos, A. R. S. (2003). Anti-inflammatory compounds of plant origin: Part 1-Action on Arachidonic acid pathway, nitric oxide and Nuclear factor kappa B. *Planta Medica*, **69**: 973-983.

Calviello, G., Filippi, G.M., Toesca, A., Palozza, P., Maggiano, N., Nicuolo, F.D., Serini, S., Azzena, G.B., and Galeotti, T. (2005). Repeated exposure to pyrrolidine-dithiocarbamate induces peripheral nerve alterations in rats. *Toxicol Lett.*, **158**: 61-71.

Cao, G., Sofic, E., and Prior, R.L. (1997). Antioxidant and prooxidant behavior of flavonoids: structure-activity relationships. *Free Radic Biol Med.*, **22**: 749-60.

Carey, M., W., Rao, N., V., Kumara, B., R., and Mohana, G., K. (2010). Anti-inflammatory and analgesic activities of methanolic extract of *Kigelia pinnata* DC flower. *J Ethnopharmacology*, **130**: 179–182.

Chaaib, F., Queiroz, E.F., Ndjoko, K., Diallo, D., and Hostettmann, K. (20030. Antifungal and antioxidant compounds from the root bark of *Fagara zanthoxyloides*. *Planta Med.*, **69**: 316-20.

Chacha, M., Bojase-Moleta, G., and Majinda, R.R. (2005). Antimicrobial and radical scavenging flavonoids from the stem wood of *Erythrina latissima*. *Phytochemistry*, **66**: 99-104.

Chanda S., and Dave, R. (2009). *In vitro* models for antioxidant activity evaluation and some medicinal plants possessing antioxidant properties: An overview. *African Journal of Microbiology Research*, **3**(13): 981-996.

Chen, Y.J., Dai, Y.S., Chen, B.F., Chang, A., Chen, H.C., Lin, Y.C., Chang, K.H., Lai, Y.L., Chung, C.H., and Lai, Y.J. (1999). The effect of tetrandrine and extracts of *Centella asiatica* on acute radiation dermatitis in rats. *Biol Pharm Bull.*, **22**: 703-6.

Chen, Z.Y., Chan, P.T., Ho, K.Y., Fung, K.P., and Wang, J. (1996). Antioxidant activity of natural flavonoids is governed by number and location of their aromatic hydroxyl groups. *Chem Phys Lipids*, **79**: 157-63.

Chernishov, G., Arragie, M., and Etana, A. (1978). Preliminary pharmacological studies on mettere (*Glinus lotoides*). II. Effects upon the cardiovascular and gastrointestinal system. *Ethiop Med J.*, **16**: 105-10.

Chiang, Y.M., Chuang, D.Y., Wang, S.Y., Kuo, Y.H., Tsai, P.W., and Shyur, L.F. (2004). Metabolite profiling and chemopreventive bioactivity of plant extracts from *Bidens pilosa*. *J Ethnopharmacol.* **95**: 409-19.

Chorepsima, S., Tentolouris, K., Dimitroulis, D., and Tentolouris, N. (2013). Melilotus: Contribution to wound healing in the diabetic foot. *Journal of herbal medicine*, **3**: 81–86.

Chothani D., L., and Vaghasiya H., U. (2011). A review of *Balanites aegytiaca* Del. (desert dale): Phytochemical constituents, traditional uses, and pharmacological activities. *Pharmacognosy Review*, **5**(9): 55-62.

Chukwujekwu, J.C., van Staden, J., and Smith, P. (2005). Antibacterial, anti-inflammatory and antimalarial activities of some Nigerian medicinal plants. *S Afr J Bot.*, **71**: 316-325.

Cimanga, K., De Bruyne, T., Apers, S., Pieters, L., Totté, J., Kambu, K., Tona, L., Bakana, P., van Ufford, L.Q., Beukelman, C., Labadie, R., and Vlietinck, A.J. (1999).

Complement-inhibiting constituents of *Bridelia ferruginea* stem bark. *Planta Med.*, **65**: 213-217.

Cody, V., Middleton, E., Jr., Harborne, J.B., and Bertz, A. (1988). Plant flavonoids in Biology and Medicine II: Biochemical, Cellular and Medicinal Properties. Alan R. Liss ed., New York

Combes, S., Finet, J.-P., and Siri, D. (2002). On the optical activity of the 3-aryl-4-hydroxycoumarin isolated from *Millettia griffoniana*: molecular modelling and total synthesis. *J Chem Soc, Perkin Transactions*, **1**: 38 – 44.

Conforti, F., Statti, G., Loizzo, M.R., Sacchetti, G., Poli, F., and Menichini, F. (2005). *In Vitro* antioxidant effect and inhibition of alpha-amylase of two varieties of *Amaranthus caudatus* seeds. *Biol Pharm Bull.*, **28**: 1098-102.

Cook, J.A., VanderJagt, D.J., Dasgupta, A., Mounkaila, G., Glew, R.S., Blackwell, W., and Glew, R.H. (1998). Use of the Trolox assay to estimate the antioxidant content of seventeen edible wild plants of Niger. *Life Sci.*, **63**: 105-10.

Coombes, P.H., and Rogers, C.B. (2002). Methyl gardenolate A, a novel cycloartenoid ester from the leaves of *Combretum woodii* (Combretaceae). *Nat Prod Lett.*, **16**: 301-4.

Cotelle, N., Bernier, J.L., Catteau, J.P., Pommery, J., Wallet, J.C., and Gaydou, E.M. (1996). Antioxidant properties of hydroxy-flavones. *Free Radic Biol Med.*, **20**: 35-43.

Dagne, E., Alemu, M., and Sterner, O. (1993). Flavonoids from *Euclea divinorum*. *Bull Chem Soc Ethiop.*, **7**: 87-92.

Dagne, E., and Bekele, A. (1990). C-prenylated isoflavones from *Millettia ferruginea*. *Phytochemistry*, **29**: 2679-82.

Dagne, E., Bekele, A., Noguchi, H., Shibuya, M., and Sankawa, U. (1990). O-Geranylated and O-prenylated flavonoids from *Millettia ferruginea*. *Phytochemistry*, **29**: 2671-73.

Dagne, E., Mammo, W., Bekele, A., Odyek, O., and Byaruhanga, M.A. (1991). Flavonoids of *Millettia dura*. *Bull Chem Soc Ethiop.*, **5**: 81-86.

Dagne, E., Mammo, W., and Sterner, O. (1992). Flavonoids of *Tephrosia polyphylla*. *Phytochemistry*, **31**: 3662-63.

Dagne, E., and Steglich, W. (1984). Knipholone: a unique anthraquinone derivative from *Kniphofia foliosa*. *Phytochemistry*, **23**: 1729 –1731.

Dagne, E., and Yenesew, A. (1994). Anthraquinones and the chemotaxonomy of the Asphodelaceae. *Pure and Appl. Chem.*, **66**: 2395-2398.

Deba, F., Xuan, T., D., Yasuda, M., and Tawata, S. (2008). Chemical composition and antioxidant, antibacterial and antifungal activities of the essential oils from *Bidens pilosa* Linn. var. Radiata. *Food Control*, **19**: 346–352.

De Bruyne, T., Cimanga, K., Pieters, L., Claeys, M., Dommisse, R., and Vlietinck, A. (1997). Gallocatechin-(4.>O >7)-Epigallocatechin, a new biflavonoid isolated from *Bridelia ferruginea*. *Nat. Prod. Let.*, **11**: 47-52.

De Paola, R., Muia, C., Mazzon, E., Genovese, T., Crisafulli, C., Menegazzi, M., Caputi, A.P., Suzuki, H., and Cuzzocrea, S. (2005). Effects of *Hypericum perforatum* extract in a rat model of ischemia and reperfusion injury. *Shock*, **24**: 255-63.

Derese, S., Yenesew, A., Midiwo, J.O., Heydenreich, M., and Peter, M.G. (2003). A new isoflavone from stem bark of *Millettia dura*. *Bulletin of the Chemical Society of Ethiopia*, **17**: 113-115.

Diallo, D., Hveem, B., Mahmoud, M.A., Berge, G., Paulsen, B.S., and Maïga, A. (1999). An ethnobotanical survey of herbal drugs of Gourma District, Mali. *Pharmaceutical Biology (Formerly International Journal of Pharmacognosy)*, **37**: 80 – 91.

Diallo, D., Marston, A., Terreaux, C., Toure, Y., Paulsen, B.S., and Hostettmann, K. (2001). Screening of Malian medicinal plants for antifungal, larvicidal, molluscicidal, antioxidant and radical scavenging activities. *Phytother Res.*, **15**: 401-6.

Dilsiz, N., Sahaboglu, A., Yildiz, M.Z., and Reichenbach, A. (2006). Protective effects of various antioxidants during ischemia-reperfusion in the rat retina. *Graefes Arch Clin Exp Ophthalmol.*, **244**: 627-633.

Djeridane, A., Yousfi, M., Nadjemi, B., Boutassouna, D., Stocker, P., and Vidal, N. (2006). Antioxidant activity of some Algerian medicinal plants extracts containing phenolic compounds. *Food Chemistry*, **97**: 654-660.

Djeridane, A., Yousfi, M., Nadjemi, B., Vidal, N., Lesgards J., F., and Stocker, P. (2007). Screening of some Algerian medicinal plants for the phenolic compounds and their antioxidant activity. *Eur Food Res Technol.*, **224**: 801–809.

Djeridane A, Yousfi, M, Brunel, J., M., and Stocker, P. (2010). Isolation and characterization of a new steroid derivative as a powerful antioxidant from Cleome arabica in screening the in vitro antioxidant capacity of 18 Algerian medicinal plants. *Food and Chemical Toxicology*, **48**: 2599–2606.

Drewes, S.E., Khan, F., van Vuuren, S.F., and Viljoen, A. M. (2005). Simple 1, 4-benzoquinones with antibacterial activity from stems and leaves of *Gunnera perpensa*. *Phytochemistry*, **66**: 1812-6.

Dubois, M.A., Ilyas, M., and Wagner, H. (1986). Cussonosides A and B, two triterpene-saponins from *Cussonia barteri*. *Planta Med.*, **56**: 80-3.

Duddeck, H., Yenesew, A., and Dagne, E. (1987). Isoflavonoids from *Taverniera abyssinica*. *Bull Chem Soc Ethiop.*, **1**: 36-41.

Dufall, K.G., Ngadjui, B.T., Simeon, K.F., Abegaz, B.M., and Croft, K.D. (2003). Antioxidant activity of prenylated flavonoids from the West African medicinal plant *Dorstenia mannii*. *J Ethnopharmacol.*, **87**: 67-72.

Duwiejua, M., Woode, E., and Obiri, D.D. (2002). Pseudo-akuammigine, an alkaloid from *Picralima nitida* seeds, has anti-inflammatory and analgesic actions in rats. *J Ethnopharmacol.*, **81**: 73-9.

Egan P., A., and van der Kooy, F. (2013). Phytochemistry of the carnivorous sundew genus Drosera (Droseraceae) –Future perspectives and ethnopharmacological relevance. *Chemistry and Biodiversity*, **10**: 1774-1790.

Eldeen, I. M., E. E. Elgorashi, *et al.* (2006). Anolignan B: a bioactive compound from the roots of *Terminalia sericea*. *J Ethnopharmacol*., **103**(1): 135-8.

El Tahir, A., Ibrahim, A.M., Satti, G.M.H., Theander, T.G., Kharazmi, A., and Khalid, S.A. (1998). The potential antileishmanial activity of some Sudanese medicinal plants. *Phytotherapy Research,* **12**: 576 – 579.

El Sherbeiny, A.E., El Ansari, M.A., Nawwar, M.A., and El Sayed, N.H. (1977). Flavonol glycosides and flavonol glucoside gallates from *Peltophorum africanum*. *Planta Med.*, **32**: 165-70.

El-Sherbiny, D.A., Khalifa, A.E., Attia, A.S., and Eldenshary Eel, D. (2003). *Hypericum perforatum* extract demonstrates antioxidant properties against elevated rat brain oxidative status induced by amnestic dose of scopolamine. *Pharmacol Biochem Behav.,* **76**: 525-33.

El-Ghorab, A.H., El-Massry, K.F., Marx, F., and Fadel, H.M. (2003). Antioxidant activity of Egyptian *Eucalyptus camaldulensis* var. *brevirostris* leaf extracts. *Nahrun,g* **47**: 41-5.

El-Masry, S., Amer, M.E., Abdel-Kader, M.S., and Zaatout, H.H. (2002). Prenylated flavonoids of *Erythrina lysistemon* grown in Egypt. *Phytochemistry,* **60**: 783-7.

El-Mousallamy, A.M. (1998). Leaf flavonoids of *Albizia lebbeck*. *Phytochemistry,* **48**: 759-61.

Eloff, J.N., Famakin, J.O., and Katerere, D.R.P. (2005a). *Combretum woodii* (Combretaceae) leaf extracts have high activity against Gram-negative and Gram-positive bacteria. *African Journal of Biotechnology,* **4**: 1161-1166.

Eloff, J.N., Famakin, J.O., and Katerere, D.R.P. (2005b). Isolation of an antibacterial stilbene from *Combretum woodii* (Combretaceae) leaves. *African Journal of Biotechnology,* **4**: 1167-1171.

Emam, A.M., Elias, R., Moussa, A.M., Faure, R., Debrauwer, L., and Balansard, G. (1998). Two flavonoid triglycosides from *Buddleja madagascariensis*. *Phytochemistry,* **48**: 739-42.

Emami, S., A., Asili, J., Mohagheghi Z., and Hassanzadeh, M., K. (2007). Antioxidant activity of leaves and fruits of Iranian conifers. *eCAM.*, **4**(3)313–319.

Endale, A. (2005). Isolation, structural elucidation, quantification and formulation of the saponins and flavonoids of seeds of *Glenus lotoides.* PhD Thesis. Universität Tübingen, der Fakultät für Chemie und Pharmazie der Eberhard-Karls

Endale, A., Kammerer, B., Gebre-Mariam, T., and Schmidt, P.C. (2005). Quantitative determination of the group of flavonoids and saponins from the extracts of the seeds of *Glinus lotoides* and tablet formulation thereof by high-performance liquid chromatography. *J Chromatogr A.*, **1083**: 32-41.

Endale, A., Schmidt, P.C., and Gebre-Mariam, T. (2004). Standardisation and physico-chemical characterisation of the extracts of seeds of *Glinus lotoides*. *Pharmazie*, **59**: 34-8.

Erasto, P., Bojase-Moleta, G., and Majinda, R.R. (2004). Antimicrobial and antioxidant flavonoids from the root wood of *Bolusanthus speciosus*. *Phytochemistry*, **65**: 875-80.

Erkan, N., Ayranci, G., and Ayranci, E. (2008). Antioxidant activities of rosemary (*Rosmarinus officinalis* L.) extract, blackseed (*Nigella sativa* L.) essential oil, carnosic acid, rosmarinic acid and sesamol. *Food Chemistry*, **110**, 76–82.

Esteban, M.D., Collado, I.G., Macias, F.A., Massanet, G.M., and Luis, F.R. (1986). Flavonoids from *Artemisia lanata*. *Phytochemistry*, **25**: 1502-1504.

Etoundi C.B., Kuaté D., Ngondi J.L., and Oben J. (2010). Anti-amylase, anti-lipase and antioxidant effects of aqueous extracts of some *Cameroonian* spices. *Journal of Natural Products*, **3**: 165-171.

Ezea, J, Iwuji, T. C., and Oguike, M. A. (2014). Growth responses of pregnant rabbits and their litters fed spreading day flower (*Commelina diffusa* Burm. F.) and rock fig (*Ficus ingens* miquel) leaves. *Journal of Global Biosciences*, **3**(2): 619-625.

Ezeamuzie, I.C., Ambakederemo, A.W., Shode, F.O., and Ekwebelem, S.C. (1996). Antiinflammatory effects of *Moringa oleifera* root extract. *Pharmaceutical Biology (Formerly International Journal of Pharmacognosy)*, **34**: 207 – 212.

Ezeamuzie, I.C., Ojinnaka, M.C., Uzogara, E.O., and Oji, S.E. (1994). Anti-inflammatory, antipyretic and anti-malarial activities of a West African medicinal plant *Picralima nitida*. *Afr J Med Med Sci.*, **23**: 85-90.

Fabiyi, O. A. Atolani, O., Adeyemi, O., S., and Olatunji, G., A. (2012). Antioxidant and cytotoxicity of β-amyrin acetate fraction from *Bridelia ferruginea* leaves. *Asian Pacific Journal of Tropical Biomedicine*, S981-S984.

Fakae, B.B., Campbell, A.M., Barrett, J., Scott, I.M., Teesdale-Spittle, P.H., Liebau, E., and Brophy, P.M. (2000). Inhibition of glutathione S-transferases (GSTs) from parasitic nematodes by extracts from traditional Nigerian medicinal plants. *Phytother Res.*, **14**: 630-4.

Fakeye, T.O., Itiola, O.A., and Odelola, H.A. (2000). Evaluation of the antimicrobial property of the stem bark of *Picralima nitida* (Apocynaceae). *Phytother Res.*, **14**: 368-70.

Fall, D., Gleye, C., Franck, X., Laurens, A., and Hocquemiller, R. (2002). Cis-bullatencin, a linear acetogenin from roots of *Uvaria chamae*. *Nat Prod Lett.*, **16**: 315-21.

Farah, N., Benghuzzi, H., Tucci, M., and Cason, Z. (2005). The effects of isolated antioxidants from black seed on the cellular metabolism of A549 cells. *Biomed Sci Instrum.*, **41**: 211-6.

Farombi, E.O. (2000). Mechanisms for the hepatoprotective action of kolaviron: studies on hepatic enzymes, microsomal lipids and lipid peroxidation in carbontetrachloride-treated rats. *Pharmacol Res.*, **42**: 75-80.

Farombi, E.O. (2003). African indigenous plants with chemotherapeutic potentials and biotechnological approach to the production of bioactive prophylactic agents. *African Journal of Biotechnology*, **2**: 662-671.

Farombi, E.O., Adepoju, B.F., Ola-Davies, O.E., and Emerole, G.O. (2005). Chemoprevention of aflatoxin B1-induced genotoxicity and hepatic oxidative damage in rats by kolaviron, a natural bioflavonoid of *Garcinia kola* seeds. *Eur J Cancer Prev.*, **14**: 207-14.

Farombi, E.O., Alabi, M.C., and Akuru, T.O. (2002). Kolaviron modulates cellular redox status and impairment of membrane protein activities induced by potassium bromate (KBrO(3)) in rats. *Pharmacol Res.*, **45**: 63-8.

Farombi, E.O., and Fakoya, A. (2005). Free radical scavenging and antigenotoxic activities of natural phenolic compounds in dried flowers of *Hibiscus sabdariffa* L. *Mol Nutr Food Res.*, **49(12)**: 1120-1128.

Farombi, E.O., Moller, P., and Dragsted, L.O. (2004). *Ex-vivo* and *in vitro* protective effects of kolaviron against oxygen-derived radical-induced DNA damage and oxidative stress in human lymphocytes and rat liver cells. *Cell Biol Toxicol.*, **20**: 71-82.

Farombi, E.O., and Nwaokeafor, I.A. (2005). Anti-oxidant mechanisms of kolaviron: studies on serum lipoprotein oxidation, metal chelation and oxidative membrane damage in rats. *Clin Exp Pharmacol Physiol.*, **32**: 667-674.

Farombi, E.O., Ogundipe, O., and Moody, J.O. (2001). Antioxidant and anti-inflammatory activities of *Mallotus oppositifolium* in model systems. *Afr J Med Med Sci.*, **30**: 213-215.

Farombi, E.O., Ogundipe, O.O., Samuel Uhunwangho, E., Adeyanju, M.A., and Olarenwaju Moody, J. (2003). Antioxidant properties of extracts from *Alchornea laxiflora* (Benth) Pax and Hoffman. *Phytother Res.*, **17**: 713-6.

Farombi E. O., Shrotriya S., and Surh., Y-J. (2009). Kolaviron inhibits dimethyl nitrosamine-induced liver injury by suppressing COX-2 and iNOS expression via NF-κB and AP-1. *Life Sciences*, **84**: 149–155.

Farombi, E.O., Tahnteng, J.G., Agboola, A.O., Nwankwo, J.O., and Emerole, G.O. (2000). Chemoprevention of 2-acetylaminofluorene-induced hepatotoxicity and lipid peroxidation in rats by kolaviron, a *Garcinia kola* seed extract. *Food Chem Toxicol.*, **38**: 535-41.

Fernandes, A.C., Cromarty, A.D., Albrecht, C., and van Rensburg, C.E. (2004). The antioxidant potential of *Sutherlandia frutescens*. *J Ethnopharmacol.*, **95**: 1-5.

Ficarra, R., Ficarra, P., Tommasini, S., Carulli, M., Melardi, S., Di Bella, M.R., Calabro, M.L., De Pasquale, R., Germano, M.P., Sanogo, R., and Casuscelli, F. (1997). Isolation and characterization of *Guiera senegalensis* J.F.Gmel. active principles. *Boll Chim Farm*, **136**: 454-9.

Fratianni F., Tucci M., De Palma M., Rosa Pepe R., and Nazzaro, F. (2007). Polyphenolic composition in different parts of some cultivars of globe artichoke (*Cynara cardunculus* L. var. *scolymus* (L.) Fiori). *Food Chemistry*, **104**: 1282–1286.

Gagnier, J.J., Chrubasik, S., and Manheimer, E. (2004). *Harpagophytum procumbens* for osteoarthritis and low back pain: A systematic review. *BMC Complement Altern Med.*, **4**: 13.

Galvez, J., Crespo, M.E., Jimenez, J., Suarez, A., and Zarzuelo, A. (1993). Antidiarrhoeic activity of quercitrin in mice and rats. *J Pharm Pharmacol.*, **45**: 157-9.

Garcý´a-Lafuente A., Guillamo´ E., Villares A., Rostagno M., A., and Martý´nez J., A. (2009). Flavonoids as anti-inflammatory agents: implications in cancer and cardiovascular disease. *Inflamm. Res.*, **58**: 537–552.

Gebhardt, R. (1997). Antioxidative and protective properties of extracts from leaves of the artichoke (*Cynara scolymus* L.) against hydroperoxide-induced oxidative stress in cultured rat hepatocytes. *Toxicol Appl Pharmacol.*, 1**44**: 279-86.

Gebre-Mariam, T., Asres, K., Getie, M., Endale, A., Neubert, R., and Schmidt, P.C. (2005). *In vitro* availability of kaempferol glycosides from cream formulations of methanolic extract of the leaves of *Melilotus elegans*. *Eur J Pharm Biopharm.*, **60**: 31-8.

Gebre-Mariam, T., Neubert, R., Schmidt, P.C., Wutzler, P., and Schmidtke, M. (2006). Antiviral activities of some Ethiopian medicinal plants used for the treatment of dermatological disorders. *J Ethnopharmacol.*, 1**04**: 182-7.

Germano, M.P., D'Angelo, V., Sanogo, R., Morabito, A., Pergolizzi, S., and De Pasquale, R. (2001). Hepatoprotective activity of *Trichilia roka* on carbon tetrachloride-induced liver damage in rats. *J Pharm Pharmacol.*, **53**: 1569-74.

Germano, M.P., Sanogo, R., Guglielmo, M., De Pasquale, R., Crisafi, G., and Bisignano, G. (1998). Effects of *Pteleopsis suberosa* extracts on experimental gastric ulcers and *Helicobacter pylori* growth. *J Ethnopharmacol.*, **59**: 167-72.

Getie, M., Gebre-Mariam, T., Rietz, R., Hohne, C., Huschka, C., Schmidtke, M., Abate, A., and Neubert, R.H. (2003). Evaluation of the anti-microbial and anti-inflammatory activities of the medicinal plants *Dodonaea viscosa*, *Rumex nervosus* and *Rumex abyssinicus*. *Fitoterapia*, **74**: 139-43.

Getie, M., Gebre-Mariam, T., Rietz, R., and Neubert, R.H. (2002). Evaluation of the release profiles of flavonoids from topical formulations of the crude extract of the leaves of *Dodonea viscosa* (Sapindaceae). *Pharmazie*, **57**: 320-2.

Gidado, A., Ameh, D.A., and Atawodi, S.E. (2005). Effect of *Nauclea latifolia* leaves aqueous extracts on blood glucose levels of normal and alloxan-induced diabetic rats. *African Journal of Biotechnology*, **4**: 91-93.

Gobel, H., Heinze, A., Ingwersen, M., Niederberger, U., and Gerber, D. (2001). [Effects of *Harpagophytum procumbens* LI 174 (devil's claw) on sensory, motor und vascular muscle reagibility in the treatment of unspecific back pain]. *Schmerz*, **15**: 10-8.

Gordon, M.H. (1990). The mechanism of antioxidant action *in vitro*. In Food Antioxidants, Hudson, B. F. J. (ed.). Elsevier, London

Gouda, Y.G., Abdel-baky, A.M., Darwish, F.M., Mohamed, K.M., Kasai, R., and Yamasaki, K. (2003). Iridoids from *Kigelia pinnata* DC. fruits. *Phytochemistry*, **63**: 887-92.

Grant, L., McBean, D., E., Fyfe, L., and Warnock, A., M. (2007). A Review of the biological and potential therapeutic actions of *Harpagophytum procumbens. Phytother. Res.*, **21**: 199–209.

Grayer, R.J., Kite, G.C., Abou-Zaid, M., and Archer, L.J. (2000). The application of atmospheric pressure chemical ionisation liquid chromatography-mass spectrometry in the chemotaxonomic study of flavonoids: characterisation of flavonoids from *Ocimum gratissimum* var. *gratissimum*. *Phytochemical Analysis*, **11**: 257 – 267.

Gülçin, I, Elmastat M., and Aboul-Enein, H., Y. (2007). Determination of antioxidant and radical scavenging activity of basil (*Ocimum basilicum* L. Family Lamiaceae) Assayed by different methodologies. *Phytother. Res.*, **21**: 354–361.

Gyamfi, M.A., and Aniya, Y. (1998). Medicinal herb, *Thonningia sanguinea* protects against aflatoxin B1 acute hepatotoxicity in Fischer 344 rats. *Hum Exp Toxicol.*, **17**: 418-23.

Gyamfi, M.A., and Aniya, Y. (2002). Antioxidant properties of Thonningianin A, isolated from the African medicinal herb, *Thonningia sanguinea. Biochem Pharmacol.*, **63**: 1725-37.

Gyamfi, M.A., Hokama, N., Oppong-Boachie, K., and Aniya, Y. (2000). Inhibitory effects of the medicinal herb, *Thonningia sanguinea*, on liver drug metabolizing enzymes of rats. *Hum Exp Toxicol* **19**: 623-31.

Gyamfi, M.A., Ohtani, II, Shinno, E., and Aniya, Y. (2004a). Inhibition of glutathione S-transferases by thonningianin A, isolated from the African medicinal herb, *Thonningia sanguinea, in vitro. Food Chem Toxicol.*, **42**: 1401-8.

Gyamfi, M.A., Tanaka, T., and Aniya, Y. (2004b). Selective suppression of cytochrome P450 gene expression by the medicinal herb, *Thonningia sanguinea* in rat liver. *Life Sci.*, **74**: 1723-37.

Gyamfi, M.A., Yonamine, M., and Aniya, Y. (1999). Free-radical scavenging action of medicinal herbs from Ghana: *Thonningia sanguinea* on experimentally-induced liver injuries. *Gen Pharmacol.*, **32**: 661-7.

Habtemariam, S. (2010). Knipholone anthrone from *Kniphofia foliosa* induces a rapid onset of necrotic cell death in cancer cells. *Fitoterapia*, **81**: 1013–1019.

Hanasaki, Y., Ogawa, S., and Fukui, S. (1994). The correlation between active oxygens scavenging and antioxidative effects of flavonoids. *Free Radic Biol Med.*, **16**: 845-50.

Harborne, J.B., and Williams, C.A. (2001). Anthocyanins and other flavonoids. *Nat Prod Rep.*, **18**: 310-33.

Herold, A., Cremer, L., Calugaru, A., Tamas, V., Ionescu, F., Manea, S., and Szegli, G. (2003a). Antioxidant properties of some hydroalcoholic plant extracts with antiinflammatory activity. *Roum Arch Microbiol Immunol.*, **62**: 217-27.

Herold, A., Cremer, L., Calugaru, A., Tamas, V., Ionescu, F., Manea, S., and Szegli, G. (2003b). Hydroalcoholic plant extracts with anti-inflammatory activity. *Roum Arch Microbiol Immunol.*, **62**: 117-29.

Hertog, M.G., Feskens, E.J., Hollman, P.C., Katan, M.B., and Kromhout, D. (1993). Dietary antioxidant flavonoids and risk of coronary heart disease: the Zutphen Elderly Study. *Lancet*, **342**: 1007-11.

Hiermann, A., and Bucar, F. (1994). Influence of some traditional medicinal plants of Senegal on prostaglandin biosynthesis. *J Ethnopharmacol.*, **42**: 111-6.

Hohtola Anja. (2010). Bioactive compounds from Northern Plants: Bio-Farms for nutraceuticals, functional food and safety control for biosensors. Edited by Giardi MT, Rea G and Berra B. Landes Bioscience, New York.

Hosseinzadeh, H., Karimi, G.R., and Rakhshanizadeh, M. (2005). Anticonvulsant effect of *Hypericum perforatum*: role of nitric oxide. *J Ethnopharmacol.*, **98**: 207-8.

Houghton, P.J., Hylands, P.J., Mensah, A.Y., Hensel, A., and Deters, A.M. (2005). *In vitro* tests and ethnopharmacological investigations: wound healing as an example. *J Ethnopharmacol.*, **100**: 100-7.

Hu, Y., Xu, J., and Hu, Q. (2003). Evaluation of antioxidant potential of aloe vera (*Aloe barbadensis* Miller) extracts. *J Agric Food Chem.*, **51**: 7788-91.

Huang, T.H., Tran, V.H., Duke, R.K., Tan, S., Chrubasik, S., Roufogalis, B.D., and Duke, C.C. (2006). Harpagoside suppresses lipopolysaccharide-induced iNOS and COX-2 expression through inhibition of NF-kappa B activation. *J Ethnopharmacol.*, **104**: 149-55.

Ibewuike, J.C., Ogundaini, A.O., Ogungbamila, F.O., Martin, M., Gallard, J., Bohlin, L., and Pais, M. (1996). Piliostigmin, a 2-phenoxychromone, and c-methylflavonols from *Piliostigma thonningii*. *Phytochemistry*, **43**: 687-690.

Ibewuike, J.C., Ogungbamila, F.O., Ogundaini, A.O., Okeke, I.N., and Bohlin, L. (1997). Antiinflammatory and antibacterial activities of C-methylflavonols from *Piliostigma thonningii*. *Phytotherapy Research*, **11**: 281 – 284.

Ighodaro O., M., Agunbiade S., O., Omole, J., O., and Kuti, A., O. (2012). Evaluation of the chemical, nutritional, antimicrobial, and antioxidant-vitamin profiles of *Piliostima thonningii* leaves (Nigeria species). *Research Journal of Medicinal Plants*, **6 (7)**: 537-543.

Igoli, J.O., Igwue, I.C., and Igoli, N.P. (2003). Traditional Medicinal Practices amongst the Igede people of Nigeria. *Herbs, Spices and Medicinal Plants*, **10**: 1-10.

Igoli, J.O., Ogaji, O.G., Tor-Anyiin, T.A., and Igoli, N.P. (2005). Traditional medicine practice amongst the Igede people of Nigeria. Part II. *Afr. J. Trad. CAM.*, **2**: 134 – 152.

Imaga, N., O., A., Shaire E., A., Ogbeide S., and Samuel K., A. 2011. *In vitro* biochemical investigations of the effects of *Carica papaya* and *Fagara zanthoxyloides* on antioxidant status and sickle erythrocytes. *African Journal of Biochemistry Research*, **5(8)**: 226-236.

Inanami, O., Asanuma, T., Inukai, N., Jin, T., Shimokawa, S., Kasai, N., Nakano, M., Sato, F., and Kuwabara, M. (1995). The suppression of age-related accumulation of lipid peroxides in rat brain by administration of Rooibos tea (*Aspalathus linearis*). *Neurosci Lett.*, **196**: 85-8.

Ishani, A., MacDonald, R., Nelson, D., Rutks, I., and Wilt, T.J. (2000). *Pygeum africanum* for the treatment of patients with benign prostatic hyperplasia: a systematic review and quantitative meta-analysis. *Am J Med.*, **109**: 654-64.

Isong, E.U., and Essien, I.B. (1996). Nutrient and antinutrient composition of three varieties of *Piper* species. *Plant Foods Hum Nutr.*, **49**: 133-7.

Ito, C., Itoigawa, M., Kojima, N., Tan, H.T., Takayasu, J., Tokuda, H., Nishino, H., and Furukawa, H. (2004a). Cancer chemopreventive activity of rotenoids from *Derris trifoliata*. *Planta Med.*, **70**: 585-8.

Ito, C., Itoigawa, M., Kojima, N., Tan, H.T., Takayasu, J., Tokuda, H., Nishino, H., and Furukawa, H. (2004b). Cancer chemopreventive activity of rotenoids from *Derris trifoliata*. *Planta Med.*, **70**: 8-11.

Iwalewa E., O., Adewale I., O., Taiwo, B., J., Arogundade T., Osinowo A., Daniyan, O., M., and Adetogun, G., E. (2008). Effect of *Harungana madagascariensis* stem bark on the antioxidant amrkers in alloxan induced diabetic and carrageenan induced inflammatory disorders in rats. *Journal of Complementary and Integrative Medicine*, **5**(1): 1553.

Iwalewa E., O., McGaw L., J., Naidoo V., and Eloff J., N. (2007). Inflammation: the foundation of diseases and disorders. A review of phytomedicines of South African origin used to treat pain and inflammatory conditions. *African Journal of Biotechnology*, **6** (25): . 2868-2885.

Iwu, M.M., Okunji, C.O., Ohiaeri, G.O., Akah, P., Corley, D., and Tempesta, M.S. (1990). Hypoglycaemic activity of dioscoretine from tubers of *Dioscorea dumetorum* in normal and alloxan diabetic rabbits. *Planta Med.*, **56**: 264-7.

Jager, A.K., Hutchings, A., and van Staden, J. (1996). Screening of Zulu medicinal plants for prostaglandin-synthesis inhibitors. *J Ethnopharmacol.*, **52**: 95-100.

Jäger, A.K., and van Staden, J. (2005). Cyclooxygenase inhibitory activity of South African plants used against inflammation. *Phytochemistry Reviews*, **4**: 39-46.

Jayaprakasha, G., K., Negi, P., S., Jena, B., S., and Rao, L., J., M. (2007). Antioxidant and antimutagenic activities of *Cinnamomum zeylanicum* fruit extracts. *Journal of Food Composition and Analysis*, **20**: 330–336.

Jayashree, G., Kurup Muraleedhara, G., Sudarslal, S., and Jacob, V.B. (2003). Anti-oxidant activity of *Centella asiatica* on lymphoma-bearing mice. *Fitoterapia*, **74**: 431-4.

Jimoh, F.O., and Oladiji, A.T. (2005). Preliminary studies on *Piliostigma thonningii* seeds: Proximate analysis, mineral composition and phytochemical screening. *African Journal of Biotechnology*, **4**: 1439-1442.

Joubert, E., Richards E., S., van der merwe, J. D., de beer, D., Manley, M., and Gelderblom, W., C., A. (2008). Effect of species variation and processing on phenolic composition and *in vitro* antioxidant activity of aqueous extracts of *Cyclopia* spp. (Honeybush Tea). *J. Agric. Food Chem.*, **56**, 954–963.

Juliani, H.R., Simon, J.E., Ramboatiana, M.M.R., Behra, O., Garvey, and A., Raskin, I. (2004). Malagasy aromatic plants: essential oils, antioxidant and antimicrobial activities. *Acta Horticulturae*, **629**: 77-81.

Juma, B.F., and Majinda, R.R. (2004). Erythrinaline alkaloids from the flowers and pods of *Erythrina lysistemon* and their DPPH radical scavenging properties. *Phytochemistry*, **65**: 1397-404.

Juma, B.F., Yenese, A., Midiwo, J.O., and Waterman, P.G. (2001). Flavones and phenylpropenoids in the surface exudate of *Psiadia punctulata*. *Phytochemistry*, **57**: 571-4.

Kabiri, N., Asgary S., Madani H., and Mahzouni P. (2010). Effects of Amaranthus caudatus l. extract and lovastatin on atherosclerosis in hypercholesterolemic rabbits. *J Medicinal Plants Res.*, **4**(5): 355-361.

Kamara, B.I., Brand, D.J., Brandt, E.V., and Joubert, E. (2004). Phenolic metabolites from honeybush tea (*Cyclopia subternata*). *J Agric Food Chem.*, **52**: 5391-5.

Kamara, B.I., Brandt, E.V., Ferreira, D., and Joubert, E. (2003). Polyphenols from Honeybush tea (*Cyclopia intermedia*). *J Agric Food Chem.*, **51**: 3874-9.

Kamatou, G.P., Viljoen, A.M., Gono-Bwalya, A.B., van Zyl, R.L., van Vuuren, S.F., Lourens, A.C., Baser, K.H., Demirci, B., Lindsey, K.L., van Staden, J., and Steenkamp, P. (2005). The *in vitro* pharmacological activities and a chemical investigation of three South African *Salvia* species. *J Ethnopharmacol.*, **102**: 382-90.

Kamgang, R., Vidal Pouokam Kamgne, E., Fonkoua, M.C., Penlap, N.B.V., and Biwole Sida, M. (2006). Activities of aqueous extracts of *Mallotus oppositifolium* on *Shigella dysenteriae* A1-induced diarrhoea in rats. *Clin Exp Pharmacol Physiol.*, **33**: 89-94.

Kansci, G., Dongo, E., and Genot, C. (2003). 2, 2-diphenyl-1-picrylhydrazyl (DPPH*) test demonstrates antiradical activity of *Dorstenia psilurus* and *Dorstenia ciliata* plant extracts. *Nahrung*, **47**: 434-7.

Karabina, S.A., Lehner, A.N., Frank, E., Parthasarathy, S., and Santanam, N. (2005). Oxidative inactivation of paraoxonase: implications in diabetes mellitus and atherosclerosis. *Biochim Biophys Acta.*, **1725**: 213-21.

Kaszkin, M., Beck, K.F., Koch, E., Erdelmeier, C., Kusch, S., Pfeilschifter, J., and Loew, D. (2004). Downregulation of iNOS expression in rat mesangial cells by special extracts of *Harpagophytum procumbens* derives from harpagoside-dependent and independent effects. *Phytomedicine*, **11**: 585-95.

Katerere, D.R., and Eloff, J.N. (2004). Variation in chemical composition, antibacterial and antioxidant activity of fresh and dried *Acacia* leaf extracts. *South African Journal of Botany*, **70**: 303-305.

Katsoulis, L.C., Veale, D.J., and Havlik, I. 2000. The pharmacological action of *Rhoicissus tridentata* on isolated rat uterus and ileum. *Phytother Res.*, **14**: 460-2.

Kaviarasan S., Naik G., H., Gangabhagirathi R.,, Anuradha C., V., and Priyadarsini, K., I. (2007). *In vitro* studies on antiradical and antioxidant activities of fenugreek (*Trigonella foenum* graecum) seeds. *Food Chemistry*, **103**: 31–37.

Kim H., Moon J., Y., Kim H., Lee D-S, Cho M., Choi H-K., Kim Y., S., Mosaddik A., and Cho, S., K. (2010). Antioxidant and antiproliferative activities of mango (*Mangifera indica* L.) flesh and peel. *Food Chemistry*, **121**: 429–436.

Klimczak, I., Malecka, M., and Pacholek, B. (2002). Antioxidant activity of ethanolic extracts of amaranth seeds. *Nahrung.*, **46**: 184-6.

Koeppen, B.H., and Roux, D.G. (1966). C-glycosylflavonoids. The chemistry of aspalathin. *Biochem J.*, **99**: 604-9.

Koeppen, B.H., Smit, J.B., and Roux, D.G. (1962). The flavone C-glycosides and flavonol O-glycosides of *Aspalathus acuminatus* (rooibos tea). *Biochem J.*, **83**: 507-11.

Koffi, A.M., Kanko, C., Ramiarantsoa, H., Figueredo, G., Chalchat, J.-C., Bessière, J.-M., Koukoua, G., and N'Guessan, Y.T. (2004). Dérivés phénoliques et benzéniques des huiles essentielles de trois Uvaria (Annonaceae) de Côte-d'Ivoire : *Uvaria chamae* (P.Beauv), *Uvaria afzelii* (Sc. Elliot) et *Uvaria* sp. (Aké Assi). *Comptes Rendus Chimie.*, **7**: 997-1002.

Kokwaro, J.O. (2009). Medicinal Plants of East Africa, 3rd ed. University of Nairobi Press, Nairobi.

Kouadio, F., Kanko, C., Juge, M., Grimaud, N., Jean, A., N'Guessan, Y.T., and Petit, J.Y. (2000). Analgesic and antiinflammatory activities of an extract from *Parkia biglobosa* used in traditional medicine in the Ivory Coast. *Phytother Res.*, **14**: 635-7.

Kouam, S.F., Ngadjui, B.T., Krohn, K., Wafo, P., Ajaz, A., and Choudhary, M.I. (2005). Prenylated anthronoid antioxidants from the stem bark of *Harungana madagascariensis*. *Phytochemistry*, **66**: 1174-9.

Krenn, L., Beyer, G., Pertz, H.H., Karall, E., Kremser, M., Galambosi, B., and Melzig, M.F. (2004). *In vitro* antispasmodic and anti-inflammatory effects of *Drosera rotundifolia*. *Arzneimittelforschung*, **54**: 402-5.

Kuete, V., and Sandjo, L., P. (2012). Isobavachalcone: An overview. *Chin J Integr Med.*, **18(7)**: 543-547.

Kuete, V., Ngameni B., Mbaveng, A., T., Ngadjuib, B., Meyer, J., J., M., and Lall, M. (2010). Evaluation of flavonoids from *Dorstenia barteri* for their antimycobacterial, antigonorrheal and anti-reverse transcriptase activities. *Acta Tropica*, **116**, 100–104.

Kweifio-Okai, G. (1991a). Antiinflammatory activity of a Ghanaian antiarthritic herbal preparation: I. *J Ethnopharmacol.*, **33**: 263-7.

Kweifio-Okai, G. (1991b). Antiinflammatory activity of a Ghanaian antiarthritic herbal preparation: II. *J Ethnopharmacol.*, **33**: 129-33.

Lamien-Meda A., Lamien C., E., Compaoré M., M., Y., Meda R., M., T., Kiendrebeogo M., Zeba B., Millogo J., F., and Nacoulma, O. G. (2008). Polyphenol content and antioxidant activity of fourteen wild edible fruits from Burkina Faso. *Molecules,* **13**: 581-594.

Latté, K.-P. (1999). Phytochemische und pharmacologische Untersuchungen an *Pelargonium reniforme* Curt. PhD thesis, University of Berlin.

Latte, K.P., and Kolodziej, H. (2004). Antioxidant properties of phenolic compounds from *Pelargonium reniforme. J Agric Food Chem.,* **52**: 4899-902.

Lee, O-H., and Boo-Yong Lee, B-Y. (2010). Antioxidant and antimicrobial activities of individual and combined phenolics in *Olea europaea* leaf extract. *Bioresource Technology,* **101**: 3751–3754.

Li, H., Xia, N., Brausch, I., Yao, Y., and Forstermann, U. (2004). Flavonoids from artichoke (*Cynara scolymus* L.) up-regulate endothelial-type nitric-oxide synthase gene expression in human endothelial cells. *J Pharmacol Exp Ther.,* **310**: 926-32.

Lin, J., Opoku, A.R., Geheeb-Keller, M., Hutchings, A.D., Terblanche, S.E., Jager, A.K., and van Staden, J. (1999). Preliminary screening of some traditional zulu medicinal plants for anti-inflammatory and anti-microbial activities. *J Ethnopharmacol.,* **68**: 267-74.

Lindsey, K.L., Motsei, M.L., and Jäger, A.K. (2002). Screening of South African food plants for antioxidant activity. *Journal of Food science,* **76**: 2129-2131.

Lindsey, K.L., Viljoen, A.M., and Jäger, A.K. (2003). Screening of *Aloe* species for antioxidant activity. *South African Journal of Botany,* **69**: 599-602.

Ling, L., T., Yap S-A., Radhakrishnan A., K., Subramaniam, T, Cheng, H., M., and Palanisamy, U., D. (2009). Standardised *Mangifera indica* extract is an ideal antioxidant. *Food Chemistry,* **113**: 1154–1159.

Lis-Balchin, M., Hart, S., and Simpson, E. (2001). Buchu (*Agathosma betulina* and *A. crenulata,* Rutaceae) essential oils: their pharmacological action on guinea-pig ileum and antimicrobial activity on microorganisms. *J Pharm Pharmacol.,* **53**: 579-82.

Liu, C.Z., Murch, S.J., El-Demerdash, M., and Saxena, P.K. (2004). *Artemisia judaica* L.: micropropagation and antioxidant activity. *J Biotechnol.,* **110**: 63-71.

Lourens, A.C., Reddy, D., Baser, K.H., Viljoen, A.M., and Van Vuuren, S.F. (2004). *In vitro* biological activity and essential oil composition of four indigenous South African *Helichrysum* species. *J Ethnopharmacol,* **95**: 253-8.

Lourens, A., C., Viljoen, A., M., and van Heerdena, F., R. (2008). South African *Helichrysum* species: A review of the traditional uses, biological activity and phytochemistry. *Journal of Ethnopharmacology,* **119**: 630–652.

Luximon-Ramma, A., Bahorun, T., Soobrattee, M.A., and Aruoma, O.I. (2002). Antioxidant activities of phenolic, proanthocyanidin, and flavonoid components in extracts of *Cassia fistula. J Agric Food Chem.,* **50**: 5042-7.

Mabona, U., and van Vuuren, S., F. (2013). Southern African medicinal plants used to treat skin diseases. *South African Journal of Botany*, **87**: 175–193.

MacKay, D., and Miller, A.L. (2003). Nutritional support for wound healing. *Altern Med Rev.*, **8**: 359-77.

MacNee, W. (2005). Pulmonary and systemic oxidant/antioxidant imbalance in chronic obstructive pulmonary disease. *Proc Am Thorac Soc.*, **2**: 50-60.

Maduka, H.C., and Okoye, Z.S. (2002). The effect of *Sacoglottis gabonensis* stem bark extract, a Nigerian alcoholic beverage additive, on the natural antioxidant defences during 2,4-dinitrophenyl hydrazine-induced membrane peroxidation *in vivo*. *Vascul Pharmacol.*, **39**: 21-31.

Maduka, H.C., Okoye, Z.S., and Eje, A. (2002). The influence of *Sacoglottis gabonensis* stem bark extract and its isolate bergenin, Nigerian alcoholic beverage additives, on the metabolic and haematological side effects of 2,4-dinitrophenyl hydrazine-induced tissue damage. *Vascul Pharmacol.*, **39**: 317-24.

Maduka, H.C.C., and Okoye, Z.S.C. (2001). The Effect of *Sacoglottis gabonensis* and its Isolate Bergenin on Doxorubicin - Ferric Ions (Fe) - Induced 3+ Degradation of Deoxyribose. *The Sciences,* **1**: 316-319.

Mahomed, I.M., Nsabimana, A.M., and Ojewole, J.A.O. (2005). Pharmacological effects of *Harpagophytum procumbens* DC [Pedaliaceae] secondary root aqueous extract on isolated gastro-intestinal tract muscles of the chick, guinea-pig and rabbit. *Afr. J. Trad. Comp. Alt. Med. (AJTCAM),* **2**: 31-45.

Mahomed, I.M., and Ojewole, J.A. (2004). Analgesic, antiinflammatory and antidiabetic properties of *Harpagophytum procumbens* DC (Pedaliaceae) secondary root aqueous extract. *Phytother Res.*, **18**: 982-9.

Mahomed, I.M., and Ojewole, J.A. (2006a). Anticonvulsant activity of *Harpagophytum procumbens* DC [Pedaliaceae] secondary root aqueous extract in mice. *Brain Res Bull.*, **69**: 57-62.

Mahomed, I.M., and Ojewole, J.A.O. (2006b). Oxytocin-like effect of *Harpagophytum procumbens* DC (Pedaliaceae) secondary root aqueous extract on rat isolated uterus. *Afr. J. Trad. Complem. Alternative Med. (AJTCAM).*, **3**: 82-89.

Mahomoodally M., F. (2013). Traditional medicines in Africa: An Appraisal of ten potent African medicinal plants. *Evidence-Based Complementary and Alternative Medicine*, Article ID 617459, 14 pages

Maiga A., Malterud K., E., Drissa Diallo D., and Paulsen, B., S. (2006). Antioxidant and 15-lipoxygenase inhibitory activities of the Malian medicinal plants *Diospyros abyssinica* (Hiern) F. White (Ebenaceae), *Lannea velutina* A. Rich (Anacardiaceae) and *Crossopteryx febrifuga* (Afzel) Benth. (Rubiaceae). *Journal of Ethnopharmacology,* **104**, 132–137.

Maiga, A., Malterud, K., E., Mathisen, G., H., Paulsen, R., E., Thomas-Oates, J., Bergström, E., Reubsaet, L., Diallo, D., and Paulsen B., S. (2007). Cell protective antioxidants from the root bark of *Lannea velutina* A. Rich., a Malian medicinal plant. *Journal of Medicinal Plants Research,* **1**(4): 066-079.

Majinda, R.R., Motswaledi, M., Waigh, R.D., Waterman, P.G. (1997). Phenolic and antibacterial constituents of *Vahlia capensis*. *Planta Med.*, **63**: 268-70.

Majinda, R.R.T., Abegaz, B.M., Bezabih, M., Ngadjui, B.T., Wanjala, C.C.W., Mdee, L.K., Bojase, G., Silayo, A., Masesane, I., and Yeboah, S.O. (2001). Recent results from natural product research at the University of Botswana. *Pure Appl Chem.*, **73**: 1197 -1208.

Males, Z., Medic-Saric, M., and Bucar, F. (1998). Flavonoids of *Guiera senegalensis* J. F. GMEL. - Thin-layer chromatography and numerical methods. *Croatica Chemica Acta*, **71**: 69-79.

Mancini-Filho, J., Van-Koiij, A., Mancini, D.A., Cozzolino, F.F., and Torres, R.P. (1998). Antioxidant activity of cinnamon (*Cinnamomum zeylanicum*, Breyne) extracts. *Boll Chim Farm*, **137**: 443-7.

Manguro, L.O., Ugi, I., Hermann, R., and Lemmen, P. (2003°). Flavonol and drimane-type sesquiterpene glycosides of *Warburgia stuhlmannii* leaves. *Phytochemistry*, **63**: 497-502.

Manguro, L.O., Ugi, I., Lemmen, P., and Hermann, R. (2003b). Flavonol glycosides of *Warburgia ugandensis* leaves. *Phytochemistry*, **64**: 891-6.

Mariani, E., and Polidori, M.C., Cherubini, A., Mecocci, P. (2005). Oxidative stress in brain aging, neurodegenerative and vascular diseases: An overview. *J Chromatogr B Analyt Technol Biomed Life Sci.*, **827(1)**: 65-75.

Maroyi, A. (2013). Traditional use of medicinal plants in south-central Zimbabwe: review and perspectives. *Journal of Ethnobiology and Ethnomedicine*, **9**: 31.

Masoko P., and Eloff, J. N. (2008). Screening of twenty-four South African *Combretum* and six *Terminalia* species (combretaceae) for antioxidant activities. *Afr. J. Trad. CAM*, **4** (2): 231 – 239.

Mathisen, E., Diallo, D., Andersen, O.M., and Malterud, K.E. (2002). Antioxidants from the bark of *Burkea africana*, an African medicinal plant. *Phytother Res.*, **16**: 148-53.

Mbaveng A., T., Kuete, V., Ngameni B., Beng, V., P., Ngadjui B., T. Meyer J., J., M., and Lall, N. (2012). Antimicrobial activities of the methanol extract and compounds from the twigs of *Dorstenia mannii* (Moraceae). *BMC Complementary and Alternative Medicine*, **12**: 83.

McGregor, G., Fiebich, B., Wartenberg, A., Brien, S., Lewith, G., and Wegener, T. (2005). Devil's Claw (*Harpagophytum procumbens*): An anti-inflammatory herb with therapeutic potential. *Phytochemistry Reviews*, **4**: 47-53.

McKay D., L., and Jeffrey B. Blumberg, J., B. (2007). A review of the bioactivity of South African herbal teas: Rooibos (*Aspalathus linearis*) and honeybush (*Cyclopia intermedia*). *Phytother. Res.*, **21**, 1–16.

Menan, H., Banzouzi, J.T., Hocquette, A., Pelissier, Y., Blache, Y., Kone, M., Mallie, M., Assi, L.A., and Valentin, A. (2006). Antiplasmodial activity and cytotoxicity of plants used in West African traditional medicine for the treatment of malaria. *J Ethnopharmacol.*, **105**: 131-136.

Menche, D. (2002). Knipholon und verwandte Phenylanthrachinone: Isolierung, Strukturaufklärung und erste, stereoselektive Totalsynthese sowie Atropselektive Darstellung des AB-Biarylfragmentes von Vancomycin. Dissertation Zur Erlangung Des Naturwissenschaftlichen Doktorgrades Der Bayerischen Julius-Maximilians-Universität Würzburg. pp. 240, Würzburg

Mensah, A.Y., Houghton, P.J., Akyirem, G.N., Fleischer, T.C., Mensah, M.L., Sarpong, K., and Adosraku, R. (2004). Evaluation of the antioxidant and free radical scavenging properties of *Secamone afzelii* Rhoem. *Phytother Res.*, **18**: 1031-2.

Meral, I., Yener, Z., Kahraman, T., and Mert, N. (2001). Effect of *Nigella sativa* on glucose concentration, lipid peroxidation, anti-oxidant defence system and liver damage in experimentally-induced diabetic rabbits. *J Vet Med A Physiol Pathol Clin Med.*, **48**: 593-9.

Messmer, W.M., Tin-Wa, M., Fong, H.H., Bevelle, C., Farnsworth, N.R., Abraham, D.J., and Trojanek, J. (1972). Fagaronine, a new tumor inhibitor isolated from *Fagara zanthoxyloides* Lam. (Rutaceae). *J Pharm Sci.*, **61**: 1858-9.

Meziti, A., Meziti, H., Boudiaf, K., Mustapha, B., and Bouriche, H. (2012). Polyphenolic profile and antioxidant activities of *Nigella sativa* seed extracts *in vitro* and *in vivo*. *World Academy of Science, Engineering and Technology*, **64;** 24-32.

Middleton, E., Jr., Kandaswami, C., and Theoharides, T.C. (2000). The effects of plant flavonoids on mammalian cells: implications for inflammation, heart disease, and cancer. *Pharmacol., Rev* **52**: 673-751.

Midiwo, J.O., Owino, N.O., and Dagne, E. (1994). Flavonoids of *Polygonum senegalense* part III: Isolation of dihydrochalcone glucoside and quercetin glycosides. *Bull Chem Soc Ethiop.*, **8**: 79-84.

Midiwo, J.O., Yenesew, A., Juma, B.F., Derese, S., Ayoo, J.A., Aluoch, A.O., and Guchu, S. (2002). Bioactive compounds from some Kenyan ethnomedicinal plants: Myrsinaceae, Polygonaceae and *Psiadia punctulata*. *Phytochemistry Reviews*, **1**: 311 – 323.

Midiwo, O., Gikonyo, N.K., Wanjau, D.O., Matasi, J.J., and Waterman, P.G. (1992). Flavonoids of *Polygonum senegalense* Part II. More surface and internal tissue flavonoid aglycones. *Bull. Chem. Soc. Ethiopia*, **6**: 119-122.

Miranda, C.L., Stevens, J.F., Ivanov, V., McCall, M., Frei, B., Deinzer, M.L., and Buhler, D.R. (2000). Antioxidant and prooxidant actions of prenylated and nonprenylated chalcones and flavanones *in vitro*. *J Agric Food Chem.*, **48**: 3876-84.

Mohamed, E.M. (1998). Phytochemical investigation of *Glinus lotoides* growing in Egypt. *Egyptian J. Pharm. Sci.*, **38**: 377-390.

Möllenbeck, S., König, T., Schreier, P., Schwab, W., and Rajaonarivony, J., L., R. (1997). Chemical composition and analyses of enantiomers of essential oils from Madagascar. *Flavour and Fragrance Journal*, **12**: 63-69.

Mongalo N. I. (2013). Peltophorum africanum Sond [Mosetlha]: A review of its ethnomedicinal uses, toxicology, phytochemistry and pharmacological activities. *J Medicinal Plants Research,* 7(48): 3484-3491.

Moniruzzaman M., Rokeya B., Ahmed S., Bhowmik A., Khalil M., I., and Gan S., H. (2012). Antioxidant effects of *Aloe barbadensis* Miller extractsand the potential role of these extracts as antidiabetic and antilipidemic agents on streptozotocin-induced Type-2 diabetic model rats. *Molecules,* **17**: 12851-12867.

Moody, J.O., Bloomfield, S.F., and Hylands, P.J. (1995). *In vitro* evaluation of the antimicrobial activities of *Enantia chlorantha* Oliv. extractives. *Afr J Med Med Sci.,* **24**: 269-73.

Moody, J.O., Robert, V.A., Connolly, J.D., and Houghton, P.J. (2005). Anti-inflammatory activities of the methanol extracts and an isolated furanoditerpene constituent of *Sphenocentrum jollyanum* Pierre (Menispermaceae). *J Ethnopharmacol.,* **104(1-2)**: 87-91.

Mothana, R., A., A., Abdo, S., A., A., Hasson S., Althawab, F., M., N., Alaghbari, S., A., Z., and Lindequist, U. (2010). Antimicrobial, antioxidant and cytotoxic activities and phytochemical screening of some Yemeni medicinal plants. *eCAM,* **7(3)**: 323–330.

Munodawafa M, Chagonda L., S., and Moyo, S. R. (2013). Antimicrobial and phytochemical screening of some Zimbabwean medicinal plants. *Journal of Biologically Active Products from Nature,* **3(5-6)**: 323-330.

Mutanyatta, J., Bezabih, M., Abegaz, B.M., Dreyer, M., Brun, R., Kocher, N., and Bringmann, G. (2005). The First 6'-O-Sulfated Phenylanthraquinones: Isolation from *Bulbine frutescens,* Structural Elucidation, Enantiomeric Purity, and Partial Synthesis. *Tetrahedron,* **61**: 8475-8484.

Naidoo V., McGaw, L. J., Bisschop, S. P. R., Duncan N., and Eloff, J. N. (2008). The value of plant extracts with antioxidant activity in attenuating coccidiosis in broiler chickens. *Veterinary Parasitology,* **153**: 214–219.

Naili, M., B., Alghazeer R., O., Nabil A. Saleh N., A., and Al-Najjar, A., Y. (2010). Evaluation of antibacterial and antioxidant activities of *Artemisia campestris* (Astraceae) and *Ziziphus lotus* (Rhamnacea). *Arabian Journal of Chemistry,* **3**: 79–84.

Nair, V. D. P., Dairam, A., Agbonon, A., Arnason, J. T., Foster, B. C., and Kanfer, I. (2007). Investigation of the antioxidant activity of African potato (*Hypoxis hemerocallidea*). *J. Agric. Food Chem.,* **55**: 1707-1711.

Nakatani, N. (1997). Antioxidants from spices and herbs. In: Natural antioxidants. Chemistry, Health effects, and Applications Ed. Shahidi, F. AOCS Press, Champaign, Illinois (USA)

Nana, O., Momeni, J., Tepongning, R., N., and Ngassoum, M., B. (2013). Phythochemical screening, antioxydant and antiplasmodial activities of extracts from *Trichilia roka* and *Sapium ellipticum*. *Journal of Phytopharmacology,* **2(4)**: 22-29.

Nassar, M. I., Gaara, A. H., El-Ghorab, A. H., Farrag, A.R. H., Shen, H., Huq E., and Mabry T., J. (2007). Chemical constituents of Clove (*Syzygium aromaticum*, Fam. Myrtaceae) and their antioxidant activity. *Rev. Latinoamer. Quím.*, **35(3)**: 47-57.

Natarajan, K.S., Narasimhan, M., Shanmugasundaram, K.R., and Shanmugasundaram, E.R. (2006). Antioxidant activity of a salt-spice-herbal mixture against free radical induction. *J Ethnopharmacol.*, **105**: 76-83.

Nébié, R.H.C., Yaméogo, R.T., and Bélanger, A., Sib, F.S. (2004). Composition chimique des huiles essentielles d'*Ageratum conyzoïdes* du Burkina Faso. *Comptes Rendus Chimie*, **7**: 1019-1022

Neergheen, V. S., Bahorun T., Jen, L-S., and Aruoma, O. I. (2007). Bioefficacy of Mauritian endemic medicinal plants: Assessment of their phenolic contents and antioxidant potential. *Pharmaceutical Biology*, **45(1)**: 9–17.

Nergard, C.S., Diallo, D., Inngjerdingen, K., Michaelsen, T.E., Matsumoto, T., Kiyohara, H., Yamada, H., and Paulsen, B.S. (2005). Medicinal use of *Cochlospermum tinctorium* in Mali Anti-ulcer-, radical scavenging- and immunomodulating activities of polymers in the aqueous extract of the roots. *J Ethnopharmacol.*, **96**: 255-69.

Nergard, C.S., Diallo, D., Michaelsen, T.E., Malterud, K.E., Kiyohara, H., Matsumoto, T., Yamada, H., and Paulsen, B.S. (2004). Isolation, partial characterisation and immunomodulating activities of polysaccharides from *Vernonia kotschyana* Sch. Bip. ex Walp. *J Ethnopharmacol.*, **91**: 141-52.

Ngadjui, B., Dongo, E., Tamboue, H., Fogue, K., and Abegaz, B. (1999a). Prenylated flavanones from the twigs of *Dorstenia mannii*. *Phytochemistry*, **50**: 1401 – 1406.

Ngadjui, B.T., Abegaz, B.M., Dongo, E., Tamboue, H., and Fogue, K. (1998a). Geranylated and prenylated flavonoids from the twigs of *Dorstenia mannii*. *Phytochemistry*, **48**: 349-354.

Ngadjui, B.T., Dongo, E., Abegaz, B.M., Fotso, S., and Tamboue, H. (2002a). Dinklagins A, B and C: three prenylated flavonoids and other constituents from the twigs of *Dorstenia dinklagei*. *Phytochemistry*, **61**: 99-104.

Ngadjui, B.T., Dongo, E., Happi, E.N., Bezabih, M.T., and Abegaz, B.M. (1998b). Prenylated flavones and phenylpropanoid derivatives from roots of *Dorstenia psilurus*. *Phytochemistry*, **48**: 733 – 737.

Ngadjui, B.T., Kapche, G.W.F., Tamboue, H., Abegaz, B.M., and Connolly, J.D. (1999b). Prenylated flavonoids and a dihydro-4-phenylcoumarin from *Dorstenia poinsettifolia*. *Phytochemistry* **51**: 119-123.

Ngadjui, B.T., Kouam, S.F., Dongo, E., Kapche, G.W., and Abegaz, B.M. (2000). Prenylated flavonoids from the aerial parts of *Dorstenia mannii*. *Phytochemistry*, **55**: 915-9.

Ngadjui, B.T., Ngameni, B., Dongo, E., Kouam, S.F., and Abegaz, B.M. (2002b). Prenylated and geranylated chalcones and flavones from the aerial parts of *Dorstenia ciliata*. *Bull Chem Soc Ethiopia*, **16**: 157-163.

Ngadjui, B.T., Tabopda, T.K., Dongo, E., Kapche, G.W.F., Sandor, P., and Abegaz, B.M. (1999c). Dorsilurins C, D and E, three prenylated flavonoids from the roots of *Dorstenia psilurus*. *Phytochemistry,* **52**: 731 – 735.

Ngadjui, B.T., Watchueng, J., Keumedjio, F., Ngameni, B., Simo, I.K., and Abegaz, B.M. (2005). Prenylated chalcones, flavone and other constituents of the twigs of *Dorstenia angusticornis* and *Dorstenia barteri* var. *subtriangularis*. *Phytochemistry,* **66**: 687-92.

Ngameni, B., Ngadjui, B.T., Folefoc, G.N., Watchueng, J., and Abegaz, B.M. (2004). Diprenylated chalcones and other constituents from the twigs of *Dorstenia barteri* var. *subtriangularis*. *Phytochemistry,* **65**: 427-32.

Ngamga, D., Yankep, E., Tane, P., Bezabih, M., Ngadjui, B.T., Fomum, Z.T., and Abegaz, B.M. (2005a). Antiparasitic prenylated isoflavonoids from seeds of *Millettia griffoniana*. *Bulletin of the Chemical Society of Ethiopia,* **19**: 75-80.

Ngamga, D., Yankep, E., Tane, P., Bezabih, M., Ngadjui, B.T., Fomum, Z.T., and Abegaz, B.M. (2005b). Isoflavonoids from Seeds of *Millettia griffoniana* (Bail). *Z. Naturforsch.,* **60b**: 973 – 977.

Ngondi, J.L., Oben, J.E., and Minka, S.R. (2005). The effect of *Irvingia gabonensis* seeds on body weight and blood lipids of obese subjects in Cameroon. *Lipids Health Dis.,* **4**: 12.

N'Guessan J., D., Bidié A., P., Lenta B., N., Weniger B., André P., and Guédé-Guina, F. (2007). *In vitro* assays for bioactivity-guided isolation of antisalmonella and antioxidant compounds in *Thonningia sanguinea* flowers.

Niassy, B., Um, B.-H., Lobstein, A., Weniger, B., Koné, M., and Anton, R. (2004). Flavonoïdes de *Tephrosia deflexa* et *Tephrosia albifoliolis*. *Comptes Rendus Chimie.,* **7**: 993-996.

Nimenibo-Uadia, R. (2003). Control of hyperlipidaemia, hypercholesterolaemia and hyperketonaemia by aqueous extract of *Dioscorea dumetorum* tuber. *Trop J Pharm Res.,* **2**

Nkengfack, A.E., Vardamides, J.C., Fomum, Z.T., and Meyer, M. (1995). Prenylated isoflavanone from *Erythrina eriotricha*. *Phytochemistry,* **40**: 1803-8.

Nkengfack, A.E., Vouffo, T.W., Vardamides, J.C., Kouam, J., Fomum, Z.T., Meyer, M., and Sterner, O. (1997). Phenolic metabolites from *Erythrina* species. *Phytochemistry,* **46**: 573-578.

Noamesi, B.K., Bogale, M., and Dagne, E. (1990). Intestinal smooth muscle spasmolytic actions of the aqueous extract of the roots of *Taverniera abyssinica*. *J Ethnopharmacol.,* **30**: 107-13.

Noguchi, N., Watanabe, A., and Shi, H. (2000). Diverse functions of antioxidants. *Free Radic Res.,* **33**: 809-17.

Nunes J., de M., Pinto, P. S., Bordignon, S. A. L., Rech S. B., and von Poser, G. L. (2010). Phenolic compounds in *Hypericum* species from the Trigynobrathys section. *Biochemical Systematics and Ecology,* **38**: 224–228.

Nwaehujor C. O., Ezeja M. I., Udeh N. E., Okoye D. N., and Udegbunam, R. I. (2012). Anti-inflammatory and anti-oxidant activities of *Mallotus oppositifolius* (Geisel) methanol leaf extracts. http: / / dx.doi.org / 10.1016 / j.arabjc.2012.03.014.

Nwankwo, J.O., Tahnteng, J.G., and Emerole, G.O. (2000). Inhibition of aflatoxin B1 genotoxicity in human liver-derived HepG2 cells by kolaviron biflavonoids and molecular mechanisms of action. *Eur J Cancer Prev.*, **9**: 351-61.

Occhiuto, F., Sanogo, R., Germano, M.P., Keita, A., D'Angelo, V., and De Pasquale, R. (1999). Effects of some Malian medicinal plants on the respiratory tract of guinea-pigs. *J Pharm Pharmacol.*, **51**: 1299-303.

Odukoya, A.O., Ilori, O.O., Sofidiya, M.O., Aniunoh, O.A., Lawal, B.M., and Tade, I.O. (2005). Antioxidant Activity of Nigerian Dietary Spices. *Electronic Journal of Environmental Agricultural and Food Chemistry*, **4**

Ofori-Kwakye, K., Kwapong, A. A., and Adu, F. (2009). Antimicrobial activity of extracts and topical products of the stem bark of *Spathodea campanulata* for wound healing. *Afr. J. Trad. CAM.*, **6(2)**: 168 – 174.

Ogundipe, O.O., Moody, J.O., Houghton, P.J., and Odelola, H.A. (2001). Bioactive chemical constituents from *Alchornea laxiflora* (Benth) Pax and Hoffman. *J Ethnopharmacol.*, **74**: 275-80.

Ogwal-Okeng, J.W., Obua, C., and Anokbonggo, W.W. (2003). Acute toxicity effects of the methanolic extract of *Fagara zanthoxyloides* (Lam.) root-bark. *Afr Health Sci.*, **3**: 124-6.

Ohtani, II, Gotoh, N., Tanaka, J., Higa, T., Gyamfi, M.A., and Aniya, Y. (2000). Thonningianins A and B, new antioxidants from the African medicinal herb *Thonningia sanguinea*. *J Nat Prod.*, **63**: 676-9.

Ojewole, J.A. (2002). Antiinflammatory properties of *Hypoxis hemerocallidea* corm (African potato) extracts in rats. *Methods Find Exp Clin Pharmacol.*, **24**: 685-7.

Ojewole, J.A. (2003). Evaluation of the anti-inflammatory properties of *Sclerocarya birrea* (A. Rich.) Hochst. (family: Anacardiaceae) stem-bark extracts in rats. *J Ethnopharmacol.*, **85**: 217-20.

Ojewole, J.A. (2004). Evaluation of the analgesic, anti-inflammatory and anti-diabetic properties of *Sclerocarya birrea* (A. Rich.) Hochst. stem-bark aqueous extract in mice and rats. *Phytother Res.*, **18**: 601-8.

Ojewole, J.A. (2005a). Antiinflammatory, analgesic and hypoglycemic effects of *Mangifera indica* Linn. (Anacardiaceae) stem-bark aqueous extract. *Methods Find Exp Clin Pharmacol.*, **27**: 547-54.

Ojewole, J.A. (2005b). Antinociceptive, anti-inflammatory and antidiabetic effects of *Bryophyllum pinnatum* (Crassulaceae) leaf aqueous extract. *J Ethnopharmacol.*, **99**: 13-9.

Ojewole, J.A. (2005c). Antinociceptive, antiinflammatory and antidiabetic effects of *Leonotis leonurus* (L.) R. BR. [Lamiaceae] leaf aqueous extract in mice and rats. *Methods Find Exp Clin Pharmacol.*, **27**: 257-64.

Ojewole, J.A. (2006). Antinociceptive, anti-inflammatory and antidiabetic properties of *Hypoxis hemerocallidea* Fisch. and C.A. Mey. (Hypoxidaceae) corm ['African Potato'] aqueous extract in mice and rats. *J Ethnopharmacol.*, **103**: 126-34.

Ojo-Amaize, E.A., Kapahi, P., Kakkanaiah, V.N., Takahashi, T., Shalom-Barak, T., Cottam, H.B., Adesomoju, A.A., Nchekwube, E.J., Oyemade, O.A., Karin, M., and Okogun, J.I. (2001). Hypoestoxide, a novel anti-inflammatory natural diterpene, inhibits the activity of IkappaB kinase. *Cell Immunol.*, **209**: 149-57.

Ojo-Amaize, E.A., Nchekwube, E.J., Cottam, H.B., Bai, R., Verdier-Pinard, P., Kakkanaiah, V.N., Varner, J.A., Leoni, L., Okogun, J.I., Adesomoju, A.A., Oyemade, O.A., and Hamel, E. (2002). Hypoestoxide, a natural nonmutagenic diterpenoid with antiangiogenic and antitumor activity: possible mechanisms of action. *Cancer Res.*, **62**: 4007-14.

Okigbo R. N., Anuagasi C. L., and Amadi J. E. (2009) Advances in selected medicinal and aromatic plants indigenous to Africa Advances in selected medicinal and aromatic plants indigenous to Africa. *Journal of Medicinal Plants Research*, **3(2)**: 086-095.

Okpekon, T., Yolou, S., Gleye, C., Roblot, F., Loiseau, P., Bories, C., Grellier, P., Frappier, F., Laurens, A., and Hocquemiller, R. (2004). Antiparasitic activities of medicinal plants used in Ivory Coast. *J Ethnopharmacol.*, **90**: 91-7.

Okwu Donatus Ebere, and Iroabuchin F. (2009). Phytochemical Composition and Biological Activities of *Uvaria chamae* and *Clerodendoron splendens* http: // www.e-journals.net, **6(2)**: 553-560.

Olajide, O.A., Aderogba, M.A., Adedapo, A.D., and Makinde, J.M. (2004). Effects of *Anacardium occidentale* stem bark extract on *in vivo* inflammatory models. *J Ethnopharmacol.*, **95**: 139-42

Olajide, O.A., Awe, S.O., Makinde, J.M., Ekhelar, A.I., Olusola, A., Morebise, O., Okpako, D.T. (2000a). Studies on the anti-inflammatory, antipyretic and analgesic properties of *Alstonia boonei* stem bark. *J Ethnopharmacol.*, **71**: 179-86.

Olajide, O.A., Makinde, J.M., Okpako, D.T., and Awe, S.O. (2000b). Studies on the anti-inflammatory and related pharmacological properties of the aqueous extract of *Bridelia ferruginea* stem bark. *J Ethnopharmacol.*, **71**: 153-60.

Olaleye, S.B., Farombi, E.O., Adewoye, E.A., Owoyele, B.V., and Onasanwo, S.A. (2000). Analgesic and anti-inflammatory effects of kolaviron (a *Garcinia kola* seed extract). *Afr. J. Biomed. Res.*, **3**: 171 – 174.

Olorunnisola O., S., and Afolayan, A., J. (2013). *In vivo* antioxidant and biochemical evaluation of *Sphenocentrum jollyanum* leaf extract in *P. berghei* infected mice. *Pak. J. Pharm. Sci.*, **26**(3): 445-450.

Olorunnisola O., S., Akintola A, O., and A. J. Afolayan A., J. (2011). Hepatoprotective and antioxidant effect of *Sphenocentrum jollyanum* (Menispermaceae) stem bark extract against CCl_4- induced oxidative stress in rats. *African Journal of Pharmacy and Pharmacology*, **5**(9): 1241-1246.

Oloyede G. K, Onocha, P., A.,, Soyinka, J.,Oguntokun O., W., and Thonda E. (2010). Phytochemical screening, antimicrobial and antioxidant activities of four Nigerian medicinal plants. *Annals of Biological Research*, **1** (2): 114-120.

Omisore, N.O., Adewunmi, C.O., Iwalewa, E.O., Ngadjui, B.T., Adenowo, T.K., Abegaz, B.M., Ojewole, J.A., and Watchueng, J. (2005). Antitrichomonal and antioxidant activities of *Dorstenia barteri* and *Dorstenia convexa*. *Braz J Med Biol Res.*, **38**: 1087-94.

Omisore, N.O., Adewunmi, C.O., Iwalewa, E.O., Ngadjui, B.T., Watchueng, J., Abegaz, B.M., and Ojewole, J.A. (2004). Antinociceptive and anti-inflammatory effects of *Dorstenia barteri* (Moraceae) leaf and twig extracts in mice. *J Ethnopharmacol.*, **95**: 7-12.

Onyeyili, P.A., Nwosu, C.O., Amin, J.D., and Jibike, J.I. (2001). Anthelmintic activity of crude aqueous extract of *Nauclea latifolia* stem bark against ovine nematodes. *Fitoterapia*, **72**: 12-21.

Opoku, A.R., Maseko, N.F., and Terblanche, S.E. (2002). The *in vitro* antioxidative activity of some traditional Zulu medicinal plants. *Phytother Res.*, **16 Suppl 1**: S51-6.

Opoku, A., R., Nethengwe M., F., Dludla P., Madida K., T., Shonhai A., Smith P., and Singh M. (2011). Larvicidal and antimalarial activity of some Zulu medicinal plants. *Planta Medica*, **77**: PF51.

Orèiæ D., Z., Mimica-Dukiæ N., M., Franciškoviæ M., M., Petroviæ S., S., and Jovin E., D. (2011). *Chemistry Central Journal*, **5**: 34.

Orhan N, Orhan I., E., and Ergun, F. (2011). Insights into cholinesterase inhibitory and antioxidant activities of five *Juniperus* species. *Food and Chemical Toxicology*, **49**: 2305–2312.

Owira P., M., O.,and Ojewole, J., A., O. (2009) 'African Potato' (*Hypoxis hemerocallidea* corm): A Plant-Medicine for Modern and 21st Century Diseases of Mankind? – A Review. *Phytother. Res.*, **23**, 147–152.

Palé, É., Kouda-Bonafos, M., and Nacro, M. (2004). Caractérisation et mesure des activités anti-radicalaires d'anthocyanes de plantes du Burkina Faso. *Comptes Rendus Chimie.*, **7**: 973-980.

Papajewski, S., Vogler, B., Conrad, J., Klaiber, I., Roos, G., Walter, C.U., Sussmuth, R., and Kraus, W. (2001). Isolation from *Cussonia barteri* of 1'-O-chlorogenoylchlorogenic acid and 1'-O-chlorogenoylneochlorogenic acid, a new type of quinic acid esters. *Planta Med.*, **67**: 732-6.

Paper, D.H., Karall, E., Kremser, M., and Krenn, L. (2005). Comparison of the antiinflammatory effects of *Drosera rotundifolia* and *Drosera madagascariensis* in the HET-CAM assay. *Phytother Res.*, **19**: 323-6.

Patel D., K., Patel K., Kumar R., Gadewar M., and Tahilyani, V. (2012). Pharmacological and analytical aspects of bergenin: a concise report. *Asian Pacific Journal of Tropical Disease*, 163-167.

Paubert-Braquet, M., Cave, A., Hocquemiller, R., Delacroix, D., Dupont, C., Hedef, N., and Borgeat, P. (1994). Effect of *Pygeum africanum* extract on A23187-stimulated production of lipoxygenase metabolites from human polymorphonuclear cells. *J Lipid Mediat Cell Signal*, **9**: 285-90.

Philipov, S., Ivanovska, N., Istatkova, R., Velikova, M., and Tuleva, P. (2000). Phytochemical study and cytotoxic activity of alkaloids from *Uvaria chamae* P. Beauv. *Pharmazie*, **55**: 688-9.

Pietta, P.G. (2000). Flavonoids as antioxidants. *J Nat Prod.*, **63**: 1035-42.

Pinchuk, I., and Lichtenberg, D. (2002). The mechanism of action of antioxidants against lipoprotein peroxidation, evaluation based on kinetic experiments. *Prog Lipid Res.*, **41**: 279-314.

Pittella F., Dutra R. C., Junior, D. D., Lopes M. T. P., and Barbosa, N. R. (2009). Antioxidant and cytotoxic activities of *Centella asiatica* (L) Urb. *Int. J. Mol. Sci.*, **10**: 3713-3721.

Popat, A., Shear, N.H., Malkiewicz, I., Stewart, M.J., Steenkamp, V., Thomson, S., and Neuman, M.G. (2001). The toxicity of *Callilepis laureola*, a South African traditional herbal medicine. *Clin Biochem.*, **34**: 229-36.

Prakash, B., Shukla, R., Singh, P., Mishra, P. K., Dubey, N. K., and Kharwar, R., N. (2011). Efficacy of chemically characterized *Ocimum gratissimum* L. essential oil as an antioxidant and a safe plant based antimicrobial against fungal and aflatoxin B1 contamination of spices. *Food Research International*, 44: 385–390.

Prenesti, E., Berto, S., Daniel, P. G., and Toso, S. (2007). Antioxidant power quantification of decoction and cold infusions of *Hibiscus sabdariffa* flowers. *Food Chemistry*, **100**: 433–438.

Qhotsokoane-Lusunzi, M.A., and Karuso, P. (2001). Secondary Metabolites from Basotho Medicinal Plants. II. *Bulbine capitata*. *Australian Journal of Chemistry*, **54**: 427 – 430.

Quattara L. K. J., Zongo C., Barro N., Savadogo A., Bassole I. H. N., Quattara A.S., and Traore A. S. (2011). Antioxidant and anti-bacterial activities of three species of *Lannea* from Burkina Fasso. *J Appld Sciences*, **11**(10): 157-162.

Queiroz, E.F., Kuhl, C., Terreaux, C., Mavi, S., and Hostettmann, K. (2003). New dihydroalkylhexenones from *Lannea edulis*. *J Nat Prod.*, **66**: 578-80.

Rabe, C., Steenkamp, J.A., Joubert, E., Burger, J.F., and Ferreira, D. (1994). Phenolic metabolites from rooibos tea (*Aspalathus linearis*). *Phytochem Anal.*, **35**: 1559-65.

Ragusa, S., De Pasquale, R., Flores, M., Germano, M.P., Sanogo, R., and Rapisarda, A. (2001). Micromorphological investigations on *Entada africana* Guill. et Perr. (Mimosaceae). *Farmaco.*, **56**: 361-3.

Rajasekaran, S., Sivagnanam, K., and Subramanian, S. (2005°). Antioxidant effect of *Aloe vera* gel extract in streptozotocin-induced diabetes in rats. *Pharmacol Rep.*, **57**: 90-6.

Rajasekaran, S., Sivagnanam, K., and Subramanian, S. (2005b). Modulatory effects of *Aloe vera* leaf gel extract on oxidative stress in rats treated with streptozotocin. *J Pharm Pharmacol.*, **57**: 241-6.

Ramadan, M.F., Kroh, L.W., and Morsel, J.T. (2003). Radical scavenging activity of black cumin (*Nigella sativa* L.), coriander (*Coriandrum sativum* L.), and niger (*Guizotia abyssinica* Cass.) crude seed oils and oil fractions. *J Agric Food Chem.*, **51**: 6961-9.

Ramful D., Aumjaud, B., Neergheen V. S., Soobrattee M. A., Googoolye K., Aruoma O. I., and Bahorun, T. (2011). Polyphenolic content and antioxidant activity of *Eugenia pollicina* leaf extract *in vitro* and in model emulsion systems. *Food Research International*, **44**: 1190–1196.

Randhir, R., Lin, Y.T., and Shetty, K. (2004). Phenolics, their antioxidant and antimicrobial activity in dark germinated fenugreek sprouts in response to peptide and phytochemical elicitors. *Asia Pac J Clin Nutr.*, **13**: 295-307.

Rannilla *et al.*, 2010 (page 21) Ranilla L., G., Kwon Y-I, Apostolidis E., and Shetty, K. (2010). Phenolic compounds, antioxidant activity and *in vitro* inhibitory potential against key enzymes relevant for hyperglycemia and hypertension of commonly used medicinal plants, herbs and spices in Latin America. *Bioresource Technology*, **101**: 4676–4689.

Reddy, J.S., Rao, P.R., and Reddy, M.S. (2002). Wound healing effects of *Heliotropium indicum*, *Plumbago zeylanicum* and *Acalypha indica* in rats. *J Ethnopharmacol.*, **79**: 249-51.

Reiersen, B., Kiremire, B.T., Byamukama, R., and Andersen, M.O. (2003). Anthocyanins acylated with gallic acid from chenille plant, *Acalypha hispida*. *Phytochemistry*, **64**: 867-71.

Rennard, S.I. (2004). Anti-inflammatory therapies other than corticosteroids. *Proc Am Thorac Soc.*, **1**: 282-7.

Rice-Evans, C.A., Miller, N.J., and Paganga, G. (1996). Structure-antioxidant activity relationships of flavonoids and phenolic acids. *Free Radic Biol Med.*, **20**: 933-56.

Rojas, A., Cruz, S., Ponce-Monter, H., and Mata, R. (1996). Smooth muscle relaxing compounds from *Dodonaea viscosa*. *Planta Med.*, **62**: 154-9.

Rojas, J.J., Ochoa, V.J., Ocampo, S.A., and Munoz, J.F. (2006). Screening for antimicrobial activity of ten medicinal plants used in Colombian folkloric medicine: a possible alternative in the treatment of non-nosocomial infections. *BMC Complement Altern Med.*, **6**: 2.

Rotimi, V.O., Laughon, B.E., Bartlett, J.G., and Mosadomi, H.A. (1988). Activities of Nigerian chewing stick extracts against *Bacteroides gingivalis* and *Bacteroides melaninogenicus*. *Antimicrob Agents Chemother.*, **32**: 598-600.

Roy, S., Sehgal, R., Padhy, B.M., and Kumar, V.L. (2005). Antioxidant and protective effect of latex of *Calotropis procera* against alloxan-induced diabetes in rats. *J Ethnopharmacol.*, **102**: 470-3.

Sachdev, K., and Kulshreshtha, D.K. (1983). Flavonoids from *Dodonaea viscosa*. *Phytochemistry*, **22**: 1253-1256.

Saleh, N., El-Hawary, Z., El-Shobaki, F.A., Abbassy, M., and Morcos, S.R. (1977). Vitamins content of fruits and vegetables in common use in Egypt. *Z Ernahrungswiss.*, **16**: 158-62.

Salawu *et al.*, 2011 (page 37) Salawu O., A., Tijani A., Y., Babayi H., Nwaeze A., C., Ezeonu C., M., Igwe S., and Ndukuba M., A. (2011). Gastro-protective effect of *Crossopteryx febrifuga* in wistar rats. *Afr J Tradit Complement Altern Med.*, **8**(3): 300306.

Sanogo, R., Germanò, M.P., DAngelo, V., Guglielmo, M., and De Pasquale, R. (1998). Antihepatotoxic properties of Entada africana (Mimosaceae). *Phytotherapy Research*, **12**: S157 - S159.

Schinella, G.R., Tournier, H.A., Prieto, J.M., Mordujovich de Buschiazzo, P., and Rios, J.L. (2002). Antioxidant activity of anti-inflammatory plant extracts. *Life Sci.*, **70**: 1023-33.

Schneider, I., and Bucar, F. (2005). Lipoxygenase inhibitors from natural plant sources. Part 1: Medicinal plants with inhibitory activity on arachidonate 5-lipoxygenase and 5-lipoxygenase[sol]cyclooxygenase. *Phytother Res.*, **19**: 81-102.

Selloum, L., Arrar, L., Medani, B., Khenchouche, A., and Bisker, H. (1995). Effect of *Cleome arabica* leaves extract on inflammatory cells response in rat. *Biochem Soc Trans.*, **23**: 609S.

Selloum, L., Sebihi, L., Mekhalfia, A., Mahdadi, R., and Senator, A. (1997). Antioxidant activity of *Cleome arabica* leaves extract. *Biochem Soc Trans.*, **25**: S608.

Shai, L.J., Magano, S.R., Lebelo, S.L., and Mogale, A.M. (2011). Inhibitory effects of five medicinal plants on rat alpha-glucosidase: Comparison with effects on yeast of alpha-glucosidase. *J. Med. Plant Res.*, **5**(13): 2863-2867.

Sharma, J., and Sharma, R. (2002). Radioprotection of Swiss albino mouse by *Centella asiatica* extract. *Phytother Res.*, **16**: 785-6.

Sharma, R., and Sharma, J. (2005). Modification of gamma ray induced changes in the mouse hepatocytes by *Centella asiatica* extract: *in vivo* studies. *Phytother Res.*, **19**: 605-11.

Sharma, R.A., Gescher, A.J., and Steward, W.P. (2005). Curcumin: the story so far. *Eur J Cancer*, **41**: 1955–1968, 2005.

Shirwaikar, A., Bhilegaonkar, P.M., Malini, S., and Kumar, J.S. (2003). The gastroprotective activity of the ethanol extract of *Ageratum conyzoides*. *J Ethnopharmacol.*, **86**: 117-21.

Shukla, A., Rasik, A.M., and Dhawan, B.N. (1999a). Asiaticoside-induced elevation of antioxidant levels in healing wounds. *Phytother Res.*, **13**: 50-54.

Shukla, A., Rasik, A.M., Jain, G.K., Shankar, R., Kulshrestha, D.K., and Dhawan, B.N. (1999b). *In vitro* and *in vivo* wound healing activity of asiaticoside isolated from *Centella asiatica*. *J Ethnopharmacol.*, **65**: 1-11.

Snijman *et al.*, 2009 (page 25) Snijman, P., W., Joubert, E.,Ferreira, D., Li, X., X-C., Ding., X., Y., Green, I., R., and Gelderblom, W., C., A. (2009). Antioxidant activity of the dihydrochalcones aspalathin and nothofagin and their corresponding flavones in relation to other rooibos (*Aspalathus linearis*) flavonoids, epigallocatechin gallate, and trolox. *J. Agric. Food Chem.*, **57**: 6678–6684.

Sombie, P. A., Hilou, A., Mounier, C., Coulibaly, A.Y., Kiendrebeogo, M., Millogo, J.F., and Nacoulma, O. G. (2011). Antioxidant and anti-inflammatory activities of galls of *Guiera senegalensis* J.F. Gmel. (Combrectaceae). *Research Journal of Medicinal Plants*, **5**(4): 448-461.

Somova, L.I., Shode, F.O., Ramnanan, P., and Nadar, A. (2003). Antihypertensive, antiatherosclerotic and antioxidant activity of triterpenoids isolated from *Olea europaea*, subspecies africana leaves. *J Ethnopharmacol.*, **84**: 299-305.

Sonibare M. A., and Abegunde, R. B. (2012). *In vitro* antimicrobial and antioxidant analysis of *Dioscorea* species. *J. Appl. Biosc.*, **51**: 3583-3590.

Souza, J.S.N., Machado, L.L., Pessoa, O.D.L., Braz-Filho, R., Overk, C.R., Yao, P., Cordell, G.A., and Lemos, T.L.G. (2005). Pyrrolizidine alkaloids from *Heliotropium indicum*. *J. Braz. Chem. Soc.*, **16**: 1410-1414.

Speroni, E., Cervellati, R., Govoni, P., Guizzardi, S., Renzulli, C., and Guerra, M.C. (2003). Efficacy of different *Cynara scolymus* preparations on liver complaints. *J Ethnopharmacol.*, **86**: 203-11.

Speroni, E., Cervellati, R., Innocenti, G., Costa, S., Guerra, M.C., Dall' Acqua, S., and Govoni, P. (2005). Anti-inflammatory, anti-nociceptive and antioxidant activities of *Balanites aegyptiaca* (L.) Delile. *J Ethnopharmacol.*, **98**: 117-25.

Srinivasan, K., Sambaiah, K., and Chandrasekhara, N. (2004). Spices as beneficial hypolipidemic food adjuncts: A review. *Food Reviews International*, **20**: 187– 220.

Steenkamp, V., Gouws, M.C., Gulumian, M., Elgorashi, E.E., and van Staden, J. (2005a). Studies on antibacterial, anti-inflammatory and antioxidant activity of herbal remedies used in the treatment of benign prostatic hyperplasia and prostatitis. *J Ethnopharmacol.*,

Steenkamp, V., Grimmer, H., Semano, M., and Gulumian, M. (2005b). Antioxidant and genotoxic properties of South African herbal extracts. *Mutat Res.*, **581**: 35-42.

Steenkamp, V., Mathivha, E., Gouws, M.C., and van Rensburg, C.E. (2004). Studies on antibacterial, antioxidant and fibroblast growth stimulation of wound healing remedies from South Africa. *J Ethnopharmacol.*, **95**: 353-7.

Steenkamp, V., Stewart, M.J., and Zuckerman, M. (1999). Detection of poisoning by Impila (*Callilepis laureola*) in a mother and child. *Hum Exp Toxicol.*, **18**: 594-7.

Stoddard, F.L. (2013). Novel feed and non-food uses of legumes. Legume Futures Report 1.3. Available from www.legumefutures.de.

Sugandha S., Varsha P., and Shashi R. (2013). Medicinal uses of *Adansonia digitata*: an endangered tree species. *J Pharmceutical Scientific Innovation*, **2(3)**: 14-16.

Tai, J., Cheung, S., Chan, E., and Hasman, D. (2004). *In vitro* culture studies of *Sutherlandia frutescens* on human tumor cell lines. *J Ethnopharmacol.*, **93**: 9-19.

Tambama, P., Abegaz, B., and Mukanganyama S. (2014). Antiproliferative activity of the isofuranonaphthoquinone isolated from *Bulbine frutescens* against Jurkat T cells. *BioMed Research International*, Article ID 752941, 14 pages http: // dx.doi.org/10.1155/2014/752941.

Tan, P.V., Nyasse, B., Dimo, T., Wafo, P., and Akahkuh, B.T. (2002). Synergistic and potentiating effects of ranitidine and two new anti-ulcer compounds from *Enantia chlorantha* and *Voacanga africana* in experimental animal models. *Pharmazie*, **57**: 409-12.

Tatsimo, S. J. N., Tamokou, J. D., Havyarimana L., Csupor, D., Forgo P., Hohmann J., Kuiate J-R., and Tane P. (2012). Antimicrobial and antioxidant activity of kaempferol rhamnoside derivatives from *Bryophyllum pinnatum*. *BMC Research Notes*, **5**: 158.

Teffo, L. S., Aderogba, M. A., and Eloff, J. N. (2010). Antibacterial and antioxidant activities of four kaempferol methyl ethers isolated from *Dodonaea viscosa* Jacq. var. Angustifolia leaf extracts. *South African Journal of Botany*, **76**: 25–29.

Togola, A., Diallo, D., Dembele, S., Barsett, H., and Paulsen, B.S. (2005). Ethnopharmacological survey of different uses of seven medicinal plants from Mali, (West Africa) in the regions Doila, Kolokani and Siby. *J Ethnobiol Ethnomedicine*, **1**: 7.

Tomas-Barberan, F.A., and Hostettmann, K. (1988). A cytotoxic triterpenoid and flavonoids from *Crossopteryx febrifuga*. *Planta Med.*, **54**: 266-7.

Toyokuni, S., Tanaka, T., Kawaguchi, W., Fang, N.R., Ozeki, M., Akatsuka, S., Hiai, H., Aruoma, O.I., and Bahorun, T. (2003). Effects of the phenolic contents of Mauritian endemic plant extracts on promoter activities of antioxidant enzymes. *Free Radic Res.*, **37**: 1215-24.

Tsopmo, A., Tene, M., Kamnaing, P., Ayafor, J.F., and Sterner, O. (1999). A new diels-alder-type adduct flavonoid from *Dorstenia barteri*. *J Nat Prod.*, **62**: 1432-4.

Tsopmo, A., Tene, M., Kamnaing, P., Ngnokam, D., Ayafor, J.F., and Sterner, O. (1998). Geranylated flavonoids from *Dorstenia poinsettifolia*. *Phytochemistry*, **48**: 345-348.

Ugochukwu, N.H., and Babady, N.E. (2002). Antioxidant effects of *Gongronema latifolium* in hepatocytes of rat models of non-insulin dependent diabetes mellitus. *Fitoterapia*, **73**: 612-8.

Ugochukwu, N.H., and Babady, N.E. (2003). Antihyperglycemic effect of aqueous and ethanolic extracts of *Gongronema latifolium* leaves on glucose and glycogen metabolism in livers of normal and streptozotocin-induced diabetic rats. *Life Sci.*, **73**: 1925-38.

Ugochukwu, N.H., Babady, N.E., Cobourne, M., and Gasset, S.R. (2003). The effect of *Gongronema latifolium* extracts on serum lipid profile and oxidative stress in hepatocytes of diabetic rats. *J Biosci.*, **28**: 1-5.

Ulicna, O., Vancova, O., Bozek, P., Carsky, J., Sebekova, K., Boor, P., Nakano, M., and Greksak, M. (2005). Rooibos tea (*Aspalathus linearis*) partially prevents oxidative stress in streptozotocin-induced diabetic rats. *Physiol Res.,*

Van der Merwe, M.J., Jenkins, K., Theron, E., and van der Walt, B.J. (1993). Interaction of the di-catechols rooperol and nordihydroguaiaretic acid with oxidative systems in the human blood. A structure-activity relationship. *Biochem Pharmacol.,* **45**: 303-11.

Van Wyk, B., Van Outdshoorn, B., and Gierike, N. (1997). Medicinal plants of South Africa. Briza Publications, Pretoria

vanWyk B-E., and Albrecht, C. (2008). A review of the taxonomy, ethnobotany, chemistry and pharmacology of *Sutherlandia frutescens* (Fabaceae). *Journal of Ethnopharmacology,* **119**: 620–629.

Vasange, M., Liu, B., Welch, C.J., Rolfsen, W., and Bohlin, L. (1997). The flavonoid constituents of two *Polypodium* species (Calaguala) and their effect on the elastase release in human neutrophils. *Planta Med.,* **63**: 511-7.

Veeru, P., Kishor, M., P. and Meenakshi M. (2009). Screening of medicinal plant extracts for antioxidant activity *J Medicinal Plants Res.,* **3**(8): 608-612.

Vertuani, S., Braccioli E., Buzzoni V., and Manfredini S. (2002). Antioxidant capacity of Adansonia digitata fruit pulp and leaves. *Acta phytotherapeutica,* **(2)**: 2-7.

Vieira, R.F., Grayer, R.J., Paton, A., and Simon, J.E. (2001). Genetic diversity of *Ocimum gratissimum* L. based on volatile oil constituents, flavonoids and RAPD markers. *Biochem Syst Ecol.,* **29**: 287-304.

Violi, F., and Cangemi, R. (2005). Antioxidants and cardiovascular disease. *Curr Opin Investig Drugs,* **6**: 895-900.

Wagner, H., and Bladt, S. (1975). Coumarins from South African *Pelargonium* species. *Phytochemistry,* **14**: 2061-2064.

Wang, C.J., Wang, J.M., Lin, W.L., Chu, C.Y., Chou, F.P., and Tseng, T.H. (2000). Protective effect of *Hibiscus anthocyanins* against tert-butyl hydroperoxide-induced hepatic toxicity in rats. *Food Chem Toxicol.,* **38**: 411-6.

Wang D., Li Y., Hou G., Wang P., Zhang J., Laudon V., and Shi, B. (2010). *Pygeum africanum*: effect on oxidative stress in early diabetes-induced bladder. *Int Urol Nephrol.,* **42**: 401–408.

Wanjala, C.C., Juma, B.F., Bojase, G., Gashe, B.A., and Majinda, R.R. (2002). Erythrinaline alkaloids and antimicrobial flavonoids from *Erythrina latissima*. *Planta Med.,* **68**: 640-2.

Wanjala, C.C., and Majinda, R.R. (2001). Isoflavone glycosides from the root wood of *Erythrina latissima*. *J AOAC Int.,* **84**: 451-3.

Watson, A.R., Coovadia, H.M., and Bhoola, K.D. (1979). The clinical syndrome of Impila (*Callilepis laureola*) poisoning in children. *S Afr Med J.,* **55**: 290-2.

Watt, John Mitchell and Breyer-Brandwijk, Maria Gerdina (1962). The Medicinal and Poisonous Plants of Southern Africa – being an account of their medicinal uses, chemical composition, pharmacological effects and toxicology in man and animal. E and S Livingstone, Edinburgh.

Williams, C.A., and Grayer, R.J. (2004). Anthocyanins and other flavonoids. *Nat Prod Rep.*, **21**: 539-73.

Wube, A.A., Bucar, F., Asres, K., Gibbons, S., Adams, M., Streit, B., Bodensieck, A., and Bauer, R. (2006). Knipholone, a selective inhibitor of leukotriene metabolism. *Phytomedicine*, **13**: 452-6.

Wube, A.A., Bucar, F., Asres, K., Gibbons, S., Rattray, L., and Croft, S.L. (2005a). Antimalarial compounds from *Kniphofia foliosa* roots. *Phytother Res.*, **19**: 472-6.

Wube, A.A., Bucar, F., Gibbons, S., and Asres, K. (2005b). Sesquiterpenes from *Warburgia ugandensis* and their antimycobacterial activity. *Phytochemistry*, **66**: 2309-15.

Wojdy³o A., Oszmian´ski J., and Czemerys, R. (2007). Antioxidant activity and phenolic compounds in 32 selected herbs. *Food Chemistry*, **105**: 940–949.

Wube, A.A., Streit, B., Gibbons, S., Asres, K., and Bucar, F. (2005c). *In vitro* 12(S)-HETE inhibitory activities of naphthoquinones isolated from the root bark of *Euclea racemosa* ssp. *schimperi*. *J Ethnopharmacol.*, **102**: 191-6.

Yankep, E., Fomum, Z.T., Bisrat, D., Dagne, E., Hellwig, V., and Steglich, W. (1998). O-Geranylated isoflavones and a 3-phenylcoumarin from *Millettia griffoniana*. *Phytochemistry*, **49**: 2521-2523.

Yankep, E., Fomum, Z.T., and Dagne, E. (1997). An O-geranylated isoflavone from *Millettia griffoniana*. *Phytochemistry*, **46**: 591-593.

Yankep, E., Mbafor, J.T., Fomum, Z.T., Steinbeck, C., Messanga, B.B., Nyasse, B., Budzikiewicz, H., Lenz, C., and Schmickler, H. (2001). Further isoflavonoid metabolites from *Millettia griffoniana* (Bail). *Phytochemistry*, **56**: 363-8.

Yankep, E., Njamen, D., Fotsing, M.T., Fomum, Z.T., Mbanya, J.C., Giner, R.M., Recio, M.C., Manez, S., and Rios, J.L. (2003). Griffonianone D, an isoflavone with anti-inflammatory activity from the root bark of *Millettia griffoniana*. *J Nat Prod.*, **66**: 1288-90.

Yenesew, A., Derese, S., Irungu, B., Midiwo, J.O., Waters, N.C., Liyala, P., Akala, H., Heydenreich, M., and Peter, M.G. (2003a). Flavonoids and isoflavonoids with antiplasmodial activities from the root bark of *Erythrina abyssinica*. *Planta Med* **69**: 658-61.

Yenesew, A., Derese, S., Midiwo, J.O., Bii, C.C., Heydenreich, M., and Peter, M.G. (2005a). Antimicrobial flavonoids from the stem bark of *Erythrina burttii*. *Fitoterapia*, **76**: 469-72.

Yenesew, A., Induli, M., Derese, S., Midiwo, J.O., Heydenreich, M., Peter, M.G., Akala, H., Wangui, J., Liyala, P., and Waters, N.C. (2004). Anti-plasmodial flavonoids from the stem bark of *Erythrina abyssinica*. *Phytochemistry*, **65**: 3029-32.

Yenesew, A., Irungu, B., Derese, S., Midiwo, J.O., Heydenreich, M., and Peter, M.G. (2003b). Two prenylated flavonoids from the stem bark of *Erythrina burttii*. *Phytochemistry*, **63**: 445-8.

Yenesew, A., Kiplagat, J.T., Derese, S., Midiwo, J.O., Kabaru, J.M., Heydenreich, M., and Peter, M.G. (2006). Two unusual rotenoid derivatives, 7a-O-methyl-12a-hydroxydeguelol and spiro-13-homo-13-oxaelliptone, from the seeds of *Derris trifoliata*. *Phytochemistry*,

Yenesew, A., Midiwo, J.O., Guchu, S.M., Heydenreich, M., and Peter, M.G. (2002). Three isoflav-3-enes and a 2-arylbenzofuran from the root bark of *Erythrina burttii*. *Phytochemistry*, **59**: 337-41.

Yenesew, A., Midiwo, J.O., Heydenreich, M., and Peter, M.G. (1998a). Four isoflavones from the stem bark of *erythrina sacleuxii*. *Phytochemistry*, **49**: 247-249.

Yenesew, A., Midiwo, J.O., Heydenreich, M., Schanzenbach, D., and Peter, M.G. (2000). Two isoflavanones from the stem bark of *Erythrina sacleuxii*. *Phytochemistry*, **55**: 457-9.

Yenesew, A., Midiwo, J.O., Miessner, M., Heydenreich, M., and Peter, M.G. (1998b). Two prenylated flavonones from stem bark of *Eryhrina burttii*. *Phytochemistry*, **48**: 1439 1443.

Yenesew, A., Mushibe, E.K., Induli, M., Derese, S., Midiwo, J.O., Kabaru, J.M., Heydenreich, M., Koch, A., and Peter, M.G. (2005b). 7a-O-methyldeguelol, a modified rotenoid with an open ring-C, from the roots of *Derris trifoliata*. *Phytochemistry*, **66**: 653-7.

Yuan *et al.*, 2008 (page 19) Yuan L-P., Chena, F-H., Ling L., Dou P-F., Bob H., Zhong M-M., and Xi, L-J. (2008). Protective effects of total flavonoids of *Bidens pilosa* L. (TFB) on animal liver injury and liver fibrosis. *Journal of Ethnopharmacology*, **116**: 539–546.

Zabri *et al.*, 2008 (page 18) Zabri H., Kodjo C., Benie, A., Bekro J., M., and Bekro, Y., A. (2008). Phytochemical screening and determination of flavonoids in *Secamone afzelii* (Asclepiadaceae) extracts. *African Journal of Pure and Applied Chemistry*, **2**(8): 080-082.

Zabka M., Pavela R., and Slezakova L. (2009). Antifungal effect of *Pimenta dioica* essential oil against dangerous pathogenic and toxinogenic fungi. *Industrial Crops and Products*, **30**: 250–253.

Zapata P. J. D., Navarro, F., Guillén, S., Castillo, D., Valero M-R., D., and Serrano, M. (2013). Characterisation of gels from different *Aloe* spp. as antifungal treatment: Potential crops for industrial applications.

Zhang, X.F., Wang, H.M., Song, Y.L., Nie, L.H., Wang, L.F., Liu, B., Shen, P.P., and Liu, Y. (2006). Isolation, structure elucidation, antioxidative and immunomodulatory properties of two novel dihydrocoumarins from Aloe vera. *Bioorg Med Chem Lett.*, **16**: 949-53.

Zheng, W., and Wang, S.Y. (2001). Antioxidant activity and phenolic compounds in selected herbs. *J Agric Food Chem.*, **49**: 5165-70.

Zingue, S., Njamen D., Tchoumtchoua J., Halabalaki M., Evan Simpson E., Clyne, C., and Nde, C., B., M. (2013). Effects of *Millettia macrophylla* (Fabaceae) extracts target organs of female Wistar rat. *J Pharmacol Sci.*, **123**: 120 – 131.

Zishiri, V.K. (2005). Potentising and application of a *Combretum woodii* leaf extract with high antibacterial and antioxidant activity. University of Pretoria, M.Sc. Desertation. Paraclinical Sciences, Pretoria.

Zou, Y., Lu, Y., and Wei, D. (2004). Antioxidant activity of a flavonoid-rich extract of *Hypericum perforatum* L. *in vitro*. *J Agric Food Chem.*, **52**: 5032-9.

Natural Products: Research Reviews Vol. 3 (2016) *Pages* **305–326**
Editor: **V.K. Gupta**
Published by: **DAYA PUBLISHING HOUSE, NEW DELHI**

10

Ethanomedicinal Importances of *Rauvolfia serpentine* L. Benth. Ex Kurz in the Prevention and Treatment of Diseases

Renu Chaudhary[1], Bharat Singh[1] and Anil Kumar Chhillar[1*]

ABSTRACT

In past, tribal used medicinal plants for the treatment of various types of diseases. So far numbers of compounds have been identified from medicinal plants which were used as a source for crude medicines. Now a days, it is observed that most of the drugs used for curing diseases are not enough due to the development of resistances in pathogen and also their associated side effects on the host. Therefore, medicinal plants have a promising future in health care. One of the medicinally important plants used by the tribal for the purposes of obtaining drugs is Rauvolfia serpentine. This review provide a set of information about the Rauvolfia serpentine and how crude components of this plants are being used for the treatment of various diseases. This plant has been used against snake, insect and animal bite, mental illness, schizophrenia, hypertension, blood pressure, gastrointestinal diseases, circulatory disorders, pneumonia, fever, malaria, asthma, skin diseases, scabies, eye diseases, spleen diseases, acquired immuno-deficiency syndrome, rheumatism and body pain. The phyto-constituents which are present in the plant are mainly alkaloids, phenols, tannins and flavonoids which are responsible for the curative effects.

Keywords: Ethnomedicine, Herbal, Medicinal plant, Reserpine, *Rouvolfia serpentine*, Tribal.

1 Center for Biotechnology, Maharshi Dayanand University, Rohtak – 124 001, Haryana, India.

* *Corresponding author*: E-mail: anilchhillar@gmail.com

Introduction

As the population of the world keeps on increasing the requirement for food, shelter, cloths and medicine also increases. As millions of people suffering from various types of diseases, several pharmacological preparations available in the market can be used for the treatments of a number of diseases however; they are seems to be expensive, ineffective and cause undesirable toxic side effects. Therefore, there is need for us to use an alternative with natural or herbal medications which have no side effect or optimal side effects. Approximately 80 per cent of world population depends on herbal medicine and their associated chemical moieties such as alkaloids, hydrocarbon, volatile, nitrogen, resins, glycosides, oils, fatty acids, gums and tannins produced by various medicinal plants are being widely used for curing diseases (Kamboj, 2000). Medicinal values of plants make the scientist to work on the phytochemical analysis of the plant extracts and due to its medicinal importance they are used for treatment of various diseases (Hu *et al.*, 2008; Harisaranraj *et al.*, 2009; Mittal *et al.*, 2012). As per World Health Organization (WHO) any chemical substance released by plant or its part which has been used therapeutically or used as raw material for chemical or pharmaceutical production is known as a drug (Kumari *et al.*, 2011). Worldwide 300 species of aromatic and medicinal plants have been used as a part of food, pharmaceuticals, cosmetics and fragrance industries (Robber *et al.*, 1996; Harisaranraj *et al.*, 2009; Deshmukh *et al.*, 2012). One of the medicinally important plants used for the purposes of obtaining drugs and for treatment of many diseases is *Rauvolfia serpentina* (*R. serpentina*), as it is a rich source of vitamins, minerals and phytochemicals (Harisaranraj *et al.*, 2009; Mittal *et al.*, 2012).

Morphology and Distribution

Kingdom : Plantae

Division : Magnoliophyta

Class : Magnoliopsida

Order : Gentianales

Family : Apocynaceae

Genus : *Rouvolfia*

Species : *serpentina*

R. serpentine L. Benth. Ex Kurz commonly known as Sarpagandha, Chandrika, Chandrabagha, Snake root plant, Chotachand, and Harkaya etc. of Apocynaceae family inhabitats tropical and subtropical regions (Mallick *et al.*, 2012). The family includes 50 species, spread worldwide in the region of the Himalayas, Indian peninsula, Indonesia, Sri Lanka, and Burma. It is originally belongs to India, Bangladesh and other regions of Asia (Ghani *et al.*, 1998; Singh *et al.*, 2009). In India, it is widely distributed in the sub-Himalayan tract from Punjab to Nepal, Bhutan and Sikkim. It is also found in the Eastern Ghats, Western Ghats lower, hills of Gangetic plains and Andamans (Mabberley, 2008). It is evergreen, erect, glabrous and perennial shrub with maximum height upto 60 cm and a very good source of many important

biologically active alkaloids (Alamgir and Ahamed, 2005). The plant possess tuberous root with pale brown cork (Deshmukh *et al.*, 2012) contains a large number of alkaloid and secondary metabolites of medicinal importance and this have fascinated the attention of specialists of indigenous system of medicine (Mittal *et al.*, 2012; Poonam *et al.*, 2013). It has been used in India as a part of the ayurvedic medicines for the treatment of various diseases (Pant *et al.*, 2012). In Ayurveda, the roots of *R. serpentine* have been used as remedy for curing insomnia, hypertension, epilepsy, mental agitation, gastrointestinal disorders, traumas, excitement, schizophrenia, anxiety, excitement, sedative insomnia and insanity (Poonam *et al.*, 2013; Meena *et al.*, 2008). The root juice and extracts have been used to treat liver and abdomen pain, crushed roots and leaf buds with milk was applied externally as paste on the snakebite affected areas by the folks of Bhadra wildlife sanctuary in Karnataka (Parinitha *et al.*, 2004; Prakasha *et al.*, 2010; Singh *et al.*, 2010). According to a report of Rajendran *et al.* (2007), fruits and seeds have also been used for its medicinal or ethnobotanical purposes by the ethnic tribes of Virudhungar district, Tamil Nadu, India. It has been used as anti-helmintic and anti-hypertensive drug. The plant was also used by local people of Uttar Pradesh, Eastern Ghats, Karnataka and Bangladesh as an antidote against snake bite and bites of other poisonous insects (Sankaranarayanan *et al.*, 2010). In diarrhoea, cholera, dysentery, central epilepsy, fever, opacity of the cornea and ecbolic *R. serpentine* also played an important role (Singh *et al.*, 2009, Ghani, 1998). The other diseases such as body aches, burns, pneumonia, malaria, eczema, menstrual disorders, respiratory problems, scabies, skin cancers, asthma, eye inflammation, fever and spleen diseases can also be cured by *R. serpentine* (Anisuzzaman *et al.*, 2007; Dey *et al.*, 2011; Rai, 2004; Itoh *et al.*, 2005; Mohanta *et al.*, 2006; Bhattarai *et al.*, 2009a; Dey *et al.*, 2010; Azmi *et al.*, 2013). Therefore, the present review deals with the exhaustive research work undertaken by various groups in recent year's world wide to explore the use of *R. serpentine* for the treatment of different ailments by the tribal and aboriginals as part of their ethnomedical system.

Phytochemical Constituents

Over periods major research has been going on *R. serpentine* due to its phytochemical properties and production of various secondary metabolites such as alkaloids, phenols, tannins and flavonoids which provide defense against various pathogens (Mittal *et al.*, 2012; Singh *et al.*, 2009; Mallick *et al.*, 2012; Poonam *et al.*, 2013; Tiwari, 2008).

Alkaloids are complex organic molecules containing a heterocyclic nitrogen ring, which have been widely exploited for their diverse pharmacological properties. Such compounds are produced by many different organisms, including microbes, animals but a particularly different group of alkaloids are produced by plants.

Approximately 10 per cent of plant species are believed to produce alkaloids as secondary metabolites, where they work predominantly in providing defense against herbivores and pathogens. Alkaloids often isolated from plants used as pure compounds, these purely isolated alkaloids and their synthetic derivatives are used as medicinal agents for their analgesic, antispasmodic and bactericidal effects (Okwu *et al.*, 2004). On the basis of the structure there are three types of alkaloids namely, weak basic indole alkaloids, intermediatly basic alkaloids and strong anhydronium bases. The state of India (Assam) and December month is best respective place and time for collection of plant with the high abundance of alkaloids. As a matter of fact percentage of alkaloid, does not depend on the plant age. The major alkaloids identified in *Rauvolfia* samples collected from India includes reserpine, ajmaline, ajmalimine, rescinnamine, rescinnamidine, reserpiline ajmalicine, yohimbine, deserpidine, indobine, indobinine, serpentinine and serpentine etc.

Reserpine

Reserpine the most important alkaloid isolated in 1950s from roots of *R. serpentine*. It is a weak tertiary base present in the oleoresin fraction of the roots and is the most prominent alkaloids mainly used as a natural tranquillizer (Pullaiah, 2002; Banerjee *et al.*, 2010). It is also use in the treatment of hypertension, cardiovascular diseases and neurological diseases (Howes *et al.*, 1990; Weiss *et al.*, 2000). It is responsible for antihypertensive action and is now being utilized as a tool in physiologic studies of body functions and in pharmacological studies.

Reserpine inhibits the function of postganglionic adrenergic neurons, thus inhibiting sympathetic function. It binds to catecholamine storage vesicles in central and peripheral sympathetic nerve terminals. Storage vesicles become nonfunctional as a result of interacting with reserpine and lose the ability to store and concentrate catecholamines. Catecholamine from vesicles are enzymatically destroyed in the cytoplasm and as a consequence little transmitter is released upon nerve ending depolarization. Depletion of transmitter in both the central and peripheral nervous system suggest that both sites are important mediators of the antihypertensive response (Ellenhorn *et al*., 1988; Gilman *et al*., 1990; Nammi *et al*., 2005). These substances are mostly involved in controlling heart rate, cardiac contraction and peripheral resistance. It also helps in sedation and lowering of blood pressure, especially in cases of hypertension exacerbated by stress and sympathetic nervous system activity.

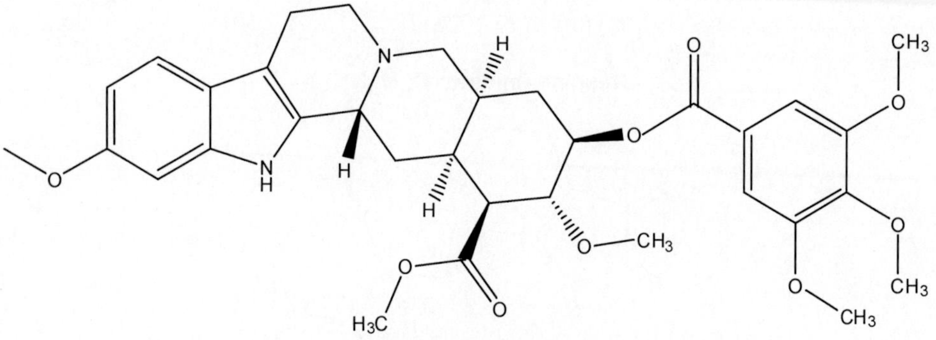

Reserpine ($C_{33}H_{40}N_2O_9$)

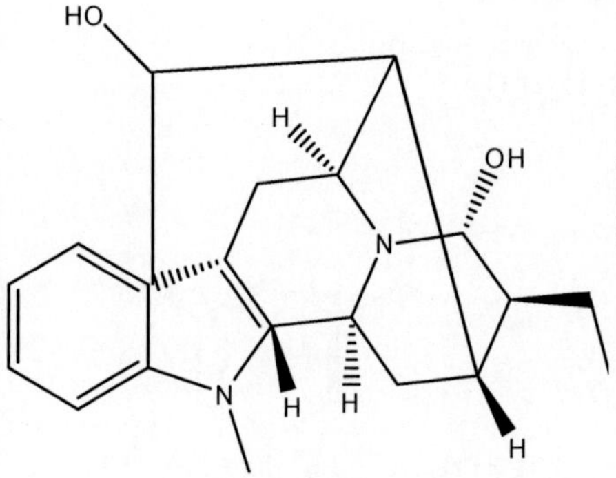

Ajmaline ($C_{20}H_{26}N_2O_2$)

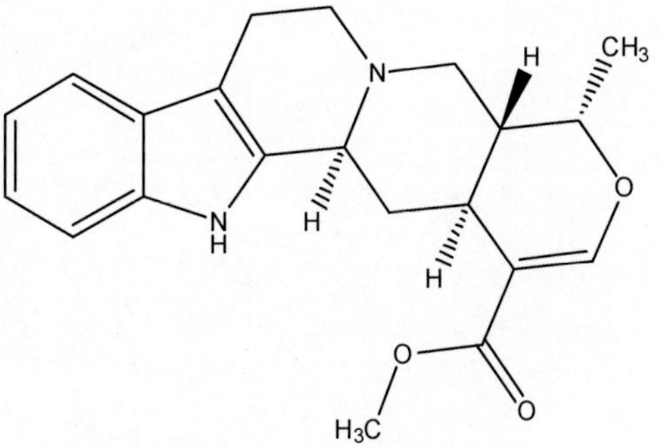

Rescinnamine ($C_{35}H_{42}N_2O_9$)

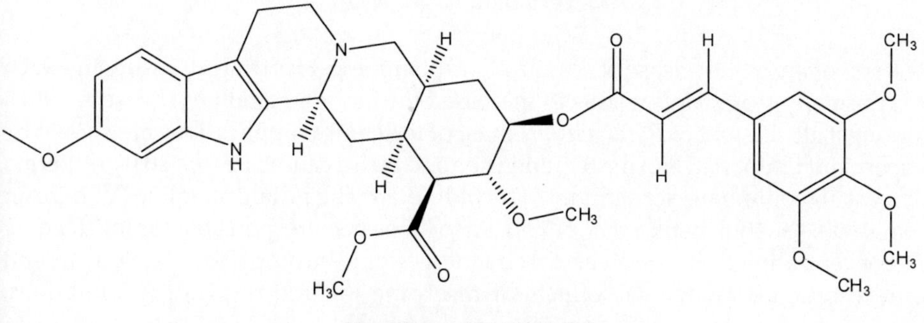

Ajmalicine ($C_{21}H_{24}N_2O_3$)

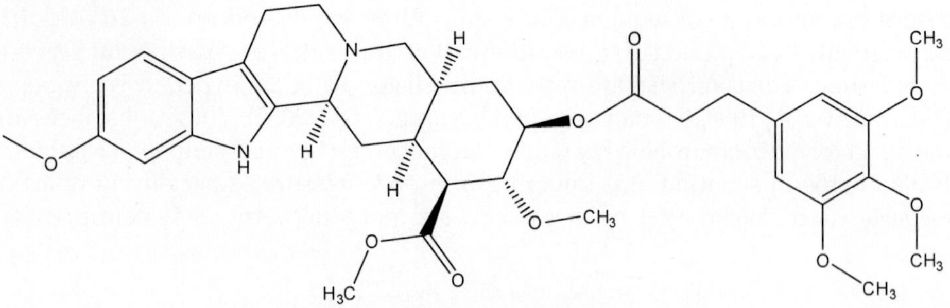

Rescinnamidine ($C_{35}H_{44}N_2O_9$)

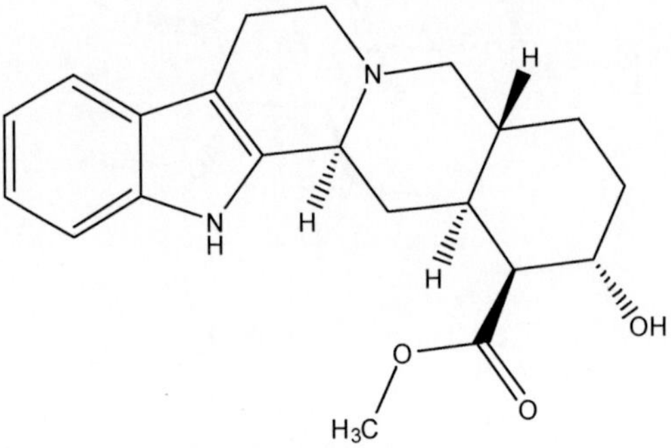

Yohimbine ($C_{21}H_{26}N_2O_3$)

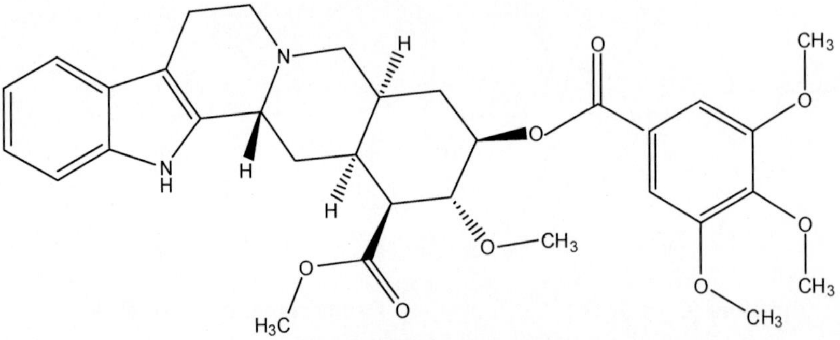

Deserpidine ($C_{32}H_{38}N_2O_8$)

Reserpine causes the release of 5-hydroxytryptamine (5-HT) from all tissues in which it is normally stored and results in increase of urinary metabolites (Prusoff, 1961). Amongst all alkaloid of *R. serpentine*, most of the clinical applications are shown by resperine. It is reported that the hypertension and breast cancer are treated by resperine along with yohimbine, serpentine, deserpidine, ajmalicine and ajmaline (Von-Poser *et al.*, 1990; Klushnichenko *et al.*, 1995). Also it has been responsible for influencing the concentration of glycogen, acetyl choline, g-amino butyric acid, nucleic acids and anti-diuretic hormone. The effects of reserpine include respiratory inhibition, stimulation of peristalsis, myosis and relaxation of nictating membranes and also influence temperature regulating centre. It increases the volume and free acidity of gastric secretion (Mittal *et al.*, 2012).

Ajmaline

In 1931, Salimuzzaman Siddiqui isolated a compound ajmaline from the roots of *R. serpentine* and named it after Hakim Ajmal Khan an eminent experts of Unani medicine in South Asia (Siddiqui *et al.*, 1985; Siddiqui, 2013).

Ajmaline "class Ia" anti-arrhythmic drug obtained from roots of *R. serpentine* used in several countries as first-line treatment for ventricular tachyarrhythmia, (Ito *et al.*, 1990; Schuchert and Meinertz, 2000). In the treatment of sustained Ventricular tachycardia, ajmaline was reported to be more effective than lidocaine (Manz *et al.*, 1992) and is highly useful in diagnosis of Brugada Syndrome,(a genetic disease characterized by abnormal electrocardiogram results in high risk of sudden cardiac death. It helps in distinguishing patients subtypes with that of disease (Rolf *et al.*, 2003). On the basis of mechanism of action these agents are mainly categorized into four major groups *i.e.* calcium channel blockade, sodium channel blockade, beta-adrenergic blockade, repolarization prolongation. Ajmaline a sodium channel blocker when given intravenously shows instant action, and used for diagnostic purposes. The administration of *Rauvolfia* alkaloid to patients with this type of arrhythmia is known as the "Ajmaline Test" (Kostin *et al.*, 1986; Brugada *et al.*, 2000). It has been reported to stimulate respiration and intestinal movements. The mode of action of ajmaline on systemic and pulmonary blood pressure is similar to serpentine (Gawade *et al.*, 2012).

Ajmalicine

The ajmalicine is derived from tryptophan which is converted to tryptamine through secologanin, strictosidine and cathenamine. Cathenamine reduced to ajmalicine by enzyme NADPH and tryptophan decarboxylase (TDC). Most important enzyme involved in the synthesis of ajmalicine in *Rauvolfia* is decarboxylase (Liu *et al.*, 2012). The synthetic pathway starts with geraniol through iridodial and irdotrial by the synthesis of loganin, which on oxidation converts loganin into secoloanin. This helps the tryptamine to form corynan the type nucleus that results in the formation of ajmalicine (O'Connor *et al.*, 2006; Li *et al.*, 2004). Serpentine, a type II topoisomerase inhibitor, exhibits antipsychotic properties. (Dassonneville *et al.*, 1999; Costa-Campos *et al.*, 2014). The enzyme peroxidase is responsible for oxidation of ajmalicine to serpentine by catalyzing bis indole alkaloid localized in the vacuole (O'Connor *et al.*, 2006). An estimated 3500 kg of ajmalicine is isolated annually from either *Rauvolfia* or *Catharanthus* spp. by pharmaceutical industries for the treatment of circulatory diseases.

Ajmalicine have a large number of applications it involved in the treatment of circulatory diseases, especially in providing relief to normal cerebral blood flow. It affects the function of smooth muscle prevent strokes and helps in lowering blood pressure.

Rescinnamine

A purified ester alkaloid of alseroxylon fraction was isolated from *Rauvolfia* in 1950's. It is clinically less potent alkaloid than reserpine in lowering blood pressure but it is chemically and pharmacologically related to reserpine with similar uses. Rescinnamine binds and inhibits angiotensin converting enzyme, peptidyl dipeptidase which inhibits the conversion of angiotensin I to angiotensin II that stimulates the secretion of aldosterone from adrenal cortex. Inhibition of angiotensin converting enzyme results in decreased angiotensin II concentration in plasma. As angiotensin II is a negative-feedback mediator for renin and vasoconstrictor, its results in lowering down the blood pressure and stimulation of baroreceptor reflex mechanisms, which eventually results in decreased aldosterone secretion and vasopressor activity (http://www.drugbank.ca/drugs/DB01180).

Deserpidine

An ester alkaloid deserpidine isolated from *Rauvolfia* is almost similar to respirine except methoxy group at C-1. Deserpidine and angiotensin I competes with one other for binding with angiotensin-converting enzyme, deserpidine when binds with angiotensin converting enzyme inhibits its action resulting in the blockage of angiotensin I conversion to angiotensin II (Varchi *et al.*, 2005) and due to antipsychotic and antihypertensive properties, deserpidine act on the heart, blood vessels and lower down high blood pressure by controlling nerve impulses along various nerve pathways.

Yohimbine A

Pharmacologically categorized alkaloid Yohimbine is used as alpha-adrenergic antagonist or alpha-blocker for the treatment of erectile dysfunction by dilating blood vessels and increasing blood flow in the penis (Morales, 2000b; Andersson, 1993; Andersson, 2001; Goldberg *et al.*, 1983). Yohimbine has also been used in animal and human models as therapy for diabetes as it carries polymorphisms in α2A-adrenergicreceptor gene. Antagonism at these receptors results in low blood pressure and relaxation of smooth muscle (Rosenren *et al.*, 2009).

Phenols

Phenols also known as phenolics are class of chemical compounds consisting of hydroxyl group bounded directly to an aromatic hydrocarbon group. Phenolic compounds have been found in the plant kingdom mainly in herbs, shrubs, vegetables and trees as secondary plant metabolites (Bonilla *et al.*, 2003; Naira *et al.*, 2013). The presence of phenolic compounds in plants has been deliberated toxic for the growth and development of several pest and pathogens (Singh *et al.*, 1988). *R. serpentine* has important antidiabetic and hypolipidemic properties due to the presence of high quantity of total polyphenolic compounds (Azmi *et al.*, 2013; Qureshi *et al.*, 2009) and can also use as an expectorant and emulsifying in medicine.

Tannins

Gallic and Diagallic acid present in tanni are responsible for the inhibition of oxidation (Ihekoronye *et al.*, 1985). Stringent properties of tannins accelerate the healing of wounds and inflamed mucous membranes. Due to these reasons *R. serpentine* has been used as traditional medicine healers in south eastern India for the treatment of many diseases (Harisaranraj *et al.*, 2009; Agoha, 1974).

Flavonoids

Flavonoids are effective water-soluble antioxidants and free radical scavengers, which stop oxidative cell damage and have strong anticancerous activity (Salah *et al.*, 1995; Del-Rio *et al.*, 1997). In herbal medicine flavonoids used as antioxidants, provide anti-inflammatory activity against the treatment of various diseases. Flavonoids also lower the risk of heart diseases when present in intestinal tract (Mittal *et al.*, 2012; Okwu, 2004).

Saponins

In over 70 families of plants saponins have been identified as glycoside of both triterpenes and sterols. Several features of saponins include cholesterol binding properties, formation of foams in aqueous solutions, hemolytic activity, bitterness and coagulating red blood cells (Sodipo *et al.*, 2000). Due to the high saponin content in *R. serpentine* confirms the use of its extracts in stop bleeding and wounds treatment (Harisaranraj *et al.*, 2009; Basu *et al.*, 1967).

Mineral Composition

Rauvolfia contains large number of macro and micro-nutrients. Calcium is the most abundant macronutrient and because of high content of calcium. *R. serpentine* has been used in blood coagulation, stop bleeding and treating wounds. *R. serpentine* has low sodium content and because of direct relationship of sodium intake with hypertension, low sodium content have advantageous effects (Dahl, 1972). Presence of zinc in *R. serpentine* has been used in the controlling diabetes, which cause due to insulin malfunction (Agoha,1974).

R. serpentine is also an outstanding source for ascorbic acids, riboflavin, thiamin and niacin. Ascorbic acid plays an important role in normal wound healing and lack of it damages the normal formation of intercellular substances throughout the body including collagen, bone matrix and tooth dentine (Okwu, 2004).

Medicinal Uses

In India, since the prevedic period, roots of *R.serpentine* have been used as drug for the treatment of fever and snake bite. It is shown in the ancient literature that plant has been used against snake and insect bite (pandey, 1984). *R. serpentine* regarded as a typical drug of Ayurveda belongs to small group of plants deserving attention as a traditional medicinal plant since 3000 years (Ruppert *et al.*, 2005b). The ayurvedic preparations of *R. serpentine* are Sarpagandha ghanavati, Sarpagandha yoga, Sarpagandha churna and Mheshvari vati (Vaidya, 2006) and roots are used as a valuable remedy for snake, insect and animal bite, insomnia, anxiety, excitement, mental illness, schizophrenia, high blood pressure, insanity, epilepsy, hypochondria and other diseases of the central nervous system (Monachino, 1954; Kirtikar and Basu, 1993). Alkaloids of this plant have a great medicinal importance to treat cardiovascular diseases (Anitha and kumari., 2006), high blood pressure (vakil, 1955), hypertension (Von Poster *et al.*, 1990), arrhythmia (Kirillova *et al.*, 2001), various psychiatric diseases (Bhatara *et al.*, 1997), mental disorders (Noce *et al.*, 1954), breast cancer (Stanford *et al.*, 1986), human promyelocytic leukemia (Itoh *et al.*, 2005) like disease. The Unani formulation "Pitkriya capsuls" contain arsole of *R. serpentine* which acts as musakkin- wo -munawwim (sedative and hypnotic), Mudir (diuretic), Musakkin-e-Asa (nervine sedative) and Mukkhaddir (anesthetic) (Shamsi *et al.*, 2006).

Ethnobotanical Uses

Snake, Insect and Animal Bite

Peoples have many folk-lonistic about *R. serpentine*. Some believes that mongoose before fighting with cobra first chew the leaves of *R. serpentine* to gain power. Some other believe, when freshly ground leaves of *R. serpentine* applied on the toes can serve as an antidote for snake poison. A third one is that, mentally challenged person is relieved of his insanity if he eats the roots (Pandey, 1984). It has been reported that *R. serpentine* has been used in different parts of India, Bangladesh and Nepal against insect, snake and animal bite (Sankaranarayanan *et al.*, 2010; Singh *et al.*, 2010; Rahamatullah *et al.*, 2010b,2010a; Mollik *et al.*, 2010; Prakasha *et al.*, 2010; Pattanaik *et al.*, 2009; Bhattarai *et al.*, 2009a; Arjariya and Chaurasia, 2009; Singh, 2008; Mohapatra *et al.*, 2008; Behera *et al.*, 2007; Srivastava and Pandey, 2006; Jeeva *et al.*, 2006; Pattanaik *et al.*, 2006; Ghorbani *et al.*, 2006; Nayak *et al.*, 2004; Parinitha *et al.*, 2004). Kurichayas of Kannur district, Western Ghats, Kerala, India used a paste of fresh root of *R. serpentine* grounded with the roots of *Aristolochia indica* taken twice daily for three days against snake bite (Rajith and Ramachandran, 2010). Khamptis of Arunachal Pradesh, India used juice of leaves extract twice a day for three days or apply a paste of 15 g of *R. serpentine* roots along with roots of *Cassia tora* and *Holarrhena pubescens* twice a day for two days for snake-bite (Sen *et al.*, 2008). Roots and leaf buds crushed with milk was applied externally as paste on the snakebite affected areas by the folks of Bhadra wildlife sanctuary in Karnataka (Parinitha *et al.*, 2004).

Mental Illness, Schizophrenia, Hypertension, Blood Pressure

It has been reported that roots and the other parts of *R. serpentine* has been used in the treatment of mental illness, schizophrenia, hypertension and blood pressure (Singh *et al.*, 2010; Biswas *et al.*, 2010; Mollik *et al.*, 2010; Meena *et al.*, 2009; Barbhuiya *et al.*, 2009; Tomar, 2009; Das *et al.*, 2009; Arjariya and Chaurasia, 2009; Silja *et al.*, 2008; Roy *et al.*, 2008; Mohapatra *et al.*, 2008; Rahman *et al.*, 2007; Behera *et al.*, 2007; Jeeva *et al.*, 2006; Behera *et al.*, 2006; Acharya *et al.*, 2006; Yusuf *et al.*, 2006; Kumar *et al.*, 2004). The roots of *R. serpentine,* a medicinal plants of Karnataka forest, India has been used in mental disorders, nervous disorders and psychosis (Shiddamallayya *et al.*, 2010). While studying the medicinal plants in the graveyards of Rajshahi city, Bangladesh it has been reported that roots of *R. serpentine* has been used in blood pressure and tonic (Rahman *et al.*, 2008). In Unani system of medicine *Pitkriya* formulation contains as arsol of this plant which act as diuretic, nervine sedative, hypnotic and anesthetise (Shamsi *et al.*, 2006). Plant has been used by the Jaunsari tribe of Garhwal Himalaya Uttarakhand for anxiety, epilepsy and nervous disorders

(Bhatt and Negi, 2006). Fabricant and Farnsworth(2001) reported that plant has been used as a traditional antihypertensive and tranquilizer.

Gastrointestinal Diseases

It has been reported that plant has been used in the treatment of dysentery, intestinal disorder and stomach ache (Mollik *et al.*, 2010; Rao *et al.*, 2010; Rajith and Ramachandran, 2010; Rahman *et al.*, 2008; Nayak *et al.*, 2004; Shiddamallayya *et al.*, 2010; Yesodharan and Sujana, 2007; Bhatt and Negi, 2006; Rout and Thatoi, 2009; Roy *et al.*, 2008; Sen *et al.*, 2008; Mohapatra *et al.*, 2008; Udayan *et al.*, 2007; Jeeva *et al.*, 2006). The Ethnic groups in the Eastern Ghats of Andhra Pradesh, India use mixture of *R. serpentine* root, Cassia fistula stem bark, stem barks of Mango, and *Piper longum* for jaundice (Naidu and Khasim, 2010). People of Nawalparasi district, Central Nepal used a mixture of 100g each root paste of *Cyperus rotundus, Polygala arillata, R. serpentina, Cissampelos pareira*, three leaves of Citrus lemon and one bird (Chatail bird in Gurau) mixed in two bottle of local alcohol, cooked for some time and for any stomach diseases (Bhattarai *et al.*, 2009a). To expel intestinal worms in children the tribes of Chhatarpur district, Madhya Pradesh, India take root decoction (3g) and black pepper (Arjariya and Chaurasia, 2009). *R. serpentina* root juice has been used by local people of Madhupur, Tangail, Bangladesh to treat liver pain (Anisuzzaman *et al.*, 2007). Chakma tribe in hill tracts districts of Bangladesh drink leaf and fruit extract of *R. Serpentina* and *Ixora villosa* for abdominal pain (Rahman *et al.*, 2007).

Prostate Cancer

Prostate cancer is considered to be major causes of cancer-related deaths in men. Modern therapies such as radiotherapy and chemotherapy available in the medical centers for treatments of prostate cancer have not provided significant persistence benefits to patients as compared to chemotherapy and radiotherapy (American Cancer Society, 2006). Natural products have proven to be a major treatment for variety of ailments and diseases (www.pdrhealth.com). Previously it has been reported that the tumour cell growth reduced in mice inoculated with YC8 lymphoma cells or Ehrlich ascetic cells by a compound known as ß-carboline and the extracts obtained from the root bark of *R. serpentina* are enriched with compounds ß-carboline of alkaloid family with main constituent alstonine (Beljanski *et al.*, 1986). The root bark extract has been reported to be endowed with activity against prostate cancer, demonstrated by using *in vitro* as well as *in vivo* model systems, it may be suggested that the components of root bark extract affecting gene expression pattern of cancer cells, reflected by alternations in important cellular pathways such as DNA damage and signaling control of cell cycle (Bemis *et al.*, 2006).

Circulatory Disorders

Fabricant and Farnsworth (2001) reported that the plant has being used as traditional medicines to treat circulatory disorders.

Pneumonia

People of Jhapa District, Eastern Nepal (Rai, 2004) use the young shoot extract of this plant to cure pneumonia at early stages of infection.

Malaria

A paste made up of *R. serpentine* root and black pepper has been used by Garo tribe inhabiting the Madhupur forest region of Bangladesh to treat malaria (Mia *et al.*, 2009). Kol tribes in Similipal Biosphere Reserve, Orissa, India have been using roots of this plant ground with equal quantities of *Cissampelos pareira* roots *in* water and take it for 5 days to treat malaria (Rout and Thatoi, 2009). Khamptis of Arunachal Pradesh, India used Juice extract of leaves of *R. serpentina* along with the juice of *Andrographis paniculata* and *Azadirachta indica* with honey to cure malaria (Sen *et al.*, 2008). Local people of Madhupur, Tangail, Mandi of Bangladesh make pills of *R. serpentina* leaf and root which is further sun dried and used in malarial fever.

Skin Diseases and Scabies

Scabies can be treated from *R. serpentine* root juice when mixed with leaves juice of *A. paniculata* and *Nyctanthes arbortristis* (Mohanta *et al.*, 2006). In Unani system of medication this plant has been used to treat urticarial patients (Shamsi *et al.*, 2006). It has been reported, that Kandhas of Kandhamal district of Orissa use root paste of this plant and *A. paniculata* for itches, boils and eczema (Behera *et al.*, 2006). A paste prepared by mixing seeds of *Datura fastusa,* roots of *Rauvolfia,* lime juice and boric acid has been used by people of Bhadra wildlife sanctuary in Karnataka to treat scabies (Parinitha *et al.*, 2004).

Asthma and Respiratory Problems

While scrutinizing the Kani tribal Botanical knowledge in Agasthiayamalai Biosphere Reserve, South India, Britto and Mahesh (2007) reported that flowers and the leaves of this plant has been used for treatment of asthma. It has been reported that folkloric groups of Betbunia locality of Rangamati district, Bangladesh recommended fresh root juice or dried root powder of this plant soaked in water for respiratory problem (Yusuf *et al.*, 2006).

Eye Diseases

It has been reported that eye inflammation can be prevented by fresh leaf juices of *R. serpentine* as used by Local people of Madhupur, Tangail, Bangladesh (Anisuzzaman *et al.*, 2007).

Headache

Plant has been used by folk medicinal practitioners in three villages of Natore and Rajshahi districts, Bangladesh for headache (Rahamatullah *et al.*, 2010b).

Fever

R. serpentine roots have been used by the peoples of different district of India and Bangladesh for fever as reported by (Arjariya and Chaurasia, 2009; Rahman *et al.*, 2008; Nayak *et al.*, 2004).

Spleen Diseases

Plant has been used by Garo tribe inhabiting the Madhupur forest region of Bangladesh for the treatment of spleen diseases (Mia *et al.*, 2009).

Acquired immuno-deficiency syndrome. It has been reported that plant is being used by folk medicinal practitioners in three villages of Natore and Rajshahi districts, Bangladesh for acquired immuno-deficiency syndrome (Rahamatullah *et al.*, 2010b).

Conclusion

Versatility of medicinal plants (about half million in diversity) around the world, make them a promising future because most of their medical activities have not investigate yet, and their medical properties could be significant in the treatment of disease in present or future. The traditional and ethnomedicinal literatures showed that *R. serpentine* is very effective and safe for medicinal uses. The compiled data of *R. serpentine* may be helpful for the leading investigators to focus on the priority areas of research yet to be discovered. More attention is needed to isolate the constituents responsible for the biological actions. By using the reverse pharmacological approaches in natural drug discovery a potent and safe drug can be investigated from the plant for various diseases.

References

Acharya (Siwakoti) E., and Pokhrel, B. (2006). Ethno-Medicinal Plants Used by Bantar of Bhaudaha, Morang, Nepal. *Our Nature*, **4**: 96-103.

Agoha, R.C. (1974). Medicinal plants of Nigeria, offset Drakkerij, Faculfcitder Wiskunde in Naturwetenschappen, the Netherlands, 41-33.

Alamgir, A.N.M., and Ahamed, M. (2005). Growth and phytochemical investigation of *Rauvolfia serpentina* Benth. *Bangladesh J. Bot.*, **34**: 7-10.

American Cancer Society, Cancer Facts and Figures (2006). Atlanta: American Cancer Society,

Andersson, K.E. (1993). Pharmacology of lower urinary tract smooth muscles and penile erectile tissues. *Pharmacological Reviews*, **45**: 254-308.

Andersson, K.E. (2001). Pharmacology of penile erection. *Pharmacological Reviews*, **53(3)**: 417-450.

Anisuzzaman, M., Rahman, A.H.M.M., Harun-Or-Rashid, M., Naderuzzaman, A.T.M., and Islam, A.K.M.R. (2007). An ethnobotanical study of Madhupur, Tangail. *J. Appl. Sci. Res.*, **3(7)**: 519-530.

Anitha, S., and kumarii, B. (2006). Stimulation of reserpine biosynthesis in the callus of *Rouvolfia tetraphyla* L by precursor feeding. *Afr.J. Biotechnol.*, **5**: 659-66.

Anitha, S., and Kumari, B.D.R. (2006). Stimulation of reserpine biosynthesis in the callus of *Rauvolfia tetraphyla* L. by precursor feeding. *African Journal of Biotechnology*, **5**: 659-661.

Arjariya, A., and Chaurasia, K. (2009). Some medicinal plants among the tribes of Chhatarpur district (M.P.). *Ind. Ecoprint.*, **16**: 43-50.

Arts, I.C., and Hollman, P.C. (2005). Polyphenols and disease risk in epidemiologic studies, *American Journal of Clinical Nutrition*, **81**: 317S-325S

Azmi, M.B., Qureshi, S.A. (2013). Methanolic root extract of *Rauvolfia serpentina* Benth. improves the glycemic, antiatherogenic, and cardioprotective indices in alloxan-induced diabetic mice. *Journalof Applied Pharmaceutical Science*, **3(7)**: 136-141.

Banerjee, M., and Modi, P. (2010). A novel protocol for micropropagation of *Rauvolfia serpentina*: In low concentration of growth regulators with sucrose and phenolic acid. *International Journal of Plant Sciences,* **5(1)**: 93-97.

Barbhuiya, A.R., Sharma, G.D., Arunachalam. A., and Deb, S. (2009). Diversity and conservation of medicinal plants in Barak valley, Northeast, India. *Ind. J. Trad. Knowl*, **8(2)**: 169-175.

Basu, N., and Rastogi, R.P. (1967). Triterpenoid, saponins and sapogenins. *Photochemistry,* **6**: 1249-1270.

Behera, K.K., Sahoo, S., and Mohapatra, P.C. (2007). Medicinal plant resources for bioprospecting and drug development in tribal rich district of Orissa, India. *Ethnobotanical Leaflets*, **11**: 106-112.

Behera, S.K., Panda, A., Behera, S.K., and Misra, M.K. (2006). Medicinal plants used by the Kandhas of Kandhamal district of Orissa. *Ind. J. Trad. Knowl*., **5(4)**: 519-528.

Beljanski, M., and Beljanski, M.S. (1986). Three alkaloids as selective destroyers of cancer cells in mice, synergy with classic anticancer drugs. *Oncology,* **43**: 198-203.

Bemis, D.L., Capodice, J.L., Gorroochurn, P., Katz, A.E., and Buttyan, R. (2006). Antiprostate cancer activity of a beta-carboline alkaloid enriched extract from *Rauvolfia vomitoria*. *International Journal of Oncology*, **29(5)**: 1065-1073.

Bhatara, V.S., Sharma, J.N., Gupta, S., and Gupta, Y.K. (1997). Images in psychiatry. *Rauwolfia Serpentina*: The first antipsychotic. *Am.J. Psychiatry*., **154**: 894-894.

Bhatara, V.S., Sharma, J.N., Gupta, S., and Gupta, Y.K. (1997). Images in psychiatry *Rauvolfia serpentina*: The first antipsychotic. *American Journal of Psychiatry*, **154**: 894-894.

Bhatt, V.P., and Negi, G.C.S. (2006). Ethnomedicinal plant resources of Jaunsari tribe of Garhwal Himalaya, Uttaranchal. *Ind. J. Trad. Knowl*., **5(3)**: 331-335.

Bhattarai, S., Chaudhary, R.P., and Taylor, R.S.L. (2009a). Ethno-medicinal plants used by the people of Nawalparasi District, Central Nepal. *Our Nature*, **7**: 82-99.

Biswas, A., Bari, M.A., Roy, M., and Bhadra, S.K. (2010). Inherited folk pharmaceutical knowledge of tribal people in the Chittagong Hill tracts, Bangladesh. *Ind. J. Trad. Knowl.*, **9(1)**: 77-89.

Bonilla, E.P., Akoh, C.C., Sellappan, S., and Krewer, G. (2003). Phenolic content and antioxidant capacity of muscadine grapes. *Journal of Agriculture and Food Chemistry*, **51**: 5497-5503.

Britto, J.D., and Mahesh, R. (2007). Exploration of Kani Tribal Botanical Knowledge in Agasthiayamalai Biosphere Reserve - South India. *Ethnobotanical Leaflets*, **11**: 258-265.

Brugada, R., Brugada, J., Antzelevitch, C., Kirsch, G.E., Potenza, D., Towbin, J.A., and Brugada, P. (2000). Sodium channel blockers identify risk for sudden death in patients with ST-segment elevation and right bundle branch block but structurally normal hearts. *Circulation*, **101(5)**: 510-515.

COMSATS newsletter: Biographies of eminent scientists: Dr. Salimuzzamman Siddiqui (2013). COMSATS Secretariat, Islamabad– Pakistan 5(3),.

Costa-Campos, L., Dassoler, S.C., Rigo, A.P., Iwu, M., and Elisabetsky, E. (2004). Anxiolytic properties of the antipsychotic alkaloid alstonine. *Pharmacology Biochemistry and Behaviour*, **77**: 481-489.

Dahl, L.K. (1972). Salt and hypertension. *American Journal of Clinical Nutrition*, **25**: 231-238.

Das, S., Khan, M.L., Rabha, A., and Bhattacharjya, D.K. (2009). Ethnomedicinal plants of Manas National Park, Assam, Norteast India. *Ind. J. Trad. Knowl.*, **8(4)**: 514-517.

Dassonneville, L., Bonjean, K., Pauw-Gillet, M.C.D., Colson, P., Houssier, C., Quetin-Leclercq, J., Angenot, L., and Bailly, C. (1999). Stimulation of topoisomerase II-mediated DNA cleavage by three DNA intercalating plant alkaloids: Cryptolepine, matadine, and serpentine. *Biochemistry*, **38**: 7719-7726.

Del-Rio, A., Obdululio, B.G., Casfillo, J., Marin, F.G., and Ortuno, A. (1997). Uses and properties of citrus flavonoids. *Journal of Agriculture and Food Chemistry*, **45**: 4505-4515.

Deshmukh, S.R., Dhanashree, S.A., and Patil, B.A. (2012). Extraction and evaluation of indole alkaloids from *Rauvolfia serpentina* for their antimicrobial and antiproliferative activities. *International Journal of Pharmacy and Pharmaceutical Sciences*, **4(5)**: 329-334.

Dey, A., and De, J.N. (2011). Ethnobotanical aspects of *Rauvolfia serpentina* (L). Benth. Ex Kurz. in India, Nepal and Bangladesh. *Journal of Medicinal Plant Research*, **5(2)**: 144-150.

Dey, A., and De, J.N. (2010). *Rauvolfia serpentina* (L). Benth. Ex Kurz. A Review. *Asian Journal of Plant Sciences*, **9(6)**: 285-298.

Ellenhorn, M.J., and Barceloux, D.G. (1988). Medical Toxicology, New York, NY, Elsevier Science Publishing Company, Inc, 644-659.

Ezeigbo, I., Ezeja, M., Madubuike, K., Ifenkwe, D., Ukweni, I., Udeh, N., and Akomas, S. (2012). Antidiarrhoeal activity of leaf methanolic extract of *Rauvolfia serpentine. Asian Pacific Journal of Tropical Biomedicine*, **2(6)**: 430-432.

Fabricant, D.S., and Farnsworth, N.R. (2001). The value of plants used in traditional medicine for drug discovery. *Environ. Health Perspect.*, **109(1)**: 69-75.

Gawade, B.V., and Fegade, S.A. (2012). *Rouvolfia* (reserpine) as a potential antihypertensive agent – A review. *International Journal of Pharmaceutical and Phytopharmacological Research*, **2(1)**: 46-49.

Ghani, A. (1998). Medicinal plants of Bangladesh chemical constituents and uses. *Asiatic Society of Bangladesh*, second edition, **2**: 36.

Ghorbani, A., Naghibi, F., and Mosaddegh, M. (2006). Ethnobotany, ethnopharmacology and drug discovery. *Iran. J. Pharmaceut. Sci.,* **2(2)**: 109-118.

Gilman, A.F., Rall, W.T., Nies, A.D., and Taylor, P. (1990). Goodman and Gilman's: The Pharmacologic Basis of Therapeutics, 8th ed, Pergamon Press, New York, New York, 795.

Goldberg, M.R., Robertson, D., and Yohimbine. (1983). A pharmacological probe for study of the α2-adrenoceptor. *Pharmacological Reviews,* **35**: 143-180.

Harisaranraj, R., Suresh, K., Babu, S.S., and Achudhan, V.V. (2009). Phytochemical based strategies for pathogen control and antioxidant capacities of *Rauvolfia serpentina* extracts. *Recent Research in Science and Technology*, **1**: 67-73.

Harisaranraj, R., Suresh, K., and Saravanababu, S. (2009). Evaluation of the chemical composition *Rauvolfia serpentina* and *Ephedra vulgeris. Advances in Biological Research,* **3(5-6)**: 174-178.

Howes, L.G., and Louis, W.J. (1990). *Rauvolfia* alkaloids (Reserpine), pharmacology of antihypertensive therapeutics. *Handbook of Experimental Pharmacology*, **93 (1)**: 263-285.

http: //www.drugbank.ca/drugs/DB01180.

Hu, X.J., He, H.P., Zhou, H., Di, Y.T., Yang, X.W., Hao, X.J., and Kong, L.Y. (2008). New indole alkaloids from *Rauvolfia verticillatae, Helvetica Chimica Acta,* **89**: 1344-1350.

Ihekoronye, A.I., and Ngoddy, P.O. (1985). Integrated Food Science and Technology for the Tropics, Macmillam Education Ltd,.

Ito, M., Onodera, S., Noshiro, H., Odakura, H., Yasuda, S., Nagashima. M., Shinoda, S., and Suzuki, H. (1990). Eect of class Ia antiarrhythmic agents on fasciculoventricular bers. *J. Electrocardiol.*, **23**: 323–329.

Itoh, A., Kumashiro, T., and Yamaguchi, M. (2005). Indole alkaloids and other constituents of *Rauvolfia serpentine. Journal of Natural Products*, **68(6)**: 848-852.

Jeeva, S., Kiruba, S., Mishra, B.P., Venugopal, N., Dhas, S.S.M., Regini, G.S., Kingston, C., Kavitha, A., Sukumaran, S., Raj, A.D.S., and Laloo, R.C. (2006). Weeds of Kanyakumari district and their value in rural life. *Ind. J.Trad. Knowl.*, **5(4)**: 501-509.

Kamboj, V.P. (2000). Herbal medicine. *Current Science,* **78(1)**: 35-51.

Kirillova, N.V., Smirnova, M.G., and Komov, V.P. (2001). Sequential isolation of superoxide dismutase and ajmaline from tissue culture of *Rauvolfia serpentina* Benth. *Applied Biochemistry and Microbiology,* **37**: 181-185.

Kirtikar, K.R., and Basu, B.D. (1993). Indian Medicinal Plant. 2Edn., Dehra Dun Publisher Culcutta, India., 289.

Klushnichenko, V.E., Yakimov, S.Y., Tuzova, T.P., Syagailo, Y.V., Kuzovkina, I.N., Vulf'son, A.N., and Miroshnikov, A.I. (1995). Determination of indole alkaloids from *R. serpentina* and *R. vomitoria* by HPLC and TLC methods. *Journal of Chromatography*, **704**: 357-362.

Kostin, Y.V., Melokhova, E.I., Gendenshtein, E.I., Volkova, N.D., Astakhova, T.V., and Savel'eva, E.K. (1986). Antiarrhythmic activity of the total alkaloids from a *Rauvolfia serpentina* tissue culture. *Pharmaceutical Chemistry Journal*, **20(3)**: 214-217.

Kumar, R., Suman, N.R., and Dash, S.S. (2004). Traditional uses of plants by the tribals of Amarkantak region, Madhya Pradesh. *Ind. J. Trad. Knowl.*, **3(4)**: 383-390.

Kumari, S., Shukla, G., and Rao. A.S. (2011). The present status of medicinal plants – aspects and prospects. *International Journal of Research in Pharmaceutical and Biomedical Sciences*, **2(1)**: 19-23.

Li, S., Long, J., Ma, Z., Xu, Z., Li, J., and Zhang, Z. (2004). Assessment of the therapeutic activity of a combination of almitrine and raubasine on functional rehabilitation following ischaemic stroke. *Current Medical Research and Opinion*, **20**: 409-415.

Liu, W., Chen, R., Chen, M., Zhang, H., Peng, M., Yang, C., Ming, X., Lan, X., and Liao, Z. (2012). Tryptophan decarboxylase plays an important role in ajmalicine biosynthesis in *Rauvolfia verticillata. Planta*, **236(1)**: 239-250.

Mabberley, D.J. (2008). A portable dictionary of plants, their classification and uses. Cambridge University Press, Oxford.

Mallick, S.R., Jena, R.C., and Samal, K.C. (2012). Rapid *in vitro* multiplication of an endangered medicinal plant sarpgandha (*Rauvolfia serpentina*). *American Journal of Plant Sciences*, **3**: 437-442.

Mamgain, S.K., Goel, A.K., and Sharma, S.C. (1998). Conservation of assessment of some important threatened medicinal plants of India. *J. Non-Timber Fore. Prod.*, **5**: 1-9.

Manz, M., Mletzko, R., Jung, W., and Luderitz, B. (1992). Electrophysiology and haemodynamics effects of Iidocaine and ajmaline in the management of sustained ventricular tachycardia. *Eur. Heart. J.*, **13**: 1123-1128

Meena, A.K., Bansal, P., and Kumar, S. (2009). Plants-herbal wealth as a potential source of ayurvedic drugs. *Asian Journal of Traditional Medicines*, **4(4)**: 152-170.

Mia, M.M.K., Kadir, M.F., Hossan, M.S., and Rahmatullah, M. (2009). Medicinal plants of the Garo tribe inhabiting the Madhupur forest region of Bangladesh. *Am.-Eurasian J. Sustain. Agric.*, **3(2)**: 165-171.

Mittal, B., Meenakshi, Sharma A., and Gothecha, V.K. (2012). Phytochemical and pharmacological activity of *Rauvolfia serpentina* - a review. *International Journal of Ayurvedic and Herbal Medicine*, **2(3)**: 427-434.

Mohanta, R.K., Rout, S.D., and Sahu, H.K. (2006). Ethnomedicinal plant resources of simlipal biosphere reserve, Orissa, India. *Zoo's Print Journal*, **21(8)**: 2372-2374.

Mohapatra, S.P., Prusty, G.B., and Sahoo, H.P. (2008). Ethnomedicinal observations among forest dwellers of the Daitari range of hills of Orissa, India. *Ethnobotanical Leaflets*, **12**: 1116-1123.

Mollik, M.A.H., Hossan, M.S., Paul, A.K., Taufiq-Ur-Rahman, M., Jahan, R., and Rahmatullah, M. (2010). A comparative analysis of medicinal plants used by folk medicinal healers in three districts of Bangladesh and inquiry as to mode of selection of medicinal plants. *Ethnobot. Res. Appl.*, **8**: 195-218.

Monachino, J. (1954). *Rauwolfia serpentina*: Its history, botany and medical use. *Econ.Bot.*, **8**: 349-365.

Morales, A. (2000).Yohimbine in erectile dysfunction: the facts. *International Journal of Impotence Research*, **12(1)**: 70-74.

Naidu, K.A., and Khasim, S.M. (2010). Contribution to the floristic diversity and ethnobotany of Eastern Ghats in Andhra Pradesh, India. *Ethnobotanical Leaflets*, **14**: 920-941.

Naira, V.D., Panneerselvama, R., Gopia, R., and Hong-bob, S. (2013). Elicitation of pharmacologically active phenolic compounds from *Rauvolfia serpentina* Benth. Ex. Kurtz. *Industrial Crops and Products*, **45**: 406-415.

Nammi, S., Boini, K.M., Koppula, S., and Sreemantula, S. (2005). Reserpine-induced central effects: pharmacological evidence for the lack of central effects of reserpine methiodide. *Canadian Journal of Physiology and Pharmacology*, **83(6)**: 509-15.

Nayak, S., Behera, S.K., and Misra, M.K. (2004). Ethno-medico-botanical survey of Kalahandi district of Orissa. *Ind. J. Trad. Knowledge*, **3(1)**: 72-79.

Noce, R.H., Williams, D.B., and Rapaport, W. (1954). Reserpine (Serpasil) in the management of mentally ill and mentally retarded. *Journal of American Medical Association*, **156**: 821-824.

O'Connor, S.E., and Maresh, J. (2006).Chemistry and biology of monoterpene indole alkaloid biosynthesis. *Natural Product Reports*, **23**: 532-547.

Okwu, D.E. (2003). The potentials *Ocimum gratissimum*, *Pengluria extensa* and *Tetrapleura tetraptera* as spice and flavouring agents. *Nigeria Agricultural Journal*, **34**: 143-148.

Okwu, D.E. (2004). Phytochemicals and vitamin content of indigenous spices of Southeastern, Nigeria. *Journal of Sustainable Agriculture and Environment*, **6(1)**: 30-37.

Okwu, D.E., and Okwu, M.E. (2004).Chemical composition of *Spondias mombin* Linn. plant parts. *Journal of Sustainable Agriculture and Environment*, **6(2)**: 140-147.

Pandey, B.P. (1984). Economic Botany. Janta Vedic College, Baraut, India.

Pant, K.K., and Joshi, S.D. (2008). Rapid multiplication of *Rauvolfia serpentine* Benth. Ex. Kurz through tissue culture. *Scientific World*, **6**: 58-62.

Parinitha, M., Harish, G.U., Vivek, N.C., Mahesh, T., and Shivanna, M.B. (2004). Ethnobotanical wealth of Bhadra wildlife sanctuary in Karnataka. *Ind. J. Trad. Knowl.,* **3(1)**: 37-50.

Partha, P., and Hossain, A.B.M.E. (2007). Ethnobotanical investigation into the Mandi ethnic community in Bangladesh. *Bangladesh J. Plant Taxon.,* **14(2)**: 129-145.

Pattanaik, C., Reddy, C.S., Dhal, N.K., and Das, R. (2006). Some Phytotherapeutic claims by tribals of Rayagada district, Orissa, India. *Ethnobotanical Leaflets,* **10**: 189-197.

Pattanaik, C., Reddy, S., and Reddy, K.N. (2009). Ethno-medicinal survey of threatened plants in Eastern Ghats, India. *Our Nature,* **7**: 122-128.

PDRHealth.com. *Rauvolfia*. Available at: www.pdrhealth.com.

Poonam, Agrawal S., and Mishra, S. (2013). Physiological, biochemical and modern biotechnological approach to improvement of *Rauvolfia serpentine. Journal of Pharmacy and Biological Science,* **6(2)**: 73-78.

Prakasha, H.M., Krishnappa, M., Krishnamurthy, Y.L., and Poornima, S.V. (2010). Folk medicine of NR Pura taluk in Chikmagalur district of Karnataka. *Indian Journal of Traditional Knowledge,* **9(1)**: 55-60.

Prusoff, W.H. (1961). Effect of reserpine on the 5-hydroxytryptamine and adenosinetriphosphate of the dog intestinal mucosa. *BritishJournal of Pharmacology,* **17**: 87-91.

Pullaiah, J. (2002). Medicinal plants in India, New Delhi, Regency Publ, 2: 441-443.

Qureshi, S.A., and Udani, S.K. (2009). Hypolipidaemic activity of *Rauvolfia serpentina* Benth. *Pakistan Journal of Nutrition,* **8(7)**: 1103-1106.

Rahman, A.H.M.M., Anisuzzaman, M., Haider, S.A., Ahmed, F., Islam, A.K.M.R., and Naderuzzaman, A.T.M. (2008). Study of medicinal plants in the graveyards of Rajshahi city. *Res. J. Agric. Biol. Sci.,* **4(1)**: 70-74.

Rahman, M.A., Uddin, S.B., and Wilcock, C.C. (2007). Medicinal plants used by Chakma Tribe in Hill Tracts districts of Bangladesh. *Ind. J. Trad. Knowl.,* **6(3)**: 508-517.

Rahmatullah, M., Jahan, R., Azad, A.K., Seraj, S., Rahman, M.M., Chowdhury, A.R., Begum, R., Nasrin, D., Khatun, Z., Hossain, M.S., Khatun, M.A., and Miajee, Z.U.M.E. (2010b). Medicinal plants used by folk medicinal practitioners in three villages of Natore and Rajshahi districts, Bangladesh. *Am. Eurasian J. Sustain. Agric.,* **4(2)**: 211-218.

Rahmatullah, M., Mollik, M.A.H., Harun-or-Rashid, M., Tanzin, R., Ghosh, K.C., Rahman, H., Alam, J., Faruque, M.O., Hasan, M.M., Jahan, R., and Khatun, M.A. (2010a). A comparative analysis of medicinal plants used by folk medicinal healers in villages adjoining the Ghaghot, Bangali and Padma rivers of Bangladesh. *Am.Eurasian J. Sustain. Agric.,* **4(1)**: 70-85.

Rai, S.K. (2004). Medicinal plants used by Meche people of Jhapa district, Eastern Nepal. *Our Nature,* **2**: 27-32.

Rajendran, S.M., and Agarwal, S.C. (2007). Medicinal plants conservation through sacred forests by ethnic tribals of Virudhunagar district, Tamil Nadu. *Indian Journal of Traditional Knowledge,* **6(2)**: 328-333.

Rajith, N.P., and Ramachandran, V.S. (2010). Ethnomedicines of Kurichayas, Kannur district, Western Ghats, Kerala. *Ind. J. Nat. Prod. Resour.*, 1**(2)**: 249-253.

Rao, V.L.N., Busi, B.R., Rao, C.S., Bharathi, K., and Venkiah, M. (2010). Ethnomedical study among Savaras of Srikakulam district, Andhra Pradesh. *Ind. J. Trad. Knowledge,* **9(1)**: 166-168.

Rathi, P., Kumari, R., Chatrasal, S., and Rajput, Sawhney S.S. (2013). Therapeutic characteristics of *Rauvolfia serpentine. International Journal of Pharmacy and Pharmaceutical Sciences,* **2(2)**: 1038-1042.

Robber, J.M., and Tyler, V.S. (1996). Pharmacognosy, Pharmacobiotechnology, William and Wilkins, Baltimore, 1-14.

Rolf, S., Bruns, H.J., Wichter, T., Kirchhof, P., Ribbing, M., Wasmer, K., Paul. M., Breithardt, G., Haverkamp, W., and Eckardt, L. (2003). The ajmaline challenge in Brugada syndrome: diagnostic impact, safety, and recommended protocol. *European Heart Journal,* **24(12)**: 1104-1112.

Rosenren, A.H., Jokubka, R., Tojjar, D., Granhall, C., Hansson, O., Li, D.Q., Nagaraj, V., Reinbothe, T.M. (2009). Overexpression of alpha2A-adrenergic receptors contributes to type 2 diabetes. *Science,* 217-220.

Rout, S.D., and Thatoi, H.N. (2009). Ethnomedicinal practices of Kol tribes in Similipal biosphere reserve, Orissa, India. *Ethnobotanical Leaflets,* **13**: 379-387.

Roy, S., Zashim Uddin, M., Hassan, M.A., and Rahman, M.M. (2008). Medicobotanical report on the Chakma community of Bangladesh. *Bangladesh. J. Plant Taxon.,* **15(1)**: 67-72.

Ruppert, M., and Stockigt, J. (2005b). Alkaloid biosynthesis in Rauwolfia-CDNA Cloning of major enzymes of ajmaline Pathway. *Curr. Organic Chem.,* **9**: 1421-1444.

Salah, N., Miller, N.J., Pagangeg, G., Tijburg, L., Bolwellg, P., Rice, E., and Evans, C. (1995). Polyphenolic flavonoils as seavenger of aqueous phase radicals as chain breaking antioxidant. *Archives of Biochemistry and Biophysics,* **2**: 339-346.

Sankaranarayanan, S., Bama, P., and Ramachandran, J. (2010). Ethnobotanical study of medicinal plants used by traditional users in Villupuram district of Tamil Nadu. *Indian Journal of Medicinal Plants,* **4(12)**: 1089-1101.

Scalbert, A., Manach, C., Morand, C., Remesy, C., and Jimenez, L. (2005). Dietary polyphenols and the prevention of diseases. *Critical Reviews in Food Science and Nutrition,* **45**: 287-306.

Schlittler, E., Saner, H., and Muller, J.M. (1954). Reserpinin, ei neues alkaloid aus *Rauvolfia serpentina. Experientia,* **10**: 109-133.

Schuchert, A., and Meinertz, T. (2000). Acute eects of a class IA antiarrhythmic drug on the ventricular evoked response amplitude in patients with cardiac pacemakers. *Cardiology*, **94**: 103–105.

Sen, P., Dollo, M., Duttachoudhury, M., and Choudhury, D. (2008). Documentation of traditional herbal knowledge of *Khamptis* of Arunachal Pradesh. *Ind. J. Trad. Knowl.*, **7(3)**: 438-442.

Shamsi, Y., Kumar, H., Tamanna, S.A., and Khan, E.A. (2006). Effect of a polyherbal Unani formulation on chronic urticaria. *Ind. J. Trad. Knowl.*, **5(2)**: 279-283.

Shiddamallayya, N., Yasmeen, A., and Gopakumar, K. (2010). Hundred common forest medicinal plants of Karnataka in primary healthcare. *Ind. J. Trad. Knowl.*, **9(1)**: 90-95.

Siddiqui, S., Ahmad, S.S., Haider, S.I., and Siddiqui, B.S. (1985). Isolation and structure of a new alkaloid from the roots of *Rauvolfia serpentina* Benth. *Heterocycles*, **(3)**: 617-622.

Silja, V.P., Varma, K.S., and Mohanan, K.V. (2008). Ethnomedical plant knowledge of the Mullu kuruma tribe of Wayanad district of Kerala. *Ind. J. Trad. Knowl.*, **7(4)**: 604-612.

Singh, H. (2008). Importance of local names of some useful plants in ethnobotanical study. *Ind. J. Trad. Knowl.*, **7(2)**: 365-370.

Singh, P., Singh, A., Shukla. A.K., Singh, L., Pande, V., and Nailwal, T.K. (2009). Somatic embryogenesis and in vitro regeneration of an endangered medicinal plant sarpgandha (*Rauvolfia serpentina*. L). *Life Science Journal*, **6(3)**: 74-79.

Singh, P.K., Kumar, V., Tiwari, R.K., Sharma, A., Rao, C.V., and Singh, R.H. (2010). Medico-Ethnobotany of 'Chatara' Block of District Sonebhadra, Uttar Pradesh, India. *Advances in Biological Research*, **4(1)**: 65-80.

Singh, R., and Sawhney, S.K. (1988). Advances in frontier areas of Plant Biochemistry, Prentice Hall in India Private Ltd, New Delhi, 487.

Sodipo, O.A., Akiniy, J.A., and Ogunbamosu, J.U. (2000). Studies on certain characteristics of extracts of bark of *Pansinystalia macruceras* (K. schemp) pierre Exbeille. *Global Journal of Pure Applied Science*, **6**: 83-87.

Srivastava, S.K., and Pandey, H. (2006). Traditional knowledge for Agroecosystem management. *Ind. J. Trad. Knowl.*, **5(1)**: 122-131.

Stanford, J.L., Martin, E.J., Brintin, L.A., and Hoover, R.N. (1986). Rauvolfia use and cancer: A case-control study. *Journal of the National Cancer Institute*, **76**: 817-822.

Tiwari, S. (2008). Plant: a rich source of herbal medicine. *Journal of Natural Products*, **1**: 27-35.

Tomar, A. (2009). Folk medicinal uses of plant roots from Meerut district, Uttar Prdesh. *Ind. J. Trad. Knowl.*, **8(2)**: 298-301.

Udayan, P.S., Tushar, K.V., George, S., and Balachandran, I. (2007). Ethnomedicinal information from Kattunayakas tribes of Mudumalai Wildlife Sanctuary, Nilgiris district, Tamil Nadu. *Ind. J. Trad. Knowl.*, **6(4)**: 574-578.

Vaidya, A.D.B (2006). Reverse pharmacological correlates of Ayurvedic drug actions. *Ind. J. Pharmacol.*, **38**: 311-315.

Vakil, R.J. (1955). *Rauvolfia serpentina* in the treatment of high blood pressure. *Circulation*, **12**: 220-229.

Varchi, G., Battaglia, A., Samori, C., Baldelli, E., Danieli, B., Fontana, G., Guerrini, A., and Bombardelli, E. (2005). Synthesis of deserpidine from reserpine. *Journal of Natural Products*, **68**: 1629-1631.

Von-Poser, G., Andrade, H.H., Da-Silva, K.V., Henriques, A.T., and Henriques, J.A. (1990). Genotoxic, mutagenic and recombinogenic effects of *Rauvolfia* alkaloids. *Mutation Research Journal*, **232**: 37-43.

Weiss, R.F., and Fintelmann, V. (2000). Herbal medicine, 2nd ed. Thieme, Stuttgart, 229-230, 387-416.

Yesodharan, K., and Sujana, K.A. (2007). Ethnomedicinal knowledge among Malamalasar tribe of Parambikulam wildlife sanctuary, Kerala. *Ind. J. Trad. Knowl.*, **6(3)**: 481-485.

Yu, J., Yan, M., Jeanne, D., and Qi-Chen. (2013). Antitumor activities of *Rauvolfia vomitoria* extract and potentiation of carboplatin effects against ovarian cancer. *Current Therapeutic Research*,**75**: 8-14.

Yusuf, M., Wahab, M.A., Chowdhury, J.A., and Begum, J. (2006). Ethno-medicobotanical knowledge from Kaukhali proper and Betbunia of Rangamati district. *Bangladesh J. Plant Taxon.*, **13(1)**: 55-61.

Natural Products: Research Reviews Vol. 3 (2016) *Pages* **327–349**
Editor: **V.K. Gupta**
Published by: **DAYA PUBLISHING HOUSE, NEW DELHI**

11

Plant Remedies: How they Can Overcome *Candida* Infections

Valeriya Gyurkovska[1], Petya Dimitrova[1] and Nina Ivanovska[1*]

ABSTRACT

The use of natural products with therapeutic properties is as old as human civilization and for a long time plant products were the main source of drugs. Today, plants contribute approximately 25 per cent of currently used crude drugs and another 25 per cent is derived from chemically altered natural products. The plant based remedies can be used to limit or cure infections as well as to function synergistically with current therapies. Increased antibiotic resistance has become a global medical problem and together with the microbial persistence it raised the need to develop novel microbial agents active against drug resistant microbes and capable to kill persistent microorganisms. The high cost of important conventional therapeutics and/or inaccessibility to adequate health care facilities has led to reliance on traditional medicine since it is affordable and available to people. Infections associated with bacterial and fungal pathogens are among some of the indications treated using traditional remedies. Many studies are carried out on the plants in order to isolate, identify, characterize, and elucidate the structure of the bioactive compounds. Also, toxicological studies should be carried out. This review is trying to summarize data on the use of plants and their constituents in Candida infections based on random selection or on ethnodirected approach.

Keywords: Active constituents, *Candida* infections, Medicinal plants, Plant extracts.

Introduction

The plant kingdom provides an enormous source of medicinal plants initially used in their crude forms as herbal teas, syrups, infusions, ointments and powders.

---

1 Department of Immunology, Institute of Microbiology, 1113 Sofia, Bulgaria.

* *Corresponding author*: E-mail: nina@microbio.bas.bg

The development of chemistry and Western medicine lead to the isolation of active compounds from many species and in some cases duplicated in the form of synthetic drugs. Herbal-derived substances remain the basis for production of many commercial medications used today for the treatment of heart disease, inflammation, pain, allergy and other illnesses. They belong to four general basic systems: Traditional Chinese herbalism, Ayurveda, Western herbalism, which originally came from Greece and Rome and then spread to Europe and America, and Arab traditional medicine. The Indian sub-continent is considered to be one of the biggest sources of traditional medicinal practices in the whole world. Homeopathic, Ayurvedic, Unani, and the Siddha systems of medicine are well recognized with their clearly-defined formulations and selection of medicinal plants. Overall, alternative medicine in India describes the use of more of 7500 plant species (Mukherjee *et al.*, 2006).

There is a strong belief that plant remedies are safer and also less damaging to the human body than synthetic drugs. Therefore, many laboratories all over the world are doing wide screening of plants for biological activities with therapeutic potential. One of the major criterions for such plant selection is traditional healer's information for their therapeutic application. Before being used as a remedy any plant herb or ingredients isolated from plants should be tested combining various chemical, cell biological, biochemical, molecular biological, *in vitro* cell culture techniques and *in vivo* models.

Fungal diseases have emerged as important public health problems owing to an ever-expanding population of immuno-compromised patients (Low *et al.*, 2011; Miceli *et al.*, 2011). The antifungal active principles are different and limited classes of antifungal agents are currently available to treat yeast infections due to the high toxicity of many of them (Spampinato *et al.*, 2013). These concern poleynes, azoles and echinocadins. Strains less susceptible to the available antifungals are spreading, complicating therapeutic approaches and medical management of the infected patients. (White *et al.*, 2002). Alternative microbiological control measures that are suitable for use in humans are of the utmost importance. The need for new antimicrobial substances reinforces the growing investigation of the therapeutic potential of medicinal plants as a source of compounds with antifungal properties (Benko-Iseppon *et al.*, 2010; Dionisi *et al.*, 2012).

Brief Pathophysiology of *Candida* Infections

Candida is rounded and oval-shaped yeast (3-30 μm in diameter) that reproduces asexually through a budding process. The new daughter cells occasionally do not detach and form chains of cells called pseudohyphae. The hyphae are composed of a row of elongated cells enveloped by a cell wall (mycelium). In solid culture media, the yeast grows, forming compact colonies that are macroscopically visible after 24-48 hours of incubation. Over time nutritional and environmental variations direct its conversion to the mycelial or invasive form which is able to induce macrophage phagocytic action (Hermann *et al.*, 2001).

Candida albicans (*C. albicans*) lives in both intestinal and oral areas as a component of the normal flora. Under special circumstances the yeast causes an acute or chronic, superficial or deep infection with a very wide clinical spectrum (Figure 11.1). *Candida*

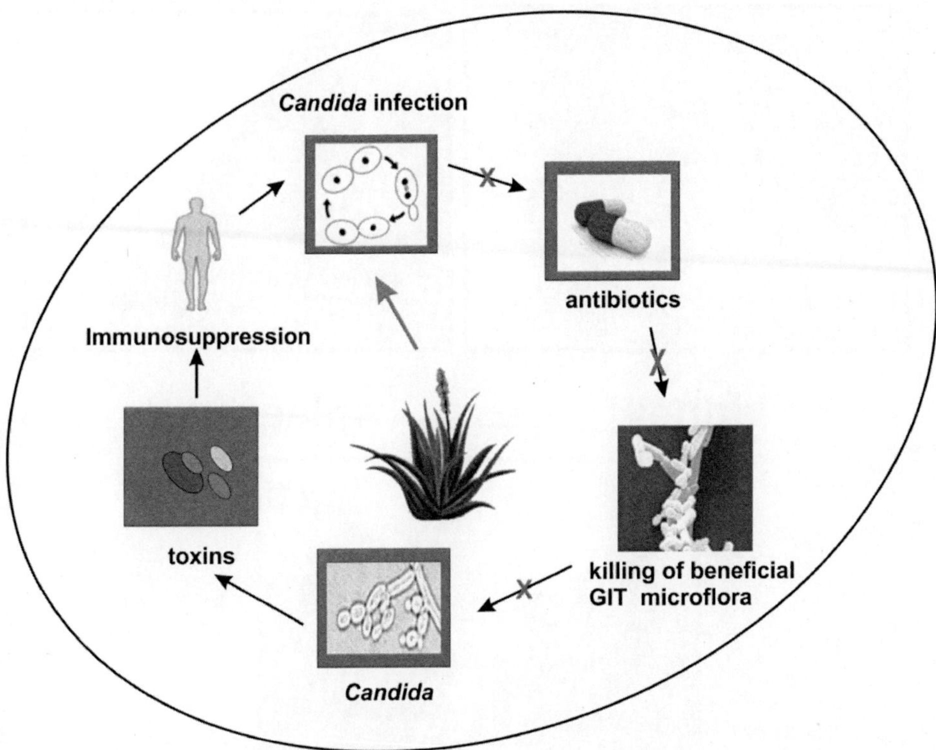

Figure 11.1: Plant Remedies can Break the Vicious Circle of *Candida* infection. *Candida* species produce substances called endotoxins, which are released after the cells break-up. They all have one thing in common: they are dangerous to the host organism. After the treatment, cells of the fungus, which have not been destroyed by the medication or poorly functioning immune system, circulate in the blood for a long period and during that time have the opportunity to cause the re-infection. In addition, antifungal drugs kill the friendly bacteria in gastrointestinal tract (GIT).

species are the most frequent cause of mucosal and invasive fungal infections (Viudes *et al.,* 2002). *C. albicans* is the leading cause of candidiasis. The yeast resides as a commensal of the human mucosae and the gastrointestinal tract. There are two different forms of the fungus. The first is an easily treatable yeast infection, while the second is a hyphal form which is very hard to eradicate. When the fungus moves from an yeast infection to the hyphal stage of growth, filament-like combinations of cells appear. They can more powerfully spread into the different organs of the body (Figure 11.2). In systemic infection, the pathogen can be detected in the tissues forming abscesses that consist inflammatory cells together with yeast and hyphal forms of the organism (Papadimitriou *et al.,* 1986). The brain and kidneys are common targets of disseminated *Candida* infections in humans (Parker *et al.,* 1976), and these organs are also major foci of infection in mice increased use of broad-spectrum antimicrobials and immunosuppressive agents, aggressive chemotherapy and other medical interventions generate patients, highly susceptible to infections caused by

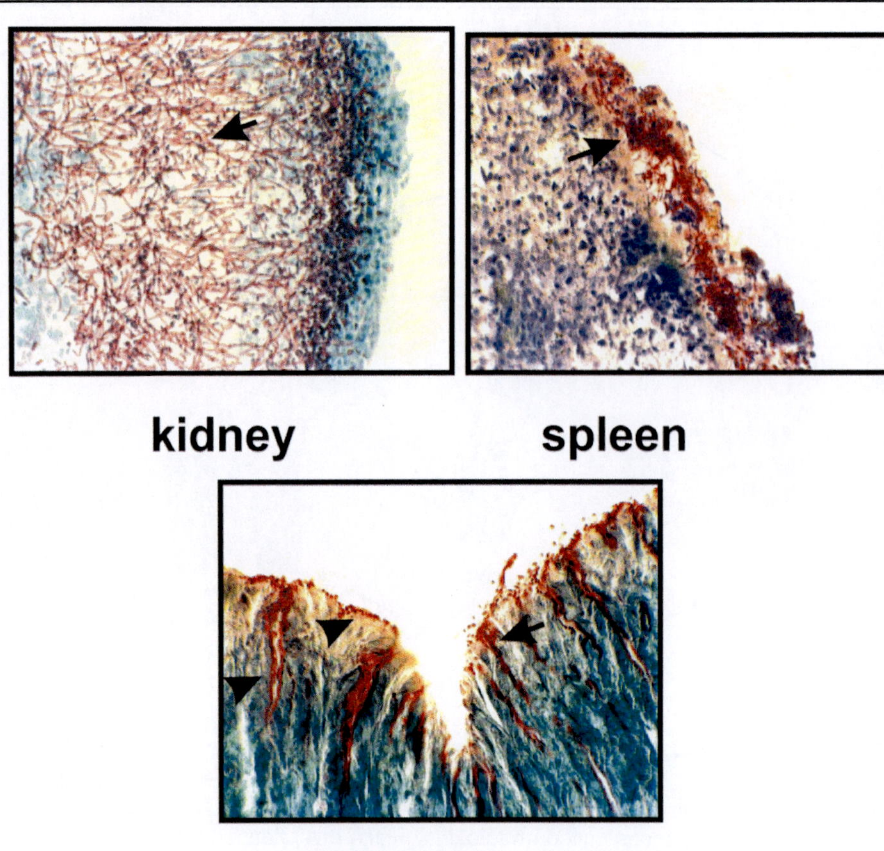

Figure 11.2: Organ Colonization in Experimental Mouse Model of *Candida albicans* Infections. Cells, which are embedded in the internal organs, germinate to form mycelium, which spreads to the whole body in a short period of time. Arrows indicate *Candida* accumulation shown by Schiff staining.

opportunistic organisms such as fungi, often rapidly progressive, and difficult to diagnose or treat (Fidel, 2002; Garber, 2001). Diseases associated with *C. albicans* infection occur as systemic (invasive or disseminated), and mucosal candidiasis, which most often affects the oropharyngeal (oral), oesophageal and vaginal mucosal barriers. These infections appear to be significant clinical and economical problems. Systemic infection is life-threatening and occurs mostly in patients with reduced immune systems like those suffering from diabetes mellitus, people who have had organ transplants, those receiving antibiotic treatment and those infected with human immunodeficiency virus (HIV) 1 or HIV 2 (Pugliese *et al.*, 2000; Sims *et al.*, 2005). Oropharyngeal candidiasis affects up to 90 per cent of HIV-positive individuals and is associated with high morbidity (Samaranayake *et al.*, 2002). Vaginal candidiasis has been estimated to occur in approximately 75 per cent of women at least once (Pfaller *et al.*, 2007). Mucosal colonisation can develop to systemic disease (Martino *et al.*, 1994), where the mortality rate may be as high as 60 per cent (Sims *et al.*, 2005). Among *Candida* species, with a significant prevalence appears to be *C. albicans* but with a growing prominence of non-*albicans Candida* species, such as *C. glabrata*, *C. krusei*, *C. parapsilosis*, *C. tropicalis* (Pfaller Diekema, 2007; Richardson, 2005). The high morbidity and mortality due to haematogenous disseminated or invasive candidiasis is a consequence of lack of early and accurate diagnostic methods, limited arsenal of antifungal drugs and the emergence of resistant strains.

Screening of Plant Extracts, Fractions and Pure Constituents for Anti-*candida* Activity

Plant Extracts

About 80 per cent of individuals from developed countries use traditional medicine, which has compounds derived from medicinal plants (Ellof, 1998). Therefore, such plants should be investigated to better understand their properties, safety and efficiency. The agar diffusion and micro-dilution methods are most used for screening purposes. When comparing the minimal inhibitory concentrations (MICs) it should be considered that the inocolum dose is not always mentioned by the authors and also reference drug is often missed. Here we have summarized some studies, including mainly a large number of plants. Thailand is a region known with its great diversity of medicinal plants. Fifty two species used in the folk medicine were screened *in vitro* for antioxidant, lipase inhibitory and antimicrobial activities (Kaewpiboon *et al.*, 2012). Only two extracts, from *Sonneratia alba* Sm. and *Sonneratia caseolaris* L. possessed anti-*C. albicans* activity (Al-Bayati *et al.*, 2008). Organic and aqueous extracts from fruits, leaves and roots of *Tribulus terrestris* L. demonstrated strong antifungal activity against *C. albicans* with a MIC value of 0.15 mg/ml as well as chloroform, methanol and aqueous extracts of *Ficus deltoidea* Jack (Mas Cotek) at 10 mg/ml, 20 mg/ml and 50 mg/ml, respectively (Abdsamah *et al.*, 2012).

In support that synergism between different compounds resulted in stronger effect than the extracts or fractions are the data on anti-*Candida* effect of *Eriosema robustum* Baker. The ethanol extract showed higher effect than the fractions and the isolated eight compounds (Awouafack *et al.*, 2013). Extracts of the stem bark of *Terminalia spinosa* Engl., a traditional medicine of Kenya were investigated for antifungal activity with the lowest MIC concentration of 62.5 μg/ml (Fabry *et al.*, 1996), while the methanol extracts of *Jatropha elliptica* (Pohl) Oeken (roots), *Physalis angulata* L. (leaves), and *Tephrosia cinerea* (L.) Pers. (roots) demonstrated moderate antifungal properties at MIC >1000 mg/ml (Faria *et al.*, 2006). Different extracts with pronounced anti-candidal effect *in vitro* was found for dichloromethane and methanol extracts of *Punica granatum* L., *Syzygium cumini* L. and *Rosmarinus officinalis* L., dichloromethane extract from *Arrabidaea chica* Verlot and methanol extract of *Mentha piperita* and *Tabebuia avellanedae* Lorentz ex Griseb with MIC varying from 60 to 1 μg/ml (Hofling *et al.*, 2010). Inhibitory effects of *P. granatum* were also observed by other authors (Duraipandiyan *et al.*, 2006) and of *S. cumini* and *R. officinalis* (Portillo *et al.*, 2001). The methanol extract of *Vernonia cinerea* Less (Asteraceae) was active against *C. albicans* (MIC value of 1.56 mg/ml) (Latha *et al.*, 2011). According to data about the use of Indian traditional medicines ethanol extracts from 78 plants were tested in agar diffusion method against *C. albicans* at a concentration of 50 mg/ml. Only 13 per cent of the plant extracts showed inhibitory activity (Valsaraj *et al.*, 1997). Fourteen Jordanian plants (51 extracts) were tested for antimicrobial activity. Butanol extract from the aerial parts of *Narcissus tazetta* and aqueous extract of *Rosa damascena* Mill were moderately active against *C. albicans* (MIC of 125 µg/ml) (Talib *et al.*, 2010). *Andrographis paniculata* (Burm. f.) Nees (Acanthaceae) is prescribed for the treatment of skin infections in India, China, and Malaysia by folk medicine practitioners. Its extract was active against *C. albicans* and *C. tropicalis* at MIC value of 250 µg/ml (Sule *et al.*, 2012).

Essential Oils

Among the different groups of plant products, essential oils are especially recommended as one of the most promising groups of natural sources for the formulation of safer antifungal agents (Varma *et al.*, 2001). These products are known by their active substances, for example, the phenolic compounds and tannins which are part of the essential oils (Saxena *et al.*, 1994). They have been applied in natural remedies, perfumes and make-up products, in dentistry, in agriculture, as food preservatives, and as additives (Bakkali *et al.*, 2008). The most potent upside of essential oils is that they do not cause resistance like drugs do. Generally, whole essential oils have greater antifungal activity due to a synergistic effect with some active components, thus they are more promising in commercial application than single compounds.

A large study on anti-*Candida albicans* activity was conducted with ethanol extracts and essential oils from the leaves and/or roots of 35 medicinal Brazilian plants (Duarte *et al.*, 2005). Essential oils from 13 plants showed anti-fungal activity, including *Aloysia triphylla* (L'Her.) Britton, *Anthemis nobilis* L., *Cymbopogon martinii* (Roxb.) Wats., *Cymbopogon winterianus* Jowitt ex Bor, *Cyperus articulatus* L., *Cyperus rotundus* Linn., *Lippia alba* (Miller) N.E.Brown, *Mentha arvensis* L., *Mikania glomerata* Spreng, *Mentha piperita* L., *Stachys byzantina* C. Koch., and *Solidago chilensis* Meyen, while ethanol extract was not effective at any of the concentrations tested. Chemical analyses showed the presence of compounds with known antimicrobial activity, including 1,8-cineole, geranial, germacrene-D, limonene, linalool, and menthol. Essential oils from five Iranian medicinal plants *Trachyspermum copticum* (L.) Link, *Zataria multiflora* Boiss, *Nigella sativa* L., *Ziziphora clinopodiodes* Boiss. and *Heracleum persicum* Desf. expressed anti-*Candida zeylanoides* activity (Shokri). As a result of an ethnopharmacological survey on a number of Caribbean plants, the anti-*Candida albicans* effect was established for the leaf essential oil of *Gossypium barbadense* (Linn.). It is claimed that α-pinene, which is a terpene that comprises 12.8 per cent of the oil is responsible for its antimicrobial properties (Luciano-Montalvo *et al.*, 2013). Essential oils of *Eucalyptus sideroxylon* A.Cunn. ex Wools and *Eucalyptus torquata* Luehm exhibited moderate to high antifungal activities against *C. albicans* (Ashour, 2008). Ten crude extracts from edible plants were investigated for antifungal activity. The phenol crude extract group exhibited the highest effect (Suwanmanee *et al.*, 2014). Among them, most prominent action showed leaf extracts from *Psidium guajava* L. (active constituents essential oils, flavonoids and triterpenoids), *Piper betel* L. (active constituents polyphenols, alkaloids, steroids, saponins, and hydrolysable tannins) and *Schefflera leucantha* R. Vig crude (active constituents was a mixture of saponins). Essential oils from *Abies holophylla* Fr. Schmidt and *Abies koreana* Wilson showed high antifungal activity against *C. glabrata* (MIC values in the range of 0.5-2.2 mg/ml) (Lee *et al.*, 2009a). Essential oils from *Physalis angulata* L. were active against *C. albicans, C. stellatoidea* and *C. torulopsis* (Osho *et al.*, 2010).

Chemical Groups and Individual Constituents Supposed to be Associated with Anti-Candida Activity

Five medicinal plant species from Cameroon traditional medicine that were found to contain alkaloids, flavonoids, polyphenols, triterpens, sterols, tannins, coumarins, glycosides and reducing sugars expressed anti-*C. albicans* effect in micro dilution method (Assob *et al.*, 2011). More detailed study was conducted on root extracts of *Morella serata* (Lam.) showing that its anti-*Candida* effect was due to the presence of tannins, saponins, flavonoids, terpenoids and steroids (Ashafa, 2013) and to flavonoids of *Lantana camara* L. (Sharma *et al.*, 2009). Hydroalcohol extracts of *Cassia fistula* Linn. showed significant antifungal activities (Bhalodia *et al.*, 2011). Leaves and roots ethanolic extracts from *Cajanus cajan* (L.) Millsp. demonstrated *in vitro* inhibitory effect against *C. albicans, C. krusei* and *C. tropicalis* (Brito). Plumbagin a major naphthoquinone component isolated from roots of *Plumbago scandens* L. showed anti-*C. albicans* activity at a concentration of 0.78 µg/ml (de Paiva *et al.*, 2003). A large group of 124 extracts of Egyptian plants were evaluated for antibacterial and anti-*C. albicans* activity. The results showed that *Lagerstroemia indica* L. leaf extract was effective against all of them and chemical analyses proved that '4-methoxy apigenin-8-C-beta-D-glucopyranoside expressed minimum lethal concentration (MLC) of 32 µg/ml against *C. albicans* (Diab *et al.*, 2012). Aqueous *Allium sativum* (garlic) extract inhibited *C. albicans* growth with MIC of 0.8 and 1.6 mg/ml (Ghannoum, 1988). The antifungal activity of six fractions derived from garlic decreased the growth of *C. albicans* the most active compounds being ajoene and allicin at less than 20 µg/ml (Yoshida *et al.*, 1987) (Table 11.1). Antifungal activity of 2 benzophenanthridine alkaloids 8-hydroxydihydrosanguinarine and 8-hydroxydihydrochelerythrine from *Chelidonium majus* Linn against clinical yeast isolates (Meng *et al.*, 2009) and lupane triterpenes from *Acacia mellifera* extracts (Mutai *et al.*, 2009) was observed. A flavonoidd isolated from the flower extract of *Retama raetam* Forssk. Webb identified as derrone possessed eminent growth inhibition activity against *C. albicans, C. krusei, C. parapsilosis* and *C. glabrata* at a concentration of 7.81 µg/ml (Edziri *et al.*, 2012). A group of Brazilian plants selected according to their use in traditional medicine as antileishmanial remedies were analyzed *in vitro* against fluconazole-resistant yeast *C. krusei*, azole-susceptible *C. parapsilosis and C. albicans.* Among the most effective tested extracts, *Plectranthus neochilus* Schlechter demonstrated strong activity against *C. kruse* and *Albizia inundata* (Mart.) Barneby and *Calamintha adscendens* Willk. and Lang., showed an antifungal activity against *C. albicans. Hydrocotyle bonariensis* Lam. presented a marked antifungal activity (EC_{50} of 31.63 µg/ml) against the fluconazole-resistant strain *C. krusei* (Tempone *et al.*, 2008). A huge survey including 327 plant species (92 families and 251 genera) from seven Latin American countries compared were screened for antifungal activity (Svetaz *et al.*, 2010). The authors tried to compare

Table 11.1: Some Active Ingredients Inhibiting *C. albicans* Growth Assayed by Agar Diffusion and Micro-dilution Tests

Active Ingredient	*Plant Species*	*Part Used*	*Concentration*	*Reference*
Derrone	*Retama raetam* Forssk. Webbb	Flowers	7.81 μg/ml	Edziri *et al.*, 2012
Flavonoids	*Lantana camara* L.	Root	120.5 mg/ml	Sharma *et al.*, 2009
Allicin	*Allium sativum (garlic)*	Whole plant	MIC_{95} 17.3 μg/ml	Yoshida *et al.*, 1987
Ajoene	*Allium sativum (garlic)* L.	Whole plant	MIC_{95} 7.6 μg/ml	Yoshida *et al.*, 1987
Catechins and theaflavins	*Black tea*	Leaves	6.25 mg/ml	Sitheeque *et al.*, 2009
18-beta glycyrrhetinic acid	*Glycyrrhiza glabra* L.	Root	6.2 μg/ml	Pellati *et al.*, 2009
Glabridin	*Glycyrrhiza glabra* L.	Root	31.25-250 μg/ml.	Fatima *et al.*, 2009
Punicalagin	*Punica granatum* L.	Fruit peel	3.9 μg/ml	Endo *et al.*, 2010
Ellagic acid	*Punica granatum* L.	Fruit peel	75 μg/ml	Endo *et al.*, 2012; Bakkiyaraj *et al.*, 2013
Tanins	*Stryphnodendron adstringens* (Barbatimão)	Stem bark	7.80 μg/ml	Ishida *et al.*, 2006
Xanthorrhizol	*Curcuma xanthorrhiza* Roxb	Rhizome	1.0-15.0 mg/l	Rukayadi *et al.*,
Pogostone	*Pogosteman cablin* (Blanco) Benth	Leaves	3.1 μg/ml	Yi *et al.*, 2013
Proanthocyanidins	*Vaccinium macrocarpon Ait.*	Fruit	6.25 μg/ml	Feldman *et al.*, 2012
Quercetin	*Croton menyhartii* Pax	Leaves	< 1 mg/ml	Aderogba *et al.*, 2013
Baicalein	*Scutellaria baicalensis* Georgi	Root	2.0 μg/ml	Fu *et al.*, 2011
Plumbagin	*Plumbago scandens* L.	Root	0.78 μg/ml	de Paiva *et al.*, 2003
Licochalcone A	*Glycyrrhiza glabra* L.	Root	0.2 μg ml	Pellati *et al.*, 2009
Essential oils	*Equisetum arvense* L.	Stems	50 mg/ml	de Oliveira *et al.*, 2013
Spirostatic saponin SC-2	*Solanum chrysotrichum* (Schldl.)	Leaves	200-800 μg/ml	Herrera-Arellano *et al.*, 2012
Berberine	*Coptis rhizoma* Franch and *Phellodendron amurense* Rupr.	Whole plant	250 μg/ml	*Park et al., 1999*

Contd...

Table 11.1–*Contd...*

Active Ingredient	*Plant Species*	*Part Used*	*Concentration*	*Reference*
Palmitine	*Coptis rhizoma* Franch and *Phellodendron amurense* Rupr.	Whole plant	500 µg/ml	*Park et al., 1999*
α-rhamnoisorobin	*Bryophyllum pinnatum* (Lank.) Oken	Whole plant	1 µg/ml	Tatsimo *et al.*, 2012
Afzelin	*Bryophyllum pinnatum* (Lank.) Oken	Whole plant	8 µg/ml	Tatsimo *et al.*, 2012
Kaempferol 3-*O*-α-D-glucopyranoside-7-*O*-α-L-rhamnopyranoside	*Bryophyllum pinnatum* (Lank.) Oken	Whole plant	16 µg/ml	Tatsimo *et al.*, 2012
Casuarinin	*Plinia cauliflora* (Mart.)	Leaves	1-580-µg/ml	Souza-Moreira *et al.*, 2013
Indirubin	*Wrightia tinctoria* R. Br.	Leaves	0.75-25 µg/ml	Ponnusamy *et al.*, 2012010
TTS-12[a]	*Tribulus terrestris* L.	Whole plant	4.4 µg/ml	*Zhang et al.*, 2005
TTS-15[b]	*Tribulus terrestris* L.	Whole plant	9.4 µg/ml	*Zhang et al.*, 2005

[a] TTS-12 tigogenin-3-O-beta-D-xylopyranosyl(1-2)-[beta-D-xylopyranosyl(1-3)]beta-D-glucopyranosyl(1-4)-[alpha-L-rhamnopyranosyl(1-2)]-beta-D-galactopyranoside.

[b] TTS-15 tigogenin-3-O-beta-D-glucopyranosyl(1–>2)-[beta-D-xylopyranosyl(1-3)]-beta-D-glucopyranosyl(1-4)-beta-D-galactopyranoside.

Table 11.2: Targets of Anti-*Candida albicans* Action

Plant	*Active Principle*	*Activity*	*Reference*
Allium sativum L.	Aqueous extract	Lipid synthesis	Adetumbi *et al.*, 1986
Coriaria nepalensis Wall.	Essential oil	Ergosterol biosynthesis and membrane integrity	Ahmad *et al.*, 2011
Fucus spiralis L.	Phlorotannins	Formation of pseudohyphae	Lopes *et al.*, 2013
Aloe vera L.	Leaves extract	Inhibition of germ tube formation	Bernardes *et al.*, 2012
Plinia cauliflora (Mart.)	Causarin	Changes in cell wall architecture	Souza-Moreira *et al.*, 2013
Coptis rhizoma Franch	Berberine and palmitine	Affection of sterol and cell wall biosyntheses	Park *et al.*, 1999
Phellodendron amurense Rupr.	Berberine and palmitine	Affection of sterol and cell wall biosyntheses	Park *et al.*, 1999
Larrea divaricata		Increased phagocytosis	Martino *et al.*, 2012
Tabernanthe iboga Baillon	Root bark extract	Protective effect in cutaneous, systemic and intestinal infections, reduction of organ colonization and inhibition of the extracellular aspartyl proteinase and lipase activities	Yordanov *et al.*, 2005; Yordanov *et al.*, 2008
Glycyrrhiza glabra L.	Liquiritigenin	Influence on CD4+ Th1 immune response (IL-2 and IFN-γ)	Lee *et al.*, 2009b
Swietenia mahogany (Linn.) Jacq.	Methanolic seed extract	Protective effect in a model of systemic infection	Sahgal *et al.*, 2011

a group of plants chosen on the base of their ethnomedical uses with a group selected at random. The authors considered that the probability to predict antifungal effect is higher for the first group meanwhile, this group included species with lower MICs.

Glycyrrhiza glabra L. (licorice), has been used since ancient times as a traditional herbal remedy in south-east Europe and south-west Asia because of its anti-inflammatory and expectorant properties. The root and leave extracts showed strong activity against *C. albicans* (Irani *et al.*, 2010). Glabridin, an active constituent of *G. glabra* roots inhibited the growth of drug resistant strains of *C. albicans* at a MIC of 31-250 μg/ml (Fatima *et al.*, 2009). Glabridin and licochalcone A at a high concentration of 100 μg/ml strongly suppressed hyphal formation and licochalcone A decreased biofilm formation (Messier *et al.*, 2011). Compound 18-beta glycyrrhetinic acid obtained from *G. glabra* root markedly inhibited the *in vitro* growth of *C. albicans* strains by a relatively low dose of 6.2 μg/ml (Pellati *et al.*, 2009).

The polyphenols catechins and theaflavins in black tea showed anti-*Candida* activity against all tested *Candida* species and demonstrated a MIC of 6.25 mg/ml for *C. albicans*. And C. *glabrata* was found to be the most sensitive species followed by *C. parapsilosis*, *C. albicans*, *C. krusei* and *C. tropicalis* (Sitheeque *et al.*, 2009). Leaf extract from *Wrightia tinctoria* R. Br. showed promising activity against *C. albicans*. Also, the major constituent indirubin expressed inhibitory action against the yeast within a MIC range of 0.75-25 μg/ml (Ponnusamy *et al.*, 2010).

An effort to find structure-activity relationships was made by Tasimo *et al.*, (2012) through evaluation of antioxidant and antifungal properties of seven kaempferol rhamnosides from *Bryophyllum pinnatum* (Lank.) Oken. The antimicrobial activity was more concentrated in ethanolic fraction, while the antioxidant activity was more concentrated in methanolic extract. The highest anti-fungal effect was demonstrated by α-rhamnoisorobin (MIC 1 μg/ml) against *C. albicans* and *C. parapsilosis* equal to those of nystatin. Less active were kaempferol 3-*O*-α-D- glucopyranoside-7-*O*-α-L-rhamnopyranoside (MIC 16 μg/ml) and afzelin (MIC 8 μg/ml).

Breaking a *Candida* Biofilm

Candida spp. are commonly identified in dental prostheses and oral candidiasis colonizing the oral mucosa, tongue and palate (Kojic *et al.*, 2004). Pathogen resides inside a protective coat made of extracellular polymeric substances. Biofilm formation appears to be an important virulence factor for *Candida* species because it increases resistance to antifungal therapy (Chandra *et al.*, 2001). Researchers found that the drugs were initially effective against *Candida*, but as the biofilm developed they became less and less effective. Hence, it is optimal to look for new agents that effectively inhibit the formation of biofilm.

Pomegranate is a common food and is also used traditionally to treat various ailments because of its chemical composition and biological properties. The bioactive potential of *Punica granatum* L. is related to its antioxidant, anticancer, antibacterial, and antifungal properties. Crude hydro alcoholic extract prepared from the fruit peel of *P. granatum* expressed strong anti-*Candida* spp. (*C. albicans, C. glabrata* and *C. tropicals*) action *in vitro* at concentrations of 50 mg/ml (Oliveira 2013). The use of

P. granatum extract as an antifungal agent against candidosis associated with denture stomatitis was reported by Vasconcelos *et al.*, (Vasconcelos *et al.*, 2003). Activity-guided analyses yielded a compound named punicalagin, effective against *C. albicans* and *C. parapsilosis* with MICs of 3.9 and 1.9 µg/ml, respectively (Table 11.1). The synergistic activity of pomegranate extract with antibiotics has shown improved efficacies against pathogens which are resistant to antibiotics. MIC for fluconazole decreased twofold when combined with the extract (Endo *et al.*, 2010). Hence, the combination of pomegranate extract with antibiotics could help to overwhelm drug-resistant infections in humans. Apart from inhibiting the formation of biofilm, extract destroyed pre-formed biofilms and inhibited germ tube formation of *C. albicans* (Endo *et al.*, 2012). Ellagic acid is another pure component identified in the methanolic pomegranate extract which inhibited growth of the yeast at a concentration of 75 µg/ml (Bakkiyaraj *et al.*, 2013; Endo *et al.*, 2012) and biofilm formation at lower concentrations of 40µg/ml (Bakkiyaraj *et al.*, 2013). The mode of extract's action might be further focused on cytokine involvement, since available data showed that the production of IL-1β was significantly higher in cultures of macrophage RAW 264.7 cells treated with *it (Endo et al., 2010)*. A special attention should be paid on the use of microparticles containing pomegranate extract by the spray-drying technique, utilizing alginate or chitosan as encapsulating agents. These particles possessed anti-*C. albicans* activity *in vitro* and point on an additional possibility to treat topically fungal infections by a spray form (Endo *et al.*, 2012).

To *P. grantum* might be added another plant extracts perspective for application in oral candidiasis. Seed extract from *Brucea javanica* (L.) Merr. was effective against several oral species such as *C. albicans, C. glabrata C. lusitaniae, C. parapsilosis* and *C. tropicalis* (Nordin *et al.*, 2013). At a concentration of 50 mg/ml, an essential oil from *Equisetum arvense* L. was effective against *C. albicans oral isolates which* can be attributed to the presence of various compounds like the phenolic monoterpene thymol (de Oliveira *et al.*, 2013). The extract of *Stryphnodendron barbatimam* Mart. produced in aqueous and ethyl-acetate fractions showed significant antifungal activity against clinical oral isolates of *C. albicans, C. parapsilosis* and *C. tropicalis*. This effect might be associated with the presence of tannins which destroy the integrity of the cell wall of *Candida leading to* decreased adhesion to host cells and germ-tube formation (de Oliveira *et al.*, 2013). Extracts and essential oils from *Allium tuberosum* Rotter, *Coriandrum sativum* L., *Cymbopogon martinii* (Roxb.) J. F. Watson, *Cymbopogon winterianus* Jowitt ex Bor, and *Santolina chamaecyparissus* L. were capable to inhibit biofilm formation when tested against clinical isolates of *C. albicans* from patients with periodontal disease (Furletti *et al.*, 2011). The best effect was registered for the total oil extract and fractions containing 2-hexen-1-ol, 3-hexen-1-ol and cyclodecane. A-type proanthocyanidins isolated from *Vaccinium macrocarpon* Ait. prevented biofilm formation and inhibited adherence of *C. albicans* to oral epithelial cells (Table 11.1). This effect was due to a decreased production by these cells of IL-8 and IL-6 and also the anti-inflammatory effect was associated with suppression of NF-kappaB p65 activation and phosphorylation of specific signal intracellular kinases (Feldman *et al.*, 2012). Recently, *C. dubliniensis* has been isolated in patients with human immunodeficiency virus (HIV) infection. This species is important, since it is associated with very severe candidiasic mucositis of the oral cavity, esophagus and other locations, and is moreover resistant to antifungal treatment. Leaf aqueous extract from *Azadirachta indica* (A. Juss) influenced the adhesion, cell surface hydrophobicity and biofilm formation, which may reduce the colonisation by *C. albicans* (Polaquini *et al.*, 2006).

Vaginal Candidiasis

Vulvovaginal candidiasis (VVC) is one of the most common vaginal infections. Almost 90 per cent of isolates from vaginal fluid are identified as *C. albicans* (Ishida *et al.*, 2006) while *C. glabrata* represents the most common cause of nonalbicans *Candida vaginitis* (Sobel, 2007). Polyene agents used in VVC are highly nephrotoxic and hepatotoxic. However, with the emergence of azole agents, possessing lower toxic effects, many fungal infections can now be successfully treated. Although, few strains resistant to antifungal agents have been isolated, VVC continues to be a common problem in immunocompetent or healthy women. The discovery of new antifungal agents thus remains an important challenge for the scientific community, and plants may supply new promising antifungal drugs. Pogostone is a natural product isolated from the leaves of *Pogostemon cablin* (Blanco) Benth tested *in vitro* and *in vivo* against *C. albicans* (Yi *et al.*, 2013) (Table 11.1). Importantly, the substance has been equally effective with the reference drug voriconazole with MIC of 3.1μg/ml against fluconazole-resistant strains (Li *et al.*, 2012). Importantly, the metabolic profile of pogostone was evaluated *in vivo* in rats. Furthermore, it reduced the fungal load in vagina of vulvovaginal candidiasis in mouse models when administered through oral and topical route (Li *et al.*, 2012).

Hydroalcoholic and n-BuOH extracts of *Sapindus saponaria* L. was effective against azole-susceptible and azole-resistant human vaginal *Candida* spp. *In vivo*, these extracts eliminated vaginal infection in rats induced with azole-suspectible and azole-resistant *C. albicans* and *C. glabrata* (Damke *et al.*, 2011). Also, in women with acute vulvovaginal infection cream containing *Zataria multiflora* Boiss extract exhibited well expressed therapeutic effect (Khosravi *et al.*, 2008). Crude extract, fractions and subfractions from the bark of *Stryphnodendron adstringens* (Barbatimão) were able to inhibit the growth of *C. albicans* isolated from vaginal fluid and this was associated with reduced phagocytosis of yeast by macrophages (Ishida *et al.*, 2006). Beneficial effect was observed in vaginitis using plants due to its components such as carvacrol, 1,8-cineole, geranial, germacrene-D, limonene, linalool, menthol, terpinen-4-ol and thymol (Azimi *et al.*, 2011; Duarte *et al.*, 2005). One study demonstrated that topical application of saponin identified in methanolic extract of *Solanum chrysotrichum* (Schldl.) in women with cervical or vaginal *C. albicans* infection showed effect equal to that of ketoconazole (Herrera-Arellano *et al.*, 2007). This spirostatic saponin SC-2 expressed fungistatic and fungicidal effect *in vitro* on *C. albicans* and other *Candida* species such as *C. lusitaniae*, *C. glabrata* and *C. krusei*, which are susceptible or resistant to fluconazole (Herrera-Arellano *et al.*, 2013) (Table 11.1). The essential oil of a Moroccan plant *Mentha suaveolens* Ehrh. deserves attention for clinical testing based on the results in experimental model of *C. albicans* vaginal infection (Pietrella *et al.*, 2011).

Animal Experimental Models and Human Clinical Trials

One recent study investigated *in vitro* and *in vivo* the effect of methanolic extracts of fruit and bark of *Ficus glomerata* Roxb (Heroor *et al.*, 2013). The results in a mouse model of cyclophosphamide-induced immunosuppression showed that these extracts to a great extent ameliorated the impairment in blood parameters. Moreover, they enhanced the phagocytic capability of human neutrophils when tested *in vitro* against *C. albicans.* Oral administration of water extract fractions obtained from a constituent herb of Juzen-taiho-to (TJ-48, Si-Quan-Da-Bu-Tang), rhizomes of *Atractylodes lancea* (Thunb.) DC. expressed protective effect in a lethal *C. albicans* infection at a dose of 140 mg/kg/day (Inagaki *et al.*, 2001). The methanolic extract did not possess such effect. A fraction containing acidic pectic polysaccharides represented mainly by galacturonic acid was responsible for its antiinfectious properties.

Very promising results were received with seed extracts from *Cassia fistula* Linn. Full inhibition of *C. albicans* growth was achieved at a concentration of 6.25 mg/ml (Jothy *et al.*, 2012). Further investigations proved that the i.p. treatment of mice at a dose of 2.5 g/kg body weight significantly reduced the recovery of *C. albicans* in kidney and blood in systemic infection. *Swietenia mahogani* (Linn.) Jacq. methanolic seed extract was investigated *in vitro* for the antifungal activity against *C. albicans* with MIC value of 12.5 mg/ml. The *in vivo* antifungal activity of seed extract was evaluated in intravenous *C. albicans* infection in mice. Significant reduction of yeast recovery in kidney and blood was observed (Sahgal *et al.*, 2011). Liquiritigenin isolated from licorice (the root of *Glycyrrhizae* plant) protects mice against disseminated *C. albicans* in mice by the influence on CD4+ Th1 immune response. The compound enhanced IL-2 and IFN-γ production (Lee *et al.*, 2009b).

A study on cutaneous candidiasis included patients between the ages of 3 months to 58 years treated with various plant extracts, free and bound flavonoids of *Piper betle* Linn. A great part of the patients were cured completely (64 per cent), 24 per cent showed significant improvement and 12 per cent showed little improvement from the disease after allopathic treatment (Bhadauria *et al.*, 2012).

Supposed Mechanisms of Anti-*candida* Action

The number of antibacterial agents on the market is large in comparison to the number of antifungal drugs available. The limited number of treatment options, coupled with the increasing frequency of cross-resistance, makes it necessary to develop new therapeutic strategies, but it is often difficult to identify a mechanism of action. The three major groups of antifungal agents in clinical use, azoles, polyenes and allylamine/thiocarbamates, all of them are claimed to derive their activity from interaction with the biosynthesis of ergosterol, the key sterol in most pathogenic fungi. In respect to that, identifying specific mechanisms of anti-*Candida* action is of great importance, although a limited number of studies report such data. Some works report on the affection of cell wall structure by plant extracts. The effect of aqueous garlic extract inhibited *C. albicans* growth *in vitro* (MIC of 6.25 mg/ml) was associated with an influence on lipid synthesis (Adetumbi *et al.*, 1986). Antifungal activity of *Coriaria nepalensis* Wall. essential oil is realized by disrupting ergosterol biosynthesis and membrane integrity of *Candida* spp. isolates (Ahmad *et al.*, 2011). Three brown seaweeds, *Cystoseira nodicaulis* (Withering) M. Roberts, *Cystoseira usneoides* (Linnaeus) M. Roberts and *Fucus spiralis* Linnaeus were tested for their antifungal activity against human pathogenic *C. albicans*. Phlorotannins from *F. spiralis* inhibited the dimorphic transition of the yeast, resulting in the formation of pseudohyphae with decreased capacity to adhere to epithelial cells (Lopes *et al.*, 2013). Similarly, *Aloe vera* (L.) Burm. f. fresh leaves extract showed that it can inhibit both the growth and the germ tube formation by *C. albicans* (Bernardes *et al.*, 2012). Tannin-rich fraction from *Plinia cauliflora* (Mart.) Kausel leaves induced changes in cell wall architecture of *C. albicans*, since it has the ability to interfere with the outer glycoprotein-rich layer of the yeast. The major component isolated from the most active antifungal fraction was the hydrolysable tannin casuarinin which evidently was responsible for the anti-*Candida* effect (Souza-Moreira *et al.*, 2013) (Table 11.1). Methanol extracts from *Coptis rhizoma* Franch and *Phellodendron amurense* Rupr. inhibited *in vitro* growth of *C. albicans, C. glabrata, C. krusei* and *C. parapsilosis*. This effect was related to protoberberines berberine and palmatine. The most potent inhibition of growth was exhibited by berberine on *C. krusei* (MIC 4 µg/ml) and palmatine on *C. parapsilosis* (MIC 16 µg/ml) through affection of sterol and cell wall biosyntheses (Park *et al.*, 1999).

Except direct action on the pathogen, another possibility to achieve a beneficial effect is through influence on the host. Recognition of β-glucan stimulates the anti-fungal immune response that leads to phagocytosis and killing of the fungus. Fractions isolated from *Larrea divaricata* Cav were injected into mice and than infected with *C. albicans*. Protective effect was observed associated with macrophage activation leading to increased phagocytosis and yeast elimination (Martino *et al.*, 2012). Different leaf extracts, flavonols and indole alkaloids isolated from *Croton menyharthii* Pax possessed anti-*C. albicans* effect *in vitro*. Further, individual compounds were also identified and their inhibition of cyclooxygenase (COX-2), alpha-glucosidase and acetylcholinesterase activities were proven (Aderogba *et al.*, 2013).

Ibogaine is a naturally occurring psychoactive indole alkaloid extracted from the root bark of the African shrub *Tabernanthe iboga* Baillon (Mash *et al.*, 2000). This substance was investigated in different models of mouse *C. albicans* infections and its mechanism of action was studied. Ibogaine enhanced the survival of mice in systemic infection and reduced organ colonization in disseminated and gastrointestinal *C. albicans* infections while the development of a specific immune response was not inuenced. The substance improved the course of cutaneous *C. albicans* infection through the inhibition of the extracellular aspartyl proteinase and lipase activities. The combined use of amphotericin B plus ibogaine in the treatment of mice with gastrointestinal infection reduced organ colonization more strongly than each substance alone (Yordanov *et al.*, 2005; Yordanov *et al.*, 2008).

Combined Treatment of Plant Constituents with Antifungal Drugs

One of the greatest obstacles related to antifungal drug application is their severe side effects. Combined treatment appears a perspective way to reduce the dose and to lesson undesirable effects. Baicalein, a component of *Scutellaria baicalensis* Georgi was investigated against thirty clinical isolates of *C. albicans* in combination with amphotericin B. Baicalein enhanced the effect of amphotericin in regard to *C. albicans* apoptosis and release of reactive oxygen species (ROS). Moreover, this combination resulted in an increase of the caspase activity and expression of the corresponding gene CaMCA1 in *C. albicans* (Fu *et al.*, 2011; Wong *et al.*, 2009). Synergistic anti-Candida action was observed when *Ocimum sanctum* essential oil was combined with fluconazole or ketoconazole (Amber *et al.*, 2010).

Plants against Drug Resistance

The indiscriminate use of antimicrobials by population in general has been causing serious public health problems all over the world due to the fact that the microorganisms have the ability to develop resistance to these therapeutic agents. There are increasing reports of resistance in *Candida* species (White *et al.*, 2002) and an on-going need to develop new antifungal agents that are required to compensate for the increase in resistance to standard antifungal agents. A number of 42 plant species of 29 families used in Iranian folk medicine were screened for anticandidal activity. It was established that 19 plant species representatives to 16 families were effective against clotrimazole-resistant *C. albicans*. The lowest MIC of 0.62 mg/ml was registered for two of them, *Terminalia chebula* Retz. and *Thymus vulgaris* L. (Bonjar, 2004). Two steroid saponins from *Tribulus terrestris* L. sowed significant *in vitro* antifungal activity against fluconazole-resistant *C. albicans* (4.4 and 9.4 μg/ml, respectively) (Zhang *et al.*, 2005). The ethanolic fraction of *Prosopis spicigera* L. was found to strongly inhibit growth of multidrug resistant strains of *C. albicans, C. krusei, C. tropicalis* and *C. glabrata* at a MIC of 4.88 μg/ml (Khan *et al.*, 2010).

Conclusion

The application of plant products for the cure and control of fungal diseases is considered as a possible alternative to synthetic fungicides. It is associated with their lower negative impact, reduced cost, and adverse reactions of plant preparations compared to conventional therapeutics. The need for new antimicrobial substances reinforces the growing investigation of the therapeutic potential of medicinal plants, where some of its compounds with the antimicrobial property such as flavonoids, alkaloids, sterols and saponins have been the subject of interest in the treatment of various human diseases. This brief review show that a great part of investigations present data on screening of total extracts. However, one should note that the phytochemical properties may vary depending on ecological and geographical conditions, age of plant, and time of harvesting. The exact level of active compound concentration in each extract also requires further study, as this information may provide valuable data for large-scale production of crude plant extracts with antifungal properties. In addition, extracts may be more beneficial than isolated constituents, since a bioactive individual component can change its properties in the presence of other compounds present in the extracts. However, the *in vitro* and *in vivo* data might serve as a base: 1) for the development of new bioactive compounds useful for a better therapeutic outcome; 2) to compensate for the increase of resistance to standard antifungal agents; 3) for combination therapies leading to lower dosage and side effects; 4) for elucidation of the mechanisms of action of promising extracts and pure compounds.

References

Abdsamah, O., Zaidi, N.T., and Sule, A.B. (2012). Antimicrobial activity of *Ficus deltoidea* Jack (Mas Cotek). *Pak J Pharm Sci.*, **25**: 675-678.

Aderogba, M.A.,Ndhlala, A.R.,Rengasamy, K.R., and Van Staden, J. (2013). Antimicrobial and selected *in vitro* enzyme inhibitory effects of leaf extracts, flavonols and indole alkaloids isolated from *Croton menyharthii*. *Molecules*, **18**: 12633-12644.

Adetumbi, M.,Javor, G.T., and Lau, B.H. (1986). *Allium sativum* (garlic) inhibits lipid synthesis by *Candida albicans*. *Antimicrob Agents Chemother.*, **30**: 499-501.

Ahmad, A.,Khan, A.,Kumar, P.,Bhatt, R.P., and Manzoor, N. (2011). Antifungal activity of *Coriaria nepalensis* essential oil by disrupting ergosterol biosynthesis and membrane integrity against *Candida*. *Yeast*, **28**: 611-617.

Al-Bayati, F.A., and Al-Mola, H.F. (2008). Antibacterial and antifungal activities of different parts of *Tribulus terrestris* L. growing in Iraq. *J Zhejiang Univ Sci B*, **9**: 154-159.

Amber, K.,Aijaz, A.,Immaculata, X.,Luqman, K.A., and Nikhat, M. (2010). Anticandidal effect of *Ocimum sanctum* essential oil and its synergy with fluconazole and ketoconazole. *Phytomedicine*, **17**: 921-925.

Ashafa, A.O. (2013). Medicinal potential of *Morella serata* (Lam.) Killick (Myricaceae) root extracts: biological and pharmacological activities. *BMC Complement Altern Med.*, **13**: 163.

Ashour, H.M. (2008). Antibacterial, antifungal, and anticancer activities of volatile oils and extracts from stems, leaves, and flowers of *Eucalyptus sideroxylon* and *Eucalyptus torquata*. *Cancer Biol Ther.*, **7**: 399-403.

Assob, J.C., Kamga, H.L., Nsagha, D.S., Njunda, A.L., Nde, P.F. *et al.* (2011). Antimicrobial and toxicological activities of five medicinal plant species from Cameroon traditional medicine. *BMC Complement Altern Med.*, **11**: 70.

Awouafack, M.D., McGaw, L.J., Gottfried, S., Mbouangouere, R., Tane, P. *et al.* (2013). Antimicrobial activity and cytotoxicity of the ethanol extract, fractions and eight compounds isolated from *Eriosema robustum* (Fabaceae). *BMC Complement Altern Med.*, **13**: 289.

Azimi, H.,Fallah-Tafti, M.,Karimi-Darmiyan, M., and Abdollahi, M. A. (2011). Comprehensive review of vaginitis phytotherapy. *Pak J Biol Sci.*, **14**: 960-966.

Bakkali, F.,Averbeck, S.,Averbeck, D., and Idaomar, M. (2008). Biological effects of essential oils–a review. *Food Chem Toxicol.*, **46**: 446-475.

Bakkiyaraj, D.,Nandhini, J.R.,Malathy, B., and Pandian, S.K. (2013). The anti-biofilm potential of pomegranate (*Punica granatum* L.) extract against human bacterial and fungal pathogens. *Biofouling*, **29**: 929-937.

Bernardes, I., Felipe Rodrigues, M.P., Bacelli, G.K., Munin, E., Alves, L.P. *et al.* (2012). *Aloe vera* extract reduces both growth and germ tube formation by *Candida albicans*. *Mycoses*, **55**: 257-261.

Bhadauria, S., and Kumar, P. (2012). Broad spectrum antidermatophytic drug for the control of tinea infection in human beings. *Mycoses*, **55**: 339-343.

Bhalodia, N.R., and Shukla, V.J. (2011). Antibacterial and antifungal activities from leaf extracts of *Cassia fistula* l.: An ethnomedicinal plant. *J Adv Pharm Technol Res.*, **2**: 104-109.

Bonjar, G.H. (2004). Inhibition of Clotrimazole-resistant *Candida albicans* by plants used in Iranian folkloric medicine. *Fitoterapia*, **75**: 74-76.

Chandra, J., Kuhn, D.M., Mukherjee, P.K., Hoyer, L.L., McCormick, T. *et al.* (2001). Biofilm formation by the fungal pathogen Candida albicans: development, architecture, and drug resistance. *J Bacteriol.*, **183**: 5385-5394.

Damke, E., Tsuzuki, J.K., Cortez, D.A., Ferreira, I.C., Bertoni, T.A. *et al.* (2011). *In vivo* activity of *Sapindus saponaria* against azole-susceptible and -resistant human vaginal *Candida* species. *BMC Complement Altern Med.*, **11**: 35.

de Oliveira, J.R., de Castro, V.C., das Gracas Figueiredo Vilela, P., Camargo, S.E., Carvalho, C.A. *et al.* (2013). Cytotoxicity of Brazilian plant extracts against oral microorganisms of interest to dentistry. *BMC Complement Altern Med.*, **13**: 208.

de Paiva, S.R.,Figueiredo, M.R.,Aragao, T.V., and Kaplan, M.A. (2003). Antimicrobial activity *in vitro* of plumbagin isolated from *Plumbago* species. *Mem Inst Oswaldo Cruz.*, **98**: 959-961.

Diab, Y.,Atalla, K., and Elbanna, K. (2012). Antimicrobial screening of some Egyptian plants and active flavones from *Lagerstroemia indica* leaves. *Drug Discov Ther.*, **6**: 212-217.

Duarte, M.C.,Figueira, G.M.,Sartoratto, A.,Rehder, V.L., and Delarmelina, C. (2005). Anti-*Candida* activity of Brazilian medicinal plants. *J Ethnopharmacol.*, **97**: 305-311.

Duraipandiyan, V.,Ayyanar, M., and Ignacimuthu, S. (2006). Antimicrobial activity of some ethnomedicinal plants used by Paliyar tribe from Tamil Nadu, India. *BMC Complement Altern Med.*, **6**: 35.

Edziri, H., Mastouri, M., Mahjoub, M.A., Mighri, Z., Mahjoub, A. *et al.* (2012). Antibacterial, antifungal and cytotoxic activities of two flavonoids from *Retama raetam* flowers. *Molecules*, **17**: 7284-7293.

Ellof, J.N. (1998). Which extractant should be used for the screening and isolation of antimicrobial components from plants? *J Ethnopharmacol.*, **60**: 1.

Endo, E.H.,Cortez, D.A.,Ueda-Nakamura, T.,Nakamura, C.V., and Dias Filho, B.P. (2010). Potent antifungal activity of extracts and pure compound isolated from pomegranate peels and synergism with fluconazole against *Candida albicans*. *Res Microbiol.*, **161**: 534-540.

Endo, E.H.,Ueda-Nakamura, T.,Nakamura, C.V., and Filho, B.P. (2012).Activity of spray-dried microparticles containing pomegranate peel extract against *Candida albicans*. *Molecules*, **17**: 10094-10107.

Fabry, W.,Okemo, P.,Mwatha, W.E.,Chhabra, S.C., and Ansorg, R. (1996).Susceptibility of *Helicobacter pylori* and *Candida* spp. to the east African plant *Terminalia spinosa*. *Arzneimittelforschung*, **46**: 539-540.

Faria, M.H.,Carvalho, T.G.,Rabenhorst, S.H.,Sidrim, J.J., and Moraes-Filho, M.O. (2006). Cytotoxic and antifungal properties of medicinal plants from Ceara, Brazil. *Braz J Biol.*, **66**: 1133-1135.

Fatima, A., Gupta, V.K., Luqman, S., Negi, A.S., Kumar, J.K. *et al.* (2009). Antifungal activity of *Glycyrrhiza glabra* extracts and its active constituent glabridin. *Phytother Res.*, **23**: 1190-1193.

Feldman, M.,Tanabe, S.,Howell, A., and Grenier, D. (2012). Cranberry proanthocyanidins inhibit the adherence properties of *Candida albicans* and cytokine secretion by oral epithelial cells. *BMC Complement Altern Med.*, **12**: 6.

Fidel, P.L., Jr. (2002). Immunity to *Candida*. *Oral Dis.*, **8 Suppl 2**: 69-75.

Fu, Z., Lu, H., Zhu, Z., Yan, L., Jiang, Y. *et al.* (2011). Combination of baicalein and Amphotericin B accelerates *Candida albicans* apoptosis. *Biol Pharm Bull.*, **34**: 214-218.

Furletti, V.F., Teixeira, I.P., Obando-Pereda, G., Mardegan, R.C., Sartoratto, A. *et al.* (2011). Action of *Coriandrum sativum* L. essential oil upon oral *Candida albicans* biofilm formation. *Evid Based Complement Alternat Med.*, **2011**: 985832.

Garber, G. (2001). An overview of fungal infections. *Drugs*, **61 Suppl 1**: 1-12.

Ghannoum, M.A. (1988). Studies on the anticandidal mode of action of *Allium sativum* (garlic). *J Gen Microbiol.*, **134**: 2917-2924.

Hermann, P.,Berek, Z.,Nagy, G.,Kamotsay, K., and Rozgonyi, F. (2001). Pathogenesis, microbiological and clinical aspects of oral candidiasis (candidosis). *Acta Microbiol Immunol Hung.*, **48**: 479-495.

Heroor, S.,Beknal, A.K., and Mahurkar, N. (2013). Immunomodulatory activity of methanolic extracts of fruits and bark of *Ficus glomerata* Roxb. in mice and on human neutrophils. *Indian J Pharmacol.*, **45**: 130-135.

Herrera-Arellano, A., Lopez-Villegas, E.O., Rodriguez-Tovar, A.V., Zamilpa, A., Jimenez-Ferrer, E. *et al.* (2013). Use of antifungal saponin SC-2 of *Solanum chrysotrichum* for the treatment of vulvovaginal candidiasis: *in vitro* studies and clinical experiences. *Afr J Tradit Complement Altern Med.*, **10**: 410-417.

Herrera-Arellano, A., Martinez-Rivera Mde, L., Hernandez-Cruz, M., Lopez-Villegas, E.O., Rodriguez-Tovar, A.V. *et al.* (2007). Mycological and electron microscopic study of *Solanum chrysotrichum* saponin SC-2 antifungal activity on *Candida* species of medical significance. *Planta Med.*, **73**: 1568-1573.

Hofling, J.F., Anibal, P.C., Obando-Pereda, G.A., Peixoto, I.A., Furletti, V.F. *et al.* (2010). Antimicrobial potential of some plant extracts against *Candida* species. *Braz J Biol.*, **70**: 1065-1068.

Inagaki, N., Komatsu, Y., Sasaki, H., Kiyohara, H., Yamada, H. *et al.* (2001). Acidic polysaccharides from rhizomes of *Atractylodes lancea* as protective principle in *Candida*-Infected mice. *Planta Med.*, **67**: 428-431.

Irani, M.,Sarmadi, M.,Bernard, F.,Ebrahimi Pour, G.H., and Shaker Bazarnov, H. (2010). Leaves antimicrobial activity of *Glycyrrhiza glabra* L. *Iran J. Pharm Res.*, **9**: 425-428.

Ishida, K., de Mello, J.C., Cortez, D.A., Filho, B.P., Ueda-Nakamura, T. *et al.* (2006). Influence of tannins from *Stryphnodendron adstringens* on growth and virulence factors of *Candida albicans*. *J Antimicrob Chemother.*, **58**: 942-949.

Jothy, S.L.,Zakariah, Z.,Chen, Y., and Sasidharan, S. (2012). *In vitro, in situ* and *in vivo* studies on the anticandidal activity of *Cassia fistula* seed extract. *Molecules*, **17**: 6997-7009.

Khan, R.,Zakir, M.,Afaq, S.H.,Latif, A., and Khan, A.U. (2010). Activity of solvent extracts of *Prosopis spicigera, Zingiber officinale* and *Trachyspermum ammi* against multidrug resistant bacterial and fungal strains. *J Infect Dev Ctries.*, **4**: 292-300.

Khosravi, A.R.,Eslami, A.R.,Shokri, H., and Kashanian, M. (2008). *Zataria multiflora* cream for the treatment of acute vaginal candidiasis. *Int J Gynaecol Obstet.*, **101**: 201-202.

Kojic, E.M., and Darouiche, R.O. (2004). *Candida* infections of medical devices. *Clin Microbiol Rev.*, **17**: 255-267.

Latha, L.Y.,Darah, I.,Jain, K., and Sasidharan, S. (2011). Effects of *Vernonia cinerea* less methanol extract on growth and morphogenesis of *Candida albicans*. *Eur Rev Med Pharmacol Sci.*, **15**: 543-549.

Lee, J.H., and Hong, S.K. (2009a). Comparative analysis of chemical compositions and antimicrobial activities of essential oils from *Abies holophylla* and *Abies koreana*. *J Microbiol Biotechnol.*, **19**: 372-377.

Lee, J.Y., Lee, J.H., Park, J.H., Kim, S.Y., Choi, J.Y. *et al.* (2009b). Liquiritigenin, a licorice flavonoid, helps mice resist disseminated candidiasis due to *Candida albicans* by Th1 immune response, whereas liquiritin, its glycoside form, does not. *Int Immunopharmacol.*, **9**: 632-638.

Li, Y.C., Liang, H.C., Chen, H.M., Tan, L.R., Yi, Y.Y. *et al.* (2012). Anti-*Candida albicans* activity and pharmacokinetics of pogostone isolated from *Pogostemonis* Herba. *Phytomedicine*, **20**: 77-83.

Lopes, G.,Pinto, E.,Andrade, P.B., and Valentao, P. (2013). Antifungal activity of phlorotannins against dermatophytes and yeasts: approaches to the mechanism of action and influence on *Candida albicans* virulence factor. *PLoS One*, **8**: e72203.

Luciano-Montalvo, C.,Boulogne, I., and Gavillan-Suarez, J. A. (2013). Screening for antimicrobial activities of Caribbean herbal remedies. *BMC Complement Altern Med.*, **13**: 126.

Martino, P., Girmenia, C., Micozzi, A., De Bernardis, F., Boccanera, M. *et al.* (1994). Prospective study of *Candida* colonization, use of empiric amphotericin B and development of invasive mycosis in neutropenic patients. *Eur J Clin Microbiol Infect Dis.*, **13**: 797-804.

Martino, R.F., Davicino, R.C., Mattar, M.A., Sasso, C.V., Casali, Y.A. *et al.* (2012). Macrophages activation by a purified fraction, free of nordihydroguaiaretic acid (NDGA), from *Larrea divaricata* Cav. as a potential novel therapy against *Candida albicans*. *Immunopharmacol Immunotoxicol.*, **34**: 975-982.

Mash, D.C., Kovera, C.A., Pablo, J., Tyndale, R.F., Ervin, F.D. *et al.* (2000). Ibogaine: complex pharmacokinetics, concerns for safety, and preliminary efficacy measures. *Ann N Y Acad Sci.*, **914**: 394-401.

Meng, F., Zuo, G., Hao, X., Wang, G., Xiao, H. *et al.* (2009). Antifungal activity of the benzo[c]phenanthridine alkaloids from *Chelidonium majus* Linn against resistant clinical yeast isolates. *J Ethnopharmacol.*, **125**: 494-496.

Messier, C., and Grenier, D. (2011). Effect of licorice compounds licochalcone A, glabridin and glycyrrhizic acid on growth and virulence properties of *Candida albicans*. *Mycoses*, **54**: e801-806.

Mukherjee, P.K., and Wahile, A. (2006). Integrated approaches towards drug development from Ayurveda and other Indian system of medicines. *J Ethnopharmacol.*, **103**: 25-35.

Mutai, C.,Bii, C.,Vagias, C.,Abatis, D., and Roussis, V. (2009). Antimicrobial activity of *Acacia mellifera* extracts and lupane triterpenes. *J Ethnopharmacol.*, **123**: 143-148.

Nordin, M.A.,Wan Harun, W.H., and Abdul Razak, F. (2013). Antifungal susceptibility and growth inhibitory response of oral *Candida* species to *Brucea javanica* Linn. extract. *BMC Complement Altern Med.*, **13**: 342.

Osho, A.,Adetunji, T.,Fayemi, S.O., and Moronkola, D.O. (2010). Antimicrobial activity of essential oils of *Physalis angulata*. L. *Afr J Tradit Complement Altern Med.*, **7**: 303-306.

Papadimitriou, J.M., and Ashman, R.B. (1986). The pathogenesis of acute systemic candidiasis in a susceptible inbred mouse strain. *J Pathol.*, **150**: 257-265.

Park, K.S., Kang, K.C., Kim, J.H., Adams, D.J., Johng, T.N. *et al.* (1999). Differential inhibitory effects of protoberberines on sterol and chitin biosyntheses in *Candida albicans*. *J Antimicrob Chemother.*, **43**: 667-674.

Parker, J.C., Jr.,McCloskey, J.J., and Knauer, K.A. (1976). Pathobiologic features of human candidiasis. A common deep mycosis of the brain, heart and kidney in the altered host. *Am J Clin Pathol.*, **65**: 991-1000.

Pellati, D.,Fiore, C.,Armanini, D.,Rassu, M., and Bertoloni, G. (2009). *In vitro* effects of glycyrrhetinic acid on the growth of clinical isolates of *Candida albicans*. *Phytother Res.*, **23**: 572-574.

Pfaller, M.A., and Diekema, D.J. (2007). Epidemiology of invasive candidiasis: a persistent public health problem. *Clin Microbiol Rev.*, **20**: 133-163.

Pietrella, D., Angiolella, L., Vavala, E., Rachini, A., Mondello, F. *et al.* (2011). Beneficial effect of *Mentha suaveolens* essential oil in the treatment of vaginal candidiasis assessed by real-time monitoring of infection. *BMC Complement Altern Med.*, **11**: 18.

Polaquini, S.R.,Svidzinski, T.I.,Kemmelmeier, C., and Gasparetto, A. (2006). Effect of aqueous extract from Neem (*Azadirachta indica* A. Juss) on hydrophobicity, biofilm formation and adhesion in composite resin by *Candida albicans*. *Arch Oral Biol.*, **51**: 482-490.

Ponnusamy, K.,Petchiammal, C.,Mohankumar, R., and Hopper, W. (2010). *In vitro* antifungal activity of indirubin isolated from a South Indian ethnomedicinal plant *Wrightia tinctoria* R. Br. *J Ethnopharmacol.*, **132**: 349-354.

Portillo, A.,Vila, R.,Freixa, B.,Adzet, T., and Canigueral, S. (2001). Antifungal activity of Paraguayan plants used in traditional medicine. *J Ethnopharmacol.*, **76**: 93-98.

Pugliese, A., Torre, D., Baccino, F.M., Di Perri, G., Cantamessa, C. *et al.* (2000). *Candida albicans* and HIV-1 Infection. *Cell Biochem Funct.*, **18**: 235-241.

Richardson, M.D. (2005). Changing patterns and trends in systemic fungal infections. *J Antimicrob Chemother*, **56 Suppl 1**: i5-i11.

Sahgal, G., Ramanathan, S., Sasidharan, S., Mordi, M.N., Ismail, S. *et al.* (2011). *In vitro* and *in vivo* anticandidal activity of *Swietenia mahogani* methanolic seed extract. *Trop Biomed.*, **28**: 132-137.

Saxena, G.,McCutcheon, A.R.,Farmer, S.,Towers, G.H., and Hancock, R.E. (1994). Antimicrobial constituents of *Rhus glabra*. *J Ethnopharmacol*, **42**: 95-99.

Sharma, B., and Kumar, P. (2009). Bioefficacy of *Lantana camara* L. against Some Human Pathogens. *Indian J Pharm Sci.*, **71**: 589-593.

Sims, C.R.,Ostrosky-Zeichner, L., and Rex, J.H. (2005). Invasive candidiasis in immunocompromised hospitalized patients. *Arch Med Res*, **36**: 660-671.

Sitheeque, M.A., Panagoda, G.J., Yau, J., Amarakoon, A.M., Udagama, U.R. *et al.* (2009). Antifungal activity of black tea polyphenols (catechins and theaflavins) against *Candida* species. *Chemotherapy*, **55**: 189-196.

Sobel, J.D. (2007). Vulvovaginal candidosis. *Lancet*, **369**: 1961-1971.

Souza-Moreira, T.M., Severi, J.A., Lee, K., Preechasuth, K., Santos, E. *et al.* (2013). Anti-*Candida* targets and cytotoxicity of casuarinin isolated from *Plinia cauliflora* leaves in a bioactivity-guided study. *Molecules*, **18**: 8095-8108.

Spampinato, C., and Leonardi, D. (2013). *Candida* infections, causes, targets, and resistance mechanisms: traditional and alternative antifungal agents. *Biomed Res Int.*, **2013**: 204237.

Sule, A., Ahmed, Q.U., Latip, J., Samah, O.A., Omar, M.N. *et al.* (2012). Antifungal activity of *Andrographis paniculata* extracts and active principles against skin pathogenic fungal strains *in vitro*. *Pharm Biol.*, **50**: 850-856.

Suwanmanee, S.,Kitisin, T., and Luplertlop, N. (2014). *In vitro* screening of 10 edible thai plants for potential antifungal properties. *Evid Based Complement Alternat Med.*, **2014**: 138587.

Svetaz, L., Zuljan, F., Derita, M., Petenatti, E., Tamayo, G. *et al.* (2010). Value of the ethnomedical information for the discovery of plants with antifungal properties.

A survey among seven Latin American countries. *J Ethnopharmacol.*, **127**: 137-158.

Talib, W.H., and Mahasneh, A.M. (2010). Antimicrobial, cytotoxicity and phytochemical screening of Jordanian plants used in traditional medicine. *Molecules,* **15**: 1811-1824.

Tatsimo, S.J., Tamokou Jde, D., Havyarimana, L., Csupor, D., Forgo, P. *et al.* (2012). Antimicrobial and antioxidant activity of kaempferol rhamnoside derivatives from *Bryophyllum pinnatum*. *BMC Res Notes,* **5**: 158.

Tempone, A.G., Sartorelli, P., Teixeira, D., Prado, F.O., Calixto, I.A. *et al.* (2008). Brazilian flora extracts as source of novel antileishmanial and antifungal compounds. *Mem Inst Oswaldo Cruz.,* 1**03**: 443-449.

Valsaraj, R.,Pushpangadan, P.,Smitt, U.W.,Adsersen, A., and Nyman, U. (1997). Antimicrobial screening of selected medicinal plants from India. *J Ethnopharmacol.,* **58**: 75-83.

Varma, J., and Dubey, N.K. (2001). Efficacy of essential oils of *Caesulia axillaris* and *Mentha arvensis* against some storage pests causing biodeterioration of food commodities. *Int J Food Microbiol.,* **68**: 207-210.

Vasconcelos, L.C.,Sampaio, M.C.,Sampaio, F.C., and Higino, J.S. (2003). Use of *Punica granatum* as an antifungal agent against candidosis associated with denture stomatitis. *Mycoses,* **46**: 192-196.

White, T.C.,Holleman, S.,Dy, F.,Mirels, L.F., and Stevens, D.A. (2002). Resistance mechanisms in clinical isolates of *Candida albicans*. *Antimicrob Agents Chemother.,* **46**: 1704-1713.

Wong, K.S., and Tsang, W.K. (2009). *In vitro* antifungal activity of the aqueous extract of *Scutellaria baicalensis* Georgi root against *Candida albicans*. *Int J Antimicrob Agents,* **34**: 284-285.

Yi, Y.Y., He, J.J., Su, J.Q., Kong, S.Z., Su, J.Y. *et al.* (2013). Synthesis and antimicrobial evaluation of pogostone and its analogues. *Fitoterapia,* **84**: 135-139.

Yordanov, M., Dimitrova, P., Patkar, S., Falcocchio, S., Xoxi, E. *et al.* (2005). Ibogaine reduces organ colonization in murine systemic and gastrointestinal *Candida albicans* infections. *J Med Microbiol,* **54**: 647-653.

Yordanov, M.,Dimitrova, P.,Patkar, S.,Saso, L., and Ivanovska, N. (2008). Inhibition of *Candida albicans* extracellular enzyme activity by selected natural substances and their application in *Candida* infection. *Can J Microbiol.,* **54**: 435-440.

Yoshida, S., Kasuga, S., Hayashi, N., Ushiroguchi, T., Matsuura, H. *et al.* (1987). Antifungal activity of ajoene derived from garlic. *Appl Environ Microbiol.,* **53**: 615-617.

Zhang, J.D., Cao, Y.B., Xu, Z., Sun, H.H., An, M.M. *et al.* (2005). *In vitro* and *in vivo* antifungal activities of the eight steroid saponins from *Tribulus terrestris* L. with potent activity against fluconazole-resistant fungal pathogens. *Biol Pharm Bull.,* **28**: 2211-2215.

Natural Products: Research Reviews Vol. 3 (2016) *Pages* **351–360**
Editor: **V.K. Gupta**
Published by: **DAYA PUBLISHING HOUSE, NEW DELHI**

12

A Review on Tissue Culture and Phytochemical Aspects of *Gloriosa superba* L.

Rathod Dipika[1*] and Patel Illa[1]

ABSTRACT

Gloriosa superba L. is a valuable medicinal plant belongs to family Liliaceae. Seeds and tubers of this plant contain medicinally potential alkaloid colchicine and have been traditionally used for treatment of various ailments. The phytochemicals present in this plant have analgesic, anti-inflammatory, antithrombotic, anticoagulant, enzyme inhibitory, antivenom and chemotherapeutic potential. Due to medicinal potential of this valuable plant, it is highly exploited for the production of natural drugs by pharmaceutical industries So, to fulfill the demand of pharmaceutical industries in the form of raw material, this plant is needed to be multiply rapidly. For the fast and excellent production, plant tissue culture is a useful tool for the mass production of this plant which will provide various secondary metabolites from its plant parts.

The current review deals with the enormous information on scientific research and reports available in different aspects of this plant involving plant tissue culture and phytochemistry. The review also includes reports on morphology and medicinal uses of this plant.

Keywords: *Gloriosa superba* L., Phytochemistry, Colchicine, Morphology.

1 Department of Life Sciences, Hemchandracharya North Gujarat University, Patan – 384 265, Gujarat, India.

* *Corresponding author:* E-mail: rathod.dipika7@gmail.com

Introduction

Medicinal plants are an important part of human life as it is helpful in various communities for making of drugs and also used for food. The traditional people have vast knowledge, which is used as a possible tool for production of novel cost effective plants for food as well as medicine (Lal and Mishra, 2011). *Gloriosa superba* L. belonging to family Liliaceae is an important endangered medicinal plant (Badola, 2002). This plant contains valuable alkaloid colchicine in its plant parts like seeds, tuber and leaves. It is one of the oldest ingredients of species from ancient time. This plant is a native of India especially Southern India. It is known as glory lily and climbing lily- in English; Karihari- in Hindi; Langli- in Sanskrit, Agnishike in Kannada, Nabhikkodi in Tamil, Karianag in Marathi (Nadkarni, 1996; Kirthikar and Basu, 1935; Khare, 2007). Traditionally this plant is used as an Ayurvedic medicinal plant to cure various diseases like gout, arthritis, cancer and bleeding. Most recently, it has now been used also in modern system of medicine in India. They are native in tropical and southern Africa to Asia and naturalized in Australia and the Pacific as well as being widely cultivated. The antioxidant activity, antimicrobial activity, snake bite potential, hapatoprotective activity, Antithrombotic/Anti coagulant potential and Analgesic and anti-inflammatory potential of *G. superba* Linn. has already been reported (Amudha and Shanthi, 2011; Hemaiswarya *et al.*, 2009; Mohandass and Indhumathi, 2011; Nalise *et al.*, 2008; Jomy *et al.*, 2009). Tuber of this plant is widely used by many pharmaceutical industries as raw material for making drugs. Usually *G. superba* is propagated by corm and seeds but due to low germination capability and dormancy of tubers it restricts the regeneration. One seed give one plant and similarly one tuber give one plant that is also during only rainy season. Consequently, in order to maintain and conserve this important plant, biotechnological approaches would be very useful (Sivakumar and Krishnamurthy 2002). To meet the demand for raw material to pharmaceutical industries, this plant should be conserved. The biotechnological tools play a crucial role in conservation of rare and endangered medicinal plants. The present review is mainly focused on micropropagation and phytochemical aspects.

Taxonomy (Singh *et al.*, 2013)

Scientific Classification

Kingdom	: Plantae
Subkingdom	: Tracheobionta
Division	: Spermatophyta
Class	: Liliopsida
Subclass	: Liliidae
Order	: Liliales
Family	: Liliaceae
Genus	: Gloriosa
Species	: *Gloriosa superba* Linn.

Figure 12.1: *G. superba* Flowering Twig.

Botanical Description

It is a herbaceous, climbing perennial, growing plant between 3.5 to 6 m in length. the leaves spirally arranged or sub opposite, the upper ones with cirrhose tips; flowers solitary, large, borne on long, spreading pedicels, usually yellow or red in colour, actinomorphic, hermaphrodite; perianth segments 6, free, lanceolate, keeled within at base, long-persistent; stamens 6, hypogynous, the anthers extrorse, medifixed and versatile, opening by longitudinal slits; ovary superior, 3-celled, the carpels joint

Figure 12.2: *G. superba* Tubers.

only by their inner margins, the ovules numerous, the style deflected at base and projecting from the flower more or less horizontally; fruit a loculicidal capsule with many seeds (Kavithamani *et al.*, 2013). Stem annual, glabrous and sparsely branched; tuber perennial, horizontal, abruptly bent in a V or L shape, roots fibrous (Badwaik *et al.*, 2011).

Chemical Constituents

G. superba is mostly known for colchicine, a valuable alkaloid. Tubers and seed are two main parts which contains colchicine. Studies reveal that all parts of the plant, especially the tubers are extremely toxic due to the presence of a highly active alkaloid, colchicine. The species also contains another toxic alkaloid, gloriosine. Other compounds such as lumicolchicine, 3-demethyl-N-deformyl-N-deacetyl-colchicine,3-demethylcolchicine, N-formyldeacetylcolchicine have been isolated from the plant (Chulabhorn *et al.*, 1998).

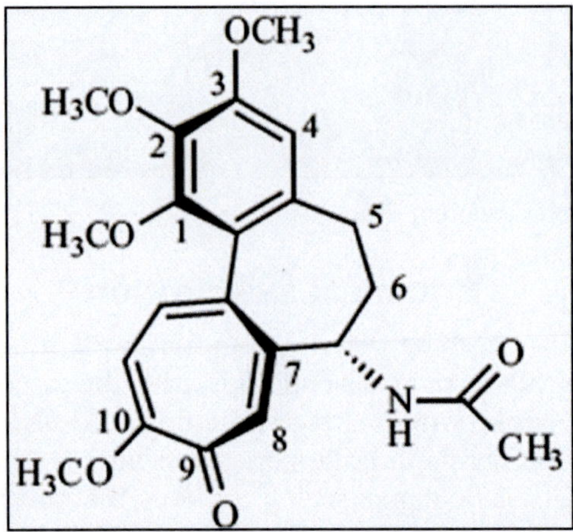

Figure 12.3: Structure of Colchicine (N-[(7*S*)-1,2,3, 10- tetramethoxy-9-oxo-5,6,7,9-tetrahydrobenzo [*a*]heptalen- 7-yl]acetamide) (Cerquaglia *et al.*, 2005).

The tubers or dried roots contain colchicine, benzoic and salicylic acid, sterols and resinous substances-colchicine, 3-demethyl colchicine, 1,2-didemethyl colchicine, 2,3- didemethyl colchicine, N-formyl, N-deacetyl colchicines, colchicocide, gloriosine, tannins and superbine (Jain and Suryavanshi, 2010). Flowers contain Luteolin, its Glucoside, N-Formyl-de-Me-colchicine, its Glucoside and 2-de-Me-colchicine Content of Colchicine 0.25 per cent (Singh *et al.*, 2013). Hence in the world market seed, tubers and flowers are considered as rich sources of colchicine and gloriosine (Srivastava and Chandra, 1975; Prajapati *et al.*, 2003).

Medicinal Uses

G. superba is used as a traditional medicine as well as Ayurvedic medicine in India and in most of the parts of the world. It contains precious alkaloid colchicine, which has been used effectively to treat acute gout, intestinal worms, infertility, wounds and other skin problems (Kavithamani *et al.*, 2013). Colchicine, an alkaloid extracted from the tubers and seeds gives high price in the market and used in scientific research. The tuber is used for the treatment of colic, bruises and sprains, chronic ulcers, haemorrhoids, impotence, cancer, nocturnal seminal emission, and leprosy and also for labour pains and abortions (Kala, 2011). *Gloriosa superba* also used in skin related problems, wounds, fever, inflammation, piles, blood disorders, uterine contractions, general body toner, and poisoning (Shrivastava and Chandra, 1977). Roots are acrid, anthelmintic, antipyretic, bitter, digestive, expectorant, highly poisonous and promoting expulsion of the placenta. Root paste is effective against paralysis, rheumatism; snake bite and insect bites (Chitra and Rajamani, 2009).This plant has gained great importance in herbal medicine in recent years for the production of colchicine in large scale (Kokate *et al.*, 2004). The sap from the leaf tip is used as a smoothening agent for pimples and skin eruptions (Anonymous, 1971; Airaj *et al.*, 1999; Badola *et al.*, 2002).

Current Status of *Gloriosa*

Gloriosa superba L. is a valuable plant of Asia and Africa. It is used to cure gout, cancer and also used for treating intestinal worms, bruises, infertility, impotence and skin problem. Tubers and seeds of this plant have colchicine and colchicosides, both are very costly substances, and highly demanded by pharmaceutical industries. Due to excessive collection of this plant for various medicinal purposes by tribal people and pharma companies has put this plant into threatened and included in red data book (Ade, 2009; kavina *et al.*, 2011).

Tissue Culture Studies

Tissue culture is the technique which helps in multiplication of plant from very limited source by using small part as an explant. It is also used for improving the quality of plants for food, fodder, cosmetic, health and medicinal purpose etc. This technique required a complete aseptic condition for *in vitro* development of plants. Tissue culture technique has been successfully used for the regeneration of the plants which are difficult to propagate through conventional methods. Sivakumar *et al.*(2003), reported callus formation from the node explants when inoculated on SH (Schenk and Hildebrand's) medium supplemented with 2,4-D (2,4-Dichlorophenoxyacetic acid) combined with 2iP (N6-(2-isopentyl) adenine) and kept in darkness. The calli were yellowish in colour and compact. Sivakumar and Krishnamurthy (2004), achieved 98 per cent of callus induction in non dormant corm bud explants of *Gloriosa superba* L. Rishi, 2011 reported that the best results for callus formation obtained in B5 (Gamborg. Medium) medium supplemented with Kin. (Kinetin) along with NAA (α-Naphthalene acetic acid). The greatest numbers of multiple shoots (43) were observed in corm-derived callus. The greatest number of multiple shoots was observed in corm derived calluses. Sivakumar and Krishnamurthy (2002) reported *in vitro* organogenetic responses of *G. superba.* They used MS medium supplemented with BA (Benzyle adenine) and ADS (Adenine sulphate). The callus induction was reported in non-dormant corm bud explants. The maximum number of multiple shoots was obtained in corm-derived calluses. Samarjeeva *et al.* (1993) studied clonal propagation of *G. superba* from apical bud and node segment of shoot tip, cultured on solidified agar (0.8 per cent w/v) Gamborg's B5 medium containing BA, Kinetin, IAA (Indole-3-acetic acid), NAA, IBA (Indole-3-butyric acid) or 2,4-D. Custers and Bergervoet (1994) reported tissue culture studies of *G. superba* from node, internodes, leaves, flowers, pedicels and tuber explants. *In vitro* propagation and corm formation in *G. superba* was reported by Somani *et al.*(1989). The dissected propagules with shoot and root primordia of *G. superba* were placed on MS basal medium containing 3 per cent sucrose and 0.6 per cent agar. Explant germinated on the MS medium producing shoot and root, which formed new corm within one month. Hassan and Roy (2005) recorded shoot regeneration from apical and axillary buds of *G. superba* in MS medium supplemented with BA and NAA. Sivakumar and Krishnamurthy (2000), reported that the best root formation was obtained in *Gloriosa superba* L. when transferred to medium containing IBA and NAA. Singh *et al.* (2013) studied *in vitro* propagation from apical bud elongation. Microtubers and shoots developed on MS medium with BAP alone or combination with NAA. Best Roots were developed on ½ MS with IBA and NAA.

Much research has not been carried out on *Gloriosa* due to some problems like, the plant is monocot having very low germination capacity, and life cycle is also very short just 2-3 months, conventional propagation is very limited and also slow since one tuber produces only one plant at a time, besides this the plant is the richest source of colchicine, which has very high demand in pharmacological industries from all over the world. The whole plant is used for the medicinal purposes for production of natural drugs. There are many positive reports already available on micropropagation and secondary metabolites production of *G. superba* (Sivakumar and Krishnamurthy, 2004; Jha *et al.,* 2005; Ade and Rai, 2009).

Phytochemical Studies

Earlier studies on analysis of important phytochemicals from *Gloriosa* gives an idea about medicinal value of this plant. A new colchicine glycoside, 3-0-demethyl colchicine 3-O-alpha-D-glucopyranoside in *G.superba* seeds was reported by Suri *et al.*(2001). Kaur *et al.* (2007) studied purification of 3 – monomeric monocot mannose – binding lectins and their evaluation for antipoxviral activity isolated from *G.superba.* Thakur *et al.*(1975) reported the substances from plant of the sub family

Wurmbaeoideae and their derivatives along with alkaloids from the *Gloriosa*. Jha *et al.*(2005) reported production of forskolin, withanolides, colchicine and tylophorine from plant source by using biotechnological approach.

Muthukrishnan and Subramaniyan (2012), reported that Phytochemical analysis of the *Gloriosa* seeds showed the presence of Alkaloids, Carbohydrates, Glycosides, Steroids, Flavanoids, Terpenoids and Phenolic compound in it; *Gloriosa* tuber showed the presence of Alkaloids, Carbohydrates and Flavanoids; *Gloriosa superba* leaves showed the presence of Alkaloids, Carbohydrates, Steroids, Flavanoids and Terpenoids. The Preliminary phytochemical screening revealed that the extracts contain Alkaloids, Glycosides, Steriods, Terpenoids and Tannins in Seeds and Tuber (Senthilkumar, 2013). Primary phytochemical analysis of ethanolic and methanolic extracts of tuber of *Gloriosa* showed the presence of tannins, flavonoids, glycosides, steroid and alkaloids and TLC profiling of methanolic extract of tuber of *Gloriosa* showed maximum 9 bands (Dadsena *et al.*, 2013). Basak *et al.* (2012) reported TLC and HPLC study for estimation of colchicine in Tubers of *Gloriosa* originated from different zones of Orrisa, India. They reported appreciable difference in colchicine content found in both wild and cultivated tubers. Sarin *et al.* (2010), reported the qualitative and quantitative estimation of identified Alkaloid (Colchicine) from *Gloriosa in vivo* and *in vitro* samples. The *In vitro* studies showed the maximum amount of colchicine in 6 week old cultures. Singh *et al.* (2012) reported on thin layer chromatographic studies of *Gloriosa superba* L. TLC profiling of the extracts give an idea about the various phytochemicals like Alkaloids and flavonoids. Kavina *et al.* (2011) studied HPLC analysis of seed and tuber of *Gloriosa superba* L. and also recorded that treatment of GA_3 and *Psedomonas aerugenosa* enhance the content of colchicine in seeds and tubers. Colchicine content was higher in Seeds compared to the tubers sample.

Conclusion

The literature shows that the medicinal plant *Gloriosa* has been widely used for curing various diseases. Phytochemical studies shows that a number of medicinally important phytochemicals such as gloriosine and colchicines have been isolated from this plant. It has been affirmed as endangered plant by IUCN and hence there is an urgent need to conserve the plant by multiplication. Plant tissue culture can be used for the multiplication and also used for *in vitro* production of Colchicine, so as to meet the increasing demand from the pharmaceutical industries. This report gives an idea about herbal uses of *Gloriosa,* further efforts are required in order to get a better understanding of the biological activities reported, and to isolate, purify and chemically characterize the active principle of *Gloriosa superba*. Present review on Tissue culture and Phytochemical studies gives brief information about the conservation strategy and medicinally potent active compounds of this valuable plant.

References

Ade, R., and Rai, M. K. (2009). Current advances in *Gloriosa superba* L. *Biodiversity,* **10(4)**: 210-214.

Airaj, P.P., Khoohaswan, Y., Wongkrajang, P., and Peungrisha, P. (1999). Anticough and antimicrobial activities of *Psidium guajava* Linn. Leaf extract. *J. Ethnopharmacol.*, **67**: 203-21.

Amudha, P., and Shanthi, P. (2011). Antioxidant activity of some rare medicinal plants. *Journal of Pharmacy Research*, **4(3)**: 698-699.

Anirudha, Rishi. (2011). *In vitro* callus induction and regeneration of healthy plants of *Gloriosa superba* Linn. *Indian Journal of Fundamental and Applied Life Sciences*, **1(1)**: 64-65.

Anonymous. (1971). National Committee for Clinical Laboratory Standards. NCCLS, Performance Standards for antimicrobial susceptibility testing. 9th International Supplement. M100-S9.

Badola, H.K. (2002). Endangered medicinal plant species in Himachal Pradesh. A report on the International Workshop on "Endangered Medicinal Plant Species in Himachal Pradesh', organized by G.B. Pant Institute of Himalayan Environment and Development at Himachal Unit, Mohal - Kullu during 18 -19 March 2002. *Curr. Sci.*, **83**: 797-798.

Basak, U.C., Dash, D., and Mahapatra, A.K. (2012). Estimation of colchicine in tubers of *Gloriosa superba* L. originated from different agro-climatic zones of Odisha, India. *International Journal of Pharmacognosy and Phytochemical Research*, **4(3)**: 157-161.

Cerquaglia, C., Diaco, M., Nucera, G., Regina, M.L., Montalto, M., and Manna, R. (2005). Pharmacological and clinical basis of treatment of Familial Mediterranean Fever (FMF) with colchicine or analogues: An update. *Current Drug Targets - Inflammation and Allergy*, **4**: 117-124.

Chitra, R., and Rajamani, K. (2009). Perise performance and correlation studies for yield and its quality characters in Glory lily *Gloriosa superba* (L). *Acad. J. Plant Sci.*, **2**: 39-43.

Chulabhorn, Somsak., Hunsa, Somcliai., Surang, Phannipha., Tasanee, Stitaya., and Pichas (1998). Biodiversity and natural product drug discovery. *Pure and Appl. Chem.*, **70 (11)**: 2065-2072.

Custers, J.B.M., and Bergervoet, J.H.W. (1994). Micropropagation of *Gloriosa*: Towards a practical protocol. *Sci. Hortic.*, **57**: 323-334.

Dadsena, R., Sahu, N.K., Agrawal, S., and Kumar, A. (2013). Phytochemical analysis of three endangered plants (*Costus specious, Gloriosa superba* L. and *Rauvolfia serpentine* (Linn.) Benth) from Kanker District of Chhattisgarh, India.

Hassan, S.A. and Roy, S.K. (2005). Micropropagation of *Gloriosa superba* L. through high frequency shoot proliferation. *Plant Tiss. Cult.*, **15(1)**: 67-74.

Hemaiswarya, S., Raja, R., Anbazhagan, C., and Thiagarajan, V. (2009). Antimicrobial and mutagenic properties of the root tubers of *Gloriosa superba* Linn. (kalihari). *Pak. J. Bot.*, **41(1)**: 293-299.

Hemant Badwaik, Mukesh Kumar Singh, Deepa Thakur, Tapan Kumar Giri and Tripathi, D.K. (2011). The botany, chemistry, pharmacological and therapeutic application of *Oxalis corniculata* Linn.: A review. *International Journal of Phytomedicine*, **3**: 1-8.

Jain, A.P., and Suryavanshi, S. (2010). *Gloriosa superba* Linn. – A pharmacological review. *International journal of Pharma Research and Development*, **2(8)**: 24-26.

Jha, S., Bandypadhyay, M., and Chaudhary, K.N. (2005). Biotechnological approaches for the production of farskolin, withanolides, colchicine and tylophorine. *Plant Genetic Resource*, **3**: 101-115.

Jomy, J., Jennifer, F., Tanaji, N., Samir, N., Alok, S., and Pradeep, D. (2009). Analgesic and anti-inflammatory activities of the hydro alcoholic extract from *Gloriosa superba* Linn. *International Journal of Green Pharmacy*, 215-219.

Kala, C.P. (2011). Indigenous uses and sustainable harvesting of trees by local people in the Pachmarhi Biosphere Reserve of India. *Int. J. Med. Arom. Plants*, **1(2)**: 153-161.

Kaur, A., Kamboj, S.S., and Singh, J. (2007). Purification of 3 Monomeric monocot mannose – binding lectins and their evaluation for antipoxviral activity: potential application in multiple viral diseases caused by enveloped viruses. *Biochemistry and Cell Biology*, **85**: 88-95.

Kavina, J., Gopi, R., and Panneerselvam, R. (2011). Quantification of colchicine in seed and tuber samples of *Gloriosa superba* by high performance liquid chromatography method. *Journal of Applied Pharmaceutical Science*, **01 (07)**: 116-119

Kavithamani, D., Umadevi, M. and Geetha, S. (2013). A review on *Gloriosa superba* L. as a medicinal plant. *Indian Journal of Research in Pharmacy and Biotechnology*, **1(4)**: 554-557.

Khare, C.P. (2007). 'Indian medicinal plants: an illustrated dictionary', Springer Verlag Berlin/Heidelberg. 293-294.

Kirthikar, K.R., and Basu, B.D. (1935). 'Indian medicinal plant', 2nd edition, allahbad popular publication. 2525-2526.

Kokate, C.K., Purohit, A.P., and Gokhale S.B. (2004). *Pharmacognosy*, Nirali Prakashan, Pune, pp. 506.

Lal, H. S., and Mishra, P. K. (2011). *Gloriosa superba* – An endangered plant spotted for the first time from forest of Tpchanchi, Hazaribag (Jharkhand) India. *Science Research Reporter*, **1(2)**: 61-64.

Mohandass, S., and Indhumathi, T. (2011). Hepato protective efficacy of *Gloriosa superba* against paracetamol treated experimental rats-An *in vivo* study. *Golden Research Thoughts*, **1(3)**: 1-4.

Muthukrishnan, S., and Subramaniyan, A. (2012). Phytochemical constituents of *Gloriosa superba* seed, tuber and leaves. *Research Journal of Pharmaceutical, Biological and Chemical Sciences*, **3(3)**: 111-117.

Nadkarni, K.M. (1996). *'Indian Materia Medica'*, 3rd edition, Popular Prakashan, Mumbai, p. 579.

Nalise, L.K., Nandipha, M., Hajierah, D., Ryno, J. Naudé, and Carminita, L.F. (2008). Antithrombotic/anticoagulant and anticancer activities of selected medicinal plants from South Africa. *African Journal of Biotechnology*, **7(3)**: 217-223.

Prajapati, N.D., Purohit, S.S., Sharma, A.K., and Kumar, T. (2003). *A hand book of Medicinal Plants*: Agro bios publishers (India); Jodhpur. 221-234.

Renu Sarin, Anirudha Rishi and Ajit Kumar. (2010). *In vivo* and *In vitro* Estimation of Colchicine in *Gloriosa superba* L. by High Pressure Liquid Chromatography. *J. Exp. Sci.*, 1**(4)**: 1-2.

Samarajeewa, P. K., Dassanayake, M. D., and Jayawardena, S.D.G. (1993). Clonal propagation of *Gloriosa superba. Indian Journal of Experimental Biology,* **31**: 719-720.

Senthilkumar, M. (2013). Phytochemical screening and antibacterial activity of *Gloriosa superba* Linn. *International Journal of Pharmacognosy and Phytochemical Research,* **5(1)**: 31-36.

Singh, D., Mishra, M., and Yadav, A. (2013). *Gloriosa superba* Linn: An important endangered medicinal plant and their conservation strategies. *International Journal of Botany and Research*, **3(1)**: 19-26.

Sivakumar, G., and Krishnamurthy, K.V. (2002). *Gloriosa superba* L.–A very useful medicinal plant. In: Singh, V.K., J.N. Govil, S. Hashmi, and G. Singh (eds). Recent Progress in Medicinal Plants, Vol. 7. Ethnomedicine and Pharmacognosy, Part II Texas: Series Sci Tech Pub, Texas, USA.

Sivakumar, G., and Krishnamurthy, K.V. 2004. *In vitro* organogenetic responses of *Gloriosa superba. Russian Journal of Plant Physiology*, **51**: 790-798.

Sivakumar, G., Krishnamurthy, K.V., and Rajendran, T.D. (2003). Embryoidogenesis and plant regeneration from leaf tissue of *Gloriosa superba. Planta Med.*, **69**: 479-481.

Somani, V.J., John, C.K., and Thengane, R.J.(1989). *In vitro* propagation and corm formation in *Gloriosa superba. Indian Journal of Experimental Biology*, **27**: 578-579.

Srivastava, U.C., and Chandra, V. (1977). *Gloriosa superba* Linn. (kalihari) - An important colchicine producing plant. *J. Res. Indian Med.*, **10**: 92-95.

Suri, O.P., Gupta B. D., and Suri K. A. (2001). A new glycoside, 3-O- demethylcolchicine-3-O-alpha-d- glucopyranoside from *Gloriosa* seeds. *Natural Product Letters,* **15**: 217-219.

Thakur, R. S., Potesilova, H., and Santavy, F. (1975). Substances from plants of the subfamily Wurmbaeoideae and their derivaties. Part LXXIX. Alkaloids of the plant *Gloriosa superba* L. *Planta Medica*, **3**: 201-209.

Natural Products: Research Reviews Vol. 3 (2016) *Pages* **361–387**
Editor: **V.K. Gupta**
Published by: **DAYA PUBLISHING HOUSE, NEW DELHI**

13

Biological Activity of Extracts from *Artemisia arborescens* (Vaill.) L.: An Overview about Insecticidal, Antimicrobial, Antifungal and Herbicidal Properties

Marcello Militello[1] and Alessandra Carrubba[1*]

ABSTRACT

Following their evolutionary process, many plants have oriented their secondary metabolism towards the production of complex molecules, which often show an apparent protective function against pests (insects, fungi, viruses, but also competing plants). The identification and the study of such compounds could play a significant role in the frame of the development of new techniques for sustainable plant protection, where they could help in achieving the goal of a reduction in the use of the most highly-impact chemicals. The Mediterranean basin, with its very large biodiversity, represents a huge reservoir of highly promising plant material,which is scarcely explored as far. Indeed, preliminary trials have allowed to state that a large number of Mediterranean plants possess definite biological activities, that give them a great interest as potential raw matters for many industrial sectors. Among these, Artemisia arborescens (Vaill.) L., a common shrub from the family Asteraceae, was found to exert a wide range of biocidal activities, which may represent the starting point for the best exploitation of this plant.In this paper, a thorough review is made about the studies that have been carried out worldwide about the composition and some of the biological activities of A. arborescens. Original data are included and discussed, concerning trials about

1 D/SAF (Department of Agricultural and ForestrySciences, Università degli Studi di Palermo), viale delle Scienze ed. 4 ingr. L, 90128 Palermo, Italy

* *Corresponding author*: E-mail: alessandra.carrubba@unipa.it

the evaluation of the activity of essential oil and water extracts against selected mealybugs, fungal and microbial strains, as well as against the germination capacity of the seeds of several common weeds.

Keywords: *Artemisia arborescens*, Insecticidal, Antimicrobial, Antifungal, Herbicidal properties.

Introduction

The term "biopesticides" identifies a huge number of molecules, naturally produced by plants, microorganisms or other biological sources, that are able to express a control activity towards phytopathogens such as viruses, bacteria, fungi or arthropods, or exert some control effect against noxious weeds. These activities may range from the simple repression of a special development stage of the target organism (sporulation, vegetative growth, reproduction), to the induction of reactions and metabolic processes generating its death.

As a matter of fact, during the history of agriculture, many plant products have been suggested and successfully used as bio-based pesticides. Although the market is dominated, as far, by chemically-derived substances, a long series of bio-pesticides are currently listed and authorized at the EPA (Environmental Protection Agency) in the US (EPA, 2010), but in other countries as well, their use is rapidly increasing, also encouraged by laws and regulations that open the way to the application of non-chemical items in organic agriculture (Cavoski *et al.*, 2011; Hillocks 2012). Among these, some compounds of natural origin, such as extracts from *Azadirachta indica* (neem), *Chrysantemum cinerariaefolium* (pyrethrum), *Derris elliptica*, *Cymbopogon winterianus*, *Capsicum frutescens* and *Cyperus rotundus*, are largely and traditionally used as insecticides all over the world. Other examples are offered by D-limonene and pelargonic acid, used as herbicides (Dayan *et al.*, 2009); isoflavonoids such as pisatin and phaseollin, useful against fungi and bacteria (Wyman and VanEtten, 1978); alkaloids, traded in commercial products suggested for fungal diseases in field and horticultural crops (Castillo *et al.*, 2012).

A strong interest in these new products comes from the emerging sectors of organic agriculture and, more generally, from all environmentally-friendly production techniques (Dayan *et al.*, 1999 and 2009). Although there is no real guarantee that a "natural" product is also innocuous, such molecules are usually non-persistent in the environment (Isman, 2000; Gahukar, 2014). This characteristic, traditionally considered a constraint, may become a positive issue when the healthiness and safety of fresh water and soil are of concern.

In many cases, the study concerning new formulations from plants is developed from species chosen into the wild flora; hence, biodiversity represents a considerable basin onto which to identify and use new active molecules. Two approaches are possible: a general, randomly-performed, screening of plants, aiming to identify as many chemical compounds as possible, followed by the evaluation of their biological activity (if any); and the validation of information coming from ethnobotany and ethnopharmacology, by one side, or plant ecology, by another (Veerpoorte, 1998). The first approach, although in former years leading to brilliant successes, also with the help of combinatorial chemistry (Cragg and Newman, 2013), has been progressively abandoned by industrial research in favor of the implementation of synthetic compound libraries, in which the synthesis and evaluation procedures were undoubtedly much faster and easier (Lam, 2007; Li and Vederas, 2009). The second approach, requiring a high level of interdisciplinary expertise, may allow to narrow the range of chemical compounds to be tested, paying a greater attention to those reported among folk or traditional uses, or expressing some significant *in vivo* effect (Veerpoorte, 1998; Cragg and Newman, 2013).

A first big constraint is that very often, the efficiency of one plant extract is different from that of its components, singularly considered. The approach aiming to the isolation, purification and massive use of the so-called "active compound" is therefore not always successful, since in many cases, the given biological effect is due to the simultaneous, synergistic, action of many compounds (Einhellig, 1995).

Actually, plants have a high ability to synthesize secondary metabolites, generally produced in response of specific environmental inputs such as thermal, water or salt stress, pests or disease attack, or competition with other plants (Selmar, 2008; Pichersky and Gang, 2000; Neilson *et al.*, 2013). Chemically, secondary metabolites belong to many different classes: flavonoids, anthocyanins, terpenes, alkaloids, cyanogenic glucosides and others (Nugroho *et al.*, 1999; Verpoorte and Memelink, 2002; Dias *et al.*, 2012).

The level of biological activity of plant extracts, being related to their chemical composition, is often subjected to large variations between the different plant populations; this justifies the splitting of the starting plant populations into groups (chemotypes), botanically identical and differentiable only on the basis the chemical makeup of their extracts (Lahlou, 2004; Franz and Novak, 2010). In many plant species, it was stated that the differentiation of chemotypes, mostly under a fairly strict genetic control, has a geographical basis (Selmar, 2008; Franz and Novak, 2010). Much work must still be done, however, to determine how much this variation may be affected by environmental features.

Among the botanical families that deserve greater attention, an important role is played by the *Asteraceae* (formerly *Compositae*), dealing with a high number of species with a recognized biological activity; inside this family, *Artemisia arborescens* (Vaill.) L. is one of the most promising species.

Artemisia arborescens (Vaill.) L.: Morphological, Botanical and Phytochemical Aspects

Artemisia arborescens (Vaill.) L. is a shrub, 1-2.5 m high, with erect, woody and whitish branches, and partitioned, strongly-smelling leaves, white-silvery, silky and almost shining on the upper surface. The small yellow flowers are tubular, gathered in characteristic head inflorescence. The fruit is an achene (Tutin *et al.*, 1964-80; Pignatti, 1982). It is a spring-flowering (April-May) nanophanerophyte (or phanerophyte), heliophilous, xerophilous and thermophilous, which fits to a variety of substrates, however preferring open areas with arid, often degraded and turned over soils, sometimes characterized by a marked nitrophily.

The entity denotes an ancestral anthropic character, such as element of basiphilous and synanthropic maquis (*e.g.*, archaeological sites, old ruins, etc.). Under the phytosociological aspect it is considered as a typical ("character") species of the alliance *Oleo sylvestris-Ceratonion siliquae* (Brullo andMarcenò, 1985; Minissale *et al.*, 2005; Brullo *et al.*, 2009); it falls to be part of more xeric vegetation aspects, ascribed to the alliance *Artemision arborescentis* (Biondi *et al.*, 1988; Minissale *et al.*, 2005), in turn framed in the order *Salsolo vermiculatae-Peganetalia harmalae* (class *Pegano harmalae-Salsoletea vermiculatae*). The species is widespread along the coastal belt of the western Mediterranean, west up to the Iberian Peninsula, east to Greece and along the coasts of North Africa, from Morocco to Tunisia (Tutin *et al.*, 1964-80; Pignatti, 1982). In Sicily it is more or less common throughout the island, up to 500-600 m a.s.l. (Lojacono, 1888-1909; Giardina *et al.*, 2007), and reported for many inner and coastal areas (Marcenò *et al.*, 1985; Gianguzzi *et al.*, 2005, 2006, 2011; Caldarella *et al.*, 2009) and circum-Sicilian small islands (Ferro and Furnari, 1970).

The major phytochemical aspects of *A.arborescens* have been elucidated with a satisfactory degree of detail, and many works have been concerned in the chemical composition of its essential oil (Table 13.1). One of the most characteristic and interesting compounds is thujone, an oxygenated monoterpene commonly found in spice and medicinal plants such as sage (*Salvia officinalis* L.), clary sage (*Salvia sclarea* L.), wormwood (*Artemisia absinthium* L.) and tansy (*Tanacetum vulgaris* L.). Thujone is present in some alcoholic beverages, and occurs in essential oils both as α- and β- stereoisomers. The isomer α -thujone, however, is claimed to be toxic for human consumption (EC, 2003; Lachenmeier *et al.*, 2006; NTP, 2011),with a higher toxicity level than β-thujone or than a racemic mixture of both isomers (NTP, 2011). Thus, the use of thujone undergoes several restrictions, although the average intake by consumers is about 100 times lower than the NOEL (no observed-effect level) derived from a 14-week study in rats (EC, 2003). In many genotypes of *A. arborescens*, α - and β -thujone may be detected at very high amounts (up to 1.3 and 68 per cent, respectively), but many populations have shown a very low content of both isomers, or nothing at all.

Chamazulene, found in amounts higher than 49 per cent in plants from Sicily (Militello *et al.*, 2012), is considered the second most important compound in the essential oil of *A. arborescens*. It is a sesquiterpene hydrocarbon, characteristic of several plants of *Asteraceae* family, such as chamomile and achillea. Actually, it is an artifact molecule obtained during steam distillation from the precursor prochamazulene contained in plants (Franz and Novak, 2010; Wagner *et al.*, 2005), and is responsible for the dark blue colour of the essential oil (Sacco *et al.*, 1983). Chamazulene possesses interesting and well-known anti-inflammatory properties (Salamon, 2009), and for this reason it has relevant industrial and pharmaceutical importance.

In some genotypes, another largely represented compound is camphor, sometime more abundant than β -thujone and chamazulene.

Notwithstanding this high variability, attempts have been made to class the species on the basis of the composition of essential oil, and the occurrence of high-thujone, high-chamazulene and intermediate (β-thujone/chamazulene) chemotypes,

Table 13.1: Composition (per cent) of the Essential Oil of *Artemisia arborescens* (Vaill.) L. as Reported by Several Bibliographic Sources, Listed in Order of Decreasing β-thujone

Compounds	*El Beyrouthy et al., 2011 (Lebanon)*	*Biondi et al., 1993 (Italy) Liguria*	*Biondi et al., 1993 (Italy) Apulia 1*	*Militello et al., 2012 (Italy) Sicily 3*	*Ornano et al., 2013 (Italy) Sardinia*	*Militello et al., 2012 (Italy) Sicily 2*	*Militello et al., 2011 (Italy) Sicily*	*Militello et al., 2012 (Italy) Sicily 4*	*Biondi et al., 1993 (Italy) Ischia*	*Militello et al., 2012 (Italy) Sicily 5*
Monoterpene hydrocarbons	4,5	2,7	3,9	11,2	2,9	6,2	8,3	12,2	5,6	15,8
α-thujene	0,7	0,3	0,2	t		t		t	t	t
sabinene	0,8	0,6	0,6	1,4	1,3	t	2,3	2,0	2,5	3,0
myrcene	1,1	0,6	1,4	5,1	0,1	2,5	2,0	5,4	0,6	6,8
others	1,9	1,2	1,7	4,7	1,5	3,7	4,0	4,8	2,5	6,0
Oxygenated monoterpenes	73,1	60,3	60,6	64,6	57,2	54,8	56,6	61,9	59,0	56,6
1,8-cineole	0,4	0,2	0,2	0,2	0,2	0,2		0,2	1,8	0,2
α-thujone	0,9	1,0	0,9	1,5	1,3	1,0	0,8	0,8	0,8	0,5
β-thujone	68,5	58,4	58,1	55,9	53,2	49,5	45,0	44,0	42,9	41,7
linalool	1,0	0,1	0,1	t		t	0,2	t	3,4	t
camphor	0,7	0,1	t	2,3		1,3	6,8	10,7	5,4	6,5
terpinen-4-ol	1,5	0,4	1,0	1,8	1,0	1,2	2,2	1,5	0,1	3,9
others	0,1	0,1	0,3	2,9	1,5	1,6	1,7	4,7	4,6	3,8
Sesquiterpene hydrocarbons	13,6	22,4	21,8	19,6	6,1	30,0	27,1	19,3	25,8	21,8
α-copaene	0,4	0,3	0,1		0,4		0,2		0,1	
β-caryophyllene	0,2	1,1	1,8	0,8	1,3	0,9	0,9	0,7	1,4	0,8
germacrene-D	0,7	3,4	0,2	3,1	4,3	3,4	3,3	2,4	0,5	2,3
chamazulene	12,3	17,3	19,2	15,2		24,9	22,7	15,7	23,4	18,5
others	0,0	0,3	0,5	0,5	0,1	0,8	0,0	0,5	0,4	0,2

Contd...

Table 13.1–*Contd...*

Compounds	*El Beyrouthy et al., 2011 (Lebanon)*	*Biondi et al., 1993 (Italy) Liguria*	*Biondi et al., 1993 (Italy) Apulia 1*	*Militello et al., 2012 (Italy) Sicily 3*	*Ornano et al., 2013 (Italy) Sardinia*	*Militello et al., 2012 (Italy) Sicily 2*	*Militello et al., 2011 (Italy) Sicily*	*Militello et al., 2012 (Italy) Sicily 4*	*Biondi et al., 1993 (Italy) Ischia*	*Militello et al., 2012 (Italy) Sicily 5*
Oxygenated sesquiterpenes	0,5	0,5	0,3	0,5	0,0	0,4	0,7	0,4	0,2	0,4
caryophillene oxide	0,3	0,2	0,2	0,3		0,3	t	0,3	0,1	0,2
geranyl isovalerate	0,2	0,3	0,1				0,7		0,1	
others	0,0	0,0	0,0	0,2	0,0	0,1	0,0	0,1	0,0	0,2
Phenylpropanoids	0,0	0,0	0,0	0,0	0,2	0,0	0,0	0,0	0,0	0,0
methyl eugenol				t	0,2	t	t	t		t
isochavicol isobutyrate										
Other compounds	**0,7**	**0,4**	**0,8**	**0,9**	**0,1**	**0,3**	**1,0**	**0,6**	**0,8**	**0,8**
Total Compounds	**92,4**	**86,3**	**87,4**	**96,8**	**66,5**	**91,7**	**93,7**	**94,4**	**91,4**	**95,4**

Contd...

Table 13.1–*Contd...*

Compounds	*Biondi et al., 1993 (Italy) Sardinia (1)*	*Marongiu et al., 2006 (Italy) Sardinia (2)*	*Pappas and Sheppard-Hanger, 2000 (Morocco)*	*Abderrahim et al., 2010 (Algeria)*	*Lai et al., 2007 (Italy) Sardinia*	*Militello et al., 2012 (Italy) Sicily 1*	*Marongiu et al., 2006 (Italy) Sardinia (1)*	*Lo Presti et al., 2007 (Italy) Lipari isl.*	*Biondi et al., 1993 (Italy) Apulia 2*	*Lo Presti et al., 2007 (Italy) Sicily*
Monoterpene hydrocarbons	4,3	7,7	28,2	0,0	6,6	7,7	2,6	4,3	19,2	4,5
α-thujene	0,4		0,3			0,2			2,8	
sabinene	1,0	1,8	4,0			1,3	0,8	0,3	0,2	0,1
myrcene	1,3	1,1	9,1			1,6	0,2	2,1	8,7	2,0
others	1,6	4,8	14,8	0,0	6,6	4,6	1,6	1,9	7,5	2,4
Oxygenated monoterpenes	51,1	49,0	59,6	30,8	61,8	36,6	22,9	33,5	45,7	26,9
1,8-cineole	0,2					0,4			0,1	
α-thujone	t	1,3	1,0	0,6		0,9	0,4		t	
β-thujone	42,1	33,7	30,1	27,8	23,9	20,5	14,0	6,6	3,8	1,5
linalool	0,1	1,5	2,1			1,5	0,3	1,2	0,1	0,7
camphor	4,9	6,7	21,7		35,7	8,4	6,2	20,1	35,5	21,4
terpinen-4-ol	1,3	3,9	3,3	1,8	2,2	1,0	0,1	3,2	4,2	2,0
others	2,5	1,8	1,4	0,6	0,0	3,9	1,9	2,4	2,0	1,3
Sesquiterpene hydrocarbons	25,1	32,1	4,4	32,1	11,7	53,9	9,3	44,6	18,3	46,5
α-copaene	0,2	0,2	0,3	0,1		0,4	0,4	0,3	t	0,4
β-caryophyllene	0,3	0,8	0,4	0,6		0,6	1,3	4,8	0,7	3,5
germacrene-D	0,3	1,6	2,1	1,2		2,8	0,9	3,9	0,7	4,5
chamazulene	24,2	26,6	1,5	30,2	7,7	49,4	6,0	34,6	16,2	37,6
others	0,1	2,8	0,2	0,0	4,1	0,7	0,8	1,0	0,7	0,5

Contd...

Table 13.1–*Contd...*

Compounds	*Biondi et al., 1993 (Italy) Sardinia (1)*	*Marongiu et al., 2006 (Italy) Sardinia (2)*	*Pappas and Sheppard-Hanger, 2000 (Morocco)*	*Abderrahim et al., 2010 (Algeria)*	*Lai et al., 2007 (Italy) Sardinia*	*Militello et al., 2012 (Italy) Sicily 1*	*Marongiu et al., 2006 (Italy) Sardinia (1)*	*Lo Presti et al., 2007 (Italy) Lipari isl.*	*Biondi et al., 1993 (Italy) Apulia 2*	*Lo Presti et al., 2007 (Italy) Sicily*
Oxygenated sesquiterpenes	0,4	2,1	0,0	8,7	0,0	0,2	7,3	3,3	0,4	1,0
caryophillene oxide	0,1	0,4		0,6		0,2	0,3	1,5	0,2	0,9
geranyl isovalerate	0,3							0,8	0,2	
others	0,0	1,7	0,0	8,1	0,0	0,0	7,0	1,0	0,0	0,1
Phenylpropanoids	0,0	0,0	0,0	0,0	0,0	0,0	0,0	0,2	0,0	0,1
methyl eugenol						t		0,2		0,1
isochavicol isobutyrate										
Other compounds	0,5	0,0	0,0	0,0	0,0	0,0	0,0	0,1	1,4	0,1
Total Compounds	**81,4**	**90,9**	**92,1**	**71,6**	**80,1**	**98,4**	**42,1**	**86,0**	**85,0**	**79,1**

Contd...

Table 13.1–*Contd...*

Compounds	*Younes et al., 2012 (Algeria) Bidar*	*Younes et al., 2012 (Algeria) Chetouane*	*Younes et al., 2012 (Algeria) Beni Snous*	*Biondi et al., 1993 (Italy) Tremiti*	*Baykan Erel et al., 2012 (Turkey) Anatolia*	*Pappas and Sheppard-Hanger, 2000 (USA)*	*Bouzenna and Krichen, 2013 (Tunisia)*	*Lo Presti et al., 2007 (Italy) Calabria*	*Biondi et al., 1993 (Italy) Sardinia 2*
Monoterpene hydrocarbons	15,9	10,5	5,2	20,0	8,8	16,8	0,0	4,0	8,9
α-thujene	0,1	0,1	0,1	1,8		0,4			0,9
sabinene	0,7	0,1	0,9	0,1		0,2			0,1
myrcene	7,4	2,7	2,0	9,3	1,5	5,1		1,0	1,8
others	7,7	7,6	2,2	8,8	7,3	11,1	0,0	3,0	6,1
Oxygenated monoterpenes	61,8	50,9	78,4	43,8	41,4	22,5	49,2	45,9	37,8
1,8-cineole	0,2	0,3	0,1	0,2					0,1
α-thujone	0,3	0,4	0,2	t	1,3				
β-thujone	1,1	1,0	0,9	0,5	0,1				
linalool				1,6		0,7	5,3	0,8	0,2
camphor	50,3	32,8	72,2	32,5	33,4	16,7	25,8	39,5	32,5
terpinen-4-ol	6,1	8,9	2,1	7,3	4,8	2,1		3,7	4,0
others	3,8	7,5	2,9	1,7	1,8	3,1	18,1	1,9	1,0
Sesquiterpene hydrocarbons	6,4	14,8	2,6	18,8	23,1	51,0	34,3	31,7	39,1
α-copaene	0,2	0,2				0,4		0,2	
β-caryophyllene	0,3	0,7	0,4	0,7	2,1	3,6	1,2	2,6	0,4
germacrene-D	2,1	3,5	1,5	1,3		7,2	1,2	1,3	0,9
chamazulene	3,2	8,7	0,1	16,5	21,1	39,6	31,9	27,1	37,5
others	0,6	1,7	0,6	0,3	0,0	0,2	0,0	0,5	0,3

Contd...

Table 13.1–*Contd...*

Compounds	*Younes et al., 2012 (Algeria) Bidar*	*Younes et al., 2012 (Algeria) Chetouane*	*Younes et al., 2012 (Algeria) Beni Snous*	*Biondi et al., 1993 (Italy) Tremiti*	*Baykan Erel et al., 2012 (Turkey) Anatolia*	*Pappas and Sheppard-Hanger, 2000 (USA)*	*Bouzenna and Krichen, 2013 (Tunisia)*	*Lo Presti et al., 2007 (Italy) Calabria*	*Biondi et al., 1993 (Italy) Sardinia 2*
Oxygenated sesquiterpenes	2,9	4,1	3,0	0,1	12,0	0,0	1,8	3,2	0,4
caryophillene oxide	0,1	0,3	0,1	0,1	4,4		1,1	0,6	0,1
geranyl isovalerate	0,2	0,3	0,5	t			0,8	1,7	0,3
others	2,6	3,5	2,4	0,0	7,7	0,0	0,0	0,9	0,0
Phenyl propanoids	0,2	0,4	0,3	0,0	0,0	0,0	0,6	0,3	0,0
methyl eugenol	0,2	0,3	0,2				0,6	0,3	
isochavicol isobutyrate		0,1	0,1						
Other compounds	1,0	1,1	0,7	0,8	0,0	0,0	0,0	0,1	0,3
Total Compounds	**88,2**	**81,8**	**90,2**	**83,5**	**85,3**	**90,2**	**85,9**	**85,2**	**86,5**

is acknowledged by many Authors (Marongiu *et al.*, 2006; El Beyrouthy *et al.*, 2011; Militello *et al.*, 2012; Ornano *et al.*, 2013). Some utility should have, in this sense, the study of other chemical compounds, such as lignans, sesquiterpene lactons and flavonoids, that could maybe be used as intra-specific markers (Betto *et al.*, 1988).

Of course, since the diverse compounds have very different industrial and pharmaceutical targets, a more detailed characterization of such chemotypes could be very useful in addressing breeding efforts. It is expected for the future a better development of knowledge as concerns the sources of variability in essential oil composition, that are probably many. A variation in essential oil composition has been found with varying the collection season, with an increase in β-thujone and a decrease in chamazulene when the plants pass from vegetative to flowering stage (Militello *et al.*, 2012; Ornano *et al.*, 2013). Furthermore, it is worth noticing that, to some extent, the essential oil composition may undergo modifications according to the extraction method (Marongiu *et al.*, 2006).

Many trials carried out all over the world allowed to state that *A. arborescens* exerts a wide range of activities (antiviral, anti-inflammatory, antiallergenic, antihistamine, anti-catarrhal, choleretic, mucolytic) and is therefore largely appreciated in aromatherapy (Sheppard-Hanger, 1995; Zarga *et al.*, 1995; Saddi *et al.*, 2007).

Ethnobotanical studies have recorded the use of its leaves against dermatitis and skin inflammations, as vermifuge and antihelmintic, and in decoction against malaria fever (Amico and Sorce, 1997; Lentini, 2000).

Insecticidal Activity

The bio-based insecticides exert their activity through a series of actions ranging from the bare repression of a special development stage of the phytophagous, to the induction of reactions and metabolic processes that may ultimately lead it to death. Hence, many different mechanisms may be involved, *e.g.* a repellant activity (that modifies the response of the phytophagous to the host-plant); a general modification of the substrate, that becomes less suitable for ovideposition or feeding; the transmission of information about an unsuitable host composition; an interference with the adsorption of nutrients; or, finally, a direct toxicity effect (Gahukar, 2014).

For their best exploitation, besides being effective, these compounds furthermore must not have any noxious effect on photosynthesis, growth or further aspects linked to the metabolism and to the physiological performance of the crops to be protected, must show selectivity for natural enemies and lack of toxicity for mammalians and "useful" organisms, and be degradable in the environment (Jacobson *et al.*, 1978; Islam, 1984; Di Ilio *et al.*, 1999; Zehnder *et al.*, 2007; Tsolakis and Ragusa, 2004, 2008). The occurrence of all these constraints poses a serious limit to a wider diffusion of natural products in farms. Nevertheless, many studies have been carried out about this topic, and further developments will surely solve such crucial issues.

The essential oil of *A. arborescens* demonstrated a high variability in insecticidal/ nematicidal effect, according to the target species. Some nematicidal effect of the essential oil (13.5 per cent mortality) was stated against the pine wood nematode,

Bursaphelenchus xylophilus (Kim *et al.*, 2008). A low effectiveness was shown against the onion thrips, *Thrips tabaci* (van Tol *et al.*, 2007). By contrast, other studies highlighted a good activity level of *A. arborescens* extracts against *Aphis gossypii* (a pest of citrus fruits), adult and neanids of *Bemisia tabaci* and adults of *Lymantria dispar* (a pest of *Quercus* spp.) (Lai *et al.*, 2003), whereas the extracts from the hybrid *A. arborescens x absinthium* were effective against young larvae of codling moth, *Cydia pomonella* (Durden *et al.*, 2009).

It is often assumed, however, that an effective pest control strategy should preferably involve the simultaneous recourse to several different control means, in an "integrated" overall management (Zehnder *et al.*, 2007). Hence, in all cases when the use of chemicals is not strictly forbidden, an attractive idea is to use natural compounds in addition to chemically-derived pesticides, in order to reduce the use of the most toxic ones. With this purpose, Militello (2012) tested the possibility to use *A. arborescens* essential oil as a synergistic agent for mineral oil in the control programs against some mealybugs, so avoiding the use of very toxic products such as phospho-organics. Three treatments were put into comparison: mineral oil, mineral oil plus *A. arborescens* essential oil (in three different formulations), and a control with water and an emulsifying agent (Table 13.2). The treatments were arranged by plunging in each test solution 10 *Citrus lemon* fruits of lemon for 3 seconds and then exposing the fruits to air for drying. After the treatment, the fruits were put in a thermoregulated chamber

Table 13.2: Insecticidal Activity of the Essential Oil of *A. arborescens* L. against *Aoenidiella aurantii;* Mortality Expressed in per cent Over 500 (50 x 10) Mealybugs Individuals (Militello, 2012)

Composition of the Tested Extracts				
Tests	*Essential Oil of A. arborescens (per cent)*	*Mineral Oil (per cent)*	*Mortality per cent*	*Statistical Significance (Fisher's Test)*
C (control)	–	–	44.9	
M_{all} (mineral oil)				
M_2	–	2	+++	***
M_1	–	1	+++	***
$M_{0,5}$	–	0,5	–	n.s.
MA_{all} (mineral oil + *A. arborescens* essential oil)				
M_2A_2	2	2	+++	***
M_1A_1	1	1	+++	***
$M_{0,5}A_1$	1	0,5	+	***
Orthogonal contrasts analysis				
M_{all} *vs* C				***
MA_{all} *vs* C				***
MA_{all} *vs* M_{all}				***
$M_{0,5}A_1$ *vs* C				***
$M_{0,5}A_1$ vs $M_{0,5}$				***

(T 28 °C, R.H. 70 per cent, day/night 18/6 hh). 15 days after, the mortality of 50 mealybugs individuals for each fruit was counted, considering "dead" all individuals that had not completed the turning from the first development stage to the second. The addition of essential oil of *A. arborescens* enhanced the effect of the treatment with mineral oils: the species seems therefore able to substitute the phospho-organicals even when the essential oil is added in the formulation treatment at very low doses.

Antimicrobial Activity

A great interest is devoted to study the antimicrobial effects of naturally-derived products, and a number of plant derivatives, above all essential oils, have been tested against many microbial species responsible for food alterations, as well for human and animal diseases (Burt, 2004). The application fields of such researches, indeed, are many and diverse, involving agro-food and pharmaceutical industry, alternative medicine and natural therapies (Hammer *et al.*, 1999).

The antimicrobial activity of the essential oil of *A. arborescens* was tested by several Authors (Baykan Erel *et al.*, 2012; Militello *et al.*, 2011;Younes *et al.*, 2012) by means of the disc diffusion method (Table 13.3); with this method, discs of sterile filter paper are wet with fixed amounts (20-30µL) of the substance to be tested, and put on the surface of substrate (generally liquid agar), where the fungal strains had been previously inoculated. After 24 hours incubation, the development areas of the fungal colonies are measured and expressed in mm. The controls are normally represented by a paper disc where 10 ml of sterile water and acetone had been added (negative control) and one with streptomycin (positive control). Although the comparison between different tests on antimicrobial activity is often problematic, due to the numerous differences that show up between the working protocols of every research group (Smith-Palmer *et al.*, 1998; Hammer *et al.*, 1999), it is possible to see that *A. arborescens* exhibits a great potential as an antimicrobial agent. A deep attention, however, must be paid to the choose of the proper plant genotype, since a chemotype-based variability in the response is evident, as found in other similar works (Pauli and Schilcher, 2010).

Fungicidal Activity

Fungicides are largely used all over the world. They exert their action by means of the fungal growth inhibition, or through a damage to fungal cells, or preventing the fungal development (Martinez, 2012).The fungicidal activity of *A. arborescens* (Table 13.4) was evaluated *in vitro* against some colonies of *Fusarium oxysporum* f. sp. *dianthi* (isolated from clove); *Verticillium dahliae* (isolated from tomato), *Phythophtora citrophtora* (isolated from *Citrus* spp.) and *Botritys cinerea*, all of them pathogenic fungi largely represented both in field and in greenhouses.

The efficacy of *A. arborescens* essential oil against fungal strains was tested at four concentrations (1x10; 1x10^2, 5x$10^2$and 1x10^3µg ml^{-1}), compared with an untreated control prepared by adding to the substrate only ethanol and sterile distilled water. The essential oil was diluted in ethanol (C_2H_5OH) and 100 µl were distributed in the center of Petri dishes(Ø 9 cm). Each treatment was repeated four times, and every test was repeated twice. After 24 hr. from 6-7 days-old fungal colonies previously grown

Table 13.3: Antimicrobial Activity of Essential Oils (EO) and Methanolic Extract (ME) of *Artemisia arborescens* L.[1]

Microbial Strains	*Baykan Erel et al., 2012*[2]		
	EO	*ME*	
SA	–	–	
SE	+	–	
ST	–	–	
PA	++	++	
EF	–	–	
ECL	–	–	
ECO	–/+	–/+	
	Militello et al., 2011[3]		
	EO		
EB	–		
LY	+/++		
SAL	–		
		Younes et al., 2012[4]	
	EO Tipe I	*EO Type II*	*EO Type III*
SA	+	++	++
BC	–	++	++
EF	+	++	++
LY	–	–	–
PA	+	++	++
EC	–	–	–
ST	–	++	++
AB	+	++	++
CF	+	++	++
PM	+	++	++
KP	–	++	++

1: No inhibition; +: 8-10 mm inhibition; ++: > 10 mm inhibition.

2: SA: *Staphylococcus aureus* (ATCC 6538/P), SE: *Staphylococcus epidermidis* (ATCC 12228), ST: *Salmonella typhimurium* (CCM5445); PA: *Pseudomonas aeruginosa* (ATCC 27853); EF: *Enterococcus faecalis* (ATCC29212); ECL: *Enterobacter cloacae* (ATCC13047); ECO: *Escherichia coli* (x 2 strains).

3: EB: *Enterobacter* (x 4 species; 10 strains total); LY : *Listeria monocytogenes* (x 16 strains); SAL : *Salmonella* x 17 species.

4: SA: *Staphylococcus aureus* (ATCC 25923); *Bacillus cereus* (ATCC 10876); EF: *Enterococcus faecalis* (ATCC 49452); LY: *Listeria monocytogenes* (ATCC 15313); PA: *Pseudomonas aeruginosa* (ATCC 27853); EC: *Escherichia coli* (ATCC 25922); ST: *Salmonella typhimurium* (ATCC 13311); AB: *Acinetobacter baumanii* (ATCC 19606); CF: *Citrobacter freundii* (ATCC 8090); PM: *Proteus mirabilis* (ATCC 35659); KP: *Klebsiella pneumoniae* (ATCC 700603).

Table 13.4: Fungicidal Activity of the Essential Oil of *Artemisia arborescens* L. (Militello, 2012)

Fungi	*10 µl/ml*	*100 µl/ml*	*500 µl/ml*	*1000 µl/ml*
		Inhibition [1]		
		BC_1 [2]		
Fusarium oxysporum f. sp. *dianthii*	+	+	++	++
Verticillium dahliae	+	+	+	++
Phytophthora citrophthora	+	++	+	++
Botrytis cinerea	+	+	–	++
		BC_2		
Fusarium oxysporum f. sp. *dianthii*	–	+	–	+
Verticillium dahliae	–	–	–	+
Phytophthora citrophthora	–	++	–	–
		BTHU		
Fusarium oxysporum f. sp. *dianthii*	–	+	+	+
Verticillium dahliae	–	+	–	++
Phytophthora citrophthora	–	+	+	+
		CHAM		
Fusarium oxysporum f. sp. *dianthii*	–	–	+	++
Verticillium dahliae	–	+	+	+++
Phytophthora citrophthora	–	++	+	++

1: Low inhibition (≤ 5 per cent); + to +++: increasing inhibition (+ 6 per cent to 25 per cent; ++ 26 per cent to 50 per cent; +++ >50 per cent).

2 BC_1, BC_2: essential oil of two *Artemisia arborescens* L. chemotypes dealing with equal percentages of β-thujone and chamazulene; BTHU: essential oil with prevailing β-thujone; CHAM: essential oil with prevailing chamazulene.

in PDA (Potato Dextrose Agar), a 8 mm-diameter mycelium segment was picked up, and transferred at the center of each plate. The plates were finally sealed with Parafilm® and put in the dark at 22±1°C.

The inhibitory effect of the *A. arborescens* essential oil was evaluated by measurement of the radial growth of the colonies (average of two orthogonal diameters) after 3, 6 and 9 days.

The best inhibitory activity was achieved with the highest concentration ($1x10^3$µg ml^{-1}) and using the CHAM(high-chamazulene) oil, that caused 65.2 per cent, 40.4 per cent and 34.5 per cent inhibition of fungal growth in colonies of *V. dahliae, F. oxysporum* and *P.citrophthora* respectively, whereas at lower concentration ($1x10^2$µg·ml^{-1}), 31.0 per cent inhibition was measured only on *P. citrophthora.* The treatments with the BTHU oil (with prevailing β-thujone) gave the highest inhibition against colonies of *V. dahliae* (35 per cent), *P. citrophthora* (24.6 per cent) and *F. oxysporum* (22.3 per cent), always at the maximum concentration. As a matter of fact, some Authors (Kordali *et al.*, 2005; Giordani *et al.*, 2008) have found that β-thujone, although toxic, does not always show a fungicidal activity, and if present, such activity is however highly specific. Other compounds, such as camphor or 1.8-cineole, seem to exhibit a wider activity range (Kordali *et al.*, 2005; Farzaneh *et al.*, 2006; Giordani *et al.*, 2008; Zuzarte *et al.*, 2011). β-caryophillene as well showed a fair fungicidal activity, mostly against *Fusarium* sp.(Kordali *et al.*, 2005; Zuzarte *et al.*, 2011). Hence, the differences in chemical composition should explain the higher fungicidal activity of the high-chamazulene essential oil, dealing with a higher amount of camphor, and of the BC1 type with respect to BC2 (5.5 per cent camphor in the first type and 3.7 per cent in the second). However, the inhibitory activity is rarely attributable to the effect of one compound only, rather to the synergic action of several compounds (Farzaneh *et al.*, 2006; Giordani *et al.*, 2008).

Herbicidal Activity

The studies about the possibility to use plant extracts and natural molecules as herbicides are rather more recent than those addressed to the exploitation of other plants' biocidal effects; this is probably due to the high incidence in agricultural practice of synthetic herbicides, that are usually highly effective, selective toward crops and low-costing (Dayan *et al.*, 2009). Nevertheless, the expansion of organic agriculture, the increasing resistance phenomena in weeds, and environmental concerns about the fate of chemicals in the ecological chains gave a strong impulse also to this field of use of natural substances (Scheepens *et al.*, 2001; Dayan *et al.*, 2009; Choi and Wang, 2013).

A great deal of references holds about the study of various aspects of allelopathic effects. In a proper sense, the term "allelopathy" relates to the effect observed on all those molecules which are emitted into the environment by living plants, and somehow affect the surrounding flora (Einhellig, 1995). Such a phenomenon is well known for several species, such as some *Salvia* species, in which many volatile compounds emitted by the roots (mainly terpenes such as camphor, camphene,1,8-cineole and β-pinene), inhibit germination and growth of the neighbor plants (Kegge and Pierik, 2009). Hence, those studies may be considered as opening the way to more detailed researches, addressed to the possibility to use plant extracts as directly-acting herbicides. Many experiments have been conducted taking into consideration plant phytocomplexes, *i.e.*, extracts of plants (essential oils, or water and/or alcohol extracts) (Dudai *et al.*, 1999; Tworkoski, 2002; Angelini *et al.*, 2003; Verdeguer *et al.*, 2012; Gitsopoulos *et al.*, 2013). Several efforts aimed to isolate the single compound responsible for the specific herbicidal action. Some have been identified in the class of terpenes, the commonest being α- and β-pinene, camphene, limonene, α-phellandrene, *p*-cymene, borneol and camphor (Angelini *et al.*, 2003), whereas the most active belong to the chemical group of cineols (Duke *et al.*, 2002).

It must be observed, however, that (1) the most abundant compounds is not necessarily the "active" one; and (2) interferences are always possible among compounds acting in mutual synergism or antagonism. These constraints set some additional difficulty to the identification of single molecules to point out for industrial development; anyway, the studies about "whole" natural extracts are still far to have exhausted their potentialities.

Studies about the herbicidal activity of *A. arborescens* were made using both essential oils (Militello, 2012) and water extracts (unpublished data), modifying the procedure already adopted by Verdeguer *et al.* (2012)on extracts from *Cistus ladaniferus*. The inhibitory activity of the essential oils was evaluated taking into account the germination of the seeds of a few species, some of them very common weeds in the cropping systems of Mediterranean basin: *Portulaca oleracea, Araujia sericifera* and *Conyza australis*. Samples of 20 (*Portulaca* and *Conyza*) and 10 seeds (*Araujia*) were put to germinate in Petri dishes (10 cm diameter), between two layers of filter paper (Whatman No. 3) wet with 4 ml of distilled water. The inhibition of the germinative ability was tested by putting on the paper disc covering the seeds fixed amounts of mixtures that had been previously prepared adding the expected essential oil quantity on 10 ml of acetone: water 1:1 solution, in order to guarantee a proper emulsion of the Artemisia oil. Similar set-up but without essential oil served as control. Each treatment was replicated 5 times for *Portulaca* and *Conyza* and 10 times for *Araujia*. Petri dishes were closed and sealed with a stretched Parafilm® to prevent oil volatilization and were incubated under controlled conditions (25°C, 70 per cent relative humidity, 16-h photoperiod).After 3, 6 and 10 days the number of germinated seed was counted and, when present, the length of sprouts and roots was measured(Dudai *et al.*, 1999; Verdeguer *et al.*, 2012).

As shown in Figure 13.1, the essential oil of *A. arborescens* showed a selective herbicidal activity depending on the weed: *P. oleracea* showed a high tolerance to the treatment, whereas the germination of seeds of *A. sericifera* and *C. australis* was greatly affected by the addition of essential oil. In particular, the doses of 0.5 and 1 ppm totally inhibited germination in *Araujia sericifera,* and *C. australis* experienced a decrease of germination of 81 per cent and 98 per cent, respectively.

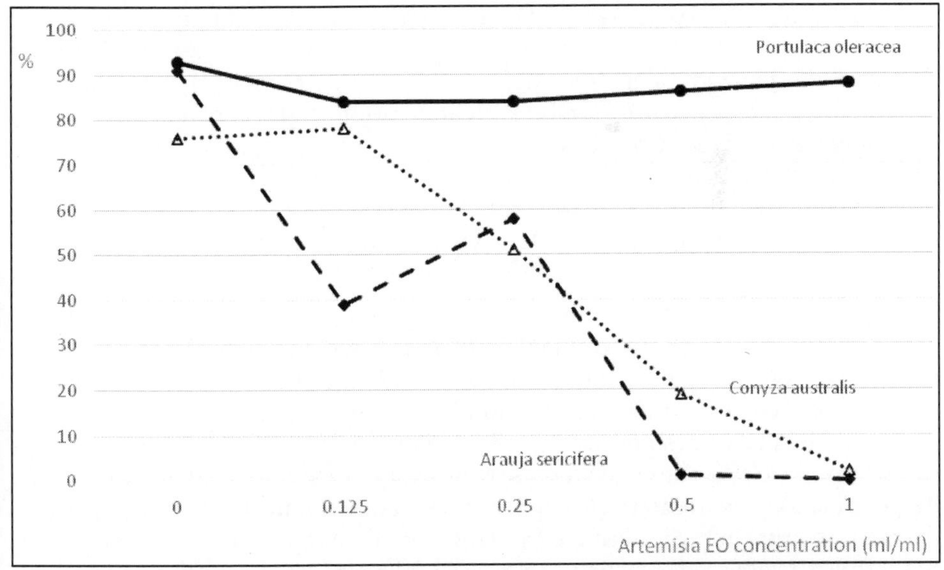

Figure 13.1: Effects of different Concentrations of *A. arborescens* Essential Oil on Seeds Germination (Per cent) in different Herbaceous Species (Militello, 2012).

The preliminary results of the evaluation of the antigerminative effects of *A. arborescens* water extracts are reported instead in Figure 13.2. For these evaluations, a solid/liquid extraction was performed, starting from plant material collected from the wild at full blooming time.

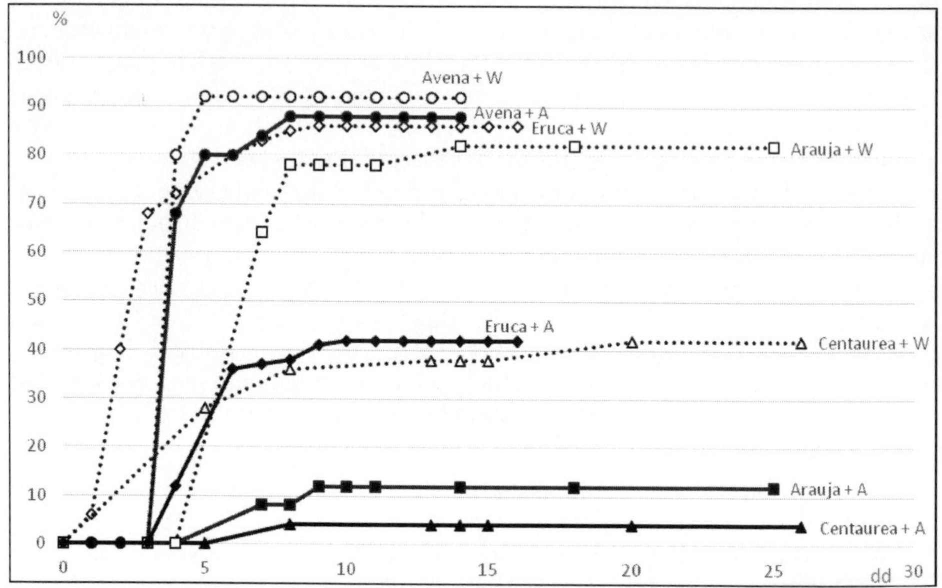

Figure 13.2: Effects of *A. arborescens* Water Extracts on Seeds Germination Rate (Per cent) in different Herbaceous Species (A: Artemisia extract; W: Water).

The extracts were prepared by putting 100 g of ground plant dry material in 1 l distilled water, and shakering the mixture overnight at room temperature. Thereafter they were filtered with filter paper and stored in refrigerator (4-6 °C) until use. The seeds of the plant species used for the evaluation of effectiveness of the extracts were distributed, 25 seeds each, on discs of filter paper (Whatman no. 4) inside Petri dishes (10 cm diameter). Each dispositive was replicated four times. The seeds used for the experiments came from weeds very common in the Mediterranean environments (*Centaurea diluta* Aimon, *Eruca sativa* L., *Avena fatua* L.), compared with an herbaceous species that had been previously used for evaluation of herbicidal effect of Artemisia essential oil, *i.e., Arauja sericifera.*

5 ml of pure water extract (containing 1.03 per cent of dry extract) were put on the Petri dishes containing the seeds, thereafter the dishes were sealed with Parafilm® and stored at controlled temperature until seeds germination. The controls were made with 4 Petri dishes per each species, watered with 5 ml of distilled water. As shown in Figure 13.2, the effectiveness of the extracts was variable with the species; the germination rate was found to decrease in *Arauja, Eruca* and *Centaurea*. Oppositely, the treatment was not very effective on *Avena*, where, although lower than the control, the germination of seeds remained on very high levels.

Similar results had been obtained by Araniti *et al.* (2012), who found that aqueous extract of *A. arborescens* exhibited the highest phytotoxicity with respect to the extracts from other 16 wild Mediterranean species, causing after 24 and 48 hours an average inhibition of root elongation in seedlings of lettuce of 70 per cent and 84 per cent, respectively, compared to control.

Constraints and Perspectives for Application

The narrow host range of biocontrol products limits their market size (Schepeens *et al.*, 2001).Generally speaking, phytochemicals (secondary metabolites and essential oils) have a shorter shelf-life in comparison with synthetic pesticides; *e.g.*, in essential oils, their feeble chemical stability to air, light, moisture and high temperatures causes the quick evaporation or degradation of some active compounds. Such a character, above all linked to the complexity of involved molecules, is one of the most typical traits of natural products.

Unfortunately sometimes, common agricultural practices still tend to consider this trait as a defect, and pesticides, in order to be economically suitable, are required to show persistence. Another important constraint to use is given by the strong variability in concentration and biological activity of the active principles that are responsible of the activity against pests and weeds. With the purpose to achieve better safety and effectiveness of pesticides, special systems for controlled release have been pointed. As an example, a constant and controlled release of the active substances may be obtained by means of antioxidants or various protective agents, or by means of the adsorption of the active compounds on nanostructured substrates (Ahola *et al.*, 2000; Liu *et al.*, 2001; Parris *et al.*, 2005; Perlatti *et al.*, 2013). Nanotechnology is becoming increasingly important for the agricultural sector allowing to address the urgent issues of environmental protection and pollution. In comparison to micronic particles (≥ 1000 nm), nanoparticles (< 1000 nm) offer the advantage of effective loading due to the larger surface area, easy attachment and fast mass transfer. The active ingredient is adsorbed, attached, encapsulated or entrapped unto or into the nano-matrix. The use of nanomaterials for pesticides may reduce the doses ensuring a controlled and slow release of the active ingredient, reducing the pollution risks, preventing their quick evaporation and degradation and therefore increasing their stability, also maintaining the minimum effective dosage per application (Perlatti *et al.*, 2013).

Some interesting application examples have regarded the essential oil from garlic, that was loaded on polymer nanoparticles (240 nm) to evaluate their insecticidal activity against adult *Tribolium castaneum* (Yang *et al.*, 2009).The oil-loading efficiency reached 80 per cent and remained unchanged after five months, while the free garlic essential oil has an efficiency of 11 per cent. Nanoparticles based on polymeric materials were successfully experimented for controlled release of fungicides for solid wood preservation (Liu *et al.*, 2001) and pesticides ivermectin and acetamiprid (Zheng and Shang, 2005).

In several studies carried out to evaluate the possibility to incorporate the *A. arborescens* essential oil into nanoparticles, formulations were obtained that showed a high physical stability and a good capability to reduce the essential oil evaporation, resulting highly promising as carriers in order to use of this essential oil in agriculture. (Lai *et al.*, 2003, 2006, 2007).

Conclusions

Many environmental claims, *e.g.*, about the increasing of the resistant strains of different phytophagous species, and the chronic toxicity caused by the residues of chemicals in the agricultural products, call attention of market, industry and research towards the development of new molecules, that must be at the same time environmentally friendly and effective against the chosen pest. The active ingredients present in some bio-based pesticides seem to accomplish these conditions.

A. arborescens has shown in many cases a satisfactory activity level which allow to consider it as a resource deserving further studies. The use of *A. arborescens* extracts in food items is limited by the occurrence inside them of highly toxic compounds, such as thujone. An interesting application would be, however, the use of extracts in the storage and packaging materials. The use of controlled release substrates and nanoparticles seems a promising tool to enhance the field efficiency of this natural substance.

References

Abderrahim, A., Belhamel, K.,Chalchat, J.-C., and Figuérédo, G. (2010). Chemical composition of the essential oil from *Artemisia arborescens* L. growing wild in Algeria. *Rec. Nat. Prod.*, **4 (1)**: 87-90.

Ahola, M., Kortesuo, P., Kangasniemi, I., Kiesvaara, J., and Yli-Urpo, A. (2000). Silica xerogel carrier material for controlled release of toremifene citrate. *Int J Pharmaceutics*, **195**: 219–227.

Amico, F.P., and Sorce, *E.G.* (1997). Medicinal plants and phytotherapy in Mussomeli area (Caltanissetta, Sicily, Italy). *Fitoterapia*, **63 (2)**: 143-159.

Angelini, L.G., Carpanese, G., Cioni, P-L., Morelli, I., Macchia, M., and Flamini G. (2003). Essential oils from Mediterranean lamiaceae as weed germination inhibitors. *J. Agric. Food Chem.*, **51**: 6158-6164.

Araniti, F., Sorgona, A., Lupini, A., and Abenavoli, M.R. (2012). Screening of Mediterranean wild plant species for allelopathic activity and their use as bio-herbicides. *Allelopathy Journal*, **29 (1)**: 107-124.

Baykan Erel, S., Reznicek, G., Senol, S.G., KarabayYavasogulu, N.U., Konyalioglu, S., and Zeybek A.U. (2012). Antimicrobial and antioxidant properties of *Artemisia* L. species from western Anatolia. *Turk J Biol.*, **36**: 75-84.

Betto, P., Casinovi, C.G., Gabriele, R., Grandolini, G., and Menichini, F. (1988). Estudio sistematico de los componentes menores de la *Artemisia arborescens* L. (Compositae) de Calabria. *Rev. Latinoamericana Quim.*, 19 **(1)**: 40-42 (In Spanish).

Biondi, E., Allegrezza, M., and Filigheddu, R. (1988). Su alcune formazioni ad *Artemisia arborescens* L. della Sardegna settentrionale. *Boll. Soc. Sarda Sci. Nat.*, **26**: 177-185 (In Italian).

Biondi, E., Valentini, G., Bellomaria, B., and Zuccarello, V. (1993). Composition of essential oil in *Artemisia arborescens* L. from Italy. *Acta Horticulturae*, **344**: 290-304.

Bouzenna, H., and Krichen, L. (2013). *Pelargonium graveolens*L'Her. and *Artemisia arborescens* L. essential oils: Chemical composition, antifungal activity against *Rhizoctonia solani* and insecticidal activity against *Rhysopertha dominica*. *Natural Product Research*, **27 (9)**: 841–846.

Brullo, S., Gianguzzi, L., La Mantia, A., and Siracusa G. (2009). La classe *Quercetea ilicis* in Sicilia. *Boll. Accad. Gioenia Sci. Nat. Catania*, **41 (369)**: 1-77. (In Italian)

Brullo, S., and Marcenò C. (1985). Contributo alla conoscenza della classe *Quercetea ilicis* in Sicilia. *Not. Fitosoc.*, **19 (1)**: 183-229. (In Italian).

Burt, S. (2004). Essential oils: their antibacterial properties and potential applications in foods–a review. *International Journal of Food Microbiology*, **94**: 223– 253.

Caldarella, O., Gianguzzi, L., Romano, S., and Fici S. (2009). The vascular flora of Nature Reserve "Pizzo Cane, Pizzo Trigna and Grotta Mazzamuto" (NW Sicily) – *Webbia*, **64(1)**: 101-151.

Castillo, F., Hernández, D., Gallegos, G., Rodríguez, R., and Aguilar, C.N. (2012). Antifungal Properties of Bioactive Compounds from Plants. In: "Fungicides for Plant and Animal Diseases": 81-106, Dhanasekaran D. (Ed.) (InTechopen.com).

Cavoski, I., Caboni, P., and Miano, T. (2011). Natural Pesticides and Future Perspectives. In: "Pesticides in the Modern World - Pesticides Use and Management": 169-189, Stoytcheva M. (Ed.) (InTechopen.com).

Choi, J.-S., and Hwang, I.-T. (2013). New Natural Herbicide Candidate for *Sicyon angulatus* Control. In: "Herbicides - Current Research and Case Studies in Use": 315-327, Price A.J., Kelton J.A.(Eds.) (InTechopen.com).

Cragg, G.M., and Newman, D.J. (2013). Natural products: A continuing source of novel drug leads. *Biochimica et Biophysica Acta*, **1830**: 3670–3695.

Dayan, F.E., Romagni, J., Tellez, M., Rimando, A., and Duke, S.O. (1999). Managing weeds with natural products. *Pesticide Outlook*, **October**: 185-188.

Dayan, F.E., Cantrell, C.L., and Duke, S.O. (2009). Natural products in crop protection. *Bioorganic and Medicinal Chemistry*, **17**: 4022–4034.

Dias, D.A., Urban,S., and Roessner, U. (2012). A historical overview of natural products in drug discovery. *Metabolites*, **2**: 303-336.

Di Ilio, V., Cristofaro, M., Marchini, D., Nobili, P., and Dallai, R. (1999). Effects of a neem compound on the fecundity and longevity of *Ceratitis capitata* (Diptera: Tephritidae). *J. Econ. Entomol.*, **92(1)**: 76-82.

Dudai, N., Poljakoff-Mayber, A., Mayer, A.M., Putievsky, E., and Lerner, H.R. (1999). Essential oils as allelochemicals and their potential use as bioherbicides. *Journal of Chemical Ecology*, **25 (5)**: 1079-1089.

Duke, S.O., Rimando, A.M., Baerson, S.R., Scheffler, B.E., Ota, E., and Belz, R.G. (2002). Strategies for the use of natural products for weed management. *J. Pesticide Sci.*, **27**: 298-306.

Durden, K., Sellars, S., and Pszczolkowski, M.A. (2009). Preventing fruit infestation by codling moth neonates with *Artemisia* extracts. *Pestycydy/Pesticides*, **1-4**: 51-56.

Einhellig, F.A. (1995). Allelophaty: current status and future goals. In: "Allelopathy: organisms, processes, and applications": 1-24,Dakshini K.M.M.,Einhellig F.A (Eds.), Am. Chem. Soc., Washington DC, USA.

El Beyrouthy, M., Arnold-Apostolides, N., Labaki, M., Cazier, F., Najm, S. and AbouKais, A. (2011). Chemical composition of the essential oil of the *Artemisia arborescens* L. growing wild in Labanon. *Lebanese Science Journal*, **12(1)**: 71-78.

EPA, (2010). Pesticides: Regulating Pesticides. New Biopesticide Active Ingredients – 2010. U.S. Environmental Protection Agency. http: //www.epa.gov/ oppbppd1/biopesticides/product_lists/new_ai_2010.html (last accessed July 12th, 2014).

European Commission, (2003). Opinion of the scientific committeeon food on Thujone. Expressed on 2nd Dec 2002. EuropeanCommission, Health and Consumer Protection Directorate. General, Scientific Committee on Food–SCF/CS/FLAV/ FLAVOUR/23 ADD2 Final, 6 February 2003.

Farzaneh, M., Ghorbani-Ghouzhdi, H., Ghorbani, M., and Hadian, J. (2006). Composition and antifungal activity of essential oils of *Artemisia sieberi*. *Pakistan Journal of biological science*, **9 (10)**: 1979-1982.

Ferro, G., and Furnari, F. (1970). Flora e vegetazione di Vulcano (Isole Eolie). Pubbl. Ist. Bot. Univ. Catania (In Italian).

Franz, C., and Novak, J. (2010). Sources of essential oils. In: "Handbook of essential oils: science technologyand applications": 39–81, Baser K.H.C., Buchbauer G. (Eds.),CRC Press, Boca Raton:

Gahukar, R.T. (2014).Potential and Utilization of PlantProducts in Pest Control. In: "Integrated Pest Management": 125-139, Abrol D.P. (Ed).

Gianguzzi, L., D'Amico, A., Caldarella, O., and Romano S. (2011). La flora vascolare delle Rocche di Entella (Entroterra della Sicilia). *Il Naturalista Siciliano*, **35 (3-4)**: 363-405 (In Italian).

Gianguzzi, L., La Mantia, A., Ottonello, D., and Romano, S. (2005). La flora vascolare della Riserva naturale di Monte Cofano (Sicilia occidentale). *Il Naturalista Siciliano*, **29 (3-4)**: 107-152 (In Italian).

Gianguzzi, L., Scuderi, L., and Pasta, S. (2006). La flora vascolare dell'Isola di Marettimo (Arcipelago delle Egadi, Canale di Sicilia): aggiornamento ed analisi fitogeografica. *Webbia*, **61 (2)**: 359-402 (In Italian).

Giardina, G., Raimondo, F.M., and Spadaro, V. (2007). A catalogue of plants growing in Sicily. *Bocconea*, **20**: 1-582.

Giordani, R., Hadef, Y., and Kaloustian, J. (2008). Compositions and antifungal activities of essential oils of some Algerian aromatic plants. *Fitoterapia*, **79**: 199-203.

Gitsopoulos, T.K., Chatzopoulou, P., and Georgoulas, I. (2013). Effects of essential oils of *Lavandula x hybrida* Rev., *Foeniculum vulgare* Mill and *Thymus capitatus* L. on the germination and radical length of *Triticum aestivum* L., *Hordeum vulgare* L., *Lolium rigidum* L., and *Phalaris brachystachys* L. *TEOP*, **16 (6)**: 817-825.

Hammer, K.A., Carson, C.F., and Riley, T.V. (1999). Antimicrobial activity of essential oils and other plant extracts. *Journal of Applied Microbiology*, **86**: 985–990.

Hillocks, R.J. (2012). Farming with fewer pesticides: EU pesticide review and resulting challenges for UK agriculture. *Crop Protection*, **31**: 85-93.

Islam, B.N. (1984). Pesticidal action of neem and certain indigenous plants and weeds of Bangladesh.- Proc. 2nd Int. Neem Conf., Rauischholzhausen 1983: 263-290.

Isman, M.B. (2000). Plant essential oils for pest and disease management. *Crop Protection*, **19**: 603–608.

Jacobson, M., Reed, D.K., Crystal, M.M., Moreno, D.S., and Soderstrom, E.L. (1978). Chemistry and biological activity of insect feeding deterrents from certain weed and crop plants. *Ent. Exp. Appl.*, **24**: 248-257.

Kegge, W., and Pierik, R. (2009). Biogenic volatile organic compounds and plant competition. *Trends in Plant Science*, **15 (3)**: 126-132.

Kim, J., Seo, S.-M., Lee, S.-G., Shin, S.-C., and Park, I.-K., 2008. Nematicidal activity of plant essential oils and components from coriander *(Coriandrum sativum)*, oriental sweet gum*(Liquidambar orientalis)*, and valerian*(Valeriana wallichii)* essential oils against pine wood nematode *(Bursaphelenchus xylophilus). Journ.Agric. Food Chem.*, **56**: 7316–7320.

Kordali, S., Cakir, A., Mavi, A., Kilic, H., and Yildirim, A. (2005). Screening of chemical composition and antifungal and antioxidant activities of the essential oils from three Turkish *Artemisia* species. *J AgricFood Chem.*, **53**: 1408–1416.

Lachenmeier, D.W., Emmert, J., Kuballa, T., and Sartor, G. (2006). Thujone–Cause of absinthism? *Forensic Science International*, **158**: 1–8.

Lahlou, M. (2004). Essential oils and fragrance compounds: bioactivity and mechanisms of action. *Flavour and Fragrance Journal*, **19**: 159–165.

Lai, F., Sinico, C., De Logu, A., Zaru, M., Müller, R.H., and Fadda, A.M. (2007). SLN as a topical delivery system for *Artemisia arborescens* essential oil: *In vitro* antiviral activity and skin permeation study. *International Journal of Nanomedicine*, **2 (3)**: 419–425.

Lai, F., Sinico, C., Valenti, D., Casu, L., Loy, G., and Fadda, A.M. (2003). *Artemisia arborescens* L essential oil-loaded beads: preparation and characterization. Proceedings of the 30th Annual Meeting and Exposition of the Controlled Release Society (CRS); July 19-23, 2003; Glasgow, Scotland, Minneapolis: Controlled Release Society; 2003: 45.

Lai, F., Wissing, S.A., Müller, R.H., and Fadda, A.M. (2006). *Artemisia arborescens* L. essential oil–loaded solid lipid nanoparticles for potential agricultural application: preparation and characterization. *AAPS Pharm Sci Tech.*, **7**: 1–9.

Lam, K.S. (2007). New aspects of natural products in drug discovery. *Trends in Microbiology*,**15 (6)**: 279-289.

Lentini, F. (2000). The role of ethnobotanics in scientific research. State of ethnobotanical knowledge in Sicily. *Fitoterapia*, **71**: S83-S88.

Li, J.W.-H., and Vederas, J.C. (2009). Drug Discovery and Natural Products: End of an Era or an Endless Frontier? *Science*, **325(5937)**: 161-165.

Liu, Y., Yan, I., Heiden, P., and Laks, P. (2001). Use of nanoparticles for controlled release of biocides in solid wood. *J. Appl. Polym. Sci.*, **79**: 458–65.

Lojacono, M. (1888-1909). Flora Sicula, o Descrizione delle Piante vascolari spontanee o indigenate in Sicilia. 3 voll. Tip. Virzì, Palermo, Italy (In Italian).

Lo Presti, M., Crupi, M.L., Zellner, B. d'A., Dugo, G., Mondello, L., Dugo, P. and Ragusa, S. (2007). Characterization of *Artemisia arborescens* L. (Asteraceae) leaf-dervied essential oil from Southern Italy. *J. Essent. Oil Res.*, **19**: 218-224.

Marcenò, C., Colombo, P., and Princiotta, R. (1985). La Flora. – In: AA.VV., Ricerche climatologiche e botaniche sui Monti Sicani (Sicilia centro-occidentale). *Naturalista siciliano*, **8 (suppl.)**: 69-133 (In Italian).

Marongiu, B., Piras, A., and Porcedda, S. (2006). Comparative analysis of the oil and supercritical CO_2 extract of *Artemisia arborescens* L. and *Helichrysum splendidum* (Thunb.) Less. *Natural Product Research*, **20 (5)**: 421–428 (InTechOpen.com).

Martínez, J.A. (2012). Natural Fungicides Obtained from Plants. In: "Fungicides for Plant and Animal Diseases": 3-28, DhanasekaranD. (Ed.).

Militello, M., Carrubba, A., and Blázquez, M.A. (2012). *Artemisia arborescens* L.: essential oil composition and effects of plant growth stage in some genotypes from Sicily. *Journal of Essential Oil Research*, **24 (3)**: 229–235.

Militello, M., Settanni, L., Aleo, A., Mammina, C., Moschetti, G., Giammanco, G. M., Blazquez, M. A., and Carrubba, A. (2011). Chemical composition and antibacterial potential of *Artemisia arborescens* L. essential oil. *Current Microbiology*, **62 (4)**: 1274-1281.

Militello, M. (2012). Studi fitochimici e agronomici su *Artemisia arborescens* L. *(Asteraceae)* della flora spontanea siciliana e attività biocida degli oli essenziali. Ph.D. dissertation, University of Palermo.

Minissale, P., Sciandrello, S., and Spampinato, G. (2005). Analisi della biodiversità vegetale e relativa cartografia della Riserva Naturale Orientata Isola Bella e del territorio circostante (Taormina - ME - Sicilia). *Quaderni di Botanica Ambientale Applicata*, **16**: 175-208 (In Italian).

Neilson, E.H., Goodger, J.Q.D., Woodrow, I.E., and Lindberg Møller, B. (2013). Plant chemical defense: at what cost? *Trends in Plant Science*, **18 (5)**: 250-258.

Nugroho, B.W., Edrada, R.A., Wray, V., Witte, L., Bringmann, G., Gehling, M., and Proksch, P. (1999). An insecticidal rocaglamide derivatives and related compounds from *Aglaia odorata* (Meliaceae). *Phytochemistry*, **51**: 367-376.

NTP, (2011). National Toxicology Program Technical Report on the toxicology and carcinogenesis studies of α,β-thujone (CAS NO. 76231-76-0) in F344/N rats and B6C3F1 mice (gavage studies). NIH Publication No. 12-5912: 260 pp.

Ornano, L., Venditti, A., Ballero, M., Sanna, C., Quassinti, L., Bramucci, M., Lupidi, G.,Papa, F., Vittori, S.,Maggi, F., and Bianco, A. (2013). Chemopreventive and antioxidant activity of the chamazulene-rich essential oil obtained from *Artemisia arborescens* L. growing on the Isle of La Maddalena, Sardinia, Italy. *Chemistry and Biodiversity*, **10**: 1464-1474.

Pappas, R., and Sheppard-Hanger, S. (2000). *Artemisia arborescens* essential oil of the Pacific Northwest: a high-chamazulene, low-thujone essential oil with potential skin-care applications. *Aromatherapy J.*, **10**: 30-33.

Parris, N., Cooke, P.H., and Hicks, K.B. (2005). Encapsulation of Essential Oils in Zein Nano spherical Particles. *J. Agric. Food Chem.*,**53**: 4788-4792.

Pauli, A., and Schilcher, H. (2010). *In Vitro* Antimicrobial Activities of Essential Oils Monographed in the European Pharmacopoeia 6th Edition. In: "Handbook of essential oils: science technology and applications": 350–547, Baser K.H.C., Buchbauer G. (Eds.). CRC Press, Boca Raton.

Perlatti, B., de Souza Bergo, P.L., das Graças Fernandes da Silva, M.F., Fernandes, J.B., and Rossi Forim, M. (2013). Polymeric Nanoparticle-Based Insecticides: A Controlled Release Purpose for Agrochemicals. In: "Insecticides - Development of Safer and More Effective Technologies": 523-550, Trdan S. (Ed.) (InTechOpen.com).

Pichersky, E., and Gang, D.R. (2000). Genetics and biochemistry ofsecondary metabolites inplants: an evolutionaryperspective. *Trends in Plant Science*, **5 (10)**: 439-445.

Pignatti, S. (1982). Flora d'Italia. Voll. 1-3. Edagricole, Bologna (In Italian).

Sacco, T., Frattini, C., and Bicchi, C. (1983). Constituents of essential oil of *Artemisia arborescens*. *Planta Medica*, **47**: 49-51.

Saddi, M., Sanna, A., Cottiglia, F., Chisu, L. Casu, L., Bonsignore, L., and De Logu, A. (2007). Antiherpevirus activity of *Artemisia arborescens* essential oil and inhibition of lateral diffusion in Vero cells. *Annals of Clinical Microbiology and Antimicrobials*, **6**: 10.

Salamon, I. (2009). Chamomile biodiversity of the essential oil.Qualitative–quantitative characteristics. In: "Innovations in chemical biology": 83–90,Sener B. (Ed.), Springer Science and Business Media,Dordrecht.

Scheepens, P.C., Müller-Schärer, H., and Kempenaar, C. (2001). Opportunities for biological weed control in Europe. *Bio Control*, **46**: 127–138.

Sheppard-Hanger, S. (1995). The Aromatherapy Practitioner Reference Manual. Tampa, FLA: Atlantic Institute of Aromatherapy, Chart 197

Selmar, D. (2008). Potential of salt and drought stress to increase pharmaceutical significant secondary compounds in plants. *Landbauforschung - vTI Agriculture and Forestry Research*, **1/2 (58)**: 139-144.

Smith-Palmer, A., Stewart, J., and Fyfe, L. (1998). Antimicrobial properties of plant essential oils and essences against five important food-borne pathogens. *Letters in Applied Microbiology*, **26**: 118–122.

Tsolakis, H., and Ragusa, S. (2004). Laboratory evaluation of plant extracts on Tetranychus urticae Koch (Acariformes, Tetranychidae). In: "Acarine Biodiversity in the Natural and Human Sphere": 539-548, Weigmann G., Alberti G., Wohltmann A., Ragusa S. (Eds.), Phytophaga, 16.

Tsolakis, H., and Ragusa, S. (2008). Effects of a mixture of vegetable and essential oils and fatty acid potassium salts on *Tetranychus urticae* and *Phytoseiulus persimilis*. *Ecotoxicology and Environmental Safety*, **70**: 276-282.

Tutin, T.G., Heywood, V.H., Burges, N.A., Valentine, D.H., Walters, S.M., and Webb, D.A. (eds.), 1964-1980. Flora Europaea. Cambridge Univ. Press, Cambridge, London, New York, Melbourne, Vol. 1, Lycopodiaceae to Platanaceae: xxxii + 464 pp. (1964);2, Rosaceae to Umbelliferae: xxvii + 475 pp. (1968); 3, Diapensiaceae to Myoporaceae: xxix + 370 pp. (1972); 4, Plantaginaceae to Compositae (and Rubiaceae): xxix + 505 pp. (1976); 5, Alismataceae to Orchidaceae (Monocotyledones): xxxiv + 452 pp. (1980).

Tworkoski, T. (2002). Herbicide effects of essential oils. *Weed Science*, **50**: 425–431.

Van Tol, R.W.H.M., James, D.E., Jan De Kogel, W., and Teulon, D.A.J. (2007). Plant odours with potential for a push–pull strategy to control the onion thrips, *Thripstabaci*. *Entomologia Experimentalis et Applicata*, **122**: 69–76.

Verdeguer, M., Blázquez, M. A., and Boira, H. (2012). Chemical composition and herbicidal activity of the essential oil from a *Cistus ladanifer* L. population from Spain. *Natural Product Research: Formerly Natural Product Letters*, **17 (26)**: 1602-1609.

Verpoorte, R. (1998). Exploration of nature'schemodiversity: the role of secondary metabolites as leads in drug development. *Drug Discovery Today*, **3 (5)**: 232-238.

Verpoorte, R., and Memelink, J. (2002). Engineering secondary metabolite production in plants. *Current Opinion in Biotechnology*, **13**: 181–187.

Wagner, C., Friedt, W., Marquard, R.A., and Ordon, F. (2005). Molecular analyses on the genetic diversity and inheritance of (-)-α-bisabolol and chamazulene content in tetraploid chamomile (*Chamomilla recutita* (L.) Rausch.). *Plant Science*, **169**: 917–927.

Wyman, J.G., and VanEtten, H.D. (1978). Antibacterial activity of selected isoflavonoids. *Phytopathology*, **68**: 583-589.

Yang, F.L., Li, S.G., Zhu, F., and Lei, C.L. (2009). Structural characterization of nanoparticles loaded with garlic essential oil and their insecticidal activity against *Tribolium castaneum* (Herbst) (Coleoptera: Tenebrionidae). *J Agric Food Chem.*, **57**: 10156–10162.

Younes, K., Merghache, S., Djabou, N., Merghache, D., Muselli, A., Tabti, B., and Costa, J. (2012). Chemical composition, antibacterial and antioxidant activities of a new essential oil chemotype of Algerian *Artemisia arborescens* L. *African Journal of Pharmacy and Pharmacology*, **6(42)**: 2912-2921.

Zarga, M.A., Qauasmeh, R., Sabri S., Munsoor, M., and Abdalla, S. (1995). Chemical constituents of *Artemisia arborescens* and the effect of the aqueous extract on rat isolated smooth muscle. *PlantaMedica*, **61(3)**: 242-245.

Zehnder, G., Gurr, G.M., Kuhne, S., Wade, M.R., Wratten, S.D., and Wyss, E. (2007). Arthropod Pest Management in Organic Crops. *Annu. Rev. Entomol.*, **52**: 57.80.

Zheng, H., and Shang, Q. (2005).Water suspension acetamiprid nano capsule preparation and its reparing method. CN1491558. *ChemicalAbstracts*, **143**: 73729.

Zuzarte, M., Gonçalves, M.J., Cavaleiro, C., Canhoto, J., Vale-Silva, L., Silva, M.J., Pinto, E., and Salgueiro, L. (2011). Chemical composition and antifungal activity of the essential oils of *Lavandula viridis* L'Hér. *Journal of Medical Microbiology*, **5(60)**: 612-618.

Natural Products: Research Reviews Vol. 3 (2016) *Pages* **389–406**
Editor: **V.K. Gupta**
Published by: **DAYA PUBLISHING HOUSE, NEW DELHI**

14

Review of Phytochemical and Pharmacological Properties and the Antioxidant Activity Studies of *Thespesia populnea*

Beena Joy[1*] and Lekshmi Devi[1]

ABSTRACT

Thespesia populnea, belonging to Malvaceae family, is used in folk-medicine for the treatment of liver diseases, cutaneous affections such as scabies, psoriasis, eczema, herpetic diseases etc. in India. Earlier phytochemical investigations revealed that Thespesia populnea appears to be rich in secondary metabolites and it is widely used in traditional medicine to combat and cure various ailments. The anti inflammatory, astringent, asthma and diarrhoea effects of the plant can be attributed to its high flavonoids, steroids, alkaloids, tannins, terpenoids and saponins. Therefore it is necessary to exploit its maximum potential in the field of medicinal and pharmaceutical sciences for novel and fruitful application as antioxidant activity can be further related with other pharmacological activities. Exploitation of these pharmacological properties involve further investigation and identification of active ingredients by implementation of techniques like extraction, separation, crystallization and characterization of the secondary metabolites. This chapter summarizes a review of phytochemical and pharmacological properties of Thespesia populnea. This is the first report on the antioxidant activity studies of Thespesia populnea heart wood.

Keywords: *Thespesia populnea*, Biochemical studies, Pharmacological properties, Phytochemical properties.

1 Agroprocessing and Natural Products, NIIST, Thiruvananthapuram, India.

* *Corresponding author*: E-mail: bjoy1571@gmail.com

Introduction

Since ancient times, people have been exploring the nature particularly plants in search of new drugs. Plants have been used in treating human diseases for thousands of years. The use of medicinal plant is not just a custom of the distant past. This has resulted in the use of large number of medicinal plants with curative properties to treat various diseases (Verpoorte, 1998) Nearly 80 per cent of the world's population relies on traditional medicines for primary health care, most of which involve the use of plant extracts (Sandya *et al.*, 2006). In India, almost 95 per cent of the prescriptions were plant based in the traditional systems of Unani, Ayurveda, Homeopathy and Siddha (Satyavati *et al.*, 1987). *Thespesia populnea* is a large tree belongs to the family Malvaceae, found in tropical regions and coastal forests of India. Various parts of this plant are found to possess useful medicinal properties. *Thespesia populnea* is an evergreen tree. The leaves are alternate, simple, with petioles of length 5-10cm long. The flowers Hibiscus like single at upper leaf axils, corolla yellow with a red center. The Fruits are globose. The seeds are black, hairy (Saravana kumar *et al.*, 2009). The main chemical constituents reported are kaempferol, quercetin and its glycosides, herbacetin and its glucoside, populneol, populnin, populnetin, rutin, gossipetin, gossypol, lupeol, sesquiterpenoidal quinones *viz.*, thespeson, thespone, mansonones C,D,E and F, amino acids and carbohydrates (Patill *et al.*, 2012).The phytochemical activity is due to the presence of phenolic compounds, terpenoids, flavanoids and tannins etc.

Traditional Uses

Thespesia populnea is commonly seen in tropical regions and coastal forests of India is a medicinal plant and found to possess useful medicinal properties (Allen, 2002; Fosberg,1972) such as antifertility, anti-inflammatory, antioxidant, purgative and heptoprotective (Shirwaikar kumar *et al.*, 1995) activity etc. In India it is used in folk-medicine in India for the treatment of liver diseases. The decoction of the bark is used by Ayurvedic physicians for the treatment of skin and liver diseases (Varier,1994). The bark and flower of *Thespesia populnea* possess pharmacological properties such as hepatoprotection, antioxidant activity, anti-inflammation, memory enhancement property and hypocholesterolemic activity (Yuvaraj *et al.*, 2009; Ilvarasan *et al.*, 2003; Vasudevan *et al.*, 2006; Sathyanarayana, 2004; Shirwaikar *et al.*, 1995). Bark and fruits posses more curative properties. The plant is astringent, cooling, depurative, anti-inflammatory, haemostatic, anti-diarrhoeal and anti-bacterial. It is useful in dematopathy such as scabies, psoriasis, ringworm, leprosy, arthritis, haemorrhages wounds, ulcer, cholera, diabetes, as cites, dispsia, cough, asthma, catarrh (Indian medicinal plants, 1996). It acts against dysentery, piles, diabetes and haemorrhoids. It cures ulcers, itching; scabies, other skin diseases and urinary disorders. Leaves are beneficial for painful joints. Fruits,leaves are applied externally to scabies, psoriasis and other skin diseases (Joshi, 2000). *Thespesia populnea* fruit and bark have antihelmintic and antidiarrheal activity (shirwaikar *et al.*, 2011; Viswanatha *et al.*, 2011). The bark, fruits, leaves; flowers are reported to be useful in cutaneous infection such as scabies, psoriasis, and ringworm. The juice of fruits was employed in treating certain disease like herpetic disease etc. Seed yields a deep red, thick fatty oil which is also used in cutaneous diseases. The astringent bark, root and fruits were reported to be used in dysentery, cholera, haemorrhoids. The mashed bark is employed as a politice or hot fomentation for wounds (wealth of India,2003).

Pharmacological Activity

Antibacterial Activity

The methanolic extract of *Thespesia populnea* exhibited moderate significant antibacterial activity against *R.terrae, B. licheniformis, M. flavum* and *B. leuteum* at lower concentration range (Sharavana kumar *et al.*, 2012). Thc methanolic extract of *Thespesia populnea* leaves and its corresponding callus were studied against effect of *Escherichia coli, Staphylococcus aureus, Kiebsiella pneumonia* and *Salmonella typhii* by Bauer-kirby method. Both the leaf of the plant and its corresponding callus showed great potential as source of antibacterial agents (Archana Moon *et al.*, 2010).

Hepatoprotective Activity

The ethanol extract fraction of *Thespesia populnea* administered orally to different groups of rats was evaluated using CCl_4 induced hepatotoxic model of liver injury, showed significant activity (Shirwaikar *et al.*, 1995). Yuvaraj and Subramoniam had reported the hepatoprotective activity of *Thespesia populnea* in carbon tetrachloride (CCl_4)-induced liver injury in rats. The water suspension of leaf, flower and stem bark of *Thespesia populnea* showed varying levels of protective action against CCl_4- induced liver damage as evidenced from significant reduction in the activities of serum marker enzymes for liver damage (alanine transaminase, aspartate transaminase, and alkaline phosphatase), and bilirubin levels when compared with CCl_4-intoxicated control rats. The stem bark suspension showed maximum hepatoprotection compared with leaf and flower. An ethanol extract of the stem bark was more active than n-hexane and water extracts. The hepatoprotective effect of this extract was almost comparable to that of silymarin, a reference herbal drug (Yuvaraj *et al.*, 2009).

Antidiabetic Activity

Thespesia populnea leaves are widely used in traditional medicines to treat diabetes and hyperglycemia generated reactive radical species as it has been proved to be a potential source of bioactives. The tannins and flavanoids present in the extracts may be responsible for the antidiabetic activity. Ethyl acetate extract of *Thespesia populnea* leaves showed highest α-glucosidase inhibition capacity as compared to methanol extract. Thus the ethyl acetate extract of *Thespesia populnea* leaves is the most active extract in terms of its antidiabetic properties (Beena Joy *et al.*, 2012)· The hypoglycaemic activity of methanolic extract of flowers and leaves of *Thespesia populnea* was evaluated in alloxan induced diabetic rats. The result indicates that the methanolic extract of *Thespesia populnea* flower and leaf exhibited significant antihyperglyceamic property compared to standard Glibenclamide (Haja Shereif *et al.*, 2011).The hypoglycemic activity of ethanolic extract of bark and leaves of *Thespesia populnea* was evaluated in STZ induced diabetic rats. Ethanolic extract of *Thespesia populnea* in diabetic rats showed a significant reduction in lipid peroxidase in hepatic and renal tissues as compared to diabetic group. The ethanolic extract of *Thespesia populnea* bark and leaf exhibited significant antihyperglycemic activity in streptozotocin (STZ)-induced diabetic rats.This extracts has also showed the effect due to enhancing effect on cellular antioxidant defenses to protect against oxidative damage (Parthasarathy *et al.*, 2009).

Antihelmintic Activity

Ethanol and aqueous extract of *Thespesia populnea* (bark) showed paralysis as well as death of worm in lesser time in comparison with the standard piperazine citrate (Yuvaraj *et al.*, 2013). The stem bark extract of the *Thespesia populnea* showed significant anthelmintic activity. The ethyl acetate fraction and phenolic acid fraction of the stem bark showed paralysis and death of earth worms in comparison with the standard drug Albendazole (Yuvaraj *et al.*, 2013).

Antipsoriatic Activity

Screening for anti-psoriatic activity was carried out by topical application of different extracts and isolated compounds of *Thespesia populnea* bark in the form of a cream using the Perry's scientific mouse tail model. Successive pet-ether extract showed maximum antipsoriatic activity (Siddharth Shrivastav *et al.*, 2009).

Memory Enhancing Activity

Elevated plus-maze and Hebb-Williams maze served as the exteroceptive behavioral models for testing memory. Diazepam-, scopolamine-, and ageing-induced amnesia served as the interoceptive behavioral models. The ethanol extract of *Thespesia populnea*.was administered orally in three dose for seven successive days to different groups of young and aged rats. *Thespesia populnea* resulted in significant improvement in memory of young and aged rats. *Thespesia populnea* also reversed the amnesia induced by scopolamine and diazepam. Cholesterol-lowering, anticholinesterase, anti-inflammatory, and antioxidant properties of *Thespesia populnea* may favorably contribute to its memory-enhancement effect. Therefore, *Thespesia populnea* bark appears to be a promising candidate for improving memory, and it would be worthwhile to explore the potential of this plant in the management of Alzheimer patients (Mani Vasudevan *et al.*, 2007).

Anti Ulcer Activity

The plant *Thespesia populnea* used for the treatment of ulcer, in the folk medicine. Patil studied about the anti-ulcer potential of the terpenoid fraction from the leaves of *Thespesia populnea*. Terpenoid fraction decreased ulcer index in dose dependent manner. Terpenoid fraction from *Thespesia populnea* leaves showed significant antiulcer activity in experimentally induced ulcer in rat model by decreasing the gastric secretions and by enhancing glycoprotein levels. Thus the results of the study substantiate the folk claims that the use of *Thespesia populnea* is beneficial in gastric ulcer (Patil *et al.*, 2010).

Anti Inflammatory Activity

The studies claimed to have significant anti-inflammatory activity against experimentally induced paw oedema in rats. This is indicated due to the presence of active phytoconstituents and their influence on the prostaglandins pathway. The ethanolic extract of leaf of *Thespesia populnea* was screened for anti-inflammatory activity on carrageenan induced paw oedema in rats. The acute toxicity studies of oral doses of ethanolic extract of *Thespesia populnea* leaf in rats revealed that it has a high safety profile, as the extract was well tolerated by the animals. The results of the study suggest that the extract possesses anti-inflammatory activity (Elakkiya S *et al.*, 2011).

Immunomodulatory Activity

Studies have been carried out to find out immunomodulatory potential of methanolic extract of leaves of *Thespesia populnea*.The measurement of immunomodulatory property was carried out by delayed type hypersensitivity, humoral antibody titre response to SRBC, and Cyclophosphamide induced myelosuppression. The increasing doses of methanolic extract have shown the augmentation of antibody titre. Cyclophosphamide induced immunosuppression was counteracted by methanolic extract of *Thespesia populnea*. Phytochemical screening showed the presence of flavonoids, triterpenoids, proteins, amino acids, phenolic and steroidal compounds. The immunomodulatory activity of plant is attributed to these phytoconstituents. In the study they concluded that methanolic extract of *Thespesia populnea* stimulate both cellular and humoral immunity (Switi Gaikwad *et al.*, 2011).

Anti-oxidant Activity

Antioxidant potential of methanolic extract root was evaluated by *in vitro* antioxidant studies like free radical scavenging activity by DPPH method, nitric oxide method, anti- lipid peroxidation study, and reducing power assay and expressed as per cent scavenging and IC_{50}. Ascorbic acid was served as standard. The phytochemical screening and TLC profilingof the extract showed the presence of phenolic compounds, terpenoids and tannins. The extracts showed the significant antioxidant activity as compared to control, but less than that of ascorbic acid. The antioxidant potential was increased in dose dependent manner. The methanolic extract of root of *Thespesia populnea* exhibited better antioxidant activity due to prominent presence of phenolic compounds and terpenoids (Patil1 *et al.*, 2012).

Antioxidant Studies of Heartwood of *Thespesia populnea.*

Materials and Methods

Material Collection

Heartwood of *Thespesia populnea* were collected from the outskirts of Trivandrum, Kerala, South India. The specimen was identified by botanist and a voucher specimen is deposited in the herbarium of NIIST, Trivandrum, Kerala.

Extraction Procedures

About 250g of Heartwood of *Thespesia populnea* was subjected to hot extraction using various solvents like acetone and methanol.

Chemicals and Reagents

Chemicals and reagents used were Gallic acid, sodium carbonate, Folin-ciocalteu reagent 2,2-diphenyl-2-picryl hydrazyl (DPPH), methanol,2,2′-Azino-bis-ethyl benzothiazoline-6-sulfonic acid(ABTS), Trolox,potassium persulphate, acetone,ethanol,aluminium chloride, Quercetin, Sodium nitrate and sodium hydroxide. All solvents used were of analytical grade.

Estimation of Total Phenolic Content

Total content of phenolics was determined according to the method of Folins-Ciocalteu method (Shingh *et al.,* 2009).In this Gallic acid is used as a reference compound. The sample solution volume was made up to 3.5 mL with distilled water. Then 0.5 mL of Folins-Ciocalteu reagent was added to each tube and mixed well by swirling. The mixture was allowed to stand for 3 min, followed the addition of 1 mL of 20 per cent sodium carbonate. All the tubes were incubated at room temperature for 90 min. The absorbance of the resulting solution was measured colorimetrically at 760nm in a Schimadzu UV-160 Spectrophotometer. The total phenolic content was expressed as mg/g of gallic acid equivalent.

Total Flavonoid Content

Total flavonoids content was measured by aluminum chloride colorimetic assay (Marinova *et al.,* 2005). 1ml of extracts or standard solution of quercetin was added to 10 ml volumetric flask containing 4 ml of distilled water. To above mixture, 0.3 ml of 5 per cent $NaNO_2$ was added. After 5 min, 0.3 ml of 10 per cent $AlCl_3$ was added. At 6^{th}min, 2 ml of 1M NaOH was added and the total volume was made up to 10 ml with distilled water. The solution was mixed well and the absorbance was measured against prepared reagent blank at 510nm.

Antioxidant Activity

Antioxidant activity was evaluated by a number of methods as the plant extracts are complex in nature and involves diverse mechanism for their therapeutic properties (Verhagen *et al.,* 2003).

DPPH Radical Scavenging Activity

DPPH radical scavenging activity was determined according to the Burits and Bucar (Burits *et al.,* 2000).A total of 1 ml of a 0.3 mM DPPH in ethanol solution was added to 2.5 ml of sample solution of different concentrations and allowed to react at room temperature. A mixture of 2.5mL ethanol and 1mL of 0.3 mM DPPH solution were used as negative control and 2.5 mL of the standard antioxidant and 1mL of 0.3 mM DPPH solution was taken as positive control. The absorbances were measured colorimetrically at 518 nm in a Shimadzu-UV-160 Spectrophotometer. All extracts were analysed in triplicates.

ABTS Radical Cation Decolorisation Assay

The experiments were carried out using an improved ABTS decolourisation assay (Pellegrini *et al.,* 1999), which involves the generation of the $ABTS^{+\bullet}$ chromophore by the oxidation of ABTS with potassium per sulphate. It is applicable for both hydrophilic and lipophilic compounds. The ABTS+• solution was diluted to an absorbance of 0.7 ± 0.05 at 734 nm (Shimadzu UV–Visible Spectrophotometer, Model 2100). Absorbance was measured at 7 min after the initial mixing of different concentrations of the extracts with 1 ml of ABTS+• solution. Trolox was used as a reference standard. A standard curve was plotted by measuring the reduction in absorbance of the ABTS+• solution at different concentrations of trolox over a period of 7 min. The trolox equivalent antioxidant capacity of an extract represents the concentration of trolox solution that has the same antioxidant capacity as the extract. All determinations were carried out in triplicate.

Results and Discussion

Total Phenolic Content (TPC)

The total phenolic contents were examined using the Folin-Ciocalteu's reagent. The values obtained for the concentration of total phenols are expressed as mg of GA/g of extract (Table 14.1 and Figure 14.1). Among the extracts analysed both acetone and ethanol have comparatively equal quantity of TPC. The acetone extract of heartwood yielded 2.5±0.13mg/g GAE whereas the ethanolic fraction yielded 4.3±0.1547mg/g GAE. The results showed that ethanolic extract of *Thespesia populnea* has promising activity than acetone extract.TPC gives a measure of the phenolic hydroxyl group present in various compounds like flavanoids, alkaloids and carotenoids, tannins present. However it is obvious that the TPC measured by the Folin-Ciocalteu procedure do not give a full picture of the quantity or quality of the phenolic constituents (Katsube *et al.*, 2004). But it is a measure of activity gradient.

Table 14.1: Total Phenolic Content of Heartwood Extract of *Thespesia populnea*

Sl.No.	*Sample*	*Extract*	*TPC (mg/g of GAE)*
1.	*Thespesia populnea*	Acetone	2.5±0.13
		Ethanol	4.3±0.1547

Total Flavanoid Content (TFC)

The acetone and ethanol extract of heartwood were analysed for its total flavanoid content. The acetone extract of *Thespesia populnea* exhibit 12.37±0.087mg/g QE and ethanol extract exhibit 13.323±0.54mg/g QE. The result showed both the extract have more or less same flavanoid content.

Table 14.2: Total Flavanoid Content of Extracts of *Thespesia populnea.*

Sl.No.	*Sample*	*Extract*	*TFC (mg/g of GAE)*
1.	*Thespesia populnea*	Acetone	12.37±0.087
		Ethanol	13.323±0.54

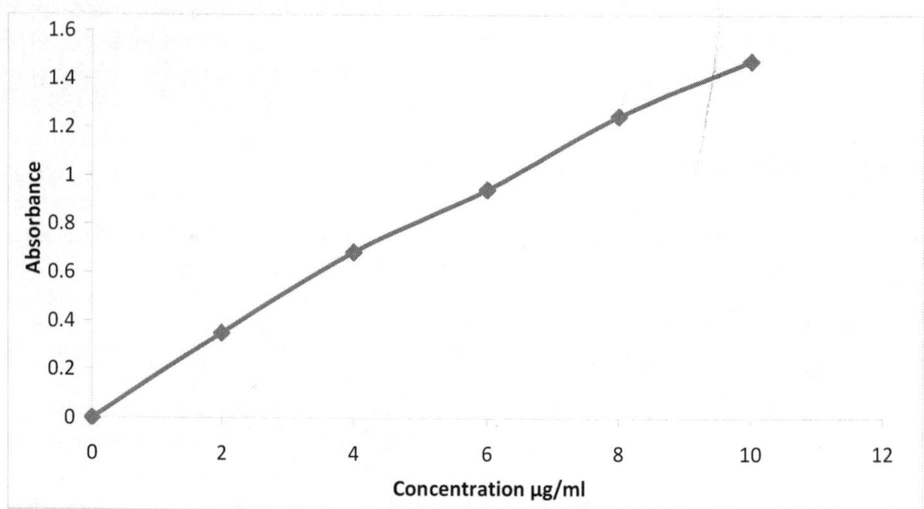

Figure 14.1: Total Phenolic Content for Standard Gallic Acid.

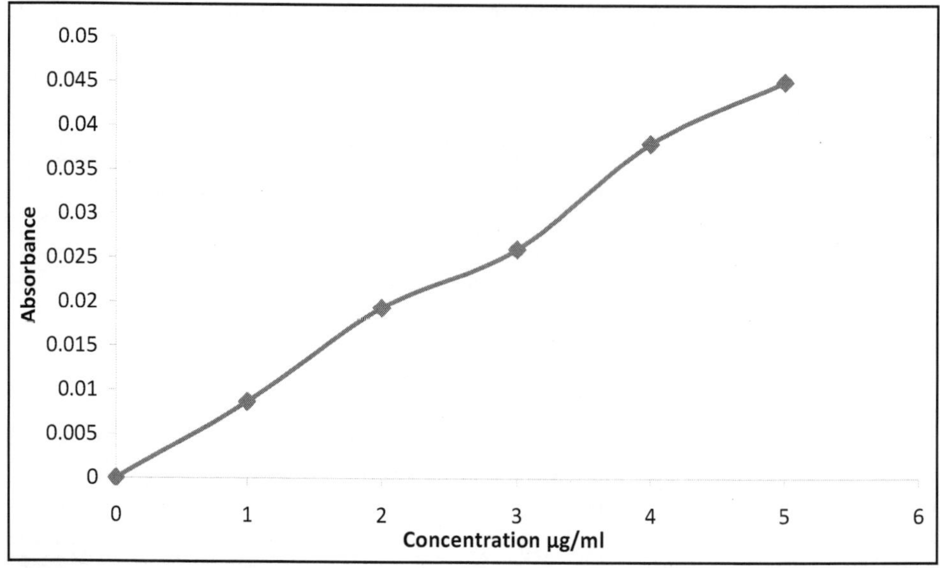

Figure 14.2: Total Flavanoid Content for Standard Quercetin.

Antioxidant Activity

DPPH Scavenging Activity

DPPH is one of the powerful free radical which is used to evaluate the electron donating capacity of antioxidants. Antioxidant compounds either transfer electrons or hydrogen atoms to DPPH radical and neutralizes the free radical character which is visualized by the color change from purple to yellow due to reduction of DPPH there will be a decrease in absorbance value. The decrease in absorbance depends on the amount of antioxidant compounds. The DPPH radical strongly absorbs at 518nm and upon reduction the absorption decreases. The ethanolic extract of heartwood showed highest radical scavenging activity (IC_{50} value 31.35±0.78 µg/ml) than acetone extract (IC_{50} value 62.52±1.5 µg/ml).A comparison of the scavenging activity with the standard Gallic acid was (3.01±0.82µg/ml) summarized in Table 14.2 and Figure 14.2. Low IC_{50} value of ethanol extract indicates high radical scavenging activity due to more phenolic antioxidants in the extract.

Table 14.3: DPPH Evaluation of different Extract of *Thespesia populnea*

Sl.No.	*Sample*	*Extract*	*IC_{50} Value (µg/ml)*
1.	*Thespesia populnea*	Acetone	62.52±1.5
		Ethanol	31.35±0.78

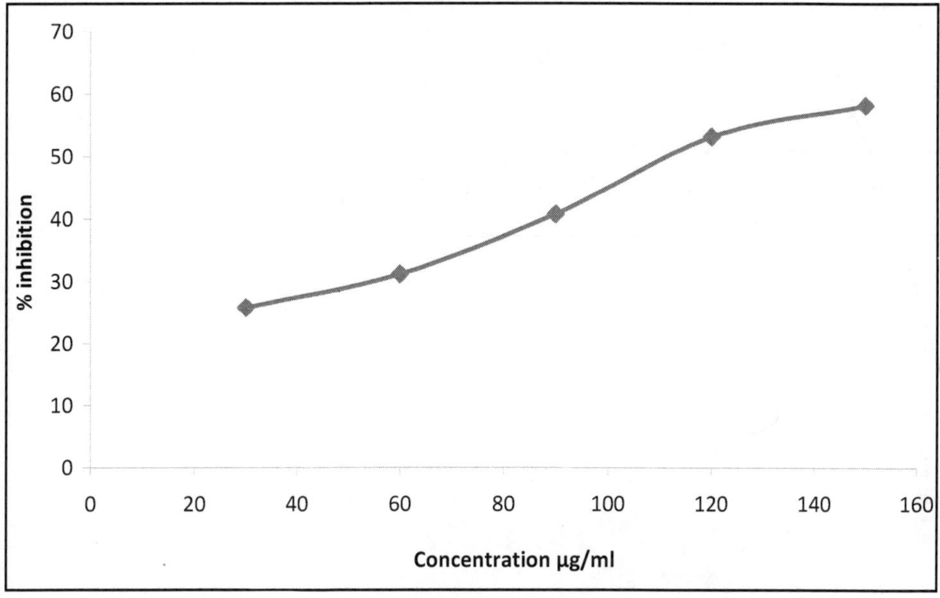

Figure 14.3: DPPH Assay for Acetone Extract o *Thespesia populnea.*

ABTS Cation Decolorisation Capacity

ABTS radical assay is often used in evaluating total antioxidant power of single compounds and complex mixtures of various plants. The acetone extract of *Thespesia populnea showed a* less cation scavenging effect than ethanol extract.

Table 14.4: Evaluation of ABTS Cation Decolourisation Capacity of Extracts of *Thespesia populnea*

Sl.No.	*Sample*	*Extract*	*IC_{50} Value (µg/ml)*
1.	*Thespesia populnea*	Acetone	4.38±0.659
		Ethanol	22.082±0.431

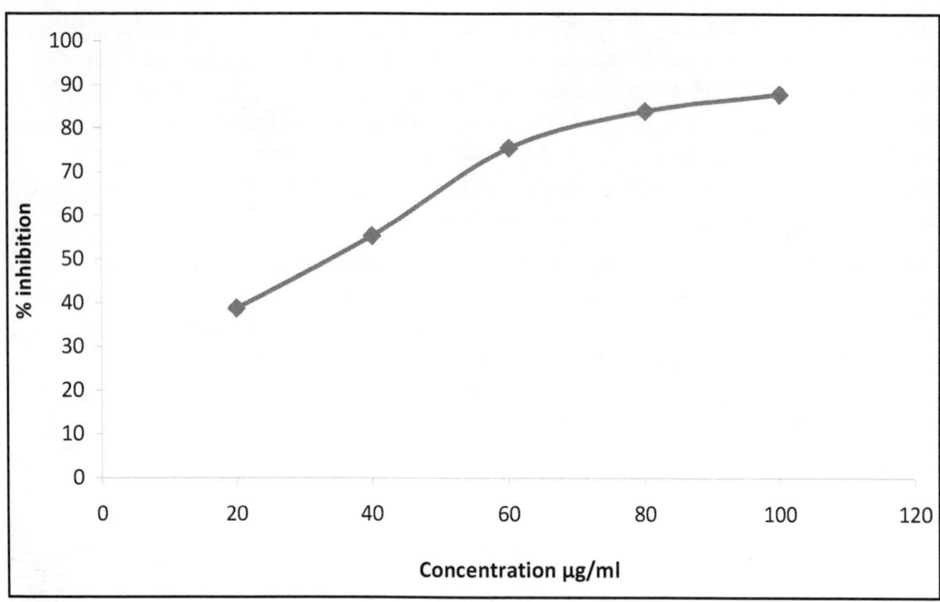

Figure 14.4: Ethanol Extract of *Thespesia populnea.*

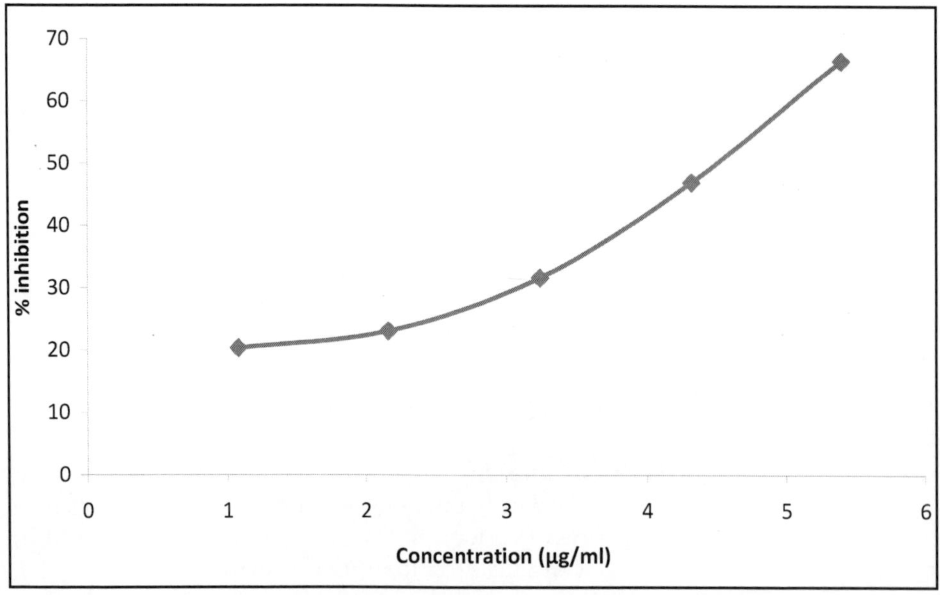

Figure 14.5: ABTS Assay for the Acetone Extract of *Thespesia populnea.*

Phytochemical Properties

Leaves contain lupeol, β-sitosterol, lupenone, quercetin, ferulic, syringic, melilotic acid (Figure 14.7) (Daniel, 2006).Gossypol is the main color pigment present in this plant (Akhila *et al.*, 1993). Four naturally occurring quinines, *viz.* thespone, mansonone-

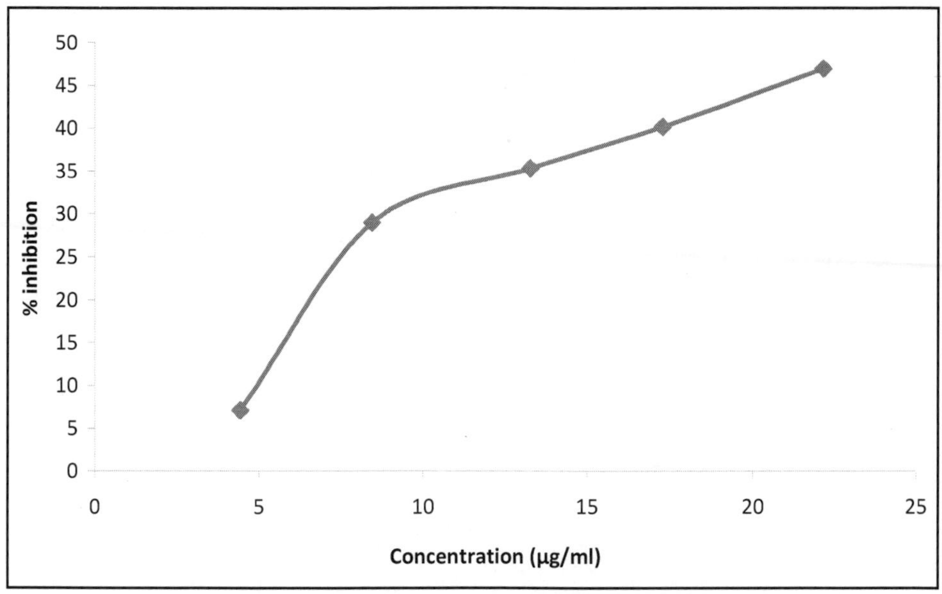

Figure 14.6: ABTS Assay for the Ethanol Extract of *Thespesia populnea*.

D, mansonone-H, and thespesone, have also been extracted from heartwood of *Thespesia populnea* (Johnson *et al.*, 1999). Five sesquiterpenoid quinones were isolated from the aerial parts of *Thespesia populnea*. Their structures were determined on the basis of spectral data. The Mansonones were identified as: Mansonones C (1), G (2), E (3), H (4) and 7-hydroxy-2,3,5,6-tetrahydro-3,6,9-trimethylnaphtho[l,8-b,c]pyran-4,8-dione (Abd ELnasser *et al.*, 2013). The various chemical constituents were isolated from the *Thespesia populnea are* Gossypol (Datta *et al.*, 1972), 7 Hydroxy-2,3,5,6-tetrahydro-3,6,9-Trimethylnaphto [1,8-B,C] Pyran-4,8-Dione (Milbrodt *et al.*, 1997), Kaempferol, Quercetin (Kasim *et al.*, 1975), Kaempferol 3-glucoside,Quercetin 3-glucoside,rutin (Datta *et al.*, 1973), Nonacosane, lupenone, myricyl alcohol, lupeol, β-sitosterol and β-sitosterol-β-D-glucoside (Seshadri *et al.*, 1975), 5, 8-dihydroxy-7-methoxyflavone, 7-1ydroxyisoflavone and Thespone (Neelakantan *et al.*, 1983). Mansonones D, E and F, Populneol, Thespesin (Valderrama *et al.*, 2008).

The study of aqueous and methanolic extracts of *Thespesia populnea* (Linn) bark showed significant anti oxidant activity against CCl_4 induced liver injury in rat. The study paved way for the isolation of a rare flavanoid-qucertin-7-*O*-rhamnoglucoside. The compound was identified and confirmed by spectral studies (Ilavarasan *et al.*, 2003).

The investigation of chemical constituents report ten compounds. They were isolated and purified with silica column chromatography, Sephadex LH-20 column chromatography and HPLC methods. Mansonone H (1), Populene A (2), 1,8-Dihydroxy-3-methoxy-6-methyl-9,10-anthraquinone (3), 1, 4, 5-Trihydroxy-3-methoxy-6-methyl-9,10-anthraquinone (4), 1, 8-Dihydroxy-3-carboxy-9, 10-anthraquinone (5), Catacerebroside B (6), beta-Sitosterol (7), beta-Daucosterol (8),

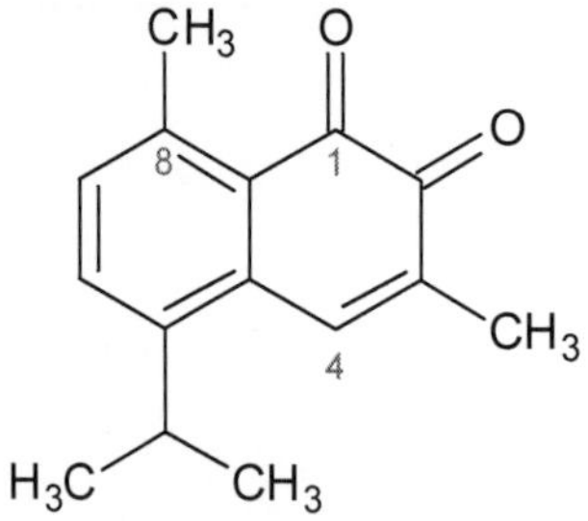

Mansonone 1

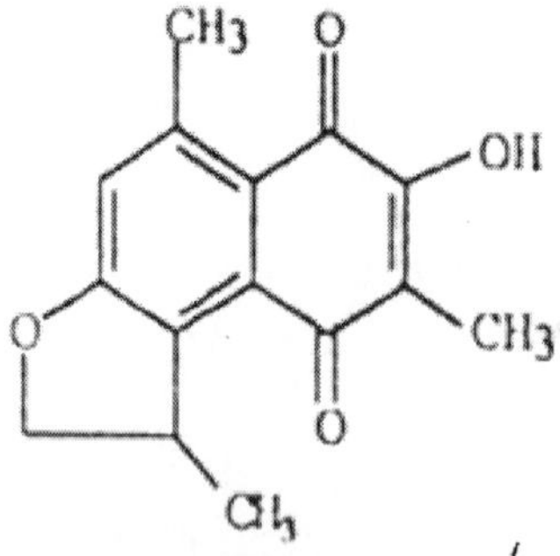

Thespesone

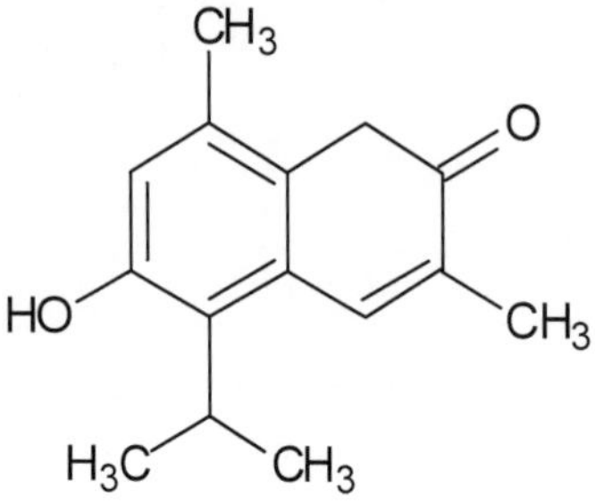

Lupeol

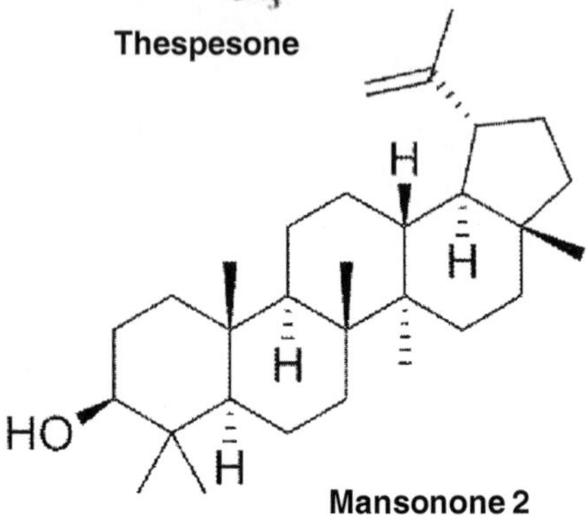

Mansonone 2

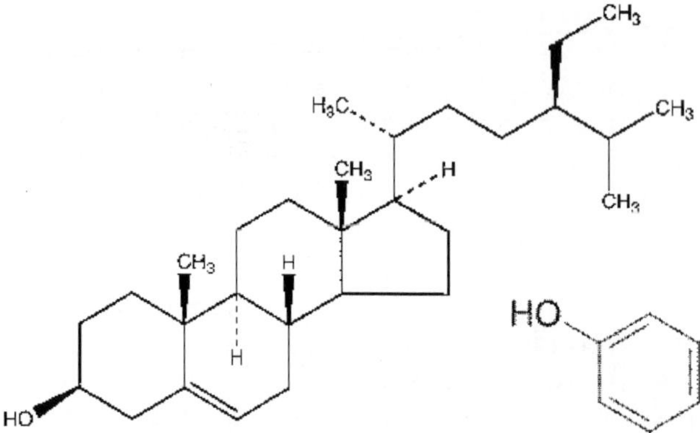

β-sitosterol

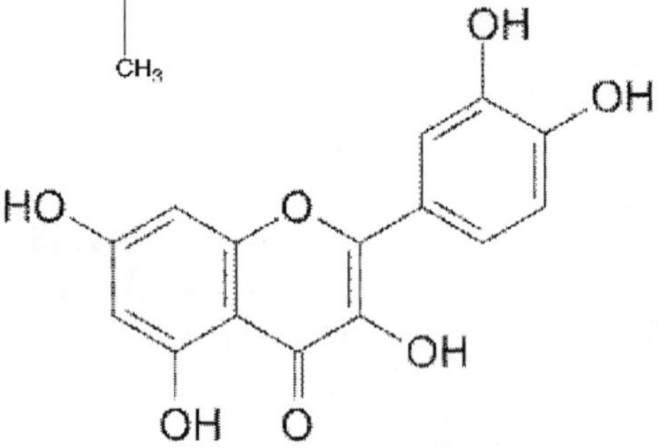

Quercetin

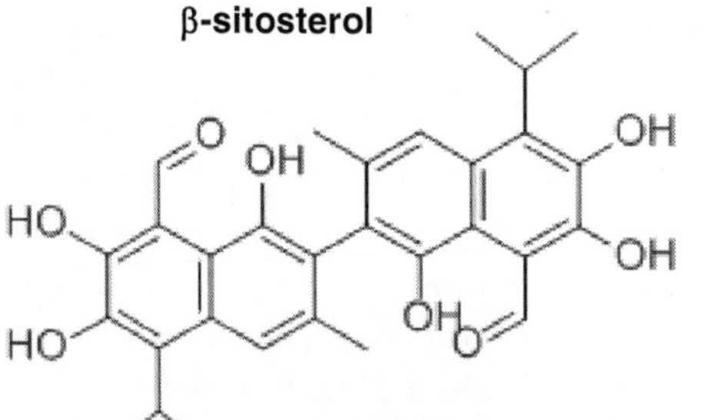

Gossypol

Figure 14.7: Structures Reported from Leaves of *Thespesia populnea.*

Palmitic acid (9), 3-Hydroxy-13 (18) -ene-28-oleanolic acid (10), respectively. Compounds 3-5, 10 are obtained from *Thespesia populnea* for the first time. Their structures were identified by physico-chemical analysis, spectroscopic analysis (^{1}H-NMR, ^{13}C-NMR and MS), and comparison with the data of literatures (Xu *et al.*, 2012).

Eight new sesquiterpenoids, named populene A-H, were isolated from dichloromethane extracts of the wood and dark heartwood of *Thespesia populnea*, together with 11 known compounds. The cytotoxic activity of isolated compounds was evaluated against four cancer cells. Mansonone E and (+)-gossypol showed significant activities. Their antibacterial properties are also studied (Sompong Boonsri *et al.*, 2008).

The isolation and structure elucidation of biologically active compound, (+)-gossypol from the bark were reported recently (Annamalai *et al.*, 2013).

The phytochemical and pharmacognostic studies were conducted on the root (Patil *et al.*, 2012) and leaf (Sai Koteswar Sarma *et al.*, 2011) is helpful for further standardization of the drugs. A new protease populenein was purified from the leaves of *Thespesia populnea* (Linn.) and characterized (Ishwarya *et al.*, 2013).The studies showed that the aqueous and ethanol extracts of leaf have both analgesic and anti-inflammatory activity (Ilavarasan *et al.*, 2012).

The essential oil analysis of Thespesia leaves reported that the principal chemical constituents are 1,2,3,5,6,8-hexahydro-4,7-dimethyl-1-(1-methylsethyl)-,(1S-cis)-Naphthalene (16.14 per cent), α-Farnesene (13.16 per cent), 1-methyl-4-(5-methyl-1-methylene-4-hexenyl)-,(S)-Cyclohexene (8.12 per cent) and n-Hexadecanoic acid (8.52 per cent) (YUAN Ting *et al.*, 2012).

Phytochemical studies of leaves revealed the presence of flavanoids, tannins, steroids, glycosides, saponins, phenols, terpenoids, alkaloids, anthocyanins, carbohydrates and proteins; and absence of reducing sugars, anthraqouinones leucoanthrocyanins and emodins (Savithramma *et al.*, 2011).

Study showed that the leaves were effective in inhibiting α-amylase, thereby proving to be potential hyperglycemic agents (Sangeetha *et al.*, 2012).The work has been carried out to find out immunomodulatory potential of methanolic extract of leaves of *Thespesia populnea.* (Switi *et al.*, 2011). Terpenoid fraction from *Thespesia populnea* leaves showed significant antiulcer activity in experimentally induced ulcer in rat model by decreasing the gastric secretions and by enhancing glycoprotein levels (Patil *et al.*, 2010).The yield of crude seed oil from the kernels of *Thespesia populnea* (Linn) was compared to the oil yield obtained from SG 747 cottonseed kernels. Studies showed that its seed oil has 2 per cent of fatty acids while later present in minor quantities (Michael, 2012).

Gossypol was reported to be present in the bark and fruit of the plant and it was found to be optically active (Datta *et al.*, 1972).The fruit kernels were reported to contain β-sitosterol, ceryl alcohol and a yellow pigment, thespesin (Srivastava *et al.*, 1963; Datta *et al.*, 1973). The ethanolic extract of the fruits were reported to have activity against Ranikhet disease virus and also anticancer against Lewis lung-carcinoma in the mice model (Dhawan, 2012).

Flavonoids, the major composition of *Thespesia populnea* flowers and phenolic contents were evaluated. The antibacterial activity of methanolic extract was investigated by agar well diffusion method (Saravanakumar *et al.*, 2009). The flowers contained kaempferol, kaempferol-7-glucoside and gossypetin. The floral extract of *T. populnea* exhibited anti steroidogenic activity in mouse ovary (Kavimani *et al.*, 1999).

Conclusion

Thespesia populnea is widely used in traditional system of medicine to combat and cure various ailments due to the presence of active secondary metabolites. The extensive literature survey revealed that *Thespesia populnea* is a important medicinal plant with diverse pharmacological spectrum. This is the first report on the antioxidant activity of *Thespesia populnea* heartwood. The pharmacological and phytochemical review shows that the plant is worth looking deeply. Further studies are needed to exploit the pharmacological activity of plant in combination with its composition, detailed chemical investigation are under progress at Agro Processing and Natural Products.

Acknowledgements

The authors are thankful to the Director, National Institute for Interdisciplinary Science and Technology for providing all the facilities and CSIR for the financial support to carry out the investigations.

References

Abd ELnasser, B., Sengab, Mohamed, R., Elgindi and Mahmoud, A.,(2013). Sesquiterpenes quinones from *Thespesia populnea* and their biological studies. *Journal of Pharmacognosy and Phytochemistry*, **2 (3)**: 136-139.

Akhila, A., and Rani, K.(1993). Biosynthesis of gossypol in *Thespesia populnea*. *Phytochemistry*, **33**: 335-40.

Allen, J.A. (2002).Use of plants, *Cordia subcordata*. Tropical Tree Seed Manual pp. 18–419. Agriculture Handbook 721. USDA Forest Service, Washington, DC. In: Vozzo,J.A. (ed.)

Anonymous (1996). Indian medicinal plants (Compendium of 500 species), 5: 281.

Anonymous, Wealth of India. (2003).First supplementary Series, Vol-X, (Sp-w), Raw materials. Published by Niscom, 10: 223-225.

Archana Moon, Aqueel Khan and Bharat Wadher (2010). Antibactrial potential of *Thespesia populnea*(Linn.) Sol.ex Corr. leaves and its corresponding callus against drug resistant isolates. *Indian journal of Natural products and resources*, **1(4)**: 444-449

Beena Joy, Kavitha Sankar, and Anjali (2012). *In-vitro* antioxidant/antidiabetic activity of *Thespesia populnea* leaves extract. *International Journal of Current Pharmaceutical Research*, **4(3)**: 138-143.

Burits, M., and Bucar, F.(2000). Antioxidant activity of *Nigella sativa* essential oil. *Phytotheraphy Research*, **14**: 323-328.

Daniel, M. (2006). Medicinal plants- Chemistry and properties. Science publishers, New Hampshire, 184.

Datta, S.C., Murti, V.V.S., and Seshadri, T.R. (1972). Isolation and study of (+)–Gossypol from *Thepesia populnea*. *Indian Journal of Chemistry*, **10**: 263-266.

Datta,S.C., Murti, V.V.S., and Seshadri,T.R.(1972). Isolation and study of (+) - Gossypol from *Thespesia populnea*. *Indian J Chem.*, **10**: 263-6.

Datta, S.C., Murti, V.V.S., Seshadri, T.R., and Sharma, N.N. (1973). Glycosidic component of *Thespesia populnea* flowers. *Indian Journal of Chemistry*, **11**: 506-507.

Dhawan, B.N. (2012). Anti-viral activity of indian plants. *Proc. Natl. Acad. Sci. Sect B. Biol. Sci.*, **82(1)**: 209–224.

Elakkiya, S., and Ananthi, T. (2011).Studies on anti-inflammatory activity of *Thespesia populnea* Linn on the drug induced male albino rats. *Journal Chem Pharm Res.*, **3(5)**: 473-477.

Fosberg, F.R., and Sachet, M.H. (1972). *Thespesia populnea* (L.) Solander ex Correa and *Thespesia populneoides* (Roxburgh) Kosteletsky (Malvaceae). *Smithsonian Contributions to Botany*, **7**: 1–13.

Haja Shereief, S., Moideen, K., Sengottuvelu, S., and Sivakumar, T. (2011). Antidiabetic effect of methanolic extract of *Thespesia populnea* flower and leaf in normal and alloxan- induced diabetic rats. *IJRAP.*, **2(3)**: 936-939.

Ilavarasan, R., Mohideen, S., and Venkataraman, S. (2012). Analgesic and anti-inflammatory properties of *Thespesia populnea* leaf extracts. *Nat Prod Res.*, **26(17)**: 1616-20.

Ilavarasan, R.,Vasudevan, M., Anbazhagan, S., and Venkataraman, S. (2003). Antioxidant activity of *Thespesia populnea* bark extracts against carbon tetrachloride-induced liver injury in rats. *J Ethnopharmacology*, **87(2-3)**: 227-30

Ishwarya, S., and Sangeetha, R.(2013). A new serine protease from the leaves of *Thespesia populnea. Prep Biochem Biotechnol.*, **43(1)**: 95-107.

Johnson, J.I., Gandhidasan, R., and Murugesan, R. (1999). Cytotoxicity and superoxide anion generation by some naturally occurring quinines. *Fr Rad Bio Med.*, **26**: 1072–78.

Jopsy, T., Simon, 1., and Karthikeyan, D.(2012). Antibacterial Activity of methanolic extract of *Thespesia populnea* (Malvaceae) flowers. *J. Environ. Nanotechnol.*, **1(1)**: 50-52.

Joshi, S.G.(2000). Medicinal plants, Oxford and IBH Publishing Co. Pvt. Ltd, New Delhi,259.

Kasim, S.M., Neelakantan, S., and Raman, P.V. (1975). Chemical components of Indian medicinal plants [*Thespesia populnea* wood and Floribunda flowers]. *Current Science*, **44**: 888-889.

Katsube,T.,Tabata, H.,Ohta, Y.,Yamasaki, Y., Anuurad, E., Shiwaku, K., and Yamane, Y. (2004). Screening for antioxidant activity in edible plant products: Comparison of low-density lipoprotein oxidation assay, DPPH radical scavenging assay and Folin–Ciocalteu assay. *Journal of Agri Food Chem.*,**52**: 2391-2396.

Kavimani, S., Ilango, R., Karpagam, S., Suryaprabha, K., and Jaykar, B. (1999). Antisteroidogenic activity of floral extract of *Thespesia populnea* Corr. in mouse ovary. *Indian Exp Biol.*, **37**: 1241-2.

Mani Vasudevan and Milind Parle. (2007). Memory-enhancing activity of *Thespesia populnea* in rats. *Pharmaceutical Biology*, **45(4)**: 267-273.

Marinova, D., Ribarova, F., and Atanassova, M. (2005).Total phenolics and total flavonoids in Bulgarian fruits and vegetables. *Journal of the University of Chemical Technology and Metallurgy*, **40**: 255-260.

Michael. K., and Dowd. (2012). Identication of the unsaturated heptadecyl fatty acids in the seed oils of *Thespesia populnea* and *Gossypium hirsutum. J Am Oil Chem Soc.*, **89**: 1599–1609.

Milbrodt, M., Konig, W.A., and Hausen, B.M. (1997). 7 Hydroxy-2,3,5,6-tetrahydro-3,6,9-Trimethylnaphto [1,8-B,C] Pyran-4,8-Dione from *Thespesia populnea. Phytochemistry*, **45**: 1523-1525.

Neelakantan, S., Rajagopalan, V., and Raman, P.V. (1983). Thespesone and thespone, two new mansonones of heartwoods of *Thespesia Populnea*sol. ex. corr (Fam: Malvaceae). *Indian Journal of Chemistry*, **22B**: 95-96.

Parthasarathy, R., Ilavarasan, R., and Karrunakaran, C.M.(2009). Antidiabetic activity of *Thespesia Populnea* bark and leaf extract against streptozotocin induced diabetic rats. *International Journal of Pharm Tech Research*, **1(4)**: 1069-1072.

Patil, P.H., Patel, J.B., and Surana, S.J.(2010). Evaluation of antiulcer activity of the terpenoid fraction from the leaves of *Thespesia populnea* in albino rats. *Research Journal of Pharmaceutical and Chemical Sciences*, **1(4)**: 495-513.

Patil, P.H., Patil, J.Y., Mahale, J.N., Patel, J.B., and Surana, S.J.(2010). Evaluation of antiulcer activity of the terpenoid fraction from the leaves of *Thespesia populnea* (L) (Malvaceae) in albino rats. *Research Journal of Pharmaceuticals, Biological and Chemical Sciences*, **1(4)**: 495.

Patil, P.S., Venkatanarayanan, R., Argade, P.D., and Shinde, P.R.(2012). Assessment of pharmacognostic and phytochemical standards of *Thespesia populnea* (L.) root. *Asian Pacific Journal of Tropical Biomedicine*, 1212-1216.

Patill,P.S., Venkatnarayanan, R., Argade, P.D., Ghule, R.S., and Shinde, P.R. (2012). Phytochemical screening and antioxidant potential of *Thespesia polpunea* (L.)root extract using *in –vitro* models. *International Journal of Reasearch in Pharmaceutical and Biomedical Sciences*, **3**: 1294-1299.

Re, R., Pellegrini, N., Proteggente, A., Pannala, A., Yang, M., and Rice-Evans, C. (1999). Antioxidant activity applying an improved ABTS radicalcation decolorization assay. *Free Radical Biology and Medicine*, **26**: 1231–1237.

Sai Koteswar Sarma, D., and Venkata Suresh Babu, A. (2011). Pharmacognostic and phytochemical studies of *Thespesia populnea* Linn. *J. Chem. Pharm. Res.*, **3(4)**: 237-244.

Sandya, B., Thomas, S., Isabel, W., and Shenbagathai, R. (2006). *Complementary and alternative medicines*, **3**: 101-114.

Sangeetha, R., and Vedasree, N. (2012). *In Vitro* α-amylase inhibitory activity of the leaves of *Thespesia populnea*. *International Scholarly Research Network*, Article ID 515634, 4 pages.

Saravanakumar, A., Venkateshwaran, K., Vanitha, J., Ganesh, M., Vasudevan, M., and Sivakumar, T. (2009). Evaluation of antibacterial activity, phenol and flavonoid contents of *Thespesia populnea* flower extracts. *Pak. J. Pharm. Sci.*, **22(3)**: 282-286.

Sathyanarayana, T., Sarita, T., Balaji, M., Ramesh, A., and Boini, M.K. (2004). Antihyperglycemic and hypoglycemic effect of *Thespesia populnea* fruit in normal and alloxan-induced diabetes in rabbits. *Saudi Pharmacol. J.*, **12**: 107-111.

Satyavati, G.V., Gupta, A.K., and Tandon, N. (1987). Medicinal plants of India. Indian Council of Medical Research, New Delhi, India.

Savithramma, N., Linga Rao, M., and Bhumi, G.(2011). Phytochemical screening of *Thespesia populnea* (L.) Soland and *Tridax procumbens* L. *J. Chem. Pharm. Res.*, **3(5)**: 28-34.

Seshadri, R., and Sharma, N.N. (1975). Neutral component of *Thespesia populnea* flowers. *Current Science*, **44**: 109-110.

Sharavana Kumaar, A.P., Vanitha, J., Venkateshwaran, K., Srikanth Reddy, K., Jopsy, T., and Simon, Karthikeyan. D. (2012). Antibacterial activity of methanolic extract of *Thespesia populnea* (Malvaceae) flowers. *J. Environ. Nanotechnol.*, **1(1)**: 50-52

Shingh, J., Upadhyay, A.K., Singh, S., and Rai, M. (2009). Total phenolics content and free radical scavenging activity of brassica vegetables. *Journal of Food Science Technology*, **46**: 595–597.

Shirwaikar, A., Devi, S., Premalatha, K., and Siju, E.N. (2011). *In Vitro* anthelmintic activity of *Thespesia Populnea*. *Asian J.Biochem.Pharmaceutic.Res.*, **1(3)**: 1.

Shirwaikar, A., Kumar, A.V., Krishnanand, B. R., and Sreenivasan, K. K. (1995). Chemical investigation and antihepatotoxic activity of *Thespesia populnea*. *International Journal of Pharmacogonosy*, **33(4)**: 305-310.

Siddharth Shrivastav, Rakesh Sindhu,.K., Sanjeev Kumar, and Pradeep Kumar.(2009). Anti-psoriatic and phytochemical evaluation of *Thespesia populnea* bark extracts. *International Journal of Pharmacy and Pharmaceutical Sciences*, **1(1)**: 176-185.

Siju, E.N., Arun Shirwaikar, Sarala Devi, Premalatha.(2011). *In vitro* anthelmintic activity of *Thespesia populnea*. *Asian Journal of Biochemical and Pharmaceutical Research*, **3(1)**: 1-5.

Sompong Boonsri, Chatchanok Karalai, Chanita Ponglimanont Suchada Chantrapromma and Akkharawit Kanjana-opas. (2008).Cytotoxic and antibacterial sesquiterpenes from *Thespesia populnea*. *J. Nat. Prod.*, **71**: 1173–1177.

Srivastava, S.N., Bhakuni, D.S., and Sharma, V.N.(1963).Chemical investigation of *Thespesia populnea* Soland. *Indian J Chem.*, 1: 451.

Switi, B., Gaikwad, and Krishna Mohan, G.(2011).Immunomodulatory activity of methanolic extract of *Thespesia populnea* leaves in Wistar albino rats. *Asian J Pharm Clin Res.*, **4(4)**: 99-101.

Valderrama, J.A., Leiva, H., Jaime, A., Guez, R., Theoduloz, C., and Schmeda-Hirshmann, G. (2008). Studies on quinones. Part 43: Synthesis and cytotoxic evaluation of polyoxyethylene-containing 1,4-naphthoquinones. *Bioorganic and Medicinal Chemistry*, **16**: 3687–3693..

Varier, P.V.S. (1994). Indian medicinal plant. Orient Longman Limited, **(5)**: 282.

Vasudevan, M., and Parle, M. (2006). Pharmacological actions of *Thespesia populnea* relevant to Alzheimer's disease. *Phytomedicine*, **13**: 677-687.

Vasudevan, M., Gunnam, K.K., and Parle, M. (2007). Antinociceptive and anti-inflammatory effects of *Thespesia populnea* bark extract. *J. Ethnopharmacol.*, **109**: 264-270.

Verhagen, H., Aruoma, O.I., Van Delft, J.H., Dragsted, L.O., Freguson, L.R., and Knasmüller, S. (2003).The 10 basic requirements for a scientific paper reporting antioxidant, antimutagenic or anticarcinogenic potential of test substances in *in vitro* experiments and animal studies *in vivo*. *Food and Chemical Toxicology*, **41(5)**: 603-610.

Verpoorte, R. (1998). Chemodiversity and the biological role of secondary metabolites, some thoughts for selecting plant material for drug development. *Proc Phytochem Soc Europe*, **43**: 11-14.

Viswanatha, G.L., Hanumanthappa, S., Krishnadas, N., and Rangappa, S.(2011). Antidiarrheal effect of fractions from stem bark of *Thespesia populnea* in rodents: Possible antimotility and antisecretory mechanisms. *Asian Pac.J.Trop.Med.*, **4(6)**: 451.

Xu, H.W., Cheng, F., Wu, J., and Zhou, Y. (2012). Chemical constituents from wood of the Chinese Hainan mangrove-associated plant, *Thespesia populnea*. *Zhong Yao Cai.*, **35(12)**: 1953-6.

Yuan Ting, Wang Cheng-fang, Fei Chao, Wang Dong, and Du Shu-shan. (2012). Studies on the essential oil of *Thespesia populnea* leaf. *Chinese Journal of Experimental Traditional Medical Formulae* 03.

Yuvaraj, P., and Ann Shine, Paul. (2013). Investigation of *in vitro* anthelmintic activity of phenolic acid fraction from *Thespesia populnea* (Stem Bark). *Asian Journal of Biochemical and Pharmaceutical Research*, **3 (3)**: 13-16.

Yuvaraj, P., and Subramoniam, A. (2009). Hepatoprotective property of *Thespesia populnea* against carbon tetrachloride induced liver damage in rats. *J Basic Clin Physiol Pharmacol.*, **20(2)**: 169-77.

Natural Products: Research Reviews Vol. 3 (2016) *Pages* **407–424**
Editor: **V.K. Gupta**
Published by: **DAYA PUBLISHING HOUSE, NEW DELHI**

15

Importance of Toxic Plants: Review of the *Palicourea* Genus

Ana Claudia F. Amaral[1*], Aline de S. Ramos[1], José Luiz P. Ferreira[1,2], Deborah Q. Falcão[2], Thamiris de A. Souza[1], Arith R. de Souza[1], Priscila E. Alves[1] and Jefferson Rocha de A. Silva[3]

ABSTRACT

The Palicourea genus (Rubiaceae family) includes about 200 species of shrubs and small trees, and is widely distributed in the tropics. It is notable for its toxicity to animals, and "sudden death" syndrome in cattle and rodents is frequently attributed to species of the genus. The geographic distribution of Palicourea longiflora is mainly restricted to the Amazon region, but it has the same characteristics of toxicity. Monofluoroacetic acid and methyl salicylate are some of the toxic substances present in the leaves and stems of P. longiflora, which explains the reported deaths of animals after ingesting the plant in the Amazon region. There is still little information about the genus Palicourea. The research already carried out on the toxic species of the genus Palicourea occurring in Brazil is presented in this chapter.

Keywords: Rubiaceae, *Palicourea*, *Palicourea longiflora*, Chemical constituents, Morphology, Taxonomy, Toxic plants, Toxicity.

---

1 [1]Laboratório de Plantas Medicinais e Derivados, Depto de Produtos Naturais, Farmanguinhos – FIOCRUZ – Rua Sizenando Nabuco, 100 – Manguinhos/RJ, 21041-250, Brazil

2 Faculdade de Farmácia – UFF- Rua Mario Viana, 523-Niterói, 24241-000, RJ, Brazil

3 Laboratório de Cromatografia – Depto. de Química - UFAM – Setor Sul/Av. Rodrigo Otávio, 3000 - Japiim - Manaus/AM, 69077-000, Brazil

* *Corresponding author*: E-mail: acamaral@fiocruz.br

Introduction

There are an estimated 250 thousand species of higher plants worldwide. Many of these species produce compounds that are toxic to living organisms. Recent hypotheses suggest that plants could form secondary metabolites as defense against predators. Many of these species are cited as toxic. Over time, toxic plants have been used in very different ways. In ancient times, and also in the twentieth century, in some regions, toxic plants have played an important role in providing food, through their use as poison to fishes. Toxic plants were widely used in the Middle Ages for political, military and personal purposes, as resources in intentional poisoning. More recently, in the late nineteenth century and beginning of the twentieth century, exotic plants have played an important role as sources of active substances capable of providing molecular models for drug development (Schenkel *et al.*, 1999).

The genus *Palicourea* belongs to the Rubiaceae family and is known to cause poisoning in cattle and wild animals at concentrations of 0.5-0.75 g/kg (Pessoa *et al.*, 2013). The species *P. marcgravii*, popularly known as "cafezinho", "café bravo", "erva de rato", "roxa", "roxinha", "roxana", "vick" and "cambará", is the most important toxic plant to herbivores in the Amazon region, causing "sudden death" syndrome in cattle, resulting in severe economic losses (Tokarnia *et al.*, 2002; Pessoa *et al.*, 2013). The syndrome is characterized by a peracute poisoning: after eating the leaves, the animals suddenly drop down to the ground, dying within a few minutes. Other species of the genus, *P. grandifolia*, *P. aeneofusca* and *P. juruana*, are also known to cause sudden death (Tokarnia *et al.*, 2002). The species *P. longiflora*, found in the Brazilian Amazon, has more restricted distribution than *P. marcgravii*, but the same characteristics of toxicity (Coelho *et al.*, 2007).

On the whole, there is still little information about the genus *Palicourea* and a review of *Palicourea* species, mainly *P. longiflora*, an important species of the Amazon region, are presented in this chapter.

Taxonomy

History of taxonomic classification

Taylor (1997), in his "Conspectus of the Genus *Palicourea*" describes important facts for taxonomic classification of this genus:

- ☆ In 1891, Schumann was one of the first authors to distinguish the genus *Palicourea* of the genus *Psychotria*. In his classification of the genus, he separated *Palicourea* by the corolla curved and gibbous at the base. However, he included in this genus the *Psychotria* species (*P. domingensis*) and excluded the *Palicourea* species (*P. aurantiaca*). Thus, his genus classification is not entirely consistent.
- ☆ In 1936, even using the most outstanding characteristics ("more, or less curved elongated tubular corolla, gibbous at the base; inflorescence branch usually clear, red or yellow"), Standley's classification appears to be based on a series of individual species classifications rather than a characterization of the genus as a whole, and considers as *Palicourea* several

species with these characteristics, but with clumps of green branches and white or off-white corolla (*P. andrei* Standl, *P. andaluciana* Standl and *P. garciae* Standl). Meanwhile, he considers many similar species as *Psychotria* (*Psychotria bella* Standl. [=*P. bella* (Standl.) Dwyer], *Psychotria brenesii* Standl. [=*P. standleyana* C. M. Taylor].

- ☆ The general concept of *Palicourea* by Steyermark of 1972 and 1974 is more consistent and more strongly grounded in characteristic corollas. In practice, the species included by Steyermark as *Palicourea*, usually present relatively well-developed tubular corolla.

According to Taylor (1989, 1993) the genus *Palicourea* is distinguished by its colorful flowers with relatively well-developed tubular corolla, with swollen base and internal pubescence ring. These characteristics appear to represent adaptations for pollination by hummingbirds.

Taxonomy

The Rubiaceae family, despite its cosmopolitan distribution, occurs mainly in tropical regions and includes about 550 genera and 9000 species. In the Brazilian territory, it is represented by approximately 1500 species included in 130 genera. It is considered one of the largest plant families in Brazil and is easily recognizable by its opposite leaves, and occasionally, verticillata and interpetiolar leaves (Campos *et al.*, 1999; Souza and Lorenzi, 2005). The inflorescences are essentially cymose, highly variable in type, shape and size. The flowers are usually hermaphroditic, actinomorphic, white, yellow and blue, with a pleasant odor, and unisexual in the *Alibertia*, *Amaaioua*, *Duroia* and *Kotchubaea* genera. The family is pollinated by moths, butterflies and bees. The fruits may be dry or fleshy, and the seeds, which are often winged, are scattered by the wind. The associations with ants are common occurring in hollow branches that serve as shelter for several species (Cronquist, 1981; Joly, 1993; Campos *et al.*, 1999).

Recent works on phylogeny transfer the Rubiaceae family from the order Rubiales to the order Gentianales, evidencing its monophyletic character, and emphasizing the inclusion of Dialypetalanthaceae and Henriqueziaceae. The Angiosperm classification system (APGIII) places the family in the following taxonomic position:

Eudicots

Core eudicots

Asterids

Euasterids I (Lamiids)

Gentianales

Rubiaceae

The family is divided into three subfamilies: Cinchonoideae, Ixoroideae and Rubioideae (Bremer *et al.*, 1995; Rova *et al.*, 1997) and Psychotrieae tribe is subordinated to Rubioideae and encompasses many genera, including *Palicourea* (Aubl.) Rich.

Morphology

Anatomic studies are highly significant, especially when associated with ecological, physiological and comparative aspects, supporting taxonomic works (Metcalfe and Chalk, 1972). The authenticity of a plant sample is given by the parameters of botanical identity, through macro- and microscopic assays, as well as the presence of active chemical compounds and/or characteristics of the species (Farias, 1999).

For example, the leaves of *Palicourea longiflora* (Aubl.) Rich., species of Amazon region, collect in June/2002 at the Experimental Farm of UFAM (Manaus – voucher specimen INPA 200639) show in transversal sections the uniseriate adaxial epidermis, with rectangular cells and a thick cuticle. Uni- or bicellular tector trichomes, some short and conical and others long and uni- or multicellular, with warty walls, can be observed more frequently in the abaxial epidermis. The dorsiventral mesophyll contains one, or more rarely, two palisade cells layers and 4 to 5 layers of spongy parenchyma, containing bundles of acicular crystals of calcium oxalate and vascular bundles wrapped by a parenchymatous sheath. The midrib is biconvex, narrower on the adaxial side, and presents an epidermis with rounded cells. The same trichomes described for the mesophyll region are also seen in this region. Internally, there is one angular collenchyma, a developed fundamental parenchyma of rounded cells, and collateral vascular bundles arranged in an open arch. A group of sclerenchymatous cells, with thick and lignified walls, can be seen wrapping the vascular bundles (Figure 15.1).

Calcium oxalate crystals, in the form of raphides or acicular crystals, are found in the leaves of *P. longiflora*. This type of crystal is long and needle-like, sharp at both sides, and generally gathered in bundles. They are easily found in the parenchymatic cells of the pith, cortex and secondary phloem (Cutter, 1986; Coelho *et al.*, 2007). According to Metcalfe and Chalk (1972) calcium oxalate crystals in the form of raphides are characteristic of the *Palicourea* species. The plants invest carbon and calcium sources into the formation of crystals, indicating that this is an important basic process to their growth and development.

The crystals can also contribute to the toxicity of the species. In a few toxic species of the *Dieffenbachia* (Araceae) genus, injury is due to the conjugate action of one mechanical factor allied with one toxic lipid substance. The former is present under the form of myriads of very thin calcium oxalate needles that can be found inside the cell. These crystals penetrate the oropharynx mucosa, causing severe irritation. Furthermore, they enable the performance of lipids, generating glottis constriction and lethal asphyxia. The leaves of some *Palicourea* species lead to "sudden death" in animals, and the same mechanism may be involved in this genus, monofluoracetic acid being the harmful substance that is introduced by the calcium oxalate crystals.

Geographic Distribution

The genus *Palicourea* has about 200 species, including shrubs and small trees, widely distributed in the tropics. In the Brazilian rainforest there are seven species:

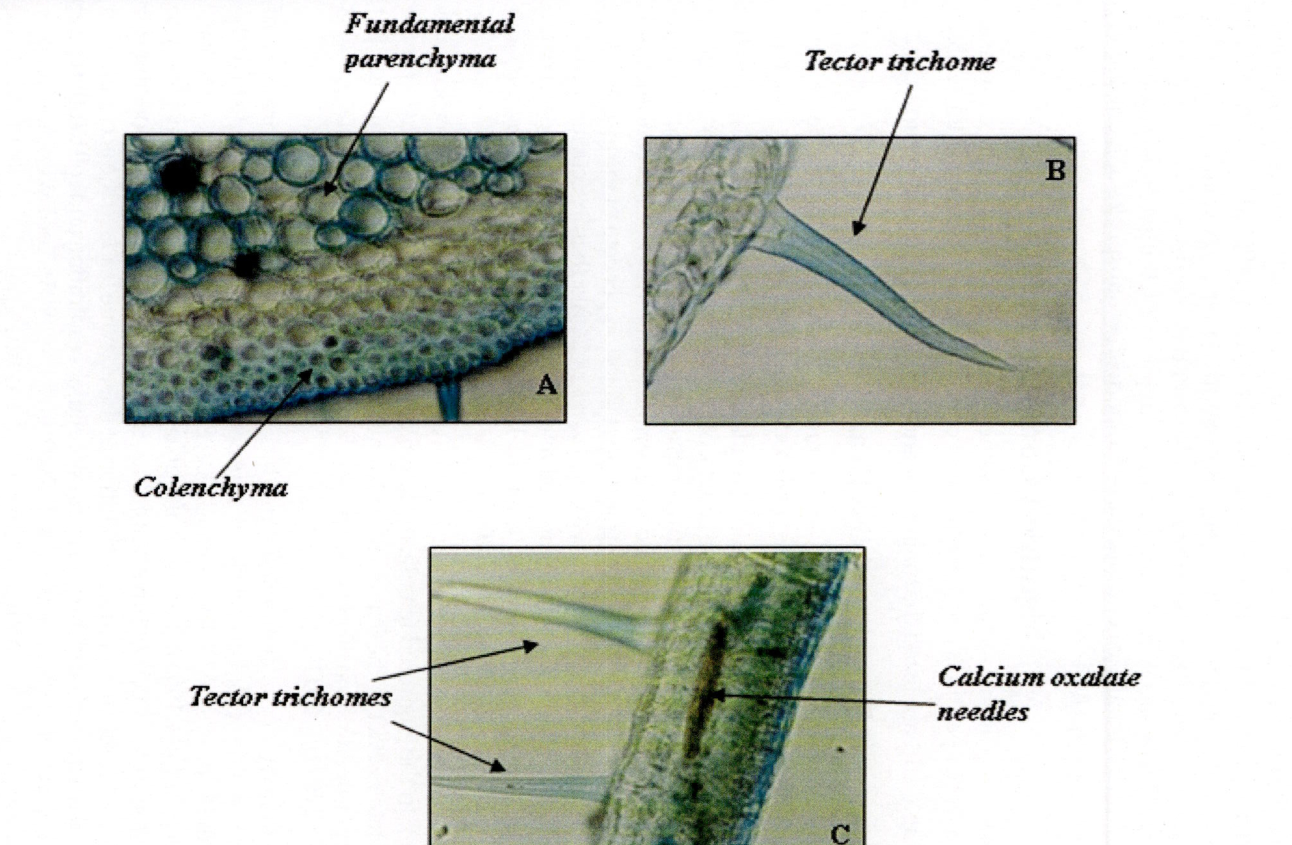

Figure 15.1: Anatomy of the Leaf of *P. longiflora* (Transverse section): (a) Abaxial midrib region (50 µm), (b) Trichomes of the lower region of the mesophyll (25 µm), (c) Mesophyll with acicular crystals of calcium oxalate (50 µm).

Palicourea australis C.M. Taylor, *P. brasiliensis* Wawra, *P. fulgens* (Müll.Arg.) Standl., *P. longipedunculata* Gardner, *P. radians* (Müll.Arg.) Standl., *P. rudgeoides* (Müll.Arg.) Standl. and *P. veterinariorum* Kirkbride (Zappi and Taylor, 2010). The name *Palicourea* originates in the Guianas, where is widespread (Barroso, 1991). A large number of species is concentrated in tropical regions of South America, particularly Ecuador, Colombia, Venezuela, Peru and Brazil. It is also found in North and Central America, particularly Costa Rica, Mexico and Panama (Taylor, 1989; Burger and Taylor, 1993; Taylor, 1997). In Brazil the genus includes 55 species, six subspecies and one variety (Taylor, 1997; Zappi and Taylor, 2010).

Identified Compounds

Palicourea genus is a source of triterpenes, alkaloids, organic salts, coumarins and macrocyclic peptides and some of these compounds are shown in Figure 15.2.

At Table 15.1 are listed the main substances related to *Palicourea* species.

Toxicity of the Brazilian Species

References on the genus toxicity are frequently found in literature. The species of the genus *Palicourea* toxic for cattle in Brazil are described in Table 15.2.

Due to its toxicity and insecticide potential, the *Palicourea marcgravii* species has been studied as an alternative for the control of crop pests. The aqueous extract of the plant was assayed against the brown citrus aphid *Toxoptera citricida* Kirkaldy (Homoptera: Aphididae) and caused the mortality of all the insects at concentration above 50 mg/mL. The lethal concentration that killed 50 per cent of the insects (LC_{50}) was 10.61mg/mL (Gonzaga *et al.*, 2008). *Palicourea rigida* also has insecticidal activity: the peptide fraction obtained from the methanol/dichloromethane (1:1) extract of the aerial parts can kill 60 per cent of larvae of Lepidoptera (*Diatraea saccharalis*) at a concentration of 1 μM (Pinto *et al.*, 2012).

P. marcgravii has acaricidal activity on the cattle tick *Rhipicephalus* (*Boophilus microplus*). In the engorged female, the lethal concentration 50 per cent (LC_{50}) of the ethyl acetate extract of the leaves, the most active extract, is 30.08 mg/mL, and the inhibitory concentration 50 per cent (IC_{50}) is 5.79 mg/mL. On larvae, the LC_{50} is 4.38 mg/mL (Silva *et al.*, 2011).

The toxicity of the genus *Palicourea* has been associated with the presence of monofluoroacetic acid. This substance interferes with the energy metabolism of the cells, *i.e.*, the Krebs Cycle (Tokarnia *et al.*, 2002; Eckschmidt *et al.*, 1989; Moraes-Moreau, 1995; Krebs *et al.*, 1994). The monofluoroacetic acid is converted *in vivo* into fluorocitrate, which competitively inhibits the aconitase enzyme responsible for conversion of citrate to isocitrate. Inhibition of the enzyme leads to the accumulation of citrate, reduction of power supply, loss of cellular respiration and cell death (Lee *et al.*, 2012; Krebs *et al.*, 1994; Tokarnia *et al.*, 2002; Gribble, 1973). The hydropic-vacuolar degeneration associated with nuclear pyknosis of the epithelial cells of the distal convoluted tubules in the kidney is a typical sign of intoxication by monofluoracetic acid in some animal species (Nogueira *et al.*, 2011). These findings are related to the sudden death caused by the ingestion of *Palicourea* species, and suggest that

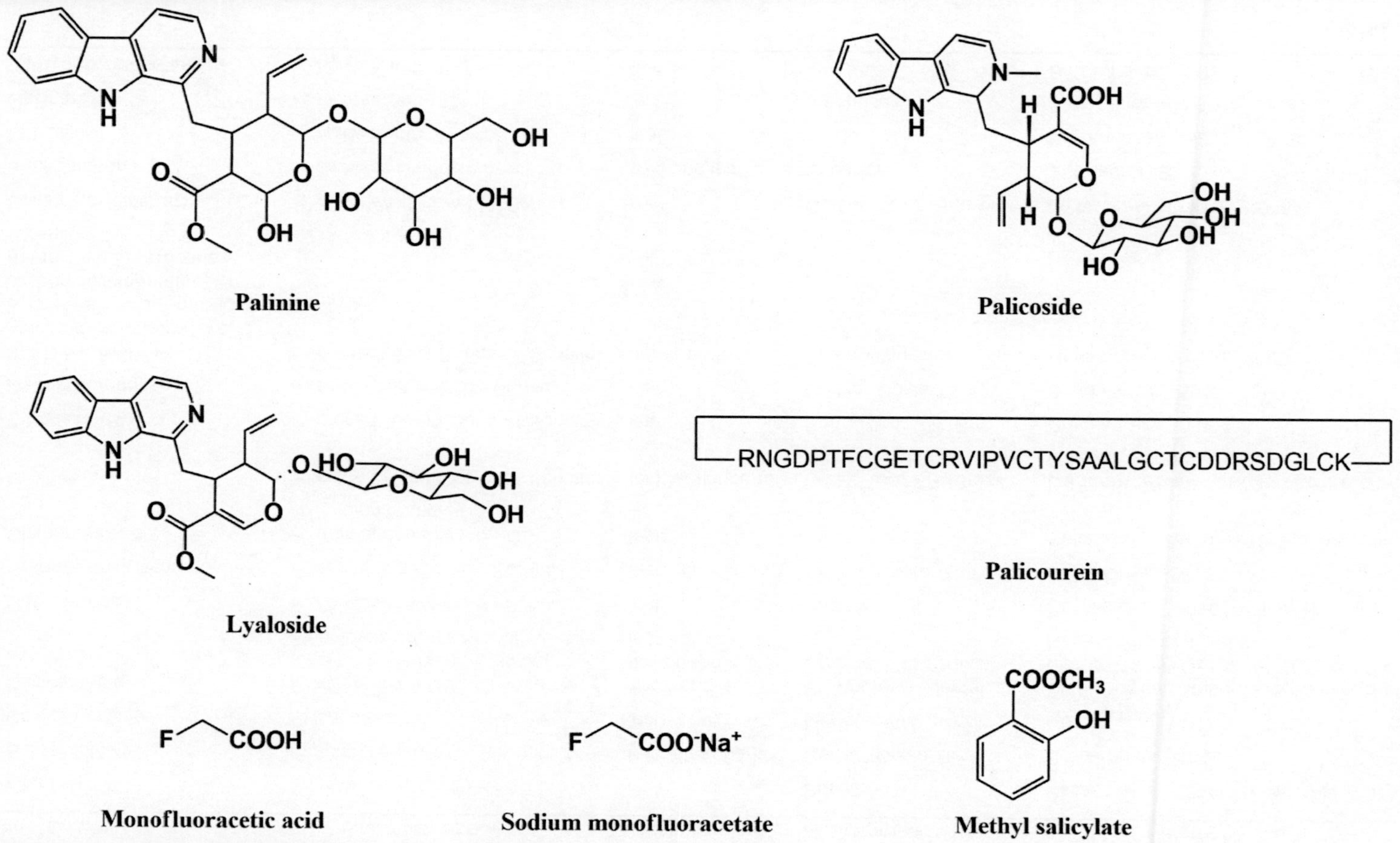

Palinine

Palicoside

Lyaloside

RNGDPTFCGETCRVIPVCTYSAALGCTCDDRSDGLCK

Palicourein

Monofluoracetic acid

Sodium monofluoracetate

Methyl salicylate

Figure 15.2: Some Constituents of *Palicourea* Genus.

Table 15.1: Substances Identified in *Palicourea* Species

Constituent	*Palicourea Species*	*Plant Material*	*Geographic Origin*	*Reference*
Allantoin	*P. coriacea* (Cham.) K. Schum	leaf	Brazil)	Passos *et al.*, 2010; Freitas *et al.*, 2011
Bahienoside B	*P. acuminata* (Benth.) Borhidi	leaf; stem bark	Golfito (Costa Rica)	Berger *et al.*, 2012
Calcium oxalate	*P. longiflora*	leaf	Manaus, AM (Brazil)	Coelho *et al.*, 2007
Calycanthine	*P. alpina* (Sw.) DC.; *P. coriacea* (Cham.) K. Schum; *P. fendleri* Standley; *P. ovalis* Standl.	Stem and leaf; flower and fruit; leaf	St. Andrews (Jamaica); Goiânia, GO (Brazil)	Woo-Ming and Stuart, 1975; Nakano and Martin, 1976; Garcia *et al.*, 1997; do Nascimento *et al.*, 2006
Calycantrine	*P. coriacea* (Cham.) K. Schum	leaf	(Brazil)	Passos *et al.*, 2010; Freitas *et al.*, 2011
Carboxystrictosidine-5α	*P. acuminata* (Benth.) Borhidi	stem bark	Golfito (Costa Rica)	Berger *et al.*, 2012
Chimonanthine	*P. domingensis* (Jacq.) DC.; *P. fendleri* Standley	leaf		Nakano and Martin. 1976; Ripperberger, 1982
Croceaine A	*P. crocea* (Sw.) Roem. et Schultes	leaf; aerial parts	Porto Rico, PR (Brazil); Arima (Trinidad)	Düsman *et al.*, 2004; Narine and Maxwell, 2009;
Croceaine B	*P. crocea* (Sw.) Roem. et Schultes	leaf	Porto Rico, PR (Brazil)	Düsman *et al.*, 2004
Desoxycordifoline	*P. acuminata* (Benth.) Borhidi	leaf;	Golfito (Costa Rica)	Berger *et al.*, 2012
3,4-Dihydro-1-(1-β-D-glucopyranosyloxy-1,4a, 5,7a-tetrahydro-4-methoxy-carbonylcyclopenta[c] pyran-7-yl)-β-carboline N2-oxide	*P. crocea* (Sw.) Roem. et Schultes	aerial parts	Arima (Trinidad)	Narine and Maxwell., 2009
Epi-strictosidinic acid	*P. coriacea* (Cham.) K. Schum	root	Goiânia, GO (Brazil)	do Nascimento *et al.*, 2006;
3-friedelanone	*P. demissa* Standl.	twig and leaf	Ecuador)	El-Seedi, 1999
Friedelin	*P. rigida* Kunth	leaf		Bolzani *et al.*, 1992
Harman	*P. alpina* (Sw.) DC.		(Jamaica)	Stuart and Woo-Ming, 1974
3β-hydroxyfriedelane	*P. rigida* Kunth	leaf		Bolzani *et al.*, 1992

Contd...

Table 15.1–*Contd...*

Constituent	*Palicourea Species*	*Plant Material*	*Geographic Origin*	*Reference*
30-hydroxyfriedelan-3-one	*P. rigida* Kunth	leaf		Bolzani *et al.*, 1992
Isorhamnetin 3-glucoside	*P. rigida* Kunth	leaf	Goiânia, GO (Brazil)	Rosa *et al.*, 2010
Isoscopoletin	*P. demissa* Standl.	twig and leaf	(Ecuador)	El-Seedi, 1999
Isovanillic acid	*P. demissa* Standl.	twig and leaf	(Ecuador)	El-Seedi, 1999
Logamboside	*P. acuminata* (Benth.) Borhidi	leaf	Golfito (Costa Rica)	Berger *et al.*, 2012
Loganin	*P. rigida* Kunth	leaf	Minas Gerais (Brazil); Goiás (Brazil); Minas Gerais (Brazil); São Paulo (Brazil); São Paulo (Brazil); Minas Gerais (Brazil); Goiás (Brazil)	Lopes *et al.*, 2004; Morel *et al.*, 2011; Silva *et al.*, 2013
Lyaloside	*P. adusta* Standley	leaf	Cartago (Costa Rica)	Valverde *et al.*, 1999; 13-Dusman *et al.*, 2004
Lyaloside-(*E*)-*O*-cinnamoyl-4''-hydroxy-3'',5''-dimethoxy	*P. adusta* Standley	leaf	Cartago (Costa Rica)	Valverde *et al.*, 1999; 13-Düsman *et al.*, 2004
Lyaloside-(*E*)-*O*-cinnamoyl-4''-hydroxy-3''-methoxy	*P. adusta* Standley	leaf	Cartago (Costa Rica)	Valverde *et al.*, 1999; 13-Düsman *et al.*, 2004
Methyl salicylate	*P. longiflora* (Aubl.) A. Rich	leaf	Manaus, AM (Brazil)	Coelho *et al.*, 2007
Palicoside	*P. marcgravii* A. St.-Hil.	leaf	São Paulo (Brazil)	Morita *et al.*, 1989; Düsman *et al.*, 2004
Palicourein	*P. condensata* Standl.	stem bark	Pastaza (Ecuador)	Bokesch *et al.*, 2001
Palidimine	*P. alpina* (Sw.) DC.		(Jamaica)	Stuart and Woo- Ming, 1974
Palinine	*P. alpina* (Sw.) DC.		(Jamaica)	Stuart and Woo Ming, 1974; Düsman *et al.*, 2004
Parigidin-br1	*P. rigida* Kunth	Leaf, inflorescence and peduncle	Brasília (Brazil)	Pinto *et al.*, 2012

Contd...

Table 15.1–*Contd...*

Constituent	*Palicourea Species*	*Plant Material*	*Geographic Origin*	*Reference*
Protocatechuic acid	*P. demissa* Standl.	twig and leaf	(Ecuador)	El-Seedi, 1999
Psychollatine	*P. crocea* (Sw.) Roem. et Schultes	aerial parts	Arima (Trinidad)	Narine and Maxwell, 2009
Quercetin 3-*O*-sophoroside	*P. rigida* Kunth	leaf	Goiânia, GO (Brazil)	Rosa *et al.*, 2010
Quercetin 3-*O*-β-D-glucoside	*P. rigida* Kunth	leaf	Goiânia, GO (Brazil)	Rosa *et al.*, 2010
Scopoletin	*P. demissa* Standl.	twig and leaf	(Ecuador)	El-Seedi., 1999
Sitosterol-β	*P. demissa* Standl.	twig and leaf	(Ecuador)	El-Seedi., 1999
Sodium monofluoroacetate/ monofluoroacetic acid	*P. aeneofusca* (Müll. Arg.) Standl.; *P. aff. Juruana*; *P. aff. longiflora;* *P. amapaenses* Steverm.; *P. grandiflora* (Kunth) Standl.; *P. longiflora* (Aubl.) A. Rich.; *P. macarthurorum* C.M. Taylor; *P. marcgravii* A. St.-Hil.; *P. nigricans* K. Krause; *P. vacillans* Standl.	leaf	Manaus, AM (Brazil); Areia, PB (Brazil); herbarium specimens; Goiás, Paraíba (Brazil)	Coelho *et al.*, 2007; Silva *et al.*, 2011; Lee *et al.*, 2012; de Oliveira *et al.*, 2013; Cook *et al.*, 2014
Stigmasterol	*P. demissa* Standl.	twig and leaf	(Ecuador)	El-Seedi, 1999
Strictosidine	*P. elata* (Sw.) Borhidi			Berger *et al.*, 2012
Strictosidinic acid	*P. coriacea* (Cham.) K. Schum	root	Goiânia, GO (Brazil)	do Nascimento *et al.*, 2006
Strictosidinic acid-4'-*O*-3''-sucrose	*P. coriacea* (Cham.) K. Schum	root	Goiânia, GO (Brazil)	do Nascimento *et al.*, 2008
Strictosidinic ketone	*P. coriacea* (Cham.) K. Schum	root	Goiânia, GO (Brazil)	do Nascimento *et al.*, 2006
Tectoquinone-7-hydroxy	*P. rigida* Kunth	leaf	Goiânia, GO (Brazil)	Rosa *et al.*, 2006

Contd...

Table 15.1–*Contd...*

Constituent	*Palicourea Species*	*Plant Material*	*Geographic Origin*	*Reference*
Umbelliferone	*P. demissa* standl.	twig and leaf	(Ecuador)	El-Seedi., 1999
Ursolic acid	*P. coriacea* (Cham.) K. Schum	root and leaf; leaf; leaf and stalk	Goiânia, GO (Brazil)	do Nascimento *et al.,* 2006; Passos *et al.,* 2010; Freitas *et al.,* 2011
Vallesiachotamine	*P. rigida* Kunth	leaf	Goiânia, GO (Brazil)	Vencato *et al.,* 2006; Soares *et al.,* 2012
Vanillic acid	*P. demissa* Standl.	twig and leaf	(Ecuador)	El-Seedi, 1999
Vomifoliol	*P. alpina* (Sw.) DC.	leaf, twigs and bark	St. Andrews (Jamaica)	Stuart and Woo- Ming, 1975

Table 15.2: Toxic Species of the Genus *Palicourea* in Brazil

Species	*Vernacular Name*	*Geographic Distribution*	*Reference*
Palicourea aeneofusca (Müll. Arg.) Standl.	Erva de rato, papaconha	Northeast region	de Oliveira *et al.*, 2013; Lee *et al.*, 2012; Tokarnia *et al.*, 2002
Palicourea barbiflora H. B. K Standl	Erva de rato graúda	Paranaiba Valley	Coelho, 2003
Palicourea coriacea (Cham.) K. Schum	Folha de couro, congonha dura, congonhão, douradinha, douradinha do campo, congonha do campo	Midwestern Region	Passos *et al.*, 2010; Freitas *et al.*, 2011
Palicourea corymbifera (Müll. Arg.)	Without information	Amazon forest and Rainforest	Ribeiro de Assis *et al.*, 2006
Palicourea crocea (Sw.) Roem. et Schultes	Cachimbo, palicourea amarela, palicourea vermelha	Southern Mexico to Southern Brazil and Paraguay	Düsman *et al.*, 2004
Palicourea cuiabense	Without information	Without information	Coelho, 2003
Palicourea grandiflora (Kunth) Standl.	Pau jacaré	Amazon region	Tokarnia *et al.*, 2002
Palicourea juruana Krause	Roxa, Roxinha	Minas Gerais e Paraná, Amazon rainforest	Cook *et al.*, 2014; Tokarnia *et al.*, 2002
Palicourea longiflora Aubl.) A. Rich	Without information	Amazonas	Coelho *et al.*, 2007
Palicourea longipedunculata Gardner	Congonha do mato, Congonha	Minas Gerais	Coelho, 2003
Palicourea marcgravii H. B.K. St. Hill	"Cafézinho", "erva-de-rato", "erva-café", "café bravo", "erva de rato verdadeira", "erva de rato palha", "erva de rato da mata", "roxa", "roxinha", "roxona"	São Paulo, Mato Grosso, Rio Grande do Sul, Amazonas, Pernambuco	Pessoa *et al.*, 2013; Moraes-Moreau *et al.*, 1995; Tokarnia *et al.*, 2002, Eckschmidt *et al.*, 1989
Palicourea nicotianifolia Cham e Schltd	Falsa erva de rato	Rio de Janeiro, São Paulo e Minas Gerais	Coelho, 2003
Palicourea rígida Kunth	Douradão, douradinha do campo, gritadeira.	Midwestern Region	Rosa *et al.*, 2010
Palicourea squarrosa (Müll. Arg.) Standl.	Congonha escura, escurinha	Minas Gerais	Coelho, 2003

monofluoracetic acid is the main constituent responsible for the deaths of these animals (Tokarnia *et al.*, 2002). This hypothesis is supported by the protective effect of acetamide when administered as an antidote against monofluoroacetic acid in several animal species prior to consumption of *P. marcgravii* (Gorniak *et al.*, 1994; Peixoto *et al.*, 2011; Peixoto *et al.*, 2012).

Other substances may contribute to the toxicity of the genus *Palicourea*. Methyl salicylate, identified in *P. longiflora*, is considered another cause of toxicity. This substance is soluble in lipids and can cause convulsions and skin eruptions, also affecting the acid-base balance (Coelho *et al.*, 2007). The oral LD_{50} in rats is 1110 mg/kg (Davison *et al.*, 1961). Calcium oxalate needles, also found in *P. longiflora*, can penetrate the pharynx causing irritation, glottis constriction and lethal asphyxia (Coelho *et al.*, 2007).

Some ways of controlling plant poisoning by plants containing monofluoroacetic acid have been studied. One of them is the induction of resistance to *Palicourea* sp. poisoning by continuous administration of the dry plant. The aerial parts of *P. aeneofusca*, containing 0.09 ± 0.05 per cent monofluoroacetic acid, was administered at 0.06 g/kg/day to goats (Alpine and Saanen). It was observed that the treatment increased the resistance to poisoning, probably due to the proliferation of monofluoracetic acid degrading bacteria in the rumen (de Oliveira *et al.*, 2013). Another way of controlling the poisoning is to introduce monofluoroacetic acid degrading bacteria directly into the rumen of the animals (Camboim *et al.*, 2012).

Pharmacological Potential of Brazilian Species

According to reports in the literature, some species exhibit important biological activities and have potential therapeutic uses.

The ethanol extract of the aerial parts of *Palicourea coriaceae* has diuretic activity. The administration of a single oral dose (20, 40 and 80mg/kg) to rats (Wistar) increased the urinary flow in 24h and the excretion of both ions sodium and potassium, without toxicity signals (Freitas *et al.*, 2011). *P. coriaceae* aqueous extract at 5, 10, or 15 mg extract/mL protects somatic cells of *Drosophila melanogaster* against the genotoxic effect induced by doxorubicin (Passos *et al.*, 2010).

The ethyl acetate fraction of *Palicourea rigida*, which is rich in flavonoids, has moderate antioxidant activity, with IC_{50} of 192.77 µg/mL (Rosa *et al.*, 2010). The alkaline fraction of the ethanol extract of *Palicourea rigida* can inhibit the proliferative effect on human melanoma cells on in *in vitro* assays (IC_{50} = 14.7 ± 1.2 µM). The vallessiachotamine isolated from this extract can cause cytotoxicity and necrosis at 50 µM (Soares *et al.*, 2012).

References

Barroso, G. M. (1991). Sistemática das Angiospermas do Brasil, UFV, Viçosa, Brazil, v. 3, 326 pp.

Berger, A., Hannes Fasshuber, H., Schinnerl, J., Lothar Brecker, L., and Greger, H. (2012). Various types of tryptamine-iridoid alkaloids from *Palicourea acuminate* (=*Psychotria acuminata*, Rubiaceae). *Phytochemistry Letters*, **5**: 558–562.

Bokesh, H. R., Pannell, L. K., Cochran, P. K., Sowder, R. C., Mckee, T. C., and Boyd, M. R. (2001). A novel anti-HIV macrocyclic peptide from *Palicourea condensata*. *Journal of Natural Products*, **64**: 249–250.

Bolzani, V. S., Trevisan, L. M. V., and Young, M. C.M. (1992). Triterpenes of *Palicourea rigida* H.B.K. *Revista Latinoamericana de Quimica*, **23**: 20-21.

Bremer, B., Andreasen, K., and Olsson, D. (1995). Subfamilial and tribal relationships in the Rubiaceae based on rbcL sequence data. *Annals of the Missouri Botanical Garden*, **82**: 383–397.

Burger, W. C., and Taylor C. M. (1993). Flora Costaricensis: Family #202. Rubiaceae. *Fieldiana*: *Botany, New Series*, **33**: 1-333.

Camboim, E. K. A., Almeida, A. P., Tadra-Sfeir, M. Z., Junior F. G., Andrade P. P., McSweeney, C. S., Melo, M. A., and Riet-Correa, F. (2012). Isolation and identification of sodium fluoroacetate degrading bacteria from caprine rumen in Brazil. *The Scientific World Journal*, article ID 178254, 6 pp.

Campos, M. T. V. do A., Brito, J. M., and Taylor, C. Rubiaceae. (1999). In: Ribeiro, J. E. L. *et al.*, Flora de Reseva Ducke – Guia de identificação das plantas vasculares de uma floresta de terra-firme na Amazônia Central, p. 625-647.

Coelho, E. G. (2003). Contribuição à química da família Rubiaceae: *Palicourea longiflora* (Aubl.) A. Rich (Dissertation). UFAM, Instituto de Ciências Exatas, Manaus, Brazil, 52 pp.

Coelho, E. G., Amaral, A. C. F., Ferreira, J. L. P., dos Santos, A. G., Pinheiro, M. L. B., and Silva, J. R. de A. (2007). Calcium oxalate crystals and methyl salicylate as toxic principles of the fresh leaves from *Palicourea longiflora*, an endemic species in the Amazonas state. *Toxicon*, **49**: 407-409.

Cronquist, A. (1981). An integral system of classification of flowering plants. Columbia University Press New York, 1262 pp.

Cook, D., Lee, S. T., Taylor, C. M., Bassüner, B., Riet-Correa, F., Pfister, J. A., Dale R., and Gardner, D. R. (2014). Detection of toxic monofluoroacetate in *Palicourea* species. *Toxicon*, **80**: 9–16.

Cutter, E. G. (1986). Anatomia Vegetal. Roca, São Paulo.

Davison, C., Zimmermanm E. F., and Smith, P. K. (1961) On the metabolism and toxicity of methyl salicylate. *Journal of Pharmacology and Experimental Therapeutics*, **132**: 207-211.

de Oliveira, M. D., Franklin Riet-Correa, F., Carvalho, F. K. L., Silva, G. B., Pereira, W. S., and Medeiros, R. M. T. (2013). Indução de resistência à intoxicação por *Palicourea aeneofusca* (Rubiaceae) mediante administração de doses sucessivas não tóxicas. *Pesq. Vet. Bras.*, **33**(6): 731-734.

do Nascimento, C. A., Gomes, M. S., Lião, L. M., Oliveira, C. M. A., Kato, L., Silva, C. C., and Tanaka, C. M. A. (2006). Alkaloids from *Palicourea coriacea* (Cham.) K. Schum. *Zeitschrift für Naturforschung B, Journal of Biosciences*, **61b**: 1443–1446.

do Nascimento, C. A., Lião, L. M., Kato, L., da Silva, C. C., Tanaka, C. M. A., Schuquel, I. T. A., and de Oliveira, C. M. A. (2008). A tetrahydro β-carboline trisaccharide from *Palicourea coriacea* (Cham.) K. Schum. *Carbohydrate Research,* **343**: 1104–1107.

Düsman, L. T., Jorge, T. C. M., de Souza, M. C., Eberlin, M. N., Meurer, E. C., Bocca, C. C., Basso, E. A., and Sarragiotto, M. H. (2004). Monoterpene indole alkaloids from *Palicourea crocea*. *Journal of Natural Products,* **67** (11): 1886-1888.

Eckschmidt, M., Brizolla, M., Tarragá D. P., and Palermo-Neto J. (1989). Is monofluoroacetic acid the active neurotoxic principle in *Palicourea marcgravii* (St. Hill) leaves? *Brazilian Journal of Medical and Biological Research,* **22**: 975-977.

El Seedi, H. R. (1999). Coumarins, benzoic acids and triterpenoids from *Palicourea demissa*. *Revista Latinoamericana de Quimica,* **27**: 13–16.

Farias, M. R. (1999). Avaliação da qualidade de matéria-prima vegetal. In: Simões C.M.O. *et al.,* Farmacognosia: da planta ao medicamento. Ed. Universidade/ UFRGS, Porto Alegre, p. 263-288.

Freitas, P. C. M., Pucci, L. L., Vieira, M. S., Lino Jr., R. S., Oliveira, C. M. A., Cunha, L. C., Paula, J. R., M. C., and Valadares, M. (2011). Diuretic activity and acute oral toxicity of *Palicourea coriacea* (Cham.) K. Schum. *Journal of Ethnopharmacology,* **134**: 501–503.

Garcia, L. A., Tobón, C. F., and Mora, C. E. (1997). Citotoxicidad de los componentes de *Palicourea ovalis* CE. *Revista Colombiana Ciencias Quimico-farmaceuticas,* **26**: 55–57.

Gonzaga, A. D., Garcia, M. V. B., Sousa, S. G. A., Py-Daniel, V., Correa, R. S., and Ribeiro, J. D. (2008) Toxicity of cassava manipueira (*Manihot esculenta* Crantz) and erva-de-rato (*Palicourea marcgravii* St. Hill) to adults of *Toxoptera citricida* Kirkaldy (Homoptera: Aphididae). *Acta Amazonica,* **38**: 101-106.

Gorniak, S. L., Palermo-Neto J., and Spinosa, H. S. (1994) Effects of acetamide on experimentally induced *Palicourea marcgravii* (St Hill) poisoning in rats. *Veterinary and Human Toxicololy,* **36**: 101-102.

Gribble, W.G. (1973) Fluoroacetate Toxicity. *Journal of Chemical Education,* **50**: 460-462.

Joly, A. B. (1993) Botânica: Introdução à Taxonomia Vegetal, Nacional Ed., São Paulo, Brasil, 777 p.

Krebs, H. C., Kemmerling, W., and Habermehl, G. (1994). Qualitative and quantitative determination of fluoroacetic in *Arrabidea bilabiata* and *Palicourea marcgravii* by ^{19}F-NMR spectroscopy. *Toxicon,* **32**: 909-913.

Lee, S. T., Cook, D., Riet-Correa, F., James A. Pfister, J. A., Anderson, W. R., Lima, F. G., and Gardner, D. R. (2012). Detection of monofluoroacetate in *Palicourea* and *Amorimia* species. *Toxicon,* **60**: 791–796.

Lopes, S., Poser, G. L., Kerber, V. A., Farias, F. M., Konrath E. L., Moreno, P., Sobral M. E., Zuanazzi J. A. S., and Henriques, A.T. (2004). Taxonomic significance of alkaloids and iridoid glucosides in the tribe Psychotrieae (Rubiaceae). *Biochemical Systematics and Ecology,* **32**: 1187-1195.

Metcalfe, C. R., and Chalk, L. (1972) Anatomy of the dicotiledons: leaves, stem and wood in relation to taxonomy whit notes on economic uses; Oxford, v. II,

Moraes-Moreau, R. L., Haraguchi, M., Morita, H., and Palermo-Neto, J. (1995) Chemical and biological demonstration of the presence of monofluoroacetate in the leaves of *Palicourea marcgravii* St. Hil. *Brazilian Journal of Medical and Biological Research,* **28**: 685-692.

Morel, L. J. F., Baratto, D. M., Pereira, P. S., Contini, S. H. T., Mom, H. G., Bertoni, B. W., França, S. C., and Pereira, A. M. S. (2011). Loganin production in *Palicourea rigida* HBK (Rubiaceae) from populations native to Brazilian Cerrado. *Journal of Medicinal Plants Research,* **5**: 2559–2565.

Morita, H., Ichihara, Y., Takeya, K., Watanabe, K., Itokawa, H., and Motidome, M. (1989). A new indol alkaloid from the leaves of *Palicourea marcgravii. Planta Medica,* **55**: 288-289.

Nakano, N., and Martin, A. (1976). Studies of the alkaloids of *Palicourea fendleri. Planta Medica,* **30**: 186–189.

Narine, L. L., and Maxwell, A. R. (2009). Monoterpenoid indole alkaloids from *Palicourea crocea. Phytochemistry Letters,* **2**: 34–36.

Nogueira V.A., Peixoto T.C., França T.N., Caldas S.A., and Peixoto P.V. (2011). Intoxicação por monofluoroacetato em animais. *Pesquisa Veterinária Brasileira,* **31 (10)**: 823-838.

Passos, D. C. S., Ferreira, H. D., Vieira, I. L. F. B., Nunes, W. B., Felício, L. P., Silva, E. M.,Vale, C. R., Duarte, S. R., Silva, E. S., and Carvalho, S. (2010). Modulatory effect of *Palicourea coriacea* (Rubiaceae) against damage induced by doxorubicin in somatic cells of *Drosophila melanogaster. Genetics and Molecular Research,* **9 (2)**: 1153-1162.

Peixoto, T. C., Oliveira, L. I., Caldas, S. A., Catunda Junior, F. E. A., Carvalho, M. G., França, T. N., and Peixoto, P. V. (2011). Efeito protetor da acetamida sobre as intoxicações experimentais em ratos por monofluoroacetato de sódio e por algumas plantas brasileiras que causam morte súbita. *Pesquisa Veterinária Brasileira,* **31 (11)**: 938-952.

Peixoto, T. C., Nogueira, V. A., Caldas, S. A., França, T. N., Anjos, B. L., Aragão A. P., and Peixoto, P. V. (2012). Efeito protetor da acetamida em bovinos indica monofluoroacetato como princípio tóxico de *Palicourea marcgravii* (Rubiaceae). *Pesquisa Veterinária Brasileira,* **32 (4)**: 219-328.

Pessoa, C. R. M., Medeiros, R. M. T., and Riet-Correa, F. (2013) Importância econômica, epidemiologia e controle das intoxicações por plantas no Brasil. *Pesquisa Veterinária Brasileira,* **33 (6)**: 752-758.

Pinto, M. F. S., Fensterseifer, I. C. M., Migliolo, L., Sousa, D. A., Capdville, G., Arboleda-Valencia, J. W., Colgrave, M. L., Craik, D. J., Magalhães, B. S., Simoni C. Dias, S.

C., and Franco, O. L. (2012). Identification and structural characterization of novel cyclotide with activity against an insect pest of sugar cane. *The Journal of Biological Chemistry*, **287(1)**: 134-147.

Ribeiro de Assis, J. C. S., Suffredini, I. B., Moreno, P. R. H., Young, M. C., Varella, A. D., Younes, R. N., and Bernardi, M. M. (2006). Analysis of the toxic potential of *Palicourea corymbifera* (Müll. Arg.) Standl. In laboratory animals. *Research in Veterinary Science*, **80**: 209-217.

Ripperberger, H. (1982). Chimonanthin aus *Palicourea dominguensis*. *Pharmazie*, **37**, H.

Rosa, E. A., Silva1, C. C., Tanaka, C. M. A., Oliveira, C. M. A., Lucília Kato, L., and Ferreira, H. D. (2006). Antraquinona de *Palicourea rigida* (Rubiaceae). *In*: 29ª Reunião Anual da Sociedade Brasileira de Química, Águas de Lindóia, Brasil.

Rosa, E. A., Silva, B. C., Silva, F. M., Tanaka, C. M. A., Peralta, R. M., de Oliveira, C. M. A., Kato, L., Ferreira, H. D., and da Silva, C. C. (2010). Flavonoides e atividade antioxidante em *Palicourea rigida* Kunth, Rubiaceae. *Brazilian Journal of Pharmacognosy*, **20 (4)**: 484-488.

Rova, J. H. E., Andersson, L., Delprete, P. G., and Albert, V. A. (1997). Macrophylogeny of the Rubiaceae from trnL-F sequence data. In: Abstract of the AIBS meetings, Canada. *American Journal of Botany*, **84 (suppl.1)**: 227-227.

Schenkel, E. P., Zannin, M., Mentz, L. A., Bordignon, S. A. L., Irgang, B. (1999). Plantas Tóxicas. In: Simões, C. M., Schenkel, E. P., Gosmann, G., Mello, J. C. P., Mentz, L. A., Petrovick, P. R. Farmacognosia - Da planta ao medicamento. 1ª. Ed. Porta Alegre/Florianópolis, Brasil, p. 755-780. Ed. Universidade do Rio Grande do Sul e Ed. UFSC.

Silva, W. C., Martins, J. R. S., Cesio, M. V., Azevedo, J. L., Heinzen, H., and de Barros, N. M. (2011). Acaricidal activity of *Palicourea marcgravii*, a species from the Amazon forest, on cattle tick *Rhipicephalus* (*Boophilus*) *microplus*. *Veterinary Parasitology*, **179**: 189–194.

Silva, M. S.; Pereira, A. N. S., Morel, L. J. F. França, S. C. and Bertoni, B. W. (2013). Association of loganin contents with the genetic characterization of natural populations of *Palicourea rigida* Kunth determined by AFLP molecular markers. *Biochemical Systematics and Ecology*, **51**: 189–194.

Soares, P. R. O. de Oliveira, P. L., de Oliveira, C. M. A., Kato, L., and Guillo, L. A. (2012). *In vitro* antiproliferative effects of the indole alkaloid vallesiachotamine on human melanoma cells. *Archiv Pharmacal Research*, **35 (3)**: 565-571.

Souza, V. C., and Lorenzi, H. (2005). Botânica Sistemática. Guia Ilustrado para Identificação das Famílias de Angiospermas da Flora Brasileira Baseado em APG III. Instituto Plantarum de Estudos da Flora, Nova Odessa, Brasil, p. 468-483.

Stuart, K. L., and Woo-Ming, R. B. (1974). *Palicourea* alkaloids: The structure of palinine. *Tetrahedron Letters*, **44**: 3853-3856.

Stuart, K. L., and Woo-Ming, R. B. (1975). Vomifoliol in *Croton* and *Palicourea* species. *Phytochemistry,* **14**: 594-595.

Taylor, C. M. (1989). Revision of *Palicourea* (Rubiaceae) in Mexico and Central America. *Systematic Botany Monographs,* **26**: 1-102.

Taylor, C. M. (1993). Revision of *Palicourea* (Rubiaceae: Psychotrieae) in the West Indies. *Moscosoa,* **7**: 201-241.

Taylor, C. M. (1997). Conspectus of the genus *Palicourea* (Rubiaceae: Psychotrieae) with the description of some new species from Ecuador and Colombia. *Annals of the Missouri Botanical Garden,* **84 (2)**: 224-262.

Tokarnia, C. H., Döbereiner, J., and Peixoto, P. V. (2002). Poisonous plants affecting livestock in Brazil. *Toxicon* **40**: 1635-1660.

Valverde, J., Tamayo, and G., Hesse, M. (1999). Beta-carboline monoterpenoid glucosides from *Palicourea adusta*. *Phytochemistry,* **52**: 1485-1489.

Vencato, I., Silva, F. M., Oliveira, C. M. A., Kato, L., Tanaka, C. M. A., Silva, C. C., and Sabino, J. R. (2006). Vallesiachotamine. *Acta Crystallographica Section E,* **E62**: 429-431.

Woo-Ming, R. B., and Stuart, K. L. (1975). Calycanthine from *Palicourea alpina*. *Phytochemistry,* **14**: 2529-2529.

Zappi, D., and Taylor, C. (2010). <http: / /floradobrasil.jbrj.gov.br/2010/FB014133, *Palicourea in* Lista de Espécies da Flora do Brasil. Jardim Botânico do Rio de Janeiro. Access in 14/05/2014.

Natural Products: Research Reviews Vol. 3 (2016) *Pages* **425–443**
Editor: **V.K. Gupta**
Published by: **DAYA PUBLISHING HOUSE, NEW DELHI**

16

Neurodegenerative Disorders and Natural Products: A Contemplation

Yogesh A. Kulkarni[1*], Mayuresh S. Garud[1] and Kalyani H. Barve[1]

ABSTRACT

The incidence and prevalence of neurodegenerative disorders are rising throughout the globe. There is an imperative need for development of a better way of treatment and management of these diseases which will also help in alleviating the social and financial burden associated with such disorders. It is hypothesized that neurodegeneration is progressed via excitotoxicity due to excessive activation of excitatory amino acid receptors, associated apoptosis, oxidative stress induced by free radicals and deposition of abnormal proteins like β-amyloid and tau proteins. Protection against the neurodegeneration has been proved as promising approach in management of neurodegenerative disorders and hence promotes use of neuroprotective agents in treatment of these disorders. Ayurveda, an ancient system of medicine, is now gaining the greater attention despite of the evidences that are being generated by the scientific research. Use of the natural products for neuroprotection is now well accepted approach for management of neurodegenerative disorders. These agents act by different mechanisms like free radical scavenging activity, anti-excitotoxicity, apoptosis inhibition etc. Medicinal plants like Bacopa monniera, Curcuma longa, Emblica officinalis, Ocimum sanctum and Withania somnifera have been studied experimentally for their neuroprotective properties. Natural products involve the use of crude drug as well as extract of the part of the plant and also the isolated phytoconstituents like phenolic. In this text we have made an attempt to compile the literature providing the scientific support for the use of natural products as neuroprotective agents for neurodegenerative disorders.

Keywords: Amyloid, Alzheimer's, Curcumin, Excitotoxicity, Neuroprotection, Parkinson's, Resveratrol.

1 SPP School of Pharmacy and Technology Management, SVKM's NMIMS, V.L. Mehta Road, Vile Parle (W), Mumbai – 400 056, Maharashtra, India.

* *Corresponding author*: E-mail: yogeshkulkarni101@yahoo.com

Introduction

Patients with neurodegenerative disorders are clinically characterized by cognitive, motor, and/or behavioral dysfunction (Woulfe, 2008). These involve degeneration of functionally or anatomically connected constrained group of neurons. Among the different factors involved in neurodegenerative diseases, aging is one of the major risk factor (Hung *et al.*, 2010). The worldwide demographic trends that cause rapidly aging populations across the globe are also causing the rise in incidence and prevalence of these disorders (Trojanowski and Hampel, 2011). According to the World Health Organization (WHO), by 2040, neurodegenerative diseases will become a major cause of death (Ip *et al.*, 2012).

Neurodegenerative diseases which are wide spread include Alzheimer's disease (AD) and Parkinson's disease (PD). In the elderly, AD is the most frequent cause of dementia which accounts for approximately 70 per cent of dementia cases (Silva *et al.*, 2014). Other less common conditions include amyotrophic lateral sclerosis (ALS), Huntington's disease (HD) and vascular dementia (VaD) (Trojanowski and Hampel, 2011).

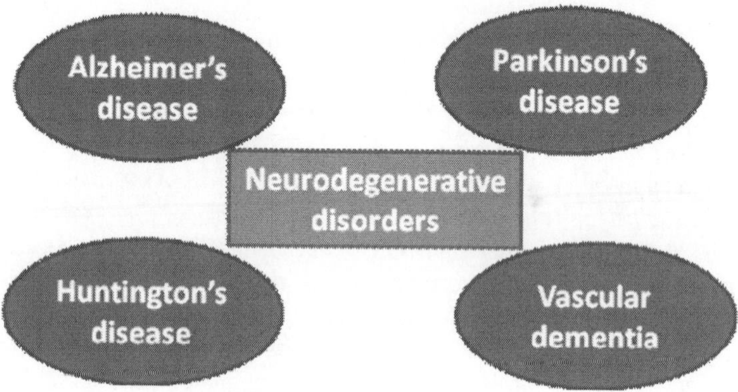

Figure 16.1: Common Neurodegenerative Disorders.

Understanding Neurodegeneration

It is necessary to understand the biological processes involved in neurodegeneration to design an effective therapy for treatment and management of the neurodegenerative disorders. An extensive work has been done on this and researchers have come up with following suggestions.

Abnormal protein aggregation is characteristic biochemical and histomorphological feature of neurodegenerative diseases. In most of the cases, 'inclusion bodies' are formed due to microscopically-visible cellular deposits of these protein aggregates (Woulfe, 2008). These inclusion bodies are specific for particular neurodegenerative conditions. These inclusion bodies include β-amyloid peptides and tau/phosphorylated tau proteins in AD, α-synuclein in PD, superoxide dismutase in amyotrophic lateral sclerosis, and mutant huntingtin in Huntington's diseases

(Hung *et al.*, 2010). It is believed that, the neurodegeneration progresses slowly for many years before it becomes symptomatically evident (Paine and Lowe, 2011).

Following factors are supposed to be responsible for the development and progression of the neurodegeneration.

Excitotoxicity

The term 'excitotoxin' was first used by Olney in 1969 which describe the neurotoxic effects of excitatory amino acids responsible for degeneration of neurons (Hall, 1995). The term excitotoxicitycan be defined as neuronal cell death caused by activation of excitatory amino acid receptors. Glutamate mediates most excitatory synaptic transmission in mammalian brain. Although glutamate is important in normal neuronal functioning, it can lead to cell death via excitotoxic mechanism when present excessive amounts (Lipton and Rosenberg, 1994).Activation of glutamate receptors triggers excessive influx of calcium into neurons through ionic channels which may mediate neuronal injury in various neurodegenerative diseases (Lipton and Rosenberg, 1994).

Apoptosis

Most of the time, apoptosis is associated with excitotoxicity which can be identified through changes in nucleus *viz.* chromatin aggregation, DNA fragmentation and cell shrinkage and can be initiated by various cell signals. Apoptosis finally ends with activation of caspases which in turn activates various inactive proteins (Steller, 1995).

Amyloid Cascade

Other hypothesis proposed by scientist is amyloid cascade. According to this hypothesis the neuronal cells get removed due to abnormal deposition of β-amyloid. It is reported that lethal chemical cascade in neurons is initiated in patients having abnormality of DNA coding Amyloid Precursor Protein (APP). Misrepresented APP is formed due to abnormal DNA which in turn gets accumulated as β amyloid deposits (Vyawahare *et al.*, 2008).

Oxidative Stress

Now it is well accepted that generation of free radicals and related oxidative stress are involved in pathogenesis of number of disorders. Evidences provide the proof that oxidative stress can lead to cell and tissue injury (Pandit, 2011). *In vivo* formation of oxygen radicals takes place as by-product of enzymatic reactions. These free radicals results in a cytotoxic consequence causing oxidative stress.Generation of free radicals is usually observed during neurodegeneration and ageing process. During oxidative stress the pro-oxidant/antioxidant balance shift towards the pro-oxidant that can occur as an output of an increase in oxidative metabolism (Sing *et al.*, 1997).

Sodium Nitrite

The nitrite salt of sodium (Sodium nitrite) is capable of converting hemoglobin to methemoglobin. These conversion causes reduction in oxygen carrying capacity of blood, leading to memory impairment and plays critical role in progression of neurodegenerative disorders (Zhang *et al.*, 2006).

Role of Neuroprotection in Neurodegenerative Disorders

Neuroprotection refers to the therapeutic strategieswhich prevent damage to the nerve cells and hence the death of neurons. It is one of the approach for protection of neuronal injury or degeneration of CNS following acute disorders. The neuroprotection is aimed to limit neuronal dysfunction caused by injury and to maintain the integrity of cellular interactions which in term keep the neural function undisturbed. A wide range of neuroprotective agents of natural origin are being clinically used and are being tested for their effectiveness in prevention of neurodegeneration. These agents can be classified as free radical scavengers, anti-excitotoxic agents, apoptosis inhibitors, neurotrophic factors etc. as per their mechanism of action.

Different aspects of neuroprotection are being examined, concentrating on different elements leading to loss of nerve cells (Halliwell, 1994). It was found that, agents with antioxidant property plays important role in neuroprotection. It is well evident that phytochemicals having antioxidant properties have ability to reduce the symptoms of neurodegeneration (Pandit, 2011). These antioxidants may be of endogenous or exogenous in origin. The enzyme superoxide dismutase (SOD), catalase (CAT) and glutathione peroxidase (GSHPx) constitute the major intracellular antioxidant protection systems by removing superoxide anion and hydrogen peroxide and become most prominent defenses against oxidative stress induced damage.

Neuroprotective Herbal Drugs

Ayurveda, an ancient system of medicine in India, is now gaining greater attention despite of the evidences that are being generated by the scientific research. Ayurveda makes the use of medicinal plants and is known for its disease preventive and health promotive approach, which is now getting increased acceptability. Ayurveda has mentioned the 'Rasyana chikitsa' which includes revitalization and rejuvenation therapy. These drugs have been found to be a rich source of antioxidants and to modulate the nuero-endocrino-immune systems (Brahma and Debnath, 2003; Pushpangadan, 2005).Researchers have also isolated the phytoconstituents from the medicinal plants which are supposed to be active constituent of it. A growing interest in the traditional medicines may be due to the lack of effective and widely applicable pharmacological treatments in the modern therapy for neurodegenerative disorders (Nadkarni, 1976). Some available scientific literature revealing neuroprotective action of such medicinal plants are described below in this article.

Artemisia absinthium

Artemisia absinthium is one of the widely used anthelmintic, which is traditionally used to treat patients with lost or declining cognitive function. Crude alcohol extract of *A. absinthium* was found to have good antioxidant activity which may have its role in the treatment of AD (Gomes *et al.*, 2009; Lachenmeier, 2010).

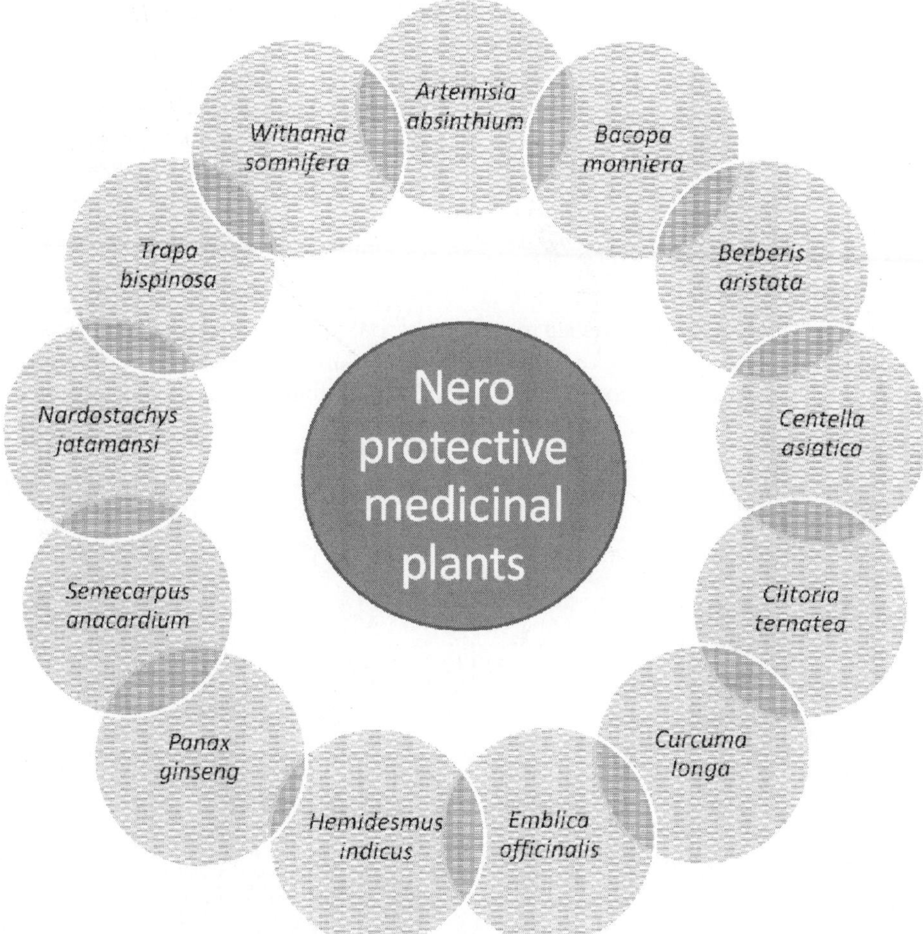

Figure 16.2: Neuroprotective Medicinal Plants.

Bacopa monniera

Ayurveda has mentioned *Bacopa monniera*, commonly known as 'Brahmi', as nerve tonic, diuretic, cardiotonic as well as effective in treatment of epilepsy, insomnia, asthma and rheumatism. It is suggested that extract of this plant may improve acetylcholine (Ach) levels and cerebral blood flow. Extracts of *B. monniera* showed nootropic, antiamnestic, antistress, anticonvulsive, antidepressant, and anxiolytic effects in experimental studies. *B. monniera* was also found to be effective in preventing $AlCl_3$-induced learning and memory deficits (Gomes *et al.*, 2009; Uabundit *et al.*, 2010; Thippeswamy *et al.*, 2013). Gajare and co-workers have proved the neuroprotective effect of *B. monniera* on lipo fuscinogenesis and fluorescence product in brain of D-Galactose induced aging accelerated mice. The study demonstrated that reaction between D-Galactose and amino groups side chains of protein result in to advanced glycation end product (AEGs) (Gajare *et al.*, 2007)and cause increased oxidative stress by chemical oxidation of AEGs (Mullarkey *et al.*, 1990). Another study reported that *B. Monniera* prevents aluminum neurotoxicity in experimental animals (Jyoti *et al.*, 2007). Damage of proteins and lipid macromolecules induced by aluminium interfere with functioning of cell organelles including mitochondria and lysosomes and decrease their efficiency(Stahl, 1998). *B. monniera* was found to protect these macromolecules and hence showed neuroprotection. The treatment with *B. monniera* conserves the endogenous antioxidant enzymes which indicated the neuroprotection from the free radical induced toxicity.

Berberis aristata

Ayurveda uses *Berberis aristata* in various herbal formulations which are used to treat different diseases like malaria, bleeding, fever, skin and eye infections, jaundice, diarrhea and hepatitis (Mahomoodally *et al.*, 2013). Berberine, an active component of the berberis was found to exert anti-depressant-like effect in various behavioral disorders.

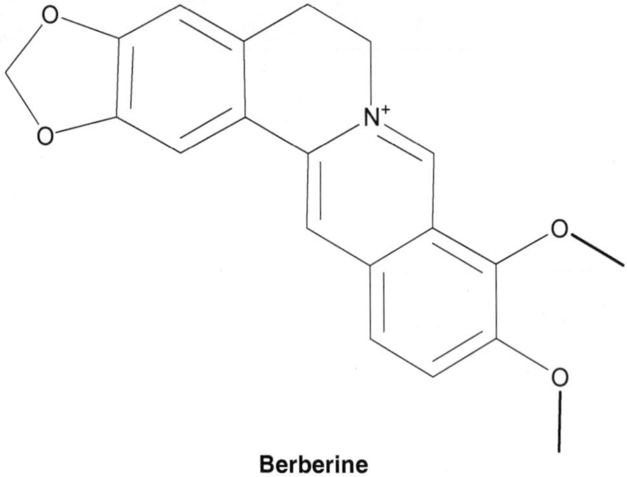

Berberine

The proposed mechanism of action of this activity is modulation of brain biogenic amines (norepinephrine, serotonin and dopamine) which can play role in management of neurodegenerative disorders (Potdar *et al.*, 2012).

Centella asiatica

The whole plant of *Centella asiatica* is used as a nervine tonic and its syrup is used as memory booster in children. Traditional systems of medicine in India have suggested its use to improve memory against dementia and aging. Asiatic acid, a triterpene present in the plant showed protection against H_2O_2-induced cell death and also lowered intracellular free radical concentration. Asiatic acid and its derivatives also showed protection from glutamate-induced excitotoxicity *in vitro* (Adams *et al.*, 2007).The aqueous extract of *Centella asiatica* showed significant improvement in learning and memory MDA is end product of lipid peroxidation and is a measure of a free radical generation which is prime reason for neurodegeneration caused by oxidative stress.

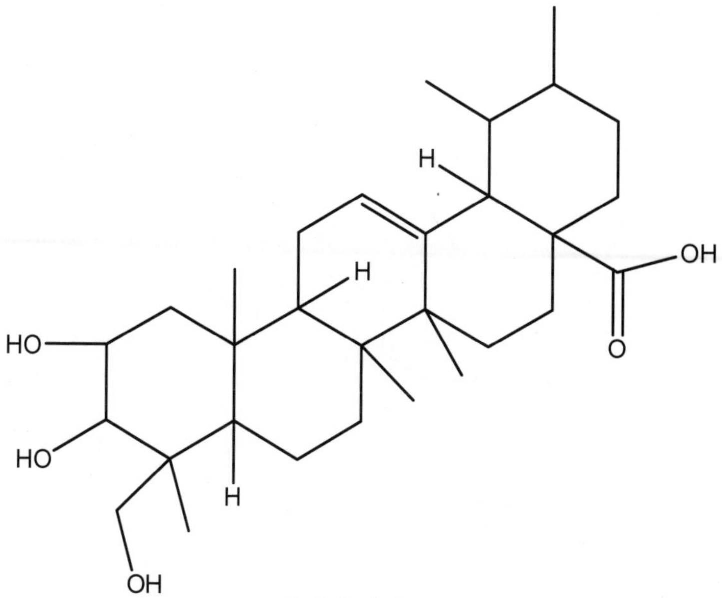

Asiatic Acid

Glutathione is major anti oxidant and free radical scavenging enzyme. The researchers reported significant decrease in the levels of malondialdehyde (MDA) after treatment with aqueous extract of *C. asiatica* with simultaneous significant level of glutathione (Joshi and Parle, 2006). The increased levels of MDA and glutathione support that *C. asiatica* act via antioxidant mechanism by augmenting the endogenous antioxidant enzymes in the brain (Joshi and Parle, 2006).

Clitoria ternatea

Clitoria ternatea is one of the important drugs mentioned in Ayurvedic medicine (Wang *et al.*, 2006). Vyawahare and co-workers reported that alcoholic extract of roots of *C. ternatea* shows protection against scopolamine induced memory disruption in experimental animals. The study reported significant prevention of scopolamine induced memory disruption, and thereby supports the use of it in Ayurveda.

Curcuma longa

Curcuma longa is extensively used herb in Ayurvedic system for variety of disorders. Curcumin, an active constituent of *C. longa* which may improve cognitive function in the elderly. This activity is demonstrated clinically by better cognitive abilities in elderly populations who consume turmeric regularly. Curcumin may play a dual role in protecting neuron against oxidative damage involved in AD (Goel *et al.*, 2008; Villaflores *et al.*, 2012).

Emblica officinalis

Emblica officinalis (amla) is described as a tonic for heart and brain in many of the literatures. Many herbal preparations such as Triphala and Chyawanprash use amla

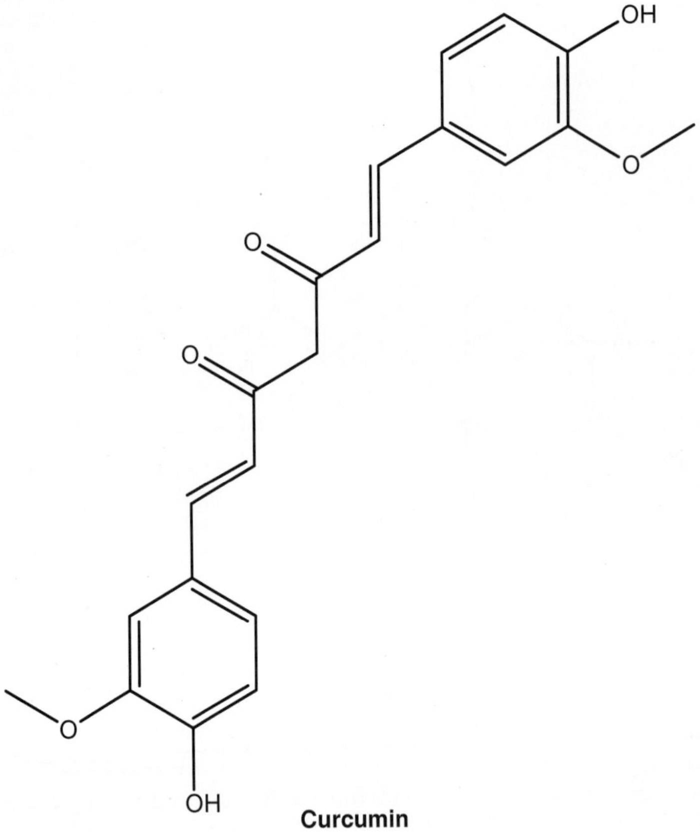

Curcumin

as their important ingredient. Ayurvedic herbal preparation, Amala churna was found to elevate ACh level in brain and ultimately improved memory of both young and aged mice, through the antioxidant properties of *E. officinalis* (Vasudevan and Parle, 2007).

Hemidesmus indicus

Hemidesmus indicus, Indian sarsaparilla or Anantmool, is widely recognized in folk medicine and as ingredient in Ayurvedic and Unani preparations against various diseases (Singh, 1997). The root of the plant has antioxidant property and scientific studies showed that it also have the nootropic (memory enhancer) potential when tested in experimental animals.

Treatment with ethanolic extract of *Ocimum sanctum* was found to ameliorate scopolamine as well as aging induced memory deficit in experimental animals. It is proposed that *O. sanctum* modulates cholinergic transmition and hence indicate possible utility in the management of AD and age associated dementias (Shukla *et al.*, 2000).

Panax ginseng

Panax ginseng is another plant that has been used since ancient time to treat ailments including neurodegeneration associated with aging (Vyawahare *et al.*, 2008). Animals treated with ginseng extract or its composite preparations is claimed to improve learning and memory (Limpeanchob *et al.*, 2008).

Semecarpus anacardium and *Nardostachys jatamansi*

Semecarpus anacardium and *Nardostachys jatamansi* were observed to inhibit AchE (acetylcholine esterase). *S. anacardium* was found to prevent stress induced neurodegeneration (Kirtikar and Basu, 1993b; Shukla *et al.*, 2006).

Trapa bispinosa

Trapa bispinosa a floating herb is one of the medicinal plants that have been used as nerve tonic from many years (Kirtikar and Basu, 1993). It is reported to have neuroprotective activity which may be due to reduction in oxidative stress induced by D- galactose by activating glutathione peroxidase and catalase consequently reducing lipid peroxidation (Ambikar *et al.*, 2009).

Withania somnifera

Withania somnifera, popularly known as Ashwagandha, has been in use for more than 2500 years for diverse clinical condition especially for improvement of memory and cognition enhancement (Vinutha *et al.*, 2007). Sitoindosides and withaferin-A, isolated from the roots, augmented learning acquisition and memory in young as well as old rats.

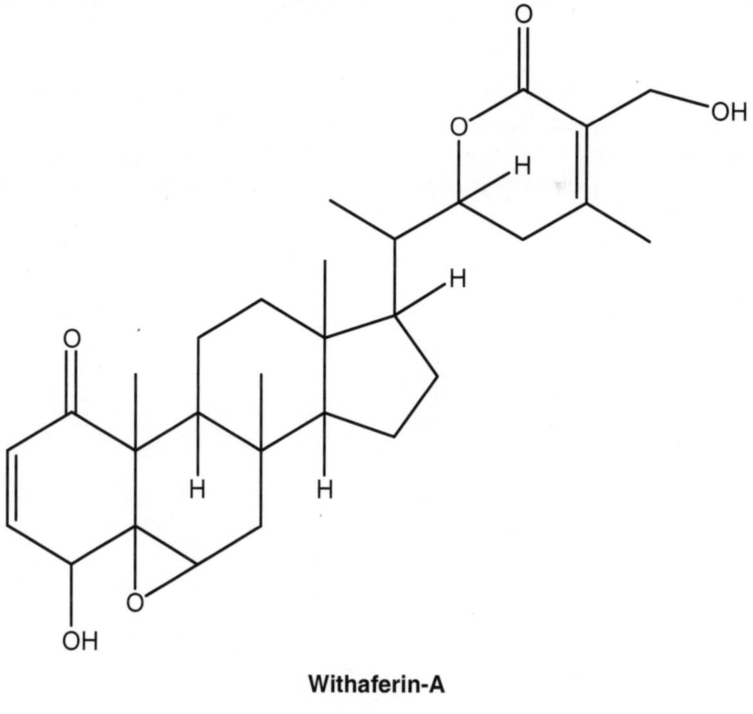

Withaferin-A

The proposed mechanism behind this activity is through modulation of cholinergic neurotransmission. Withanolide A modulates APP (amyloid precursor protein) metabolism by down regulating BACE-1 (beta secretase I). The glycowithanolides have shown anxiolytic and antidepressant activities in experimental animals. It also showed calcium antagonistic ability which makes the withanolides as possible drug candidates for further study to treat AD (Gomes *et al.*, 2009; Liu *et al.*, 2012).

W. somnifera root extract exhibits nootropic effects in mice and causes inhibition of acetylcholinesterase (AchE), suggesting that it indirectly facilitates cholinergic transmission which may have important role in preventing neurodegeneration associated with cholinergic deficiencies (Dhuley, 2001; Trigunayat *et al.*, 2007).

Biochemical investigation reflected significant increase in the lipid peroxidation, superoxide dismutase activity and fall in total suphahydral (T-SH) levels which suggests antioxidant properties of *W. somnifera* in neuroprotection (Veerendrakumar and Gupta, 2002). Another study with methanolic extract of *W. somnifera* reported significant inhibition of AchE activity and thereby suggested its role in AD prominently associated with cholinergic depletion (Vinutha *et al.*, 2007).

Phytochemicals Effective in Neurodegenerative Disorders

Scientific study on the isolated active phytoconstituents of medicinal plants is one of the fields of research which is gaining popularity among the scientists as it is giving promising and improved result than the parent herb. Here we have tried to collect the facts about the phytochemicals which have shown significance in treatment and management of neurodegenerative disorders.

Phenolics

Phenolic compounds have greater attraction due to growing evidence of their beneficial effect on human health. *In vitro, in vivo* studies and clinical trials have made it evident that polyphenols from different sources possess a stronger neuroprotection than antioxidant vitamins. The phenolic compounds like catechins

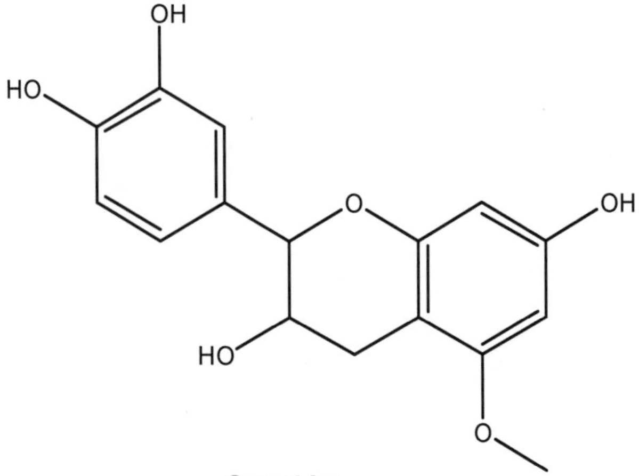

Catechins

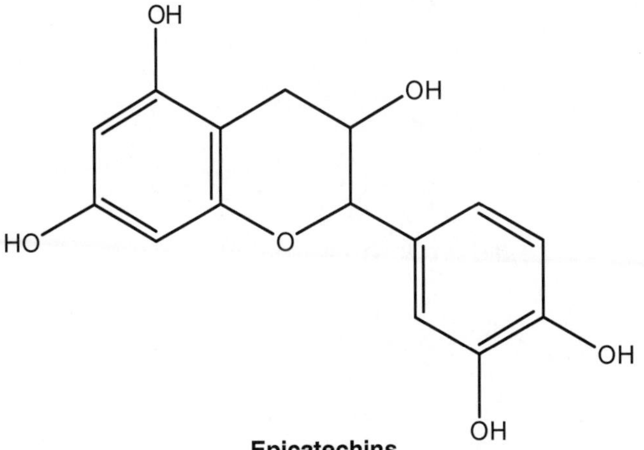

Epicatechins

and epicatechins present in green tea are capable of protecting neurons against a range of oxidative and metabolic insults.

It was found that these phenolic compounds protect dopaminergic neurons from damage induced by 6-hydroxydopamine in a rat model of Parkinson's disease (Guo *et al.*, 2007). They also protect retinal neurons against ischemia reperfusion injury (Zhang *et al.*, 2007) as well as reduces mutant huntingtinmisfolding and neurotoxicity in Huntington's disease models (Ehrnhoefer *et al.*, 2006).

Neuroprotective activities of catechins are proposed to be due to activation of protein kinase C (PKC) and transcription factors that induce the expression of cell-survival genes (Mandel *et al.*, 2005). Scientific studies have revealed that catechins and their metabolites activate multiple signaling pathways that are involved in cell-survival and anti-inflammatory actions which also include alteration of the expression of pro-apoptotic and anti-apoptotic proteins, and upregulating antioxidant defenses (Sutherland *et al.*, 2005).

Resveratrol

Resveratrol, a phytophenol isolated from its rich source red grapes, enters the CNS (central nervous system) rapidly after peripheral administration.

Resveratrol can protect neurons in the brain and spinal cord against ischemic injury (Huang *et al.*, 2001). Resveratrol was found to protect cultured neurons against nitric oxide-mediated oxidative stress induced death. Resveratrol also protected dopaminergic neurons in midbrain slice cultures against metabolic and oxidative insults which show its significance in treatment of Parkinson's disease (Huang *et al.*, 2001). *In vitro* studies showed that resveratrol can also protect cells against the toxicity of mutant huntingtin (Parker *et al.*, 2005). *In vitro* models relevant to Alzheimer's disease showed that resveratrol prevented neuronal cell death induced by amyloid β-peptide and also promoted the removal of amyloid β-peptide from cultured cells (Marambaud *et al.*, 2005).

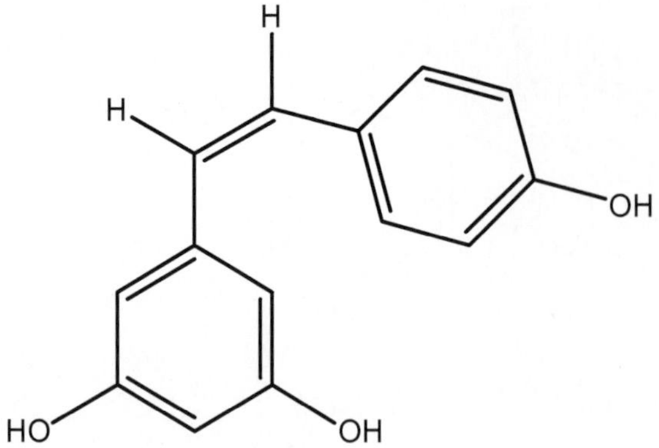

Resveratrol

Flavonoids

Researchers are showing an intense interest in the flavonoids because of their potential to modulate neuronal function and prevent age-related neurodegeneration. Studies in experimental animals as well as humans, proved that flavonoid rich plant or food extracts are capable of improving memory and learning, by protecting vulnerable neurons, enhancing existing neuronal function or by stimulating neuronal regeneration (Kumar and Khanum, 2012). Isolated flavonoids, such as the citrus flavanone tangeretin, were reported to maintain nigro-striatal integrity and functionality following lesioning with 6-hydroxydopamine. This suggest that this flavonoids may act as a potential neuroprotective agent against the basic pathology linked with Parkinson's disease (Youdim *et al.*, 2004).

Hesperetin, naringenin and their *in vivo* metabolites have been shown to traverse the blood brain barrier (BBB) which is an important step required to show the CNS activity (Youdim *et al.*, 2002; Youdim *et al.*, 2004).

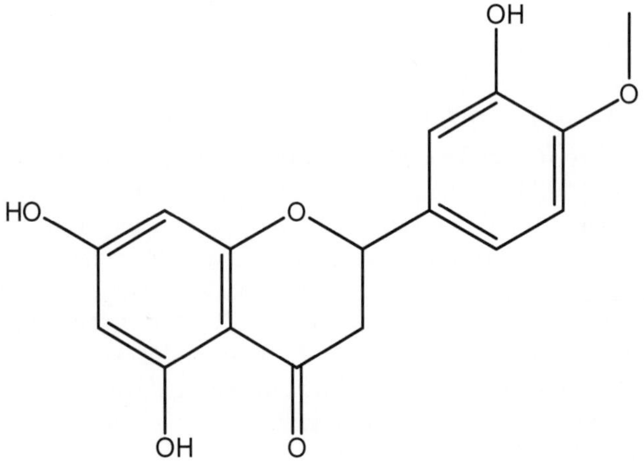

Hesperetin

Naringenin

Alkaloids

Among the natural products, alkaloids are a enormous group of compounds containing nitrogen, carbon, hydrogen and usually oxygen. Alkaloids may affect the CNS, includes nerve cells and spinal cord. Indole alkaloids may interfere or compete with the action of serotonin in the brain (Pearson and Galantamine, 2001). Ergot alkaloids have shown potential effects on blood flow, which is considered to be the mechanism of action. Vinpocetine, an alkaloid obtained from *Vinca minor* is potent vasodilator.

Vinpocetine

Clinical studies reported selective enhancement of cerebral blood flow and metabolism, including enhanced glucose uptake, which may protect against the effects of hypoxia and ischaemia after treatment with vinpocetine (Halliwell, 2007). Galantamine is a tertiary alkaloid having cholinesterase inhibitory activity and which can stimulate nicotinic receptors which further enhance cognition and memory (Halliwell, 2007).

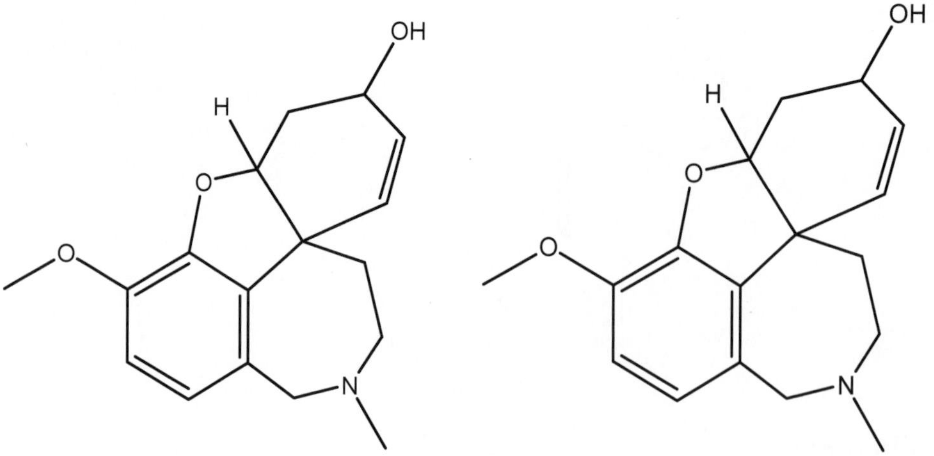

Galantamine

Terpenoids and Saponins

Brahminoside, a triterpenoid and monoterpenes, β-pinene and γ-terpinene isolated from *Centella asiatica* are active chemical substances in revitalizing and strengthening the nervous function (Sakina and Dandiya, 1990; Nalini *et al.*, 1992).

Fatty Acids

Consuming monounsaturated fatty acids and polyunsaturated fatty acids (PUFAs) was shown to slow cognitive decline in animals and in humans. Structural integrity of neurons is crucial for its normal functioning and fatty acid composition of the neuronal membrane plays important role in it (Yehuda *et al.*, 2002). Decline in the fatty acid composition of neuronal membranes is observed during aging, while diet containing essential fatty acids was reported to improve membrane fluidity and PUFA content. PUFAs in the form of phospholipids in neuronal membranes also plays role in signaling cascades which promote neuroprotection (Yehuda *et al.*, 2002).

Sulforaphane

Sulforaphane is an isothiocyanate found in high amounts in cruciferous vegetables. There are several reports on neuroprotective effects of sulforaphane in acute and chronic neurodegenerative conditions when studied in animal models., Sulforaphane administration reduced the amount of brain damage and brain edema in a rodent model of stroke (Zhao *et al.*, 2006). In vitro studies on cultured neurons showed that sulforaphane protect against oxidative stress and dopaminergic neurons against mitochondrial toxins (Han *et al.*, 2007).

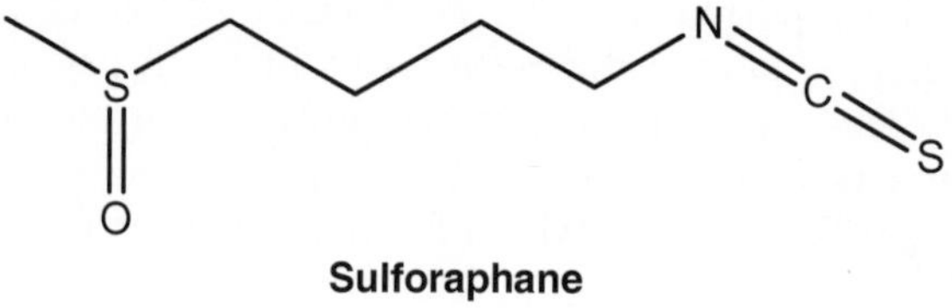

Sulforaphane

Curcumin

Curcumin, which is a widely used and studied phytoconstituent also has several beneficial effects on the nervous system. Curcumin treatment of experimental animals was found to protect neurons against ischemic cell death and ameliorated behavioral deficits (Xu *et al.*, 2007). *In vitro* and *in vivo* studies have revealed that curcumin is having potential to prevent or treat major disabling age-related neurodegenerative diseases including Alzheimer's and Parkinson's.

Allium and *Allicin*

Garlic and onions are rich in organosulfur compounds, such as the allium and allicin. These compounds reported to have neuroprotective activity. Along with antioxidant activities, allium and allicin is supposed to activate stress-response pathways, which results in the upregulation of neuroprotective proteins which includes mitochondrial uncoupling proteins (Oi *et al.*, 1999).

Conclusion

Natural products are a promising alternative to conventional therapy for neurodegenerative disorders due to their much lesser side effects and affordable cost. Various plant extracts, isolated phytochemicals have been screened for their effects in animal models of neurodegenerative disorders. There is unmet need of mechanistic and clinical studies of these natural products. Formulation into specific drug delivery system is also required to further augment the effectiveness of plant extracts or phytochemicals for treatment/management of these disorders.

References

Adams, M. Gmünder, F., and Hamburger, M. (2007). Plants traditionally used in age related brain disorders-A survey of ethnobotanical literature. *J Ethnopharmacol.*, **113**: 363–381.

Ambikar, D.B., Harle, U.N., Khandare, R.A., Bore V.V., and Vyawahare, N.S. (2009). Neuroprotective effect of hydroalcoholic extract of dried fruits of *Trapa bispinosa* Roxb on lipofuscinogenesis and fluorescence product in brain of D-galactose induced ageing accelerated mice. *Ind J Expt Biol.*, **48**: 378-382.

Brahma, S.K., and Debnath P.K. (2003).Therapeutic importance of Rasayana drugs with special reference to their multi-dimensional actions. *Aryavaidyan*, **16**: 160-163.

Dhuley, J.N. (2001). Nootropic-like effect of ashwagandha (*Withania somnifera* L.) in mice. *Phytotherapy Res.*, **15**: 524–528.

Ehrnhoefer, D.E., Duennwald, M., Markovic, P., Wacker J.L., and Engemann S. (2006). Green tea (")-epigallocatechin-gallate modulates early events in hunting tinmisfolding and reduces toxicity in Huntington's disease models. *Hum Mol Genet.*, **15**: 2743–51.

Gajare, K.A., Deshmukh, A.A., and Pillai, M.M. (2007). Neuroprotective effect of *Bacopa monniera* Linn. extract on lipofuscinogenesis and fluorescence product in brain of d-galactose induced ageing accelerated mice. *J Cell Tissue Res.*, **7(2)**: 1167-1172.

Goel, A., Kunnumakkara, A.B., and Aggarwal. B.B. (2008). Curcumin as "Curecumin": From kitchen to clinic. *BiochemPharmacol.*, **75**: 787-809.

Gomes, N.G.M., Campos, M.G., Órfão, J.M.C., and Ribeiro, C.A.F. (2009). Plants with neurobiological activity as potential targets for drug discovery. *Prog Neuropsychopharmacol Biol Psychiatry*, **33**: 1372-1389.

Guo, S., Yan, J., Yang, T., Yang, X., Bezard, E., and Zhao, B. (2007). Protective effects of green tea polyphenols in the 6- OHDA rat model of Parkinson's disease through inhibition of ROS-NO pathway. *Biol Psychiatry*, **62**: 1353-62.

Hall, E.D. (1995) Inhibition of lipid peroxidation in central nervous system trauma and ischemia. *J.Neurol Sci.*, **134**: 79-83.

Halliwell, B. (1994). Free radicals, antioxidants and human disease: curiosity or consequence? *The Lancet*, **344**: 721-724.

Halliwell, B. (2007). Oxidative stress and cancer: Have we moved forward? *Biochem J.*, **401**: 1–11.

Han, J.M., Lee, Y.J., Lee, S.Y., and Kim, E.M. (2007). Protective effect of sulforaphane against dopaminergic cell death. *J Pharmacol Exp Ther.*, **321**: 249-56.

Huang, S.S., Tsai, M.C., Chih, C.L., Hung L.M., and Tsai SK. (2001). Resveratrol reduction of infarct size in Long- Evans rats subjected to focal cerebral ischemia. *Life Sci.*, **69**: 1057-65.

Hung, C.W., Chen, Y.C., Hsieh, W.L., Chiou, S.H., and Kao, C.L. (2010). Ageing and neurodegenerative diseases. *Ageing Research Reviews*, **9S**: S36–S46.

Ip, P.S.P., Tsim, K.W.K., Chan, K., and Bauer, R. (2012). Application of complementary and alternative medicine on neurodegenerative disorders: Current status and future prospects. *Evidence-Based Complementary and Alternative Medicine*, **2012**: 1-2.

Joshi, H., and Parle, M.(2006). Cholinergic basis of memory improving effect of *Ocimum tenuiflorum*linn. *Ind J. Pharma Sci.*, **68(3)**: 364-365.

Jyoti, A., Sethi, P., and Sharma, D. (2007). *Bacopa monniera* prevents from aluminium neurotoxicity in the cerebral cortex of rat brain *J.Ethnopharmacol.*, **111(1)**: 56-62.

Kirtikar K.R., and Basu B.D. (1993). Indian medicinal plants II edition, International Book Distributors, Deharadun.

Kirtikar, K.R., and Basu, B.D. (1993). Indian medicinal plants, Vol. III., International Book Distributors, Deharadun.

Kumar, G.P., and Khanum, F. (2012). Neuroprotective potential of phytochemicals. *Pharmacogn Rev.*, **6 (12)**: 81-90.

Lachenmeier, D.W. (2010). Wormwood (*Artemisia absinthium* L.)–A curious plant with both neurotoxic and neuroprotective properties?, *J Ethnopharmacol.*, **131**: 224–227.

Limpeanchob, N., Jaipan, S., Rattanakaruna, S., Phrompittayarat, W., and Ingkaninan, K. (2008). Neuroprotective effect of *Bacopa monnieri* on beta-amyloid-induced cell death in primary cortical culture. *J Ethnopharmacol.*, **120**: 112-117.

Lipton, S.A., and Rosenberg, P.A. (1994). Excitatory amino acids a final common pathway for neurologic disorders *N. Engl. J. Med.*, **330**: 626-632.

Liu, L.F. Durairajan, S.S.K. Lu, J.H. Koo, I., and Li. M. (2012). *In vitro* screening on amyloid precursor protein modulation of plants used in Ayurvedic and Traditional Chinese medicine for memory improvement, *J Ethnopharmacolo.*, **141**: 754–760.

Mahomoodally, M.F., Bhugun, V., and Chutterdharry, G. (2013). Complementary and alternative medicines use against neurodegenerative diseases. *Advances in Pharmacology and Pharmacy*, **1(3)**: 103-123.

Mandel, S.A, Avramovich-Tirosh, Y., Reznichenko, L., Zheng, H., Weinreb, O., and Amit, T. (2005). Multifunctional activities of green tea catechins in neuroprotection. Modulation of cell survival genes, irondependent oxidative stress and PKC signaling pathway. *Neurosignals*, **14**: 46–60.

Marambaud, P., Zhao, H., and Davies, P. (2005). Resveratrol promotes clearance of Alzheimer's disease amyloid- beta peptides. *J Biol Chem.*, **280**: 37377-82.

Mullarkey, C.J., Edelstein, D., and Brownlee, L. (1990). Free radical generation by early glycation products: A mechanism for accelerated atherogenesis in diabetes. *Biochem Biophys Res Commun.*, **173**: 932–939.

Nadkarni, K.M. (1976). Indian Materia Medica Vol. 1 and II, Popular Prakashan Pvt. Ltd, Mumbai.

Nalini, K., Aroor, A., Karanth, K., and Rao, A. (1992). *Centella asiatica* fresh leaf aqueous extract on learning and memory and biogenic amine turnover in albino rats. *Fitoterapia*, **63**: 232-7.

Oi, Y.M., Imafuku, C., Shishido, Y., Kominato, S., and Nishimura, K. (1999). Allyl-containing sulfides in garlic increase uncoupling protein content in brown adipose tissue, and noradrenaline and adrenaline secretion in rats. *J Nutr.*, **129**: 336-42.

Paine, S.M.L., and Lowe, J.S. (2011). Approach to the post-mortem investigation of neurodegenerative diseases: from diagnosis to research. *Diagnostic Histopathology*, **17**: 211-216.

Pandit, M.K. (2011). Neuroprotective properties of some Indian medicinal plants. *International Journal of Pharmaceutical and Biological Archives*, **2(5)**: 1374-1379.

Parker, J.A., Arango, M., Abderrahmane, S., Lambert, E., Tourette, C., and Catoire, H. (2005). Resveratrol rescues mutant polyglutamine cytotoxicity in nematode and mammalian neurons. *Nat Genet.*, **37**: 349-50.

Pearson, V.E. (2001). Galantamine: A new Alzheimer drug with a past life. *Ann Pharmacother.*, **35**: 1406–13.

Potdar, D. Hirwani, R.R., and Dhulap, S. (2012). Phyto-chemical and pharmacological applications of *Berberis aristata*. *Fitoterapia*, **83**: 817–830.

Pushpangadan, P. (2005). Antioxidant approach to disease management and the role of 'Rasayana' herbs of Ayurveda. *J Ethnopharmacol.*, **99**: 165-178.

Sakina, M.R., and Dandiya, P.C. (1990). A psycho-neuropharmacological profile of *Centella asiatica* extract. *Fitoterapia*, **61**: 291-6.

Shukla, P.K., Khanna, V.K., Ali,M.M., Maurya, R., Khan,M.Y., and Rikhab C.S.(2006). Neuroprotective effect of *Acorus calamus* against middle cerebral artery occlusion–induced ischaemia in rat. *Human ExptToxicol.*, **25**: 187- 194.

Shukla, S.D., Jain, S., Sharma, K., and Bhatnagar,M. (2000).Stress induced neuron degeneration and protective effects of *Semecarpus anacardium* Linn. and *Withania somnifera* Dunn. in hippocampus of albino rats: an ultrastructural study. *Ind J. Expt. Bio.*, **38**: 1007–1013.

Silva, A.R., Santos, A.C., Farfel, J.M., Grinberg, L.T., Ferretti, R.E., Campos, A.H., Cunha, I.W., Begnami, M.D., Rocha, R.M., Carraro, D.M. de Bragança Pereira, C.A., Jacob-Filho, W., and Brentani, H. (2014). Repair of oxidative DNA damage, cell-cycle regulation and neuronal death may influence the clinical manifestation of Alzheimer's disease. *PLoS One*, **17**: **9(6)**: e99897

Sing. S., Melkani, G.C., Gaur, S.P.S., Agrawal, V. and Agrawal, C.G. (1997). Oxidative stress and metabolid control in non-insulin dependent diabetes mellitus. *Ind. J. Biochem. Biophys.*, **34**: 512-517

Stahl, S.M. (1998). Essential Psychopharmacology, Cambridge University Press.

Steller, H.(1995). Mechanisms and genes of cellular suicide. *Science, 267*: 1445-1449.

Sutherland, B.A., Rahman, R.M., and Appleton, I. (2005). Mechanisms of action of green tea catechins, with a focus on ischemia-induced neurodegeneration. *J Nutr Biochem.*, **17**: 291–306.

Thippeswamy, A.H. Rafiq, M. Viswantha, G.L.S. Kavya, K.J. Anturlikar, S.D., and Patki. P.S. (2013). Evaluation of *Bacopa monniera* for its synergistic activity with rivastigmine in reversing aluminum-induced memory loss and learning deficit in rats. *J Acupunct Meridian Stud.*, **6(4)**: 208-13.

Trigunayat, A., Raghavendra, M., Singh, R.K., Bhattacharya, A.K., and Acharya, S.B. (2007). Neuroprotective effect of *Withania somnifera* (WS) in cerebral ischemiareperfusion and long-term hypoperfusion induced alterations in rats. *J. Nat. Rem.*, **7(2)**: 234-246.

Trojanowski, J.Q., and Hampel, H. (2011). Neurodegenerative disease biomarkers: Guideposts for disease prevention through early diagnosis and intervention, *Progress in Neurobiology*, **95**: 491-495.

Uabundit, N. Wattanathorn, J. Mucimapura, S., and Ingkaninan. K. (2010). Cognitive enhancement and neuroprotective effects of *Bacopa monnieri* in Alzheimer's disease model. *J Ethnopharmacol.*, **127**: 26–31.

Vasudevan, M., and Parle. M. (2007). Memory enhancing activity of amala churna (*Emblica officinalis* Gaertn.): An ayurvedic preparation. *Physiology and Behavior*, **91**: 46–54.

Veerendrakumar, M.H., and Gupta, Y.K. (2002). Effect of different extracts of *Centella asiatica* on cognition and markers of oxidative stress in rats. *J. Ethnopharmacol.*, **79**: 253–260.

Villaflores, O.B. Chen, Y.J. Chen, C.P. Yeh, J.M., and Wu. T.Y. (2012). Curcuminoids and resveratrol as anti-Alzheimer agents. *Taiwan J Obstet Gynecol.*, **51**: 515-525.

Vinutha,B., Prashanth,D., Salma,K., Sreeja,S.L., Pratiti, D., Padmaja, R., Radhika, S., Amita, A., Venkateshwarlu, K., and Deepak, M. (2007). Screening of selected Indian medicinal plants for acetylcholinesterase inhibitory activity. *J Ethnopharmacol.*, **109**: 359– 363

Vyawahare, N.S. Ambikar, D.B., Patil, G.T. Kamble, P.N., and Chitte, N.S. (2008). Phytomedicines for neuroprotection. *Electronic J. Pharma. Therapy*, **1**: 15-23.

Wang, L.C.H., Wang, B., Sum-Yan, N., Tze-fun L., and Lawrence C.H. (2006). Effects of ginseng saponins on β-amyloid-induced amnesia in rats. *J.Ethnopharmacol.*, **103**: 103-108.

Woulfe, J. (2008). Nuclear bodies in neurodegenerative disease. *Biochimicaet Biophysica Acta*, **1783**: 2195–2206.

Xu, Y., Ku, B., Cui, L., Li, X., Barish, P.A., Foster, T.C., and Ogle, W.O. (2007). Curcumin reverses impaired hippocampal neurogenesis and increases serotonin receptor 1A mRNA and brain-derived neurotrophic factor expression in chronically stressed rats. *Brain Res.*, **1162**: 9-18.

Yehuda, S., Rabinovitz, S., and Carasso, R.L. (2002). The role of polyunsaturated fatty acids in restoring the aging neuronal membrane. *Neurobiol Aging*, **23**: 843-53.

Youdim, K.A., Qaiser, M.Z., and Begley, D.J. (2004). Flavonoid permeability across an in situ model of the blood-brain barrier. *Free RadicBiol Med.*, **36**: 592–604.

Youdim, K.A., Spencer, J.P., Schroeter, H., and Rice-Evans, C. (2002). Dietary flavonoids as potential neuroprotectants. *Biol Chem.*, **383**: 503–19.

Zhang, B., Safa, R., Rusciano, D., and Osborne, N.N. (2007). Epigallocatechingallate, an active ingredient from green tea, attenuates damaging influences to the retina caused by ischemia/reperfusion. *Brain Res.*, **1159**: 40–53.

Zhang, D., Gengtao, L., Jiangong, S., and Zhang, J. (2006). *Coeloglossum viride* var. bracteatum extract attenuates d-galactose and $NaNO_2$ induced memory impairment in mice. *J Ethnopharmacol.*, **104(1-2)**: 250-256.

Zhao, J., Kobori, N., Aronowski, J., and Dash, P.K. (2006). Sulforaphane reduces infarct volume following focal cerebral ischemia in rodents. *Neurosci Lett.*, **393**: 108-12.

Natural Products: Research Reviews Vol. 3 (2016) *Pages* **445–458**
Editor: **V.K. Gupta**
Published by: **DAYA PUBLISHING HOUSE, NEW DELHI**

17

Recent Advances in Antioxidative Potentiality of Vitamin-C as Yield Inducer

Mohammad Mazid[1]

ABSTRACT

L-ascorbic acid (AA, vitamin C) fulfils essential metabolic functions in the life of animals and plants. It is a small, water soluble, reductone sugar acid with antioxidant properties and acts as a primary substrate in the cyclic pathway for enzymatic detoxification of a number of reactive oxygen species (ROS). In a large number of studies associated with stress mitigation/ tolerance in plants by adding the different natural and synthetic compounds such as PPGs (phenypropanoid glycosides), AA is used as a reference compound. It has been shown to have important functions in photosynthesis such as in the protection of the photosynthetic apparatus against the oxygen radicals and H_2O_2 that are formed during photosynthetic activity and against photo-inactivation, since it is a factor of carotenoid-de-epoxidation. Indeed, AA plays an important role in resistance to abiotic stress such as heavy metal, saline and ultra-violet and biotic stresss. Rapidly increasing evidences indicate that AA is centrally involved in several physiological processes including photosynthesis, cell wall production and its extension by involving in a complex enigmatic array of phytohormone-mediatedsignalling networks. Due to its function as a co-factor for the biosynthesis of various plant hormones, AA appears to influence not only the endogenous level but also signaling of these plant hormones, and thus affect responses against the environmental stresses.

Keywords: Abiotic stress, Ascorbic acid, Antioxidant, Environmental stress, Photosynthesis, Plant hormone, Vitamin-C.

1 Department of Botany, Faculty of life sciences, Aligarh Muslim University, Aligarh – 202 002, India.

* *Author*: E-mail: mazidmohd@gmail.com

Introduction

L-AA serves as a co-factor of many enzymes, for example, α-ketoglutarate-dependent dioxygenases, important for formation of covalent adducts with electrophilic secondary metabolites in plants. L-ascorbic acid fulfils essential metabolic functions in the life of animals and plants. It is a small, water soluble, reductone sugar acid with antioxidant properties and acts as a primary substrate in the cyclic pathway for enzymatic detoxification of a number of ROS such as H_2O_2 and many other, harmful to normal functioning of plant metabolism. Shan *et al.* (2011) studies investigates that JA induced increase in the transcript levels and activities of APX, GR, MDHAR, DHAR, the contents of AA, GSH, ratio of AA/DHA and GSH/GSSG and reduced the GSSG/GSH. Studies of ET signaling in Arabidopsis indicate that receptor gene families in plants may function similar to many of their animal counterparts, increasing their flexibility at responding to stressful environments. Also, studies in the past few years have contributed to the accumulating evidences that ethylene (ET) signaling has a substantial role in disease resistance. The extent of the alternative pathway as a percentage of the total respiration in Arabidopsis mesophyll protoplasts is much higher in *Vtc1* than in wild type and become pronounced at high light and/or when the alternative pathway is inhibited (Hamada, 1998; Rao *et al.*, 1996).

As well as, acknowledging the importance of the cytochrome-C oxidase pathway and hypothesize that AA and the alternative pathway may complementary to each other to protect photosynthesis against photo- inhibition. Treatment with plant hormone (BS, brassinolide) as foliar spray mitigating salt stress by inducing enzyme activities responsible for antioxdation and detoxification as well as by elevating content of AA (Dionisio-Sese-Tobita,1998). Abiotic stress factors such as metal action, salinity and extremes of light and temperature have long been known as major limitations to crop yield. Oxidative stress, which frequently accompanies high temperature, salinity or heavy metal stress may cause denaturation of functional and structural proteins.

Interactive Approaches and Ascorbic Acid

Molecular and genetic studies present the notion that the cross talk between AA and various plant hormones exists. It includes alteration in the expression of hormones biosynthetic genes and/or signaling intermediates. The exogenous treatment of AA serves as an important factor in biosynthetic regulation of tissue levels and therefore the signaling of various plants hormones including gibberellic acid (GA_3), ET, ABA (Abscissic acid), SA (Salicylic acid), cytokinin (CK) and JA (Jasmonic acid). GA_3 also play a vital role in the detoxification of heavy metal and in tolerance to salt stress by improving plant growth, chlorophyll synthesis and activities of antioxidant enzymes and by preventing lipid peroxidaton. ABA represents a good example of a combination of genetic, molecular and biochemical approaches can lead to the elucidation of a complex biosynthetic pathway. Moreover, phenols act as a substrate for many antioxidants enzymes, so, therefore, it mitigates the saline induced injuries. Similarly, Rivero *et al.* (2001) suggested that an accumulation of phenolic compounds (*e.g.*, SA) in response to abiotic stress would be attributed to activation of phenyl

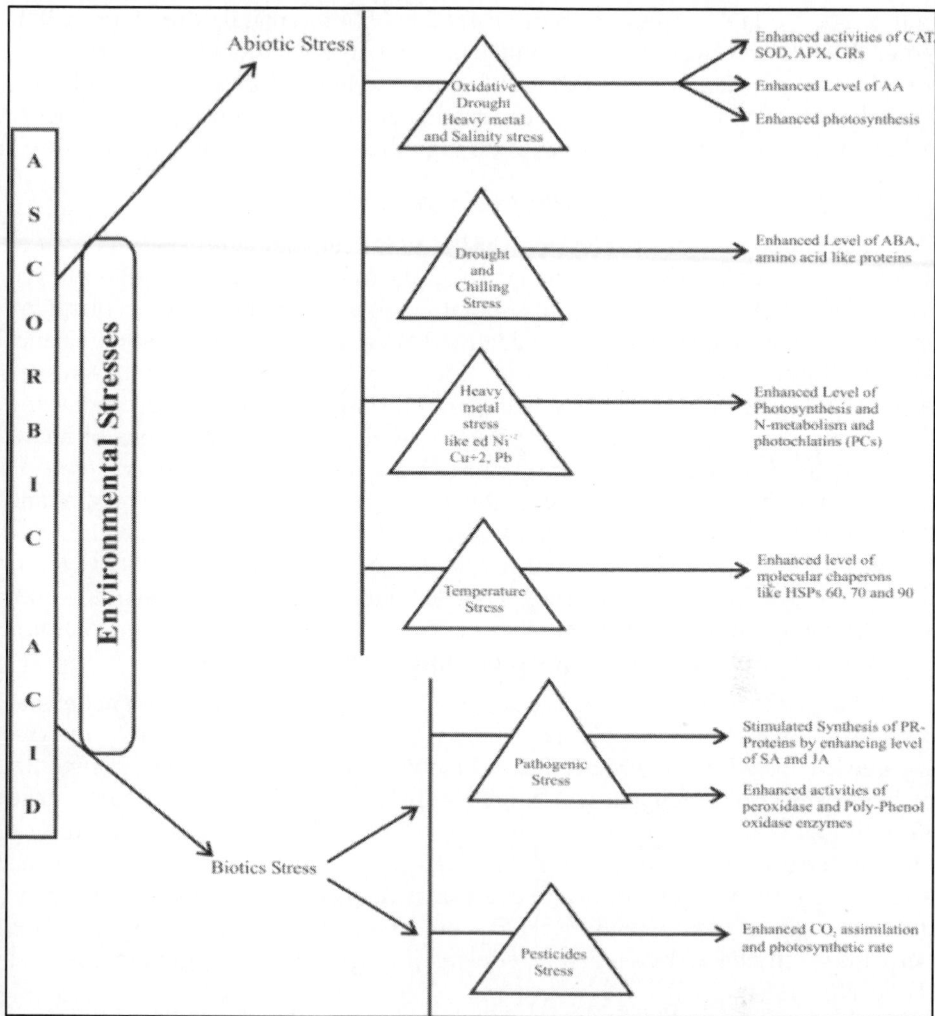

Figure 17.1: Effect of Ascorbic Acid on Plant in Terms of Protection provide through Gene Expression and Inducing Responses.

alanine ammonia lyase (PAL). The SA, and other ubiquitous plant phenolic compounds, is recognized as an endogenous regulator in many plant physiological processes (Yalpini and Raskin, 1993). More direct evidence for the role of AA in regulating the expression of SAGs is provided by a study that revealed an up-regulation of select SAG transcripts in the AA-deficient Arabidopsis mutant *vtc1* when grown under a long-day photoperiod, suggesting that AA-deficiency induces a senescent phenotype. However, GA can also promote cell death in this system, a protectant involves ROS, and notably H_2O_2 which is produced in glyoxysomes (a cell organelle, microbody) by the activity of a flavin-containing acyl CoA oxidase. ABA can present cell death by promoting high activities of enzymes that destroy ROS induced by oxidative stress produced by action of heavy metal. Moreover, APX is synergistically induced by ABA and oxidative stress while other peroxidases areoxidatively induced by wounding and photo-oxidation.This increase may be due to total phenols play a significant role in mechanism of regulation of plant metabolic processes and consequently overall plant growth as well as lignin biosynthesis.

Metal Stress

The antioxidant defense system in the plant cells includes both enzymatic antioxidants such as SOD, CAT (catalase), APX (ascorbate-peroxidase) and non-enzymatic antioxidants like AA, GSH (glutathione) and tocopherol. When plants are subjected to environmental stresses, oxidative damage may results because the balance between the production of ROS and their detoxification by the antioxidative system is altered (Gomez *et al.*, 1999). Since, water pH is an important factor directly affecting the toxicity of metals in algae. The presence of elevated concentrations of heavy metals in the growth medium of the germinating seeds suppresses mobilization and translocation of reserve materials from the reserve tissue to the growing regions and their subsequent utilization there. In conditions of metal toxicity, higher activities of non-enzymatic antioxidative substances like AA are important for plants to tolerate the metal stress. These are originally thought to function as osmotic buffers. However, apart from the osmotic adjustment, they also seem to play a key role in maintaining the natural state of macromolecules, probably by scavenging ROS.

Strategies to minimize oxidative damage are a universal feature of plant defence responses. Several lines of evidence suggest that plants posses various strategies capable to cope with dangerous effect of heavy metals or of altering environmental mobility of metal contaminants with subsequent effects on the potential for root uptake. Aquatic plants are primary producers of most aquatic food chains and account for much of the production base of fresh water and manure ecosystems. Heavy metals toxicity can elicit a variety of adoptive responses in plants. The oxidative damage to different cellular components by H_2O_2 could be minimized either by CAT and peroxidase activities or by a reaction sequence known as ascorbate-glutathione cycle (Foyer *et al.*, 1994).

Response of Saline Stress

Today, soil salinity has become a serious environmental problem which affects the growth and productivity of many crops. The impairment of productivity observed for many plant species subjected to excess salinity is often associated with a reduction in photosynthetic capacity. Photosynthesis may decrease due to stomatal closure or by the direct effect of salt on the photosynthetic apparatus. MDA content, a product of lipid per-oxidation, has been considered an indicator of oxidative damage. The capacity to scavenge ROS and to reduce their damaging effects on macromolecules such as proteins and DNA appears to represent and important stress tolerance trait. For instance, an Arabidopsis mutant vst 1 which exhibit increased tolerance to salt stress is found to have an increased capacity to scavenge ROS. Elimination of ROS is mainly achieved by antioxidant compounds such as AA, GSH, thioredoxin and carotenoids and by ROS scavenging enzymes. The expression of genes encoding many of these enzymes is also regulated by other stresses as well as by ABA. Upon ABA treatment, there is an increase in H_2O_2 levels in maize cultured cells and in Arabidopsis guard cells. Transgenic plants with reduced CAT activity showed increased sensitivity to salt stress. Studies of Hassanein *et al.* (2009) and Abd-El Hamid (2009) suggested that AA increase the content of IAA, which stimulates cell division and/cell enlargement and this, in turn, improve plant growth. One possibility is that additional AA would inhibit stress induced increase in the leakage of essential electrolytes following peroxidative damage to plasma membranes. However, additional AA did not inhibit increases in leakage of electrolytes from roots of salt stressed tomato seedlings.

The remarkable protective effect of exogenous AA appeared to be specially related to its antioxidant activity, rather than it's possible utility as an organic substrate for energy respiratory metabolism (Shalata and Neumann, 2001). Meanwhile, addition of AA and 400mg/l at normal conditions or under salinity levels caused significant increase in cellulose per cent and cellulose/lignin per cent and decrease in lignin per cent as compared with the control plants or the corresponding salinity levels. This evidence points that AA being an important part of the plant defence machinery maintaining the integrity and normal function of the photosynthetic apparatus and act directly to neutralize $O_2^{\cdot-2}$ or $^1O_2$ in plant cells. The adverse effects of salt stress treatments on cellulose and lignin contents in shoots, roots are partially or completely alleviated by soaking grains in AA that could perhaps alleviate the inhibitory effects of salinity on glucose incorporation to cell wall polysaccharides. The parallel increase in AA found resembles the response of maize to water deficit ROS produced as a result of abiotic environmental stresses needs to be scavenged for maintenance of normal growth. AA can scavenge $O_2^{\cdot-}$ and H_2O_2 non-enzymatically, and also takes part in APX mediated scavenging of H_2O_2 (Asada, 1992) as well as also involved in the regulation of another important non-enzymatic antioxidant α-tocopherol. Higher H_2O_2 accumulation and lipid per-oxidation in sensitive cultivars of pea and rice have been reported earlier (Booth and Beardall, 1998). Constitutive level provide protection from oxidative stress arising from normal oxidative metabolism, the salinity induced increase in activities of anti-oxidants in response to increase in oxidative stress actually decide the level of tolerance of a plant.

UV Raditions

UV-B is a key environmental signal regulates diverse responses in plants and promotes UV protection, survival in sunlight and influences metabolism, development and defense responses. Also, the direct effects of UV radiation are mostly damaging, because UV photons have enough energy to create lesions in important UV-absorbing biomolecules such as nucleic acids and proteins. In addition, many of the effects of UV-B involve the differential regulation of gene expression. Even though a number of studies have shown that UV-B radiation inhibits plant growth and regulates cell cycle progress but however, little is known about the molecular and cellular mechanisms involved. In the leaves of tropical tress, the ambient UV-B radiation might contribute to the reversible decline in potential PS-ÉÉ efficiency observed upon exposure to full and direct sunlight. In the combined conditions of high light and low temperature where metabolism is slowed relative to photochemistry, the Mehler reaction may occur at increased rates. Enhanced operation of the Mehler reaction may thus, diminish the extent of photo-inhibition, the slowly reversible reduction in photosynthetic efficiency and capacity which occurs when light energy is in excess.

In the APX reaction, H_2O_2 is reduced to water and ascorbate is oxidized to MDHA, the univalent product of ascorbate oxidation. In the chloroplast, MDHA is rapidly reduced to ascorbate by reduced ferredoxin. In addition, MDHA reductases (MDHAR) rapidly reduce MDHA to ascorbate using NADPH. The presence of these membranes for recycling MDHA has cause doubt on the requirement for GSH-dependent reduction of DHA in ensuring ascorbate regeneration. When MDHA is not reduced, it rapidly disproportionate to AA and DHA, the divalent oxidized product. Data obtained by addition of oxidants to enzymes removed from the light and membrane dependent thioredoxin reduction system lack any relevance whatsoever to *in vivo* condition, where the activation state of thiol-regulated enzymes will reflect the differences between reductive and oxidative fluxes. The abundance of a putative cytosolic MDAR mRNA from tomato (*Lycopersicon esculentum* Mill) is shown to be inversely proportional to total AA and greatly elevated in response to UV radiation. Under UV-B stress, the inhibition in electron transport chain due to the degradation of the D1 protein in PS-ÉÉ might promote the energy transfer from triplet chlorophylls to oxygen to form $O_2^{\cdot-}$. The exogenous addition of ascorbate to the culture of green algae *Chlamydomonas reinhardtii* prevented the ROS increase and therefore, the ROS-mediated down-regulation of large subunits of rubisco translation induced by excess light stress.

In addition, a study of Bowler *et al.* (2002) suggests that AA also exhibited a significant protective effect on lipid peroxidation and DNA strand break. Moreover, the bleaching of chl-a andphycobilins leads to a drop in the photosynthetic rate. Quenching of ROS by antioxidants cannot completely reverse the negative effect of UV-B radiation in terms of photosynthetic yield and non-photochemical quenching.Also, a plethora of evidence indicates that ROS are involved in the DNA damage, cell death and signal transduction. ROS generated under UV stress especially OH^-, $^1O_2$, react with sugars, purines and pyrimidines.Under moderate UV-B radiation, cyanobactaria thrived to survive and the damage caused which did not result in lethal effects on all of the organisms, although a significant loss in survival occurred. In particular, AA did not demonstrate a significantly higher ROS scavenging efficiency than other antioxidants during the starts of the irradiation.

Biotic Stress

Abiotic stress factors such as metal action, salinity and extremes of light have long been known as major limitations to crop productivity. Plants have evolved direct and indirect mechanism for sensing and responding to pathogenic stimuli to prevent damage caused by pathogenic attack and it contributes to the detoxification of ROS. This antioxidant activity of AA is associated with resistance to oxidative stress and longevity in plants. Moreover, the endogenous level of AA has been suggested to be important in the regulation of developmental senescence and plant defense against pests.

Disease Registance

Plant responses to stress can be viewed as being orchestrated through a network that integrates signalling pathways characterized by the production of ET, JA, SA, ABA and GA3. The identified regulatory step in the network involves transcription, protein interaction and targeted protein damage. In plants, the mitogen activated protein kinase (MAPK) cascade plays a key role in various abiotic and biotic stress responses and in phyto-hormone responses that include ROS signalling. The Combined synergistic effect of AA induced ABA and GA3 in signal transudation has been extensively studied using the cereal aleurone layer. Evidence suggests that these hormones are perceived at the plasma membrane and that phospholipases and G-proteins are involved in the early signaling events. Down streams are Ca^{2+} and calmodulin, which, in turn, target various ion channels, protein kinases and phosphatases. Through these systems, GA_3 and ABA regulate the expression of a number of proteins in the aleorone layer, including α-amylase. Studies of ET signaling in Arabidopsis indicate that receptor gene families in plants may function similar to many of their animal counterparts, increasing their flexibility at responding to stressful environments. Also, studies in the past few years have contributed to the accumulating evidence that ET signalling has a substantial role in disease tolerance.

Pathogenesis

Plants are attacked by many diseases-causing organisms including bacteria, fungi, viruses and nematodes. Plant immunity is based on a complex response that is highly flexible in its capacity to recognize and counteract different invaders. To combat invasion by microbial pathogens and herbivores insects effectively, plants make use of pre-existing physical and chemical barrier as well as inducible defense mechanism that become activated upon attack. These pathogens cause large crop losses and probably since the agriculture have contributed to human hunger and malnutrition. Plant resist pathogens attack both with performed defenses such as antimicrobial secondary compounds and by including defense responses. Inducible defences can be activated upon recognition of general elicitors such as bacterial flagellin and even host cell fragments released by pathogen damage. This protein referred to as effectors are recognised by plant disease resistance plants in a highly specific manner (Breen and Murphy, 1995).The identification of many R genes and in many cases their corresponding effectors proteins, has accelerated research into the molecular bases of gene-for-gene disease resistance. The activation of plant defense state is correlated with a stimulation of three pathogenesis related proteins, such as chitinase, β-1, 3-glucanase and peroxidase which are molecular markers of SAR, which is triggered by necrotizing pathogens, induced systemic resistance, which is activated upon colonization of roots by selected strains of non-pathogenic rhizobacteria and wound induced defense, which is typically elicited upon tissue damage such as that caused by feeding insects.

HR is a form of PCD and is considered as a part of a complex defense response to microbial pathogens in which death of host cells at the site of a virulent pathogen entry occurs within a few hours of pathogen attacks. Pathogen-induced HR is a rapid oxidative burst at the site of microbial infection. Moreover, ROS production during HR are mediated by a NADPH oxidase localized in the plasma membrane and induced signal transduction network appear to involve a MAPK signaling cascade pathways (Cakmak *et al.*, 1993). Furthermore, Chang and Shockey (1999) reported that the effect of ascorbigen and 1-methyl ascorbigen on the disease resistance in bean against fungal pathogen, *Uromyces phaseoli* and also suggests that effectiveness of protection depended on the dosage of the applied 1-methylascorbigen and on the time interval between the chemical pre- treatment and inoculation. Studies of Coles *et al.* (1999) established that AA along with glycerol found to be most effective in increasing the phytoremediating potential of *Spirodela polyrrhiza* L. In addition Belide *et al.* (2011) also suggested that hyperhydration and necrosis of *Agrobacterium*- infected cotyledons found to be effectively controlled by using AA along with L-cysteine and carrageenan.

Now it has become increasingly clear that AA function is intertwined in a complex network that meshes the plant response to pathogens and the onset of senescence. In a large number of studies associated with stress mitigation in plants by adding the different natural and synthetic compounds, AA is used as a reference compound (Lopez-Munguia *et al.*, 2011). Generally, ROS formed during HR activate ethylene, SA and JA signaling pathways; serve to induce defense gene expression to counteract the invading pathogen and to minimize lesion formation in plants exposed to oxidative stress like (ozone) O_3-ET synthesis and emission increase in plants exposed to ozone. ET triggers PCD in the accelerated cell death mutant (*acd1*), and is involved in regulating PCD in plant pathogen interaction. SA signaling is regulated for ozone-induced cell death responses, induction of pathogens related proteins, and SAR. JA biosynthesis is induced in ozone treated Arabidopsis and hybrid poplar plants within several hour of treatment (Koch *et al.*, 2000). Treatment with JA has been shown to reduce the extent of cell death in tobacco. Arabidopsis mutants constitutively expressing JA, such as *CEV1* are mere resistant to powdery mildew. These plant hormones do not act independently in response to O_3 and/or pathogens but rather in a complex signaling network (Dong, 1998).

Moreover, in a study of Pastori *et al.* (2003), *vtc1* (AA deficient mutant of Arabidopsis) is subjected to a large-scale microarray analysis to identify genes that are differentially expressed in the AA-deficient mutant relative to the wild type in the absence of added ROS. One hundred and seventy one transcripts that were either increased or decreased in *vtc1* versus wildtype were identified. Effectors are often pathogen proteins that probably evolved to subvert various host processes for promotion of the pathogen life cycle. The function of various R proteins requires phosphorylation, protein degradation or specific localization within the host cell. Some signaling components are shared by many R gene pathways whereas others appear to be pathway specific. New technologies arising from the genomics and proteomics revolution will greatly expand our ability to investigate the role of R proteins in plant disease tolerance. In contrast, major antioxidant enzymes such as CAT and APX are largely unaffected in vtc1. Interestingly, when the AA content is artificially elevated by feeding with 10 mM ascorbate resulting in AA levels similar to wild type treated equally, the transcript abundance is reversed in vtc1, indicating that low AA induces defense response and high ascorbate suppresses the induction of defense genes (El-Hariri *et al.*, 2010). Thus, one would predict that constitutively induced defense gene expression in vtc 1 might correlate with resistance to pathogens. In addition, hyphal development and fungal conidiophores production is significantly reduced in *vtc1* and also *vtc 2*, a second non-allelic AA-deficient mutant. PR-1 and PR-5 proteins are of higher abundance in vtc 1 than in wild type, particularly within the first 24h post- inoculation with *Pseudomanas syringae* as revealed by western blot analysis (Arrigoni and De Tullio, 2000). Many studies suggest the specific requirement of AA as a cofactor for the activity of 2-oxo acid-dependent dioxygenases, a class of enzymes that includes those regulating the synthesis of hydroxyl proline containing proteins and hormones in plants (El-Mashad and Mohamed, 2011).

Cell Cycle

In plants, the evidence is based on the increased proportion of cells progressing to from G1 to S phase in onion root meristems and pericycle in response to exogenous ascorbate. More evidence for the role of ascorbate in controlling the transition from G1 to S phase is provided by studies on maize root quiescent centers. Barth *et al.* (2006) assumed that AA affects phytohormone-mediated signaling processes during the transition from the vegetative to reproductive phase and the final stage of development. This leads to internodal growth rates up to 5 mm/hr. Therefore, it has been proposed that the primary action of GA in the intercalary meristems of rice is on cell elongation and that entry in to the cell cycle is a function of cell size, a phenomenon that has been well documented in yeast. The addition of exogenous L-AA stimulates the cells of the quiescent centre to re-enter the cell cycle in both maize and in Allium (El-Saidi,1997). Moreover, in synchronized tobacco BY-2 suspension culture, L-AA content is found to increase during cell division, and it is concluded that the concentration of DHA during G1 phase may contribute to a shortening of the cell cycle and to cell elongation (Kato and Esaka 1999). lycorine, a alkaloid, decreases the ascorbate content of tissues and also inhibits cell division and cell elongation in Avena Coleoptiles and Pea internodes (De Leo *et al.*, 1973) and in onion roots, it induces the disappearance of cells in S-phase. Furthermore, there are several lines of evidence indicates that ascorbate or more likely, MDA stimulates cell proliferation in cultures of animals cells by shortening the cell cycle. Another suggestion is that ascorbate increases deoxyribonucleotide reductase activity, required for DNA replication. In Arabidopsis root, however, L-AA GSH and non-specific reducing agents such as dithiothreitol are all found to stimulate cell division in the root primary meristems and to extend the range of meristematic divisions. Cell wall development Growth in higher plants is the results of two different processes, cell proliferation and cell elongation. Growth of the plant cells is driven by water uptake which, in turn, results from stress relaxation of the cell wall (Cosgrove, 1993, 1997; Fahmy *et al.*, 1998).

Cell Wall Extension

It has been suggested that apoplastic oxalate could be responsible for the sequestering of Ca^{+2}, important for the cross-linking of pectin's and cell wall strengthening. Although, it is difficult to assess the relative contributions of these various mechanisms to cell wall expansion, it is clear that L-AA is able to modulate fundamental aspects of cell wall metabolism. In this respect, the redox status of apoplastic L-AA and the balance between L-AA and H_2O_2 level will directly or indirectly influence the degree of lignifications and cross-linking of cell wall components. By involvement of ascorbate oxidase, which have consistently been found to influence plant cell expansion by a number of proposed mechanisms. These includes the co-substrate requirements of the iron-deoxygenases required for the post-translational modification of cell wall proteins and possibly the direct reaction of DHA with lysine and arginine side chains to prevent cross-linking (Lin and Varner, 1991). High AO activity correlates with the beginning of rapid expansion in germinating seeds and can be localized to regions where cells are expanding. High AO activity is found in rapidly expanding cucurbits fruits. This developmental regulation of AO expression is supported by the effect of light and auxin. Continous far-red light stimulates AO activity. The wall therefore has a system to regenerate MDA via AO activity and a means of reducing MDA to ascorbate. Ascorbate is transported in to the apoplasts by a carrier. DHA, formed by disproportionation of any MDA which escapes reduction by cyto-b, can be transported by a high affinity carrier in to the cytosol where it is reduced to ascorbate by GSH-dependent DHAR. The net result of this system is electron transport across the PM with NAD (P) H as the reductant and extracellular oxygen is the ultimate electron acceptor. It has been suggested that electron transport stimulates the plasma-membrane H+-ATPase, which lead to increased cell expansion and solute uptake (Rayle and Cleland, 1992). This suggests the relationship between AO and growth but a number of other ways in which ascorbate and AO could interact with wall function (Foyer *et al.*, 1993).

Lignifications are associated with high wall peroxidase (POD) activity and formation of H_2O_2 in the wall. POD uses H_2O_2 as oxidant to form monolignol radicals from monolignol precursors such as coniferyl alcohol. The wall ascorbate pool is smaller and more oxidized in lignifying spruce needle (more DHA) than in mature needles/(less DHA). However the average combination of ascorbate in the apoplasts of lignifying needles is still high enough to inhibit POD. It is hypothesized that the low levels of AA cause GA deficiency and that the GA flowering pathway is favored under short-day conditions, leading to delayed senescence. Also, GA induces stem growth in many rosette plants and dwarf mutants. This growth response can be quite dramatic and is the combined result of enhanced cell division activity in the sub-apical meristems and increased cell elongation. This suggests that AA is involved in the transcriptional regulation of some SAGs during oxidative-stress-induced senescence. More direct evidence for the role of AA in regulating the expression of SAGs by a study that revealed an up-regulation of select SAG transcripts in the AA-deficient Arabidopsis mutant vtc1 when grown under a long day photoperiod suggesting that AA-deficiency induces a senescent phenotype.

Photosynthesis

AA has a critical role in photosynthesis, as the high concentration in stroma of chloroplast and it has been shown to have important functions in photosynthesis, such as in the protection of the photosynthetic apparatus against the oxygen radicals and H_2O_2 that are found during photosynthetic activity and against photo-inactivation; Since it is a cofactor of carotenoid-de- epoxidation. Also, AA in photosynthesis works in its three biochemical modes. Firstly, it acts as an antioxidant by removing H_2O_2 formed by oxygen photoreduction in PS-I, catalysed by APX, and some of which is bound to thylakoids where it can scavenge H_2O_2 as it forms. Secondly, MDA, formed by APX can act as a direct electron acceptor to PS-I. Thirdly, it is a cofactor for violaxanthin de-epoxidase. The de-epoxidase, which is bound to the lumen side of the thylakoid membrane, is dependent on ascorbate as a cofactor. However, alterations in photosynthetic activity, antioxidant capacity and the amount of ROS produced have been reported neither under long nor under short-day conditions. Hence, it appears that the flowering and senescence prototypes linked to the endogenous AA content are not connected to changes in photosynthetic capacity. It is possible that AA deficiency causes very subtle and gradual changes in these processes that may be difficult to detect. Moreover, it is hypothesized that the low levels of AA cause GA deficiency and that the GA flowering pathway is favoured under short-day conditions, leading to delayed senescence. Also, GA induces stem growth in many rosette plants and dwarf mutants. This growth responsecan be quite dramatic and is the combined result of enhanced cell division activity in the sub-apical meristems and increased cell elongation. Furthermore, under short days, senescence is delayed. By contrast, under long days, flowering and senescence are accelerated in *vtc1*. High SA content and decreased levels of GA may, in a photoperiod-dependent manner, result in early flowering and senescence under low AA conditions. In addition, during senescence, cell undergoes highly co-ordinate changes in cell structure, metabolism, and gene expression. In the early stages of senescence, chlorophyll degrades and photosynthetic activity decreases due to a decrease in the expression of, for example, the small subunit of Rubisco and chlorophyll a/b-binding genes (Fujita *et al.*, 2006). These genes are referred to as senescence-down-regulated genes. Thus, senescence is characterized by a decrease in chlorophyll as well as in the content of total RNA and protein. The early phase of senescence is followed by an up- regulation of senescence-associated genes that aid in the remobilization of biomolecules. More direct evidence for the role of AA in regulating the expression of SAGs is provided by a study that revealed an up-regulation of select SAG transcripts in the AA-deficient Arabidopsis mutant vtc1 when grown under a long day photoperiod, suggesting that AA-deficiency induces a senescent phenotype. The abundance of AA modifies the threshold for activation of plant defence responses via redox mechanisms that are independent of the natural senescence program. The endogenous level of AA has recently been suggested to be important in the regulation of developmental senescence. Plants have several L-AA biosynthetic pathways including routes via L-galactose and L-gulose, but the contribution of each one is varies between different species, organs and developmental stages (Cruz-Rus *et al.*, 2011).

At this final stage of senescence, cells undergo lipid peroxidaton, DNA degradation and the nuclei, mitochondria, membranes and vacuoles distingrates. Abiotic stress conditions such as drought heat or salinity induced dehydration constitute direct osmotic stress and thus, causes extensive losses to agricultural production worldwide. Plants respond to dehydration and low temperature with a number of physiological and developmental changes. Plants have evolved a high capacity to synthesize and accumulate non-toxic solutes.

Conclusion

Ascorbate is implicated in plant responses to abiotic environmental stresses and regulates stress response as a result of a complex sequence of biochemical reactions such as activation or suppression of key enzymatic reactions, induction of stress responsive proteins, and the production of various chemical defense compounds. In addition, an attempt has been made to connect some very intriguing observations that have been reported for the AA-deficient mutant *vtc1* in terms of some factors such as temperature, salinity and ultra-violet radiations etc to cause oxidative stress. Due to its essential function as a co-factor for the biosynthesis of various plant hormones, AA appears to influence not only the endogenous level but also signaling of these plant hormones, and thus affect responses against the environmental stresses.

Refrences

Abd–El Hamid, E.K. (2009).Physiological effects of some phytoregulators on growth, productivity and yield of wheat plant cultivated in new reclaimed soil.Ph. D. thesis, Girls College, Ain Shams Univ. Cairo, Egypt.

Arrigioni, O., and De Tullio, M.C. (2000). The role of ascorbic acid in cell metabolism: between gene directed functions and unpredictable chemical reactions. *Journal of plant physiology,***157**: 481-488.

Asada, K. (1992).Ascorbate peroxidase – a hydrogen peroxide scavenging enzyme in plants.*Physiologia Plantarum,***55**: 235-241.

Bala, R., and Thukral, A.K.(2011).Phytoremediation of Cr(VI) by Spirodelapolyrrhiza (L.)Schleiden employing reducing and chelating agents.*International Journal of Phytoremediation,***13(5)**: 465-491.

Barth, C., Moeder, W., Klessig, D.F., and Conklin, P.L. (2006). The timing of senescence and response to pathogens is altered in the ascorbate-deficient Arabidopsis mutant vitamin c-1. *Plant Physiology,***134**: 1784–1792.

Belide, S., Hac, L., Singh, S.P., Green, A.G., and Wood, C.C.(2011).Agrobacterium-mediated transformation of safflower and the efficient recovery of transgenic plants via grafting.*Plant Methods,***20**: 7-12.

Booth, W.A., and Beardall, J.(1991).Effect of salinity on inorganic carbon utilization and carbonic anhydrase activity in the halotolerant algae *Dunaliella salina* (Chlorophyta). *Phycologia,* **30**: 220–225.

Bowler,C., Van Montagu, M., and Inze, D. (1992).Superoxide dismutase and stress tolerance. *Annual Review of Plant Physiology and Plant Molecular Biology,***43**: 83–116.

Breen, A.P., and Murphy, J.A. (1995).Reactions of oxyl radicals with DNA. *Free Radical Biology and Medcine*, **18**: 1033–1077.

Cakmak,I., Strbac, D., and Marschner, H. (1993).Activities of hydrogen peroxide scavenging enzymes in germinated weed seeds.*Journal of Experimental Botany*,**44**: 127-132.

Chang, C., and Shockey, J.A. (1999).The ethylene response pathway. *Current Opinion in Plant Biology*, **2**: 352–358.

Coles, J.P., Phillips, A.L., Croker, S.J., Garcia-Lepe, R., Lewis, M.J., and Hedden, P. (1999).Modification of gibberellin production and plant development in Arabidopsis by sense and antisense expression of gibberellin 20-oxidase genes.*The Plant Journal*, **17**: 547–556.

Dionisio-Sese, M.L., and Tobita, S. (1998). Antioxidant responses of rice seedlings to salinity stress. *Plant Science*,**135**: 1-9.

Diplock, A.T., Charleux, J-L., Crozier-Willi, G., Kok, F.J., Rice-Evans, C., Roberfroid,M., Stahl, W., and Vina-Ribes, J. (1998). Functional food science and defence against reactive oxygen species. *British Journal of Nutrition*, **80**: 77-112.

El Hariri, D.M., Sadak, M.S., and El-Bassiouny, H.M.S. (2010). Response of flax cultivars to ascorbic acid and A-Tocopherol under salinity stress conditions. *International Journal of Academic Research*, **2(6)**: 101-109.

El-Mashad, A.A., and Mohamed, H.I. (2011).Brassinolide alleviates salt stress and increases antioxidant activity of cowpea plants (*Vigna sinensis*). *Protoplasma*,**10**: 23-45.

El-Saidi, T.M. (1997).Salinity and its effect on growth, yield and some physiological process of crop plant. In Jaiswal PK, Singh RP (eds.): Strategies for improving salt tolerance. Oxford and IBH Publishing Co. Ltd. New Delhi. pp. 111-127.

Fahmy, A.S., Mohamed,T.M., Mohamed, S.A., and Saker, M.M. (1998).Effect of salt stress on antioxidant activities in cell suspension cultures of cantaloupe (*Cucumismelo*).*Egyptian JournalPhysiologicalScience*,**22**: 315-326.

Foyer, C., Rowell, J., and Walker, D. (1983).Measurement of the ascorbate content of spinach leaf protoplasts and chloroplasts during illu-mination.*Planta*,**157**: 239-244.

Foyer, C.H., Descourvieres, P., and Kunert, K.J. (1994). Protection against oxygen radicals: an important defence mechanism studied in transgenic plants. *Plant Cell Environtment*,**17**: 507–523.

Fujita, M., Fujita, Y., Noutoshi, Y., Takahashi, F, Narusaka, Y., Yamaguchi-Shinozaki, K., and Shinozaki, K. (2006). Crosstalk between abiotic and biotic stress responses: a current view from the points of convergence in the stress signaling networks. *Current Opinion in Plant Biology*,**9**: 436–442.

Gomez, J.M., Hernandez, J.A., Jimez,A., Delrio, L.A., and Sevilla, F. (1999).Differential response of antioxidative enzymes of chloroplast and mitochondria to long term NaCl stress of pea Plants.*Free radical Research*, **31**: 11-18.

Hamada, A.M. (1998). Effect of exogenously added ascorbic acid, thiamine or aspirin on photosynthesis and some related activities of drought stress wheat plants.

In: Garab G (eds.). Photosynthesis: Mechanism and Effects. Kluwer academic Publishers, Dordrecht, pp. 2581-2584.

Hassanein, R.A., Bassony, F.M., Barakat, D.M., and Khalil, R.R. (2009).Physiological effects of nicotinamide and ascorbic acid on *Zea mays* plant grown under salinity stress. 1- Changes in growth, some relevant metabolic activities and oxidative defense systems. *Reserch Journal ofAgricultural and Biological Science*,**5(1)**: 72-81.

Pastori, G.M., Kiddle, G., Antoniw, J., Bernard, S., Veljovic- Jovanovic, S., Verrier, P.J., Noctor, G., and Foyer, C.H. (2003). Leaf vitamin C contents modulate plant defense transcripts and regulate genes that control development through hormone signaling. *The Plant cell*,**15**: 939–951.

Rao, M.V., Paliyath, G., and Ormrod, D.P. (1996). Ultraviolet-B and ozone induced biochemical changes in antioxidant enzymes of *Arabidopsis thaliana*. *Plant Physiology*, **110**: 125-136.

Rivero, R.M., Ruiz, J.M., Garcia, P.C., Lopez Lefebre, L.R., Sanchy, E., and Romero, L. (2001). Resistance to cold and heat stress: accumulation of phenolic compounds in tomato and water melon plants. *Plant Science*,**160**: 315-321.

Shan, C., Liang, Z., Sun, Y., Hao, W., and Han, R. (2011).The protein kinase MEK1/2 participates in the regulation of ascorbate and glutathione content by jasmonic acid in *Agropyron cristatum* leaves.*Journal of Plant Physiology*, **168(5)**: 514-518.

Natural Products: Research Reviews Vol. 3 (2016) *Pages* **459–470**
Editor: **V.K. Gupta**
Published by: **DAYA PUBLISHING HOUSE, NEW DELHI**

18

Antimicrobial and Antioxidant Activity of *Leonotis nepetifolia* (Linn.) R.Br.: A Concise Review

Sankar Narayan Sinha[1]

ABSTRACT

For the last few decades demand of medicinal plants has increased globally due to insurgence of interest in and acceptance of herbal medicines. Since the dawn of civilization people used plant materials as herbal medicines and Leonotis nepetifolia, a plant of Lamiaceae is of no exception. It is an important medicinal plant which has been used for the treatment of several diseases. It contains a number of phytochemicals that showed antioxidant and antibacterial activities. The present review critically scrutinizes the antioxidant and antibacterial properties of the plant.

Keywords: Antimicrobial activity, Antioxidant activity, Phytochemicals, *Leonotis nepetifolia*.

Introduction

The genus *Leonotis* has 12 species widely distributed in Pan Tropics and is represented by one species, *Leonotis nepetifolia* in India. *Leonotis nepetifolia* R. Br. is an aromatic erect annual herb belonging to the family Lamiaceae.

Leonotis nepetifolia is an important medicinal plant of repute in Indian traditional systems of medicine such as Ayurveda, Siddha and Unani. The plant is known as Granthiparni in Ayurveda.The plant is marketed as Barchi Buti. The plant is commonly known as Knod Grass (English), Lion's ear (English), Gathivan (Hindi),

1 Department of Botany, University of Kalyani, Kalyani – 741 235, West Bengal, India.

* *Corresponding author*: E-mail: sinhasn62@yahoo.co.in

Matijer (Gujrati), Dipmal (Marathi) and Ranabheri (Telugu). Medicinal uses of this plant are reported from many countries such as, Brazil, Canada, Kenya, Madagascar and many African countries to treat diseases like rheumatism, bronchial asthma, fever, dysmenorrhoea and diarrhoea (Pullaiah, 2010).

Geographical Distribution

This plant usually grows in Africa, tropical Asia and India particularly South India. It is originally native to tropical and subtropical Africa; it is now naturalized throughout the world. It grows along road sides in abandoned fields and in disturbed areas in hotter parts of our country. It generally grows in patches along roadside or barren agricultural wasteland during rainy season. The mature plant grows up to a height of 2 meter. The orange yellow coronated verticilaster inflorescence and distinct plant odor are amongst the unique characters of this plant.

Systematic Position in the Plant Kingdom

Domain: Eukaryota

Kingdom: Plantae

Phylum: Spermatophyta

Subphylum: Angiospermae

Class: Dicotyledonae

Order: Lamiales

Family: Lamiaceae (Labiatae)

Genus: *Leonotis*

Species: *Leonotis nepetifolia*

Taxonomic Description

The plant is an erect herb of about 2-4m length; stem straight, obtusely tetragonus, pubescent. Leaves broadly ovate, heavily crenate to serrated, acute; base truncate to cuneate. Inflorescence verticillaster, Flowers orange-scarlet, in axillary, globose spinous whorls of vertical, calyx tubular, mouth oblique, 8-10 toothed, valvate, coriaceous, spinescent, corolla tubular biliped tube with five annular rings within, upper lip hooked, lower lip 3-lobed, didynamous, exerted, filamentous, minutely bearded disc, copular. Ovary bicarpellary, tetralocular ovule, one per locule, style gynoebasic, stigma bifid, nutlets four, oblong, seeds oblong (Mukherjee, 1940).

Folklore Uses

Leaves: Leaves are used in Rheumatism, Hernia, Diabetes and Hepatitis.

Stem: Stem is used in jaundice and skeletal muscle stimulant.

Seeds: Seeds are used in Antimalaria and Diuretic.

Root: The root is used in breast inflammation, burns, pre-natal vomiting.

Flowers: Flowers are used in cut wounds, burns and jaundice.

Figure 18.1: *Leonotis nepetifolia* Showing Verticillaster Inflorescence.

Medicinal Properties

The plant is being used by the local peoples and tribal of Maharashtra as ethnomedicine on various ailments. The infusion of leaves is traditionally being used to cure the stomach pain of the children and also to cure cough and cold by tribals of Melghat (MS) India. This plant is also being used for its anti-inflammatory, anti-diarrheal properties by various communities in Indian subcontinent and also across the world (Imran *et al.*, 2012). *Leonotis nepetifolia* (Lamiaceae) is an important medicinal plant which has been used to treat malaria, asthma, diarrhoea, influenza and fever. It is also used as an analgesic. The plant showed various biological activities such as anti bacterial and antifungal activities. Phytochemical examination of this plant parts revealed the presence of various bioactive compounds. Traditionally, this plant was used to treat renal problems, rheumatism, dysmenorrheal, bronchial asthma, diarrhoea, fever and influenza. In India, the medicinal uses of the selected plant are reported for burns, breast swelling, ring worm, scalds, skin afflictions, rheumatic pain and malaria. Roots of *Leonotis nepetifolia* is recognized as the botanical sources of Granthiparna (An ayurvedic herb) which is included in various formulations such as, Nakula taila, Mritasanjivani sura Brihat Guduchi taila, and Himasagar taila (Pushpan *et al.,* 2012). The studies have reported that the plant exhibits following important constituents and activities: The antioxidant and antiproliferative potential of the plant has been reported (Veerabadran, 2013). The wound healing capabilities of drug obtained from *Leonotis nepetifolia* has also been reported (Nithya, 2011). Researchers have reported the anti bacterial properties from the drug obtained from the plant (Udaya Prakash, 2012). The anti-inflammatory and anti–rheumatic action of plant has been described by researchers (Parra-Delgado, 2004; Manocha, 2012). The analgesic and anti-cancer activities of the plant were demonstrated by the studies undertaken on selected cell lines (Gurunagarajan and Pemaiah, 2010). The works supporting the anti-bacterial activity of *Leonotis nepetifolia* have been reported (Manocha, 2012; Boalino and Tinto, 2004). The traditional utilization of the plant as a medicine, a decoction is used against fever, cough, womb prolapsed and malaria has been presented (Mendes *et al*., 1986). It has been reported that plant exhibit various biological activities such as antimicrobial, antifungal, analgesic, narcotic, anti-cancer, hypotensive, laxative (Basarkar, 2013; Narayan, 2013). The plant is used in cases of asthma and also for epilepsy in S. Africa (Burkill, 1995; Clement, 2005; Ayanwuyi, 2009). The decoction of the leaves is used to treat coughs, burns and dermatitis. Several other species of the *Leonotis* were reported to possess anti-malarial activity (Bagby, 1965; Takeda, 1999; Clarkson, 2004). The plant is used in traditional medicine in therapy of asthma, diarrhoea, fever, malaria and as an analgesic agent in menstrual pains; also to treat common cold and to alleviate cough (Clement *et al.,* 2005; Maregesi *et al.,* 2007; Cheryl, 2007). Researchers have reported the presence of antidiabetic activity of ethanolic whole plant extract of *Leonotis nepetifolia* in alloxan induced diabetic rats. (Gungurthy *et al.,* 2013). The whole plant is used for menstrual pain and unspecified female complaints (Lasserre *et al*., 1983). The plant has been evaluated for its hypotensive potential (Hortensia *et al.,* 2004), anti inflammatory activity (Gopal, 1995), anti plasmodial activity (Clarkson *et al.,* 2004), and anti bacterial activity (Kumar *et al.,* 2007). The flowers of the plant are used in case of hardly healing wounds, scars and burns. To burn healing, the seeds are used as well (Nayak *et al.,* 2004; Kala 2005; Gupta *et al.,* 2010). The plant ash, paste and leaf steam bath are used for the cure of paralysis, dermatitis, rheumatism and post natal swelling (Jain and Srivastava, 2005). The plant has been reported to posses the qualities that can regulate periods and can control diarrhoea (Basha *et al.,* 2005). Workers have reported the febrifuge properties of *L. nepetifolia* (Pawar and Patil, 2011). The role of paste from plant in controlling the body swelling has also been described by some workers (Hemadari, 2011). The decoction of the plant has inhibitory effects on the joint pain as believed by the tribes of Andhra Pradesh (Patil, 2012). Various workers reported the utilization of flowers from the plant in scalds and burns as followed in the Ethnomedicinal system followed in India (Upadhyaya, 1998; Dash and Mishra, 1999; Kumar, 2007). The paste of inflorescence of the plant with groundnut oil for wound healing has long been used by Kokani tribals of Maharastra (Dey and De, 2010; Thomas *et al.*, 2012). The possible role of root paste and flower ash of the plant for breast inflammation has been described by some researchers (Mali and Bhadane, 2010; Salave *et al.*, 2011). The potential role of plant flower juice with sugar as a cure of night blindness has been described in Ethno-pharmaceutical claims by the Vanrjari of Maharashtra State, India (Babu *et al.*, 2012). The role of plant inflorescence ash mixed with mustard oil for the post natal breast pain as followed and prescribed by the Tribals of Kota Hills in Andhra Pradesh, India (Pandey and Tripathi, 2010). The plant flowers and seed mixed with Ghee has been used for the treatment of cough cuts, wounds and burns (Anbarashan, 2011). Workers reported the possible treatment of Eczema by leaf paste of *L. nepetifolia* (Kumar and Dash, 2012). The leaf juices of *L. nepetifolia* are used for the treatment of malaria (Mendes *et al.*, 1986; Clarkson *et al.*, 2004; Pushpan *et al.*, 2012). The depurative, febrifuge and antihelmenthic properties of plant leaf infusion have been described by (Basha *et al.*, 2005). The tribes of Andhra Pradesh have been utilizing the plant leaf paste to treat the burning sensation of scorpion sting (Udaya Prakash, 2012). The possible role of decoction of plant (leaf, stem, and flower) has been reported (Basha *et al.*, 2005). The root of this plant was used to control pre-natal vomiting by tribal people of Andhra Pradesh (Kumar and Vidyasagar, 2008). The traditional knowledge of Bidar district of Karnataka state has been utilizing the plant seed paste for curing head sores (Dhawan, 2013). Workers have reported the ethno medicinal utilization of the plant roots for burn cure from the Eastern Himalayan region of India (Nayak *et al.*, 2004). Despite its luxurious growth in the tropical and sub tropical regions of South East Asia, the plant merits consideration due to its over exploitation. The National Medicinal Plant Board has given the plant immense importance and recommended this herb for cultivation (NMPB, 2004). The over exploitation of the plant duly necessitates a proper investigation of the available genotypic diversity and its conservation through commercial cultivation by employing proper agro techniques. The leaves of *Leonotis nepetifolia* R.Br. are used traditionally in the management of type-2 diabetes mellitus, pain, arthritis, coughs, indigestion, warm infestation, nasal-disorders, skin infections, malaria and flowers are used in burns, scalds, itches and skin diseases. In the Caribbean region plant was found to be used in treatment of asthma. The plant is also medicinally useful to treat rheumatism, rickets, headaches, wounds, ringworm, laxative, bitter, narcotic, fever and convulsion (Gangavati, 2011).

Phytochemical Screening

The preliminary phytochemical analysis showed presence of alkaloids, flavonoids, steroids, glycosides, phenolics, tannins and saponins. However, all these chemicals were not extractable in one solvent. Alkaloids, phenolic, flavonoids and glycosides were present in chloroform extract; tannins, steroids and saponins were found in petroleum ether extract; phenols, flavonoids and glycosides were present in methanolic extract; alkaloids, phenolic, flavonoids and saponins were found in aqueous extract while acetone extract showed presence of only steroids. The quantitative analysis indicates that the plant possesses significant level of alkaloids, phenolic, flavonoids and tannins (Imran *et al.*, 2012).

The essential oil of *L. nepetefolia* (Labiatae) showed moderate anti-bacterial activity against Gram positive and Gram negative bacteria. It also inhibited the growth of dermatophytes and other aerial fungi.

The plant *Leonotis nepetefolia* rich in alkaloids (leonurine and stachydrene), iridoid glycosides (leonuride, leonurin and leonuridine), diterpenoids (leocardin), flavonoids (rutin, quercetin, hyperoside, apigenin), volatile oil, tannins and vitamin A.

The root contains *n*-octacosanol, quercetin, *n*-octacosanoic acid, crumarin like 4,6,7-trimethoxy-5-methylchromene-2-one, campesterol and beta-sitosterol-beta-D-glucopyranoside. The seed oil contains oleic, linoleic, palmitic and stearic acids. The leaves contain neptaefolin, neptaefuran, neptaefuranol, neptaefolinol, leonitin, neptaefolinin and (")-55, 6-octadecadienoic acid. The earlier phytochemical studies carried out have reported the presence of Labdanic acid, Laballenic acid (Bagby *et al.*, 1965) nepetaefolinol and leonotinin and a coumarin, characterized as 4,6,7-trimethoxy-5-methylc-hromen-2-one (Takeda *et al.*, 1999). The predominant phytochemicals such as phenols and flavonoids were also reported by other author (Veerabadran *et al.*, 2013).

Antimicrobial Activities

Anti-microbial assays was done to analyze *in vitro* activity of the crude extracts of *Leonotis nepetifolia* against five pathogenic bacterial isolates which yielded highest inhibition zone of 23 mm against *Staphylococcus aureus* followed by 20 mm against *Streptococcus* pneumonia and 6 mm against *Escherichia coli.* No inhibition zone was noted against *Pseudomonas aeruginosa* and *Klebsiella pneumoniae.*

The alcoholic extracts of leaf of *L. nepetifolia* showed maximum antimicrobial activity against the microbial pathogens followed by inflorescence, root and stem extracts (Gnaneswari and Venkata Raju, 2012). The methanolic extract (ME) and acetone extract (AE) from aerial parts of *L. nepetifolia* were inactive towards selected strains of Gram (+), Gram (-) bacteria and dermatophytes at concentrations up to 50 mg/ml (Sobolewska, 2014).

The essential oil of *Leonotis nepetefolia* was investigated for its antibacterial activity and found to be moderately active against Gram positive and Gram negative bacteria excepting *Pseudomonas aeruginosa*. The essential oil also showed inhibition zone against dermatophytic fungi and suppressed the growth of other aerial fungi. Therefore the essential oil of this plant may be quite useful in fungal dermatitis even with secondary bacterial infections (Gopal *et al.*, 1994).

Leonotis nepetifolia showed a MIC value of 0.25 g mL^{-1} and therefore, the best activity on *Staphylococcus. Leonotis nepetifolia* may contain compounds effective against staphylococcal infections (Nalubega *et al.*, 2011).

The methanolic extract of leaf, flower and stem of *Leonotis nepetifolia* was analyzed for antibacterial activity against eight pathogenic bacterial isolates *viz., Escherichia coli, Shigella dysenteriae, Pseudomonas aeruginosa, Shigella flexneri, Vibrio cholerae, Bacillus subtilis, Micrococcus luteus* and *Staphylococcus aureus.* The methanolic extract exhibited antibacterial activity against all the bacteria tested indicating the presence of broad spectrum antibacterial compounds in the plant parts. Gram positive bacteria were found to be more susceptible than Gram negative bacteria. Extracts of leaf and flower of the plant showed higher activity than the stem (Sinha, 2012).

Methanolic extract of *Leonotis nepetifolia* of concentration 50 mg/ml showed antibacterial activity on agar plate containing *Bacillus subtilis* (Pingale *et al.*, 2013).The ethanolic leaf extracts of *L. nepetifolia* was assessed for its antidiarrheal activity and showed significant antidiarrheal activity.

In an investigation methanol, petroleum ether and ethyl acetate extract of leaves, stem, inflorescence and root of *Leonotis nepetifolia* were studied for their antibacterial activity by agar disc diffusion method against Gram positive and Gram negative bacterial strains as well as one fungal strain. It was noted that alcoholic extracts of leaf showed maximum antimicrobial activity against the pathogenic microorganisms followed by inflorescence, root and stem extracts (Gnaneswari and Venkata Raju, 2012). Among the extracts of leaves of *Leonotis nepetifolia* studied for their antibacterial potency, the ethylacetate extract showed good antibacterial activity against *Bacillus subtilis, Escherichia coli, Klebsiella pneumonia, Pseudomonas aeruginosa, Staphylococcus aureus* and *Streptococcus epidermidis* (Udaya Prakash *et al.*, 2013).

Antioxidant Activities

The DPPH assay was done for antioxidant activity which showed methanol extract of *L. nepetifolia* leaves to be more significant in scavenging free radicals with inhibition percentage of 60.57 per cent. From the data obtained, the methanol extract proved to be significant in all anti-oxidant assays and this effect was well comparable with the standard used in the study. Regarding free radical scavenging activity using DPPH, the acetone, ethanol and ethyl acetate extracts of *Leonotis nepetifolia* leaves showed better antioxidant activity when compared with that of control (Udaya Prakash *et al.*, 2013). The DPPH assay showed that methanol extract of *L. nepetifolia* leaves to be more significant in scavenging free radicals with inhibition percentage of 60.57 per cent. The methanol extract proved to be significant in all anti-oxidant assays and this effect was well comparable with the standard used in the study. The DPPH radical scavenging ability of the methanol extract of *L. nepetifolia* was significantly higher than those of α-tocopherol which revealed that the ability of proton donating capacity of the methanol extract of *L. nepetifolia* was very high. Superoxide radical can be converted into hydrogen peroxide by the action of superoxide dismutase, and the H_2O_2 can subsequently generate extremely reactive hydroxyl radicals in the presence of certain transition metal ions (*e.g.* iron and copper), or by UV photolysis. Phytochemical screening of the crude extracts revealed the presence of secondary compounds such as alkaloids, flavonoids, steroids, tannins, and phenols (Hemadari, 2011). Presence of these phytochemicals in methanol extract of *L. nepetifolia* attributed to their excellent antioxidant activity. The preliminary phytochemical analysis of *Leonotis nepetifolia* showed presence of alkaloid, phenolics, flavonoids, tannins, steroids, glycerides and saponins. However, all these chemicals were not extractable in one solvent. Chloroform extracts yields alkaloids, phenolics, flavonoids and glycosides. Tannins, steroids and saponin are found were found in petroleum ether extract. Phenols, flavonoids and glycosides were present in methanolic extract.

Alkaloids, phenolics, flavonoids and saponins were found in aqueous extract while acetone extract showed only presence of steroids (Imran *et al.*, 2012). The presence of such phytochemicals can be correlated with the medicinal potential of the plant. In another study, qualitative phytochemical analysis of alcoholic extract of leaf materials showed the presence of many secondary metabolites like flavonoids, glycosides, phenols, tannins and diterpenes (Kamalam *et al.*, 2013).Another results showed that saponins, tannins, steroids, cardiac glycoside, terpenoids and flavonoids are present in *Leonotis nepetifolia*.

Conclusion

Leonotis nepetifolia has appropriately been used in various traditional systems of medicine. Review of literature shows a plethora of pharmacological activities of the studied plant. Considering the wide range of benefits of *L. nepetifolia* against common human diseases further studies are called for the development of its agro techniques to cater the demand for the plant in various traditional systems of medicines. However, the antimicrobial potential of *L. nepetifolia* needs verification through continued experimentation and clinical trials on human populations. Considering the immense advantages of this medicinal plant, an increase in research on this plant is expected in the time to come. More works are needed to isolate and characterize other bioactive compounds responsible for antioxidant and antibacterial activities of this plant.

References

Anbarashan, K. (2011). Ethno-floriatic survey in sacred groves, Pudukottai district, Tamil Nadu, India. *Journal of Medicinal Plant Research,* **5**: 439-443.

Ayanwuyi, L.O., Yaro, A.H., and Adamu, H.Y.S. (2009). Studies on anticonvulsant activity of methanol extract of capitulum of *Leonotis nepetifolia*. *Nigerian Journal of Pharmaceutical Sciences,* **8**: 73-79.

Babu, N.C., Naidu, T. M., and Venkaiah, M. (2012). Ethnomedicinal plants of Kotia Hills of Vizianagaram District, Andhra Pradesh, India. *Journal of Phtyology,* **2**: 76-82.

Bagby, M., Smith, C.R., and Wolff, J. (1965). Laballenic Acid. A new allenic acid from *Leonotis nepetifolia* seed oil. *Journal of Organic Chemistry,* **30**: 4227–4229.

Basarkar, U.G,, Shinde, V., Gunjal, K., Aswale, N., and Adnyal, P. (2013). Antimicrobial activity of *Leonotis nepetifolia*, medicinal herb. *Proceedings of Centenary Session of Indian Science Congress, Kolkata 2013 (Part II)* Abstracts. Pp- 130

Basha, S.K., Sudarsanam, G., Parveen, D., and Verma, A. (2005). Conserving biodiversity of Yerramalais Kurnool District, Andhra Pradesh, India through people's biodiversity Registers Program. *Biodiversity Journal,* **3**: 111-118.

Boalino, D.M., and Tinto, W.F. (2004). A new diterpene of *Leonotis nepetifolia*. *Heterocyc.,* **63**: 383-387.

Burkill, H.M. (1995). The Useful Plants of West Africa. Royal Botanical Gardens, Kew, England, **2** pp. 14-16, 239.

Cheryl, L. (2007) Ethnomedicines used in Trinidad and Tobago for reproductive problems. *Journal of Ethnobiology and Ethnomedicine*, **3**: 13.

Clarkson, C., Maharaj,V.J., Crouch, N.R.,Grace, O.M., Pillay, P., Matsabisa, M.G, Bhagwandin, N., Smith, P.J., and Folb, P.I. (2004). *In vitro* antiplasmodial activity of medicinal plants native to or naturalised in South Africa. *South African Journal of Botany*, **13**: 18-26.

Clement, Y.N., Williams, A.F., Aranda, D., Chase, R., Watson, N., Mohammed, R., Stubbs, O., and Williamson, D. (2005). Medicinal herb use among asthmatic patients attending a specialty care facility in Trinidad. *BBC Complementary and Alternative Medicine*, **5**: 3.

Dash, S.S., and Mishra, M.K. (1999). Taxonomic survey and systematic census of economic plants of Narayana Patna Hills of Koraput Dist., Orissa. *Journal of Economic and Taxonomic Botany*, **23**: 473-498.

Dey, A., and De, J.N. (2010). A survey of Ethnomedicinal plants used by the tribals of Ajoydha hill region, Purulia district, India. *American Eurasian Journal of Sustainable Agriculture*, **4**: 280-290.

Dhawan, N.G., Khan, A.S., and Srivastava, P. (2013). A general appraisal of *Leonotis nepetifolia* (L) R. Br: An essential medicinal plant. *Bulletin of Environment, Pharmacology and Life Science*, **2**: 118-121.

Gangavati, S. (2011). Evaluation of anti-diabetic and wound healing activity of *leonotis nepetaefolia* R.Br leaves. M. Pharm dissertation. Rajiv Gandhi University of Health Sciences, Karnataka, Bangaluru, India.

Gnaneswari, K., and Venkata Raju, R.R. (2012). Preliminary Phytochemical screening and antimicrobial evaluation of *Leonotis nepetiifolia* (L.) R.Br. *Journal of Natural Product Plant Resources*, **2** : 689-692.

Gopal, R.H. (1995). Antibacterial activity of *Leonotis nepetifoilia*. *Fitoterapia*, **66**: 83-86

Gopal, R.H., Vasanth, S., and Vasudevan,S.V. (1994). Antimicrobial activity of essential oil of *Leonotis nepetaefolia*. *Ancient Science of Life*, **14**: 68–70.

Gungurthy, J., Sabbathi, S., Chaitanya, K.B., Ravella, A., and Ramesh, C. (2013). Antidiabetic activity of *Leonotis nepetifolia* Limn. in Alloxan induced diabetic rats. *International Journal of Preclinical and Pharmaceutical Research*, **4**: 5-9.

Gupta, A., Nagariya, A.K., Mishra, A.K, Bansal, P., Kumar, S., Gupta, V., and Singh, A.K. (2010). Ethno-potential of medicinal herbs in skin diseases: An overview. *Journal of Pharmacy Research*, **3**: 435-441.

Gurunagarajan, S., and Pemaiah, B. (2010). Anti-tumor and antioxidant potentials of ethanolic extract of *Leonotis nepetifolia* R. Br. against Ehrlich Ascites carcinoma cell lines. *Journal of Pharmacy Research*, **3**: 2990-2992.

Hemadari, K. (2011). A treatise on Treatise Medicine. 1st ed. Dr Koppula Hemadri's House of Tridal Medicine. 20 - 26.

Hortensia, P.D., Gabriela, G.R., Antonio, N.C., and Mariano, M.V. (2004). Anti-inflammatory activity of some extracts and isolates from *Leonotis nepetifolia* on TPA induced edema model. *Revista de la Sociedad Química de México,* **48**: 293.

Imran, S., Sudhakar, S.S., and Koche, D.K. (2012). Phytochemical analysis of *Leonotis nepetifolia* (L.) R. Br., A wild medicinal plant of Lamiaceae. *Bioscience Discovery,* **3**: 197-196.

Jain, S.K., and Srivastava, S. (2005). Traditional uses of some Indian plants among islander of the Indian Ocean. *Indian Journal of Traditional Knowledge,* **4**: 345-357.

Kala, C.P. (2005). Ethnomedicinal botany of the Apatani in the Eastern Himalayan region of India. *Journal of Ethnobiology and Ethnomedicine,* **1**: 12-18.

Kamalam, M., Saraswathi, C., and Umadevi, U. (2013). Evaluation of *Leonotis nepetifolia* for its phytochemical and heavy metal analysis. *International Journal of Pharmaceutical Sciences and Research,* **4**: 4591-4596.

Kumar, B., Vijayakumar, M., Govindarajan, R., and Pushpangadan, P. (2007). Ethnopharmacological approaches to wound healing-exploring medicinal plants in India. *Journal of Ethnopharmacology,* **114** : 103-113.

Kumar, P., and Vidyasagar, P. (2008). Tradicional knowledge on medicinal plants used for the treatment of skin diseases in Bidar District, Karnatka. *Indian Journal of Traditional Knowledge,* **7**: 273-276.

Kumar, S., and Dash, D. (2012). Flora of Nanadan Kanan Sancturay; Medicinal plants with their role in Health care. *International Journal of Pharmacy and Life Sciences,* **3**: 1631-1642.

Lasserre, B., Kaiser, R., and Chanh, P.H. (1983). Effects of rats of aqueous extracts of plants used in folk medicine as antihypersensitive agents. *Naturwisessenschaften,* **70**: 95-98.

Mali, P.Y., and Bhadane, V.V. (2010). Ethnomedicinal wisdom of tribals of Aurangabad district (M.S.), India. *Indian Journal of Natural Product Resource,* **2**: 102-109.

Manocha, N., Sharma, V., Samanta, K. C., Dubey, P.K., and Saluda, M. (2012). Anti-inflammatory and anti-rheumatic activity of the chemical constituents of *Leonotis nepetifolia* capitulum. *Molecular and Clinical Pharmacology,* **3**: 73-81.

Maregesi, S.M., Ngassapa, O.D, Pieters, L., and Vlietinck, A.J. (2007). Ethnopharmacological survey of the Bunda district, Tanzania: Plants used to treat infectious diseases. *Journal of Ethnopharmacy,* **113**: 457-470.

Mendes, John., Cote, ce and Cote, la, (1986). Trinidad and Tobago Dictionary. Arima, Trinidad, pp. 135.

Mukherjee, S. K. (1940). A Revision of the Labiatae of Indian Empire. *Records of the Botanical Survey of India,* **14**: 1-228.

Nalubega, R., Kabasa., J.D., Olila, D., and Kateregga, J. (2011). Antibacterial activity and phytochemical screening of eleven plants used as poultry ethnomedicine in sourthern Uganda. *Agricultural Journal,* **6**: 303-309.

Narayan, S.S. (2013). Antibacterial potential of crude methanol extract of *Leonotis nepetifolia*. *International Research Journal of Pharmacy*, **3**: 277-278.

Nayak, S., Behera, S.K., and Misra, M.K. (2004). Ethno-medico-botanical survey of Kalahandi district of Orissa. *Indian Journal of Traditional Knowledge*, **3**: 72-79.

Nithya, V., Brinda, P., and Anand K.V. (2011). Wound healing activity of *Leonotis nepetifolia* R. Br., in Wistar Albino rats. *Asian Journal of Pharmaceutical and Clinical Research*, **4**: 23-26.

NMPB, (2004). A List of Indian Medicinal Plants excluding botanical synonyms. pp. 21.

Pandey, A.K., and Tripathi, N.N. (2010). Diversity and distribution of aromatic plants in forests of Gorakhpur Division, U.P. *Indian Biological Forum*, **2**: 25-33.

Parra-Delgado, H., Garcia Ruiz, G., Camacho, A.N., and Vazquez-Martinez, M. (2004). Anti inflammatory activity of some extracts and isolates from *Leonotis nepetifolia* on TPA-induced edema model. *Revista de la Sociedad Química de México*, **48**: 293-295.

Patil, D.A. (2012). Indian Ethnomedicines: Origin in the perspective of doctrine of signature. *Life Sciences leaflets*, **1**: 6 – 15.

Pawar, S., and Patil, D.A. (2011). Ethnomedicinal plants in Jalgaon district: Current status. *Current Botany*, **2**: 15-21.

Pingale, R., Pokharkar, D., and Phadtare, S. (2013). A review on ethnopharmacolgy, phytochemistry and bioactivity of *Leonitis nepatofolia*. *International Journal of Pharm Tech Research*, **5**: 1161-1164.

Pullaiah, T. (2010). Encyclopedia world medicinal plants, Vol. 3. New Delhi. Regency Publictations, pp- 1226.

Pushpan, R., Niteshwar, K., and Kumari, H. (2012). Ethnomedicinal claims of *Leonotis nepetifolia* (L). R. Br. A Review. *International Journal of Research in Ayurveda and Pharmacy*, **3**: 783-785.

Salave, A.P., Reddy, G.P., and Diwakar, P.G. (2011). Ethnopharmaceutical claims by the Vanrjari from Pathardhi Tahasilnin, Ahmednagar District (M.S.). India. *Asain Journal of Experimental Biology*, **2**: 69 - 74.

Sinha, S.N. (2012). Antibacterial potential of crude methanolic extract of *Leonotis nepetifolia* (L) R. Br. *International Research Journal of Pharmacy*, **3**: 277-278.

Sobolewska, D., Pasko, P., Galanty, A., Was, J.M., Padlo, K. and Wasilak, W. (2012). Preliminary phytochemical and biological screening of methanolic and acetone extracts from *Leonotis nepetifolia* (L.) R. Br. *Journal of Medicinal Plants Research*, **6**: 4582-4585.

Takeda, T., Narukawa, Y., and Hada, N. (1999). Studies on the constituents of *Leonotis nepetifolia*. *Chemical Pharmaceutical Bulletein*, **47**: 284-286.

Thomas, B. Arumugan R., Veerasamy, A., and Ramamoorthy, S. (2012). Ethnomeidcinal survey of plants used by Kuruma tribe for the treatment of cuts and wounds in Kerala, India. *Asian Pacific Journal Tropical Biomedicine*, 1-3.

Udaya Prakash, N.K. (2012) A study on antibacterail activity of common weeds in northern districts of Tamil Nadu, India. *Research Journal of Medicinal Plants*, **6**: 341-345.

Udaya Prakash, N.K., Bhubaneswari, S., Balamurugan, A., Vaishnavi, S., Sugunan, S., Meena, S., Chandran, M., and Babu, H.S. (2013). Studies on antibacterial antioxidant, larvicidal, pesticidal activities and phytochemistry of *Leonotis nepetifolia* (Linn.) R. Br. *International Journal of Research in Pharmaceutical Sciences*, **4**: 303-309.

Upadhyaya Kumar, O.P., and Tiwari, K. (1998). Skin treatments of Bihar plants. *Pharmaceutical Biology*, **36**: 20-24.

Veerabadran, U., Venkatraman, A., Souprayane, A., Narayanaswamy, M., Perumal, D., Elumalai, S., Sivalingam, S., Devaraj, D., and Perumal, A. (2013). Evaluation of antioxidant potential of leaves of *Leonotis nepetifolia* and its inhibitory effects on MCF7 and Hep2 cell lines. *Asian Pacific Journal of Tropical Diseases*, **3**: 103-110.

Veerabadran, U., Venkatraman, A., Souprayane, A., Narayanasamy, M., Perumal, D., Elumalai, S, Sivalingam, S., Devaraj, V., and Perumal, A. (2013). Evaluation of antioxidant potential of leaves of *Leonotis nepetifolia* and its inhibitory effect on MCF7 and Hep2 cancer cell lines. *Asian Pacific Journal of Tropical Disease*, **3**: 103–110.

Natural Products: Research Reviews Vol. 3 (2016) *Pages* **471–500**
Editor: **V.K. Gupta**
Published by: **DAYA PUBLISHING HOUSE, NEW DELHI**

19

Natural Multifunctional Products, Drugs and Pharmaceuticals, Industrially Important Chemicals, Enzymes and Biopesticides from Microorganisms: Search and Discovery

Suresh R. Naik[1]*, Shrinivas B. Bumrela[1], Abhishek Jain[1] and Ketaki Malagi[1]

ABSTRACT

The antiquity of the art of fermentation has risen to the present time was as novel to the ancients, and with the help of microbial scientists and biotechnologists further revolutionized as a novel approach for the preparation of useful therapeutics and health related magic bullets. The review attempts to bring forth the genesis of microbial (fungi, bacteria and their near relatives) products, antibiotics, therapeutically active molecules, enzymes, biopesticides and other compounds. After the discovery of penicillin, and β-lactam antibiotics in 1940s, produced profound effect on human health by enabling rapid treatment of patients with bacterial infection that had often proved fetal. Over the next 40 years, now seen as the "golden era" of antibiotic research, the majority of antibiotic drug classes in use today were discovered. Since 1970, most newly approved antibiotics have been based on these known scaffolds with the exception of few. The development of microbial natural products have been largely for disease modifying drugs or as a research tool or getting new leads for novel chemical species to derive better therapeutic agents. Microbes have provided many potent drugs for the treatment of human diseases, infectious immune origin, cardiovascular, cancer and many more. The

1 Department of Pharmacology, Sinhgad institute of Pharmaceuticals Sciences. Kusgaon (BK), Lonavala, Pune-410401, Maharashtra, India.

* *Corresponding author*: E-mail: srnaik5@rediffmail.com

divers enzymes produced by microorganisms and their applications in different industry e.g. food, leather, environment and air pollution, chemical and pharmaceutical industries have also been described. Also included some of the microbial metabolites based new generation antibacterial that are in the pipeline and currently in active development.

***Keywords*:** Fermentation, Microorganisms, Antibiotics, Enzymes, Amino acids, Microbial metabolites, Biopesticides.

Introduction

The aging of meat resulted more pleasing to the taste than meat eaten soon after killing was discovered by the cave man. Similarly, the cave man's earliest descendants found that intoxicating drinks could be prepared from grains and fruits. This was primarily due to the ignorance that microscopic or tiny living organisms *e.g.* molds, yeast, bacteria etc. were growing on their food and producing whole range of chemicals that has changed taste as well as texture of food materials. The earlier uncontrolled haphazard fermentation process in which yeasts digested grain gave cloudy, unreliable beer was fore-runner of the present days controlled fermentation process by which microorganism produce primary and secondary metabolites considered to be the most important drugs, enzymes and valuable chemicals. The past and present gap is that modern microbiological science has acquired mastery over many organisms and also discovered numerous useful microbes. As the time passes, during the Middle Ages they learned how to improve the taste of wine bread, beer and cheese, but they had no idea that in doing so, they were actually handling tiny organisms that are responsible for fermentation and too small to be seen by the naked eye. The French scientist, Z.J. Thenard in 1803 told that the yeasts used by the wine makers were living forms of life and responsible for the fermentation of alcohol. This was subsequently established with the birth of immortal Louis Pasteur in 1857. He proved that many microorganisms were responsible for diseases of animals and humans. He studied and found that microbes were killed by some other microbes in the soil and air.

Pasteur's knowledge found more useful outside the field of medicine and sought to use microbes as a tool in the industrial production of wine and other useful chemicals and products. The mold *Aspergillus niger* transformed ordinary sugar to citric acid, and Pfizer in 1923, ventured the world's first successful citric acid fermentation plant. Similarly, Chain and Weizmann converted corn mash with the help of tetanus germ to acetone, an essential chemical in the manufacturing of cordite, an explosive, which rescued Great Britain from acute ammunition shortage during World War I. Radolf Emmerich and Oscar Low of Munich University prepared crude world's first antibiotic pyocyanase from *Pseudomonas aeruginosa*, however, subsequently abandoned due to hazardous nature and due to failure of standardization. This was followed by Alexander Fleming, a bacteriologist at St. Marry's hospital, London, discovered the mold *Penicillin* genus. The substance extracted from the mould killed *Staphylococcus* organism and named penicillin, which was many times more powerful than pyocyanase. Later Florey and Chain of oxford University with the help Pfizer, Merck and Squibb using *Penicillium notatum* prepared penicillin with higher yield and able to save the lives of wounded soldiers of World War I. Thus, the primitive brewer's magic had come a long way (Naik, 1994). With grand success of penicillin endeavor of scientist, the mold, yeast, actinomycetes, bacteria or similar organisms in the waters of lake, streams, sea and the surface ground and air were found to be the source of new antibiotics. Thus, science has developed from nature itself. Professor Walkesman and his group at Rutger University, USA, after testing thousands of soil samples using Dr. Flemings experience and approach, yielded another antibiotic, streptomycin from the strain, *Streptomyces griesus*. This antibiotic showed activity against Gram negative bacteria resistant to penicillin and also against *Mycobacterium tuberculosis*. The most important vitamins, vitamin B_2 (riboflavin), vitamin B_{12} (folic acid) were prepared by fermentation, vitamin C (ascorbic acid) derived from sorbose, was produced by a bacterium that feeds on sorbitol prepared from corn sugar, vitamin-D was produced by ultraviolet irradiation of ergo sterol, plant growth regulators, food flavors, salts, cosmetics, candy, rust remover, beverages and health care materials were also prepared by fermentation. In short, fermentation chemistry and technology contributed almost all materials of modern human life at all stages.

The antibiotics, therapeutically important enzymes and hormones, biopesticides, plant growth regulators, diagnostic enzymes, immunostimulatory products and enzymes used in pharmaceutical industries derived from microbial sources is depicted in Figure 19.1.

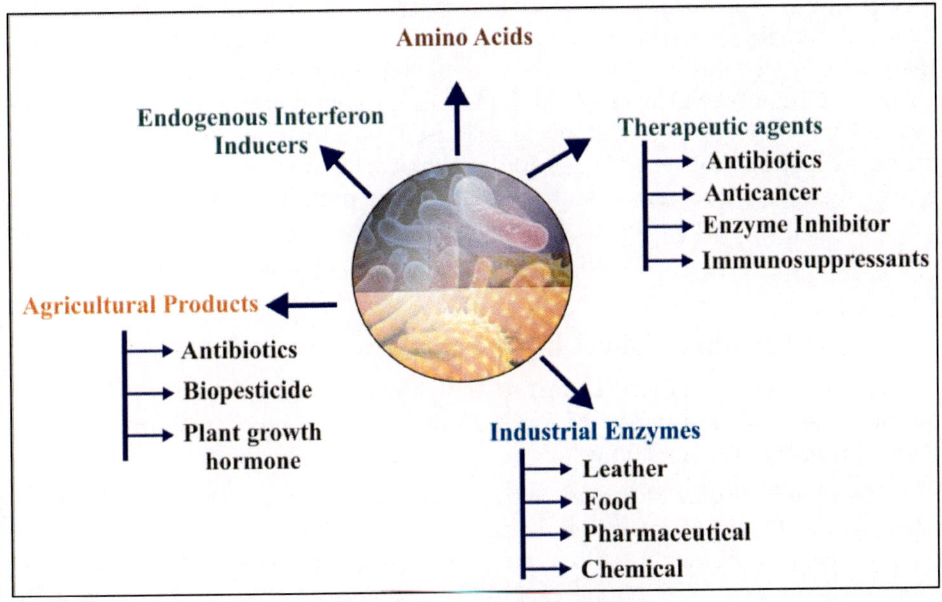

Figure 19.1: Various Products Derived from Microbial Origin.

Therapeutic and Industrial Enzymes from Microbial Sources

Actinomycetes have long been centered on their phenomenal ability to produce bioactive secondary metabolites, notably antibiotics. However, subsequently exploration of actinomycetes, *Cornybacterium, Arthrobacter, Cellulomonas* (Micromonospora and other genera like Microbiospora etc.), Streptomyces like *Streptococus lumosus* are rich sources of industrially important enzymes: Glucose isomerase (used in the manufacture of high fructose syrup),Amylase (production of maltose syrup), lignocellulose from *T. curvata* (used in degrading and preparing of trisaccaharides and solid waste compost (Ball and McCarthy, 1988), Xylanases (hemicellulases) from Chinia NCL-82-51and *Actinomycetes* as an effective alternative for biobleaching and biopulping operation in paper industry for replacing, chlorine and other toxic chemicals, Pectinases (pectinolytic activity) (Ladjama *et al.*, 1991), Chitinases (degrade the cell wall of filamentous fungi), Proteases, Lipases (lipolytic activity), chymogin (*Streptomyces and Actinomycetes)* (Seetalaxman and Srinivasan, 1993; Aramori *et al.*, 1991). Pepsin in continuous coagulation of milk in cheese, Protease I (*Achromebacter lyticus*) is used in the preparation of(semisynthetic human insulin, a biologically active peptide), steroid dehydrogenase (prednisolone), stritosidine synthase (strictosidine), carboxy peptidases (optical resolution of DL-amino acids), Alkyl sulphatases (degradation of surfactants), α-Amylases (reduction of turbidity of starch containing effluents), β-galactosidase (BOD reduction of milk industry), monooxygenase (removal of chloroform and benzopyrine), parathion hydrolase (detoxification of pesticides from waste water), peroxidase and rhodanase (removal of cyanides in effluents and air), urease (removal of urea) and many more enzymes are used in food industry in cleansing and environmental pollution, water and air purification (Shewale and Naik, 1991). A number of enzymes are also used as therapeutic agents in clinical practice for the treatment of diseases. L-Asparginase (isolated from *Serratia carotovora*) used in acute lymphoblastic leukemia in children. L-asparginase acts indirectly by catabolic depletion of serum asparagine and to some extent glutamine levels causing the inhibition of protein synthesis of neoplastic cells (Maladkar *et al.*, 1993). Deoxyribonuclease is used in the treatment of the autoimmune diseases especially, systemic lupus erythematosus, Diptheria toxin (Translocase) in cancer therapy, Fibrolysin and Streptokinase for peripheral thrombus and clot dissolution, β-glucocerebrosidase for Gauchers disease, glutathione transferase in extracorporeal shunt detoxification (Shewale and Naik, 1991) and ceredase (Gaucher's diseases).

Enzymes Used in Pharmaceutical Industry

Amino acid acylase, L-amino acid oxidases, (Aspartase, Aspartic aminotransferase) are used in L-amino acids and D-amino acids preparation. Penicillin acylases are of 3 types:

The type I, usually of fungal or *actinomycete* origin, preferentially hydrolyses (Penicillin V), the type II, of bacterial origin, preferentially hydrolyses penicillin G to 6-amino penicillanic acid (6-APA), an intermediate for semi-synthetic penicillins (Subramanian *et al.*, 1978; Borkar *et al.*, 1978; Sudhakaran *et al.*, 1992). The type III, penicillin acylase hydrolyses ampicillin. The type I and II are industrially important and obtained from microorganisms: *B. megaterium, E. coli, Micrococcus leuteus, Kluyvera citrophilia jpn, Faecacalis alcalgenas* and *Proteus retgeri. Penicillin G acyclase* also hydrolyses cephalosporin G into 7ADCA (7-amino desacetoxy cephalosporanic acid), a raw material for semisynthetic oral cephalosporins. Cephalosporin C- acyclase is produced mostly by fungi and some bacterias. *viz. Arthobacter viscous, Bacillus magaterium* (ATCC53667), *Paecilomyces* C 2106, *Penicillum griseafulvin, penicillum regulosm, Pseudomonas* sp. N176 (Aramori *et al.*, 1991; Vandamme, 1988). Cephalosporin acyclase was isolated from (*E. coli,* JM 109) by Armori *et al.* (1991), finds application in the production of 7-amino cephalosporanic acid (7-ACA) from cephalosporin C (Ceph C). Used industrially the splitting of Ceph C to 7ACA that involves 2 steps reactions:

I Step: Cephalosporin C is oxidatively deaminated by D-amino acid oxidase to form glutaryl-7 amino cephalosporinic acid (GL- 7ACA)

II Step: GL-7ACA is deacylated by glutaryl-7-ACA acylase (GL-7-ACA acyclase) to produce 7ACA (Vandamme, 1988; Kumar *et al.*, 1993).

Ceph C- acylase is produced by *P. diminuta* (P 176, V.22) (Aramori *et al.*, 1991) and GL-7-ACA acylase is produced by *Pseudomonas* sp Gk 16, sp 5E 83 (Matsuda and komatsu, 1985; Matsuda *et al.*, 1987). Amino levulinic acid dehydrogenase is used for the preparation of porphobillinogen, Enniatin synthase (glutathione), Gramicidin-s-synthase (gramcidin-s). Pimprinine (an alkaloid) produced from *Streptomyces pimprina* Thirum and *Streptomyces* CDRIL-312 under submerged fermentation (Naik *et al.*, 2001; Bhate *et al.*, 1960) showed significant anti-convulsant and anti-parkinsonian activity in animals (Naik *et al.*, 2001).

Enzymes Used in Diagnostics

β-glactosidase (*E. coli*), glucose oxidase (*Aspergillus niger*), horseradish peroxidase (*Amoracia lapathifolia*), penicillanase (*Bacillus cereus*), urease (*Canvalia ensiformis*) alkaline phosphatase and many more used in immunoassays (ELISA, enzyme labeled immunosorbent assay) for an early diagnosis of diseases and hormone (TSH, LSH, FSH, HCG) levels in biological fluids.

6MFA

A microbial metabolite, produced by *Aspergillus ochraceus* (ATCC 28706), showed antiviral activity against many arboviruses mainly by inducing endogenous interferons (Maheshwari and Gupta, 1973). Treatment with 6MFA showed amelioration of rifampicin and immunologically induced hepatotoxicity in rat model (Dhuley and Naik 1998, 1999). 6MFA treatment also inhibited the increased Gamma-glutamyltranspeptidase activity in both acute and chronic inflammation (Dhuley and Naik, 2000).

Amino Acids from Microorganisms

Japanese developed microbial fermentation process in which a number of amino acids (*e.g.* Isoleucine, histidine, arginine, lysine, proline, phenylalanine, threonine, tryptophan etc were produced initially with *Corneoforms*, subsequently using other microorganisms *viz. E. coli, B. subtilis* and yeast of the genera *Hansenula (Candidia)* and Baker's yeast *Saccharomyces cerevisiae.* Infact, *E. coli, S. marcescens, B. subtilis* are the only ones that are of potential importance from commercial point of view (Neiderberger, 1989).

Discovery of β-lactam antibiotics in 1940s had profound impact on human health by enabling better treatment of patients with bacterial infections that previously proven fetal. Over the next 50 years, now seen as "Golden Era of antibiotic research" because the majority of antibiotics classes used today were discovered during 1970-80s. A range of useful bioactive products have been isolated largely from microbial culture, genus *Streptomyces* and near relatives, that include pharmacologically, immunologically and physiologically active therapeutic agents, plant growth regulators, biopesticides and anti-metabolites (Berdy, 1988). Subsequently, professor Umezwa's Institute of microbial chemistry in Japan has discovered a number of enzymes, microbial regulatory molecules active at the receptor or enzyme level which have significant therapeutic implications (Umezawa *et al.*, 1986). It is observed glaringly that most of the newly approved antibiotics launched since 2000, are based on these scaffolds with the exceptions of linezolid, an oxazolidine, deptomycin (Walsh, 2003), pseudomenic acid and retapamulin, pleuromutilin derivative (Butler and Buss, 2006).

The range of useful bioactive microbial metabolites have been isolated largely from microbial culture genus *Streptomyces* and near relative showed pharmacological, immunological and physiological activity of therapeutic importance including plant growth regulators, bio-pesticides and anti-metabolites (Table 19.1).

Immunostimulatory Products from Fungi and Bacteria

Fungal polysaccharides have potent immunoregulatory properties and have been widely prescribed in Far East region for treating infectious diseases and cancer. Lentinan from *Lentinus edodes* was found to improve the human survival time of gastric and colorectal carcinoma (Furue *et al.*, 1981). Similarly, proteoglycan, Krestin from *Corlus versicolar* has been reported to inhibit tumor growth in mice (Mizushima *et al.*, 1982) and sold as pharmaceutical preparation in Japan. Bacterial cell wall preparations have potent immunostimulatory properties and have been investigated in the treatment of human cancers. These include Picibanil, a lyophilized preparation of *Streptococcus pyrogenes*, which has been demonstrated to activate superoxide production by macrophages and also stimulate interferon and interleukin production by lymphocytes (Saito and Tomioka, 1979; Saito *et al.*, 1982). The *Nocardia rubra* cell wall skeleton and its minimal active unit, the muramyl dipeptide, have been shown to activate macrophages and also to stimulate human monocytes to release interleukin-1. Gludapcin (FK 156) was discovered as a product of *Streptomyces olivaceogriseus* and *Streptomyces violaceus* showed stimulation of humoral antibody production and delayed type hypersensitivity, and activation of macrophages (Watanabe *et al.*, 1985). Another immunostimulant FR-900494 restored immunity induced by an immunosuppressive factor obtained from tumor bearing mice. Furthermore, FR-900494 also reduced the renal side effects of cyclosporine (Twami *et al.*, 1984). FK-506, a novel neutral macrolide found more potent than cyclosporine *in-vivo* in the suppression of mixed lymphocyte reaction, antibody production and delayed type hypersensitivity (Kino *et al.*, 1987). Florfenicol, another microbial product, found to stimulate interferon production in BCG-sensitized mice (Okura *et al.*, 1986) and ubenimex, an inhibitor of leucine aminopeptidase, enhanced the release of IL-1 and IL-2 from peritoneal macrophages and spleen cells respectively (Shibuya *et al.*, 1987).

Table 19.1: Antibiotics Discovered, Producing Strains their Mechanism of Action along with Anti-microbial and other Therapeutic Activity Profile

Antibiotics	*Producing Strains*	*Anti-microbial Activity and Mechanism of Action*	*Reference*
Penicillin G	*Penicillium crysogenum* *Penicillium notatum*	Gram positive bacteria	Fleming, 1929
Streptomycin	*Streptomycin griseus*	Gram negative enteric and mycobacterium. Irreversible inhibitors of protein synthesis	Waksman *et al.*, 1944; Busse *et al.*, 1992
Valinomycin	*Streptomyces orientalis*	Gram positive bacteria inhibits cell wall synthesis	Hunter and Schwartz, 1989; Harold and Baarda, 1967
Gentamicin	*Micromona purpurea*	Gram positive and negative bacteria, used in severe infection	Edson and Terrell, 1987
Kanamycin	*Streptomyces kanamyceticus*	Gram positive and gram negative bacteria and some mycobacteria, and protein synthesis inhibition	Edson and Terrell, 1987
Bacitracin	*Bacillus subtilis*	Active against gram positive bacteria, used for topical applications	Storm and Toscano, 1979
Tobramycin	*Streptomyces tenebrarius*	Inhibits cell wall formation activity and mechanisms similar to gentamicin	Edson and Terrell, 1987
Tetracyclines	*Streptomyces aureofaciens, Streptomyces rimosus*	Bacteriostatic for gram positive and negative bacteria, protozoans (amoebas). Binds irreversibly to 30s ribosomal unit of bacteria	Wang *et al.,* 2009; Chopra, *et al.*, 1992
Chloramphenicol	*Streptomyces venezuelae*	Gram positive and gram negative organisms of aerobic and anaerobic, protein synthesis inhibition by binding 50s submit of ribosomes bacteria	Henry and Chambers, 2009; Pongs, 1979; Feider *et al.*, 1981
Clindamycin	*Streptomyces linocolnensis*	Gram positive and gram negative organisms. It binds to bacterial 50s ribosomal RNA and inhibits protein synthesis	Henry and Chambers, 2009; Steigbigel, 1995
Erythromycin	*Streptomyces erythreus*	Gram positive and gram negative organisms. Inhibition of protein synthesis by binding ribosomal unit of 50s	Steigbigel, 1995
Bicylomycin	*Streptomycin sapporonesis* *Streptomyces aizunensis*	Active against gram positive organisms. Inhibitions of lipoprotein of the outer membrane and covalently linked to peptidoglycan cell-wall	Tanaka, 1979; Miyoshi *et al.*, 1972

Contd...

Table 19.1–*Contd...*

Antibiotics	*Producing Strains*	*Anti-microbial Activity and Mechanism of Action*	*Reference*
Framycin	*Streptomyces decaris*	Active against gram positive and gram negative organisms, but largely used for topical application. Irreversible inhibition of protein synthesis	Chambers and Deck, 2009
Rifamycin	*Nocardia mediterranel*	Active against gram positive and gram negative bacteria, Inhibition of protein synthesis via binding bacterial DNA-dependant RNA synthesis	Kaye, 2004
Capreomycin	*Streptomyces capreolus*	Inhibits mycobacteria especially drug resistant tuberculosis	Akbergenov *et al.*, 2011
Mupirocin	*Pseudomonas flurescens*	Active against gram negative *Cocci* and methicillin resistant strains. It inhibits staphylococcal isoleucylt-RNA synthase	Casewell and Hill 1983
Cycloserine	*Streptomyces orchidaceus*	Active against gram positive and gram negative organisms, and also *M. tuberculosis*. It inhibits incorporation of D-alanine into peptidoglycan pentapeptide by inhibiting alanine racemose	Salmon and Sartolli, 1998
Cephalosporin	*Cephalosporium acremonium*	Cephalosporins are more stable than penicillin to many bacterial lactamases, broader spectrum anti-microbial activity	Takeda *et al.*, 2007; Toda, 2008
Clavulanic acid	*Streptomyces clavulgerus*	β-lactase inhibitors, most active against plasmid encoded β-lactamases	Neu and Fu, 1978
Ivermectin	*Streptomyces avermitilis*	Antiparasitic- antihelminthic, microfilariae infections	Abo-Shehada and Herbert, 1984; Egerton *et al.*, 1979
Paramomycin	*Streptomycin rismosus*	Anti-ameobic, Anti- leishmaniasis	Sunder, 2007
Colistin	*Aerobacillus colistinus*	Bactericidal to many gram negative organisms but clinically used for drug resistant gonorrhea	Tascini *et al.*, 2000
Spectinomycin	*Streptomyces spectabillis*	Active against gram positive and gram negative organisms, but clinically used for drug resistant gonorrhea	Henry *et al.*, 1998
Moenomycin	*Streptomyces bamborgiensis*	Active against gram positive bacteria and nonspecifically gram negative bacteria. Inhibits bacterial cell wall synthesis	Wallhausser *et al.*, 1965

Contd...

Table 19.1–*Contd...*

Antibiotics	*Producing Strains*	*Anti-microbial Activity and Mechanism of Action*	*Reference*
Aztreonam	*Streptomyces carpreolus*	Gram negative rods, *P. aeruginosa* and gram negative organisms and anaerobes	Georgopapadakou *et al.*, 1982
Kidamycin	*Streptomyces phaeoverticillatus*	Anti-tumor, antibiotic, Ehrlich ascites, carcinoma, mouse leukemia S N- 36, L-1210, NF Sarcoma, cell-life cycle of Hela cells, inhibition gram negative bacteria	Kanda, 1971
Dactinomycin	*Streptomyces species*	Antitumor activity, dactinomycin inhibits all forms of DNA dependant RNA synthesis and affecting RNA formation	Garattini *et al.*, 1956
Phenomycin and Enomycin	*Streptomycin fervents* var. *phenomyceticus*	Antitumor activity, Ehrlich carcinoma, sarcoma-180 in mice subcutaneous solid tumors, adenocarcinoma. Inhibition of protein synthesis in mammalian system	Tanaka, 1979; Nakamura *et al.*, 1972
Pliamycin (Mithramycin) (phosphoglycolipid antibiotic)	*Streptococus plicatus*	Cytotoxic used in the disorders of bone and mineral metabolism. Also showed Antitumor activity	Núñez *et al.,* 2012
Cyclosporin A	*Actinoplanes teicomyceticus*	Immunosupressant, Human organ transplantation, selected autoimmune disorders	Cavalleri *et al.*, 1984; Pallanza *et al.*, 1984
Tacrolimus	*Streptokinase tsukubansis*	Immunosuppressant solid organ transplantation, prophylactically for graft-versus-host disease leupus nephritis	Wallemacq and Reding, 1993
Sirolimus	*Streptomyces hygroscopicus*	Anti-monocytic and inhibition of MHC antigen expression	Morelon *et al.,* 2001
Deoxyspergualin	*Bacillus laterosporus*	Antimonocytic and inhibition of MHC antigen expression.	Salmon and Sartolli, 1998
Sparomycin (suphur containing antibiotic)	*Streptomyces sparsogenes*	Inhibits bacteria, mammalian cells, yeast and mold	(Wiley and MacKellar, 1976)
Coumermycin AL	*Streptomyces rishiriensis* *Streptomyces hazeliensis*	Active against gram negative organisms. It inhibits DNA replication rather than DNA repair	Ryan, 1979
Granaticin A	*Streptomyces olivacaeus*	Active against gram positive bacteria and significant antitumor activity against P-388 lymphocytic in mice and found cytotoxic	Change *et al.*, 1975

Contd...

Table 19.1–*Contd...*

Antibiotics	*Producing Strains*	*Anti-microbial Activity and Mechanism of Action*	*Reference*
Antiamoebin I	*Emericellopsis pooriensis*	Active against *Trypansoma, Entamoeba histolytica, Trichomonas vaginalis* forms ion channels but its major membrane modifying activity appears to be as a carrier.	Snook *et al.*, 1998; Thirumalachar, 1968
Antiprotozoin	*Streptomycin* sp. *S. hygroscopicus* (Jensen) *Waksman and henrici* var. *indica* Thirum	Anti-protozoan activity	Thirumalachar, 1964

Hormones, Enzyme Inhibitors of Therapeutic Importance

Hormones, enzyme inhibitors of therapeutic importance are listed in Tables 19.2 and 19.3.

Table 19.2: Hormones from Microbial Sources (Nisbet and Porter, 1989; McMorris, 1978)

Hormone	*Organism*	*Chemical Nature*
Antheridiol	*Achlya bisexualis*	Steroid
Oogonial	*Achlya bisexualis*	Steroid
Sirenin	*Allomyces arbuscula*	Alicyclic
Trisporic acid C	*Blakesla trispora*	Alicyclic
α-Factor	*Saccharomyces cerevisiae*	Peptide
α-Factor	*Saccharomyces cerevisiae*	Peptide
Tremerogen A-10	*Tremella mesetrica*	Lipo-peptide

Table 19.3: Microbial Metabolite as Enzyme Inhibitors (Nisbet and Porter, 1989)

Microbial Metabolites	*Site of Action*
Monacolins J, L and X	HMG CoA reductase
Propioxatins	Enkephalinase B
Nucleoticidin, Melanocidin A and B	5-nucleotidase
WF-3681	Aldose reducatse
Trestatins, Mutastein	Glucosyl transferase
Plistatin	Pancreatic lipase
Plistatin, luteosporin	Phospholipase a_2
Ethylene diaminosuccinic acid	Phospholipase C
Erbstatin	EGF-kinase
MY3-469	Lipoxygenase
KF 8940, 1-681, 110	Na^+/K^+ ATPase
Gyrus oleic acid, Terferol Acylpeptides	Cyclic AMP phosphodiestrase
A-58365, L-681176, 15b2, Feroxymithine, Muracein	Angiotensin converting enzymes
Ahpatins	Renin
WB-3559A, B,C and D	Fibrinolysis
Triacsins	Acyl-CoA Synthase
Juglorin	Spermidine synthase
UCN-01, K-252	Protein Kinase C

Plant Growth Hormones/Bioregulators

Active substances mentioned here are restricted to naturally occurring plant compounds and gibberellic acid, isodureomycin, anisomycin were produced by microorganisms. *Gibberella fujikuroa,* product a new class of hormones that stimulates cell division, cell enlargement or both, Gibberallins are reported to improve growth of beans plant, grapes, sugar cane and improve tomato, coffee, chilies, peppers, pineapples, apricots, citrus, soybeans, vegetables, rice productions and improve flow latex from rubber tree and papaya plant (Nickell, 1982). Thus plant growth hormones/ bio regulators, has assured the modification of growth and development in desired direction with desired degree, hence able to enhance food production.

Bio-Pesticides

Recent biotechnological approaches towards the development of biological agents (microorganisms) as a substitute for chemical larvicides to control mosquitoes, *viz. Culex, Aedes, Anopheles* and *Psorphora, Mensonia* causing various human dreaded disease like Eastern, western Venezuelan encephalitis, yellow fever, dengue, California encephalitis, malaria, filariasis, vascular stomatitis, Wesselsbron disease, West Nile fever, Kyasanur forest disease, dysphasic menigoencephalitis, leishmaniasis, trypanosomiasis, Chagas disease etc. Among the several types of microorganisms developed, *Bacillus sphaericus, Bacillus thuringinesis* showed a strong but specific mosquito larvicidal activity and appears to be promising ones for mosquito control.

Bacillus thuringiensis was first isolated in Israel by Goldberg and Margalit (1997) and *Bacillus sphaerieus* was isolated by Kellen *et al.* (1965) and later highly toxic strain by Payne and Dividson (1993), Naik (1997). *Bacillus thuringiensis var kenyae,* a biopesticide, discovered at biological division of Bhabha Atomic Research Centre, Trombay, Mumbai during 1990 and subsequently developed with the collaboration of R and D division of Hindustan Antibiotics Ltd. Pune (Muley *et al.*, 1999; Muley *et al.*, 1997). *Bacillus thruringiensis* var *kenyae,* showed significant mosquito larvicidal activity in large scale field trials.

Antibiotic Drugs Derived from Microbial Metabolites

Twenty new antibiotics have been launched worldwide since 2000 out of which 12 are microbial metabolites derived products (Table 19.4 and Figure 19.2).

Microbial Metabolite Derived Compounds in Phase-I/II/III Clinical Trails

The metabolites derived compounds from microorganism origin which are in Phase-I, Phase-II or Phase-III clinical trials listed in Tables 19.5–19.7 along with their structures in Figures 19.3–19.5 respectively.

Conclusions

Since the discovery of penicillins, β-lactams in the early 1940, saved countless lives and alleviated human sufferings. The research and development of antibiotic drugs over the next 40 years (1940-1980) lead to the discovery of most of the current classes of antibiotics and referred as 'Golden Era of Antibiotics'. Furthermore, this is also unquestionably a landmark in microbial derived natural bioactive therapeutic agents, and industrially important enzymes. This review article also included all antibiotics that have been launched since 2000 and microbial metabolites based

Table 19.4: Antibiotic Drugs (Derived from microbial metabolites) Launched Since 2000 with Antibiotic Category, Anti-microbial Activity, Lead Molecule and Microbial Source

Year	*Name of the Antibiotic*	*Category*	*Lead and (Microbial source)*	*Reference*
2001	Telitromycin	Macrolide (ketolide)	Erythromycin(*Actinomycete)*	Johnson, 2001
2002	Biapenam	β-lactam (carbepenam)	Thienamycin(*Actinomycete)*	Projan, 2008
2002	Ertapenam	β-lactam (carbapenam)	Thienamycin(*Actinomycete)*	Silver, 2005
2003	Daptomycin	Lipopeptide	Daptomycin *(actinomycete)*	Walsh, 2003
2005	Doripenam	β-lactam (carbapenam)	Thienamycin(*Actinomycete)*	Rice and Federal, 2008
2005	Tigecycline	Tetracycline	Tetracycline(Actinomycete)	Felmingham, 2005
2007	Retapamulin	Pleuromutilin	Pleuromutilin (Fungus)	Butler and Buzz, 2006
2008	Ceftobiprole medocaril	β-lactam (cephalosporin)	Cephalosporin (Fungus)	Vergidis and Falagas, 2008
2009	Telavancin	glycopeptide	Vancomycin (*Actinomycete)*	Moellering, 2010; Guskey and Tsuji, 2010
2009	Tebipenam pivoxil	β-lactam (carbapenam)	Thienamycin*(Actinomycete)*	Talbot, 2008
2010	Ceftaroline fosamil	β-lactam (celphalosporin)	Cephalosporin (Fungus)	Livermore, 2009
2011	Fidaxomicin	Tiacumicin	Tiacumicin(*Actinomycete*)	Butler *et al.*, 2013

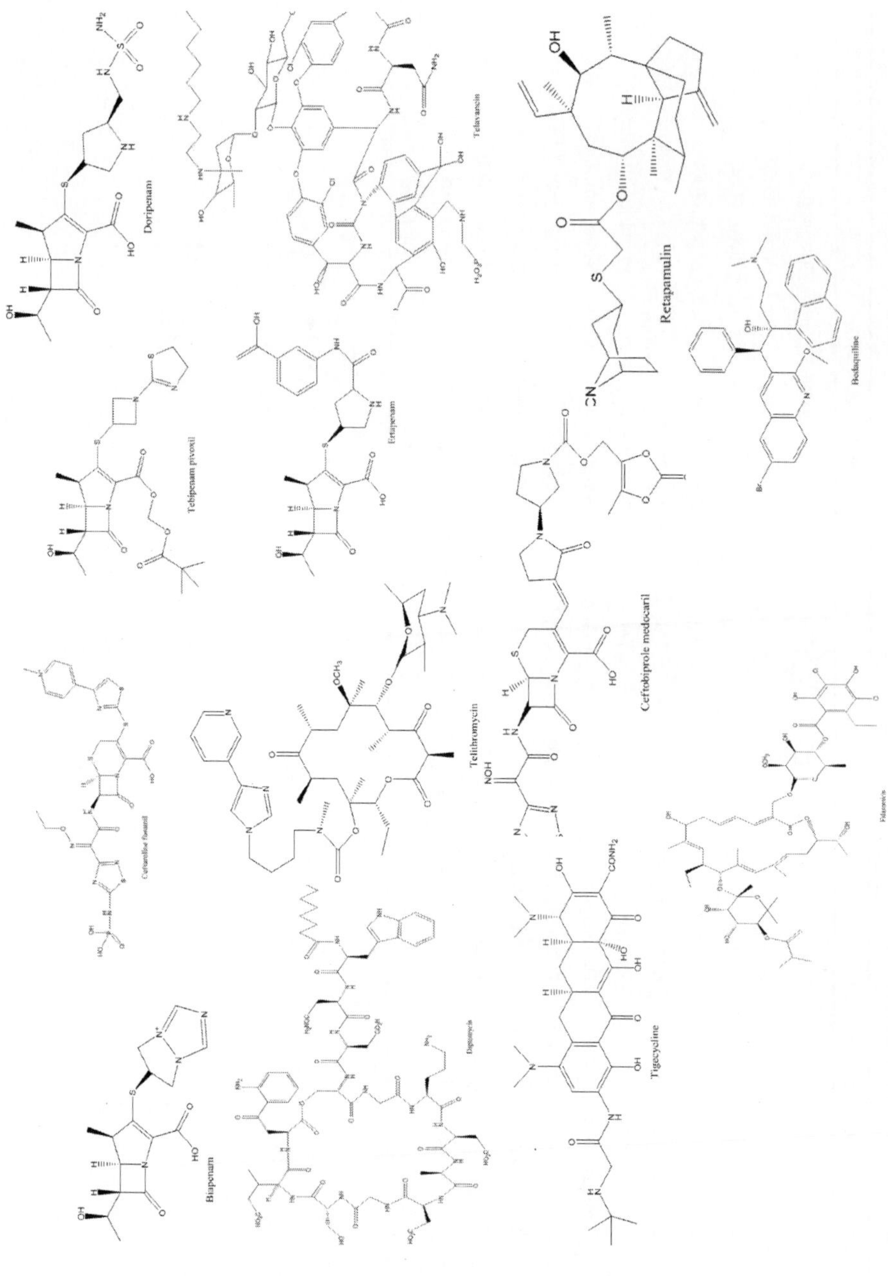

Figure 19.2: Antibiotic Drugs Derived from Microbial Metabolites.

Table 19.5: Microbial Metabolite Derived Compounds in Phase I Clinical Trails

Name of the Compound	*Lead Compound*	*Mechanism of Action*	*Status, Indication and Developing Company*
BAL 30072 Page *et al.*, 2010	Monobactum	Penicillin binding protein	Dosing studies, gram negative organisms (Basilea Pharmaceutica)
BC-7013 Biedenbach *et al.*, 2009	Pleuromutilin	Protein synthesis inhibition	Topical (Nabriva)
BC-3205 Craig *et al.*, 2009	Pleuromutilin	Protein syntheisis inhibition	Oral (Nabriva)
Lotilibcin (WAP-8294A_2) Kato, 1998	WAP-8294A_2	Phospholipid binding causing bacterial membrane damage	Interavenus formulation Methicillin resistant *S.aeurus* (Rigen)
XF-73/Exeporfinium chloride Ooi, 2009; 2010	Porphyrin	membrane perturbing activity	Topical Methicillin resistant *S.aeurus* (Density Pharma)
AZD 9742	Not-discussed	Not known	Intravenus dosing and metabolism studies
NVB 302	Deoxy-actagardine B (Type B Lantibiotic)	Lipid II binding	Active against gram positive organisms *C. difficile* associated diarrhea completed (Novacta)
GSK-2696266 Lewandowski *et al.*, 2010	Cephem	Penicillin binding protein	Gram negative infections (Shionogi/GSK)

Source: Butler and Cooper (2011, 2013), Journal of antibiotic.

Table 19.6: Microbial Metabolite Derived Compounds in Phase II Clinical Trails

Name of the Compound	*Lead Compound*	*Mechanism of Action*	*Status and Developing Company*
ACHN-490 Aggen, 2010	Aminoglycoside	Protein synthesis Inhibition	UTI and pyelonephritis (Achaogen)
BC-3781 Novak and Shlaes, 2010; Wicha *et al.*, 2010	Pleuromutilin	Protein synthesis inhibition	Acute bacterial skin and skin structure infection (Nabriva)
CB-183, 315	Daptomycin (Carpenter *et al.*, 2004)	Membrane depolarization	*C.difficile* associated diarrhea Completed (Cubist)
Ramoplanin	Ramoplanin	Cell wall production inhibition	*C.difficile* associated diarrhea. Completed (nanother-papeutics)
TP-434 Hunt, 2010	Tetracycline	Protein synthesis inhibition	Complicated intra-abdominal infection (tetraphase)
CXA-101	Cephalosporin	Penicillin binding protein	Complicated intraabdominal infection (Cubist)
GSK-1322322	Actinonin	Peptide deformylase	Complicated skin and skin structure(GSK)
PMX-300 Polymedix, 2011	Defensin	Bacterial cell membrane lysis	Acute bacterial skin and skin structure infections
NVC-422 Gottardi and Nagi, 2010	N-Chlorotaurine	Oxidation	Ophthalmic, impetigo, urinary catheter blockade and encrustation (Alcon)/Gladerma/Novabay.
LFF-571	GE2270 A (Natural product)	Elongation factor taurine	*C difficile* associated diarrhea (Norvartis)
Auricloesene	N-chlorotaurine (NP)	Oxidation	Urinary catheter blockade and encrustation adenoviral conjunctivitis, impetigo asymptomatic bacteriuria. Completed. (Novabay)
Sarecycline	Tetracycline	Protein synthesis inhibition	Acne/Rosacea (Warner Chilcott
Plazomicin	Aminoglycoside	Protein synthesis inhibition	UTI and polynephritis Completed.(Achaogen)
β-Lactam/β-Lactamase inhibitor combinations			
Imipenem/cilastatin	Carbapenem/cilastin	Penicillin binding protein/ dehydropeptidase	UTI and complicated intra-abdominal infection

Source: Butler and Cooper (2011, 2013). Journal of antibiotic.

Table 19.7: Microbial Metabolite Derived Compounds in Phase III Clinical Trials or Under New Drug Application (NDA) or Marketing Authorization Application (MAA)

Name of the Compound	*Lead Compound*	*Mechanism of Action*	*Status and Developing Company*
Fidaxomicin (tiacumicin-B) Difimycin (OPT-80)	Tiacumicin B	RNA synthesis inhibition	MAA for *Clostridium* NDA Nov, 2010 by Optimer
Omadacycline/Amadacycline (PTK-0796, MK-2764)	Tetracycline based	Protein synthesis inhibition	Phase III for complicated skin and skin structure infection by Paratek
Oritavancin Guskey and Tsuji, 2010	Glycopeptide	Cell wall production inhibition	Phase III clinical trail for acute bacterial skin and skin structure infection. The Medicines Company
Dalbavancin Malabarba and Goldstein, 2005	Glycopeptide A40926	Cell wall production inhibition	Phase III clinical trail for acute bacterial skin and skin structure infection. By Durata Therapeutics
Cethromycin Hammerschlag and Sharma 2008; Rafie *et al.*, 2010	Erythromycin	Protein synthesis inhibition	Community acquired pneumonia (NDA submitted October 2008 but rejected due to “no efficacy june, 2009 (Advanced life sciences)
Eravacycline	Tetracycline	Protein synthesis inhibition	Complicated intra-abdominal infection By Tetraphase
Solithromycin	Erythromycin	Protein synthesis inhibtion	Community acquired pneumonia By Cempra
Surotomycin	Daptomycin	Membrane depolarization	*C.difficile*-associated diarrhea. By Cubist
		β-Lactam/β-Lactamase inhibitor combinations	
CXA-201 (Ceftolozane), Tazobactam	Cephalosporin/ clavulanic acid	Penicillin binding protein/ β-lactamse inhibitor	Complicated intra-abdominal infection, and complicated urinary tract infection By Cubist
CAZ-104 (Ceftazidime, Avibactam) Donowitz, G.R., and Mandell, G.L. (1988)	Cephalosporin/ diazabicycloocatane	Penicillin binding protein/ β-lactamse inhibitor	Complicated intra-abdominal infection, Urinary tract infection By Astra Zeneca

Source: Butler and Cooper (2011, 2013), Journal of antibiotic.

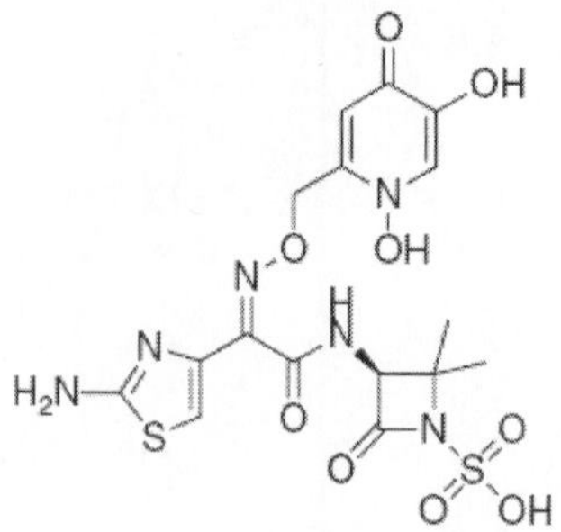

BAL30072

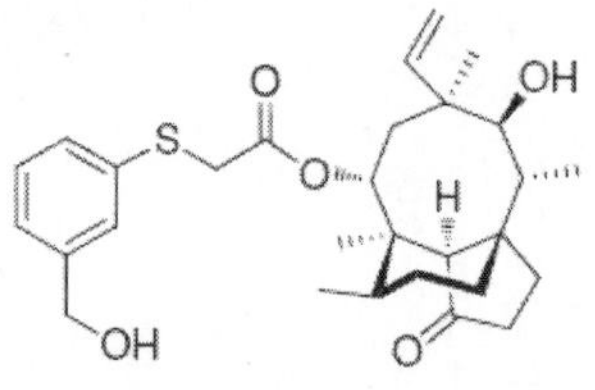

BC-7013

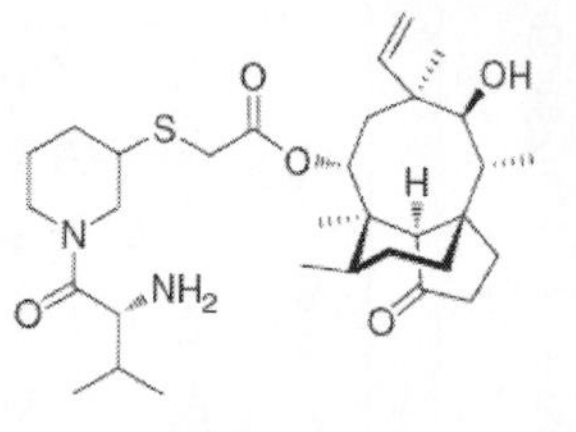

BC-3205

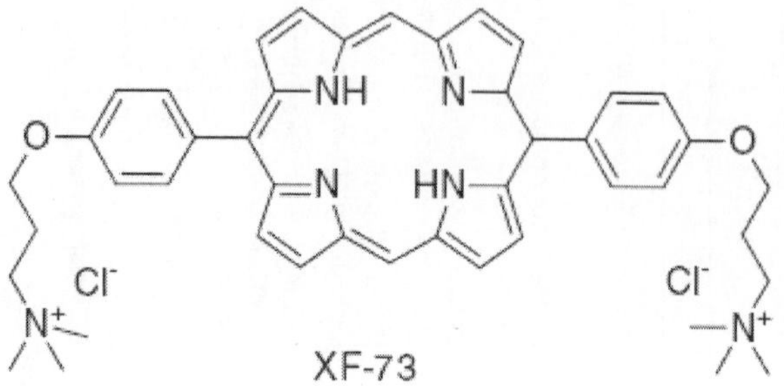

XF-73

(I)

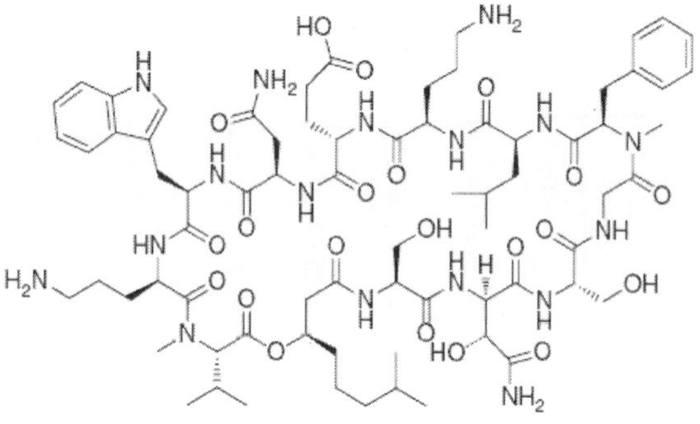

Iotilibcin

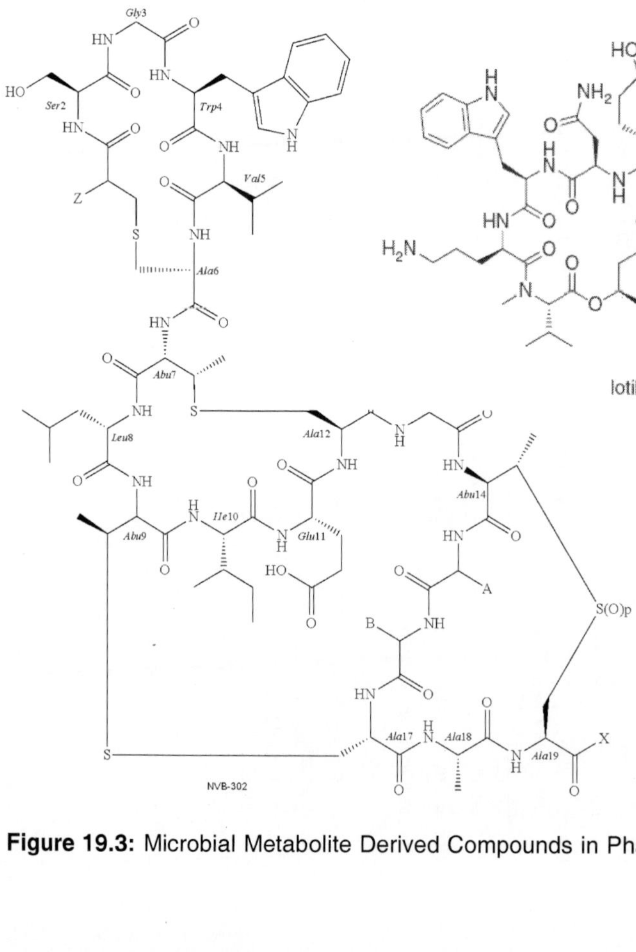

NVB-302

Figure 19.3: Microbial Metabolite Derived Compounds in Phase-I Clinical Trails.

Figure 19.4: Microbial Metabolite Derived Compounds in Phase-II Clinical Trails.

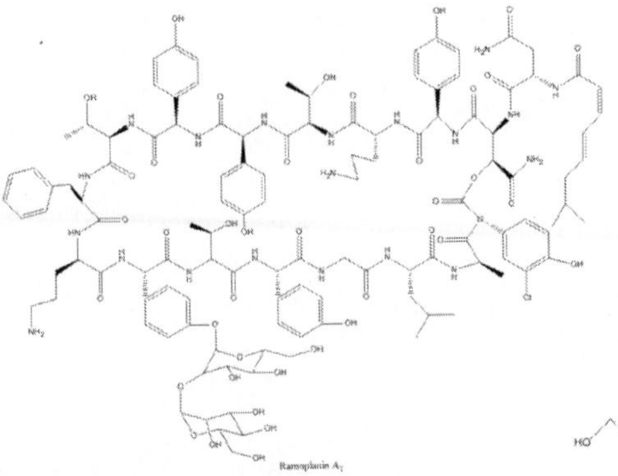

Ramoplanin A_2

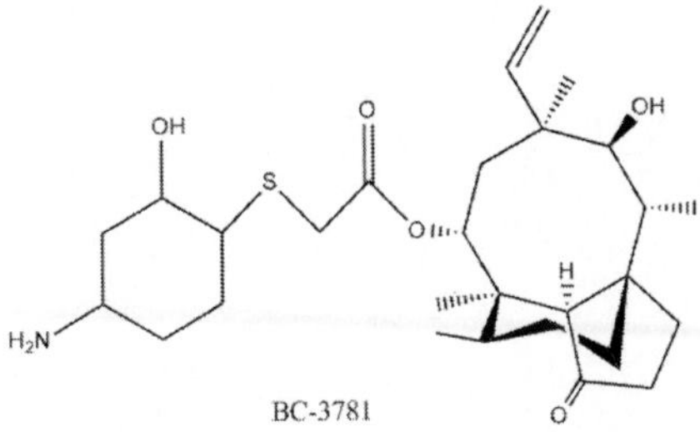

BC-3781

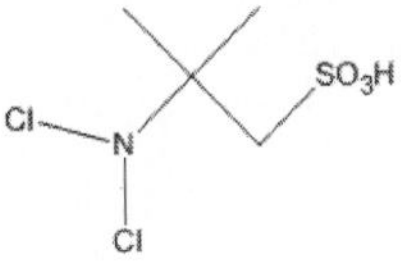

Auriclosene

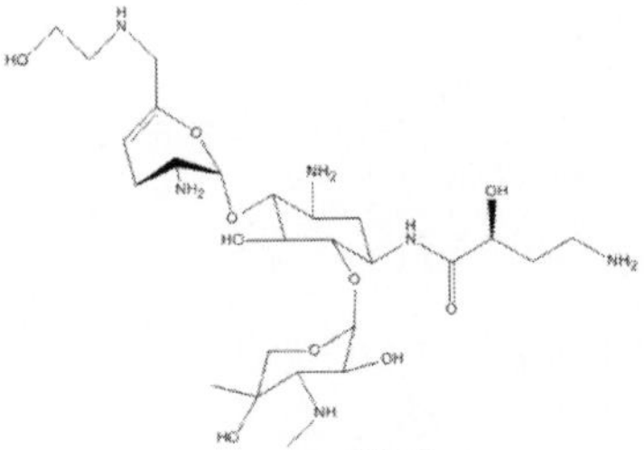

ACHN-490

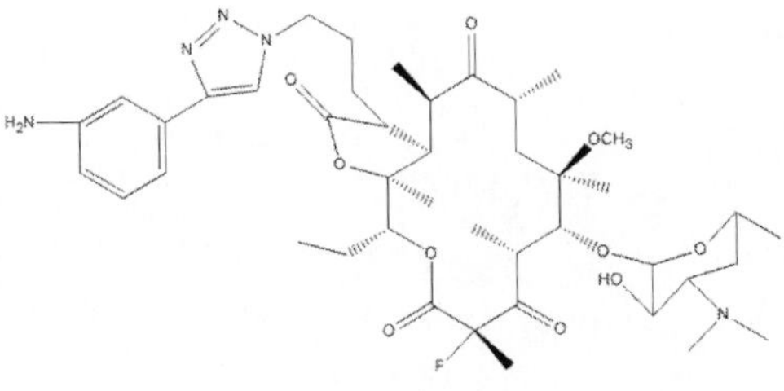

Solithromycin

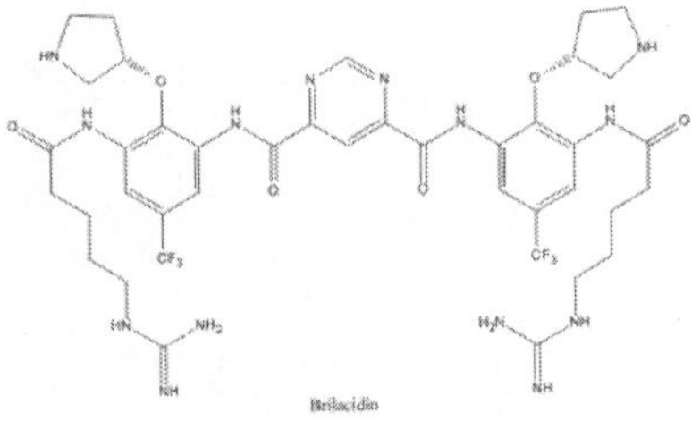

Brilacidin

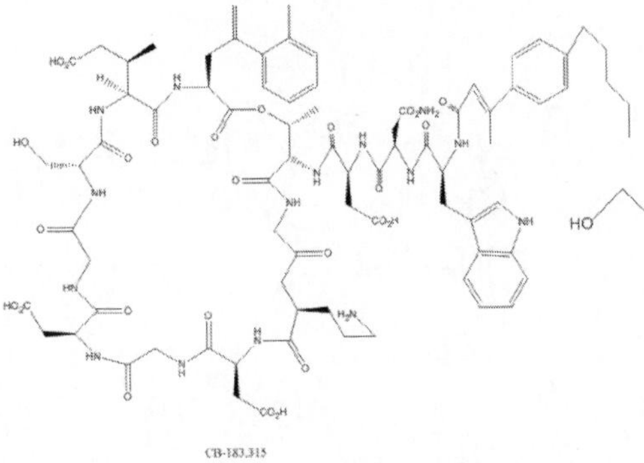

CB-183,315

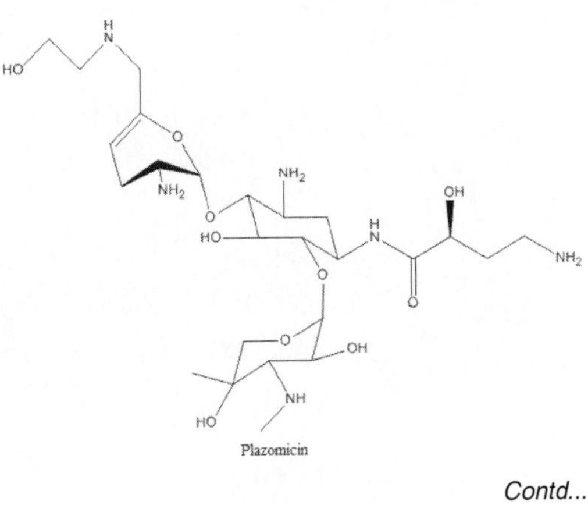

Plazomicin

Contd...

Figure 19.4–*Contd...*

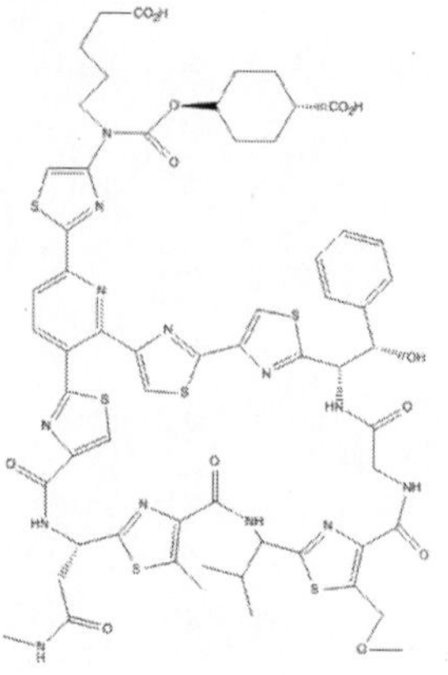

LFF-571

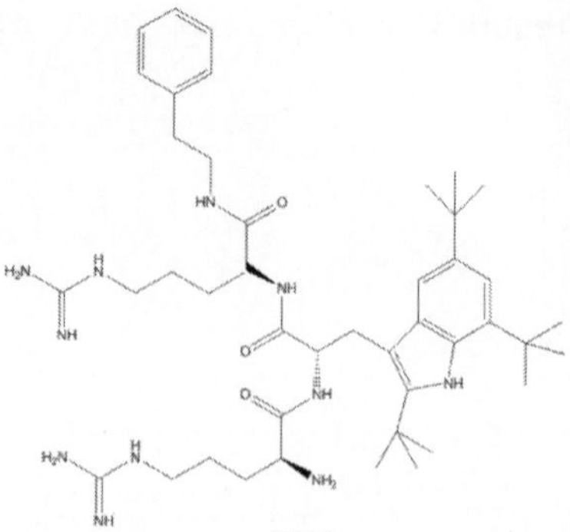

LTX-109

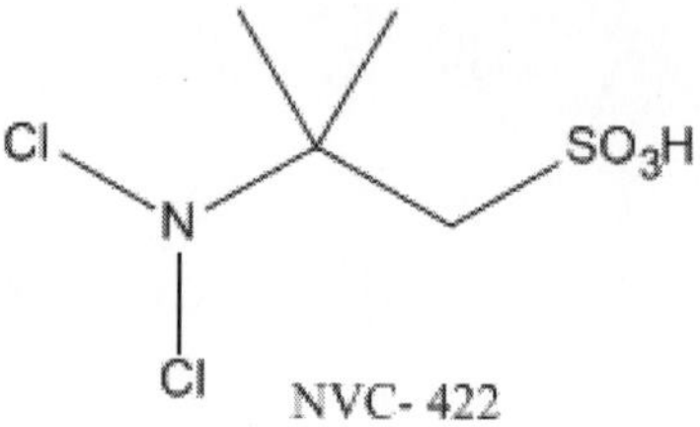

NVC- 422

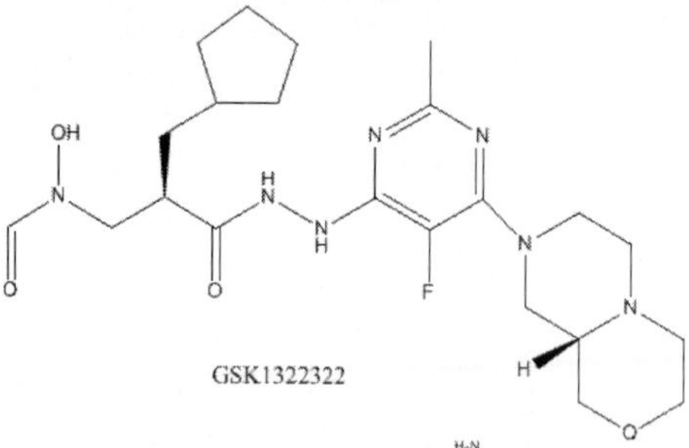

GSK1322322

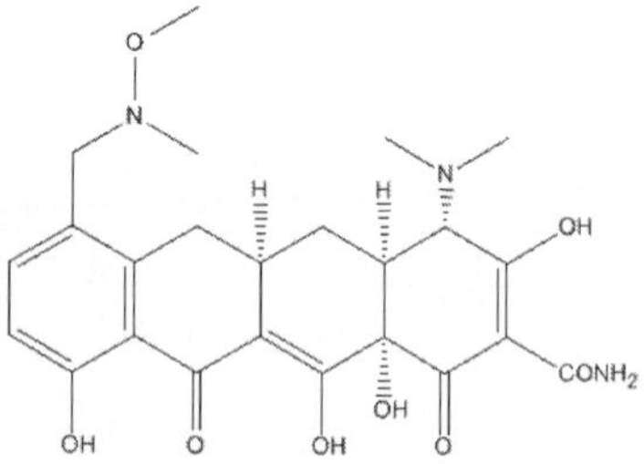

Sarecycline

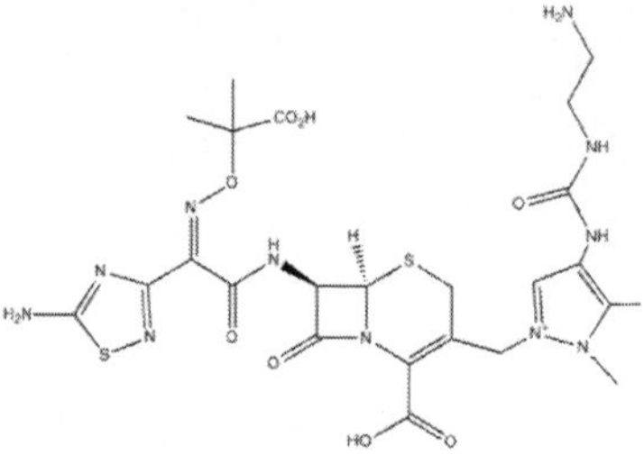

CXA-101

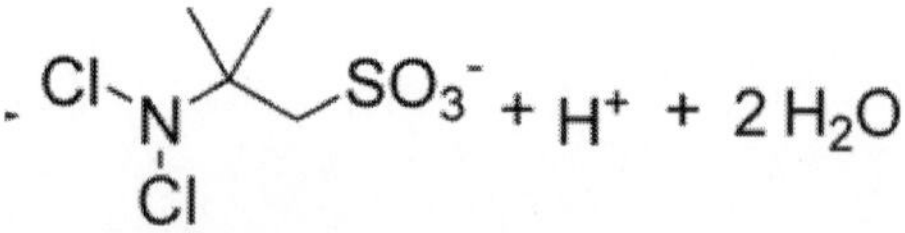

$\cdot$ Cl–N(Cl)–C(CH$_3$)$_2$–CH$_2$–SO_3^- + H^+ + 2 H_2O

NVC-422

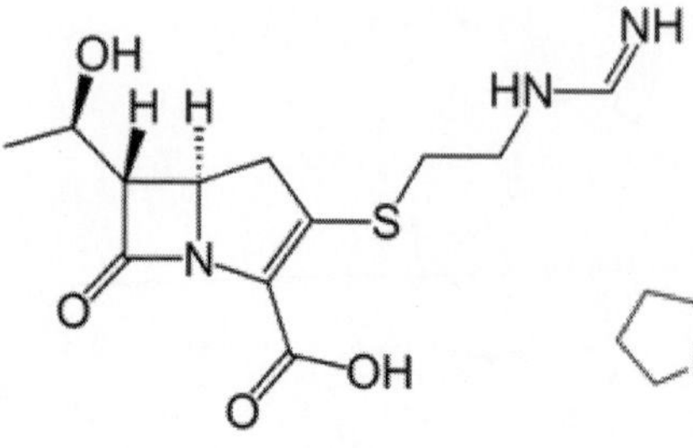

Imipenem

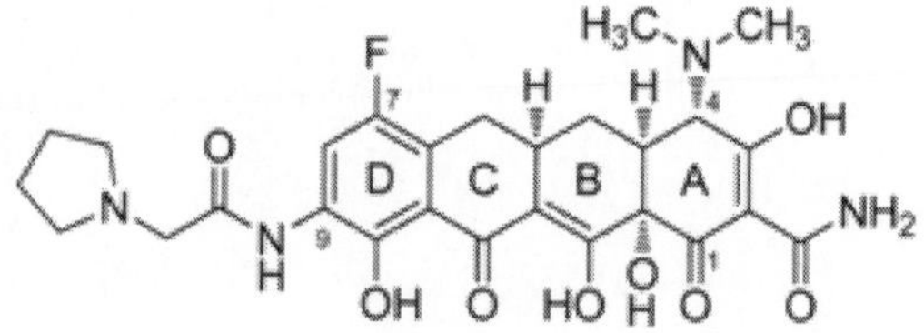

TP-434

Figure 19.5: Microbial Metabolite Derived Compounds in Phase III Clinical Trials or NDA or MAA.

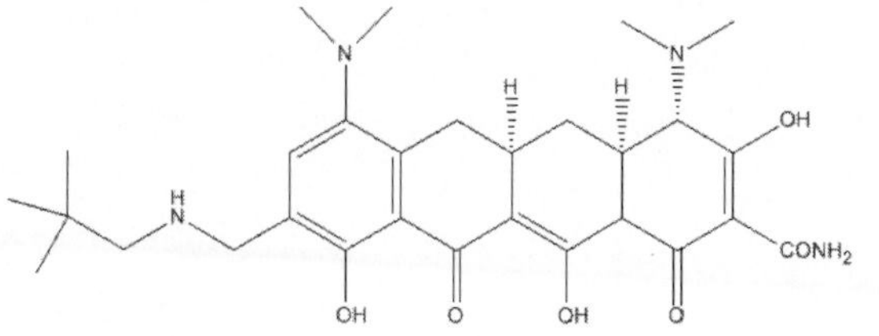

Omadacycline

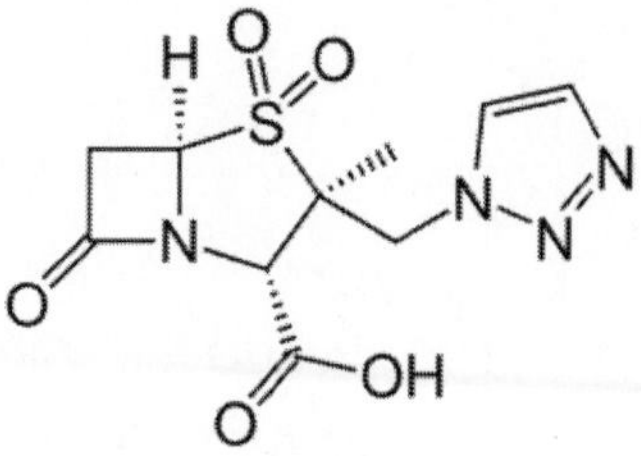

CXA-201

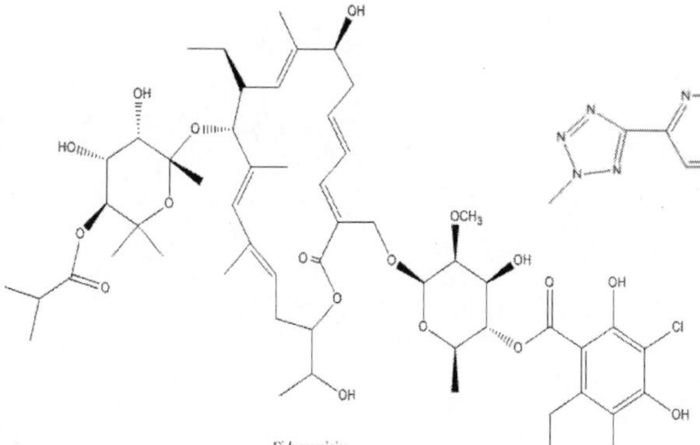

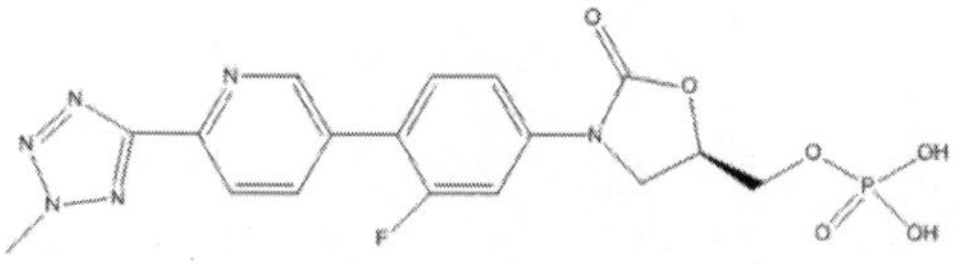

Torezolid phosphate

Fidaxomicin

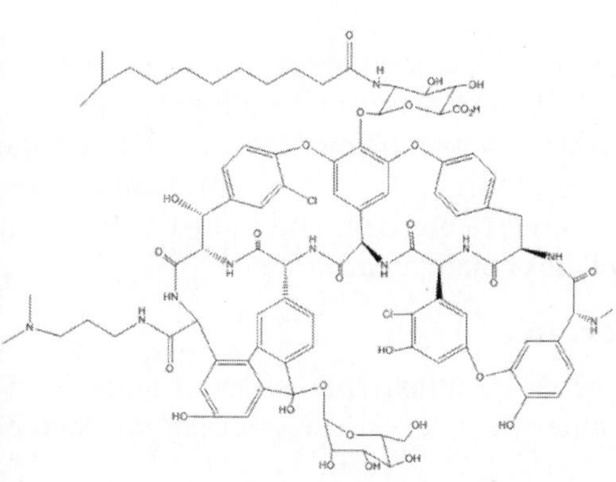

Dalbavancin

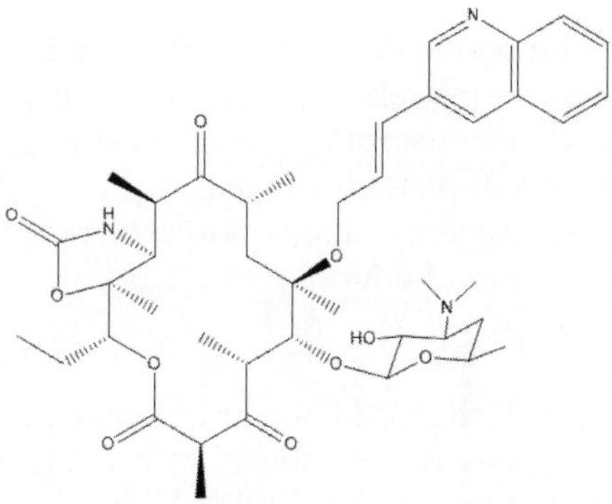

Cethromycin

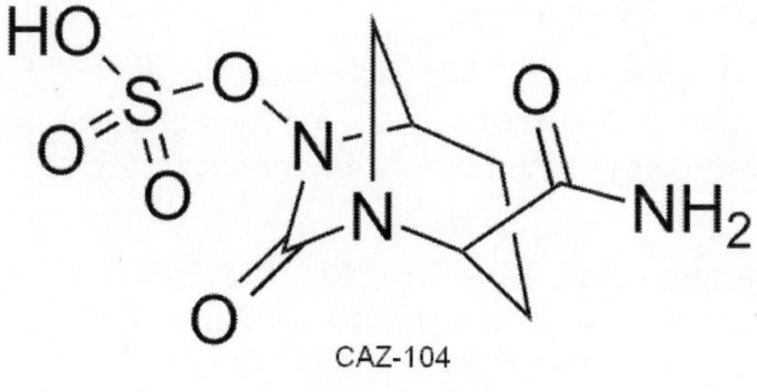

CAZ-104

Contd...

Figure 19.5–*Contd...*

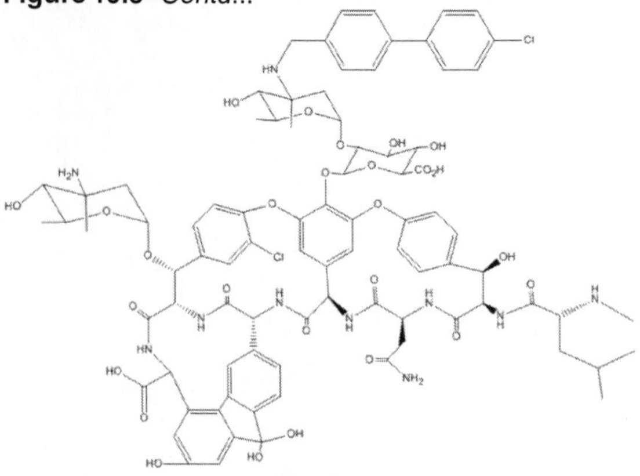

Oritavancin

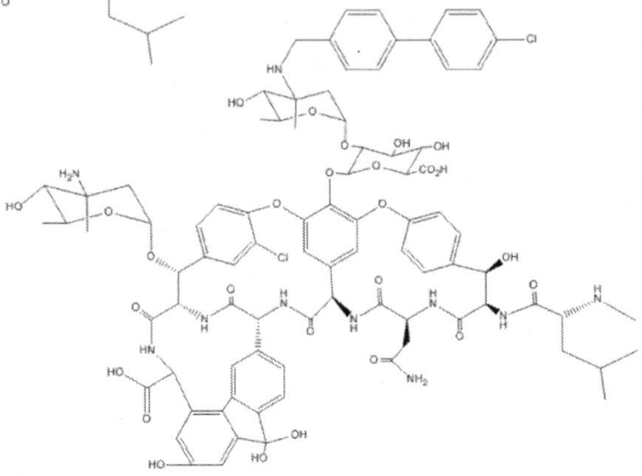

Oritavancin

compounds that are currently in clinical developmental phases. Our endeavor to explore microbial sources shall continue for better bactericidal antibiotics, enzymes and other useful chemicals through innovative screening methods, and knowledge based approaches including rational drug design for the development of new generation therapeutics to treat newly emerged superbugs and industrially useful enzymes, chemicals for better health and life of mankind at large.

References

Abo-Shehada, M.N., and Herbert, I.V. (1984). Anthelmintic effect of levamisole, ivermectin, albendazole and fenbendazole on larval *Toxocara canis* infection in mice. *Res Vet Sci.,* **36**: 87-91.

Aggen, J.B. (2010). Synthesis and spectrum of the neoglycoside ACHN-490. *Antimicrob. Agents Chemother.,* **54**: 4636-4642.

Akbergenov, R., Shcherbakov, D., Matt, T., Duscha, S., Meyer, M., Wilson, D.N., and Böttger E.C. (2011). Molecular basis for the selectivity of antituberculosis compounds capreomycin and viomycin. *Antimicrob Agents Chemother.* **55**: 4712-4717.

Aramori, I., Kukagawa, M., Tsumura, M., and Imanaka, H. (1991). A new cephalosporin C acylase. *J. Ferment. Bioeng.,***72**: 232-243.

Ball, A.S., and McCarthy, A.J. (1988). Degradation of lignin-related compounds by actinomycetes. *J. Gen. Microbiol.*, **134**: 2139-2147.

Berdy, J. (1988). The discovery of new bioactive microbial metabolites: Screening and Identiification, 2nd International Symposium on new bioactive metabolites from microorganisms Ger. GDR.

Bhate, D.S., Hulyalkar, R.K., and Menon, S.K. (1960). Isolation of iso-butyrrothine along with thiolutin and aureothricin from a *Streptomyces* species. *Experentia*, **16**: 504-505.

Biedenbach, D.J., Jones, R.N., Ivenzic, S.Z., Paukner, S., and Novak, R. (2009). *In-vitro* antibacterial spectrum of BC-7013,a novel pleumutilin derivative for topical human use in humans. *49th Interscience conference on Antimicrobial Agents and Chemotherapy Conference (San Francisco, CA)* Poster F1-1521.

Borkar, P.S., Thandani, S.B., and Ramachandran, S. (1978). 6-Aminopenillanic Acid by Immobilized Penicillin Acylase. *Hind. Antibiot.*, **30**: 81-91.

Busse, H.J., Wostmann, C., and Bakker, E.P. (1992). The bactericidal action of streptomycin: membrane permeabilization caused by the insertion of mistranslated proteins into the cytoplasmic membrane of *Escherichia coli* and subsequent caging of the antibiotic inside the cells due to degradation of these proteins. *J. Gen Micobiol.*, **138**: 551-559.

Butler, M.S., and Buss, A.D. (2006). Natural Products-the future scaffolds for novel antibiotics. *Biochem. Pharmacol.*, **71**: 919-929.

Butler, M.S., and Cooper, M.A. (2011). Antibiotics in the clinical pipeline in 2011. *J. Antibiot.*, **64**: 1-13.

Butler, M.S., and Cooper, M.A. (2013). Antibiotics in the clinical pipeline in 2013. *J. Antibiot.*, **66**: 571-591.

Carpenter, C.F., and Chamber, H.F. (2004). Daptomycin: Another novel agent for treating infections due to drug resistant gram-positive pathogens. *Clin. Infect Dis.*, **38**: 994.

Casewell, M.W., and Hill, R.L. (1983). Pseudomonic acid and control of methicillin-resistant staphylococci. *Lancet*, **2(8362)**: 1312.

Cavalleri, B., Pagani, H., Volpe, G., Selva, E., and Parenti, F. (1984). A16686, a new antibiotic from Actinoplanes I Fermentation, isolation and preliminary physico-chemical characteristics. *J. Antibiot.*, **37**: 309-317.

Chambers, H.E., and Deck, D.H. (2009). Aminoglycosides and spectinomycin: In Basic and Clinical Phamacology, 11th Edition, Katzung, B.G., Masters S.B. and Trevor, A.J. (Editors), Tata McGraw Hill Education Pvt. Ltd, pp 807-814.

Chang, C.J., Floss, H.G., Soong, P., and Chang, C. (1975). Identity of antitumor antibiotic litmomyciin with granaticin A. *J. Antibiot.*, **28**: 156-160.

Chopra, I., Hawkey, P.M., and Hinton, M. (1992). Tetracyclines, molecular and clinical aspects. *J. Antimicrob. Chemother.*, **29**: 245-277.

Craig, W.A., Andes, D., Ivenzic, S.Z., Wicha, W.W., and Novak, R. (2009). *In-vivo* pharmacodynamic activity of BC-3205,a novel pleumutilin derivative. *49th Interscience conference on Antimicrobial Agents and Chemotherapy Conference (San Francisco, CA)* Poster F1-1504.

Dhuley, J.N., and Naik, S.R. (1998). Modulation of rifampicin toxicity by 6MFA, an interferon inducer obtained from fungus *Aspergillus ocraceus. Environ. Toxicol and Pharmacol.,***5**: 237-243.

Dhuley, J.N., and Naik, S.R. (1999). Moderate effect of 6MFA a microbial metaboliteobtained from *Aspergillusochraceus* on immunological liver injury in mice. *Comp Immuno Microbiol Infect.,* **22**: 93-101.

Dhuley, J.N., and Naik, S.R. (2000). Increased γ-glutamyl transpeptidase activity during inflammation and its inhibition by 6-MFA, a metabolite obtained from *Aspergillus ochraceus. Pharm Pharmacol Commun*, **6**: 465-469.

Donowitz, G.R., and Mandell, G.L. (1988). Beta lactam antibiotics (Two parts). *N Eng. J. Medi.,* **318**: 419-423, 490-494.

Edson, R.S., and Terrell, C.L. (1987). The aminoglycosides: Streptomycin, kanamycin, gentamycin, tobramycin, aamikacin, netilmicin and sisomicin. *Mayo Clin Proc.,* **62**: 916-928.

Egerton, J.R., Ostlind, D.A., Blair, L.S., Eary, D.H., Cifelli, S., Riek, R.F., and Campbell, W.C. (1979). Avermectins, new family of potent anthelmintic agents: efficacy of the Bla components. *Antimicrob. Agents. Chemother.,* **15**: 372-378.

Feider, H.M (Jr)., Osier, C., and Maderazo, *E.G.* (1981). Chloramphenicol: A review of its use in clinical practice. *Rev Infect Dis.,* **3**: 479-482.

Felmingham D. (2005). Tigecyclin–the first glycylcycline to undergo clinical development: an overview of in vitro activity compared to tetracycline. *J. Chemother.,* **Suppl 1**: 5-11.

Fleming A. (1929). On the antibacterial action of cultures of a penicillium, with special reference to their use in the isolation of B. influenzæ. *British J Exp Pathol.,* **31**: 226–36.

Furue, H., Ito, I., Kimura, T., Kondo, T., Ogawa, N., and Taguchi, T. (1981). Phase III study on lentinan: randomized control study of lentinan on patients with advanced or recurrent gastric or colo-rectal cancer. *Gan to kagakuryoho (Cancer and Chemotherapy),* **8**: 944-946.

Garattini, S, Costa, V., Murelli, B., Palma, V., and Vegeto, N. (1956). Research on the antitumor effects of actinomycin D. *G Ital Chemioter.,* **3**: 498-511.

Georgopapadakou, N.H., Smith, S.A., and Sykes, R.B. (1982). Mode of action of azthreonam. *Antimicrob Agents Chemother.,* **21**: 950-956.

Goldberg, L.T., and Margalit, J.A. (1977). A bactericidal spore demonstrating rapid larvicidal activity against *Anapheles sergentii, Uranotaenia uniquiculata, Culex univattus, Aedes aegypti, Culex pipiens, Culex univattus. Mosq. News,* **41**: 446-459.

Gottardi, W., and Nagi, M. (2010). N-Chlorotaurine, a natural antiseptic with outstanding tolerability. *J. Antimicrob. Chemother.*, **65**: 399-409.

Guskey, M.T., and Tsuji, B. (2010). A comparative review of lipoglycopeptides: Oritavancin, Dalbavancin and Telavancin. *Pharmacotherapy*, **30**: 80-94.

Hammerschlag, M.R., and Sharma, R. (2008). Use of Cethromycin, a new ketolide, for treatment of community-aquired respiratory infections. *Expert Opin, Invest Drugs.*, **17**: 387-400.

Harold, E.M., and Baarda, J.R. (1967). Gramicidin and valinomycin and cation permeability of *Streptococcuc faecalis. J. Bacteiol.*, **94**: 53-60.

Henry, F., and Chambers, M.D. (2009). Tetracyclines, Macrolides, Clindamycin, Chloramphenicol, Streptogramins and oxazolidinones. In Lange medical book, Basic and Clinical Pharmacology, Edited by B.G. Katzung, susan B Masters and A.J. Trevor, Tata McGraw Hill Education Pvt Ltd. pp.795-804.

Henry, F., Hadley, W.K., and Jawetz, E. (1998). Aminoglycosides and Spectinomycin-In Lange Medical Book, Basic and clinical Pharmacology Edited by B.G.Katzung, *MacGraw-Hill Company*, pp.752-760.

Hunter, F.E., and Schwatz, L.S. (1989). Valinomycin In antibiotics. D.Gotleib and P.D.Shaw (Eds) Vol I, Berlin, New York, Springer 967, pp. 638-641.

Hunt, D. *et al.* (2010). TP-434 is a novel broad spectrum fluorocycline. *50th Interscience conference on Antimicrobial Agents and Chemotherapy Conference (Boston MA)* Poster F1-2157.

Johnson AP. (2001). Telithromycin. Aventis Pharma. *Curr Opin Investig Drugs*, **2(12)**: 1691-1701.

Kanda, N. (1971). A new antitumor antibiotic, kidamycin: Isolation, Purification and properties of kidamycin. *J. Antibiot.*, **25**: 592-599.

Kato, A. (1998). A new anti-MSRA antibiotic complex WAP-829A, I. taxonomy, isolation and biological activities. *J. Antibiot.*, **51**: 929-935.

Kaye, D. (2004). Current use of old antibacterial agents: Polymyxins, Rifampin and aminoglycosides. *Infect Dis Clin North Am.*, **18**: 669-689.

Kellen, W.R., Clark, T.B., Lingden, J.E., and Singer, S.Y. (1965). *Bacillus sphaericus* Neide, as a pathogen of mosquitoes. *J. Invertebr. Pathol.*,**7**: 442-448.

Kino, T., Hatanaka, H., Hashimoto, M., Nishiyama, M., Goto, T., aoki, H., and Imanaka, H. (1987). FK-506, a novel immunosuppressant isolated from a *Streptomyces* : Fermentation, isolation and physico-chemical and biological characteristics. *J. Antibiot.*, **40**: 1249-1255.

Kumar, K.K., Sudhakaran, V.K., Deshpande, B.S., Ambedkar, S.S., and Shewale, J.G. (1993). Cephalosporins acylases: Enzymes production, structure and application in the production of 7ACA. *H. Antibiot. Bull.*, **35**: 111-125.

Ladjama, A., Chardon-Loriaux, I., and Foglietti. (1991). On the pectolytic activity of two streptomyces strains. *FEMS Microbiol. Lett.*,**79**: 279-283.

Lewandowski, T., Demarsh, P., Peters, T., and Kulkarni, S. (2010). Potent activity of GSK1322322, a novel peptide deformylase inhibitor after oral dosing in a murine multi-drug resistant *Staphylococcus aureus*Infection. *50th Interscience conference on Antimicrobial Agents and Chemotherapy Conference (Boston MA)* Poster F1-2114.

Livermore, D.M. (2009). Has the era of untreatable infections arrived? *J. Antimicrob. Chemother ap.*, **64**: 29-36.

Maheshwari, R.K., and Gupta, B.M. (1973). A new antiviral agent designated 6MFA from *Aspergillus flavus*: production and assay of 6MFA against semiliki forest virus infection in mice. *Antibiotics*, **26**: 320-327.

Malabarba, A., and Goldstein, B.P. (2005). Origin, structure and activity *in-vitro* and *in-vivo* of Dalbavancin. *J. Antimicrob. Chemother.*, **55**: 1115-1120.

Maladkar, N.K., Singh, V.K., and Naik, S.R. (1993). Fermentative production and Isolation of α-Asparaginase from *Ervinacarotovara*, EC-1137. *Hind. Antibiot. Bull.*, **35**: 77-86.

McMorris, T.C. (1978). Antheridiol and the oogoniol, steroid harmones which control sexual production in Achlya. *Philosophical Transactions of the Royal Society of London Series B*, **248**: 459-470.

Miyoshi, T., Miyairi, N., Aoki, H., Sakai, M., and Imanaker, H. (1972). Bicyclomycin, a new antibiotic I Taxonomy, isolation and characterization. *J. Antibiot.*, **25**: 569-575.

Mizushima, Y.N., Yuhki, N., Hosokania, M., and Kobayshi, H. (1982). Diminution of cyclophosphamide induced suppression of antitumor immunity by an immunomodulator PS-K and combined therapeutic effect of PS-K and cyclophosphamide on transplanted tumors in rats. *Cancer Research*, **42**: 5176-5180.

Moellering, R.C. (2010). The problem of complicated skin and skin structure infections: the need for new agents. *J. Antimicrob. Chemother.* **65**: iv3-iv6.

Morelon, E., Mamzer-Bruneel, M., Peraldi, M., and Kreis, H. (2001). Sirolimus: a new promising immunosuppressive drug. Towards a rationale for its use in renal transplantation. *Nephrol Dial Transplant.*, 16: 18-20.

Muley, R.G., Sarkar, S., Bhar, B.L., Narayanan, M., and Naik, S.R. (1997). Scale-up Approach for fermentation process of production of *Bacillus thuringiensis* var *kenyae* based pesticide. *J. Sci. and Ind. Res.*, **58**: 781-784.

Naik, S.R. (1994). Historical Prospectives and Development of 'The art of fermentation". *The Eaastern Pharmacist*, 44-48.

Naik, S.R. (1997). Biocides, *Bacillus sphaericus*and *Bacillus thuringiensis* as potential Mosquito Larvicidal. *J. Sci. and Ind. Res.*,**56**: 651-656.

Naik, S.R., Harindron, J.,and Varde, A.B. (2001). Pimprinine, an extracellular alkaloid produced by *Streptomyces* CDRLL-312: Fermentation,isolation and Pharmacological activity. *J.Biotechnol.*, **88**: 1-10.

Nakamura, S., Takeuchi,T., Hori, S., Matsuzaki, M., and Umezawa, H. (1972). Phenomycin,toxicity and distribution. *J Antibiot.*, **24**: 197-199.

Neiderberger, P. (1989). Amino acid production in Microbial Eukaryotes and Prokaryotes other than Coryneforms, "In Microbial Products, New Approaches". *Cambridge University Press, N.Y. 1022 USA*, pp: 1-24.

Neu, H.C., and Fu, K.P. (1978). Clavulanic acid, a novel inhibitor of beta-lactamases. *Antimicrob Agents Chemother.*, **14**: 650-655.

Nickell, L.G. (1982). Plant growth Regulators: Agricultural uses. Springer-Verlay, Berlin Heidelberg New York,1982, pp: 142-146.

Nisbet, L.J., and Porter, N. (1989). The impact of pharmacology and molecular biology on the exploitation of microbial products. In "Microbial Products: New Approaches", Edited by Baumberg, S., Hunter, I. and Rhodes, M. *Cambridge University Press Cambridge 1989*, pp. 309-342.

Novak, R., and Shlaes, D.M. (2010). The Pleuromutilin antibiotics: a new class for human use. *CurrOpin Invest Drugs*,**11**: 182-191.

Núñez, L.E., Nybo, S.E., González-Sabín, J., Pérez, M., Menéndez, N., Braña, A.F., Shaaban, K.A., He, M., Morís, F., Salas, J.A., Rohr, J., and Méndez, C. (2012). A novel mithramycin analogue with high antitumor activity and less toxicity generated by combinatorial biosynthesis. *J Med Chem.*, **55(12)**: 5813-5825.

Okura, A., Nakadaira, M., Naito, K., and Umezanta, H. (1986). Effect of forphenical-o--interferon production in mice sensitized with BCG. *J. Anibiot.*, **39**: 569-574.

Ooi, N. (2009). XF-73, a novel antistaphylococcal membrane active agents with rapid bactericidal activity. *J. Antimicob. Chemother.*, **64**: 735-740.

Ooi, N. (2010). XF-70 and XF-73, novel antibacterial agents active against slow growing and non-dividing cultures of *Staphylococcus aureus* including biofilms. *J. Antimicro. Chemother.*, **65**: 72-78.

Page, M. G., Dantier, C., and Desarbe, E. (2010). *In-vitro* properties of BAL 30072, a novel sideophoresulfactam with activity against multiresistant Gram-negative bacilli. *Antimicrib. Agents Chemother.*, **54**: 2291-2302.

Pallanza, R., Berti, M., Scotti, R., Randisi, E., and Arioli, V. (1984). A-16686, a new antibiotic from Actinoplanes II biological properties. *J. Antibiot.*, **37**: 318-324.

Payne, M., and Dividson, E.W. (1993). Insecticidal activity of the crystalline parasporal inclusion and other components of *Bacillus sphaericus* spore complex. *J. InvertbrPathol.*, **43**: 383-388.

Pongs, O. (1979). Chloramphenicol. In: Antibiotics Mechanism of action of Antibacterial Agents, Part-I, Vol.-V (Eds) F.E. Hans Springer-Verlay, Berlin Heidelberg New York, 1979, pp26-42.

Projan, S.J. (2008). Whither antibacterial drug discovery. *Drug Discov, Today*, **13**: 279-280.

Rafie, S., MacDougall, C., and James, C.L. (2010). Cethomycin, a promising new ketolide antibiotic for respiratory infections. *Pharmacotherapy,* **30**: 290-303.

Rice, L.B. (2008). Federal funding for the study of antimicrobial resistance in nosocomial pathogens: no ESKAPE. *J. Infect. Dis.*, **197**: 1079-1081.

Ryan, M.J.(1979). Novobiocin and Coumermycin A1, In Antibiotics Mechanism of action of Antibacterial Agents, Part-I, Vol.-V (Eds) F.E. Hans Springer-Verlay, Berlin Heidelberg New York, 1979, 214-234.

Saito, H., and Tomioka, H. (1979). Enhanced hydrogen peroxide release from macrophages stimulated with streptococcal preparation OK-432. *Infection and Immunity*, **26**: 779-782.

Saito, M., Ebina, T., Koe, M., Kawale, Y., and Ishida, N. (1982). Induction of interferon on mouse spleen cells by OK432, a preparation of *Streptomyces pyogenes. Cellular Immunology*, **68**: 187-192.

Salmon, S.E., and Sartorelli, A.C. (1998). Cancer Chemotherapy In "Basic and Clinical Pharmacology" 7th Edition by Katzung B.G. (Editor), McGraw-Hill Companies. pp.881-915.

Seetalaxman, R., and Srinivasan, M.C. (1993). Actinomycetes as resource pools for Industrially useful Enzymes: An overview. *Hind. Antibiot. Bull.*, **35**: 23-32.

Shewale, J.G., and Naik, S.R. (1991). Immobilized enzymes-applications. *J. Ind. Sci. and Ind. Res.*, **50**: 947-958.

Shibuya, K., Hayashi, E., Abe, F., Takahashi, K., Takeuchi, T., and Imezawa, H. (1987). Enhancement of the interleukin-I and interleukin-II releases by ubenimex. *J. Antibiot.*, **40**: 363-369.

Silver, L.L. (2005). A retrospective on the failure and successes of antibacterial drug discovery. *IDrugs*, **8**: 651-655.

Snook, C. F., G. A. Woolley, G. Oliva, V. Pattabhi, S. P. Wood, T. L. Blundell, and B. A. Wallace. (1998). The structure and function of antiamoebin I, a proline-rich membrane-active polypeptide. *Structure*, **6**: 783–792.

Steigbigel, N.H. (1995). Erthromycin, lincomycin and clindamycin. In: principles and Practice of Infectious Diseases, 4th Edition, G.L. Mandell, J.E. Bennett and Dolin R (Editors). Churchill Livingstone 1995

Storm, D.R., and Toscano (Jr), W.A.(1979). Bacitracin, In Antibiotics Mechanism of action of Antibacterial Agents, Part-I, Vol.-V (Eds) F.E. Hans Springer-Verlay, Berlin Heidelberg New York, 1979, pp 1-17.

Subramanian, S.S., Sivaraman, H., Ratnapakshi, R.R., and Sivaraman, C. (1978). Immobilized Penicillin amidohydrolase (EC.3.5.1.11) *Hind Antibiot.*, **20**: 74-80.

Sudhakaran, V.K., Despande, B.S., and Naik S.R. (1992). Production of 6APA using packed bed Batch reactor. *Biotech. Lett.*, **14**: 913-918.

Sunder, S.S. (2007). Injectable paramomycin for visceral leishmaniasis in India. *N.Engl.J.Med.*, **356**: 2571-2574.

Takeda, S., Nakai, T., Wakai, Y., Ikeda, F., and Hatano, K. (2007). *In vitro* and *in vivo* activities of a new cephalosporin, FR264105 against *Pseudomonas aeruginosa*. *Antimicrob. Agents Chemother.*, **51**: 826-830.

Talbot, G.H. (2008).What is the pipeline for Gram-negative pathogens? *Expert Rev. Anti Infect. Ther.*,**6**: 39-49.

Tanaka, N. (1979). Bicyclomycin, In Antibiotics Mechanism of action of Antibacterial Agents, Part-I, Vol.-V (Eds) F.E. Hans Springer-Verlay, Berlin Heidelberg New York, 1979, 228-235.

Tanaka, N. (1979). Phenomycin and Enomycin. In Antibiotics Mechanism of action of Antibacterial Agents, Part-I, Vol.-V (Eds) F.E. Hans Springer-Verlay, Berlin Heidelberg New York, 1979, 235-242.

Tascini, C., Ferranti, S., Messina, F., and Menichetti, F. (2000). *In-vitro* and *in-vivo* synergistic activity of colistin, rifampin, and amikacin against a multiresistant *Pseudomonas aeruginosa* isolate. *Clin Microbiol Infect*, **6(12)**: 690-701.

Thirumalachar, M.J., Bringi, N.V., Deshmukh, P.V., and Rahalkar, P.W. (1964). Antiprotozoin, a new antifungal- antiprotozoal antibiotic. *Hind. Antibiot. Bull.*, **7**: 24-29.

Thirumalachar, M.J. (1968). Antiamoebin, a new protozoal anthelmintic antibiotic Part-1: Production and Biological studies. *Hind Antibiot. Bull.*, **10**: 287-289.

Toda, A., *et al.* (2008). Synthesis and SAR of novel parenteral anti-pseudomonal cephalosporins: Discovery of FR 264205. *Bioorg Med Chem Lett.*, **18**: 4849-4852.

Twami, M., Nakayama, O., Terano, H., Koshaka, M., Oaki, H., and Imanaka, H. (1984). A new immuno-modulator Fr-900494: Taxonomy, fermentation, isolation, physic-chemical and Biological Characteristics. *J Antibiot.*, **40**: 612-622.

Umezawa, H., Imoto, M., Sawa, T., Uchida, T., Iinuma, H., Hamada, M., and Takeuchi, T. (1986). Studies on a new epidermal growth factor- receptor kinase inhibitor, erbstatin produced by MH435-LF3. *J.Antibiot.*, **38**: 1781-1789.

Vandamme, E.J. (1988). In Mou- Young, M (Ed) Bioreactor, immobilized Enzymes and cells: Fundamentals and Applications. *Elsevier Applied Science NY*, pp. 261.

Vergidis, P.I., and Falagas, M.E. (2008). New antibiotic agents for bloodstream infections. *Int. J. Antimicrob. Agents*, **32**: S60-S65.

Wallemacq, P.E., and Reding R. (1993). FK506 (Tacrolimus), A novel immuno-suppressant in organ transplantation: Clinical, biomedical, and analytical aspects. *Clin Chem.*, **39**: 2219-2228.

Wallhausser,K.H., Nesemann, G., Prave, P., and Steigler, A. (1965). Moenomycin, a new antibiotics: Fermentation and isolation. *Antimicrob Agents Chemother.*, **64**: 734-736.

Waksman, S.A., Bugie, E., and Schatz, A. (1944). Isolation of antibiotic substances from soil microorganisms with special reference to streptothricin and streptomycin, Proc. *Staff Meetings Mayo Clinic*, **19**: 537-548.

Walsh, C. (2003). Antibiotics: Actions, Origins, Resistance (Asm Press: Washington, DC, USA, 2003).

Wang, Y., Castaner, R., Bolos, J., and Estivill, C. (2009). Amadacycline: tetracycline antibiotic. *Drugs Future,* **34**: 11-45.

Watanabe, Y., Tawara, S., Mine, Y., and Kikuchi, H. (1985). Immunoactive peptides FK156 and FK565.IV. Activation of mouse macrophages. *J Antibiot.,* **38**: 1781-1787.

Wicha, W.W., Ivezic, S.Z., and Novak, R. (2010). Efficacy of BC-3781 in murine pneumonia models. *50thInterscience conference on Antimicrobial Agents and Chemotherapy Conference (Boston MA)* Poster F1-207.

Wiley, P.F., and Mackellar, F.G. (1976). Sparomycin, structure and chemistry. *J.org. chem.,* **41**: 1858-1862.

Natural Products: Research Reviews Vol. 3 (2016) *Pages* **501–513**
Editor: **V.K. Gupta**
Published by: **DAYA PUBLISHING HOUSE, NEW DELHI**

20

Essential Oils and their Use in Skin Wound Healing

L.F.S. Gushiken[1], A.L. Rozza[1], A.J. Vieira[1], F.P. Beserra[2] and C.H. Pellizzon[1*]

ABSTRACT

Skin is a barrier that protects organisms against physical, chemical and biological agents as well as playing a role in thermoregulation and impermeability. If the skin integrity is disturbed, the process of wound healing is initiated, which involves molecular and cellular mechanisms in the inflammatory, proliferative and remodeling phases. Medicinal plants and other natural products are widely used in wound healing, especially in the pharmaceutical industry. Essential oils are secondary metabolites of aromatic plants, which have produced promising results in preclinical studies of wound healing. This manuscript reviews the use of essential oils in the skin wound healing process, with the aim of identifying the main mechanisms from which essential oils operate and promoting novel studies for plant-derived natural products that can be successfully used in the treatment of skin wounds.

***Keywords*:** Essential oil, Inflammation, Medicinal plants, Natural products, Re-epithelization, Skin wound healing, Terpenes.

Introduction

Skin is a barrier that covers the body surface and is formed by the epidermis, which is an epithelial barrier of ectodermal origin, and the dermis, which is the connective portion of mesoderm. Separating these two layers is a network of

1 Morphology Department, Biosciences Institute, UNESP – São Paulo State University, Botucatu/SP, Brazil.

2 Distrito de Rubião Junior, sn. P.O. Box 510, Zip Code 18618-970, Botucatu/SP, Brazil.

* *Corresponding author*: E-mail: claudia@ibb.unesp.br

extracellular matrix proteins known as the basal membrane. The epidermis has many layers comprised primarily of keratinocytes, melanocytes, Langerhans cells and Merkel cells. The dermis is formed by the reticular and papillary layers, with the predominance of extracellular matrix compounds such as collagen I, elastin (which promotes tissue tensile strength) and fibroblasts, which play a role in the organization of the dermis matrix (Evans *et al.*, 2013). Skin is the largest organ of human body and represents approximately 16 per cent of the total body weight. Skins serves as a vital barrier for vertebrates to protect the organism in many environments, including exposure to antigens, solvents, ultraviolet light, detergents, microorganisms, toxins, nanoparticles and local physical variations (Guo and Dipietro, 2010; Benedetto *et al.*, 2012). Moreover, the skin is impermeable to water and electrolytes because of the lipid covering produced by epidermal cells (Benedetto *et al.*, 2012).

Skin integrity is lost when an injury occurs that results in the opening or rupture of the skin, which is commonly referred to as an ulcer or lesion (Fikru *et al.*, 2012). This type of open wound can be caused by trauma, extreme heat or cold, radiation, corrosive materials or circulatory problems (Morin *et al.*, 2012). In response to this damage, the organism initiates the wound healing process, which is complex and involves a sequence of interdependent and overlapping stages: the inflammatory or exudative phase, the proliferative or reconstruction phase and the remodeling or epithelization phase. All of these stages depend on cellular, molecular and physiological processes for the structural and physiological restoration of the injured area (Martins *et al.*, 2006; Araújo, *et al.*, 2010; Agha *et al.*, 2011).

The inflammatory phase is the initial process of response after the formation of a lesion (0 to 3 days) in which the priority is the restoration of local hemostasis, with subsequent clearance of dead tissue and prevention of infection. At the edges of the wound, a hemostatic clot forms upon fibrin deposit to block blood efflux and promote platelet retention, which activates local vascular alterations (Gurtner, 2007; Profyris *et al.*, 2011). In this stage, the dilation of vessels occurs near the injured area due to the release of vasoactive amines (histamine and serotonin). There is also an increase in vessel permeability, which facilitates leukocyte migration to the injured region – a primary hallmark of the inflammatory process. Upon tissue destruction, keratinocytes in the region release interleukin 1 (IL-1) and tumor necrosis factor (TNF-α), which are essential cytokines to the inflammatory process that recruit and activate local leukocytes. Platelets release vesicles containing growth factors such as epidermal growth factor (EGF), platelet-derived growth factor (PDGF), transforming-β growth factor (TGF-β) and chemokines (Kondo and Ishida, 2010). During the first five days, neutrophils invade the lesion area, which contains abundant levels of fibrin, and promote the elimination of bacteria and degradation of dead tissue via phagocytosis and secretion of proteases. Three days after the lesion first forms, macrophages migrate to the region as part of the increasing inflammatory response, which is maintained by the presence of cellular debris and release of growth factors (FGF – fibroblast growth factor, EGF, TGF-β and PDGF), chemokines and pro-inflammatory cytokines (IL-1 and IL-6). These mitigating factors coordinate events that will manifest into the proliferative phase of wound healing (Profyris *et al.*, 2011).

The proliferative phase is characterized by the deposition of granulation tissue, which contains macrophages, fibroblasts and newly formed vessels. This phase lasts between 4 and 21 days and can be observed by the decrease in clot size (formed upon initial skin breakage). The primary mediators during this phase are PDGF, TGF-β1 and FGF, which determine fibroblast proliferation in the basal layer and initiate the synthesis of collagen III and other extracellular matrix proteins such as fibronectin and proteoglycans. ECM protein provides support for endothelial cell proliferation, which is mediated by release of VEGF, and aids in the formation of new vessels and organization of the granulation tissue (Kondo and Ishida, 2010; Profyris *et al.*, 2011). This re-epithelization process is important to establishment of the tissue integrity. The keratinocytes at the wound edges migrate towards the surface because of the cellular dedifferentiation and reorganization of adhesion molecules. The loss of adhesion with other keratinocytes and the basal membrane are critical in promoting wound closure (Profyris *et al.*, 2011).

The remodeling phase is an important process to avoid aberrations left by scarring, such as keloids. Granulation tissue formation is interrupted by cellular apoptosis, and collagen III is degraded by metalloproteinases released from macrophages, fibroblasts and endothelial cells to form collagen I, which has higher tensile strength and can better support new epithelium growth (Kondo and Ishida, 2010). Furthermore, the tissue receives a contractile response from myofibroblast activity because these cells connect to the collagen at multiple points, which facilitates the retraction of the wound area (Profyris *et al.*, 2011).

Due to the possibility of physical disability, psychosocial consequences or even death, skin injuries require significant attention in public health and would benefit from increased investments from public resources (Bayat *et al.*, 2003; Guo and Dipietro, 2010). An important aspect of wound healing by secondary intention is the decrease of the cellular response time to reconstruction of the tegmental system. For this, identifying elements that promote wound healing, such as the use of biological agents, is one of the oldest medicinal practices of humanity. Plants have been used for the treatment of diseases for thousands of years,and the utilization of natural products for the treatment of disorders has increased over the past few decades (Fabricant and Farnsworth, 2001; Veiga Jr *et al.*, 2005; Gurib-Fakim, 2006; Ganeshkumar *et al.*, 2012).

Essential oils are volatile, lipid-soluble secondary metabolites of aromatic plants that can be synthetized from all parts of the plant and are characterized by their strong odor. In nature, essential oils have important antibacterial, antifungal and antiviral activities, as well as insecticide activity, which can repel herbivores and attract insects for the dispersion of pollen and seeds. In addition to their use in the perfume industry for fragrances, essential oils have medicinal properties, including their ability to function as antimicrobial, analgesic, sedative, anti-inflammatory, spasmolytic and local anesthetic agents. They are comprised of a mixture of components and characterized by the major compounds that, generally, determine the biological properties of the essential oil. These compounds include two groups: terpenes and aromatic compounds. Terpenes are molecules that are formed by various combinations of isoprene $(C_5H_8)_n$ and are classified according to the number of carbons, including monoterpenes (10 carbons), sesquiterpenes (15 carbons) and diterpenes (20 carbons) among others. Monoterpenes ($C_{10}H_{16}$) and sesquiterpenes ($C_{15}H_{24}$) are the primary terpenes present in essential oils and are formed by two and three isoprene units, respectively. The aromatic compounds are molecules derived from phenylpropane and exist at a lower percentage in essential oils compared to terpenes (Bakkali *et al.*, 2008).

This review aims to highlight the role of essential oils in skin wound healing with the intention of promoting novel and extensive studies to identify improved treatments.

Materials and Methods

Searches for this review were performed using primary literature accessible from PubMed through April 2014. The words "essential oil", "wound healing" and "medicinal plants" were used in PubMed and Scopus searches. Moreover, we selected original articles written in English from the previous ten years without Ayurveda research. The results are expressed in Table 21.1 and discussed below.

Results and Discussion

Table 21.1 shows all of the plant species and their respective primary compounds used in skin wound healing studies described in the literature. Twenty-five species of plants belonging to eight different families were studied, and the parts of the plants used to obtain the essential oils were the leaves, trunk, cones and aerial parts.

Skin wound healing is a highly complex process that starts the moment a lesion manifests, which determines the partial or complete tissue repair and involves a series of interdependent and overlapping stages classified in three phases: inflammatory, proliferative and remodeling (Martins *et al.*, 2006).

Selection of the proper experimental model for skin wound healing studies is essential to analyze the biological mechanisms involved in the process. In the analyzed manuscripts, the most commonly implemented experimental models to quantitatively and qualitatively evaluate the healing process were excision, linear incision and wound dead space. The excision model is employed to determine lesion retraction in wound healing by secondary intention. A surgical "punch" is used to induce a lesion on the back of the animal, which allows for the measurement of daily wound retraction as well as molecular and histopathological analysis to determine the healing mechanisms during the different phases. The second model mentioned is performed by introducing paravertebral linear incisions equidistant of spinal column, which are subsequently sutured, to evaluate the primary closure. Incisions are analyzed with a densitometer to evaluate the tissue tensile strength. The wound dead space model is used to analyze the granulation tissue synthesis during the proliferative phase of wound healing. A cellulose sponge pellet is implanted under the skin in the back of animals and, after apre-determined number of days corresponding to the different phases of the healing process, the granulation tissue formed in the pellet is removed and weighed. Therefore, the skin excision and linear incision models are used to directly evaluate the three phases of skin wound healing, but the wound dead space model has emerged as an alternative to analyze the parameters of granulation tissue formation during the proliferative stage.

Table 20.1: Essential Oils Used in Skin Wound Healing Treatment

Plant Species	*Family*	*Part of the Plant*	*Majority Compound*	*Model*	*Mechanism of Action*	*Reference*
Cinnamomum zeylanicum	Lauraceae	leaves	Eugenol/trans-cinnamaldehyde	Excision	Antimicrobial/ re-epithelization	Ghosh *et al.*, 2013
Croton adamantinus	Euphorbiaceae	leaves	methyl-eugenol/ 1,8-cineole	Excision/wound dead space	Antimicrobial/ re-epithelization/collagen deposition/increase in fibroblast population	Ximenes *et al.*, 2013
Croton zehntneri	Euphorbiaceae	leaves	Trans-anethole	Excision	Increase in fibroblasts population and collagen fibers	Cavalcanti *et al.*, 2012
Eucalyptus globulus	Myrtaceae	n.i.	1,8-cineole (eucalyptol)	Excision	Anti-inflammatory/ antimicrobial	Sugumar *et al.*, 2014
Origanum majorana	Lamiaceae	n.i	Carvacrol	Linear incision/excision	Antimicrobial/anti-inflammatory/increase in connective tissue formation	Suntar *et al.*, 2011
Origanum minutiflorum	Lamiaceae	n.i	Carvacrol	Linear incision/excision	Antimicrobial/anti-inflammatory/increase in connective tissue formation	Suntar *et al.*, 2011
Salvia triloba	Lamiaceae	n.i	1,8-cineole	Linear incision/ excision	Antimicrobial/anti-inflammatory/increase in connective tissue formation	Suntar *et al.*, 2011
Juniperus virginiana	Cupressaceae	trunk	cedrol	Linear incision/excision	Ineffective	Tumen *et al.*, 2013

Contd...

Table 20.1–*Contd...*

Plant Species	*Family*	*Part of the Plant*	*Majority Compound*	*Model*	*Mechanism of Action*	*Reference*
Juniperus occidentalis	Cupressaceae	trunk	cedrol	Linear incision/excision	Re-epithelization/anti-inflammatory/increase in tensile strength/high hydroxyproline content	Tumen *et al.*, 2013
Juniperus ashei	Cupressaceae	trunk	thujopsene	Linear incision/excision	Ineffective	Tumen *et al.*, 2013
Lavandula x allardii	Lamiaceae	n.i	1,8-cineole	Excision	Ineffective	Lusby *et al.*, 2006
Lippia sidoides	Verbenaceae	leaves	thymol	Excision	Antimicrobial/amplify inflammatory response	Oliveira *et al.*, 2014
Murraya koenigii	Rutaceae	leaves	β-Caryophyllene	Excision	Ineffective	Nagappan *et al.*, 2012
Abies cilicica subsp. *Cilicica*	Pinaceae	Coniferous cones	n.i	Linear incision/excision	Increase in tensile strength/inhibition of capillary permeability	Tumen *et al.*, 2011
Abies nordmanniana subsp. equi-trojani	Pinaceae	Coniferous cones	n.i	Linear incision/excision	Re-epithelization	Tumen *et al.*, 2011
Abies nordmanniana subsp. *bornmulleriana*	Pinaceae	Coniferous cones	n.i	Linear incision/excision	Increase in tensile strength/re-epithelization	Tumen *et al.*, 2011
Abies nordmanniana subsp. *nordmanniana*	Pinaceae	Coniferous cones	n.i	Linear incision/excision	Re-epithelization	Tumen *et al.*, 2011
Cedrus libani	Pinaceae	Coniferous cones	n.i	Linear incision/excision	Increase in tensile strength/inhibition of capillary permeability/ re-epithelization	Tumen *et al.*, 2011
Picea orientalis	Pinaceae	Coniferous cones	n.i	Linear incision/excision	Ineffective	Tumen *et al.*, 2011
Pinus brutia	Pinaceae	Cones	n.i	Linear incision/excision	Ineffective	Süntar *et al.*, 2012

Contd...

Table 20.1–*Contd...*

Plant Species	*Family*	*Part of the Plant*	*Majority Compound*	*Model*	*Mechanism of Action*	*Reference*
Pinus halepensis	Pinaceae	Cones	α-Pinene	Linear incision/excision	Increase in tensile strength/ re-epithelization/ anti-hyaluronidase activity/high hydroxyproline content	Süntar *et al.*, 2012
Pinus nigra	Pinaceae	Cones	α-Pinene	Linear incision/excision	Inhibition of capillary permeability/ re-epithelization	Süntar *et al.*, 2012
Pinus pinea	Pinaceae	Cones	Limonene and β-pinene	Linear incision/excision	Increase in tensile strength/inhibition of capillary permeability/ re-epithelization/ anti-hyaluronidase activity/high hydroxyproline content	Süntar *et al.*, 2012
Pinus sylvestris	Pinaceae	Cones	α-Pinene	Linear incision/excision	Ineffective	Süntar *et al.*, 2012
Rosmarinus officinalis	Lamiaceae	Aerial parts	n.i	Excision	Increase in granulation tissue weight/increase in wound contraction	Abu-Al-Basal, M.A., 2010

n.i: not identified.

Essential oils are substances comprised of volatile and fat-soluble secondary metabolites, which can be extracted from different parts of the plant. Secondary metabolites present in essential oils can have biological activity on the healing mechanism by decreasing the retraction time and helping in wound healing. Many essential oils with reported healing activities have terpenes as their primary compounds, which are hydrocarbons whose chemical formula is derived from one or more isoprene molecules $(C_5H_8)_n$. Regarding the plant species identified in our search, the majority have either monoterpenes (a molecule derived from two isoprenes with chemical formula $C_{10}H_{16}$) or sesquiterpenes (a hydrocarbon with formula $C_{15}H_{24}$) as their primary compound, which suggests a possible relationship between this group of molecules and the mechanisms involved in skin wound healing. However, some species such as *Juniperus virginiana, Juniperus ashei* (Tumen *et al.*, 2013), *Lavandula allardii* (Lusby *et al.*, 2006), *Murraya koenigii* (Nagappan *et al.*, 2012) and *Pinus sylvestris* (Süntar *et al.*, 2011), which have the terpenes cedrol, thujopsene, 1,8-cineole, β-caryophyllene, and α-pinene, respectively, as their major compounds, had no healing activity. Therefore, it is possible that the terpene concentrations in the studied samples are a determining factor for its ability to promote healing. There is also the possibility of a synergistic effect between the different compounds in essential oils instead of only one major compound or isolated substances acting on the healing pathways. Furthermore, skin lesions treated with essential oils from *Cinnamomum zeylanicum* (Ghosh *et al.*, 2013) and *Croton adamantinus* (Ximenes *et al.*, 2013), which have the phenylpropanoids eugenol and methyl-eugenol, respectively, as their primary compounds, demonstrated wound healing activity. Phenylpropanoids are molecules consisting of phenolic compounds and derived from the aromatic amino acids phenylalanine and tyrosine; these organic compounds have been shown to exert antimicrobial and antioxidant activities and may help in the healing process (Korkina, 2007). Cavalcanti *et al.*(2012) showed that the animals treated with *Croton zehntneri* and trans-anethole essential oils presented similar results in wound healing, demonstrating a likely relationship between the primary compound present and the overall essential oil.

Skin wound healing involves molecular, cellular and physiological mechanisms that play a role in either partial or complete tissue repair (Agha *et al.*, 2011). The process may be divided into three interdependent and overlapping phases: the inflammatory, proliferative and remodeling phases. The inflammatory phase begins with the coagulation cascade to maintain hemostasis and promote of local inflammation; the proliferative phase is characterized by the deposition of granulation tissue (macrophages, fibroblasts and newly formed vessels), lesion contraction and re-epithelization; and the remodeling phase is marked by the replacement of collagen III for collagen I, which provides increased tensile strength of the tissue, complete epithelization and synthesis of the new dermis connective tissue (Rodero and Khosrotehrani, 2010; Kondo and Ishida, 2010). During the inflammatory phase after the lesion initially forms, the injured keratinocytes release IL-1 and TNF-α, which are cytokines responsible for the recruitment of neutrophils and macrophages to promote bacterial degradation and dead tissue clearance. The antimicrobial activity observed in essential oils from *Cinnamomum zeylanicum* (Ghosh *et al.*, 2013), *Croton adamantinus*

(Ximenes *et al.*, 2013), *Eucalyptus globulus* (Sugumar *et al.*, 2014), *Origanum majorana*, *Origanum minutiflorum*, *Salvia triloba* (Suntar *et al.*, 2011) and *Lippia sidoides* (Oliveira *et al.*, 2014) play an important role in wound healing, preventing infections and the consequent delay in the process. Inflammatory mediators such as IL-1 and TNF-α play a vital function in tissue reconstruction (Midwood *et al.*, 2004). Inflammation could cause destruction of the extracellular matrix, cell senescence and delays in wound healing (Oliveira *et al.*, 2014). Ahanger *et al.* (2010) showed that a decrease in TNF-α synthesis accelerated skin wound repair. Therefore, the anti-inflammatory activity of the essential oils from *Eucalyptus globulus* (Sugumar *et al.*, 2014), *Origanum majorana*, *Origanum minutiflorum*, *Salvia triloba* (Suntar *et al.*, 2011) and *Juniperus occidentalis* (Tumen *et al.*, 2013) enabled a decrease in the local edema and prevented loss of tissue function normally observed upon exacerbated inflammation. However, in cases of an attenuated inflammatory response after injury, there is a risk of infection risk due to the delayed recovery of damaged skin. There are studies demonstrating the positive modulatory effects of TNF-α and other inflammatory cytokines in healing as this pro-inflammatory cytokine synthesis influences the expression of other growth factors. Therefore, a decrease or complete abolishment of this cytokine synthesis process could impair the mechanisms involved in the proliferative and remodeling phases (Lim *et al.*, 2006). Oliveira *et al.*(2014) showed the beneficial effects amplifying inflammation in skin wounds treated with the essential oil from *Lippia sidoides* at 6 per cent and 12 per cent concentrations, suggesting the use of this oil as an alternative treatment for skin wound.

After initiation of fibrin clot synthesis and the inflammatory process, chemical mediators and growth factors such as PDGF, EGF, TGF-β1, FGF and VEGF determine specific alterations of the proliferative phase, including re-epithelization and granulation tissue synthesis (Kondo and Ishida, 2010). In the analyzed papers, the essential oils of some species from the*Pinaceae* family described by Süntar *et al.* (2011) and Tumen *et al.*(2011), as well as the *Cinnamomum zeylanicum* (Ghosh *et al.*, 2013), *Croton adamantinus* (Ximenes *et al.*, 2013) and *Juniperus occidentalis* (Tumen *et al.*, 2013) plant species, showed increased re-epithelization of the injured tissue, which occurs by the restructuring of the cell adhesion molecules near the wound edge keratinocytes. The loss of adhesion with other keratinocytes and basement membrane proteins allows for migration of these cells to aid in closing the lesion (Tziotzios *et al.*, 2012). Granulation tissue, which is comprised of macrophages, fibroblasts and newly formed vessels, plays an important role in skin wound healing because it initiates the synthesis of new extracellular matrix compounds (collagen III, proteoglycans and fibronectina) and provides support for the new tegument development (Kondo and Ishida, 2010; Tziotzios *et al.*, 2012). Subsequently, fibroblasts differentiate into myofibroblast, which are involved in the wound edge contractile response and consequently decrease the size of the injured area (Rodero and Khosrotehrani, 2010). Thus, substances that induced an increase in the number of cells in the granulation tissue could result in decreasing the healing time by promoting extracellular matrix compound synthesis, deposition of collagen fibers and retraction of the injured surface by myofibroblasts, as demonstrated by the essential oils from *Origanum majorana* (Süntar *et al.*, 2011), *Croton adamantinus* (Ximenes *et al.*, 2013) and *Croton zehntneri* (Cavalcanti *et al.*, 2012). Migration and cell proliferation are important during the proliferative phase of wound healing. Hyaluronic acid is one of the primary extracellular matrix constituents and helps in proliferation and cellular migration 6-mechanisms (Stern, 2004). Thus, compounds that have anti-hyaluronidase activity may contribute to wound healing process, such as the *Pinus pinea* and *Pinus halepensis* (Süntar *et al.*, 2012) and the *Juniperus occidentalis* essential oil (Tumen *et al.*, 2013).

During the remodeling phase, collagen III synthetized by fibroblasts is degraded by metalloproteinase and replaced with collagen I, which has greater tensile strength and can support the new epithelium (Kondo and Ishida, 2010). Collagen is the major component of the extracellular matrix, which contains the amino acid hydroxyproline–this component is critical in providing the tensile strength of collagen as well as aiding in the remodeling of injured tissue (Ganeshkumar *et al.*, 2012). Thus, hydroxyproline is used as biochemical marker for measuring collagen levels in a sample (Kumar *et al.*, 2006). Some essential oils from the *Pinaceae* family species (Süntar *et al.*, 2012) and *Juniperus occidentalis* (Tumen *et al.*, 2013) showed high hydroxyproline content, indicating higher collagen levels in the tissue samples. Increased levels of collagen, especially collagen I, correlate to improved tensile strength in the linear incision models as shown by essential oils of aforementioned species.

Conclusions and Future Directions

Most of the essential oils from the species studied in this review presented promising results in skin wound healing, by influencing the mechanisms involved in the inflammatory, proliferative and remodeling phases. Furthermore, among the studied species, the primary compounds classes identified were monoterpenes and sesquiterpenes, suggesting a possible healing effect of either these isolated compounds or a synergic effect among the secondary metabolites within the essential oils. These observations reinforce the importance of ethnopharmacological studies in skin wound healing and encourage the development of future studies in this field.

References

Abu-Al-Basal, A.M. (2010). Healing potential of *Rosmarinus officinalis* L. on full-thickness excision cutaneous wounds inalloxan-induced-diabetic BALB/c mice. *J Ethnopharmacol.*, **131**: 443–450.

Agha, R., Ogawa, R., Pietramaggiori, G., and Orgill, D.P. (2011). A Review of the role of mechanical forces in cutaneous wound healing. *Journal of Surgical Research*, **171**: 700-708.

Araújo, L.U., Grabe-Guimarães, A., Mosqueira, V.C.F., Carneiro, C.M., and Silva-Barcellos, N.M. (2010). Profile of wound healing process induced by allantoin. *Acta Cirúrgica Brasileira*, **25**: 460-466.

Bayat, A., Mcgrouther, D. A., and Ferguson, M. W. (2003). Skin scarring. BMJ., **326**: 88-92.

Bakkali, F., Averbeck, S., Averbeck, D., and Idaomar M. (2008). Biological eects of essential oils – A review. *Food and Chemical Toxicology*, **48**: 446-475.

Benedetto, A., Kubo, A., and Beck, L.A. (2012). Skin barrier disruption: A requirement for allergen sensitization? *Journal of Investigative Dermatology*, **132**: 949–963.

Cavalcanti, J.M., Leal-Cardoso, J.H., Diniz, L.R.L., Portella, V.G., Costa, C.O., Linard, C.F.B.M., Alves, K., Rocha, M.V.A.P., Lima, C.C., Cecatto, V.M., and Coelho-De-Souza, A.N. (2012). The essential oil of *Croton zehntneri* and trans-anethole improves cutaneous wound healing. *J Ethnopharmacol.*, **144**: 240–247.

Evans, N.D., Oreffo, R.O.C., Healy, E., Thurner, P.J., and Man, Y.H. (2013). Epithelial mechanobiology, skin wound healing, and the stem cell niche. *Journal of the Mechanical Behavior of Biomedical Materials*, **28**: 397-409.

Fabricant, D.S., and Farnsworth, N.R. (2001). The value of plants used in traditional medicine for drug discovery. *Environ Health Perspect.*, **109**: 69–75.

Fikru, A., Makonnen, E., Eguale, T., Debella, A., and Mekonnen, G.A. (2012). Evaluation of *in vivo* wound healing activity of methanol extract of *Achyranthes aspera* L. *Journal of Ethnopharmacology*, **143**: 469-474.

Ganeshkumar, M., Ponrasu, T., Krithika, R., Iyappan, K., Gayathri, V.S., and Suguna, L. (2012). Topical application of *Acalypha indica* accelerates rat cutaneous wound healing by up-regulating the expression of Type I and III collagen. *J Ethnopharmacol.*, **142**: 14-22.

Ghosh, V., Saranya, S., Mukherjee, A., and Chandrasekaran, N. (2013). Antibacterial microemulsion prevents sepsis and triggers healing of wound in Wistar rats. *Colloids and Surfaces B: Biointerfaces*, **105**: 152-157.

Guo, S., and Dipietro, L.A. (2010). Factors affecting wound healing. *J Dent Res.*, **89**: 219-229.

Gurib-Fakim, A. (2006). Medicinal plants: traditions of yesterday. *Mol. Aspects Med.*, **27**: 1-93.

Kondo, T. and Ishida, Y. (2010). Molecular pathology of wound healing. *Forensic Science International*, **203**: 93-98.

Korkina, L.G. (2007). Phenylpropanoids as naturally occurring antioxidants: from plant defense to human health. *Cellular and Molecular Biology*, **53**: 15-25.

Kumar, R., Katoch, S.S., and Sharma, S. (2006). Adrenoreceptor agonist treatment reverses denervation atrophy with augmentation of collagen proliferation in denervated mice gastrocnemius muscle. *Indian Journal of Experimental Biology*, **44**: 371–376.

Lim, Y., Levya, M.A., and Bray, T.M. (2006). Dietary supplementation of N-acetylcysteine enhances early inflammatory responses during cutaneous wound healing in protein malnourished mice. *The Journal of Nutritional Biochemistry*, **17**: 328-336.

Lusby, P.E., Coombes, A.L., and Wilkinson, J.M. (2006). A comparison of wound healing following treatment with lavandula x allardii honey or essential oil. *Phytother. Res.*, **20**: 755-757.

Martins, N.L.P., Malafaia, O., Ribas-Filha, J.M., Heibel, M., Baldez, R.N., Vasconcelos, P.R.L., Moreira, H., Mazza, M., and Nassif, P.A.N. (2006). Análise comparativa da cicatrização da pele com o uso intraperitoneal de extrato aquoso de *Orbignya phalerata* (babaçu). Estudo controlado em ratos. *Acta Cir Bras.*, **21**: 66-75.

Midwood, K.S., Williams, L.V., and Schwarzbauer, J.E. (2004). Tissue repair and the dynamics of the extracellular matrix. *The International Journal of Biochemistry and Cell Biology*, **36**: 1031–1037.

Morin, C., Roumegous, A., Carpentier, G., Barbier-Chassefière, V., Garrigue-Antar, L., Caredda, S., and Courty, J. (2012). Modulation of inflammation by Cicaderma ointment accelerates skin wound healing. *J Pharmacol Exp Ther*., **343**: 115-124.

Nagappan, T., Segaran, T.C., Wahid, M.E.A., Ramasamy, P., and Vairappan, C.S. (2012). Efficacy of carbazole alkaloids, essential oil and extract of *Murraya koenigii* in enhancing subcutaneous wound healing in rats. *Molecules*, **17**: 14449-14463.

Oliveira, M.L.M., Bezerra, B.M.O., Leite, L.O., Girão, V.C.C., and Nunes-Pinheiro, D.C.S. (2014). Topical continuous use of *Lippia sidoides* Cham. Essential oil induces cutaneous inflammatory response, but does not delay wound healing process. *J Ethnopharmacol*., **153**: 283–289.

Profyris, C., Tziotzios, C., and Do Vale, I. (2012). Cutaneous scarring: Pathophysiology, molecular mechanisms, and scar reduction therapeutics: Part I. The molecular basis of scar formation. *Journal of the American Academy of Dermatology*, **66**: 1-10.

Rodero, M.P., and Khosrotehrani, K. (2010). Skin wound healing modulation by macrophages. *Int J Clin Exp Pathol.*, **3**: 643-65.

Stern, R. (2004). Hyaluronan catabolism: a new metabolic pathway. *European Journal of Cell Biology*, **83**: 317–325.

Sugumar, S., Ghosh, V., Nirmala, M.J., Mukherjee, A., and Chandrasekaran, N. (2014). Ultrasonic emulsification of eucalyptus oil nanoemulsion: Antibacterial activity against *Staphylococcus aureus* and wound healing activity in Wistar rats. *Ultrasonics Sonochemistry*, **21**: 1044-1049.

Suntar, I., Akkol, E.K., Keles, H., Oktem, A., Baser K.H.C.B., and Yesilada, E. (2011). A novel wound healing ointment: A formulation of *Hypericum perforatum* oil and sage and oregano essential oils based on traditional Turkish knowledge. *J Ethnopharmacol*., **134**: 89-96.

Suntar, I., Tumen, I., Ustun, O., Keles, H., and Akkol, E.K. (2012). Appraisal on the wound healing and anti-inflammatory activities of the essential oils obtained from the cones and needles of *Pinus* species by *in vivo* and *in vitro* experimental models. *J Ethnopharmacol*., **139**: 533– 540.

Tumen, I., Akkol, E.K., Suntarb, I., and Keles, H. (2011). Wound repair and anti-inflammatory potential of essential oils from cones of Pinaceae: Preclinical experimental research in animal models. *J Ethnopharmacol*., **137**: 1215– 1220.

Tumen, I., Suntar, I., Eller, F.J., Keles, H., and Akkol, E.K. (2013). Topical wound-healing effects and phytochemical composition of heartwood essential oils of *Juniperus virginiana* L., *Juniperus occidentalis* Hook., and *Juniperus ashei* J. Buchholz. *Journal of Medicinal Food*, **16**: 48-55.

Tziotzios, C., Profyris, C., and Do Vale, I. (2012). Cutaneous scarring: Pathophysiology, molecular mechanisms, and scar reduction therapeutics: Part I. The molecular basis of scar formation. *Journal of the American Academy of Dermatology*, **66**: 1-10.

Veiga Jr, V.F., Pinto, A.C., and Maciel, M.A.M. (2005). Plantas medicinais: cura segura? *Quim Nova,* **28**: 519-28.

Ximenes, R.M., Nogueira, L.M., Cassundé, N.M.R., Jorge, R.J.B., Dos Santos, S.M., Magalhães, L.P.M., Silva, M.R., Viana, G.S.B., Araújo, R.M., De Sena, K.X.F.R., De Albuquerque, J.F.C., and Martins, R.D. (2013). Antinociceptive and wound healing activities of *Croton adamantinus* Mull. *Arg. essential oil. Journal of natural medicines*, **67**: 758-64.

Natural Products: Research Reviews Vol. 3 (2016) *Pages* **515–527**
Editor: **V.K. Gupta**
Published by: **DAYA PUBLISHING HOUSE, NEW DELHI**

21

A Specific Amino Acid Enhances Yield Under Stressful Circumstances: New Path for Productivity

Gulnaz Sabri*[1], Fiza Khan[2] and Mohammad Mazid[3]

ABSTRACT

Proline is a non-essential α-amino acid, one of the twenty DNA-encoded amino acids. It is unique among the 20 protein-forming amino acids in that the α-amino acid group is secondary. Biosynthetically, it is derived from the amino acid L-glutamate and its immediate precursor is the imino acid (s)-1-pyroline-5-carboxylate. Proline may function also as protein-compatible hydrotrope and as a hydroxyl radical scavenger. Proline considered under a class of small molecules, compatible osmolytes, also including other amino acids, quaternary ammonium compounds and the tertiary sulfinium compounds 3-dimethylsulfoniopropionate. It act as a signaling molecule to modulate mitochondrial functions, influence cell proliferation or cell death and trigger specific gene expression essential for plant recovery against various types of environmental stresses including drought, salinity, extreme temperatures, chemical toxicity and oxidative stress. In many plants, free proline accumulates in response to the imposition of a wide range of biotic and abiotic stresses. Proline is one of the most important compounds of plants defensive mixed action to environmental stresses. Although not all plants accumulate proline in sufficient amounts to help averting adverse effects of abiotic stresses. Further studies required to identification of multiple genes to enhance proline flux that could lead to new opportunities to improve plant tolerance.

Keywords: Drought, Gene expression, Proline, Resistance, Response, Stress.

1 Department of Microbiology, Hospital Laboratory, Gitam University, Visakhapatnam, India.

2 G.F.College, Rohilkhand University, Bareilly – 247 143, Uttar Pradesh, India.

3 Environmental science Division, Bareilly College, Rohilkhand University, Bareilly – 248 392, Uttar Pradesh, India.

* *Corresponding author*: E-mail: gitambotany008@gmail.com

Introduction

Proline considered under a class of small molecules, compatible osmolytes, also including other amino acids, quaternary ammonium compounds (*e.g.*, glycinebetaine, prolinebetaine, β-alaninebetaine, and choline-o-sulfate), and the tertiary sulfinium compounds 3-dimethylsulfoniopropionate (DMSP) (Miller *et al.*, 2009). Osmolytes or compatible solutes (*e.g.*, proline) are groups of low molecular weight organic compounds that accumulate in organisms in response to osmotic stress. Abiotic stress evolves multiple responses that involve a series of physiological, biochemical and molecular events. In many plants, free proline accumulates in response to the imposition of a wide range of biotic and abiotic stresses. Efficiency of proline production considered as a major stress-induced diversion of N metabolism. Although less proline accumulated in the oldest leaves, a significant amount transported from senescing to emerging leaves (Abbaspour *et al.*, 2012). Moreover, during rehydration, proline readily recycled. Proline plays a significant role in leaf N remobilization and in N use efficiency in oilseed rape also (Albert *et al.*, 2012). The purpose of this chapter is to outline recent advances and new approaches to the investigation of proline biosynthesis, accumulation and function, which are not yet completely understood.

Generalization

Proline (abbreviated as Pro) is a non-essential α-amino acid, one of the twenty DNA-encoded amino acids. It is unique among the 20 protein-forming amino acids in that the α-amino acid group is secondary (Atkinson, 1977). Biosynthetically, it is derived from the amino acid L-glutamate and its immediate precursor is the imino acid (s)-1-pyroline-5-carboxylate (P5C). Proline may function also as protein-compatible hydrotrope and as a hydroxyl radical scavenger (Ashraf and Fooland, 2007). Proline is an important osmo-protectant, and proposed to act as a compatible solute composition from dehydration injury and adjusts the osmotic potential in the cytoplasm. Proline has been shown to protect plants against damages caused by free radicals (ROS/AOS *viz.*, OH, H_2O_2) by scavenging the radicals and stabilizing of macro-molecules. Therefore, it have been seen that the more tolerant plants store more proline. In addition, proline acts as a source of energy C and N which enhances tissues recovery, and the relief of stress effects.

If the proline confined to the cytoplasm, however, the concentration of proline could exceed 200 mM in these cells and therefore, contribute substantially to cytoplasmic osmotic adjustment. Similarly, the cytosolic proline concentration of salt stressed *Distichlis spicata* cells is estimated to be ≥ 230 mM when it treated with 200 mM NaCl. Under abiotic stress proline is accumulated in cells at concentrations of a few mM, depending on the species and the extent of stresses imposed. Therefore, proline accumulation is a common biochemical indicator for assessing environmental stress in plants. Moreover, results of Dhawi and Al-Khayri (2008) showed significant two-way intervals between static magnetic field intensity and exposure duration. They studied the proline accumulation in response to magnetic fields in date palm and finally suggest that atleast lowest intensity (10 mM) of proline concentration increased in response to longer exposure durations reaching a maximum at 240 minutes (Chen and Kao, 1995).

Physiological and Biological Processes

The sensitivity of photosynthesis to all of the adverse abiotic environmental conditions capable of accusing proline accumulation is well documented. It has been seen that all of the stresses capable of eliciting proline accumulation are also likely to increase levels of the generated stress hormone ABA which is known to cause stomatal closure. However, it appears that the concentrations of proline accumulated by resistant mutant may be an order of magnitude smaller than required to produce a significant physiological effect on osmotic stress tolerance. In winter wheat the hydroxyl-proline resistant lines are significantly more frost tolerant than wildtype (Mazid *et al.,* 2010; 2011). Moreover, salt tolerant and PEG (polyethylene glycol) resistant mutants of *Nicotiana plumbaginifolia* have been derived from protoplast culture and appear to have enhanced proline accumulation in comparison to wild type. Moreover, the biocompatible feature of proline suggests that the accumulation of the end products of the biosynthetic pathway to high levels can be tolerated without deleterious effects on the stressed cell (Chinnusamy *et al.,* 2005).

A role for proline in stress recovery is consistent with the observation that the extensive accumulation of proline in stressed tissues is usually followed by its rapid disappearance when the stresses are removed. Proline differs from glycine betaine which is apparently maintained at stable levels long after relief from stress (Naidu *et al.,* 1990). A number of observations in different biochemical systems support a role for proline as a primer for TCA cycle (Tricarboxylic acid cycle/Kreb cycle) and activity in plants recovering from osmotic stress (Delauney and Verma, 1993). The mitochondrial location of proline degradation and presence of glutamate dehydrogenase (GDH) in the mitochondrial matrix suggest that this process may contribute C to the TCA cycle. In soyabean nodules, PDH detected in the cytosol suggesting that proline oxidation provides energy for bacteriods during N fixation. When P5CDH activity is limited, the P5C-proline cycle can transfer more electrons to the mitochondrial electron transport chain (ETC) and generate reactive oxygen species (ROS).

High levels of proline oxidation in the bacteriods of N-fixating root nodules of ureides-producing legumes suggests that proline may be the primary energy source used in the energy insensitive process of N-fixation (Kohl *et al.,* 1988). During stress conditions, the rate of the Calvin cycle (C3 cycle) is diminished, which prevents oxidation of NADPH and restoration of $NADP^+$. When combined with high light, electron flow in the electron transport chain is suppressed by the insufficient electron acceptor $NADP^+$ pool, leading to singlet oxygen production in the PS1 reaction centre and accumulation of ROS. Besides this, members of the small proline-rich protein family are major components of the cornified cell envelope are efficient ROS quenchers involved not only in the establishment of the skin's barrier function but also in cell migration and wound healing. In mitochondria, proline has distinct protective functions. After stress, proline pools supply a reducing potential for mitochondria through the oxidation of proline by PDH and P5CDH, provide electrons for the respiratory chain and therefore contribute to energy supply for resumed growth. An enhanced rate of proline biosynthesis in chloroplasts during stress can maintain the low NADPH: $NADP^+$ ratio, contribute to sustaining the electron flow between

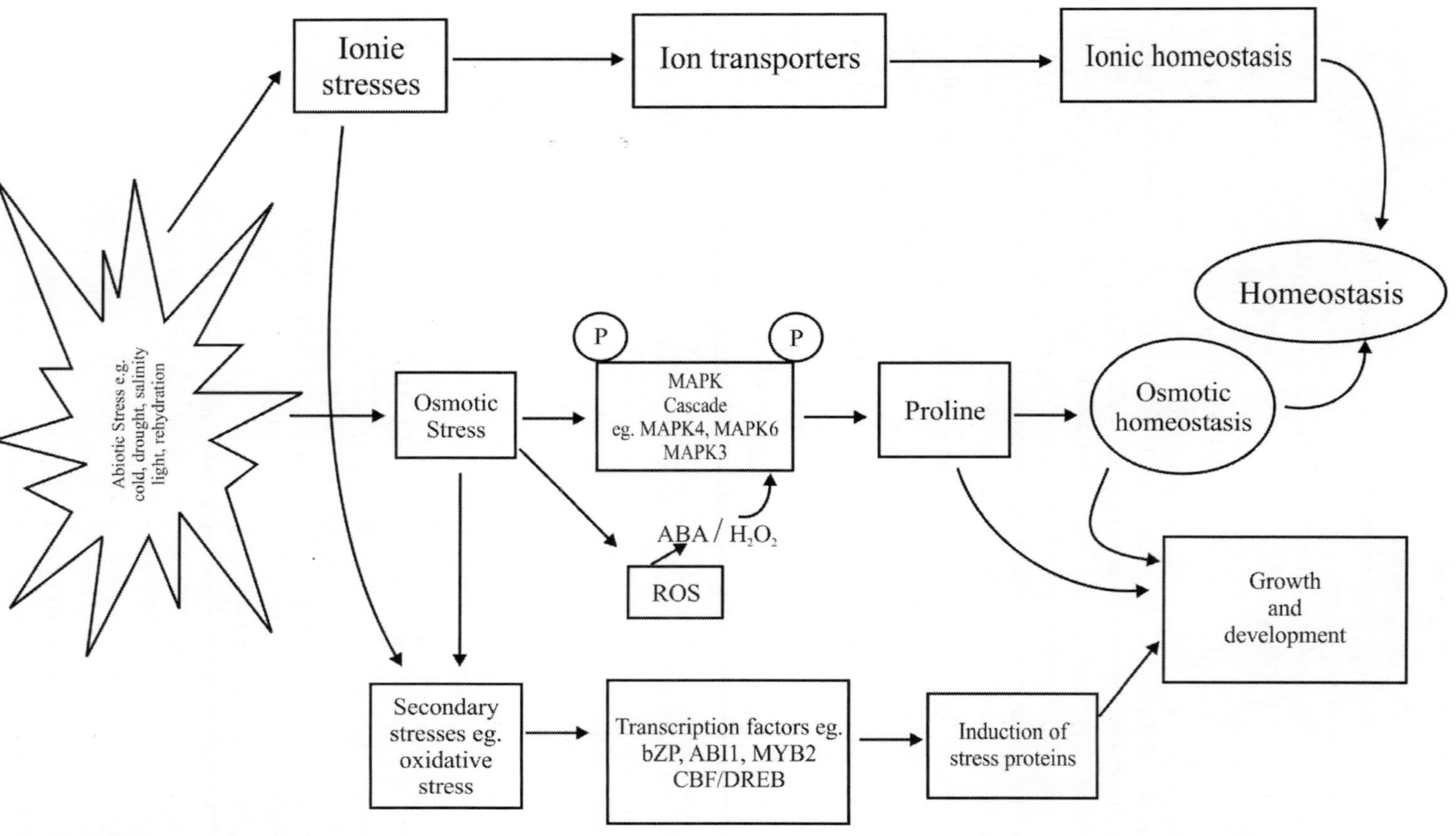

Figure 21.1: Functional Details of Proline with Interacting ROS, MAPK Cascade and Ion Transporters in Plants Cell during Influence Against Abiotic Stress Tolerance.

photosynthetic excitation centres, stabilize the redox balance, reduce photo-inhibition and damage of the photosynthetic apparatus. Therefore, proline catabolism is an important regulator of cellular ROS balance and can influence numerous additional regulatory pathways (Ghanate *et al.*, 2007).

Furthermore, proline biosynthesis is a reductive pathway and requires NADPH for the reduction of glutamate to P5C and P5C to proline, and generates $NADP^+$ that can be used further as electron acceptor. The phosphorylation of glutamate consumes ATP and produces ADP, which is a substrate for ATP biosynthesis during photosynthesis. Increasing amounts of data suggest that proline has certain regulatory function controls plant development and acts as a signal molecule (Hare and Cress, 1997). A connection between photosynthesis and proline metabolism is supported by light-dependent proline accumulation, which is regulated by the light-controlled reciprocal P5CS and PDH gene activation. Moreover, contributing C to the TCA cycle, the mitochondrial degradation of proline to 2-oxyglutarte may also provide reducing equivalents needed to support mitochondrial electron transport and the generation of ATP for recovery from stress and repair of stress-induced degree under stress conditions altered flux through certain pathway is likely to be an important mechanism used to stabilize metabolism and ensure survival when conditions are sub-optimal.

Effect of Proline on Sucrose Metabolism

Proline metabolism can also influence programmed cell death (PCD) in plants. Externally added proline was more toxic to transgenic plants expressing antisense PDH as well as to *pdh* Arabidopsis mutants than to wild type plants. Proline accumulates in plant tumours and functions as competitive antagonists of GABA-dependent plant defence, interfering with the GABA-induced degradation of quorum-sensing signal. Uncoupled induction of PDH and P5CDH in p5cdh mutants leads to P5C and ROS accumulation which can function as stress signal and cause PCD (Miller *et al.*, 2009). Besides contributing C to the TCA cycle, the mitochondrial oxidation of proline to 2-oxoglutarate may also directly provide reducing equivalents needed to support mitochondrial electron transport and the generation of ATP for recovery from stress and for their repair of stress-induced damage. Expression of genes encoding glycolytic enzymes is up-regulated in response to several environmental perturbations capable of eliciting proline accumulation (Krishnamurthy and Bhagat, 1993).

Proline catabolism in the mitochondria is connected to oxidative respiration and administers energy to resumed growth after stress. Since proline degradation generates reducing equivalents needed to during TCA cycle activity. Proline metabolism might transfer metabolic information between tissues with different metabolic requirements in the form of redox potential and activate metabolic pathways that do not involve any of the intermediates of metabolism (Kumar and Sharma, 1989). Tolerance to abiotic stresses is very complex at the whole plant and cellular levels. From a metabolic perspective, the plant is a highly integrated system, which is demonstrated by the capture of energy by photosynthesis and its use in C and N assimilation.

Proline and Abiotic Stress

In many plants, free proline accumulates in response to the imposition of a wide range of biotic and abiotic stresses. Proline accumulation may reduce stress-induced cellular acidification or proline oxidative respiration to provide energy needed for recovery. High levels of proline synthesis during stress may maintain NAD (P)$^+$/ NADP(H) ratios at value compatible with metabolism under normal conditions. Although, the regulation and function of proline accumulation are not yet completely understood, the engineering of proline metabolism could lead to new opportunities to improve plant tolerance to environmental stresses. In plants, proline is synthesized mainly from glutamate which is regulated to GSA by the P5CS enzyme, and spontaneously converted to P5C. P5C redtctase (P5CR) further reduce the P5C intermediate to proline. In most plant species, P5CS is encoded by two genes and P5CR is encoded by one (Kurkdjian and Guern, 1989).

Moreover, loss of feedback inhibition of P5CS leads to elevated proline accumulation. In Arabidopsis, P5CS splice variants has been annotated but not characterized. Conversion of P5CS to proline is not a rate limiting step in proline biosynthesis, yet the control of P5CR activity implies a complex regulation of transcription, which was shown to be under the developmental and osmotic regulation (Lee *et al.*, 1995). Over expression of P5CS1 and enhanced proline content led to the early flowering of transgenic Arabidopsis plants. The accumulation of proline under abiotic stress conditions has been studied in numerous plant species for half a century. Nevertheless, controversy has arisen as to whether an increase in free proline has any real adoptive value. Clearly, a better appreciation of the extent to which proline accumulation is of protective value to plants exposed to several different environmental extremes and an understanding of the mechanisms (s) whereby proline accumulation may exert a beneficial effect under stressful conditions, are likely to be critical pre-requisites for the success of such endeavours. These observations provide convincing evidence of a cause-and effect relationship between proline levels and abiotic stress tolerance. Important consideration in various attempts to account for a beneficial role for proline accumulation in tolerance of different stresses include whether several of the different stresses capable of eliciting the response have common negative physiological effects and assessment of whether alternations in proline metabolism may ameliorate these consequences (Llops-Tous *et al.*, 2011).

As an alternative pathway, proline can be synthesized from ornithine, which is transmitted first by ornithine-delta-aminotransferase (OAT) providing GSA and P5C, which is then converted to proline. Proline catabolism occurs in mitochondria via the sequential action of proline dehydrogenase or proline oxidae (PDH/POX) producing P5C from proline, and P5C dehydrogense (P5CDH), which converts P5C to glutamate. PDH is encoded by the two genes, whereas a single P5CDH gene has been identified in Arabidopsis and tobacco (Ribarits *et al.*, 2007). Proline accumulation induced by stress conditions is mediated both by increased synthesis and reduced oxidation of the imino acid. Metabolic labelling studies also mediate that most of the proline accumulated in plants in response to stress is the results of enhanced synthesis from glutamate. The soluble protein and free amino acid in barley organs (root and bud) increased with NaCl increasing. Central role in cellular homeostasis against environmental stress (salinity and drought stress). One mechanisms utilized by the plants for overcoming the salt stress effects might be via accumulation of compatible osmolytes, such as proline and soluble sugar. Production and accumulation of free amino acids, especially proline by plants tissue during drought, salt and water stress is an adoptive response.

Moreover, marked changes in the content of proline was observed, with some specific variations associated with the different organs of *L. albus* concerning the effects of boron deficiency in the metabolism indicate that various aspects of metabolism implicated in the amino acid accumulation affected by boron deficiency (Marin *et al.,* 2010). Also, proline increase in many other kinds of stresses and accumulated in petioles, apexes and hypocotyls. The increase in the proline concentration observed in all organs, suggests the involvement of boron with the cytoskeleton, whereas glycine decrease in leaf-blades and active growing organs like apexes and roots could be associated with the proposed role of glycine in plant signaling in processes that might be associated with the decreased growth rates observed in boron deficiency. The salt stress conditions could have effect on different stages of N metabolism, such as absorption, ionic reduction and protein synthesis. Proline content also higher in tolerant variety as compared to their lower accumulation in sensitive variety. The improved performance of the tolerant variety under high salinity accompanied by an increase in ascorbate peroxidase and catalase, though no salt-dependent increase in the activity of superoxide dismutase observed. Increased photoinhibition in sensitive variety observed by its reduced thylakoid membrane protein, D1 probably results from the greater photosynthetic damage caused by salt stress than tolerant variety (Mishra and Dubey, 2006). However, the correlation between proline accumulation and abiotic stress tolerance in plants not always apparent. For an instance, high proline levels can be characteristics of salt and cold hypersensitive Arabidopsis mutants (Chen *et al.,* 2007). During adaptation to various types of environmental stress, plants accumulate cellular solutes which include quaternary amino acids derivatives such as proline, glycine, betaine, alanine, and proline betaine (Patakas *et al.,* 2002). Still however, the relationship between this trait and stress tolerance is not clear among species. As a case, members of solanacese family, can increase their proline pool by the more than two orders of magnitude and others exhibits only a moderate increase in proline content under stress. Moreover, other researchers have challenged the value of proline accumulating potential as a positive index for resistance to osmotic stress (Heuer, 1994). It has been reported that elevated levels of proline biosynthesis in transgenic tobacco plants confer increased tolerance to hyperosmotic stress. Proline accumulations have been extensively studied as a common metabolic response of higher plants to drought and salinity stress. It is generally accepted that under conditions of drought or salinity, proline accumulation serves as a defence against osmotic challenge by acting as compatible solutes. Leaf osmotic potentials of the transgenic plants unaffected by drought and the plants displayed even less capacity for osmotic adjustment than compatible wild type tobacco. Proline appears to be the preferred organic osmoticum in many plants. Several findings highlight the need for further investigations of the mechanism whereby proline accumulation alleviates the effects of osmotic stress on plant growth (Mittal *et al.*, 2012). A number of researchers suggesting a increase in proline levels accompanying osmotic stress may be misleading since, this increase is often based on an initially small proline content. This is not always found in many glycophytes in which proline accumulation has been reported under conditions of osmotic stress. In rice plants subjected to water stress, the concentration of proline was increased in the leaves. Transgenic tobacco plants, overexpressing P5Cs have shown increased concentration of proline and resistance to both drought and salinity stresses (Kavi-Kishor *et al.*, 1995).

Genetic Recombination and Proline

Globally, approximately 22 per cent of the agricultural land is saline (FAO 2004), and areas under drought are already expanding and this is expected to increase further rapidly (Bhatnagar *et al.*, 2008). Success in breeding for better adopted varieties to abiotic stresses depend upon the concentrated efforts by various research domains including plant and cell physiology, molecular biology, genetics and breeding. Use of modern molecular biology tools for elucidating the control mechanisms of abiotic stress tolerance and for engineering stress tolerant crops is based on the expression of specific stress-related genes. Hence, genetic engineering for developing stress tolerant lands based on the genes that are known to be involved in stress response and putative tolerance might prove to be a faster track towards improving crop varieties. For beyond the initial attempts to insert single action genes engineering of the regulatory machinery involving transcription factors has emerged as a new-response genes (Pardo *et al.*, 1998). Evaluation of the transgenic plants under stress conditions and understanding the physiological effect of the inserted genes at the whole plant level remains as major challenges to overcome. In addition, the transgenic plants also showed increased expression of some drought-responding genes, and enhanced activity of antioxidant enzymes such as superoxide dismutase, peroxidase, and ascorbate peroxidase. Their membrane integrity considerably improved under water stress, as indicated by reduced MDA content and electrolyte leakage relative to control plants. These results suggest that the AtLOS5 transgenic cotton plants acquired a better drought tolerance through enhanced ABA production and ABA-induced physiological regulations (Pei *et al.*, 1998).

Transgenic approaches to improve plant stress tolerance via over-producing proline have some success. When a plant is subjected to abiotic stress a number of genes are turned on resulting in increased levels of several metabolites and proteins, some of which may be responsible for conferring a certain degree of protection to these stresses. Besides this, a key progress towards breeding better crops under stress has been to understand the changes in cellular, biochemical and molecular machinery that occur in response to stress. Moreover, modern molecular technologies involve the identification and use of molecular markers that can enhance breeding programmes (Poonlaphdecha *et al.*, 2012). Besides this, metabolic traits have been characterized genetically and appear more amenable to manipulations than structural and developmental traits. Further, genetic engineering allows controlling the timing, tissue specificity, and expression level of the introduced genes for their optimal function. The basic findings on stress promoters have led to a major shift in the paradigm for genetically engineering stress tolerant crops (Katiyar *et al.*, 1999). The most widely used promoters in generating transgenic plants are constitutively expressed *i.e.*, they are turned on all the time and throughout the plant life cycle. With an increasing number of stress genes become available and genetic transformation becoming more or less a routine procedure, characterization of stress-induced promoters.

Modulation of Proline Synthesis in Plant System

The transcription factors activate cascades of genes that act together in enhancing tolerance towards multiple stresses. Dozens of transcription factors are involved in the plant response to abiotic stress tolerance. Most of these falls into several large transcription factor families, such as AP2/ERF, bZIP, NAC, MYB, MYC, Cys2His2 zinc-finger WRKY. Individual members and of the same family often respond differently to various stress stimuli. On the other hand, some stress responsive genes may share the same transcription factors, as indicated by the significant overlap of the gene expression profiles that are induced in response to different stresses. Two families, bZIP and MYB, are involved in ABA signaling and its gene activation. Many ABA inducible genes share the (C/T) ACGTGGC consensus, cis-acting ABA-responsive element (ABRE) in their promoter regions. The use of stress inducible rd29A promoter minimized the negative effects on plant growth in these crop species. However, over-expression of DREB2 in transgenic plants did not improve stress tolerance, suggesting involvement of post-translational activation of DREB2 proteins. The DREB2 protein is expressed under normal growth conditions and activated by osmotic stress through post-translational modification in the early stages of the osmotic stress response (Smirnoff, 1993). The over expression of some drought-responsive transcription factors can lead to the expression of downstream genes and the enhancement of abiotic stress tolerance in plants. The regulatory genes/factors reported so far not only play a significant role in drought and salinity stresses, but also in submergence tolerance. Moreover, genes involved in stress signal sensing and a cascade of stress-signaling in *A. thaliana* has been of recent research interest recent. In addition, some gene products are needed in large amounts, such as LEA3, thereby necessitating the need for a very strong promoter. An ideal inducible promoter should not only be devoid of any basal level of gene expression in the absence of inducing agents, but the expression should be reversible and dose-dependent. Besides this, these promoters being constitutive in nature, by and large express the downstream transgenes in all organs and at all the stages (Verbruggen, 1993). Most of the stress promoter's conditions array of stress-specific *cis*-acting elements that are recognized by the requisite transcription cofactor; for example, the transcriptional regulation of *hsp* genes is mediated by the core the shock element (HSE) located in the promoter region of these genes 5' of the TATA box. With other gene products, such as enzymes for polyamine biosynthesis, it may be better to use an inducible promoter of moderate strength.

The promoters that have been most commonly used in the production of abiotic stress tolerant plants so far, include the CaMV 35S, ubiquitin 1 and actin promoters. Mostly promoters being constitutive in nature, by and large expression the downstream transgenes in all organs and at all the stages. An ideal inducible promoter should not only be devoid of any base at level of gene expression in the absence of inducing agents but the expression should be reversible and dose-dependent. The transcriptional regulatory region of the drought-induced and cold-induced genes have been analysed to identify several in the gene expression that is induced by abiotic stress.

Two major issues that typically need to be addressed in stress response evaluation of transgenics include; (1) means of stress imposition; (2) hard data on the response of tested material to support conclusions. An important aspect of transgenic technology is the regulated expression of transgenic. Tissues specificity of transgenes expression is also an important consideration while declining on the choice of the promoter so as to increase the level of expression of the gene. All the plant *hsp* genes sequenced so far have been shown to certain partly overlapping multiple HSEs proximal to TATA motif. The Arabidopsis *rd29A* and *rd29B* are stress responsive genes, but are differentially induced under abiotic stress conditions. The *rd29A* promoter includes both DRE and ABRE elements, where dehydration, high salinity and low temperature induce the genes, while the *rd29B* promoter include only ABREs and the induction is ABA-dependent. Although recent findings have convincingly demonstrated that proline overproduction in transgenic tobacco ameliorates the effects of hyper osmotic stress, it remains to be seen whether this phenotype alteration also increases tolerance of other commonly-encountered biotic and abiotic stresses. Despite the remarkable advances in our understanding of proline biosynthesis in plants genes over the past few years, the isolation of plant genes encoding enzymes involved in proline degradation has yet, to be reported (Wang *et al.*, 2011). Proline itself may act not only as an osmoticum but also as a substrate for the TCA cycle during recovery from stresses, while the interconversions between proline and its precursors may be involved in the regulation of cellular pH and redox potential.

Concluding Remarks

Some progress has been made in introducing genes for the production of these compounds in naturally non-accumulating or low-accumulating plant species; levels of accumulation in transgenic plants have often been low or insufficient to improve plant stress tolerance. Today's research to determine specific roles of proline in plant stress tolerance is expected to help improving their application as exogenous treatments to improve growth of plants and productivity under stress conditions. The application of system biology approaches might help to understand regulation of proline-dependent and proline mediated signaling in plants. Using a similar strategy, feedback insensitive bacterial ProBA genes were overexpressed in transgenic Arabidopsis, leading to proline hyper-accumulation and enhanced osmotolerance. Also, several other studies suggested an unchanged stress tolerance, accompanied by abnormal seed and plant development and increased proline hypersensitivity of the transgenic lines and mutants suggest that proline might not be the actual proline content but the enhanced rate of proline biosynthesis that is an important factor for stress adaptation.

References

Abbaspour, H., Saeidi-Sar, S., Afshari, H. and Abdel-Wahhab, M.A. (2012). Tolerance of Mycorrhiza infected pistachio (*Pista ciavera* L.) seedling to drought stress under glasshouse conditions. *Journal of Plant Physiology,* **169(7)**: 704-709.

Albert, B., Le Cahérec, F., Niogret, M.F., Faes, P., Avice, J.C., Leport L. and Bouchereau A (2012). Nitrogen availability impacts oilseed rape (*Brassica napus* L.) plant water status and proline production efficiency under water-limited conditions. *Planta,* PMID: 22526495.

Ashraf, M. and Foolad, M.R. (2007). Roles of glycine betaine and proline in improving plant abiotic stress resistance. *Environmental and Experimental Botany,* **59**: 206-216

Atkinson, D.E. (1977). Cellular Energy Metabolism and Its Regulation. New York: Academic Press.

Bhatnagar, P., Vadez, M.V. and Sharma, K.K. (2008). Transgenic approaches for abiotic stress tolerance in plants: retrospect and prospects. *Plant Cell Reproductio,* **27**: 411-424.

Chen, S.L., and Kao, C.H. (1995). Cd induced changes in proline level and peroxidase activity in roots of rice seedlings. *Plant Growth Regulation,* **17**: 67-71.

Chen, Z. (2007). Compatible solute accumulation and stress mitigating effects in barley genotypes contrasting in their salt tolerance. *Journal of Experimental Botany,* **58**: 4245-4255.

Chinnusamy, V., Jagendorf, A. and Zhu, J-K. (2005). Understanding and improving salt tolerance in plants. *Crop Science,* **45**: 437-448.

Delauney, A.J. and Verma, D.P.S. (1993). Proline biosynthesis and osmoregulation in plants. *Plant Journal,* **4**: 215-223.

Dhawi, F. and Al-Khayri, J.M. (2008). Proline accumulation in response to magnetic fields in date palm (*Phoenix dactylifera* L.). *The Open Agriculture Journal,* **2**: 80-83.

FAO (Food, Agriculture Organization of the United Nations) (2004). FAO production year book. FAO, Rome.

Ghanati, F., Abdolmaleki, P., Vaezzadeh, M., Rajabbeigi, E. and Yazdani, M. (2007). Application of magnetic field and iron in order to change medicinal products of *Ocimum basilicum*. *Environmentalist,* **27**: 429-34.

Hare, P. and Cress, W. (1997). Metabolic implications of stress induced proline accumulation in plants. *Plant Growth Regulator,* **21**: 79-102.

Heuer, B. (1994). Osmoregulatory role of proline in water and salt-stressed plants. In: Pessarakli M (ed). Handbook of Plant and Crop Stress, New York: Marcel Dekker, Inc, pp. 363-381.

Katiyar-Agarwal, S., Agarwal, M. and Grover, A. (1999). Emerging trends in agricultural biotechnology research: use of abiotic stress induced promoter to drive expression of a stress resistance gene in the transgenic system leads to high level stress tolerance associated with minimal negative effects on growth. *Current Science*, **77**: 1577-1579.

Kavi-Kishor, P.B., Hong, Z., Miao, G-H., Hu, C-AA. and Verma, D.P.S. (1995). Overexpression of 1-pyrroline-5-carboxylate synthetase increases proline production and confers osmotolerance in transgenic plants. *Plant Physiology*, **108**: 1387-1394.

Kohl, D.H., Schubert, K.R., Carter, M.B., Hagedorn, C.H. and Shearer, G. (1988). Proline metabolism in N2-fixing root nodules: Energy transfer and regulation of purine synthesis. *Proceedings of the National Academy of Sciences of USA*, **85**: 2036-2040.

Krishnamurthy, R. and Bhagwat, K.A. (1993). Effect of foliar application of proline on the salt stressed rice seedlings. *ActaAgronomica Hungarica*, **42**: 267-272.

Kumar, V. and Sharma, D.R. (1989). Effect of exogenous proline on growth and ion content in NaCl stressed and nonstressed cells of mungbean, *Vigna radiate* var. radiata. *Indian Journal of Experimental Biology*, **27**: 813-815.

Kurkdjian, A. and Guern, J. (1989). Intracellular pH: Measurement and importance in cell activity. *Annual Reviews of Plant Physiology*, **40**: 271-303.

Lee, J.H., Hubel, A. and Schoffl, F. (1995). Derepression of activity of genetically engineered heat-shock factor causes constitutive synthesis of heat shock proteins and increased thermotolerance in transgenic Arabidopsis. *Plant Journal*, **8**: 603-612.

Llop-Tous, I., Ortiz, M., Torrent, M. and Ludevid, M.D. (2011). The expression of a xylanase targeted to ER-protein bodies provides a simple strategy to produce active insoluble enzyme polymers in tobacco plants. *PLoS One*, **6(4)**: e19474.

Marin, J.A., Andreu, P., Carrasco, A. and Arbeloa, A. (2010). Determination of proline concentration, an abiotic stress marker, in root exudates of excised root cultures of fruit tree rootstocks under salt stress. *Revue des RégionsArides – Numérospécial – 24 (2/2010) Actes du 3ème Meeting International "Aridocultureet Cultures Oasisennes: Gestion et Valorisation des Ressources et Applications Biotechnologiquesdans les AgrosystèmesArides et Sahariens" Jerba (Tunisie)*, **1**: 16-17.

Mazid, M., Ali, B., Hayat, S. and Ahmad, A. (2010). The effect of 4-chloroindole-3-acetic acid on some growth parameters in mung bean under cadmium stress. *Turkey Journal of Biology*, **34**: 9-13.

Mazid, M., Khan, T.A. and Mohammad, F. (2011). Potential of NO and H2O2 as signaling molecules in tolerance to abiotic stress in plants. *Journal of Industrial Research and Technology*,**1(1)**: 56-68.

Miller, G. (2009). Unraveling delta1-pyrroline-5-carboxylateproline cycle in plants by uncoupled expression of proline oxidation enzymes. *Journal of Biological Chemistry*, **284**: 26482-26492.

Mishra, S. and Dubey, R.S. (2006). Inhibition of ribonuclease and protease activities in arsenic exposed rice seedlings: role of proline as enzyme protectant. *Journal of Plant Physiology*, **163**: 927-936.

Mittal, S., Kumari, N. and Sharma, V. (2012). Differential response of salt stress on *Brassica juncea*: Photosynthetic performance, pigment, proline, D1 and antioxidant enzymes. *Plant Physiology and Biochemistry,* **54**: 17-26.

Naidu, B.P., Paleg, L.G., Aspinall, D., Jennings, A.C. and Jones, G.P. (1990). Rate of imposition of water stress alters the accumulation of nitrogen-containing solutes by wheat seedlings. *Australian Journal of Plant Physiology,* **17**: 653-664.

Pardo, J.M., Reddy, M.P. and Yang, S. (1998). Stress signaling through Ca^{2+}/ Calmodulin dependent protein phosphatase calcineurin mediates salt adaptation in plants. *Proceedings of the National Academy of Sciences of USA,* **95**: 9681-9683.

Patakas, A., Nikolaou, N., Zioziuo, E., Radogluo, K. and Notisakis, B. (2002). The role of organic solute and ion accumulation in osmotic adjustment in drought-stressed grape veins. *Plant Science,* **163**: 361-367.

Pei, Z.M., Ghassemian, M., Kwak, C.M., McCourt, P. and Schroeder, J.I. (1998). Role of farnesyltransferase in ABA regulation of guard cell anion channels and plant water loss. *Science,* **282**: 287-290.

Poonlaphdecha, J., Maraval, I., Roques, S., Audebert, A., Boulanger, R., Bry, X. and Gunata, Z. (2012). Effect of Timing and Duration of Salt Treatment during Growth of a Fragrant Rice Variety on Yield and 2-Acetyl-1-pyrroline, Proline, and GABA Levels. *Journal ofAgriculture and Food Chemistr,.* **60 (15)**: 3824-3830.

Ribarits, A. (2007). Two tobacco proline dehydrogenases are differentially regulated and play a role in early plant development. *Planta,* **225**: 1313-1324.

Smirnoff, N. (1993). The role of active oxygen in the response of plants to water deficit and desiccation. *New Phytology,* **125**: 27-58.

Verbruggen, N. (1993). Osmoregulation of a pyrroline-5-carboxylate reductase gene in *Arabidopsis thaliana*. *Plant Physiology,* **103**: 771-781.

Wang, R., Huang, W., Chen, L., Ma, L., Guo, C. and Liu, X. (2011). Anatomical and physiological plasticity in *Leymus chinensis* (Poaceae) along large-scale longitudinal gradient in northeast China. *PLoS One,* **6(11)**: 254-267.

Natural Products: Research Reviews Vol. 3 (2016) *Pages* **529–571**
Editor: **V.K. Gupta**
Published by: **DAYA PUBLISHING HOUSE, NEW DELHI**

22

An Overview on Potassium Channel Openers as a Vital Target of Drugs Isolated from Plants as well as Synthetic Origin

Anil K. Tripathi[1*], Debaraj Mukherjee[1]
and Ashutosh Kumar Dash[1]

ABSTRACT

The overview depicts the voltage regulated potassium channel, which are found in CNS, and about the emerging therapeutic applications of potassium channel modulators and also it concerns about the recent developments in our understanding of the mechanisms, which control the activity of a series of channels, which are selective for potassium ions. It has been long recognized that excitability of cells is mediated by proteins, which can modulate the ability of potassium ions to travel across the cell membrane. Go back to the ten years, it has become an increasingly clear that, potassium ion channels represent an excellent target for the novel drug design (Dunne et al., 1990). Investigation of the role of potassium channels in the treatment of human diseases continues to be growing field of research. The ability of potassium channels to regulate membrane potential accords them a central role in varied cellular processes that governs excitability, action potential characteristic, stimulus secretion coupling, cell volume regulation and epithelial electrolyte transport. Attention from medicinal chemists to potassium channels as drug targets has grown with the realization that a variety of potassium channel openers offer significant therapeutic opportunities in cardiac, smooth muscle, neuronal, immune and secretary systems. Progressive improvements in molecular biology have enabled regular cloning of potassium channels of interest, and defined families of these channels have facilitated a comprehensive understanding of their function. Importantly many families of

1 CSIR-Indian Institute of Integrative Medicine, Canal Road, Jammu – 180 001, J&K State, India.

* *Corresponding author*: E-mail: tripathitripathi@rediffmail.com

increasingly selective small molecules have emerged a target validation and clinical proof of principle. Many reviews have appeared summarizing the synthesis and therapeutic potential of this channels (Roxburg et al., 1995).The scope of this report is to make you aware of potassium channel biology, which lead themselves to a more expedient identification of agents for the treatment of CNS disorders, emphasizing developments in medicinal chemistry from potassium channels where modulators would have considerable clinical potential. While every effort has been made to include all relevant reports in this discussion, any omission is inadvertent and we apologize for the same.

Keywords: K_{ATP}, Cromakalim, Benzopyran, Nicorandil.

Introduction

The diseases in man, can be treated, by plants, their extracts and plant derived compounds has been practiced for thousands of years. Numerous key compounds have been obtained and shown to be responsible for the therapeutic effects. Prevention of ailments and diseases has therefore remained a challenge and necessity for the human being. It would seem that at one time or another almost every plant at least was thought to have medicinal properties, even though a few survived into modern medicinal practice and fewer still into our present pharmacopoeias. In fact until the middle of nineteenth century they were, with a few exceptions, the only material available for the treatment and prevention of diseases. Throughout the middle ages, the material medica of plant origin seemed to be far more important and respectable. It was therefore a natural development at the beginning of the nineteenth century, when modern chemistry and pharmacology began to develop into the study of the medicinal plants.In many of the countries possessing ancient cultures, such as India and China there exist traditional system of medicine, the material medica of which in a large measure comprise plant drugs. Rig-Veda, perhaps one of the oldest repositories of human knowledge for the use of medicinal plants for curing various diseases. However, with the passage of time and advancement of sciences, an imperative need for the systematic study of herbal drugs was felt. This led to the development of natural product chemistry, synthetic organic chemistry and medicinal chemistry.

In pharmaceutical chemistry, with its roots in natural products chemistry, efforts are directed towards new biologically active compounds in plants, which are providing direction for the synthesis of newer analogues and derivatives having an equivalent or better activity than the parent molecules. Different types of molecules are present in the plants,which have aliphatic, aromatic, long chain carboxylic adopt, contain ring systems of carbons and at least one nitrogen in the ring. These compounds can be classified as nitrogen atom in the ring. These compounds can be classified as nitrogen hetero cycles, which are widely distributed in nature and are essential to life and also play a vital role in the metabolism of living cells; for examples; the pyridine and purine bases of the genetic material DNA; the vitamins and coenzymes precursors; the essential amino acids; the photosynthesizing pigment chlorophyll; the oxygen transporting pigment hemoglobin and its breakdown products like the bile pigments and hormones together with the most sugars. There are vast numbers of pharmacologically active nitrogen heterocyclic compounds, many of which are regular clinical use. Some of these natural products, for example include antibiotics such as cephalosporins and penicillin, alkaloids such as morphine, ellipticine, reserpine and vinblastine. A large majority of synthetic nitrogenous heterocycles have found wide use, for examples as insecticides, rodenticides, weed killers, pesticides, anticancer agents, analgesics, hepatoprotectives and vasodepressents.

By dint of the lone pair of electron at the apex of the pyramid on nitrogen in a cyclic system, the nitrogenous compounds possess a unique character with constant flipping from one form to another, which is responsible for unique properties of nitrogenous bearing compounds showing marked physiological activity.

When we take into account of heterocyclic compounds especially nitrogen heterocyclic there are several channel modulators, which are rapidly growing in number and are being used as drugs for the treatment of the cardiovascular and related diseases, such as, beta blockers, calcium channel blockers, sodium channel blockers and potassium channel blockers. Calcium channel blockers mainly influence myocardial cells, the cells within the specialized conducting system of the heart, and the cell of the vascular smooth muscle,while sodium channel blocker have been used as local anesthetics and anti arrhythmic for many years. Potassium channels play a key role in the maintenance of resting membrane potential and repolarization of the action potential. Fluctuations in membrane potential effects cell excitability, regulating various physiological functions in smooth muscle, cardiac tissue, and in the CNS. Potassium channels are largely regulated by voltage, cell metabolism, and calcium and receptor mediated processes (Cook *et al.*, 1989). Unlike the relatively unique the relatively unique sodium and calcium channels, there are a variety of potassium channel types, often present in the same tissue. The diversity of potassium channels modulators being unraveled by the use of biochemical and electrophysiological methods is increasing at a faster rate. There is still much to be learned about the physiological functions associated with various channel types. However, it is clear that modulation of potassium channels may form the basis of therapeutic approaches to various diseases of cardiac and smooth muscles and of neuronal tissues (Cook *et al.*, 1989;Quast *et al.*, 1989).

Sophisticated techniques are used (*e.g.*, patch and clamp) to measure the activity of various channel types has increased our ability to characterize existing pharmacological agents, these tools are very valuable for the medicinal chemist for designing novel compounds. The synthesis of new compounds still relies heavily on the structures of existing prototypes.

How does Potassium Channel Opener Works?

The potassium channels are an important class of ionic channels, whose function is regulated by changes in the intracellular level of adenosine triphosphate (ATP) levels are elevated and open, when intracellular adenosine level decline, thus linking membrane potential to the metabolic state of the cell (Gopalkrishan *et al.*, 1993), opening them allows the passage of potassium ion out of the cell, causing trans membrane hyperpolarization and repolarization. These effects reduce intracellular calcium channels and inhibiting intracellular calcium release, resulting in smooth muscle relaxation (Uchida *et al.*, 1985).

ATP-Sensitive Channel

Some ATP-sensitive K_{ATP} channels, (Noma *et al.*, 1983), which represent a family of K^+ channels inhibited by intracellular ATP, have been found in a variety of tissues including heart, pancreatic beta cells, skeletal muscles, smooth muscles, and the CNS. These K_{ATP} channels have been associated with diverse cellular functions, such as shortening of action potential duration and cellular loss of K^+ ions that occur during metabolic inhibition in heart, insulin secretion from pancreatic beta cells, smooth muscle relaxation, regulation of skeletal muscle excitability, and neurotransmitter release. Furthermore K_{ATP} channels in different tissues exhibit considerable variation in response to K^+ channel openers. For example, the pancreatic beta cell K_{ATP} channel is activated by diazoxide and only and weakly by pinacidil. The cardiac K_{ATP} channel is activated by pinacidil but not by diazoxide.The smooth muscle K_{ATP} channel is activated effectively by both of these compounds. Thus, properties of K_{ATP} channels vary among tissues, having led to the premise that this K^+ channel family may be composed of heterogeneous K^+ channel properties. K_{ATP} is known to exists in a variety of tissues, which includes cardiac muscle, (Noma *et al.*, 1983) skeletal muscle, (Spruce *et al.*, 1983) and pancreatic beta cell, (Chiu *et al.*, 1990). It is thought to exist in certain region of the brain (Ben-ari *et al.*, 2001; Bernardi *et al.*, 1988)· By making use of patch clamp methods, K_{ATP} has been shown to exist in arterial, (Standen *et al.*, 1989) and aortic, (Kovac *et al.*, 1991) smooth muscle. Recently, this channel has been shown to exist in rat liver inner mitochondrial membranes (Takehasi Inoue *et al.*, 1996). Although the exact physiological function of this channel in mitochondria is not known, it is hypothesized to control the mitochondrial membrane. The inhibition of this channel is thought to involve specific binding of ATP to the channel (Ashcroft *et al.*, 1988). The sensitivity of binding and the ratio of ATP molecules to binding site is dependent on the tissue type (Kirsch *et al.*, 1990). Although much less potent than ATP, magnesium ions and certain nucleotides (ADP, AMP) also block the activity of this channel. K_{ATP} may be regulated by adenosine receptors via G-protein coupling in rat ventricular myocytes (Wille de *et al.*, 1988). K_{ATPcos} involves three isoforms of the sulfonylurea (SUR) receptor (which forms the beta subunits of the K_{ATP} channel). The distribution of the SUR isoform provides unique targets for the development of selective K_{ATPcos} giving focused therapeutic approaches to clinical conditions for example cardiac ischemia, urinary incontinence, neuro degeneration, obesity and autoimmune diseases. Although the channel has been hypothesized to be the receptor for the sulphonylurea class of ligands definitive to demonstrate that the binding site for these compounds is the channel-forming proteins. K_{ATP} is by far the most studied channel, due largely to the availability of pharmacological agents that modulate its activity. The discovery of K_{ATP} opening as a part of the mechanism of vasodialation by a diverse group of agents has created a much excitement in studying this channel. Of all the potassium channel known to date

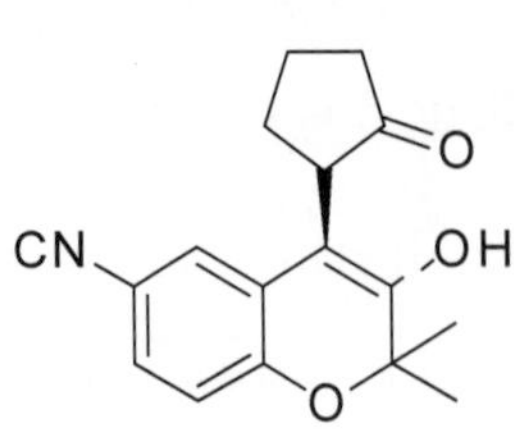

CN
OH
O
O

1

4-pyrolidine benzopyran analogues, cromakalim **1** is one of the most potent and studied compounds of this class.

The structure of other related molecules representing various structure types are depicted as under *e.g.*, 2-9.

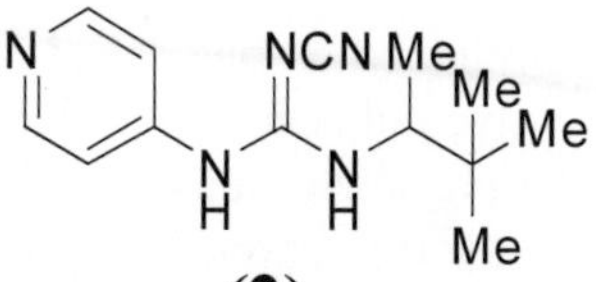

(2) 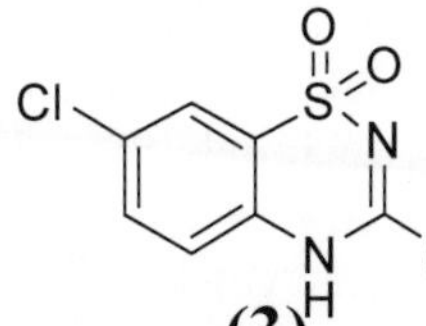 (3) 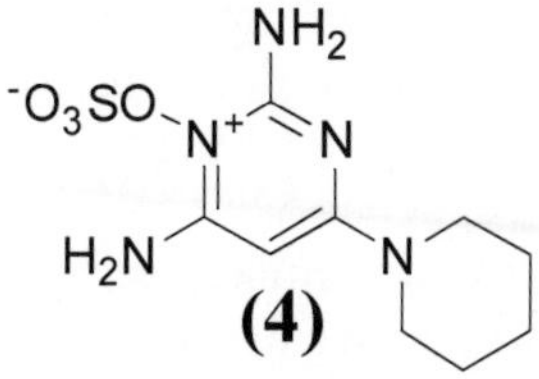 (4)

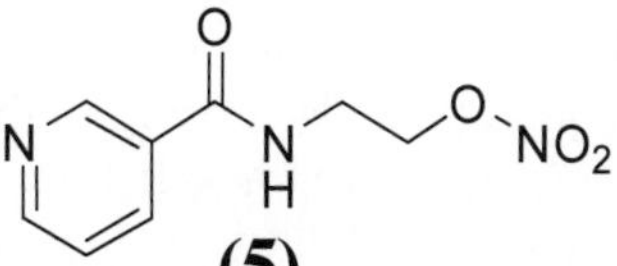

(5) 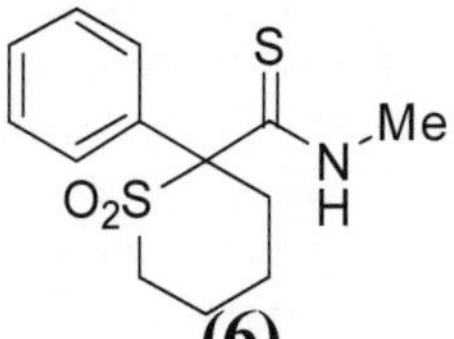 (6) 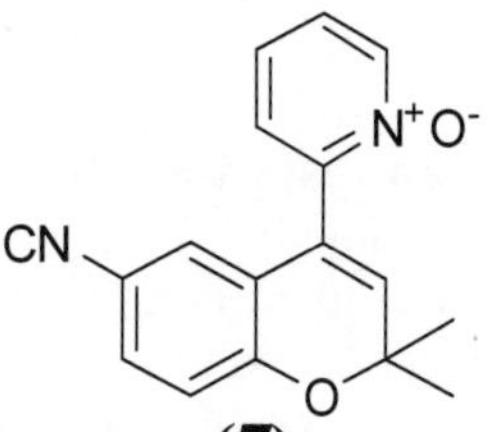 (7)

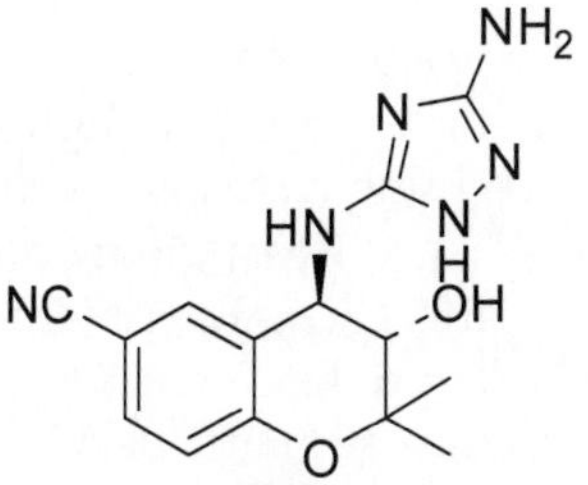

(8) 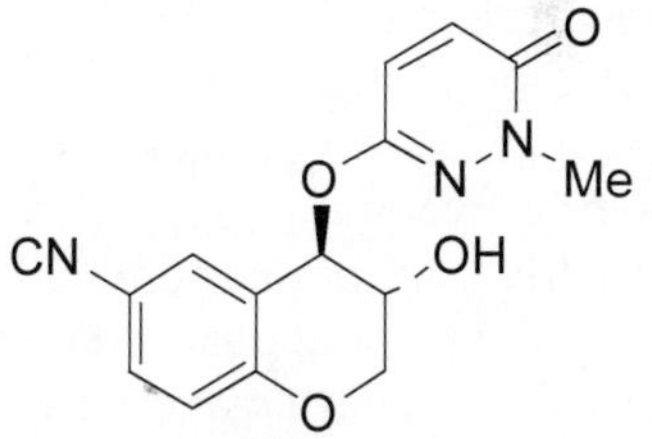 (9)

K_{ATP} Openers

Opening of K_{ATP} has been shown to play an important role in the vasodilatory action of antihypertensive drugs such as cromakalim **1**, pinacidil **2**, dizoxide **3**, and minodixil sulphate **4** (Wille de *et al.*, 1988; Quast *et al.*, 1989). Same mechanism also contributes to the vasodialator action of nicorandil **5** (Karashima *et al.*, 1982). A new structural type RP52, 891 **6** (Mondot *et al.*, 1988), has been added to this growing list of agents. Although many of these compounds may also open other types of channels, their primary mechanism of action, uncovered after the discovery of these drugs as hypotensive agents, is believed to involve opening of potassium channels, which cause membrane hyperpolarization, resulting in raising of the threshold for calcium entry through voltage sensitive calcium channels(Cook *et al.*, 1988; Quast *et al.*, 1989;Wille de *et al.*, 1988).The past few years have seen an explosive growth in the synthesis of compounds related to cromakalim **1**(Cook *et al.*, 1988; Quast *et al.*, 1989; Burrell *et al.*, 1990; Buckle *et al.*, 1991; Bergmann *et al.*, 1990; Malsumoto *et al.*, 2000; Caliendo *et al.*, 2004; Simsek *et al.*, 2004; Schou *et al.*, 2005; Calderone *et al.*, 2005; Weiping Xie *et al.*, 2005; Tyrell *et al.*, 2006; Tripathi *et al.*, 2008) and studies of their structure and activity relationship (SAR) data in terms of a hypothesis that involves binding affinity of these agents to receptor proteins. Since there is no concrete data to support receptor binding by these compounds, one can only assume that hypotensive effects of these molecules are mediated through binding to specific receptors.

Analogues of Cromakalim

Potassium channels play a crucial role in controlling the cell membrane potential. Among the different varieties of K^+ channels, the ATP- sensitive potassium channels (K_{ATP} channels) have been characterized numerous cell types, such as skeletal and smooth muscle cells, endocrine cells, cardiac cells and central neurons. Several molecules are known to activate K_{ATP} channels and have been named potassium channel openers (PCO_S).Such compounds may have a wide therapeutic potential and a few drugs are currently used as antihypertensive agents. Different chemicals series of PCO_S have been explored. This heterogenous group of organic compounds comprises the benzopyran series including potent vasorelaxant drugs, such as cromakalim. The latter compound, a typical example of potassium channel opener, exert its biological effect by activating K_{ATP} channels. This report also highlighted recent developments in the chemistry of cromakalim analogues and reports chemical aspects governing their potency and tissue selectivity. Most cromakalim analogues (Atwood *et al.*, 1989; Stemp *et al.*, 1990; Gericke *et al.*, 1990; Stemp *et al.*, 1991; Tedder *et al.*, 1989; Matsuhisa *et al.*, 1991; Tripathi *et al.*, 2003; Tripathi *et al.*, 2008;) differ with respect to substitution at C-4 of the benzopyran nucleus. The benzopyran is usually substituted at C-6 with an electron withdrawing group, although heterocyclic replacements (pyridine, thiophene, etc) of the aryl ring have also been reported to maintain potency (Evans *et al.*, 1987; Evans *et al.*, 1988). Most potent derivatives also have gem-dimethyl groups or a spirocarbocyclic ring at C-2. The presence of a trans-hydroxyl group does not appear to be critical for hypotensive activity in all series *e.g.*,**7**.The potent antihypertensive activity of the indane analogue,**11** demonstrates that the oxygen atom of the benzopyran nucleus is not a critical component of the pharmacophore. The C-4 substituent can be attached through variety of linkers, which include nitrogen **1,8,10** or oxygen **9**, or carbon **7**. Depending on the substituent at C-4 both SP^3 and Sp^2 carbon are tolerated at 3-and 4- position of benzopyran. In an interesting diversion from the benzopyran nucleus, benzoxazine derivatives *e.g.*, in the structure of **12** are reported to be potent antihypertensive agent (Matsuhisa *et al.*, 1991). The combined bioactivity of indane 11 and 12, indicates that a variety of changes are tolerated in the area of the molecular structures.

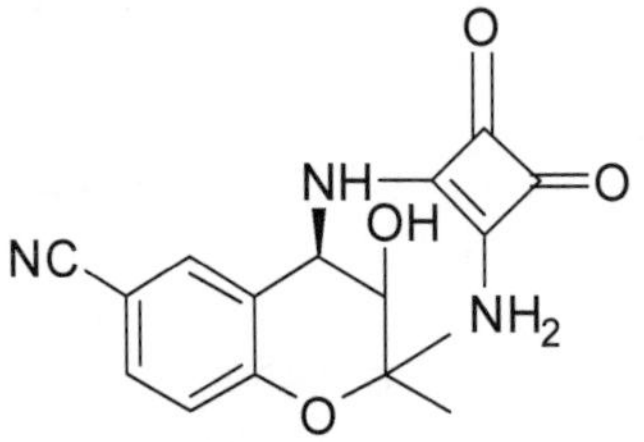

NH OH O O NC NH₂ O

(10)

N O NC OH

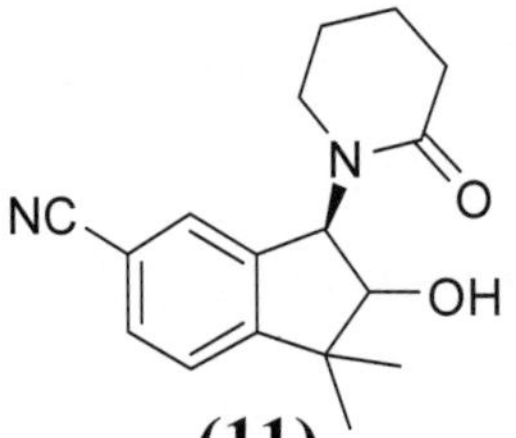

(11)

The pharmacological effect of chromakalim **1** and its analogues are stereoselective, with 3S,4R-isomer being the more potent enantiomer (Grover *et al.*, 1991). It has been well established through these studies that presence of substituent group (electron withdrawing or electron donating) in aromatic ring effect their activity to loser or greater extent.The group such as cyano,nitro, sulphonamide at C-7 potentiate the activity, while electron donating group has less effect, though they may help in modifying the activity. Similarly effect of modification at C-2 and 3and 4 positions in the benzopyran unit also result in changing the bio-activity. The presence of a heterocyclic system at C-4 may also have pronounced effect in over all biological profile on the molecule Figure 22.1.

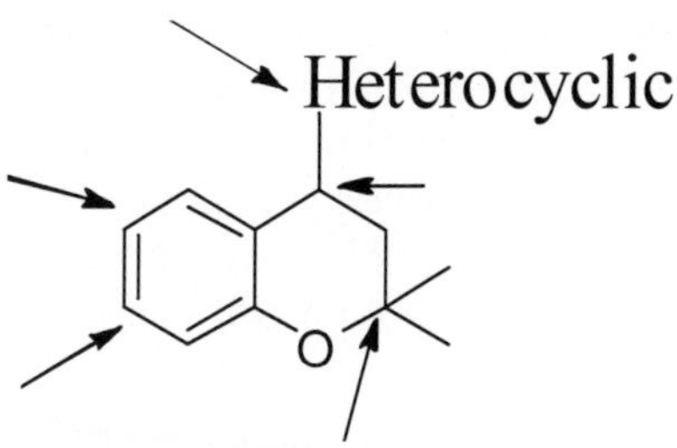

Heterocyclic

Figure 22.1

Potassium channel openers are recognized as having a therapeutic potential for the treatment of various cardiovascular and non cardiovascular diseases. However, the first generation agents open K_{ATP} in a variety of tissues that limit their potential in clinical utility. This report describes our studies aimed at the identification of a common pharmacophore among structurally diverse K_{ATP} openers and the discovery of tissue selective agents that may offer advantages over the first-generation agents for the treatment of cardiovascular diseases. A summary of structure activity relationship (SAR) for the biological activity is given in Figure 22.2.

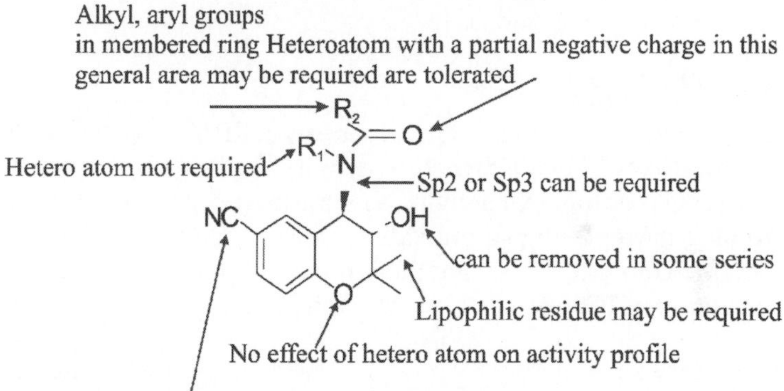

Alkyl, aryl groups
in membered ring Heteroatom with a partial negative charge in this general area may be required are tolerated
Hetero atom not required
Sp2 or Sp3 can be required
can be removed in some series
Lipophilic residue may be required
No effect of hetero atom on activity profile
withdrawing substituents is essential for better activity

Figure 22.2: Summary of SAR.

Analogues of Pinacidil

As bulk of publication has already described the structure – activity relationship of pinacidil **2**, (Peterson *et al.*, 1978) there has not been a great deal of synthetic activity in this structural class. Antihypertensive potency of aryl cyanoguanidine **13** indicates that pyridine is not a prerequist for the biological activity of pinacidil **2** (Atwal *et al.*, 1992; Atwal *et al.*, 1992). Previous studies have shown that cyanoguanidine can be replaced with a thiourea without much loss of potency

(Peterson *et al.*, 1976). A branched alkyl group on the guanidine nitrogen appears to be important for optimal activity (Peterson *et al.*, 1978; Atwal *et al.*, 1992) (Atwal *et al.*, 1992). As the cromakalim 1, the biological activity of pinacidil 2 and its analogue is reported to reside predominately in the R-(-)-enantiomer, indicating their possible interaction with a receptor protein (Steinberg *et al.*, 1990). The ethelnediamine analogue **14** of pinacidil is also reported to maintain potency (Manley *et al.*, 1991).

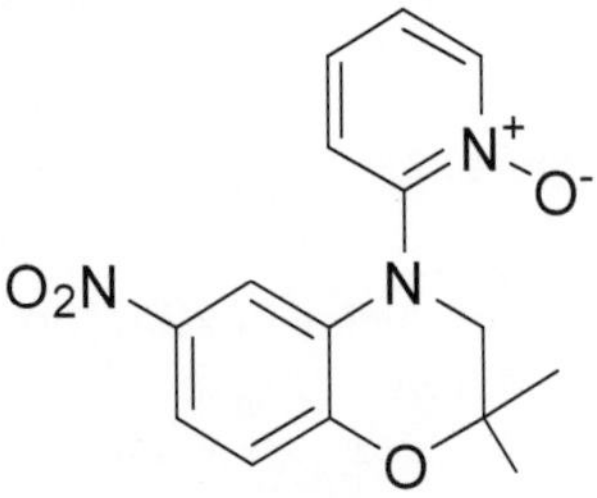

(12) 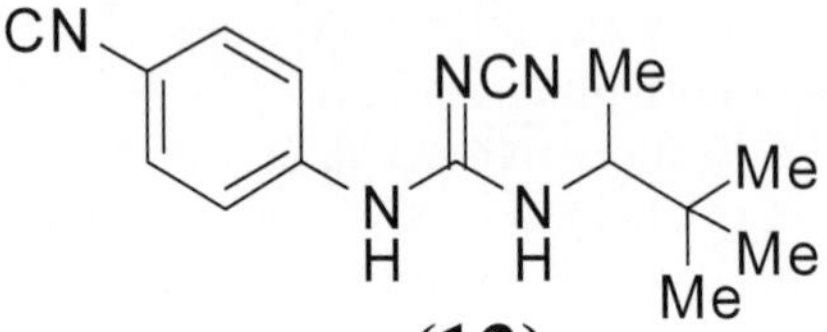 **(13)**

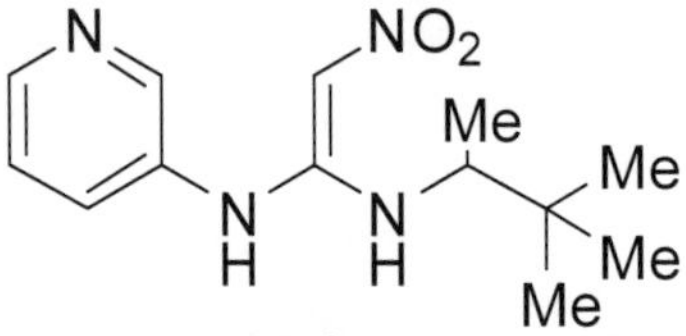

(14)

Analogues of RP 52891

An excellent manuscript describing the genesis of RP 52891(**6**) has appeared in the literature (Aloup *et al.*, 1997). Structure activity studies indicate that the tetra hydrothiopyran ring is optimal for activity, as is the methyl group on the nitrogen of thioamide. Analogues replacing sulphoxide with an olefin *e.g.*, **15** or a substituted alkyl group **16**, (Malcolm *et al.*, 1992) and those replacing pyridine with aryl **17** (Cook *et al.*, 1989) or hetero aryl (Cook *et al.*, 1988) rings have also appeared in the literature.

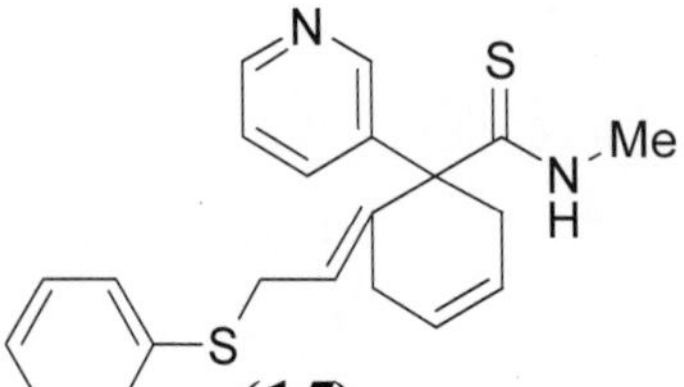

(15) 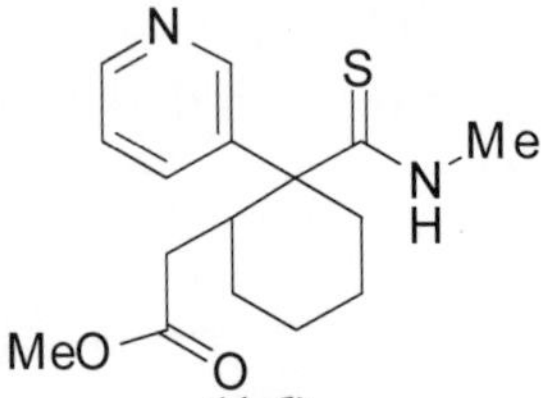 **(16)** 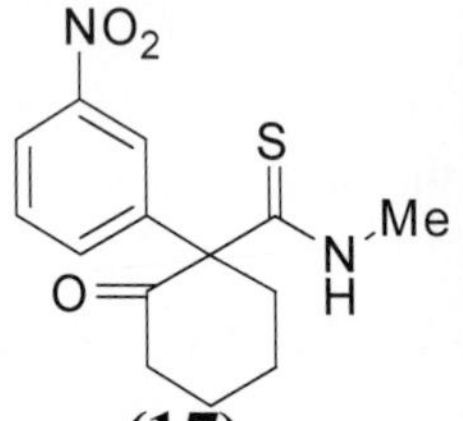 **(17)**

Other Openers

The mechanism of vasorelaxation by calcitonin gene-related peptide (CGRP) has been shown to involve, at least in part, the opening of K_{ATP} in rabbit mesenteric artery (Nelson *et al.,* 1990). A known K_{ATP} blocker glyburide **20**, which by itself had very little effect, inhibited the hyperpolarization caused by CGRP (Nelson *et al.,* 1990). As shown by patch clamp studies, the hyperpolarization caused by CGRP is due to K_{ATP} opening. Blockers of calcium-activated potassium channels had no effect on either vasorelaxation or K_{ATP} opening caused by CGRP. In a similar set of experiments, the endogenously present vasoactive intestinal peptide (VIP) and endothelium derived hyperpolarisation factor were shown to posses K_{ATP} opening as a part of their mechanism of vasodilation. By measurement of ^{86}Rb efflux in the insulinoma cells, it has been shown that the insulin –lowering hormone somastatin activates K_{ATP} (Fosset *et al.,* 1988) This effect is inhibited by the K_{ATP} blockers glyburide **20**.The effect of somatostatin was blocked by preatreatment with pertusis toxin, indicating the involvement of a G protein in mediating the effect of this hormone. The hormone gelatin, which elevates blood glucose by lowering insulin release, has also been shown to activate K_{ATP} in insulinomia cells (Wille de *et al.,* 1988). The effect of both somatostatin and gelanin depend on the intracellular concentration of cAMP (Bergmann *et al.,* 1990). In another study, 4-beta-phorbot-12-myristate-13-acetate was shown to activate K_{ATP} in rat insulinomia cells, suggesting the involvement of a second messenger system (for example, protein kinase C) in regulation of this channel (Bergmann *et al.,* 1990). Dilation of coronary arteries by adenosine in the isolated in the isolated heart has been hypothesized to involve, at least in part, opening of K_{ATP} (Daut *et al.,* 1990).

Common Structural Feature in Synthetic K_{ATP} Openers: 13

Cromakalim 1 and pinacidil 2 may share common structural features, with compounds such as**17**, (Cook *et al.,* 1989) and **18** (Atwal *et al.,* 1992), which are found to be potent vasorelaxant, antihypertensive agents. Antihypertensive activity of the piradyl analogue **19** suggest the presence of common pharmacophoric features in cromakalim **1**.It is of interest to note the most active analogues of RP 52,891 6 contain a pyridine or an aryl ring substituted with electron withdrawing groups. Thus there appears to be a general requirement of an aryl or a heteroaryl ring for optimum vasorelaxant potency. Additionally, most potassium channel openers **1-6** have a lipophilic residue and potential site for hydrogen bond formation. Therefore these compounds may express their biological effects with similar structural requirements. Since there is no evidence to support receptor binding by these molecules, it is premature to speculate whether they bind to single or multiple receptor sites.

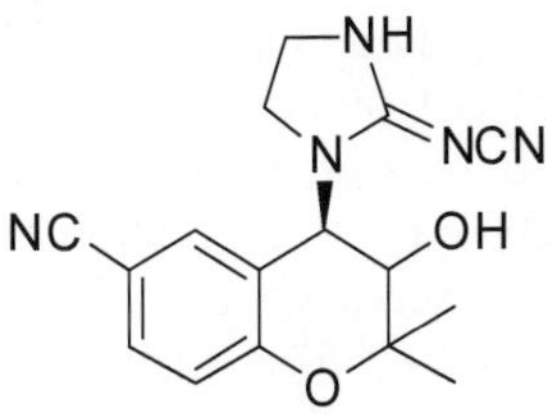

(18) 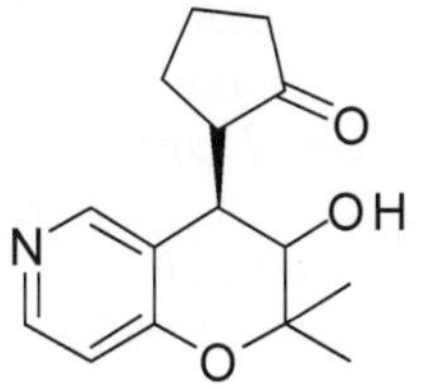 **(19)**

Therapeutic Suggestion

There is extensive clinical data to support the use pinacidil **2** for lowering blood pressure in hypertensive humans (Ahnfeit-Renne *et al.*, 1989). The reduction in blood pressure is accompanied by reflexogenic increase in heart rate, which is attenuated by co-administration of a beta-adrenergic blocker.The other side effects like edema, headache, flushing, etc, of

Pinacidil **2** appears to be typical of most general vasodialation (Goldberg *et al.*, 1988). The beneficial effects of pinacidil **2** on blood lipids seem to diminish on co-treatment with the diuretic hydrochlorothiazide to control edema. The hemodynamic profile of lemakalim (l-cromakalim) in various animal models is similar to cromakalim-**1** (Clapham *et al.*, 1991). The reduction in blood pressure by lemakalim is accompanied by reflux tachycardia and elevation in plasma- renin activity in renal hypertensive cats. (Clapham *et al.*, 1991). The blood pressure lowering in all species is a consequence of reduction in total peripheral resistance. Aside from differences in their potencies, cromakalim 1 and pinacidil **2** have similar pharmacological profiles in animal models. In a study comparing hemodynamic profiles of cromakalim1 has been reported to have a more beneficial effect on renal vascular resistance (Longman *et al.*, 1988). Based on the established intraction of diazoxide with pancreatic beta cell (Kozlowski *et al.*, 1989), there were some concerns that the compounds like cromakalim **1**, pinacidil **2**, and RP 52,891 6 may also cause hyperglycemia by opening K_{ATP} in insulin secreting beta-cells (Dunne *et al.*, 1991). Pinacidil **2** has been shown to actually inhibit through insulin release by increasing potassium efflux through K_{ATP} (Lebrum *et al.*, 1988). Studies by Pratz (Pratz *et al.*, 1991) in normal and hyperglycemic rats have shown that, while diazoxide causes glyburide reversible hyperglycemia, cromakalim**1**, RP 52,891 6, and nicorandil **5** do not effect plasma glucose level at concentrations much higher than those needed for lowering blood pressure. These studies indicate that there may be differences in regulation of K_{ATP} in various tissues and some degree of selectivity may already be built into the prototype potassium channel openers. The beneficial effects of prototype potassium channel openers on the heart are less well understood compared to their potassium channel opening properties, these agents shorten the action potential duration (Smallwood *et al.*, 1988; Osterrider *et al.*, 1988; Sanguinetti *et al.*, 1990). They can be hypothesized to have pro-or antiarrythmic effects, depending upon the type of arrhythmia (Colatsky *et al.*, 1989). Studies in vitro show cromakalim **1** and nicorandil **5** inhibit triggered activity (Lathrop *et al.*, 1990; Fish *et al.*, 1990).

Studies in globally ischemic perfused rat heart by Grover (Grover *et al.*, 1989), show this prototype K_{ATP} openers cromakalim **1**, pinacidil **5**, and RP52,891 **6** improve reperfusion function and reduce lactate dehydrogenase release. These effects of reversible by the K_{ATP} blockers glyburide **20**, indicating that K_{ATP} opening is involved in the cardioprotective effect of these agents. Advantageously, these compounds show cardioprotective activity doses that cause relatively little cardio depression compared to, for example, calcium channel blockers. In order to show efficacy in vivo, cromakalim had to administered directly into the coronary artery (Grover *et al.*, 1989). Because of their potent hypotensive and coronary dialating properties, prototype agents *e.g.*, **1, 2, 6** have a narrow marigin of safety for the treatment of myocardial ischemia, although some report demonstrating cardio protection intravenous administration have appeared in the literature (Maruyama *et al.*, 1990; Murray *et al.*, 1989).

Studies with cromakalim **1** and pinacidil **2** in animal models, (Nelson-kudsk *et al.*, 1988) and with cromakalim in isolated human bronchiole also suggest the clinical use of K_{ATP} opener for asthma (Dunne *et al.*, 1990). Cromakalim **1** was shown to be effective in inhibiting the the histamine-induced broncho- constriction in healthy humans (Baird., *et al.*, 1988; Baird *et al.*, 1988) and it was found to be effective in patients with nocturnal asthma (Williams *et al.*, 1988). Although the hypotensive activity of cromakalim does not appear to be a problem in these studies, the identification of compounds selective for the bronchial tissue is desirable for these agents to succeed in the general patient population. Since cromakalim and pinacidil relax the bladder smooth muscle, (Malmgren*et al.*, 1989; Anderson *et al.*, 1989) K_{ATP} openers may find use in the treatment of urinary incontinence (Hsin-I *et al.*, 2001).

Recently, cromakalim **1** has also been shown to prevent ischemia-induced damage in a rat skeletal muscle preparation (Hatton *et al.*, 1991).This result and the observation that cromakalim restores the membrane potential of depolarized human skeletal muscle fibres, (Quasthoff *et al.*, 1990) indicate that K_{ATP} openers may also be useful for the treatment of peripheral vascular disease. Studies in vitro and animal models indicate the potential clinical utility of K_{ATP} openers for diseases of CNS. In in-vitro experiments, K_{ATP} openers somatostatin and diazoxide **3** were shown to prevent anoxia-induced depolarization of CA **3** hippocampal neurons (Krnjevic *et al.*, 1990). These effects were inhibited by pretreatment with glyburide **20**. Cromakalim **1** and RP52,891 **6** have been shown to posses antileptic activity in a rat model of epilepsia, In another study cromakalim prevented pilocarpine induced behavioural response in mice (Tricklebank *et al.*, 1988). These studies are encouraging, but tissue selective agents that cross the blood- brain barrier are needed to realize the potential clinical applications for CNS. By virtue of their smooth muscle relaxing effects, K_{ATP} openers may also be useful for the treatment of irritable bowl syndrome, (Perkins *et al.*, 1989) and premature labour (Piper *et al.*, 1990).The use of minoxidil in treating male pattern baldness may indicate stimulation of hair growth to be a general property of K_{ATP} openers. Since the existing agents are rather promiscuous, it is clear that tissue selective compounds need to be developed to explore full clinical potential of K_{ATP} openers

K_{ATP} Blockers

The most widely known compounds that block K_{ATP} are the sulphonyl urea class of antidiabetic agents (Gylfe *et al.*, 1984). Their mechanism of hypoglycemic activity involves blocking of K_{ATP}, which causes membrane depolarization, leading to calcium entry through the voltage sensitive channel and consequently, release of insulin (Rorsman *et al.*, 1985). These compounds are useful for the treatment of non-insulin dependent diabetes. Sulphonyl ureas are also known to block K_{ATP} in cardiac (Fosset *et al.*, 1988) and brain tissues (Bemardi *et al.*, 1988). In addition to their therapeutic value, these compounds have served as important tools for studing K_{ATP} in various tissues. Glyburide **20** and its analogues have been extensively employed to study the sulphonyl urea receptor and its coupling to K_{ATP} in various tissues. There is a good correlation between the potencies of various analogues to display radio labeled glyburide from its receptor site and hypoglycemic action in beta-cell tumor, indicating the sulphonyl urea receptor may be the K_{ATP} forming proteins (Schmidt-Antomarchi *et al.*, 1987). There appears to be a difference in the potencies of sulphonyl ureas in inhibiting ^{86}Rb effluxes through K_{ATP} in insulin-secreting beta cells and Substantia nigra, suggesting the sulphonyl urea receptors may not be identical in the various tissues (Amorso *et al.*, 1990). The binding of [^{3}H]-glyburide is biphasic, depending on the tissue, and it is inhibited by certain nucleotides in the presence of magnesium and dithiotheritol (Gopalakrishnan *et al.*, 1991; Gopalakrishnan *et al.*, 1993). These studies confirm earlier observation that there may be more than one nucleotide binding site (Lederer *et al.*, 1989) and that a sulphonyl group is present at or near the K_{ATP} (Weik *et al.*, 1989). The structure of some of the sulphonyl urea analogues **20-25** are given on next page.

The potencies of various hypoglycemic agents to inhibit glyburide binding has been described by Geisen and co-workers (Geisen *et al.*, 1985). A non sulphonyl urea hypoglycemic agents linogliride **26** has also been reported to inhibit K_{ATP} in beta-cells (Ronner *et al.*, 1991). Interestingly, this compound does not have any acid functionality, usually present in glyburide **20** and its congeners. Sodium 5-hydroxyl decanoate, **27** has been reported to block K_{ATP} in guinea pig ventricular cells (Niho *et al.*, 1987). The same agent has been also demonstrated to block the anti-ischemic effect of potassium channel openers cromakalim in globally ischemic rat Langendorff heart (Mc. Cullogh *et al.*, 1991). These effects of sodium 5-hydroxydecanoate are similar to those observed in glyburide 20 (Grover *et al.*, 1989). Recent studies suggest that phentolamine **28** type alpha-adreno receptor blockers (yohimbine, antazoline, tolazoline) block K_{ATP} (Mc. Pherson *et al.*, 1989; Dunne *et al.*, 1991) Phentolamine **28** and its analogue alinidine **29** inhibit the vasorelaxation and Rb efflux is caused by K_{ATP} openers cromakalim and pinacidil in a number of smooth muscle preparation (Mc. Pherson *et al.*, 1990). However, the failure of other types of alpha-adreno receptors blockers such as prazocin, rauwolscine, and phenoxybenzamine to block the effects of cromakalim indicates that this is not a general property of alpha-adrenoreceptor blockers (Mc. Pherson*et al.*, 1989).Certain antiarrythmic drugs such as the class III agents Tedisamil **30** K (Bray *et al.*, 1991) and the class I/IV agent BRL 31660 **31**(Arch *et al.*, 1991) can also be hypothesized to have some K_{ATP} blocking activity, as they inhibit the pharmacological action of cromakalim.

Therapeutic Suggestion

The single most important clinical use of sulphonyl ureas and other compounds that block K_{ATP} in insulin secreting beta cells is for the treatment of non-insulin dependent diabetes (Dunne *et al.*, 1990). Although concerns over the development of tolerance to these agents remain, they are the most prescribed drugs for type II diabetes. Under ischemic conditions, opening of K_{ATP} Contributes to the loss of potassium from the myocardium (Kleber *et al.*, 1984; Weiss *et al.*, 1989; Wille de *et al.*, 1988).This has been thought to contribute to ischemic damage leading to derangement of electrical activity, which results in lethal arrhythmis (Harris *et al.*, 1984). Thus blocking of K_{ATP} can be hypothesized to minimize myocardial injury and prevent ischemia induced arrhythmias (Bekheit*et al.*, 1989).

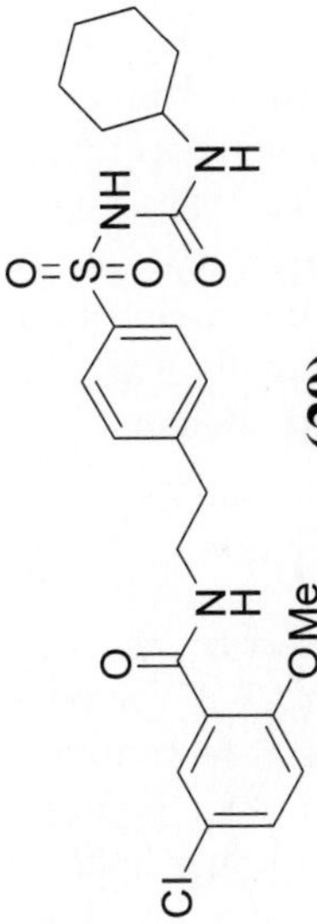

(20)

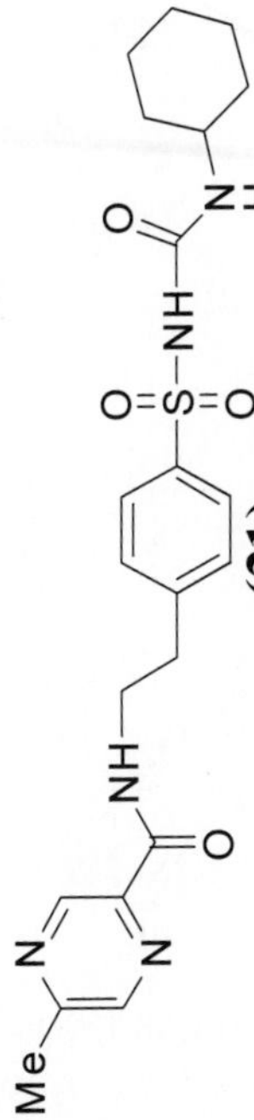

(21)

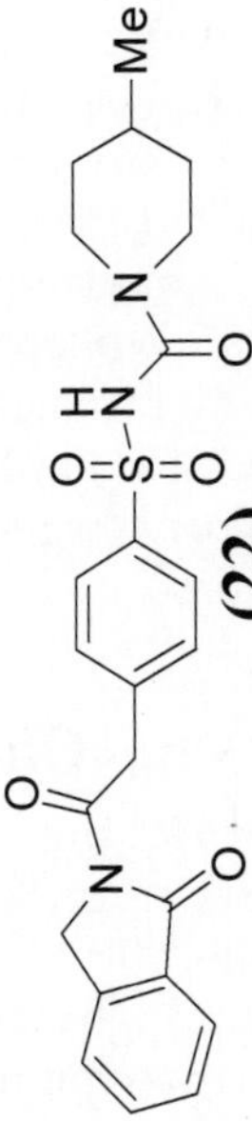

(22)

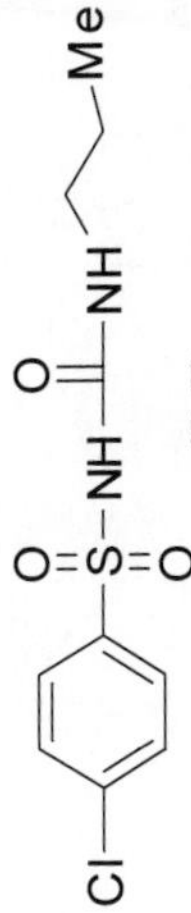

(23)

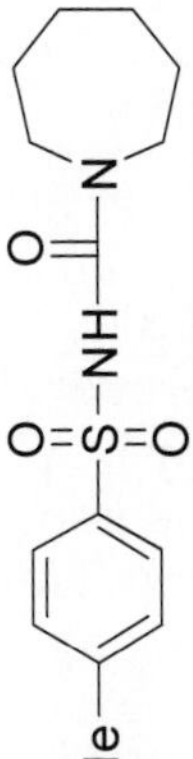

(24)

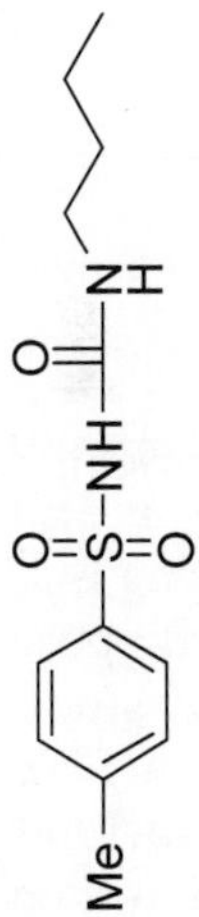

(25)

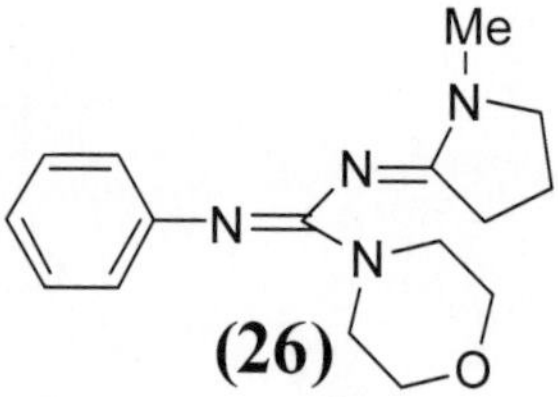

(26) 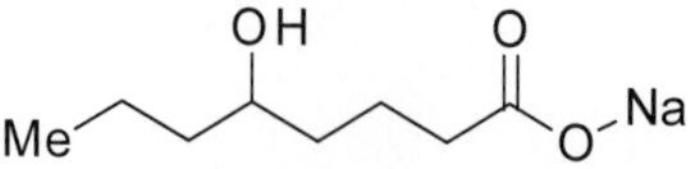 **(27)**

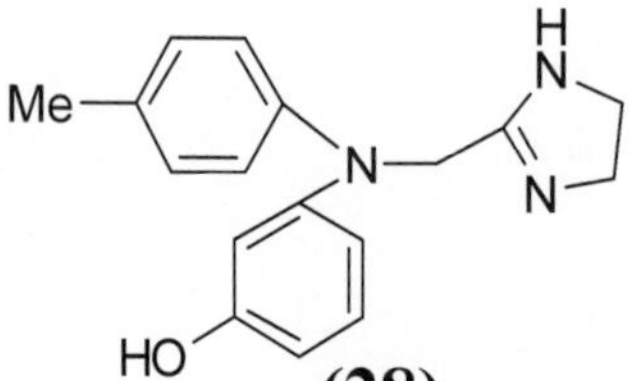

(28) 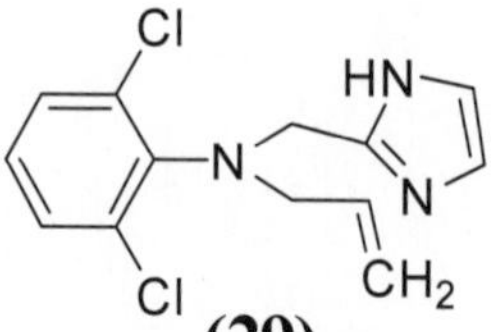 **(29)**

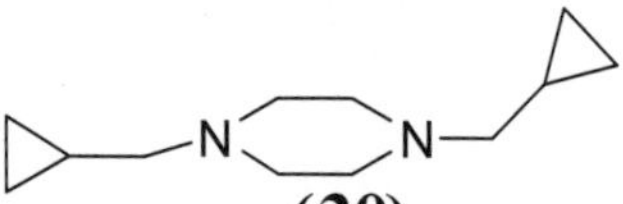

(30) 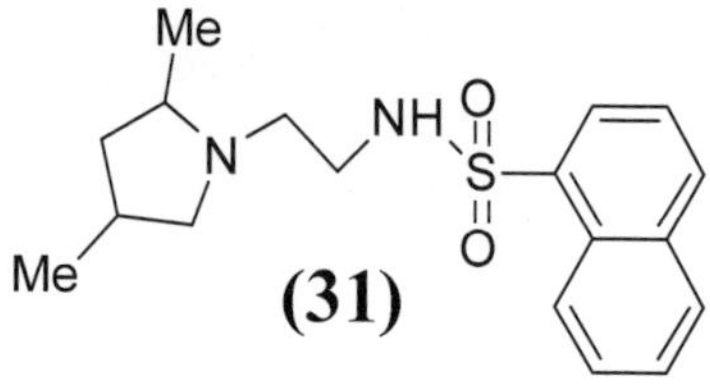 **(31)**

Studies support a beneficial role associated with K_{ATP} opening in the ischemic myocardium and more importantly they indicate that potassium efflux through the channel during ischemia may be physiological response of the heart to minimize injury. A cardioprotective role associated with K_{ATP} opening during ischemia is also supported by studies by Gross, (Auchampach *et al.*, 1991) showing K_{ATP} blockers inhibit the beneficial effects of preconditioning, an episode of brief coronary artery occluding. Experiments showing the utility of K_{ATP} blockers as potential ant arrhythmic agents in the isolated heart, (Kantor *et al.*, 1987; Wollenben *et al.*, 1989) have been difficult to reproduce *in vivo*. Recent studies by Lucchesi (Uprichard *et al.*, 1989), show glyburide **20** has no effect on ischemia-induced fibrillation in a canine model of myocardial ischemia. These data nicely compliment the results showing lack of effect of glyburide **20** on ischemia-induced damage in isolated perfused rat hearts (Grover *et al.*, 1989; Mc. Cullogh *et al.*, 1991).

Calcium- Activated Potassium Channel (K_{ca})

Calcium-activated potassium channel (K_{Ca}) is activated by membrane depolarization and by increases in intracellular calcium (Blatz *et al.*, 1987), depending on single channel conductance or maxi-K(100-250Ps), intermediate conductance (18-50Ps), or low-conductance(10-14pS) K_{Ca}. The high conductance K_{Ca} is present in neurons, cardiac cells, and various types of smooth muscles. (Singer *et al.*, 1987; Bard *et al.*, 1989; Ahmed *et al.*, 1984; Rosa *et al.*, 2000)· The intermediate conductance channel has been shown to be present in red blood cells (Hamil *et al.*, 1981)[158], and in smooth muscle (Innoue *et al.*, 1986).The low conductance channel is present in a variety of cell types (Inoue *et al.*, 1985). The main function of K_{Ca} is thought to be its involvement in repolarization of the action potentials, secretion, and volume regulation. K_{Ca} may be modulated by other regulated mechanisms. For example, in rat brain. K_{Ca} appears to be regulated by Mg-ATP via a protein kinase-catalyzed phosphorylation (Chung *et al.*, 1989). Cyclic AMP dependent kinase (A-kinase) is reported to activate K_{Ca} in tracheal myocytes (Kume*et al.*, 1989) aortic smooth muscle cells (Sadoshrina *et al.*, 1988), and in neurons (Ewald *et al.*, 1985). Existing evidence also suggests that K_{Ca} may be modulated by guanosine-5-monophosphate I vascular smooth muscle cells (Jr. Williams *et al.*, 1988), and by a combination of inositol-1,4,5-triphosphate and inositol-1,3,4,5-tetrakisphosphate in mouse lacrimal acinar cells (Morris *et al.*, 1987).

Openers of K_{Ca}

There are very few organic molecules known to specific openers of K_{Ca}. Gelbland have reported the opening by cromakalim of what appears to be a high- conductance K_{Ca} in rabit aortic smooth muscle incorporated into planar lipid bilayers (Gelband *et al.*, 1989). This effect is dose dependent and it is reversed by glyburide (Gelband *et al.*, 1989). Kloockner' s work also supports opening of K_{Ca} by cromakalim (Klockner *et al.*, 1989; Klockner *et al.*, 1989). Studies by Hermsmeyer suggest that pinacidil opens high conductance K_{Ca} in rat veins (Hermsmeyer *et al.*, 1978). Cromakalim and pinacidil analogue (P-1060) have been shown to increase the open-state probability of maxi-K chanels in smooth muscle cells enzymatically dissociated from rat portal vein (Peterson *et al.*, 1978). These effects were blocked by glyburide and charbdotoxin (toxic polypeptide), indicating that the maxi-K channel is sensitive to these blockers in the rat portal vein. Glyburide, but not charbdotoxin, inhibit the effects of cromakalim in several smooth muscle preparations (Hu *et al.*, 1990).

The calcium channel blocker niguldipine **32**, (Wickenden *et al.*, 1991) has been shown to activate K_{Ca} in vascular smooth muscle (Klockner *et al.*, 1989; Klockner *et al.*, 1989). The contribution of K_{Ca} to the vasorelaxing properties of nigludipine **32** is not clear as it also blocks voltage- sensitive calcium channel and alpha-1 adrenoreceptors. (Boer *et al.*, 1989).

Evidence also exists to support that the amino acid L-alanine **33** opens K_{Ca} in rat liver cells (Bear *et al.*, 1987). The full potential of K_{Ca} openers or blockers cannot be appreciated until tissue and channel specific are developed.

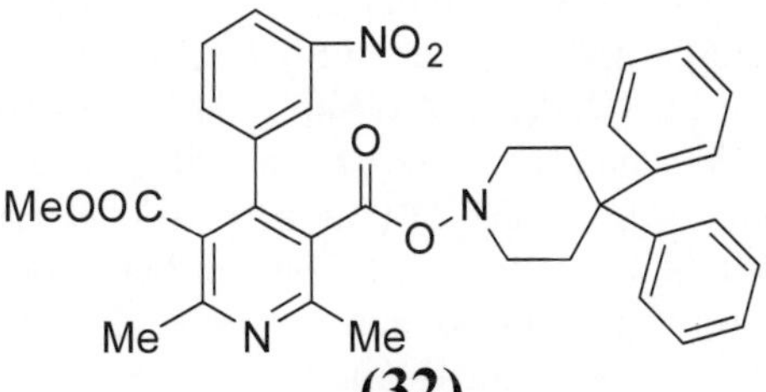

NO_2 O MeOOC N O Me N Me

(32)

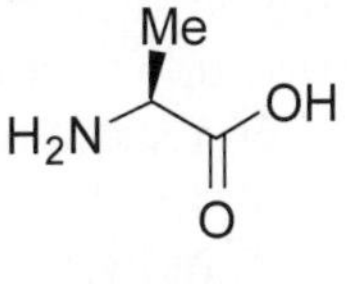

Me H_2N OH O

(33)

K_{Ca} Blockers

The most important tools to distinguish between low and high conductance K_{Ca} are the toxins (polypeptides) apamin (Marquese *et al.*, 1987) and charbdotoxin (Gimenez-Gallego *et al.*, 1988). While carbdotoxin specifically blocks maxi-K, apamin is a potent blocker of the low conductance K_{Ca}. However, in certain tissues, for example, in rat brain, charbdotoxin may block K_{Ca} of all three types (Moczydlowsky *et al.*, 1988). Although less potent than carbdotoxin, both tetraethyl ammonium (TEA) and Ba^{+2} block maxi-channel. Quinine and charbdotoxin, in addition to blocking maxi-K, weakly block the intermediate-conductance channel. Both apamin charbdotoxin have high affinity binding sites in a variety of tissue (Moczydlowsky *et al.*, 1988). The charbdotoxin block of K_{ca} has been proposed to involve electrostatic interaction between the positively charged toxin and an an acidic residue at or near the mouth of channel (Mackinnon *et al.*, 1989).It remain to be demonstrated whether a similar electrostatic interaction is involved in the inhibition of K_{ca} by other toxin. The solution structure of the 37-amino acid charbdotoxin has been described recently (Bontems *et al.*, 1991; Massefski *et al.*, 1990). The three dimensional structural of charbdotoxin and proposed importance of critical amino acids in blocking K_{Ca} (Mackinnon *et al.*, 1989), provide an excellent opportunity for the medicinal chemist to design small molecule opener/blocker for the charbdotoxin-sensitive K_{ca}. At a three fold higher dose, charbdotoxin exacerbated brain edema. The blockers of K_{ca} may find some use in reducing brain injury due to ischemia A peptide fragment, obtained by enzymatic digestion of charbdotoxin, was also found to retain protective effect against ischemic and traumatic injuries (Ohnishi *et al.*, 1991). There is no indication as to whether the effect of this peptide is still indicated through inhibition of K_{ca}. These results suggest that some of the pharmacological effect of charbdotoxin can be retained in small fragment, indicating the potential for structure -activity studies to arrive at a small peptide or non peptide modulator of K_{ca}

Delayed Rectified Potassium Channel (I_k)

Delayed rectifier potassium channel is known to exist both in cardiac (Matsura *et al.*, 1987), and non-cardiac cell types (Conti *et al.*, 1990). This channel is main function is through to be its involvement in repolarisation of the action potential (Noble*et al.*, 1969). It has been shown to be composed of two current, a rapidly activating compounds blocked by certain class III antiarrhythmic drugs and a slowly activating components (Sanguinetti *et al.*, 1990; Gerlach *et al.*, 1985). The organic compounds known to block this channel are some of the so called class III antiarrhythmic agents. Class III agents suppress arrhythmia by prolongation of the cardiac action potential duration and effective refractory period (Jr. Williams *et al.*, 1988). Examples of these compounds include clofilium **34** (Arenna *et al.*, 1988), flecainide **35** (Follmer *et al.*, 1990), sotalol **36** (Sanguinetti *et al.*, 1990), E-431 **37** (Sanguinetti *et al.*, 1990), UK-68,798 **38** (Black *et al.*, 1991), UK-66,914 **39** (Gwilt *et al.*, 1988), and **40** (Connors *et al.*, 1991). Additionaly, tetraethyl ammonium, aminopyridines, quinine, bretilium, and strychnine may also block this current. While both d- and l-enantiomers of sotalol **36** are potassium channel blockers,the beta-adrenoreceptors blocking activity resides predominately in the l-enantiomer (Bartek *et al.*, 1985; Carmeliet *et al.*, 1985; Hewawasam *et al.*, 1988).

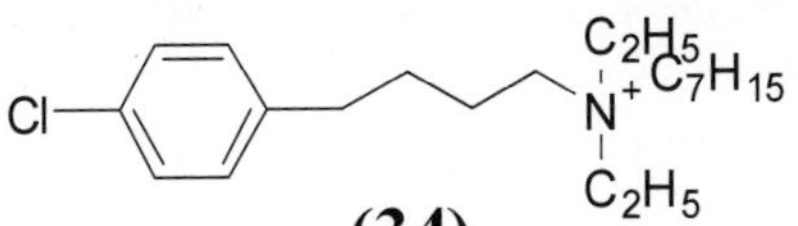

(34)

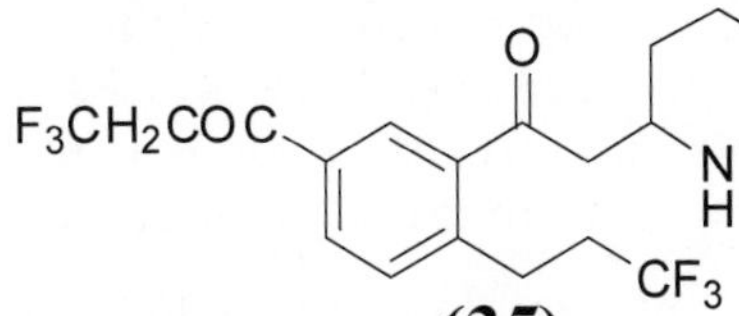

(35)

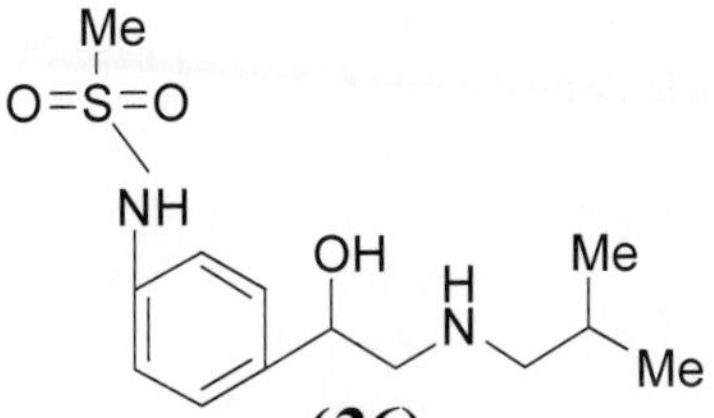

(36)

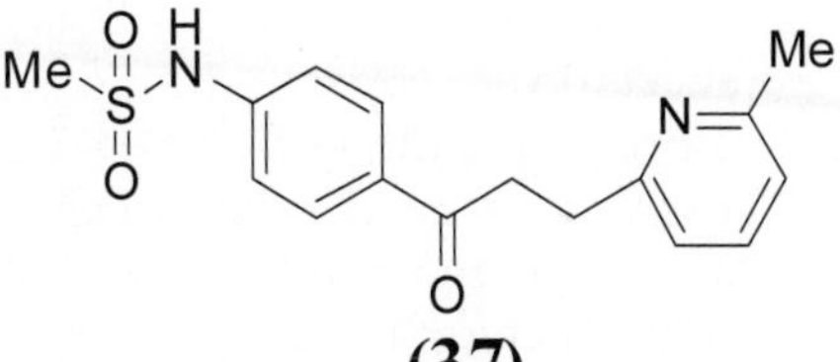

(37)

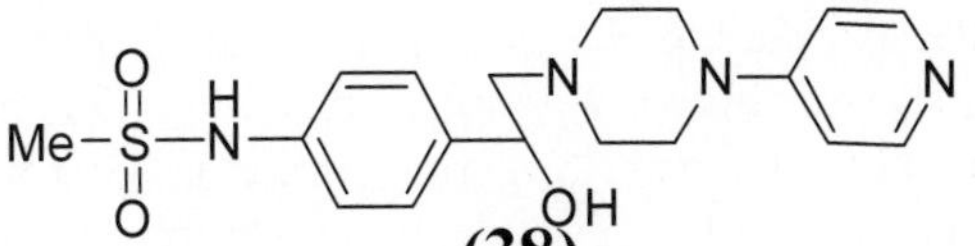

(38)

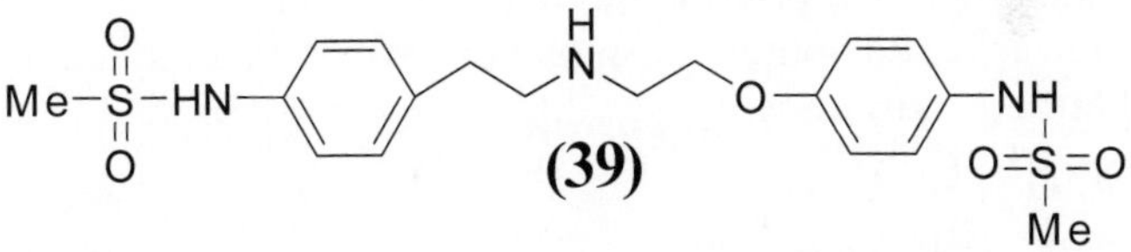

(39)

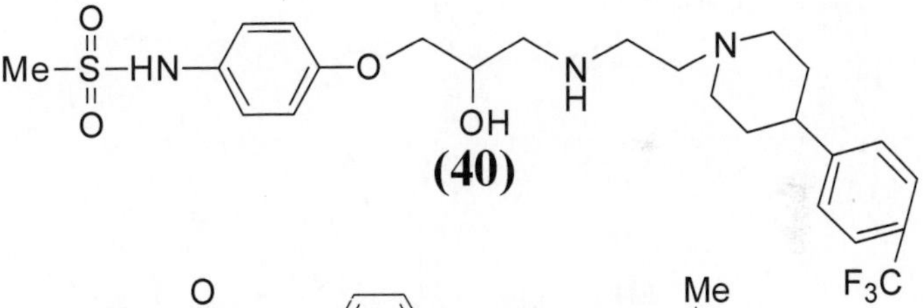

(40)

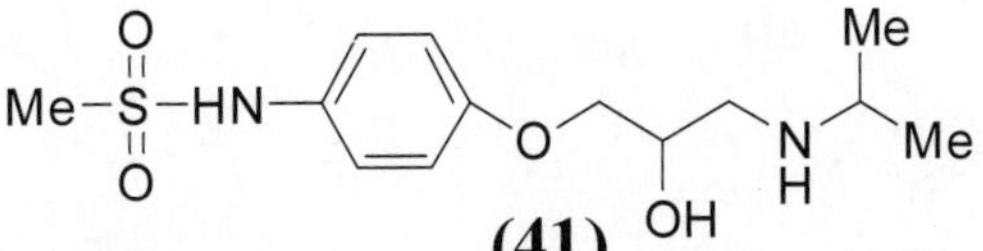

(41)

As it is clear from the structure of compounds **37-41** (Hewawasam *et al.*, 1988; Carmeliet *et al.*, 1985; Gomoll *et al.*, 1986), sotalol **36** has served as a starting point for the synthesis of most of the benzene sulphonamide type of class III agents. Studies by Baskin and co-workers show UK-68, 798, 39 to be more potent than both E-4031, **37** and sotalol **36** in increasing myocardial refractory period (Baskin *et al.*, 1991). There is a large body of structure –activity data published on the prolongation of the action potential duration by benzene sulphonamide type compounds (Lis *et al.*, 1987). Based on their structural resemblances to the benzene sulphonamide type of antiarrythmic agents **36-41**group of compounds **42-51** (Jr. Morgan. *et al.*, 1990, Lis *et al.*, 1987, Cross *et al.*, 1988, Jr. Lumma *et al.*, 1987), may also block I_K.

Fatty-Acid Modulation Potassium Channels

Arachidonic acid and certain related fatty acids can directly open potassium channel in cardiac, (Kim *et al.*, 1989) and smooth muscle cells (Ordway *et al.*, 1989). These effects do not appear to be mediated through lipoxygenase or cycloxygenase-derived products (Kim *et al.*, 1989). The effect of fatty acids on potassium channel opening is also independent of nucleotides (ATP, GTP) or calcium. These data indicate that fatty acids and the process that control their release may be important physiological determinants in control of cell excitability. In smooth muscle they may antagonize agonist induced contraction of membrane hyperpolarisation and consequently limit calcium entry (Ordway *et al.*, 1989). Since free acid levels rise during cardiac ischemia, opening of these channels may limit cell damage by inhibiting calcium entry and conserving ATP (Kim *et al.*, 1989). This protective roll may be similar to that played by cardiac K_{ATP} during ischemia (Grover *et al.*, 1989; Grover *et al.*, 1990). It has been shown that 5-lipoxygenase derived products of arachidonic acids can open pertusis toxin sensitive G-protein-gated muscarinic potassium channels in cardiac cells (Kuichi*et al.*, 1989; Kim *et al.*, 1989). Although further work is necessary to determine the physiological significance of these results, it does appear that arachidonic acid and some of its metabolites may be mediate their effects via potassium channel modulation.

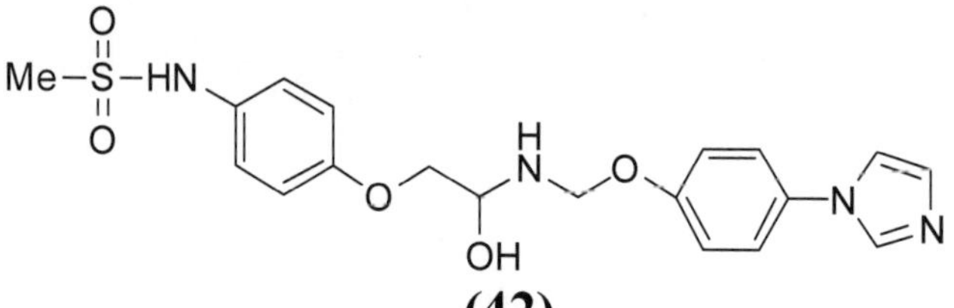

(42)

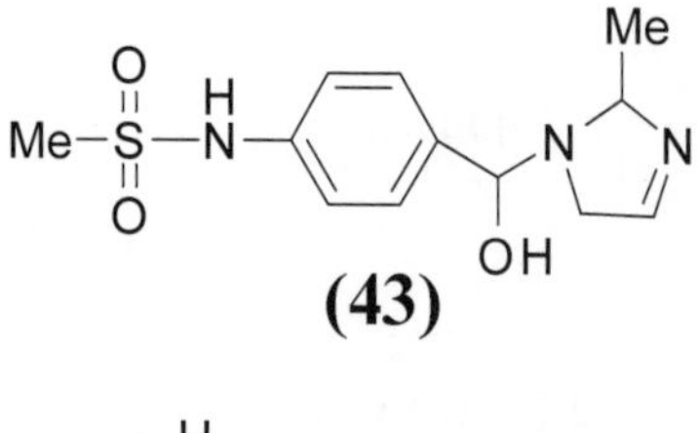

(43) 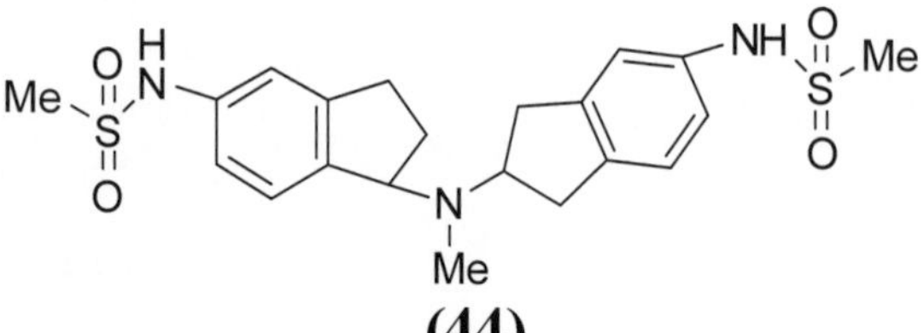 **(44)**

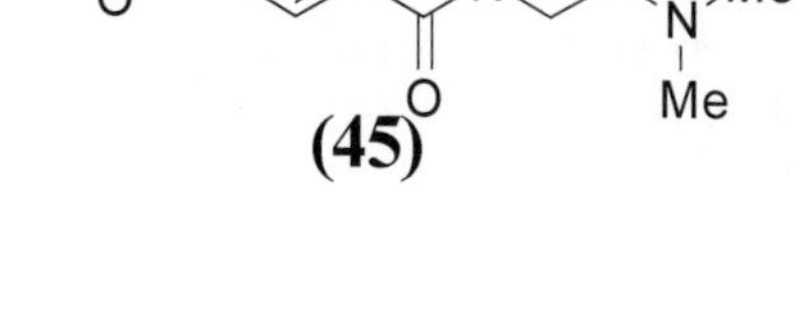

(45) 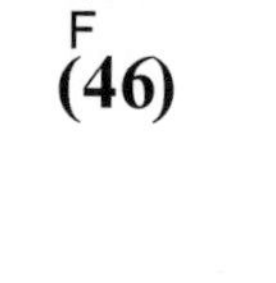 **(46)** 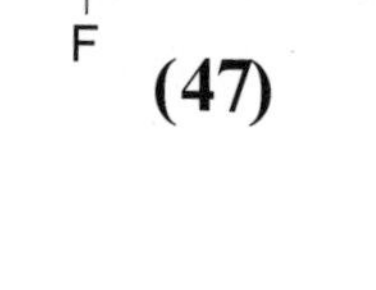 **(47)**

Anesthetic-Activated Potassium Channel

Some evidences have been found that in voltaic anesthetics such as halothane **46**, isoflurane **47**,chloroform, and diethyl ether may be mediated their effects by opening a neuronal potassium channels (Franks *et al.*, 1988). The opening of this channel is reversible. Since a good correlation could be found between channel opening and inhibition of spontaneous firing in a given neuron, potassium channel opening may be relevant to the mechanism of action of these anesthetics (Franks *et al.*, 1988). Recent studies show the (+)-Enantiomer of isoflurane **47** to be two fold more potent than the (-)-enantiomer, indicating the probable binding of these compounds to a receptor (Franks *et al.*, 1991).

However, it is not clear that whether these agents act directly on the potassium channel proteins or if their effects are mediated through secondary messenger system.

Molecular Mechanism of Potassium Channel Gating

Using technique of molecular biology, protein chemistry, and electrophysiology, the molecular mechanism of potassium channel gating are beginning to be unrevealed. Much of this progress has been made possible by cloning of the shaker gene of Drosophila (Papazian *et al.*, 1987). The cloning of this gene has allowed the identification of a number voltage-activated potassium channel proteins, which appear to be related in structure (Tempel *et al.*, 1990). Four such proteins make up a single potassium channel (Mackinnon *et al.*, 1991). In an excellent review articles by Miller *et al.* (1991) and Aldrich (Hoshi *et al.*, 1990; Caterall *et al.*, 1988), details of channel structure and gating mechanism are discussed. The inactivation process is hypothesized to obey the ball and chain model (Annstrong *et al.*, 1977). According to this model, the channel contains an amino acid region (ball) on the cytoplasmic side near the amino terminal of the monomeric proteins, which is connected to the membrane –spanning region through small stretch (chain) of amino acids. Once it is open, the ball is hypothesized to provide a physical plug to close the channel. The inactivating region is shown to interact with the voltage-sensitive region to cause charge immobilization and prevent it from returning to its original confirmation (Bezanilla *et al.*, 1991). This simple model explains the voltage sensitivity of the channel opening and its blocking by carbdotoxin and tetraethyl ammonium by binding to the pore- forming region between S5 and S6 domains (Miller *et al.*, 1991).

Current Trends

For many years potassium channel has dominated the field of ion channel research. The initial euphoria over the discovery of potassium channel opening as a mechanism for the vasodilator/antihypertensive activity of several existing agents is slowly settling down. Our knowledge of the functioning of various potassium channels is increasing at a fast rate. The diversity of potassium channels in various tissues and their controlling basic cell functions are being unraveled by interdisciplinary research. New uses are being explored, and they may determine the future of this class of agents. Since there are several types of potassium channels, often in the same tissue, channel and tissue selectivity plays a major role in advancing these agents to clinical stage. Although most of the pharmacological effects of the K_{ATP} openers (*e.g.*, cromakalim) are reversed by sulphonyl urea type K_{ATP} blockers (*e.g.*, glyburide), and other K_{ATP} openers do not effect glyburide binding at therapeutically relevant concentration (Angel *et al.*, 1991). Definitive experiments need to be performed to understand at what level the effects of potassium channel openers are reversed by sulphonyl ureas. Additional comes from the controversy over the identity of the potassium channel(s) effected by cromakalim and other prototype compounds.

The availability of organic molecules in large numbers that modulate these channels in a tissue-specific manner would be useful to explore the therapeutic potential of potassium channel modulation. Medicinal chemistry research has so far concentrated only on modification of the existing molecules. The design of molecules, which are channel and tissue specific, relies heavily on the progress in understanding the structural biology and the protein chemistry of potassium channels. At the present time, little is known about the three dimensional structure of channel protein(s) or the second messenger system that regulate their activity. Our understanding of the basic mechanism(s) of channel functioning is slowly progressing, but further work is needed to unravel how various channels differ in structure and their gating properties. This is very crucial for understanding pharmacological profile of existing potassium channel modulators. There has been a rapid extension in the synthesis of molecule related to nicorandil, cromakalim, pinacidil, in last few years. Reports of work on the pyridine regiomers or substitution of other amides for the pyrrolidinone have appeared already in the literature. Cromakalim by far remains the most mimic structure in the class of compounds.6-substituted-2,2-bis(fluoromethyl)benzopyran-4-carboxamide potassium channel openers showed a highly potent, slow and long lasting antihypertensive effect with reduced reflux tachycardia, together with the beneficial effects of potassium channel openers such as improvement in lipid metabolism. They have identified **48** as a potential candidate for PCM's (Naoki *et al.*, 2000).

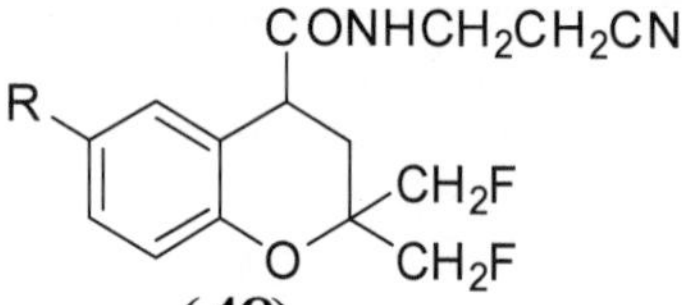

$CONHCH_2CH_2CN$
R
CH_2F
O
CH_2F

(48)

A publication reports an unsubstituted pyridine adduct **49** as highly active, where as substitution in the pyridine ring leads to a decrease in activity. Strongly electron-withdrawing substituents at the C-6 position are required for optimal activity. When the 2-pyridone ring is replaced by other heterocycles such as 4-pyridone, pyrimidone, pyrazinone, and 1,4-butanesultam, the activity is maintained (Bergmann *et al.*, 1990). The elimination of water from the chromanol leads to the formation of the chromenes, which are among the most potent antihypertensive known. Alkyl ester at the 3-hydroxy group with short-chain acid also,maintain their activity. The epoxidation of the chromene double bond also produces active compounds. Reduction of the ketone **50** produces both cis- and trans- isomer *e.g.*, **51** with different activity profile (Bergmann *et al.*, 1990).

In some of the novel congeners of cromakalim the pyrrolidin-2-one has been methylated *e.g.*, **54** (Englert *et al.*, 1988), or replaced with acylic amides *e.g.*, **59** (Evans *et al.*, 1986), and **60** (Quagliato *et al.*, 1990) the phenyl ring replaced with pyridine regiomers *e.g.*, **53** (Evans*et al.*, 1988), and a pyridine regiomer carrying trifluoromethyl substituent (Lang *et al.*, 1988). Furthur, structure have been synthesized where

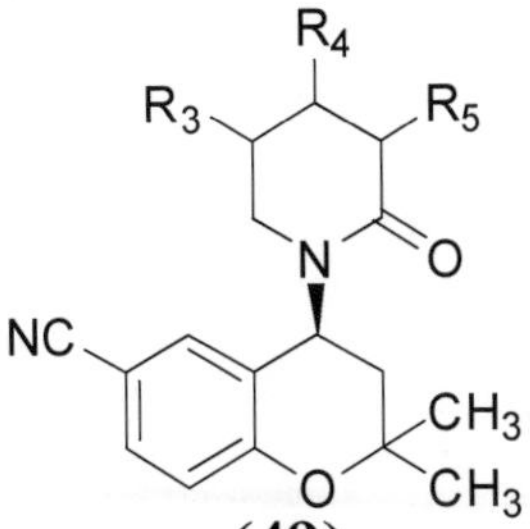

(49)

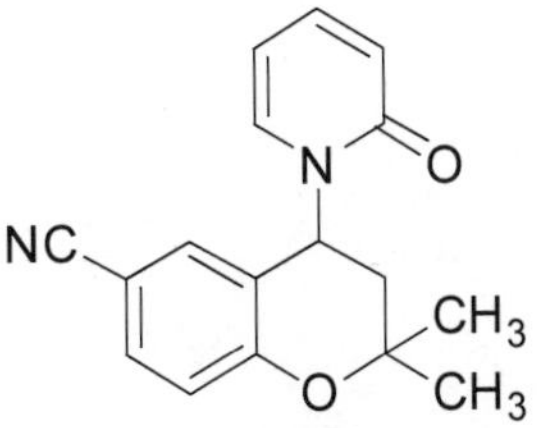

(50)

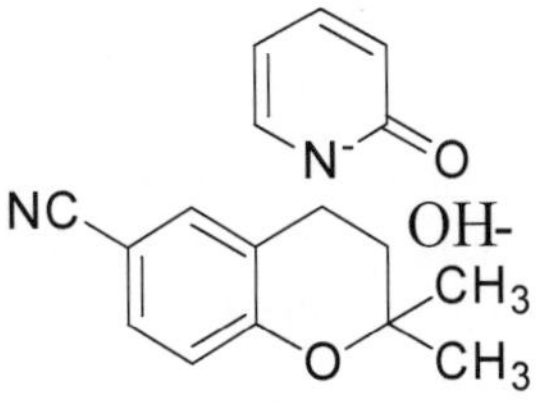

(51)

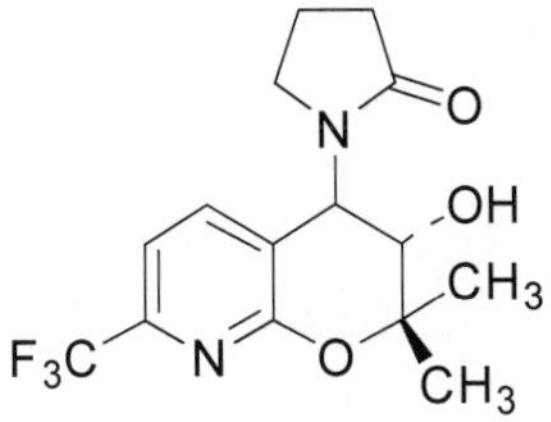

(52)

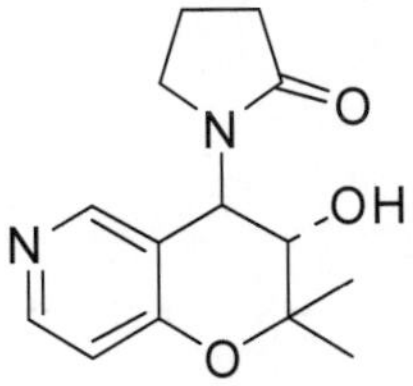

(53)

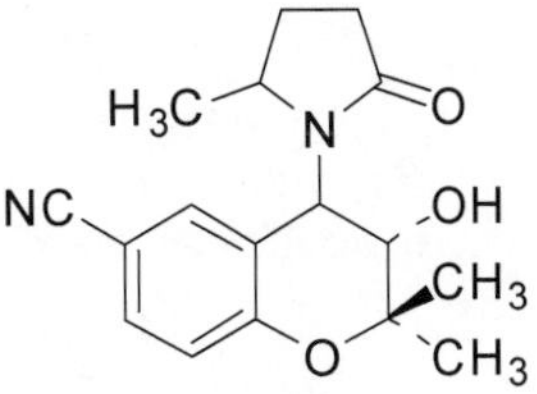

(54)

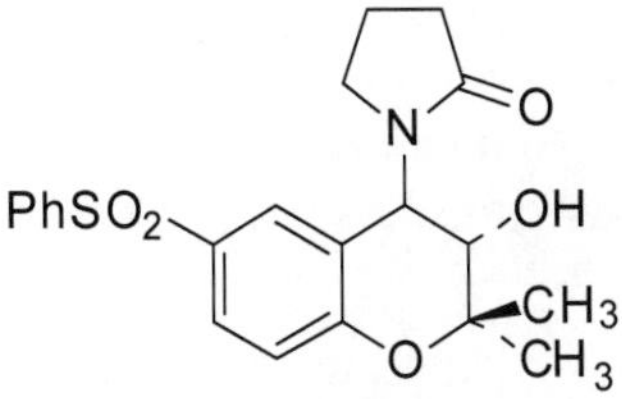

(55)

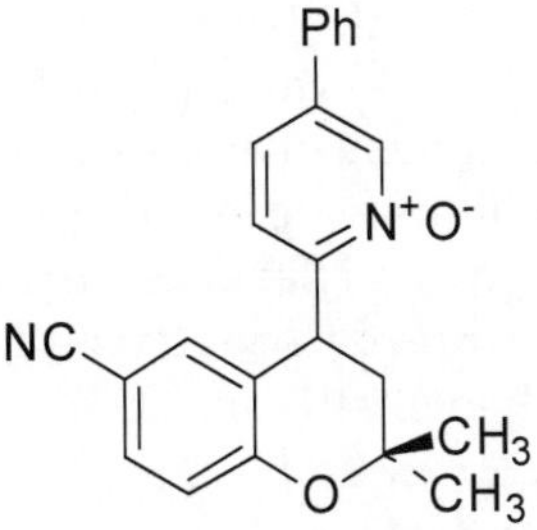

(56)

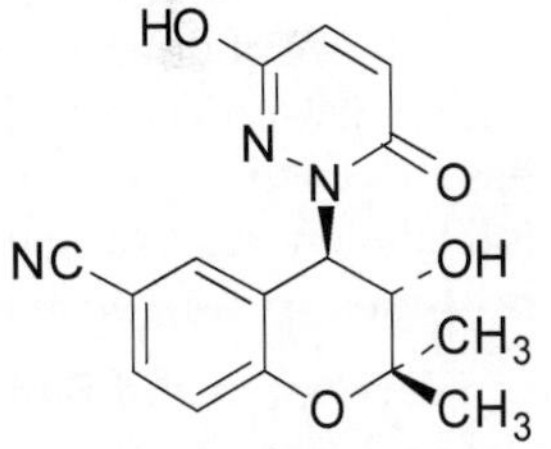

(57)

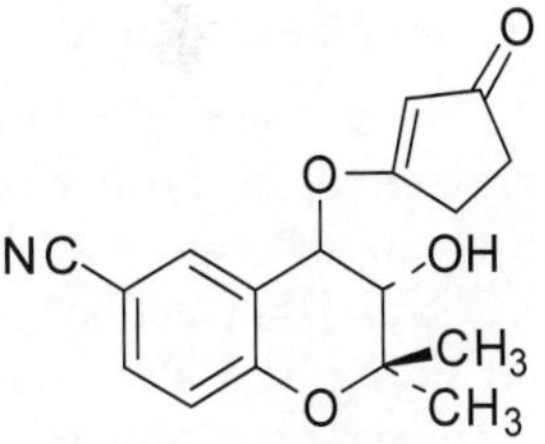

(58)

substitution in pyridine ring and 6-cyano group is replaced by a phenyl sulphonyl group *e.g.*, **55** (Englert *et al.*, 1988), the geminal dimethyl group is replaced with a spirocyclic cyclopentane moiety *e.g.*, **63** (Bergmann *et al.*, 1990), and dehydration of the chromene bimakalim was also reported (Sassen *et al.*, 1990). Compounds containing heterocycles, other than pyrrolidine, have also been synthesized such as **56** (Atwood *et al.*, 1989), and **57** (Garcia *et al.*, 1988), together with compounds containing a cyclopentenone ring, linked via an oxygen atom *e.g.*, **58** (Blarer *et al.*, 1988). In addition, the benzopyran nucleus has also been replaced with a benzofuran moiety *e.g.*, **62** (Masuda *et al.*, 1990).

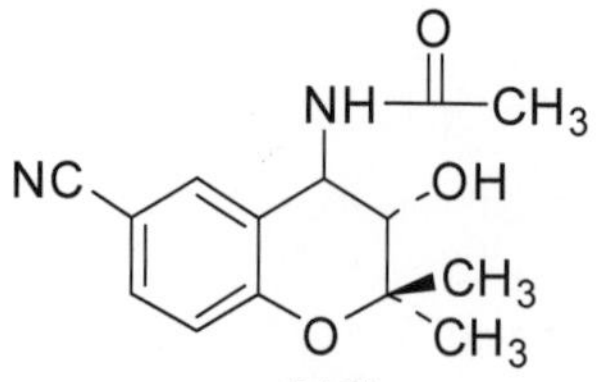

(59) 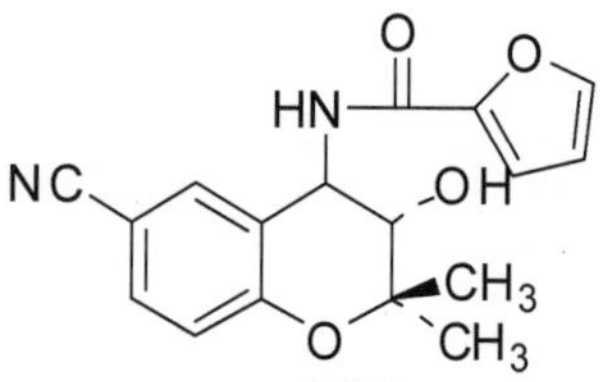 (60) 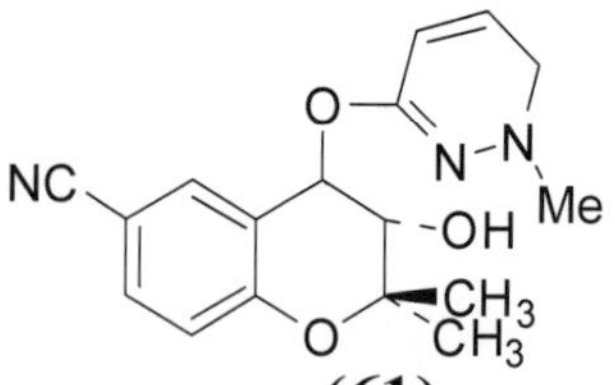 (61)

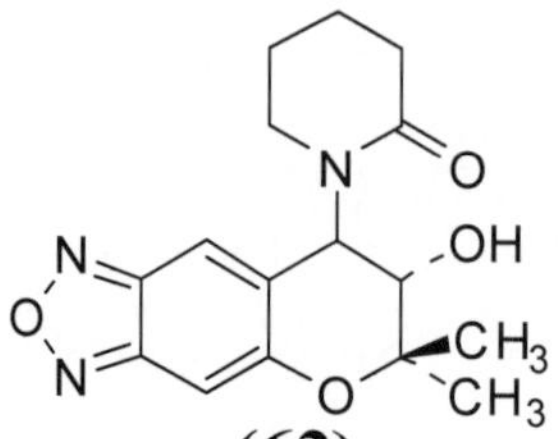

(62) 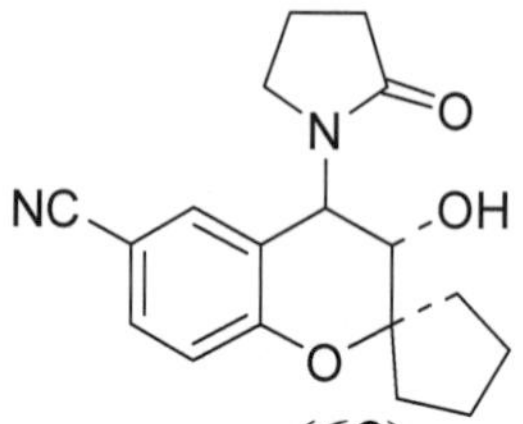 (63)

Recent Status

In the recent years a lot of works has been made with reference to potassium channel. Among them in 2013 an article published by Chang Hui showed the importance of flavonoids in treatment of breast cancer, initially molecules associated with potassium channels inhibitors were less effective towards cancer treatment. He has shown the intake of flavonols and flavones, but not other flavonoid subclasses or total flavonoids, is associated with a decreased risk of breast cancer, especially among post-menopausal women (Chang Hui *et al.*, 2013).

Derivatives of Benzopyran from the ancient age has proved to as potassium channel openers (KCOs) having antihypertensive, cardio-protective, myocardial protectors, powerful peripheral vasodilators and anti-ischemic activity have proved successively. Their usage as anti-ischemic including angina, hypertension and diabetes is thought to be due to the stimulation of KATP channels which are contemplated to produce vasorelaxation and myocardial protection.

It is observed that potassium channels are involved in mediating the cardio-protective effects of pre-conditioning in animal models and man. KCOs protect heart from an ischemic insult without contribution from vasodilatation. Characteristics of KCOs and their actions on subtypes used widely for the treatments of various diseases including hypertension, cardiac ischemia, arrhythmia, smooth muscle relaxation, diabetes, cardio-protective and anti-angiogenic activities have been shown by Mohsina Bano *et al.*(2013).

Evaluation of anti-ischemic properties of new spirocyclic benzopyran activator of the cardiac mito-KATP channel has been done in 2013. Anti-ischemic effects of the spiro-cyclic derivative on isolated rat hearts exhibited cardioprotective effect which was mainly mediated by channels expressed on mito-KATP. They have also studied this molecule more extensively using diazoxide as reference mito-KATP opener.

Protective effects of compounds under study and diazoxide were also recorded in H9c2 cells and again were inhibited by 5-HD. Compound of study and diazoxide caused swelling of cardiac mitochondria, in agreement with the profile of mito-KATP openers. These compounds evoked concentration-dependent Ca-release from Ca^{2+} release-preloaded mitochondria, prevented mitochondrial Ca^{2+} uptake and caused mitochondrial membrane depolarization. These effects were antagonized by ATP, the endogenous KATP inhibitor (Calderone *et al.*, 2010).

The synthesis and biological evaluation of 5-membered spiro heterocyclic benzopyran derivatives **66** was applied against myocardial ischemia. KATP play a key role in an endogenous "self-defence" mechanism known as ischemic pre-conditioning (IPC) which is fundamentally involved in the protection of the heart against the ischemia/reperfusion injury exogenous activation of KATP channels by pharmacological tools. It can be viewed as one of the most promising strategies for the therapy of myocardial ischemia. Apart from their synthesis and the evaluation of new cardioprotective agents, they have extensively studied several six membered spiro-heterocyclic benzopyran compounds (Rapposelli *et al.*, 2011).

X= CO_2, CH_2

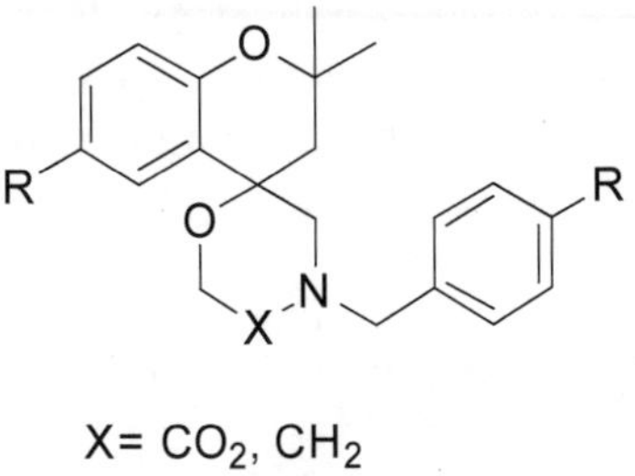

66

Synthesis and studied the electrochemical properties, cytotoxic effects and antioxidant activity of 5-amino-8-methyul-4H-benzopyran-4-one and its copper (II) complexes were done. The chromone derivative 5- amino-8-methyl 4H-benzopyran-4-one (ligand) was used to obtain a series of Cu(II) complexes as potential KCOs. The molecular structures of ligand and its Cu (II) complex have been determined by X-ray crystallography. The cytotoxicity of all obtained compounds has been evaluated on melanoma A375 cell line (Grazul *et al.*, 2012).

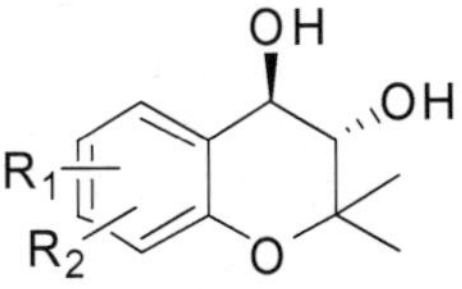

67

Ligand 5-amino-8-methyl-4H-benzopyran-4-one and its three new copper (II) complexes were synthesized, Which possess only one functional group of the two potential donors, 4-O and 5-N atoms. They are also involved in redox reactions and the reversibility of the processes is not dependent on the pH of environment. Antioxidant activity of tested compound is related to ligand. These compounds were tested for in vitro antitumor activities against human melanoma cell line A375. All of these results demonstrated that these complexes have better anti-proliferative activities in comparison with ligand. The highest anticancer activity was observed in cultures treated with copper (II) complex containing perchlorate. Its activity in A375 cell cultures was six times lower than that observed in cultures treated with cisplatin (Grazul *et al.*, 2012)

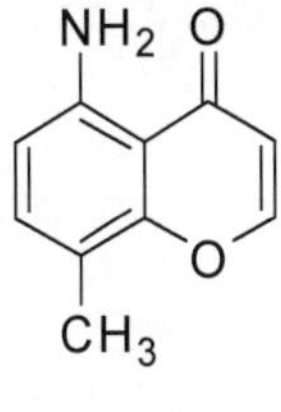

Synthesis and evaluation of a new series of cyclobutyl-6- (4-piperidyloxy) spiro [benzopyran-2, 40-piperidine] derivative 71 as high affinity and selective histamine-3 receptor (H3R) antagonists. The spirobenzopyran piperidine ether analogs demonstrated excellent

68

H3R affinity and selectivity against histamine receptor subtypes (H1R, H2R, and H4R). They were stable in liver microsomes and had selectivity against CYP P450 enzymes. Compounds under study demonstrated high H3R affinity, in vitr liver microsomal stability, selectivity against CYP isoforms (Dandu *et al.*, 2012).

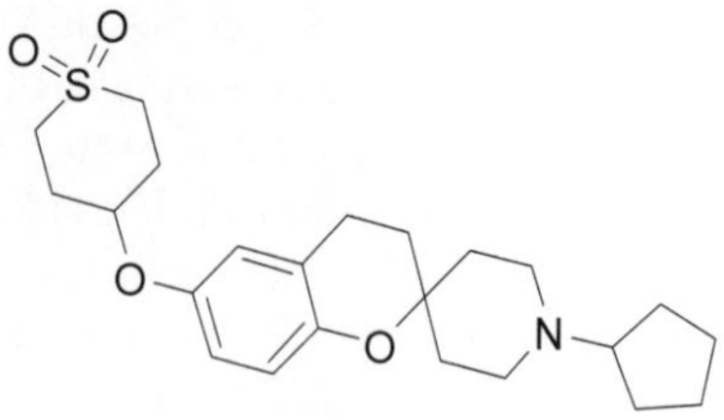

69

Pharmacological activity of N-(2,2-dimethyl-3,4-dihydro-2H-1-benzopyran-4-yl)-4H-1,2,4-Benzothiadiazine-3-carboxamides 1,1-dioxides were done after synthesis,on rat uterus, rat aorta and rat pancreatic cells. Pharmacological studies conducted on rat uterus indicated that several of these original hybrid compounds displayed a strong myorelaxant activity. The most active compounds hold a bromine atom at 6th position of the dihydrobenzopyran ring. Moreover, the compounds failed to display a marked inhibitory effect on insulin secretion and vascular myogenic activity. Finally, they suggested that the 6-bromo compounds could be relatively selective towards the uterine smooth muscle (Khelili *et al.*, 2012).

Evaluation of antiplasmodial activity of synthesized novel chromeno [2, 3-b] chromene derivatives **73** Among them 5-Methoxyflavenes and 6-methoxyflavenes were found to undergo stereoselective acid-catalyzed rearrangement to generate a range of novel chromeno [2, 3- b] chromene derivatives. When subjected to an in vitro antiplasmodial growth inhibition assay using *Plasmodium falciparum* (3D7 line), the chromene analogs were shown to display IC50 values ranging from 6.8 to 39.8 lM (Devakaram *et al.*, 2012).

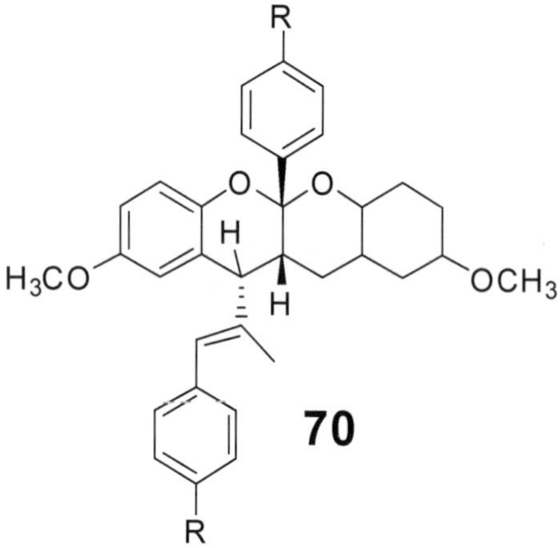

70

A series of bisindolylmethanes (BIMs) were prepared for bioevaluation using 4-keto7-alkoxy benzopyran. The results of initial antimicrobial screening of compounds showed to be the most potent inhibitors, exhibiting MIC as well as MBC values equal to or less than that of ciprofloxacin (0.5–2 lg/mL) against *Staphylococcus aureus*, MRSA and VRE. Some of the compounds were selected further to study the effect of antitumor activity. It was observed that substitution at N- position diminishes its antibacterial

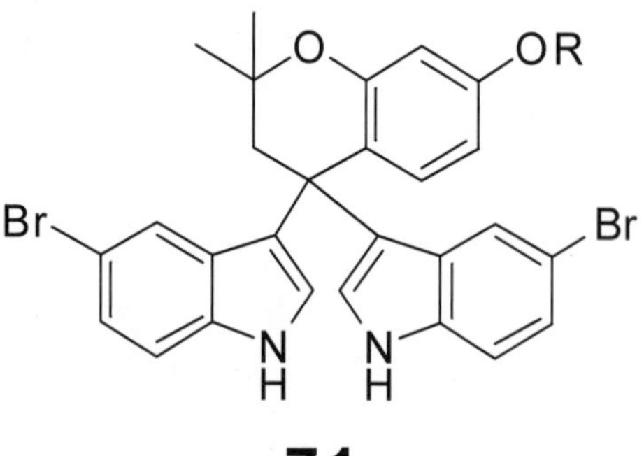

71

activity though *in vitro* antitumor activity against a panel of prostate, cervical and lung cancer cell lines remains more or less intact (Tripathi *et al.*, 2013).

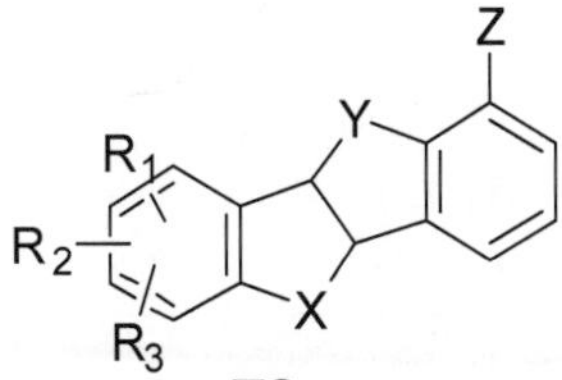

During the preparation of the present article some additional research work has appeared in the area of preparation of newer potassium channel activators. Some of the relevant synthetic research work generally covering the literature up to the latest concerning is being also reported here. A recent US patent (Antane *et al.*, 2001), revealed preparation of optical active substituted benzofuranidoles and idenoindoles **72** as novel potassium channel activators, which are found useful in the treatment of disorders associated with smooth muscle contraction via potassium channel modulation such as urinary incontinence and irritable syndrome.

72

A publication described the synthesis (Naoki *et al.*, 2000), of benzopyran-4-amides as potassium channel openers via the catalytic amidation using palladium as a catalyst in 99 per cent yield. The reaction of 4-bromo-2′2-dimethyl-6-nitro-2H-1-benzopyran with H_2NCH_2CN and CO in DMF in the presence of $Pd(OAc)_2$, potassium iodide and triphenyl phosphine gave 99 per cent N-(2-Cyanoethyl)-2,2-dimethyl-6-nitro-2H-1-benzopyran-4-carboxamide.

As a part of a program to develop new compounds through modifying cromakalim, the synthesis and potassium channel opening activity of substituted 10H-benz [4,5] fluoro[3,2-b]indole and 5,10-dihydro-indeno [1,2b]indole-1-carboxylic acids have been reported (Butera *et al.*, 2001), clamming to posses potent bladder-sensitive smooth muscle relaxant properties and thus are potentially useful for the treatment of urinary incontinence.

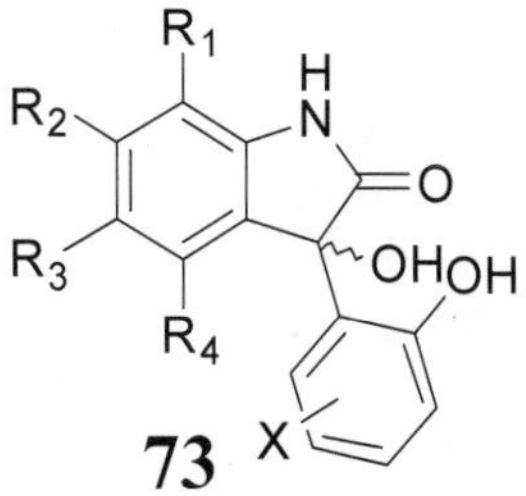

73 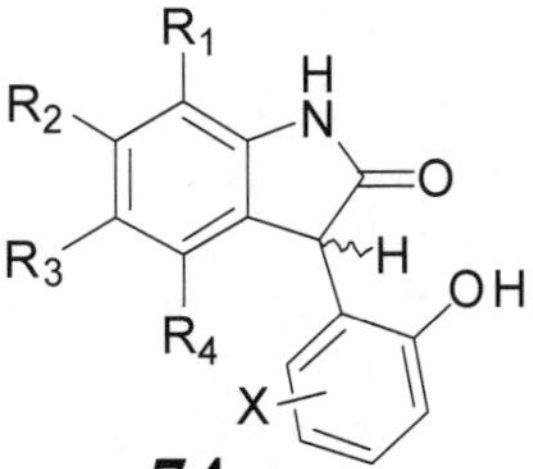 **74** 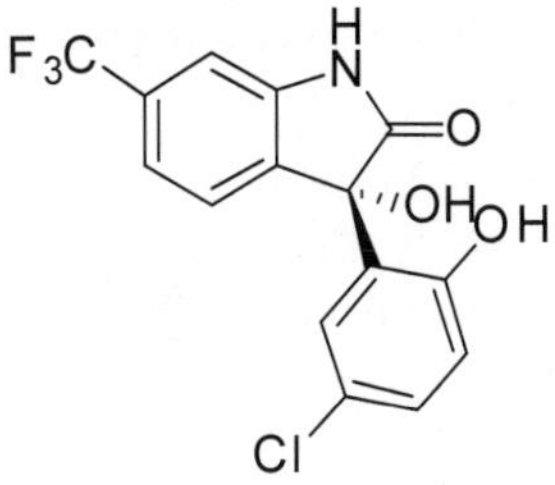 **75**

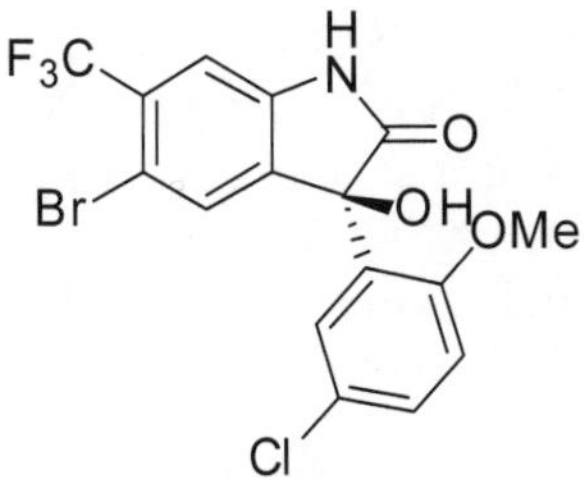

76 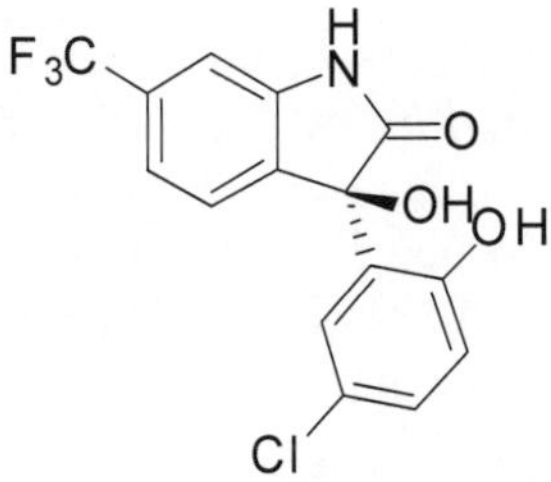 **77**

An excellent review article (Yogupoloskii *et al.*, 2001), for calcium channel blockers and/or fluorinated analogues of pinacidil, an opener of ATP sensitive potassium channels has also appeared describing the synthesis and pharmacological properties of novel type of vasorelaxant hybrid compounds.

Another paper describing the synthesis of (+/)(5chloro2methoxyphenyl)-1,3-dihydro-3-fluro-6-(triflurométhyl)-2H-indole-2-one has been reported, (Kiesewetter *et al.*, 2002) as a potent, specific and novel opener of cloned large-conductance, calcium activated (maxi-K) potassium channels. One of its enantiomer, *i.e.*, BMS-204352(Maxi Post), in undergoing clinical evaluation for its efficacy in patients with suspected acute stroke.

Another paper (Hewawasam *et al.*, 1988), in the series has been published for the synthesis and structural activity relationship of 3-aryloxindoles, a new class of calcium dependent, large conductance potassium (Maxi K) channel openers 64b-64 with neuro protective properties.

In a recent report (Hewawasam *et al.*, 2002), synthesis and characterization of BMS-204352(Maxi Post) and related 3-fluorooxindoles as openers of maxi-K potassium channels have been made. These compounds can be synthesized in two steps by the addition of aryl Grignard reagent to an isatin, followed by treatment with DAST. It can also be isolated using chiral HPLC.

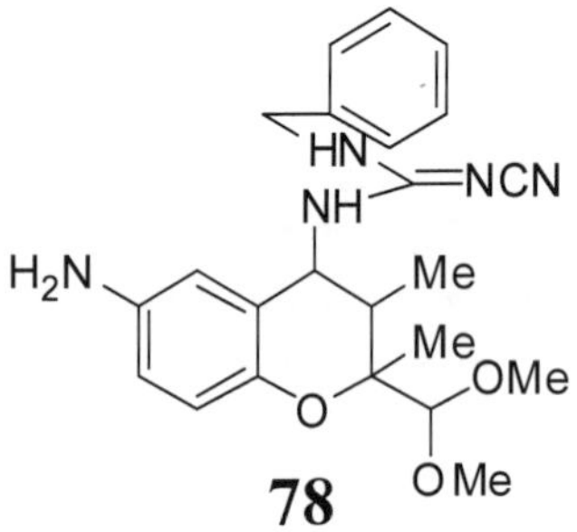

A novel anti-ischemic ATP-sensitive potassium channel opener like N-(6-Aminobenzopyranyl)-N′-benzyl-N″-cyano guanidine analogue has been developed. The design, synthesis, and biological evaluation of a novel anti-ischemic compound (Sung-eun *et al.*, 2001), (2S, 3S, 4R)-N-(6-amino-3,4-dihydro-2-dimethoxymethyl)-N′-benzyl-N″-cyanoguanidine 64g.

HN NH =NCN H$_2$N Me Me OMe O OMe

78

The new class of potassium channel openers namely (BK) channels are recently appeared in the literature (Calderone *et al.*, 2006). The large conductance calcium-activated potassium (BK) channels are expressed in excitable as well as in non-excitable cells. They control several cell function in the nervous system, BK channels contribute to the shaping of action potential and modulate the neuronal excitability and the release of neurotransmitters, also BK channels play a fundamental role in the regulation of the tone of smooth muscle cells (Calderone *et al.*, 2002; Wu *et al.*, 2003).

A publication described the synthesis (Khelil *et al.*, 2006) of some N-arylsulphonyl-N-methyl-N′-(2, 2-dimethyl-2H-1-benzopyran-4-yl) ureas and evaluated as putative potassium channel openers on the vascular and uterine smooth muscle tissue (myorelaxant effect), as well as on insulin secreting pancreatic islets(inhibition of insulin release). The pharmacological results indicated that these compounds exhibited a marketed biological activity on these three tissues.

Another recent publication discloses an excellent range of novel benzopyrans, which have been synthesized and biologically evaluated for K_{ATP} channel activity employing cromakalim as a benchmark K_{ATP} channel opener (Tyrell *et al.*, 2008).

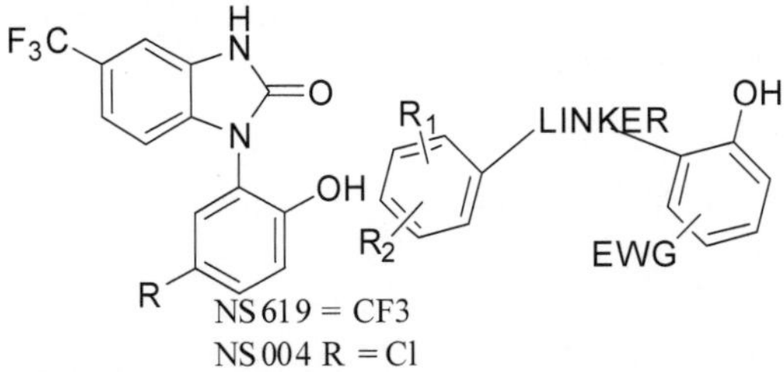

NS619 = CF3
NS004 R = Cl

A = Benzimidazole NS004. B = Generic pharmacophoric model for a BK- opener, consisting of two aryl rings, spaced by an appropiate heterocyclic or acyclic linker, EWG 1/4 electron withdrawing group; R1 and R2 represent various kinds of possible substituents.

79

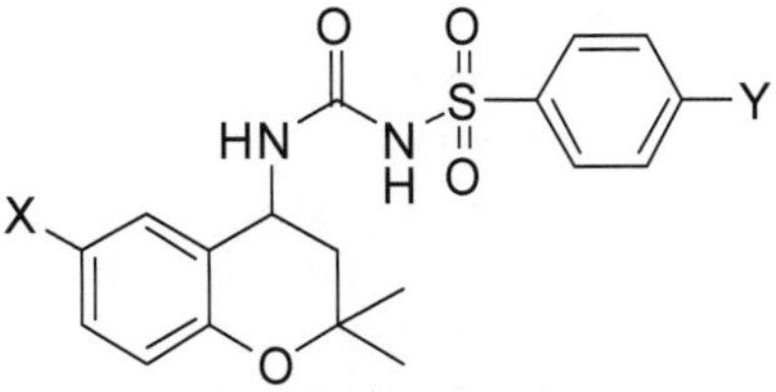

X = Br ,F, NO2
Y =H,CH3, Cl

80

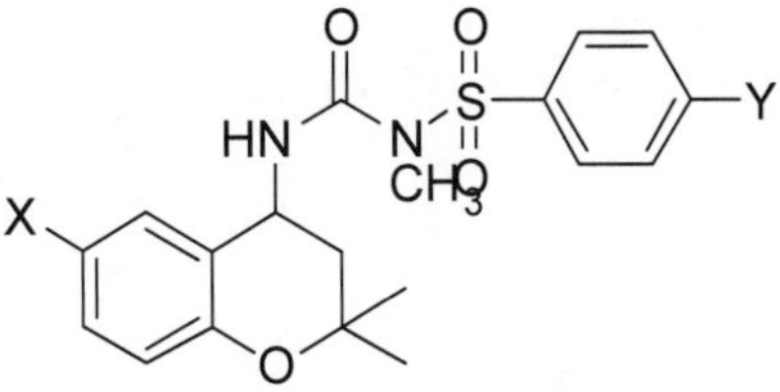

X = Br ,F,
Y =H,CH3, Cl

81

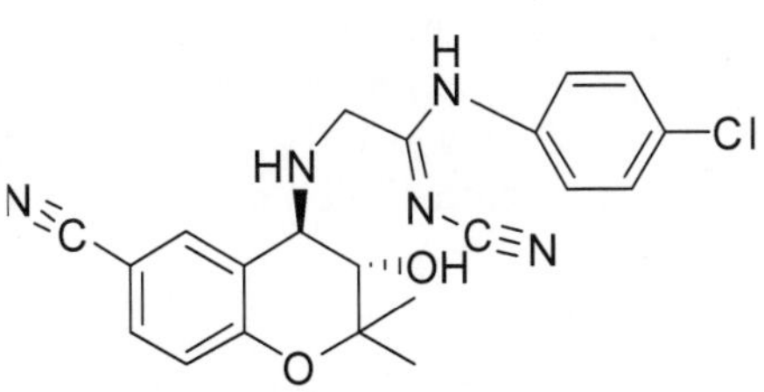

82 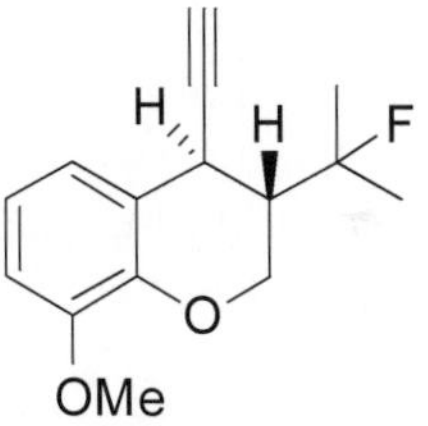 **83** 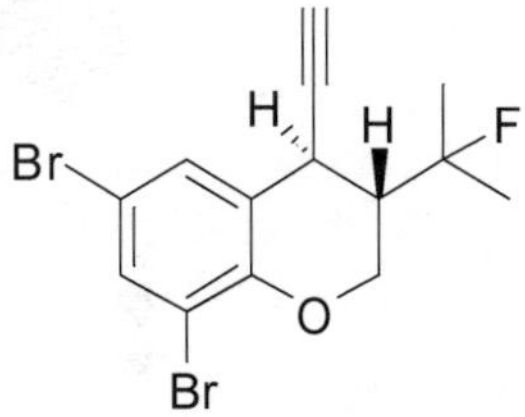 **84**

These compounds were evaluated and demonstrated a reduced ability to relax phenylephrine simulated rat thoracic tissue, with evidence that these compounds may be operating via an alternative mechanism than ATP-sensitive K^+ channel activity.

The most active analogues of benzopyrans from homochiral propargyl alcohols.

Clinical Application

Cromakalim, nicorandil and pinacidil are currently at various stages of clinical assessment. Nicorandil has been marketed in Japan as an anti anginal agent. Pinacidil was approved for use as an antihypertensive drug in Denmark. Cromakalim is in phase II/III clinical trials in hypertension.

Angina Pectoris

Recent studies in patients with efforts angina have shown favourable results with nicorandil in comparison with established therapy (Belz *et al.*, 1984; Hayata *et al.*, 1990; Araki *et al.*, 1987). Nicorandil decreases preload, but also reduces after load, as would be expected from in vitro experiments with arteries. Some recent studies have suggested that, in vasoplastic angina, the primary defect may be a decrease in the potassium ion permeability of the smooth muscle cells of coronary arteries (Uchida *et al.*, 1985; Iwki *et al.*, 1987). If this is correct, nicorandil, with a combination of guanylate cyclase stimulation and potassium channel opening properties, should be potentially useful in this condition.

Hypertension

The primary effect of cromakalim and pinacidil is dialation of peripheral arterioles with a reduction in after load and these drugs are currently being evaluated for the use in the treatment of hypertension. More than 4000 hypertensive patient have been treated till date with this drug resulting a decrease in blood pressure and increase in heart rate, effects sometimes associated with a typical "vasodialation headache". Oedema is also a prominent side effect of pinacidil, as a consequence of the reduced blood pressure and stimulation of renal compensatory mechanisms. As with vasodialator drugs such as hydralazine, the incidence of both oedema and headache is reduced by athiazide diuretic. Therefore, combination of pinacidil with a diuretic may present the optimum tereatment regime; this has been recommended for pinacidil by FDA Cardiorenal Drugs Advisory Committee in the USA. An initial investigation of cromakalim in 70 patients (active and placebo) with mild essential hypertension has found that single oral daily doses, in the range 0.5-1.5 mg, lower blood pressure (Arr *et al.*, 1987). Some incidence of headache was reported, particularly at the highest dose, but oedema was not seen. The differences in haemodynamic profile between cromakaliim and pinacidil suggest that cromakalim may posses a beneficial effect on renal blood flow in man.

Asthma

In vitro experiments have shown that cromakalim, nicorandil and pinacidil relax bronchial smooth muscle and this activity is reflected for oral cromakalim by bronchodilator activity in both the guinea pig and man (Buckle *et al.*, 1987; Baird *et al.*, 1988). Thus this drug may have potential as novel drug treatment of asthma. Such in vivo findings suggest that the opening of potassium channels, and the resultant hyperpolarisation by drugs such as cromakalim, may be as effective means of antagonizing the hyperactivity of airways smooth muscle associated with the condition.

Conclusion

Despite the tremendous interest and considerable scientific progress in potassium channel openers over the past few years, the clinical benefit of many types of these agents has yet to proven. The ever present need for potassium channel selectivity and target validation remains a key hurdle. Without doubt, the most established potassium channel openers in the clinic area the sulphonylureas, which block the potassium channel. These drugs are used for NIDDM (non-insulin type of diabetes), but their future may be under threat, to some extent, from the development of agents,which enhance the sensitivity of tissue to insulin. Compounds acting to activate, or open, potassium channels in smooth muscle have the potential to be used for a variety of disorders including hypertension, obstructive airways disease (asthma) and urinary incontinence. There is no doubt that potassium channel openers can lower blood pressure, but their use in hypertension is accompanied by an incidence of headache and edema. The success of potassium and channel openers of their hypertensive action will no doubt depend on additional properties such as beneficial effect of plasma lipids and a useful anti ischemic action. The bronchodilator and possible anti-inflammatory effects of potassium channel openers in the airway disease now obviously depends on clinical trials with the newer selective compounds. There has been a considerable search for bladder- selective potassium channel openers. Cromakalim and pinacidil are novel drugs which lower blood pressure *in vivo* and open membrane potassium channels in a variety of smooth muscle *in vitro.* Clinical experience with cromakalim, nicorandil and pinacidil is relatively limited. However, initial results in patients with cardiovascular diseases like hypertension and angina pectoris seem promising. Their use in the treatment of other disorders of smooth muscle, like asthma, is under investigation. Although work on potassium channels in muscle has been the main focus of attention in the past, studies on agents affecting potassium channels in neurons are now in progress and could lead to new types of useful drugs for various CNS disorders. Obviously the rest of the present decade and the first one of the next century should see the introduction of further type of potassium channel openers into the clinic and it will be interesting to observe how their usefulness develops.

Acknowledgements

We are thankful to the Director, Indian Institute of Integrative of Medicine, Jammu, who has given a generous support for the research in this Institute. His inspiration is the essence to formulate this article, which may be the fruit of interested explorers, may be in the recent years, those who are really interested in diversity oriented synthesis.

References

Ahmed, F., Foster, R. W., Smal, l R. C., and Weston, A. H. (1984). The spasmogenic action of oxytocin in the rat uterus. *Br. J.Pharmacol.*, **83**: 227.

Ahnfeit-Renne, I., and Jergensen, H. J. (1989). Pinacidil. *Drugs Toda.*, **25**: 65.

Aloup, J. C., Farge, D., James, C., Mondot, S., and Cavero, I. (1992). Pharmacological evidence for a role of ATP-dependent potassium channels in myocardial stunning. *Drugs Future,* **15**: 1097.

Amorso, S., Schmidt-Antomarchi, H., Fosset, M., and Lazdunski, M. (1990). Essential role of adenosine. *Science,* **287**: 852.

Anderson, K. E. (1989). Partition coefficients of low-molecular-weight volatile chemicals in various liquids and tissues. *Pharmacol. Toxicology,* **70**: 244-255.

Angel, I., and Bidet, S. (1991). The binding site for [3 H]glibenclamide in the rat cerebral cortex does not recognize K-channel agonists or antagonists other than sulphonylureas. *Fund. Clin. Pharmacol.,* **5**: 107.

Annstrong,C. M., and Bezanilla, F. (1977) Inactivation of the sodium channel. II. Gating current experiments. *Gen, J. Physiol.,***70**: 567.

Antane, S. A., Butera, J. A., and Lennox, J.R. (2001). Synthesis of 8-Chloro-10H-[1]benzofuro[3,2-b]indole-1-carboxylic acid. *US 6288099 B1 11 Sep.,* 15pp.

Araki, H., Hayata N., Matsuguchi and Nakamura, M. (1987). Effects of nicorandil on rest and effort angina unresponsive to combination therapy with a calcium antagonist and oral nitrate. *Clin Ther.,* **9**: 174-182.

Arch, J. R. S., Buckle, D. R., Carey, C., Parr-Dobrzanski, H., Faller, A., Foster, K. A., Houge-Frydrych,C. S. V., Pinto, I. L., Smith, D. G., and Taylor, S. G. (1991). Tubulin and microtubules target for anticancer drugs. *J. Med. Chem.,* **34**: 2588.

Arenna, J. P., and Kass, R. S. (1988). Block of heart potassium channels by clofilium and its tertiary analogs: relationship between drug structure and type of channel blocked. *Mol. Pharmacol.,* **34**: 60.

Ashcroft, F. M. (1988). Cloning and functional expression of the cdna encoding a novel ATP-sensitive potassium channel subunit expressed in pancreatic β-cells, brain, heart. *Annu. Rev. Neurosc.,* **11**: 97.

Atwal, K. S., Moreland, S., Mc Cullough,J. R., Relliey, B. C. O., Ahmed, S. Z., and Normandin, D. E. (1992). Modulation of potassium channels by organic molecules. *Bioorg. Med. Chem. Lett.,* **1**: 83.

Atwood, M. R., Jones, P. S., Redshaw, S. (1989). Drug Discovery Opinion.EPA 0298452, *Chem Abstract.,* **110**: 1730101b.

Auchampach, J. A. and Gross, G. J. (1991). A model of anoxic preconditioning in the isolated rat cardiac tissue. *FASEB J. S.,* A1103.

Baird A., Hamilton, T. C., Richards, D. H., Tasker, T., and Williams, A. J. (1988). Inhibition of histamine induced bronchoconstriction in normal healthy volunteers by a potassium channel activator, BRL 34915. *Br. J. Clin. Pharmacol.,* **25**: 114p.

Bard, I., and escande, D. (1989).Recent challenges in insulin delivery systems. *Plugers archive.,* **414(suppl.1)**.

Bartek, M. J., and Gomoll, A. W. (1985). Adverse effects of a single dose of (+)-sotalol in patients *Journal. Pharmacology. Exp. Ther.,* **232**: 817-823.

Baskin, E. P., Serik,C. M., Wallace, A. A., Brookes, L. M., Selnick, H. G., Claremon, D. A., Lynch, JR. J. J., and Cadiovasc, J. (1991). Effects of new and potent methanesulfonanilide class III antiarrhythmic agents on myocardial refractoriness and contractility in isolated cardiac muscle. *Pharmacol.,* **18**: 406.

Bekheit Restive, S., Henkin, M., Boutijdir,R., Henkin, M., boutjdir, R., Gooyandeh, M., and Jean-Bart, K. (1989). Hazards of electyrical shock in cardiology. *J. AM. Coll. Cardiol.*, **13**:142A.

Belz, I., Matthews, G. G., Heinrich, J., and wagner, G. (1984). Ca2+-entry blockers and atherosclerosis. *Eur. J. Clin. Pharmacy*, **26**: 681-685.

Ben-ari, Y., krnjevic, K., and Crepel, V. (1990). Activators of ATP-sensitive K^+ channels reduce anoxic depolarization in CA3 hippocampal neurons. *Neurosci.*, **37**: 557.

Bergmann, E. R. V., and Gericke, R. (1990a). Synthesis and antihypertensive activity of 4-(1, 2-dihydro-2-oxo-1-pyridyl)-2H-1-benzopyrans and related compounds, new potassium channel activators. *J. Med. Chem.*, **33**: 492.

Bergmann, R., Gerickle, R. J. (1990c). Synthesis and antihypertensive activity of KC-399, a benzopyran K channel opener with long duration of action and less tachycardia. *J. Med. Chem.*, **43**: 499.

Bernardi, H., Fosset, M., and Lazdunski, M. (1988). Potassium ion channels. *Proc. Natl. Sci. USA.*, **58**: 9816.

Bezanilla, F., Perozo, E., Papazian, D. M., and Stefani, E. (1991). Molecular basis of gating charge immobilization in Shaker potassium channels. *Science*, **254**: 679.

Black, S. C., Chi, L., Mu, D-X., and Lucchesi, B. R. (1991). The antifibrillatory actions of UK-68,798, a class III antiarrhythmic agent. *J. Pharmacol. Exp Ther Neurasci.*, **10**: 416.

Blarer, S. G. B. (1989). *Patent applications.*,2204868A.(1988).potassium channel modulators. *Chem. Abstr.*, **111**: 7227y.

Blatz, A. L., and Magleby, K. L.(1987). Cell-cycle control of a large-conductance K^+ channel in mouse early embryos. *Trends neurasci.*, **10**: 463.

Boer, R., Grassegger, A., Schudt, C., and Glossmann, H. (1989). Niguldipine binding. *Eur. J. Pharmacol.*, **172**: 131.

Bontems, F., Roumestand,C., Boyot, P., Gilquin, B., Doljansky, Y., Menez, A., and Toma, F. S. (1991). Three-dimensional structure of natural charybdotoxin in aqueous solution by 1H-NMR. Charybdotoxin possesses a structural motif found in other scorpion toxins. *Eur. J. Biochem.*, **196**:19.

Bray, K., and Quast, U. (1991). Analysis of excitation-contraction-coupling components in chronically stimulated canine skeletal muscle. *Eur.J. Pharmacol.*, **200**: 163.

Buckle, D. R., Bumstead, J., Clarke, G. D., Taylor, J. F., and Taylor,S. G. (1987). Reversal of agonist induced broncho constriction in the guinea pig by the potassium channel activator BRL 34915. *Br. J. Pharmacol.*, **92**: 744.

Buckle; D. R., Arch, J. R. S., Edge, C., Foster, K. A., Houge-frydrych, C. S. V., Pinto, I. L., Smith, D. G., Taylor, J.F., Taylor, S. G., Tedder, J. M., and Webster, R. A. B. (1991). Synthesis and smooth muscle relaxant activity of a new series of potassium channel activators: 3-amido-1,1-dimethylindan-2-ols. *J.Med.Chem.*, **34**: 919.

Burrell, G., Cassidy F., Evans, J. M., Lightowler, D., and Stemp, G. (1990). Variation in the aromatic ring of cromakalim: antihypertensive activity of pyranopyridines and 6-alkyl-2H-1-benzopyrans. *J.Med.Chem.*, **33**: 3023.

Butera, J. A., Antane,S. A., Hirth, B., Lennox, J. R, Sheldon, J. H., Norton, N. W., Werga, D., and Argentieri, T. M. (2001). Synthesis and potassium channel opening activity of substituted 10H-benzo[4,5]furo[3,2-b]indole-and 5,10-dihydro-indeno[1,2-b]indole-1-carboxylic acids. *Bioorg.Med.Chem.Lett.*, **11(16)**: 2093-2097.

Calderone, V., Coli, A., Fiamingo, F., Giorgi, F. L., Leonardi, M., Livi, O., Marteli, A., and Martinotti, E. (2006). Structural modifications of benzanilide derivatives, effective potassium channel openers. *European Journal of Medicinal Chemistry*, **41**: 1421-1429.

Calderone, V., Testai, L., Martelli, A., Rappoelli, S., Digiacomo, M., Balsamo, A., and Brech, M.C. (2010). The mitochondrial bioenergetic phenotype.Biochem. Pharmaco.,**79**: 39-47.

Calderone,V. (2002). Large conductance calcium activated potassium channel. Function, pharmacology and drugs. *Curr. Med. Chem.*, **9**: 1385-1395.

Caliendo, G., Perissutti, E., Fiorino, V., Severino, F., acirillo,B. D., Bianca, D., Lippolis, R. D. V., Pinto, A., and Sorrentino, R. (2004). Synthesis by microwave irradiation of a substituted benzoxazine parallel library with preferential relaxant activity for guinea pig trachealis. *European Journal of Medicinal Chemistry*, **39**: 815-826.

Carmeliet, E. (1985). Role of 5-HT2 receptors in serotonin-induced contractions of nonvascular smooth muscle. *J. Pharamcol. Exp. Ther.*, **232**: 817.

Caterall, W. A. (1988). Structure and function of voltage-sensitive ion channels. *Science.*, **242**: 50-61.

Chiu, Hsin-I., Lin, Yen-Chung, Yu, Chen-Yu (2001). *N*-Acyl-1,2,3,4a,5,10b-hexahydro-[1]benzopyrano-[3,4-*b*][1,4]oxazine-9-carbonitriles as bladder-selective potassium channel openers. *Bioorganic and Medicinal Chemistry*, **9**: 383.

Christie, M. J., Adelman, J. P., Douglass, J., and North, R. A. (1989). Cloning and expression of an inwardly rectifying ATP-regulated potassium channel. *Science*, **244**: 221.

Chung, S., Reinhart, P. H., Martin, B. L., Brautigan, D., and Levitan, I. B.(1991). Protein kinase activity closely associated with a reconstituted calcium-activated potassium channel. *Science*, **253**: 560.

Clapham, J. C., Hamilton, T. C., Longman, S. D., Buckingham, R. E., Campbell, C. A., and Isley, G. L. (1991). BRL 34915, a novel antihypertensive agent: comparison of effects on blood pressure and other haemodynamic parameters with those of nifedipine in animal models. *Arzenim. Forsch.*, **1**: 385.

Colatsky,T. J., and Follmer, C. H. (1989). *Cardiovasc. Drug Rev.*, **7**: 199.

Connors, S. P., Denis, P. D., Gilli, E. W., and. Terrar, D. A. (1991). Cytoprotection utilizing Aryltriazol-3-thiones *J. Med Chem.*, **34**: 1570.

Conti, F., and Neher; E. (1980). Single channel recordings of K+ currents in squid axons. *Nature.*, **285**: 140.

Cook, D. C., Hart, T. W., Mclay, I. M., Palfreyman, M. N., and Wash, R. J. A. (1989). Novel Potassium Channel Activators. III.1) Synthesis and pharmacological evaluation of 3,4- Dihydro-2H-1,4-benzoxazine Derivatives: Modification at the 2 Position.*EPA 0321274 AI.*

Cross, P. E., and Arrow smith (1988). UK-68,798: a novel, potent and highly selective class III antiarrhythmic agent which blocks potassium channels in cardiac cells.. J. E., *EP 0286277 AI.*

Cullogh, J. R. Mc., Normandin, D. E., Conder,M. L., Sleph, P. G., Dzwonczyk, S., and Grover, G. J. (1991). Heart Function Studies in Nonfibrilhating Isolated Hearts *Circular, Res.*, **69**: 949.

Dandu, R.R., Lyons, J.A., Raddatz, R., Huang, Z., and Aimone, S.D. (2012). Synthesis and evaluation of pyridazinone-phenethylamine derivatives as selective and orally bioavailable histamine H3 receptor antagonists with robust wake-promoting activity. *Bioorg. Med. Chem. Lett.*, **22:** 2151-53.

Daut, J., Mailer-Rudolph, W., Von Beckernath, N., Mehrke, G., Ginnther, K., and Meinen, Goedel. L. (1990). Do the K< sup>+</sup> channel openers relax smooth muscle by opening K< sup>+</sup> channels?. *Science*, **247**: 1341.

Devakaram, R., Black, D.S., Choomeunwaid, V., Davis, R.A., and Kumar, N. (2012), Facile Three-Component Synthesis of 7-Arylbenzo[f]chromeno[4,3-b]chromen-6(7H)-ones Catalyzed by Potassium Phthalimide in Aqueous Media. *Bioorg. Med. Chem.*, **20**: 1527-1534.

Dunne, M. J. (1991). Block of ATP-regulated potassium channels by phentolamine and other α-adrenoceptor antagonists.Br. *J. Pharmacol.*, **103**: 1847.

Dunne, M. J., Aspinall, R. J., and Peterson, O. H. (1990). Effects of pinacidil, RP 49356 and nicorandil on ATP-sensitive potassium channels in insulin-secreting cells. *Br. J. Pharmacol.*, **99**: 169.

Englert, H. C., Lang, H-J., Mania, D., and Scholkens, B. (1988). An efficient 1,3- allylic carbonyl transposition of chalcone. *Chem. Abstr.*, **10**: 57513p.

Evans, J. M. (1986). Dilator of arterial dilation. 4677116. *Chem. Abstr.*, **104**: 19511e.

Evans, J. M., and Stemp G.(1988). Variation in the aromatic ring of cromakalim: antihypertensive activity of pyranopyridines and 6-alkyl-2H-1-benzopyrans. *Synth Commun.*, **18**: 1111.

Ewald, D. A., Williams, A., and Levitan, I. B. (1985). Modulation of single Ca2+-dependent K+-channel activity by protein phosphorylation. *Nature*, **315**: 503.

Fish, F. A., Prakash, C., and Roden, D. M. (1990). Suppression of repolarization-related arrhythmias *in vitro* and *in vivo* by low-dose potassium channel activators. *Circulation*, **82**: 1362.

Follmer, C. H., and Colatsky, T. J. (1990). Block of delayed rectifier potassium current, IK, by flecainide and E-4031 in cat ventricular myocytes. *Circulation*, **82**: 289.

Fosset, M., Rweille de, J., Green, R. D., Schmid-Antomarchi, H., and Lazdunski, M. (1988). Antidiabetic sulfonylureas control action potential properties in heart cells via high affinity receptors that are linked to ATP-dependent K+ channels. *Bioi Chem.,* **263**: 7933.

Franks, N. P., and Lieb, W. R. (1988). Molecular and cellular mechanisms of general anaesthesia. *Nature*, **333**: 662.

Garcia, G., Roux, R., Nisato, D., and Gautier, P. (1989). Potassium channels and their modulators. *EPA 0296975. Chem. Abstr.,* **110**: 192654a.

Geisen, K., Hitzel, V., Okomonopoulos, R., Punter, J., Weyer, R., and Summ, H.D. (1985). Glibenclamide binding to sulfonylurea receptors by oral antidiabetics. *Arzneim Farsch*., **35**: 707.

Gelband, C. H., Lodge, N. J., and Breeman,C. V. (1989). A Ca+ activated K' channel from rabbit aorta: modulation. *Eur. J. Pharmacol*., **167**: 207.

Gericke, R., Baumgarth, M., Lues, I., Peyer, J. De., and Bergmann R. (1990). A review on Modulation of potassium channels by organic molecules *EP 400430 A2* Tedder, J. M., Pinto, I. L., and Edge, C. M. (1989). *EP 0321175 AI*.

Gerlach, T., Brende,l J., Lang, H-J., Paulus,E. F., Weidmann, K., Bruggemann, A., and Williams, E. M. V. (1985). Purinergic mechanisms in inflammation. *Eur. Heart. J.* (**suppl. D**), 145.

Gimenez-Gallego, G., Navia,M. A., Reuben, J. P., Katz, G. M., Kaczorowki, G. J., and Garcia, M. L. (1988). Purification, sequence, and model structure of charybdotoxin, a potent selective inhibitor of calcium-activated potassium channels. *Proc Natl.Acad Sci*., USA, **85:** 3329.

Goldberg,M. R., and Offen, W. W. (1988). Pinacidil with and without hydrochlorothiazide. *Drugs 36.* (**suppl.7**), 83.

Gomoll, A. W., and Bertek, M. J. (1986). Pharmacology, pharmacodynamics and pharmacokinetics of sotalol. *Eur. J. Pharmacol.,* **132**: 123.

Gopalakrishnan, M., Johnson, D.E., and Janis, R.A., Triggle, D.J. (1991). Characterization of binding of the ATP-sensitive potassium channel ligand, [3H] glyburide, to neuronal and muscle preparations. *J. Pharmacol. Exp. Ther*., **257**:1162.

Gopalkrishan, M., Janis, R. A., and Trigger, D. J. (1993). Characterization of the ATP-Sensitive Potassium Channels (K_{ATP}) Expressed in Guinea Pig Bladder Smooth Muscle Cells. *Drug Dev.Res*., **28**: 95.

Grazula, M., Kufelnicki, A., Wozniczka, M., and Lorenz, I.P. (2012). Synthesis of chromenes. *Polyhedron*., **31**: 150-158.

Grover, G. J., Dzwonczyk, S., and sleph, P. G. (1990). Reduction of ischemic damage in isolated rat hearts by the potassium channel opener, RP 52891. *Eur.J.Pharmacol*., **191**: 11.

Grover, G. J., Dzwonczyk, S., Parham, C. S., and Sleph, P. G. (1992). Role of myocardial ATP-sensitive potassium channels in mediating preconditioning in the dog heart and their possible interaction with adenosine A1-receptors. *Cardiovasc. Drugs Ther.*, **4**: 465.

Grover, G. J., Mccullough, J. R., Henry, D. E., Condor, M. L., and Sleph, P. G. (1989). Anti-ischemic effects of the potassium channel activators pinacidil and cromakalim and the reversal of these effects with the potassium channel blocker glyburide. *J. Pharmacol. Exp. Ther.*, **251:** 98.

Grover, G. J., Newburger Sleph, J., Dzwonczyk, P. S., Taylor, S. C., Ahmed, S. Z., and Atwal, K. S. (1991). Pharmacology of ATP-sensitive potassium channel (KATP) openers in models of myocardial ischemia and reperfusion. *J. Pharmacol. Exp. Ther.*, **257**: 156.

Grover, G. J., Sleph, P. G., and Dzwonczyk, S. (1990). Pharmacologic profile of cromakalim in the treatment of myocardial ischemia in isolated rat hearts and anesthetized dogs.*J. Cardiovasc. Pharmacol.*, **16**: 853.

Gwilt, M., Dalrymple, H. W., Blackbum, K. J., Burges, R. A., And Higgins, A. J. (1988). The use of rat and guinea pig cardiac action potentials to distinguish subclasses of class III antiarrhythmic drugs. *Circulations,* **78 (4)**:150.

Gylfe, E., Hellman. B., Sehlin.J., and Tliljedal, I.B. (1984). Amino acid conversion into 5-hydroxytryptamine in pancreatic ß-cells. *Experentia*, **40**: 1126.

Hamil, O. P., Marti, A., Neher, E., Sakmann, B., and Sigworth, F. J. (1981). Improved patch-clamp techniques for high-resolution current recording from cells and cell-free membrane patches. *Pflagers Archive*, **391**: 85.

Harris, A. S., Bisteni, A., Russell, R. A., Brigham, J. C., Firestone, J. E. (1984). Product of myocardial ischaemia and electrical instability of heart. *Science*, **119**: 200.

Hatton, R., Heys, C., Todd, M. H., Downing, O. A., and Wilson K. A. (1991). Synthesisd of benzopyran. *Br. Pharmacol.*, **102**: 280.

Hayata, N., Araki, H., and Nakamura, M. (1986). Effects of nicorandil on exercise tolerance in patients with stable effort angina: a double-blind study. *Am Heart J. Drugs,* **36 (Supply.7)**: 29.

Hermsmeyer, K. (1997). Reactivity based coronary vasospasm independent of atherosclerosis in rhesus monkeys. *J Am Col Cardiol.*, **29**: 671-680.

Hewawasam, P., Gribkoff, V. K., Pendri, Y., Dworretzky, S. L., Meanwell, N. A., Martinez, E., Boissard, C. G., Post-Monson, D. J., Trojnacki, J. T., Yeleswaram, K., Pajor, L. M., and Starrrett, J. E. (2002). The synthesis and characterization of BMS-204352 (maxipost). *Bioorganic and Medicinal Chemistry Letters*, **12(7)**: 1023-1026.

Hewawasam, P., Erway, M., Moon, S. L., Knipe, J., Weiner, H., Boissard, C. G., Posthafner, D. T., Berger, F., Borchard, U., Kullman, A., and Scherlitz, A. (1988). Diphenyl triazoles as potassium channel modulators. *Arzenim Forsch.*, **38**: 231.

Hoshi, T., Zagotta, W. N., and Aldrich, R. W. (1990). Biophysical and molecular mechanisms of Shaker potassium channel inactivation. *Science*, **250**: 533.

Hu, S., Kim, H. S., Okolie, P., and Weiss, G. B. (1990). Comparison of *the* actions of acetylcholine and BRL 38227. *J. Pharmacol. Express. Ther.*, **253**:771.

Innoue, R., Okabe, K., Kitamura, K., and Kuriyama, H. (1986). A newly identified Ca2+ dependent K+ channel in the smooth muscle membrane of single cells dispersed from the rabbit portal vein. *Pfleger. Archiv.*, **406**: 138.

Inoue, Takehasi, Hashimoto, Shinji, Tsuchiya, Kiyoshi, Izumi, Takeshi, Ohmori, Tetsuro, and Koyama, Tsukasa (1996). *European Journal of Pharmacology*, **311**:1-6.

Iwki, M., Mizobuchi, S., Nakaya, Y., Kawano, K., Nikki, T., and Mori, H. (1987). Tetra thylammonium induced coronary spasm in isolated perfused rabbit heart. *Cardiovasc. Res.*, **21**: 130-139.

Kantor, P.F., Coetzee, W.A., Dennis, S.C., and Opie, L.H. (1987). Effects of glibenclamide on ischemic arrhythmias. *Circulation*, **76:** 4-17.

Karashima, T., Itoh, T., and Kuriyama, H. (1982). Effects of 2-nicotinamidoethyl nitrate on smooth muscle cells of the guinea-pig mesenteric and portal veins.. *J. Pharmacol. Exp. Ther.*, **221**:472.

Khelil, S., Lebrun, P., Tullio de, P., and Pirotte, B. (2006). Synthesis and pharmacological evaluation of some N-arylsulfonyl-N-methyl-N′-(2,2-dimethyl-2H-1-benzopyran-4-yl)ureas structurally related to cromakalim, *Bioorganic and Medicinal Chemistry*, **14**: 3530-3534.

Khelili, S., Kihal, N., Yekhlef, M., Tullio-de, P., Lebrun, P., and Piorette, B. (2012). Synthesis and pharmacological activity of N-(2,2-dimethyl-3,4-dihydro-2H-1-benzopyran-4-yl)-4H-1,2,4-benzothiadiazine-3-carboxamides 1,1-dioxides on rat uterus, rat aorta and rat pancreatic β-cells *Eur. J. Med. Chem.*, **54**: 873-884.

Kiesewetter, D. O., Jagoda, E. M., Starrett, J. E. Jr., Gribkoff, V. K., Hewawasam, P., Srinivas, N., Salzar, D., and Ecckelman, W. C. (2002). Radiochemical synthesis and biodistribution of a novel maxi-K potassium channel opener. *Nuclear Medicine and Mioliogy*, **29(1)**: 55-59.

Kim, D., and Calpham, D. E. (1989). Potassium channels in cardiac cells activated by arachidonic acid and phospholipids. *Science*, **144**:1174.

Kim, D., Lewis, D. L., Graziadei, L., Neer, E. J., Bar-Sagi, D., and Clampham, D. E. (1989). G-protein Py-subunits activate the cardiac muscarinic K+-channel via phospholipase A2. *Nature*, **337**: 557.

Kirsch, G. R., Codina, J., Birnbaumer, L., and Brown, A. M. (1990). Coupling of ATP-sensitive K+ channels to A1 receptors by G proteins in rat ventricular myocytes. *Physiol, Am. J.*, **259**: H820.

Kleber, A.G. (1984). Extracellular potassium accumulation in acute myocardial ischemia., *J. Mol. Cell. Cardiac.*, **16**: 389.

Klockner, U., and Isenberg, G. (1989). The dihydropyridine niguldipine modulates calcium and potassium currents in vascular smooth muscle cells. *Br. J. Pharmacol.*, **97**: 957.

Klockner, U., Trieschman, U., and Isenberg, G. (1989). Action potential and net membrane currents of isolated smooth muscle cells (urinary bladder of the guinea-pig). *Arzneim.Forsch.*, **39**:120.

Klockner, U., Trieschman, U., and Isenberg, G. (1989). Pharmacological modulation of calcium and potassium channels in isolated vascular smooth muscle cells. *Arzneim.Forsch.*, **39**:120.

Kovac, R. J., and Nelson, M. T. (1991). *Physiol. Am. J.*, **261**:H604.

Kozlowski, R. Z., Hales, C. N., and Ashford, M. L. J. (1989). Dual effects of diazoxide on ATP-K+ currents recorded from an insulin-secreting cell line. *Br. J. Pharmacol.*, **97**: 1039.

Krnjevic, K. (1990). Adenosine triphosphate-sensitive potassium channels in anoxia. *Stroke*, **21**: 111-190.

Kuichi, Y., Ito, H., Sugimoto, T., Shimizu, T., Miki, I., and Ui, M. (1989). A central role for Stat3 in IL-6-induced regulation of growth and differentiation in M1 leukemia cells. *Nature*, **337**: 557.

Kume, H., and Takai, A. (1989). Beta-adrenergic agonists regulate kca channels in airway. *Nature*, **341**: 152.

Lang, R. W., and Wenk, P. F. (1988). Synthesis of selectively trifluoromethylated pyridine derivaties as potential antihypertensives. *Helv. Chim. Acta.*, **71**: 596.

Lathrop, D., Nanasi, P. P., and Varro, A. (1990). *In vitro* cardiac models of dog Purkinje fibre triggered and spontaneous electrical activity: effects of nicorandil. *Br. J. Pharmacol.*, **99**: 119.

Lebrum, P., Devreux, V., Hermann,M., and Herchuelz, A.(1988). Pinacidil inhibits insulin release by increasing potassium outflow from pancreatic B-cells. *Eur. J. Pharmacol.*, **156**: 283.

Lederer W. J., and Nichols, C. G. (1989). ATP dependence of KATP channel kinetics in isolated membrane patches from rat ventricle. *J. Physiol.*, **419**: 193.

Lis, R., Davey, D. D., Morgan, T. K. Jr., Lumma, W. C. Jr., Wohl, R. A., Lain, V. K., Wan, C. N., Argentieri, T. M., Sullivan, M. B., and Cantor, E. H.(1987).Synthesis and antiarrhythmic activity of novel 3-alkyl-1-[.omega.-[4-[(alkylsulfonyl)amino]phenyl]-.omega.-hydroxyalkyl]-1H-imidazolium salts and related compounds. *J. Med. Chem.*, **30**: 2303.

Longman, S. D., Clapham, J. C., Wilson, C., and Himilton, T. C. (1988). Cromakalim, a potassium channel activator: a comparison of its cardiovascular haemodynamic profile and tissue specificity with those of pinacidil and nicorandil. *J.Cardiovasc. Pharmacol.*, **11**: 535.

Lumma, W.C.Jr., Wohl, R. A., Davey, D. D., Argentier, T. M., Vitade, R. J., Gomez, R. P., Jain,V. K., Marisca, A. J., Morgan,T. K. Jr., Reiser, H. J., Sullivan, M. E., Wiggins, J., and Wong, S. S. (1987). Rational design of4-[(methylsulfonyl) amino] benzamides as class III antiarrhythmic agentsj. *Med. Chem.*, **30**:755.

Mackinnon, R. (1991). Determination of the subunit stoichiometry of a voltage-activated potassium channel. *Nature,* **350**:232-235.

Mackinnon, R., and Miller,C. (1989). Mutant potassium channels with altered binding of charybdotoxin, a pore blocking peptide inhibitor. *Science,* **245**: 1382.

Malmgren, A., Anderson, K. E., Sjogren, C., and Anderson, P. O. (1989). Facile synthesis of benzopyran. *Urol. USA.,* **142**:1134.

Manley, P. W., and Quast,U. (1991). Modulation of potassium channel by organic molecules. *Chemia.,* **45**: 88.

Marquese, B., Seagar, M. J., and Couraud, F. (1987). Characterization of Apamin binding protein associated with calcium activated potassium channel. *Eur. J. Biochem.,* **169**: 295.

Maruyama, M., and Gross, G. J. (1990). Pharmacological evidence for a role of a tp-dependent potassium channels in myocardial stunning. *FASEB J.,* **4**: A320.

Massefski, Jr. W., Redfield, A. G., and Hare, D. R. (1990). Molecular structure of carbotoxin a pore directed potassium channel inhibitor. *Science,* **249**: 521

Masuda, Y., Arakawa, C., Miyajima, M., Takeguchi, M., Yamashita, T., and Tanaka, S. (1990). Effect of an L- and *T*-Type calcium channel blocker. Jpn. *J. Pharmacol. Abstr., supply* **1**: 0-107.

Matsuhisa, S., Asano, M., Matsumoto Y., Takayama, K., Yoden, T., Tsuzuki, R., Ucida, W., and Yanagisawa, I. (1991). Modulation of potassium channels by organic molecules. *EP 0432.,* **893**, A2.

Matsumoto, Y., Tsuzuki, R., Matsuhisa, A., Yoden, T., Yamagiwa, Y.,Yanagisawa, I., Shibanuma, T., and Nohira, H. (2000). A one-pot synthesis of pyrido [2, 3-b][1, 4] oxazin-2-ones. *Biorg. Med. Chem.,* **8**: 393.

Matsura, H., Ehara, T., and Imoto,Y. (1987). Rapidly and slowly activating components of delayed rectifier K^+ current in guinea-pig sino-atrial node pacemaker cells. *Pfligers Archiv.,* **410**: 596.

Mc. Cullogh, J. R., Normandin, D. E., Conder, M. L., Sleph, P. G., Dzwonczyk, S., and Grover, G. J. (1991). Specific block of anti-ischemic effect of chromakalim bu sodium-5- hydroxydecanote. *Circular. Res.,* **69**: 949.

Mc. Pherson, G. A., and Angus, J. A. (1989). Functional and electrophysiological effects of a novel imidazoline-based k_{atp} channel blocker, IMID-4F.*Br. J. Pharmacol.,* **97**: 941.

Miller, C. (1991). Annus mirabilis of potassium channels. *Science,* **252**: 1092.

Moczydlowsky, E., Luchhesi, K., and Ravindran, A. (1988). An emerging pharmacology of peptide toxins targeted against potassium channels. *J. Membr. Biol.,* **105** : 95.

Mondo,T. S., Mestre, S. M., Cillagarh, C. G., and Cavero, I. (1988). Cardiovascular and biological effects of K channel openers, a class of drugs with vasorelaxant and cardioprotective properties.*Br. J. Pharmacol.,* **95:** 813.

Morgan, T. K.Jr., Lis, R., Lumma, W. C. Jr., Wohl, R. A., Nickisch, K., Phillip, G. B., Lind, J. M., Lampe, J. W., Meo Di, S. V., Reiser, H. J., Argentieri, T. M., Sullivan, M. E., and Cantor, E. H. (1990). Synthesis of novel (aryloxy) propanolamines and related compounds possessing both class II and class III antiarrhythmic activity. *J. Med. Chem.*, **33**: 1087.

Morris, A. P., Gallacher, D. V., Irvine, R. F., and Peterson, O. H.(1987). Synergism of inositol trisphosphate and tetrakisphosphate in activating Ca2+-dependent K+ channels. *Nature*, **330**: 653.

Murray, M. A., Boyle, J. P., and Small, R. C. (1989). Cromakalim-induced relaxation of guinea-pig isolated trachealis: antagonism by glibenclamide and by phentolamine. *Br. J. Pharmacol.*, **98**: 865.

Naoki, T., Hiroshi, K., Haruhiko, S., Takenori, I., and Tadakatsu, T.(2000). Synthesis of Benzopyran-4-amides potassium channel openers via palladium-catalyzed amidation. *Synthetic Commun.*, **30(23):** 4263-4269.

Nelson, M. T., Huang, Y., Brayden, J. E., Hescheler, J., and Standen, N. B. (1990). Arterial dilations in response to calcitonin gene-related peptide involve activation of K+ channels. *Nature*, **344**: 770.

Nelson-kudsk, J. E., Mellernkjaer, S., Siggaard, C., and Nielsen, C. B. (1988). Effects of pinacidil on guinea-pig airway smooth muscle contracted by asthma mediators. *Eur. Pharmacol.*, **157**: 221.

Niho, T., Notsu, T., Ishikawa, H., Funato, H., Yamaloaki, M., Takahashi, H., Tanaka, I., Kayamoto, M., Dabasaki, T., Ymasaki, F., Shinkawa, T., Uemura., A., and Mizota, M. (1987). *Folia Pharmacol. Jpn.*, **89**: 155.

Noble, D., and Tsien, R. W. (1969). A quantitative analysis of the slow component of delayed rectification in frog atrium. *Physiol, J.*, **200**: 205.

Noma, A. (1983). ATP-regulated K+ channels in cardiac muscle. *Nature*. **305**: 147.

Ohnishi, T. (1991).*US Patent No. 5.*, **006**:512.

Ordway, R. W., Walsh, J. V.Jr., and Singer, J. J. (1989). Arachidonic acid and other fatty acids directly activate potassium channels in smooth muscle cells. *Science*. **244**: 1176.

Osterrider, W. (1988). Modification of K + conductance of heart cell membrane by BRL 34915. *Naunyn-schmiedeberg's Arch. Pharmacol.*, **337**: 93.

Palfreyman, Malcom. N., Walsh, Roger. J. A., Brown, Thomas. J., and Chapman, Robert. F. (1992). Synthesis and biological activity of trans-(.+-.)-N-methyl-2-(3-pyridyl)-2-tetrahydrothiopyrancarbothioamide 1-oxide (RP 49356) and analogs: a new class of potassium channel openers. *Journal. Medicinal. Chemistry*, **35(20)**: 3613-3624.

Papazian, D. M., Schwarz, T. L., Tempel, B. L., Jan, L. Y., and Jan, Y. N. (1987). Sequence of a probable potassium channel component encoded at Shaker locus of Drosophila. *Science*, **237**:749.

Perkins, R. S., Paciorek, P. M., and Waterfall, J. F. (1989). Glibenclamide and procaine, but not apamin, inhibit the relaxant responses to Ro 31-6930 in guinea-pig taenia caeci. *Br. J. Pharmacol.*, **98**: 808.

Peterson, H. J. (1978). Synthesis and hypotensive activity of N-alkyl-N″-cyano-N′-pyridylguanidines. *J. Med. Chem.*, **21**: 773.

Pherson, G. A. Mc., and Angus, J. A. (1990). Characterization of responses to cromakalim and pinacidil in smooth and cardiac muscle by use of selective antagonists. *Br. J. Pharmacol.*, **100**: 201.

Piper, I., minbhall, E., Downing, S. J., Hollingsworth,M., and Sadraei, H. (1990). Effects of several potassium channel openers and glibenclamide on the uterus of the rat. *Br. J. Pharmacol.*, **101**: 901.

Pratz, J., Mondot.,S, Montier, F., and Cavero, I. (1991). Effects of the K+ channel activators, RP 52891, cromakalim and diazoxide, on the plasma insulin level, plasma renin activity and blood pressure in rats. *J. Pharmacol. Exp.*, **258**: 216.

Quagliato, D. A., and Humber, L. G. (*1990*). Benzopyran analogues as target potassium channel opener. *EPA 0314446. Chem. Abstract.*,**112**:48798z.

Quast, U., and Cook, N. S. (1989). ATP dependent potassium channel of muscle cells. *Trends Pharmacol. Sci.*, **10**: 431.

Quasthoff, S., Spuler, A., Spittelmeister, W., Lehhman-Hom, F., and Graff, P.(1990). K^+ channel openers suppress myotonic activity of human skeletal muscle *in vitro*. *Eur. J. Pharmacol.*, **186**: 125.

Rapposelli, S., Breschi, C.M., Calderone, V., Martelli, A., Tetai. L., and Vanni, M. (2011). Synthesis and biological evaluation of 5-membered spiro heterocycle-benzopyran derivatives against myocardial ischemia. *Eur. J. Med. Chem.*, **46**: 966-973.

Ronner, P., Higgins, T. J., and Kimmich, G. A. (1991). Inhibition of ATP-sensitive K+ channels in pancreatic beta-cells by nonsulfonylurea drug linogliride. *Diabetes*, **40**: 885.

Rorsman, P., and Trub, G. (1985). Synthesis and characterization of quinoline compounds. *Pflager. Arciv.*, **405**:305.

Rosa, J. C., Galanakis, D., Piergentilli, A., Bhandari,K., Ganellin, C. R., Dunn, P. M., and Jenkison, D. H.(2000). Synthesis, molecular modeling, and pharmacological testing of bis-quinolinium cyclophanes: Potent, non-peptidic blockers of the apamin-sensitive Ca^{2+}-activated K^+channel. *J. Med. Chem.*, **43**: 420.

Roxburg, C. J. (1995). The synthesis of potassium channel moducadors. *Synthesis*, 307-320.

Sadoshrina, J., Akaike, N., Kanaide, H., and Nakamura, M. (1988). Modulation of the cardiac sodium/bicarbonate cotransporter. *Am. J. Physio.*, **255**: H745.

Sanguinetti, M. C., and Jurkiewicz, N. K. (1990). A mechanistic link between an inherited and an acquird cardiac arrthytmia: *HERG* encodes the i_{kr} potassium channel *J. Gen. Physiol.*, **96**: 1995.

Sassen, L. M. A., Dunker, D. J. G. M., and Gho, B. C. G. (1990). Facile synthesis of various 2-substituted-4-(2-pyridyl) benzopyran. *Br. J. Pharmacol.*, **101**: 605.

Schmidt-Antomarchi, H., weille de, J. R., Fosset,M., and Lazdunski, M. (1987). Activation by adrenaline of a low- conductance G protein. *Biol. Chem.*, **262:** 15840.

Schou, S. C., Hansen, H. C., Tigmose, T. M., Boonen, H. C. M., Worsaae, A., Drabowski, M., Wahl, P., Arkhammar, P. O. G., Bodvarsdottir, T., Antoine, M. H., Lebrunb, P., and Hansena, J. B. (2005). Synthesis and pharmacological evaluation of 4< i> H</i>-1, 4-benzothiazine-2-carbonitrile 1, 1-dioxide and< i> N</i>-(2-cyanomethylsulfonylphenyl) acylamide. *Bioorganic and Medicinal Chemistry.*,**13:** 141-155.

Sharma, Deepak K., Tripati, Anil, K., Sharma, Rashmi., Chib, Rina., Singh, Baldev., Khan, Insad, Ali., Goswami, Anindya., Mukherjee, Debaraj. (2013). A new class of bactericidal agent against. *S.Aureus*,MRSA and VRE from bis indolylmethane. *Med Chem Res.*, DOI 10.1007/s00044-013-0764-4.

Simsek, R., Ozkan, M., Kismetli, E., Uma, S., and Safak, C. (2004). Some arylacridine derivatives possessing potassium channel opening activity. *IL FARMACO.*, **59**: 939-943.

Singer, J. J., and Steinberg, M. I. (1987). Characterization of Calcium activated Potassium Channels in Single Smooth Muscle Cells Using the Patch clamp Technique. *Pharmacol. Pflugers. Archiv.*, **408**: 98.

Smallwood, J. K., and Steinberg, M. I. (1988). Cardiac electrophysiological effects of pinacidil and related pyridylcyanoguanidines: relationship to antihypertensive activity. *Journal of Cardiovas. Pharmacol.*, **12**: 102

Spruce, A. E., Standen, N. B., and Stanfield, P. R. (1989). Multiple blocking mechanisms of ATP-sensitive potassium channels of frog skeletal muscle by tetraethylammonium ions. *Nature,* **413**: 31-48.

Standen, N. B., Quayle, J. M., Davies, N. W., Brayden, J. E., Huang,Y., and Nelson, M. T. (1989). Hyperpolarizing vasodilators activate ATP-sensitive K+ channels in arterial smooth muscle. *Science,* **245**: 177-180.

Steinberg, M. I., Wiest, S. A., Zimmerman, K. M., Ertel, P.J., Bemis, K. G., Palfreyman, M. N., Vicker N., and Walsh, R. J. (1990). Structure-activity relationships of K channel openers.*EP 0321175 AI*.

Stemp, G. S., and Burell, G. (1990). Modulation of potassium channels by organic molecules.*EPA 0399834*.

Stinson, S.C. (1991). Applications of liver acetone powders in enantioselective synthesis. *Chem. Eng. News, September,* **30**: 35.

Sung-eun, Y. Y., Yang, K., Sunkyung, L., Jeehee, S., Nakjeong, K. Ho. L. B., Ho won, S., Sun-ok, K., Dong-Ha, L., Hong, L., and Hwa-Sup, S.(2001). A Novel Anti-Ischemic ATP-Sensitive Potassium Channel (K_{ATP}) Opener without Vasorelaxation: *N-*

(6-Aminobenzopyranyl)-*N'*-benzyl-*N''*-cyanoguanidine Analogue *J. Med. Chem.*, **44(24)**: 4207-4215.

Takumi, T., Ohkubo, H., and Nakanishi, S. (1988). Cloning of a membrane protein that induces a slow voltage-gated potassium. *Current Science*, **242**: 1042.

Taylor, S. G., Bumstead, J., Morris, J. E. J., Shaw, D. J., and Taylor, J. F. (1988). Cromakalim inhibits cholinergic-mediated responses in human isolated bronchioles but not in guinea-pig airways. *Br. J. Pharmacol.*, **95(Suppl.)**: 795.

Tempel, B. L., Papazian,D. M., Schwartz, T. L., Jan, Y. N., and Jan, L. Y. (1987). Cloning of genomic and complementary DNA from Shaker, a putative potassium channel gene from Drosophila. *Science*, **237**:770

Tricklebank, M. D., Flockhart, G., and Freedman,S. G. (1988). The potassium channel activator, BRL 34915, antagonises a behavioral response to the muscarinic agonist, pilocarpine. *Eur. J. Pharmacol.*, **151**: 349.

Tripathi, A. K., Khan, K. R., and Taneja, S. C. (2003). Facile synthesis of various 2-substituted-4-(2-pyridyl) benzopyran analogues as target potassium channel opener.*Synth Commun.*, **33**: 579.

Tripathi, A. K., Koul, S., and Taneja,S. C. (2008). Microwave assisted facile and efficient synthesis of benzopyran.*Indian journal of chemistry.B.*, **48B**: 02

Tyrell, E., Tesfa, K. H., Greenwood, Iain, and Mann, A.(2008). The synthesis and biological evaluation of a range of novel functionalised benzopyrans as potential potassium channel activators.*Bioorganic and Medicinal Chemistry*, **18**: 1237-1240.

Uchida, Y. (1985). Recanalization of obstructed coronary artery by intracoronary administration of prostacyclin in patients with acute myocardial infarction. *Jap. Circulation*, **49**: 1238-1239.

Uprichard, L., Chi, A. C. G., and Lucchesi. (1989). Antifibrillatory effects of clofilium in the rabbit isolated heartb. R. *J. Mal. Cel/. Cardiol.*, **21(supply.II)**: 589.

Vandenburg, M. J. Arr., Woodward, S., Tasker, T., Stewart-Long, P., Fairhuirst, G., Stephens, J., and Wood, B. (1987).Preceeding of the third European on Hypertension. *Milan June, Abstr.*, **586.**

Wei, A., Covar-Rubias, M. A., Baker, B. A., Pak, M., and Salkoff, L. (1990). Modulation of potassium channels by organic molecules. *Science*, **248**: 599.

Weik, R., and Neumcke, B. (1989). ATP-sensitive potassium channels in adult skeletal muscle: characterization of the ATP-binding site. *J. Membr. Biol.*, **110**: 217.

Weiss, J. N., Lamp, S. T., and Shine, K. I. (1989). *Am. J. Physiol.*, **256**:H1165.

Wickenden, A. C., Grimwood, S., Grant,T. L., and Todd, M. H. (1991). Comparison of the effects of the K'-channel openers cromakalim and minoxidil sulphate on vascular smooth muscle. *Br. J. Pharmacol.*, **103**:1148.

Wille de, J. R., Schmidt-Antomarchi, H., Fosset, M., and Lazdunski, M. (1988). Structure and function of voltage sensitive potassium channels. *Proc. Natl. Acad. Sci.*, USA, **86**: 2917.

Williams, D. L. Jr., Katz, G. M., Roy-Contancin, L., and Reuben, J. (1988). Calcium activated potassium channel. *Proc. Natl. Acad. Sci*., USA, **85**:9360.

Wollenben, C. D., Sanguinetti,M. C., and Siegl, P. K. S. (1989). Influence of ATP-sensitive potassium channel modulators on ischemia-induced fibrillation in isolated rat hearts. *J. Mol. Cell. Cardiol.*, **21**:783.

Wu, S. N. (2003). Large-Conductance Ca2+-Activated K+ Channels. *Curr. Med. Chem.*, **10**: 649-661.

Xie Weiping, Wang Hong, Ding Jianhua, Wang Hai, and Hu Gang (2005). Iptakalim protects against MPP+-induced degeneration of dopaminergic neurons in association with astrocyte activation.*European Journal of Medicinal Chemistry,* **511**: 81-87.

Yogupoloski, L. M., Maletina, I. I., Petko,K. I., Fedyuk, D. V., Handrock, R., Shavaran, S. S., Klebanov, B. M., and Herzig, S. J. (2001). New fluorine-containing hypotensive preparations. *Fluorine Chem.*,**109(1)**: 87-94.

Previous Volumes Contents

— Volume 1 —

2012, xii+519p., col. plts., figs., tabls., ind., 25 cm **Rs. 2800**

ISBN 978-81-7035-775-9

— Volume 2 —

2014, xiii+534p., col. plts., figs., tabls., ind., 25 cm **Rs. 3500**

ISBN 978-81-7035-866-4

Index

A

B

D

F

G

I

J

K

L

M

N

O

P

S

T